Electrical Machinery Control Fundamentals

RIVER PUBLISHERS SERIES IN ENERGY ENGINEERING AND SYSTEMS

Series Editor

BOBBY RAUF
Semtrain LLC, USA

The "River Publishers Series in Energy Engineering and Systems" is a series of comprehensive academic and professional books focussing on the theory and applications behind various energy-related technologies and control systems. The series features handbooks for related technology, as well as manuals on the fundamentals and theoretical aspects of energy engineering projects.

The main aim of this series is to be a reference for academics, researchers, managers, engineers, and other professionals in related matters with energy engineering and control systems.

Topics covered in the series include, but are not limited to:

- Energy engineering;
- Machinery;
- Testing;
- HVAC;
- Electricity and electronics;
- Water systems;
- Industrial control systems;
- Web based systems;
- Automation;
- Lighting controls.

For a list of other books in this series, visit www.riverpublishers.com

Electrical Machinery Control Fundamentals

Dale R. Patrick
USA

Stephen W. Fardo
Department of Applied Engineering and Technology,
Eastern Kentucky University, USA

Vigyan (Vigs) Chandra
Eastern Kentucky University, USA

Brian W. Fardo
Department of Applied Engineering and Technology and
the Noel Studio for Academic Creativity,
Eastern Kentucky University, USA

River Publishers

Routledge
Taylor & Francis Group
NEW YORK AND LONDON

Published 2026 by River Publishers
River Publishers
Broagervej 10, 9260 Gistrup, Denmark
www.riverpublishers.com

Distributed exclusively by Routledge
605 Third Avenue, New York, NY 10017, USA
4 Park Square, Milton Park, Abingdon, Oxon OX14 4RN

Electrical Machinery Control Fundamentals / Dale R. Patrick, Stephen W. Fardo, Vigyan (Vigs) Chandra and Brian W. Fardo.

Routledge is an imprint of the Taylor & Francis Group, an informa business

ISBN 978-87-438-1004-9 (hardback)
ISBN 978-87-438-1006-3 (online)
ISBN 978-8-743-81005-6 (ebook master)
ISBN 978-87-438-1008-7 (epub)

DOI: 10.1201/9788743810056

Contents

List of Figures

List of Abbreviations

ACSR	aluminum conductor steel-reinforced
ALU	arithmetic logic unit
AWG	American Wire Gage
BASIC	Beginners All-purpose Symbolic Instruction Code
BCD	binary-coded decimal
BCH	binary-coded hexadecimal
BCO	binary-coded octal
BJT	bipolar junction transistor
CaS	cadmium sulfide
CEMF	counter electromotive force
CPU	central processing unit
CRM	customer relationship management
DIP	dual-inline package
DMM	digital multimeter
DP	definite purpose
DSO	digital storage oscilloscope
EAROM	electrically altered read-only memory
EMT	electrical metallic tubing
EPROM	erasable programmable read-only memory
ERP	enterprise resource planning
FET	field-effect transistor
GPIB	general-purpose interface bus
IC	integrated circuit
IDE	integrated development environment
IEEE	Institute of Electrical and Electronics Engineer
IRED	infrared-emitting diode
JFET	junction field-effect transistor
LASCR	light-activated silicon-controlled rectifier
LDR	light-dependent resistor
LSI	large-scale integration
LVDT	linear variable differential transformer
MMF	magnetomotive force

MOSFET	metal oxide semiconductor field-effect transistor
MPU	microprocessor unit
NC	normally closed
NEC	National Electrical Code
NEMA	National Electrical Manufacturers' Association
NMC	nonmetallic-sheathed cable
NO	normally open
OL	overload protection
PFET	photo-field-effect transistor
PM	preventive maintenance
PROM	programmable read-only memory
PWM	pulse-width modulation
RAM	random-access memory
RL	resistive-inductive
RISC	reduced instruction set computer
ROI	region of interest
ROM	read-only memory
SCARA	Selective Compliance Assembly Robot Arm
SSL	solid-state lamp
UJT	unijunction transistor
USB	Universal Serial Bus
USART	Universal Synchronous/Asynchronous Receiver/ Transmitter
UV	ultraviolet or infrared
VAR	voltampere reactive

UNIT ONE

Electrical Basics

To understand ***Electrical Machinery Control Systems***, you should first review electrical fundamentals. These fundamentals deal mainly with the characteristics and types of electrical systems.

Chapters 1 through 5 provide an important review of electrical basics. You should study these chapters to make sure that you are prepared in terms of the fundamentals to pursue your study of electrical machinery control for industrial automation. Control of electrical equipment has become a very comprehensive area of study. Machines are now controlled by several types of auxiliary controls, such as computer systems and sensors. You should be sure that you have a good understanding of electrical basics presented in these five chapters to improve your competence as a technician.

Chapter 1 includes topics that introduce the basic concepts of electricity. In addition, Chapter 1 discusses the Electrical Systems Model that is used as a framework to organize the textbook.

Electrical tools and safety are discussed in **Chapter 2**. Any area of electrical work requires ability to select tools, use them properly, and to always be safety conscious. You should be able to identify tools and recognize safe practices in the lab or shop.

Chapter 3 reviews basic components, symbols, and measurement methods that are fundamental for all technicians. An understanding of control systems requires an ability to quickly recognize components and symbols. In addition, using the correct measurement tools and techniques enhances the competence of any technician.

Chapter 4 covers the important concepts of direct current (DC), alternating current (AC) circuits, and Ohm's Law. Also covered are series, parallel, and

DOI: 10.1201/9788743810056-1

combination circuits, and an introduction to three-phase alternating current concepts.

A review of magnetism and electromagnetism is provided in **Chapter 5**. Many types of control devices used for electrical control apply magnetic principles. An understanding of these basics allows technicians to perform their jobs in a more confident manner.

1

Introduction to Electrical Systems

Objectives

Upon completion of this chapter, you will be able to:

- Explain the parts of an electrical system.
- Define the terms source, path, control, load, and indicator.
- Define voltage, current, and resistance.
- Define the terms insulator, conductor, and semiconductor.
- Explain electrical current flow.
- Recognize the proper measurement values–volts, amperes, and ohms.

Important Terms

ammeter
ampere (A)
conductor
control
conventional current flow
coulomb
current
electromotive force (EMF)
electron current flow
free electron
indicator
insulator
load
ohm
ohmmeter
open circuit
path
resistance
semiconductor
short circuit
source
system concept
voltage
voltmeter
volt

DOI: 10.1201/9788743810056-2

Introduction

One of the most important areas of electrical knowledge is the study of electrical control. Electrical systems supply the vast need of our country for electrical energy. Industrial automation is dependent on electrical energy. Due to our tremendous energy requirement, we must constantly be concerned with the efficient operation and control of electric machines. This chapter provides a review of the basics of electrical systems so that you can better understand control of automated industrial systems.

This chapter can be reviewed in detail before beginning your study of electrical machinery control for industrial automation, or sections can be studied as necessary for specific applications. You are encouraged to study **Sections 1.1, 1.2, and 1.3.** These sections discuss the "Electrical Systems Model," which is used as a framework for the textbook. These sections discuss the "Electrical Systems Model," which is used as a framework for the textbook.

1.1 The System Concept

Studying a complex field, such as electrical machinery control for industrial automation, by parts poses a problem that is somewhat like a puzzle. In this case, it is difficult to determine the role that each part plays in the operation of a complex system. A picture of the system divided into its essential parts, therefore, becomes an extremely important aid in understanding its operation. The **system concept** will serve as the "big picture" in the study of electrical power. In this approach, a system will first be divided into five essential blocks. The role played by each block becomes more meaningful in the operation of the overall system. After the location of each block has been established, the operation related to each block becomes more relevant. Through this approach, the way in which some of the pieces of electrical systems fit together should be more apparent.

For many years, people have worked with puzzles as a source of recreation. A puzzle contains a number of pieces that must be placed together properly to produce a picture. Each piece plays a specific role in the finished product. When a puzzle is started, it is difficult to imagine the finished product without seeing a picture of it.

1.2 Basic System Functions

The word **system** is commonly defined as an organization of parts that are connected to form a complete unit. There is a wide variety of electrical systems used today. Each system has several unique features, or characteristics,

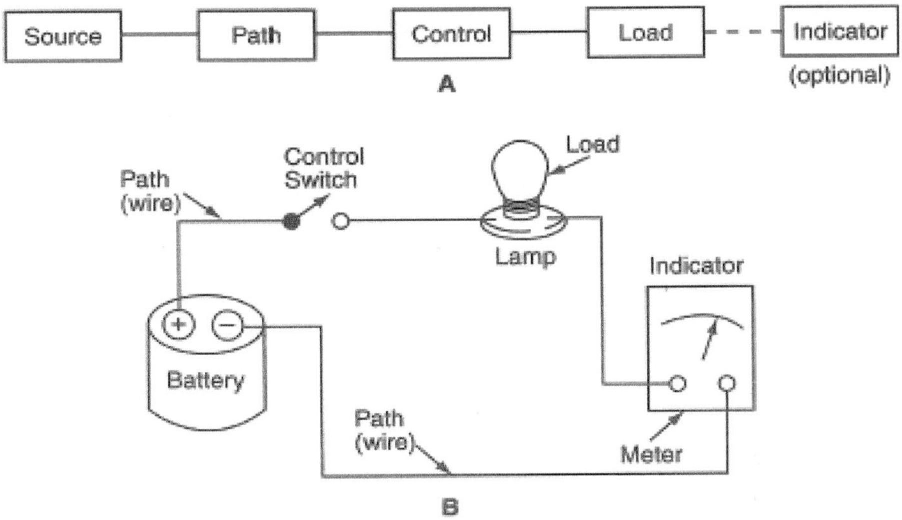

Figure 1.1 A-Block diagram of the parts of an electrical system. B-A simple circuit to show examples of each system part.

which distinguish it from other systems. More importantly, however, there is a common set of parts found in each system. Each of these parts plays the same basic role in all systems. The terms **energy source, transmission path, control, load, and indicator** are used to describe the various system parts. A block diagram of these basic parts of a system is shown in Figure 1.1, along with a simple

Each block of a basic system has a specific role to play in the overall operation of the system. This role becomes extremely important when a detailed analysis of the system is to take place. Hundreds and even thousands of discrete components can be needed to achieve a specific block function. No matter how complex the system, each block must function for the system to operate. Being familiar with these functions and being able to locate them within a complete system is a big step toward understanding the operation of the system.

The **source** for a system (See Figure 1.2) converts energy of one form into a form used by the system. Heat, light, sound, chemical, nuclear, and mechanical energy are considered primary sources of energy. A primary energy source usually goes through an energy change before it is used in an operating system.

The transmission path for a system is somewhat simple when compared with other system functions. The **path** provides a channel for the transfer of energy, see Figure 1.3. It starts with the energy source and continues through

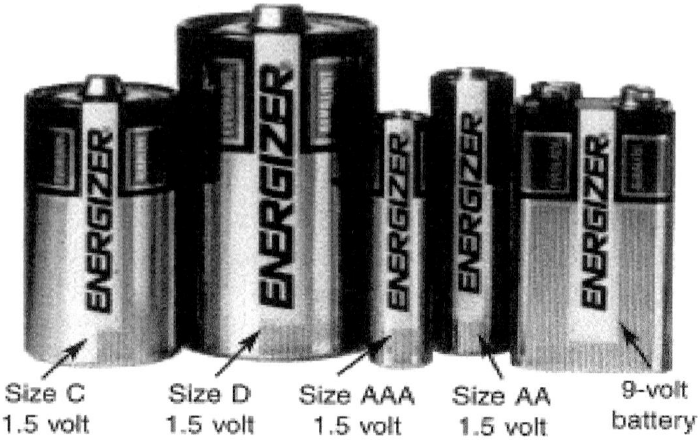

Figure 1.2 Sources of electrical energy. Batteries of various sizes used as energy sources convert chemical energy to electrical energy. (Union Carbide Corp.)

Figure 1.3 Overhead power transmission lines of an electrical power system serve as paths. (Kentucky Utilities Co.)

the system to the load. This path can be an electrical conductor, a light beam, or any other medium between the source and the load. In some systems, there is a supply line between the source and the load. In other systems, there is a supply line between the source and the load and a return line from the load to the source. There can also be several alternate paths within a complete system. These paths can be connected in series to several small load devices or connected in parallel to many independent devices.

The **control** section of a system, Figure 1.3, is by far the most complex part of the entire system. In its simplest form, **control** is achieved when

Figure 1.4 Electrical power controller. Control is made possible by various components and circuits. (Basler Electric)

a system is turned on or off. Control of this type can take place anywhere between the source and the load device. The term **full control** is commonly used to describe this operation. In addition to this type of control, a system can also employ some type of partial control. **Partial control** usually causes some type of operational change in the system other than turning it on or off. Changes in electrical current or light intensity are examples of alterations achieved by partial control.

The **load** for a system refers to a specific part or number of parts designed to produce some form of work, Figure 1.5. Work, in this case, occurs when energy goes through a transformation or change. Heat, light, chemical action, sound, and mechanical motion are some of the common forms of work produced by a load device. As a rule, a very large portion of all energy produced by the source is consumed by the load device during its operation. The load is

Figure 1.5 Electrical load. A welder that converts electrical energy into heat energy to weld metals together. (Miller Electric Mfg. Co.)

typically the most important part of the entire system because of its obvious work function.

The **indicator** of a system is primarily designed to display certain operating conditions at various points throughout the system, Figure 1.6. In· some systems, the indicator is an optional part. In other systems, it is an essential part of the operation of the system. In the latter case, system operations and adjustments are critical and are dependent on specific indicator readings. The term **operational indicator** is used to describe this application. **Test indicators** are also needed to determine different operating values. In this role, the indicator is only temporarily attached to the system to take measurements. Test lights, meters, oscilloscopes, chart recorders, and digital display instruments are some of the common indicators used in this capacity.

1.3 A Simple Electrical System Example

A flashlight is a device designed to serve as a portable light source and as a light source in an emergency. In a strict sense, flashlights can be classified as portable electrical systems. They contain the four essential parts needed to make this classification. Figure 1.7 is a cutaway drawing of a flashlight. Each component is shown associated with its appropriate system block.

The battery of a flashlight serves as the primary **energy source** of the system. Chemical energy of the battery must be changed into electrical

Figure 1.6 Chart recorder that makes a permanent record of some quantity. Courtesy of Gould Inc., Instruments Division.

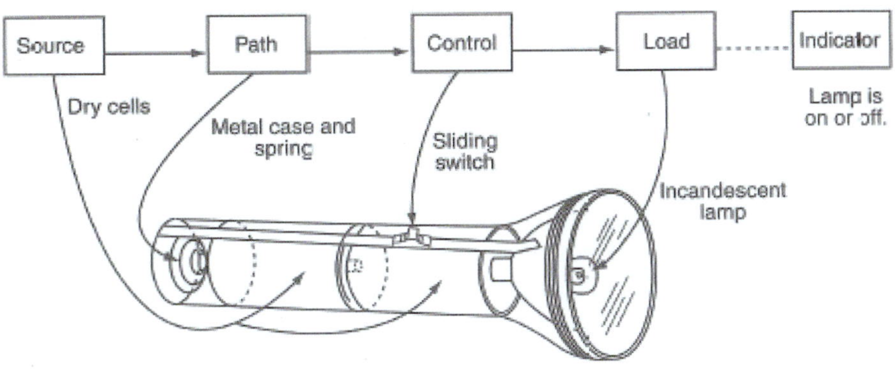

Figure 1.7 Cutaway drawing of a flashlight to show the electrical system concept.

energy before the system becomes operational. The flashlight is a synthesized system because it utilizes two distinct forms of energy in its operation. The energy source of a flashlight is an expendable item. It must be replaced or recharged periodically when it loses its ability to produce electrical energy.

The **transmission path** of a flashlight is commonly achieved through a metal case or through a conductor strip. Copper, brass, and plated steel are frequently used as paths.

The **control** of electrical energy in a flashlight is achieved by a slide switch or a push-button switch. This type of control simply interrupts the transmission path between the **source** and the **load device**. Flashlights are primarily designed to have only full control capabilities. This type of control is achieved manually by the person operating the system.

The **load** of a flashlight is a small incandescent lamp. When electrical energy from the source is forced to pass through the filament of the lamp, the lamp produces a bright glow. Electrical energy is first changed into heat and then into light energy. A certain amount of work is achieved by the lamp when this energy change takes place.

The energy transformation process of a flashlight is irreversible. This process starts at the battery when chemical energy is changed into electrical energy. Electrical energy is then changed into heat and eventually into light energy by the ***load device***. This flow of energy is in a single direction. When light is produced, it consumes a large portion of the electrical energy coming from the source. When this energy is exhausted, the system becomes inoperative.

Flashlights do not ordinarily employ a specific **indicator** as part of the system. Operation is indicated when the lamp produces light. You could say that the load of this system also serves as an indicator. In some electrical systems, the indicator is an optional part.

1.4 Review of Electrical Basics

Electrical **current**, represented by the letter I, is the movement of electrical charges from one point to another. Electrical current is produced when electrons (–) are removed from their atoms. Some electrons in the outer orbits of the atoms of certain materials are easy to remove. A force or pressure applied to these materials causes electrons to be removed. The movement of electrons from one atom to another is called **electrical current flow**.

1.4.1 Conductors

A material through which current flows is called a **conductor**. A conductor passes electrical current very easily. Copper and aluminium wire are commonly used as conductors. Conductors are said to have low **resistance** to electrical current. Many metals are electrical conductors. Each metal has a different ability to conduct electrical current. Copper is used more than any other conductor. Silver is an even better conductor than copper, but silver is too expensive to use in large amounts. Aluminium does not conduct electrical current as well as copper. Yet the use of aluminium is common, since it is cheaper and lighter than other conductors. Materials with only one electron in their outermost orbit, one *valence electron* (silver, gold, copper), are the best conductors.

1.4.2 Insulators

There are some materials that do not allow electrical current to flow easily. These materials are called **insulators**. The electrons of materials that are insulators are difficult to release. The outer orbits of some insulators are full with eight electrons. In others, the outer orbits are over half-filled with electrons. The atoms of materials that are insulators are said to be stable. Insulators have a high resistance to the movement of electrical current. Some examples of insulators are plastic and rubber.

1.4.3 Semiconductors

Materials called **semiconductors** have become very important in electronics. Semiconductor materials are neither conductors and not insulators. They will not conduct electrical current easily and do not insulate well. Their electrical characteristics, like conductors and insulators, depend on the number of electrons the atoms have in their outermost orbit. Semiconductors have four electrons in their outer orbits. Remember that conductors have outer orbits **less than** half-filled, and insulators ordinarily have outer orbits **more than** half-filled.

Figure 1.8 compares electron flow through conductors, insulators, and semiconductors. Some common types of semiconductor materials are silicon and germanium.

1.4.4 Current

Static electricity, electricity at rest, has some practical uses. Still, most of the usefulness of electricity comes from a flow of electrical current. ***Current*** is the movement of electrical charges along a conductor. Electrical current allows us to use electrical energy to do many types of work.

The movement of outer orbit electrons in conductors produces electrical current. The electrons on the outer orbit of the atoms of a conductor are called **free electrons**. Energy released by these electrons as they move allows work to be done. As more electrons move along a conductor, more energy is

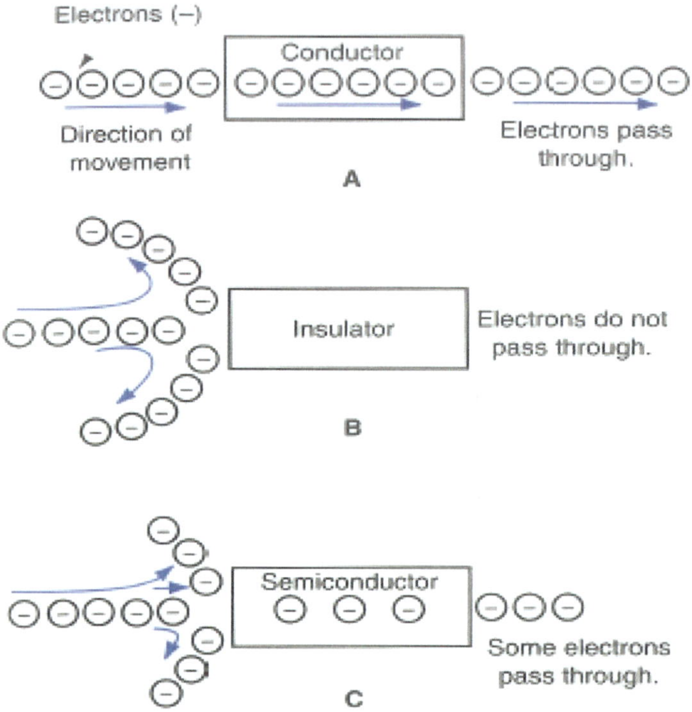

Figure 1.8 Comparison of current flowing through A – conductors, B – insulators, and C – semiconductors.

released. This increases electrical current. The movement of electrons along a conductor is shown in Figure 1.9.

To understand how electron flow takes place, it is necessary to know about the atoms of conductors. Conductors, such as copper, have atoms that are loosely held together. The atoms in copper are connected by ***metallic bonding***. The copper atoms are bonded so close together that their outer orbits overlap each other. Each copper atom has one outer orbit-electron, which is loosely held to the atom. These electrons can move from one atom to another easily. In any conductor, the outer-orbit electrons continually move in a random manner from atom to atom.

This random movement of electrons does not result in an electrical current. Electrons must move in the same direction to form current. If electrical charges are placed on each end ·of a conductor, the free electrons move in one direction. Figure 1.10 shows current flowing through a conductor caused by negative (–) and positive (+) electrical charges. Current flow takes place because there is a difference in the charges at each end of the conductor. Remember that like charges repel and unlike charges attract.

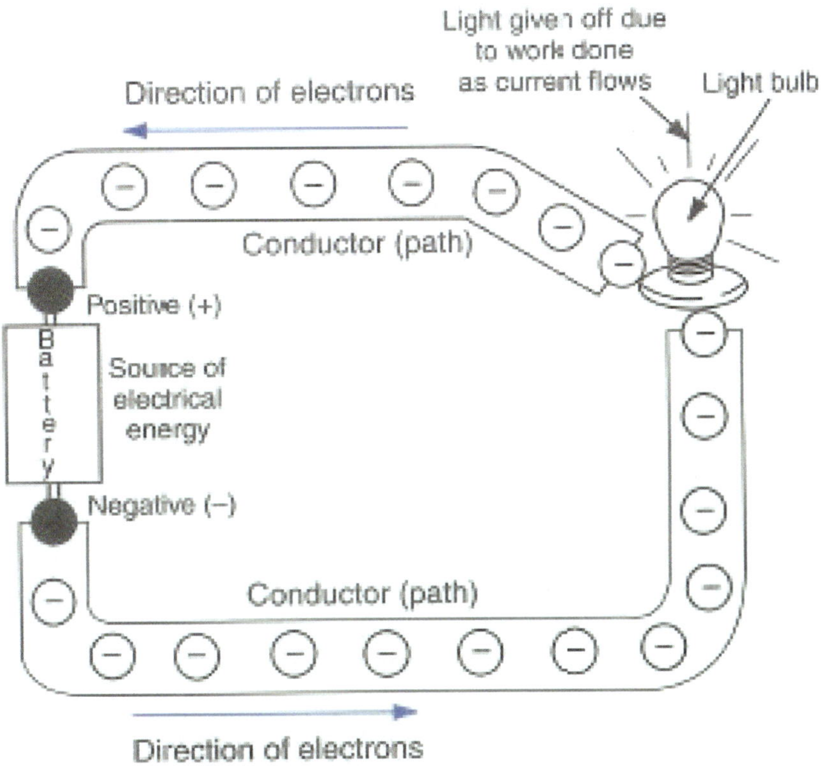

Figure 1.9 Current flowing in a closed circuit.

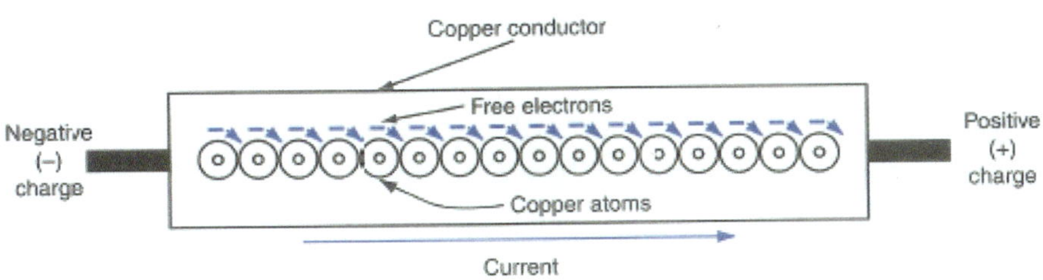

Figure 1.10 Current flowing through a conductor.

1.4.5 Direction of current

Electrical current is the movement of electrons along a conductor. Electrons are negative charges. Negative charges are attracted to positive charges and repelled by other negative charges. Electrons move from the negative terminal of a battery to the positive terminal. This is called electron current flow.

Electron current flow is in the direction of electron movement from negative to positive through a circuit.

Another way to look at electrical current is in terms of charges. Electrical charge movement is from an area of high charge to an area of low charge. A high charge can be considered positive, and a low charge negative. Using this method, an electrical charge is considered to move from a high charge (positive) to a low charge (negative). This is called **conventional current flow**.

Electron and conventional current flow should not be confusing. There are two different ways of looking at the flow of current. One deals with electron movement, and the other deals with charge movement. **In this book, electron current flow is used.**

1.4.6 Amount of current (Ampere)

The amount of electrical current that flows through a circuit depends on the number of electrons that pass a point in a certain time. The **coulomb (C)** is a unit of measurement of electrical current. In electricity, many units of measurement are used. A **coulomb** is a very large quantity of electrons. It contains 6,280,000,000,000,000,000 electrons (6.28×10^{18} in scientific notation). Since electrons are very small, it takes many to make one unit of measurement.

When one coulomb passes a point on a conductor in one second, one **Ampere (A)** of current flows in the circuit. This unit is named for A. M. Ampere, an eighteenth-century scientist who studied electricity. One ampere is a large quantity. Current is commonly measured in units called **milliamperes (mA)** and **microamperes (uA).** These are smaller units of current. A milliampere is 0.001 (1/1000) ampere, and a microampere is 0.000001 (1/1,000,000) ampere.

1.4.7 Current flow compared to water flow

Electrical current is like the flow of water through a pipe. When water flows in a pipe, some force causes it to move. It is the same with an electrical current. The pipe offers opposition, or resistance, to the flow of water. In an electrical circuit, current flows through wires (conductors), which also offer resistance to the flow of current. If the pipe is small, it is more difficult for the water to flow. If the wires are made of a material that has a **high** resistance, it is difficult for current to flow. If the wires have a **low** resistance, it is easier for current to flow in an electrical circuit. In the same way, it is easier for water to flow through a large pipe.

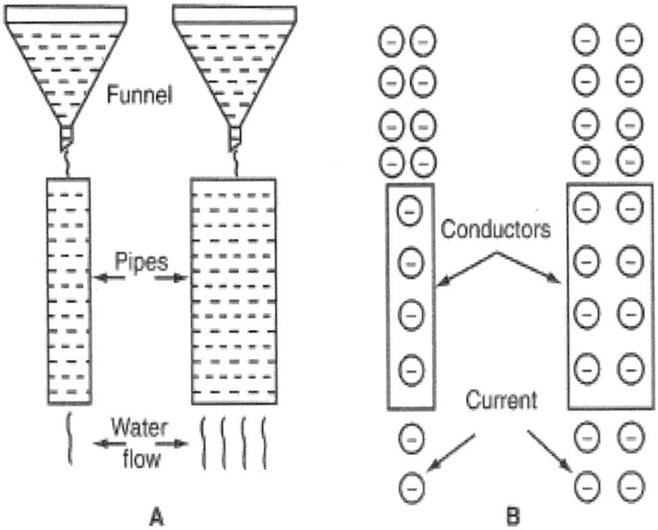

Figure 1.11 Comparison of electrical current and water flow. A-Water pipes. B-Electrical conductors.

Electrical current and water flow are compared in Figure 1.11.

1.4.8 Electrical force (voltage)

Water pressure is needed to force water along a pipe. Similarly, electrical pressure is needed to force current along a conductor. Water pressure is usually measured in pounds per square inch (lb./in.^{2}). Electrical pressure, or **voltage (V or E)**, is measured in **volts (V).** If a motor is rated at 120 V, it requires 120 volts of electrical pressure applied to the motor to force the proper amount of current through it. More pressure would increase the current and less pressure would not force enough current to flow. The motor will not operate properly with too much or too little voltage.

Water pressure produced by a pump causes water to flow through pipes. Pumps produce pressure, which causes water to flow. The same is true of an electrical energy source. A source such as a battery or generator produces current that flows through a circuit. As voltage is increased, the amount of current in a circuit is also increased. **Voltage** is also called **electromotive force (EMF)**. This term is largely responsible for the usage of *E* as an identifying letter for voltage. With the development of solid-state electronics, the letter *E* has other meanings. To avoid duplications, the letter ***V*** is used to identify voltage.

1.4.9 Resistance

The opposition to current in electrical circuits is called **resistance (R)**. Its unit of measure is the **ohm**. Resistance is not the same for all materials. The number of free electrons in a material determines the amount of resistance that material has. Atoms of some materials give up their free electrons easily. These materials offer low opposition to current, or low resistance. Other materials hold their outer electrons and offer high opposition to current, or high resistance.

Electrical current needs a source of electrical pressure to move free electrons through a material. A material will not release electrons until enough force is applied. With a constant amount of electrical force **(voltage)** and more opposition **(resistance)** to the current flow, the number of electrons flowing through a material is smaller. constant voltage, current is increased by decreasing resistance. By increasing or decreasing amount of resistance in a circuit, the amount of current can be changed.

1.4.10 Electrical Circuits

An electrical circuit is a path for an electrical current. Electrical current flows only when it has a complete, or **closed circuit** path. There must be a source of electrical energy to cause current to flow along the closed path. Examine Figure 1.9 once again. It shows a battery used as an energy source to cause ***current*** to flow through a light bulb. Notice that the **path**, or **circuit**, is complete. Light is given off by the light bulb due to the work done as electrical current flows through a closed circuit. Electrical energy produced by the battery is changed to light energy in this circuit.

Electrical current cannot flow if a circuit is ***open***. An ***open circuit*** does not provide a complete path for current to flow. If the circuit of Figure 1.9 became open, no current would flow. The light bulb would not glow. Free electrons of the conductor would no longer move in one direction. An example of an open circuit is when a light bulb burns out. The light bulb's **filament** (the part that produces light) has become open. The open filament of a light bulb stops current flow from the source of electrical energy. This causes the bulb to stop burning or producing light.

Another common circuit term is short circuit. A short circuit can be very harmful. A **short circuit** occurs when a conductor connects directly across the terminals of an electrical energy source. If a wire is placed across a battery, a short circuit occurs. For safety purposes, a short circuit should never happen. Short circuits cause too much current to flow from the source.

If a battery is short-circuited, it would probably be destroyed. In addition, the wire can get hot and possibly melt due to a short circuit.

1.4.11 Voltage, current, and resistance

It is very important to learn some of the basic electrical terms used in the study of electricity and electronics. There are three basic electrical terms that must be understood. These terms are **voltage**, **current**, and **resistance**. We have examined these terms individually; now we will look at them combined.

Voltage is best understood by looking at a battery. Look to the electrical circuit shown in Figure 1.9. The battery is a **source** of voltage. It is capable of supplying electrical energy to a light connected to it. The voltage the battery supplies should be thought of as electrical pressure. The battery has positive (+) and negative (–) terminals.

When the battery or voltage source is connected to the light bulb by conductors, a current flows. Current flows because of the electrical pressure produced by the battery. When the conductor is connected to the lamp, a current flows through the lamp. The current causes the lamp to light. Electrical current is the flow of electrons through a conductor. Electrons move because of the pressure produced by the battery. Remember that electrons have a negative charge (–).

The movement of electrons through a conductor takes place at a rate based on the resistance of a circuit. Resistance is opposition to the flow of electrical current. The conductors and lamp offer resistance to the flow of electrical current. More resistance in a circuit causes less current to flow.

Resistance can be explained by using the example of two water pipes shown in Figure 1.12. If a water pump is connected to a large pipe, such as pipe 1, water flows easily. The pipe offers a small amount of resistance to the flow of water. However, if the same water pump is connected to a small pipe, such as pipe 2, there is more opposition to the flow of water. The water flow through pipe 2 is less.

Inside a lamp bulb, the part that glows is shown in Figure 1.13. The filament offers a small amount of resistance to the flow of current. However, circuit B shows a lamp with a filament of small wire. The small wire has more resistance, or opposition to current. Therefore, less current flows in circuit B because it has a higher resistance.

The terms voltage, current, and resistance are very important. **Voltage** is the electrical pressure that causes current to flow in a circuit. **Current** is the movement of particles called electrons through a conductor. **Resistance** is opposition to current flow in a circuit.

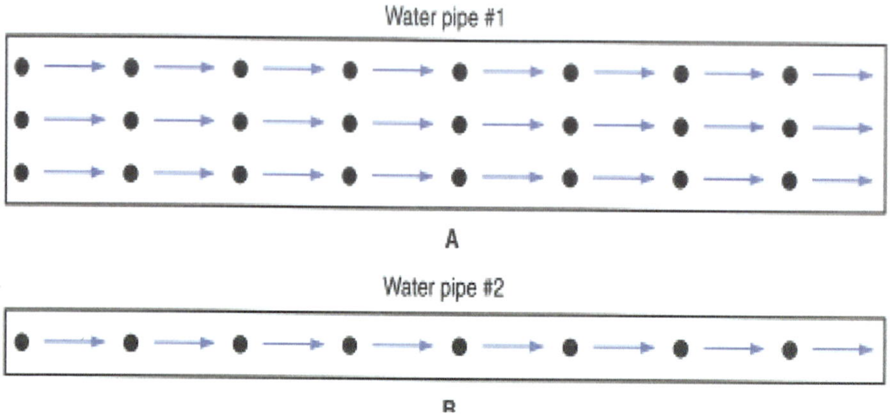

Figure 1.12 Water pipes showing the effect of resistance. A –In water pipe 1, many drops of water flow through. B – In water pipe 2, only a few drops of water flow through.

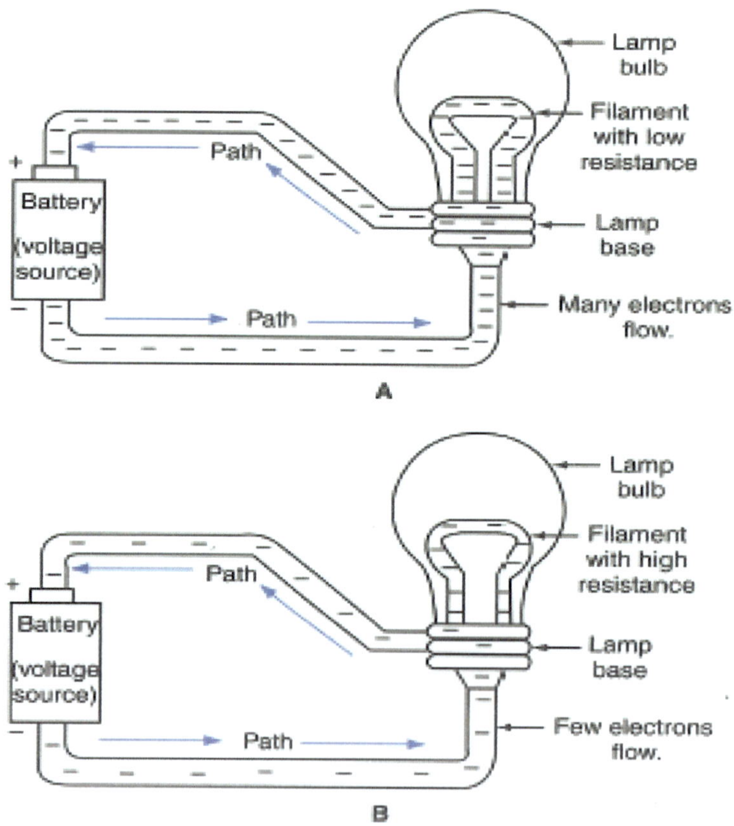

Figure 1.13 How lamp filament size affects current.

Unit	Abbreviation	Unit of Measurement	Symbol
Voltage	*V* or *E*	Volt	V
Current	*I*	Ampere	A
Resistance	*R*	Ohm	Ω
Power	*P*	'Watt	W

Figure 1.14 Basic electrical units of measurement.

1.4.12 Volts, ohms, and amperes

The similarities between water systems and electrical systems can help in understanding basic electrical quantities. The volt (unit of electrical pressure) is compared to the pressure that causes water to flow in pipes. Since the volt is a unit of electrical pressure, it is always measured across two points. An electrical pressure of 120 V exists across the terminals of electrical outlets in the home. This value can be measured with an electrical instrument called a **voltmeter**.

The **ampere** or amp is a measure of the rate of flow of electrical current or electron movement. Electrical current is similar to the rate of flow of water in a pipe. Water flow is measured in gallons per minute. An ampere is the number of electrons per unit of time flowing in an electrical circuit. An **ammeter** is used to measure the number of electrons that flow in a circuit.

The number of electrons (current) that flows is limited by the path's resistance. This resistance is similar to restrictions in a pipe carrying water. Resistance is measured with an **ohmmeter.** The unit of resistance is the ohm. Figure 1.14 summarizes the basic units of electrical measurement.

Review Questions

Answer the following questions.

1. What are the five parts of an electrical system?

2. The outer orbit of an atom contains what type of electrons?

3. Define conductors, insulators, and semiconductors.

4. Define voltage, current, and resistance.

5. Explain the difference between conventional current flow and electron current flow. Which type is used in this book?

6. What is the relationship between an ampere, a milliampere, and a microampere?

7. How does electrical current flow compare with water flow in a pipe system?

8. What factor determines the resistance of a material?

9. What are the symbols for voltage, current, and resistance?

10. Ohms are measured with a(n)_______ Voltage is measured with a(n) __________ Current is measured with a(n) ___________

2

Electrical Tools and Safety

Objectives

Upon completion of this chapter, you will be able to:

- Recognize safe and unsafe electrical practices.
- Recognize fire hazards.
- Be more "safety aware" in the laboratory or shop.
- Identify electrical hazards.
- Describe the need for using ground fault circuit interrupters (GFCIs).
- Describe the proper use of fuses and circuit breakers in electrical circuits.
- Properly identify all electrical tools and equipment used in your lab or shop.
- Provide a brief description of the proper use of each tool in your lab or shop.
- Describe proper soldering and desoldering techniques.
- Identify types of soldering and desoldering tools and equipment.
- Describe the proper grounding techniques for electrical circuits.
- Remember safety rules for the lab or shop.

Important Terms

Allen wrench
chisel
circuit breaker
cold solder joint
conduit bender
drill
fish tape
fuse
ground-fault circuit interrupter (GFCI)
hammer
heat sink
National Electrical Code (NEC)
pipe cutter

DOI: 10.1201/9788743810056-3

- pliers
- punch
- saw
- screwdriver
- socket set
- solder
- soldering
- terminal connector
- wire strippers
- wrench

Electrical tools and safety are very important in the area of electrical machinery control. The proper tools must be used when working with any electrical system.

Safety must always be kept in mind to avoid injury. Electrical technicians should always be safety-conscious. Since safety is so important, it will be discussed first. Then you will learn to identify electrical tools and equipment.

2.1 Electrical Safety

Electrical safety is very important. Many electrical dangers are not easy to see. For this reason, safety should be based on understanding basic electrical principles. Common sense is also important. The physical arrangement of equipment in the electrical lab or work area should be done in a safe manner. Well-designed electrical equipment should always be used. Electrical equipment is often improvised for the sake of economy. It is important that all equipment be made as safe as possible. This is especially true for equipment and circuits that are designed and built in the lab or shop.

Work surfaces in the shop or lab should be covered with material that is nonconducting. The floor of the lab or shop should also be nonconducting. Concrete floors should be covered with rubber tiles or linoleum. A fire extinguisher that has a nonconducting agent should be placed in a convenient location. Extinguishers should be used with caution, and their proper use should be studied.

Electrical circuits and equipment in the lab or shop should be plainly marked. Several voltage values are ordinarily used with electrical lab work. Outlets require special plugs for each voltage.

Storage facilities for electrical supplies and equipment should be neatly kept. Neatness encourages safety and helps keep equipment in good condition. Tools and small equipment should be maintained in good condition and stored in a tool panel or marked storage area. Tools that have insulated handles should be used. Tools and equipment plugged into convenience outlets should be wired with three-wire cords and plugs. The purpose of the third wire is to protect the user from electrical shocks. This is done by grounding all metal parts of the tool to the outlet.

Soldering irons are often used in the electrical shop or lab. They can be a fire hazard. They should have a metal storage rack. Soldering irons should be unplugged when not in use. Soldering irons can also cause burns if not used properly. Rosin-core solder should always be used in the electrical lab or shop. Adequate laboratory space is needed to reduce the possibility of accidents. Proper ventilation, heat, and light also provide a safe working environment. Wiring in the electrical lab or shop should conform to the specifications of the National Electrical Code. The NEC governs all electrical wiring in buildings.

2.2 Lab or Shop Practices

All activities should be done with low voltages whenever possible. Instructions need to be written, with clear directions, for performing lab activities. All lab or shop work should emphasize safety. Experimental circuits must always be checked before they are plugged into a power source. Electrical lab projects should be constructed to provide maximum safety when used.

A slight shock from an electrical circuit should be a warning that something is wrong. Equipment that causes a shock should be immediately checked and repaired, or replaced. Proper grounding is important in preventing shock.

Disconnect electrical equipment from the source of power before working on it. When testing electronic equipment, such as TV sets or other devices, an isolation transformer should be used. This isolates the chassis ground from the ground of the equipment and eliminates the shock hazard when working with equipment.

2.3 Electrical Hazards

A good **first-aid kit** should be in every electrical shop or lab. The phone number of an ambulance service or other available medical services should also be in the lab or work area. Any accident should be reported immediately to the proper school officials. Instructors need to be proficient in the treatment of minor cuts and bruises. They should also be able to apply artificial respiration. In case of electrical shock, when breathing stops, artificial respiration must be started immediately.

Extreme care should be used in removing a shock victim from the circuit that has caused the shock. If possible, turn off the power first. If the power supply cannot be reached, move the shock victim with an insulated material such as a wood pole. Care must be used so that someone else does not come in contact with the same voltage. It is not likely that a high-voltage shock will occur. However, everyone should know what to do in case of an emergency.

Normally, the human body is not a good conductor of electricity. However, when skin becomes wet, the body is a better conductor. Never work with wet hands or while standing in water.

Safety devices called **ground-fault circuit interrupters (GFCIs)** are now used for bathroom and outdoor power receptacles. They have the potential to save many lives by preventing shock. GFCIs immediately cut off power if a shock occurs. In the United States, the **National Electric Code** specifies where GFCIs should be used.

Electricity causes many fires each year. Electrical wiring with too many appliances connected to a circuit overheats the wires. Overheating can set fire to nearby combustible materials. Defective and worn equipment can allow electrical conductors to touch one another and cause a short circuit, which causes a blown fuse. It could also cause a spark or arc that might ignite insulation or other combustible materials.

2.4 Fuses and Circuit Breakers

Fuses and **circuit breakers** are important safety devices. When a fuse blows, it means that something is wrong in the circuit. Causes of blown fuses could be:

- A short circuit caused by two wires touching.
- Too much equipment on the same circuit.
- Worn insulation allows bare wires to touch grounded metal objects such as heat radiators or water pipes.

After correcting the problem, a new fuse of the proper size should be installed. Power should be turned off when replacing a fuse. Never use a makeshift device in place of a new fuse of the correct size. This destroys the purpose of the fuse. Fuses are used to cut off the power and prevent overheated wires. Some types of fuses are shown in Figure 2.1. Fuses are discussed in **Chapter 6**.

Circuit breakers are now very common. Circuit breakers operate on spring tension. They can be turned on or off like wall switches. If a circuit breaker opens, something is wrong in the circuit. Locate and correct the cause, and then reset the breaker.

Always remember to use common sense whenever working with electrical equipment or circuits. Safe practices should be followed in the electrical

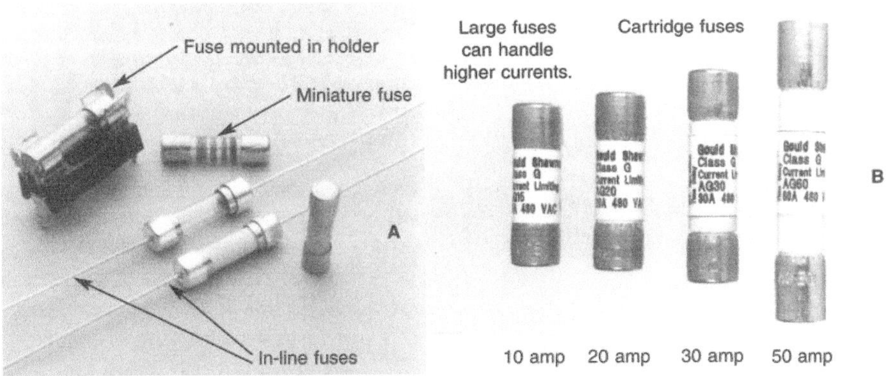

Figure 2.1 Types of fuses. A – In-line fuses. B – Cartridge fuses. (Wickmann USA, Inc., Gould, Inc.)

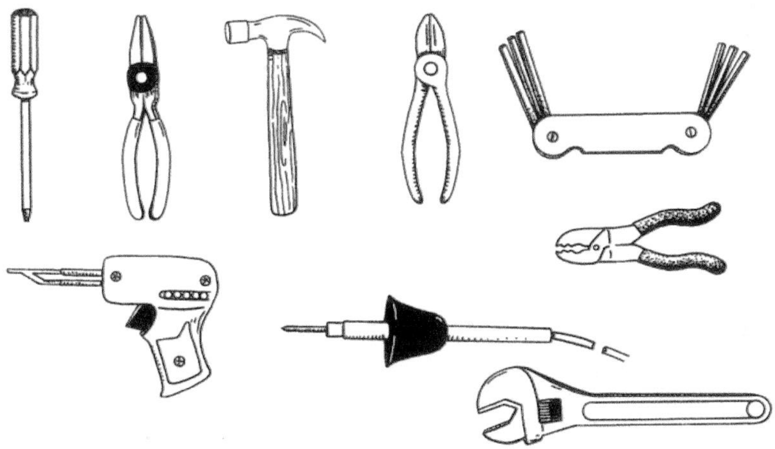

Figure 2.2 Basic hand tools. Can you identify them?

lab or shop as well as in the home. Detailed safety information is available from the National Safety Council in the United States and other organizations. It is always wise to be safe.

2.5 Electrical Tools and Equipment

Take a look at the tools shown in Figure 2.2. How many of these tools can you identify? Do you know the proper use of each tool? It is very important for all electrical technicians to be able to identify tools and know their intended use.

Screwdrivers are a commonly used fastening device. Screwdrivers come in a wide variety of sizes and lengths. They are made with a variety

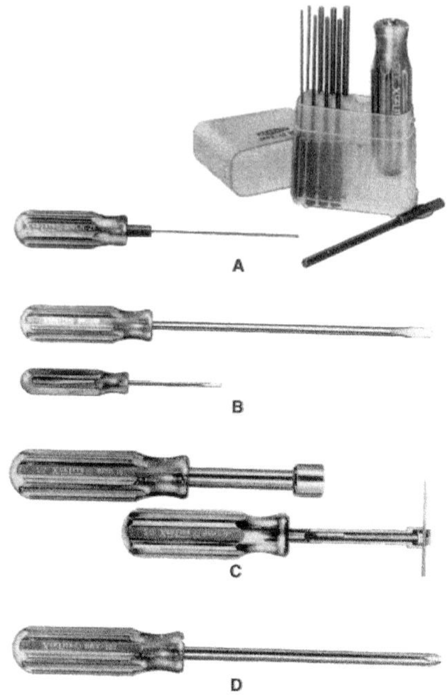

Figure 2.3 Various types of screwdrivers. They are used to tighten and loosen screws in wood, metal, or plastic. A – Allen screwdriver. B – Standard screwdrivers. C – Nut drivers. D – Phillips screwdriver. (The Cooper Group)

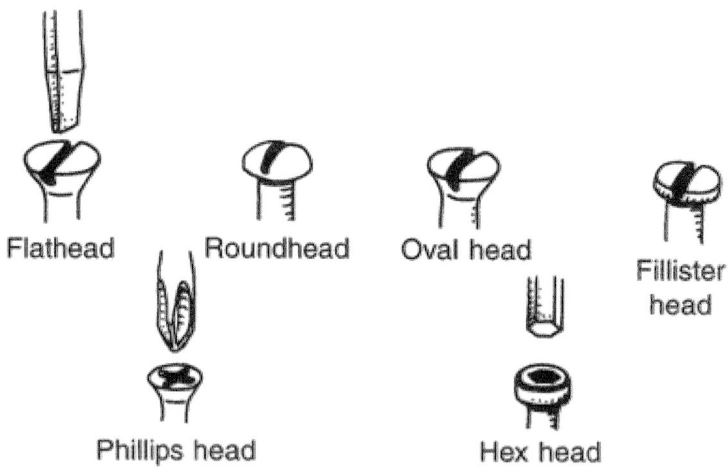

Figure 2.4 Typical head types for screws.

of different tip shapes as shown in Figure 2.3. Each different tip shape is designed to work with a different fastener. Figure 2.4 shows a close-up view of some common screw heads. Offset screwdrivers, Figure 2.5, can be used to drive in screws that are hard to reach. Always use the right size of screwdriver, Figure 2.6.

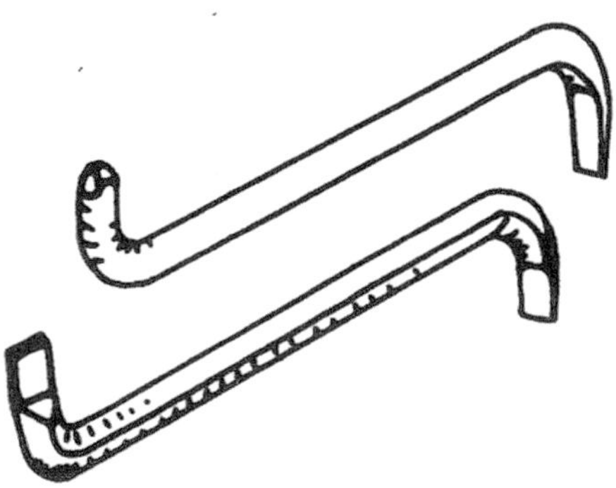

Figure 2.5 Offset screwdrivers are used for driving screws in awkward places.

Figure 2.6 Proper screwdriver head fit. Be sure that the head fits properly.

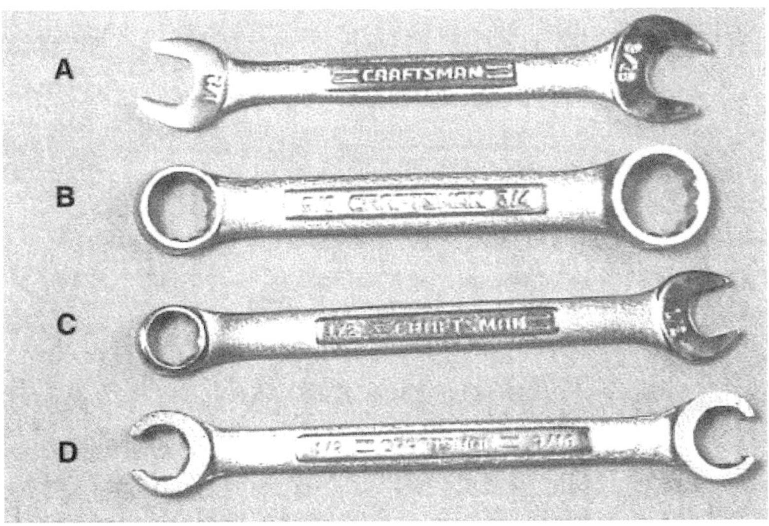

Figure 2.7 A – Open-end wrench. B – Box-end wrench. C – Combination box/open-end wrench. D – Tubing wrench.

Wrenches are tools for use in assembling and disassembling many types of threaded fasteners. As with screwdrivers, they come in a wide variety of shapes and sizes. Some common types of wrenches are shown in Figure 2.7.

An adjustable wrench is useful when a full set of wrenches is unavailable. Figure 2.8 shows the proper use of an adjustable wrench. The wrench should fit snugly on the nut or bolt and the movable jaw should face the direction of rotation. Pipe wrenches, Figure 2.9, are used to grasp and turn round objects or bolts that have had their corners rounded. Pipe wrenches will leave marks and can damage nuts and bolts. They should only be used when other wrenches cannot do the job.

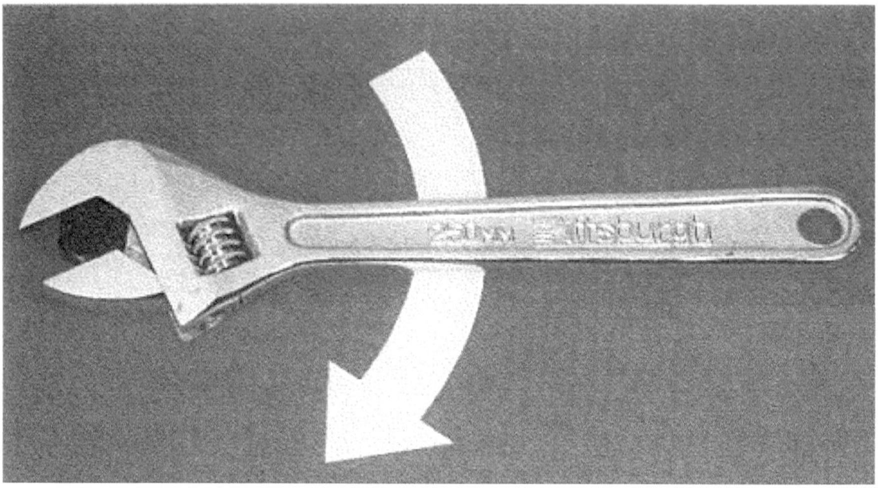

Figure 2.8 Correct way to use an adjustable wrench.

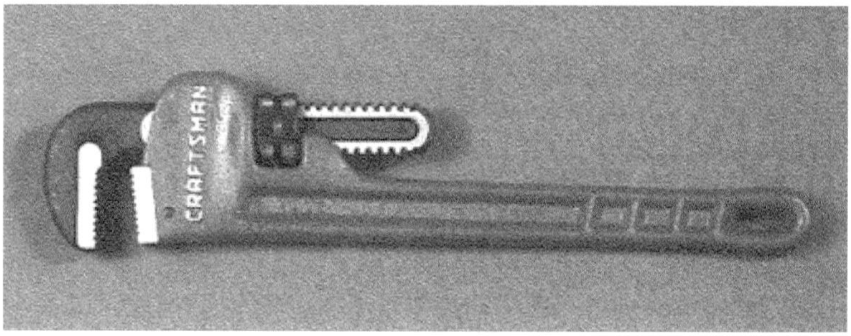

Figure 2.9 Pipe wrenches are able to grasp round objects.

A *socket set*, Figure 2.10, provides a handle and an assortment of sockets that can be used to place or remove nuts and bolts. Socket sets can come with a variety of handles (solid bar or ratchet type) and many different extensions to get to difficult areas. Figure 2.11 shows several sockets, a ratchet-type handle, and two extensions. Figure 2.12 shows several socket head types.

An **Allen wrench**, Figure 2.13, is a fastening tool commonly used with socket-headed fasteners. Allen wrenches come in both inch and metric sizes.

Pliers are tools that are used for holding, bending, and pulling. Many pliers also have cutting edges. Figure 2.14 shows a pair of slip-joint pliers. Figure 2.15 shows two types of **needle nose pliers**. They are useful for holding small and hard-to-reach objects. Round-nose pliers, Figure 2.16, have smooth inside jaws. This prevents nicking wires. **Channel-lock pliers**,

Figure 2.10 Typical socket set.

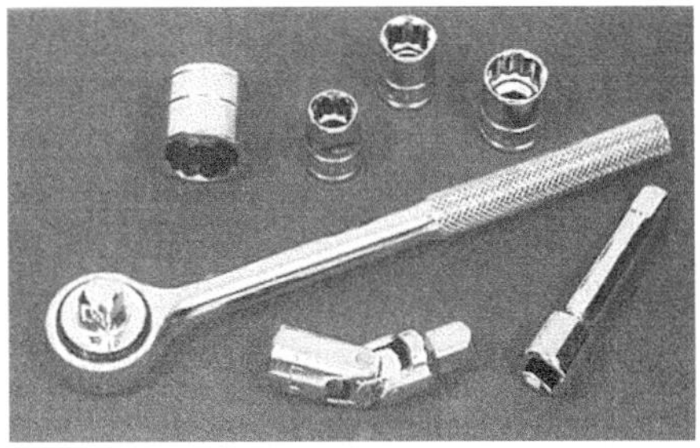

Figure 2.11 Ratchet wrench with sockets and extensions. The extension at the bottom in the center has a universal joint. This allows access to some nuts and bolts that are obstructed.

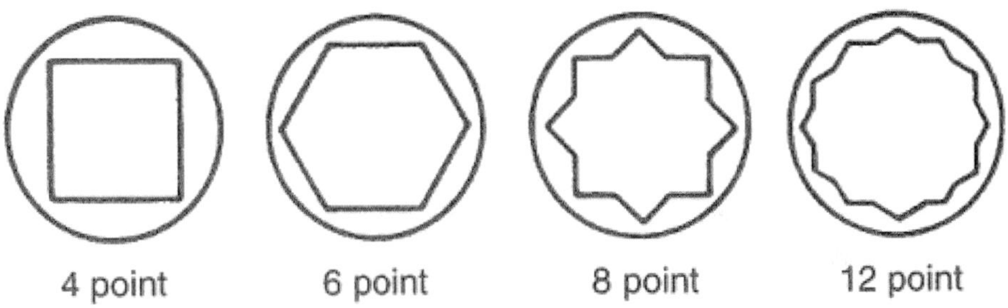

Figure 2.12 Common types of socket heads.

Figure 2.17 are designed to handle objects of various shapes. Locking pliers, Figure 2.18, have an adjustment on one side of the handle that allows them to be adjusted to grasp a wide range of sizes. Squeezing the handle after adjustment locks the pliers. The pliers can be unlocked using the relief lever on the handle. Figure 2.19 shows pliers with a cutting edge. These pliers are called side-cutting pliers.

Cutting and stripping wire are common tasks when working with electricity. Figure 2.20 shows a pair of round-nose **diagonal cutters**. These cutters will cut flush with a surface. Figure 2.21 shows a basic pair of **wire strippers**. This pair has a screw adjustment to set the strippers for various wire diameters. Figure 2.22 shows a more elaborate pair of wire strippers. In addition to stripping wire, this pair can also cut wire and bolts, and it can crimp wire connections.

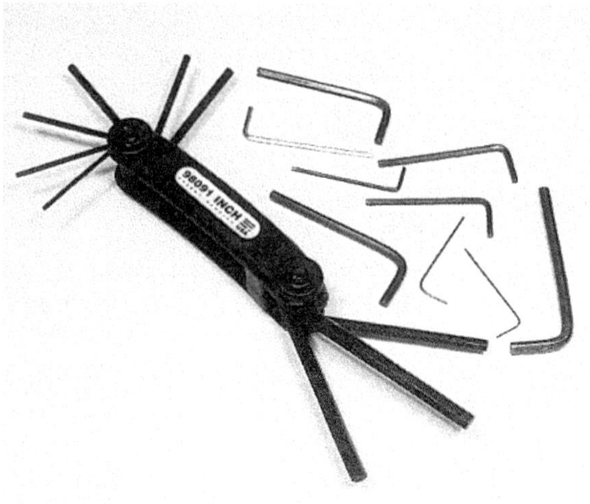

Figure 2.13 Allen (hex) wrenches can be purchased in sets or individually.

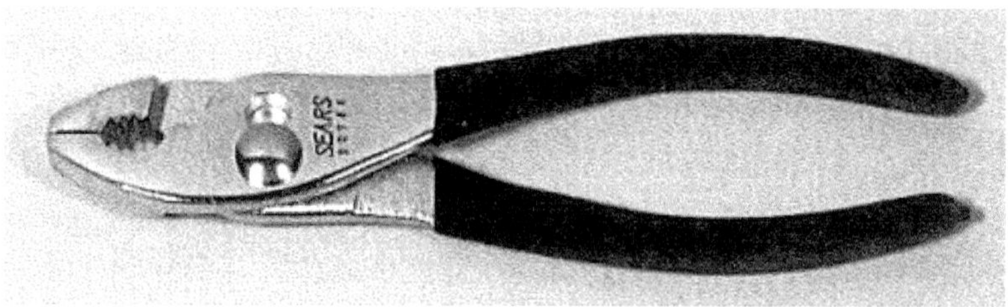

Figure 2.14 Slip-joint pliers have a second jaw position to allow the pliers to grasp larger objects.

After the wires have been cut and stripped, terminal connectors can be applied. **Terminal connectors** allow wires to be quickly connected and disconnected from machinery. Figure 2.23 shows a variety of terminal connectors. Figure 2.24 shows a cutaway of a quick disconnect terminal.

Hammers are used to drive and remove nails as well as for striking chisels and punches. The claw hammer, Figure 2.25, is usually used for nailing purposes. The ball peen hammer, Figure 2.26, is used for striking chisels and punches. Soft face hammers are also used for striking. They are used to prevent sparks and marring the surfaces being struck. Several soft face hammers are shown in Figure 2.27.

There are numerous shapes and sizes of chisels and punches used with hammers. Cold **chisels** are used to cut metal. Wood chisels are made of softer metal and should only be used for cutting wood. Figure 2.28 shows a sketch of different chisels.

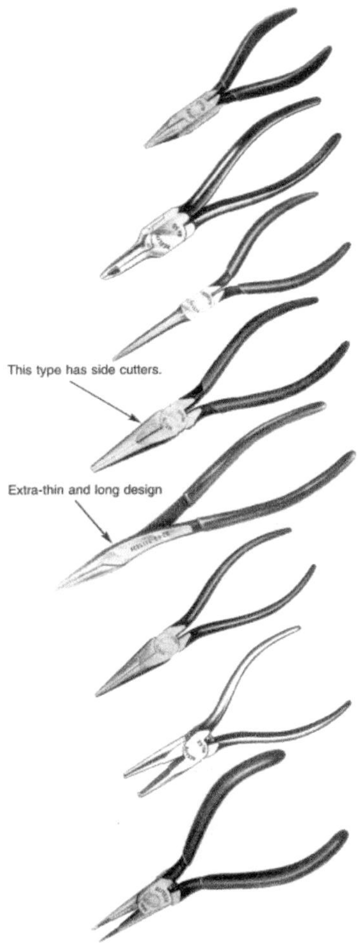

Figure 2.15 This assortment of needle-nose or long-nose pliers is designed for a wide range of applications for the electrician, mechanic, or repairman.

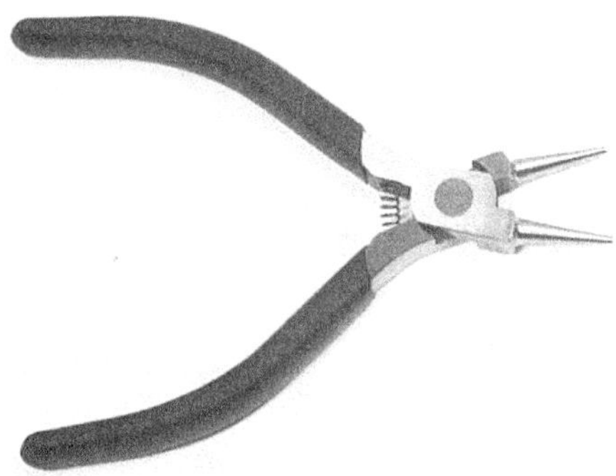

Figure 2.16 Round-nose pliers have smooth jaws so that they do not scratch the objects they grasp.

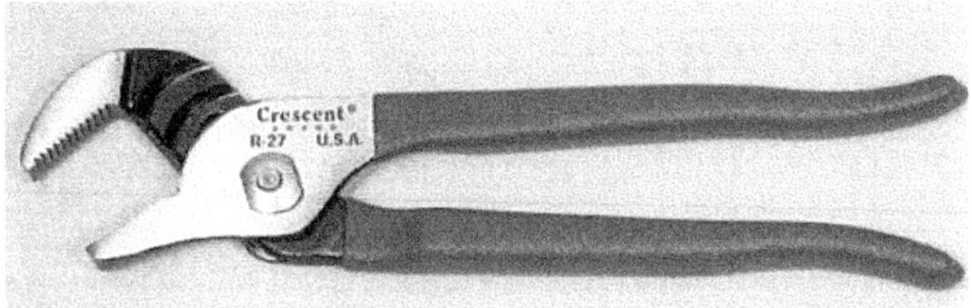

Figure 2.17 Channel-lock pliers can grasp a variety of objects, including pipes, rods, hoses and cable connections, and other mechanical devices.

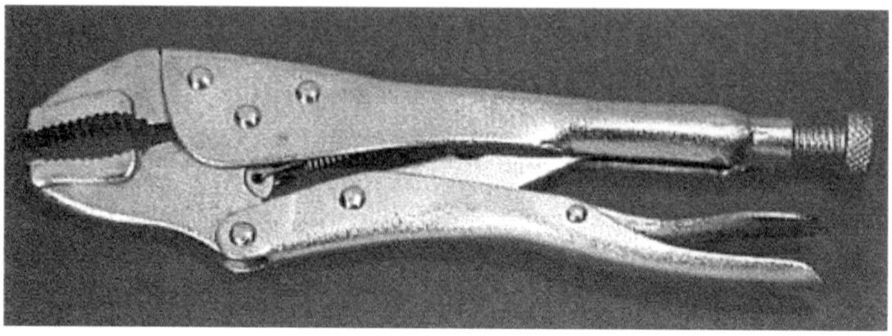

Figure 2.18 Locking pliers. Notice the adjustment on the top handle and the lock release on the bottom handle.

Figure 2.19 Side-cutting pliers are used for both holding and cutting tasks.

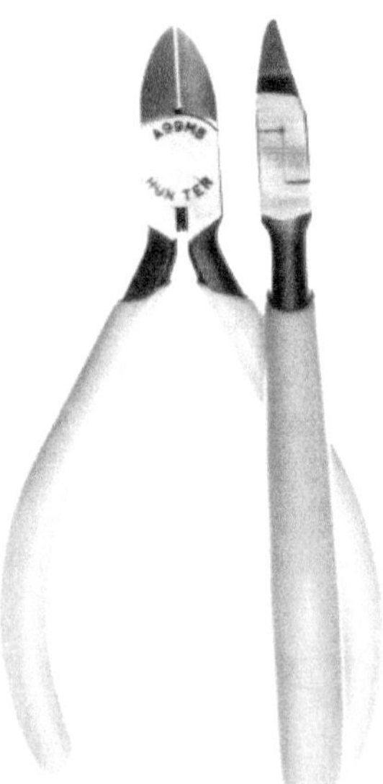

Figure 2.20 Round-nose diagonal cutters are used to cut small-diameter wire. (Hunter Tools, Division of K-D Manufacturing.)

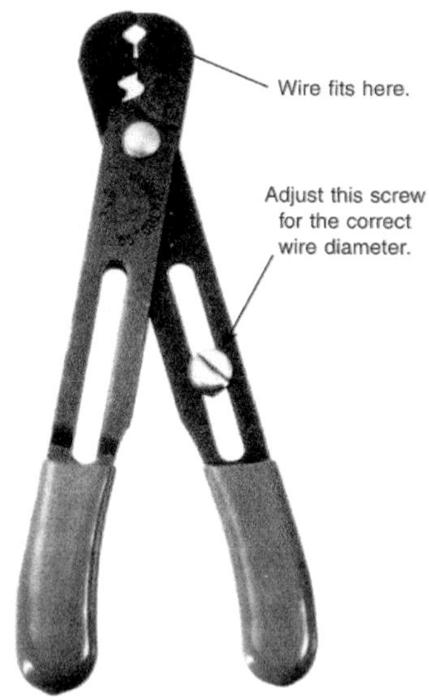

Figure 2.21 Wire strippers are used to strip insulation. (Hunter Tools, Division of K-D Manufacturing Co.)

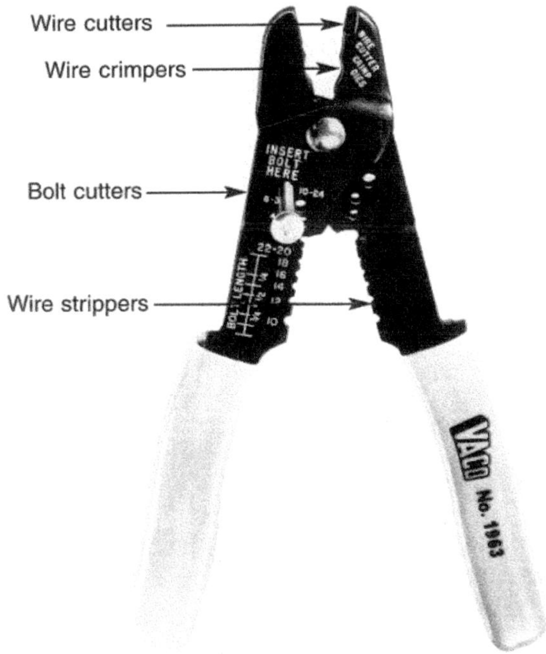

Figure 2.22 Wire crimping tool with stripping and bolt cutting capability. (Vaco Products Co.)

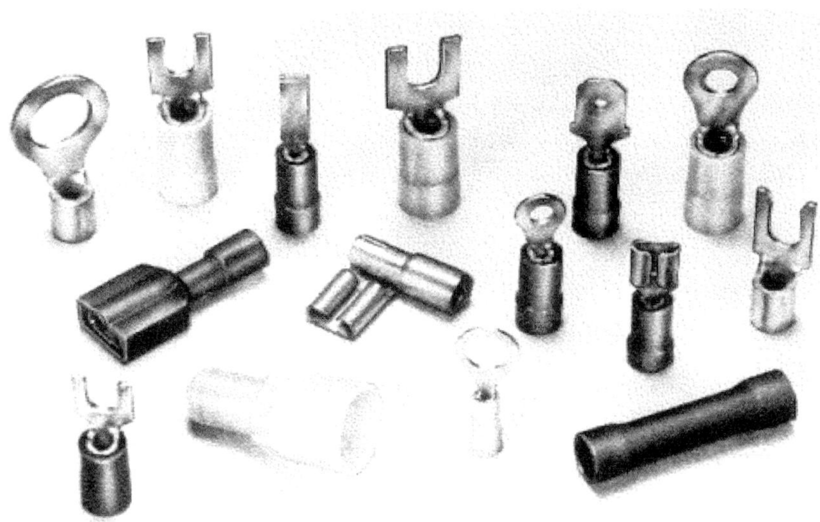

Figure 2.23 Various types of insulated and non-insulated terminal connectors. (Panduit Co.)

Figure 2.24 This quick disconnect is used to connect the wire to the machine terminals. Crimping tools are used to connect the quick disconnect and the proper size wire. (Molex-Etc., Inc.)

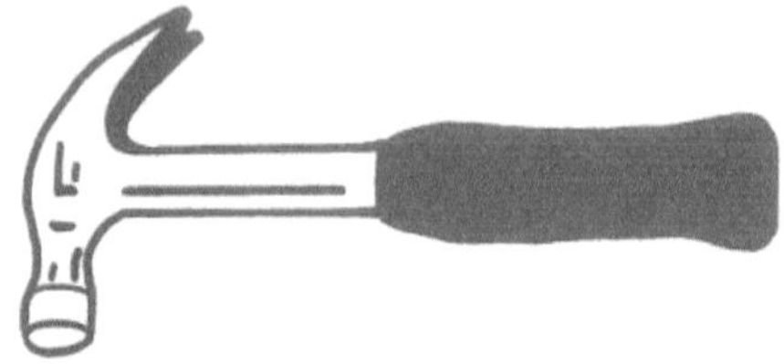

Figure 2.25 Claw hammer. Used to drive and remove nails.

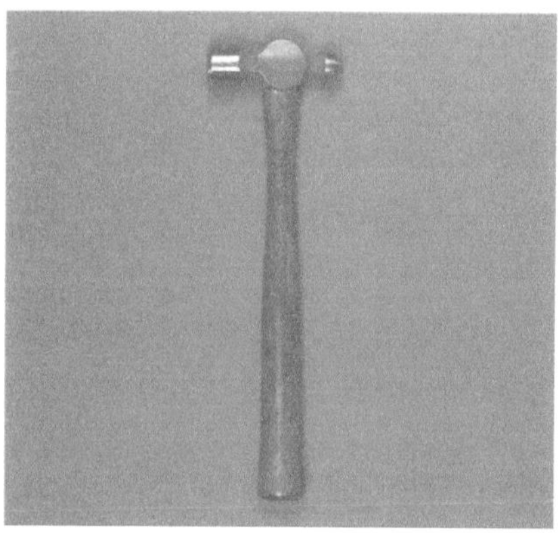

Figure 2.26 Ball peen hammer. Designed for striking chisels and punches. They may also be used for riveting, shaping, and straightening unhardened metal.

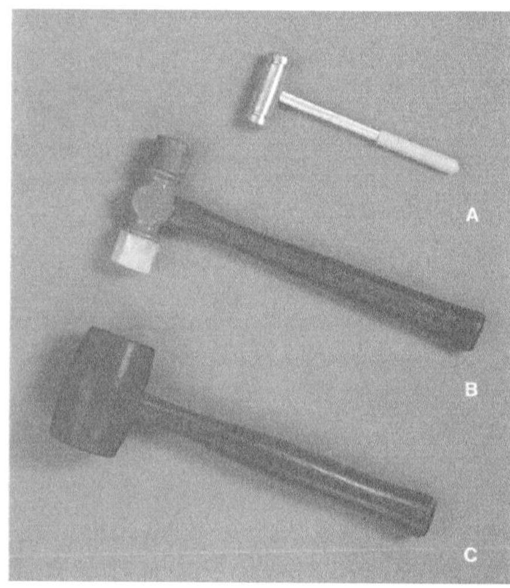

Figure 2.27 Soft face hammers. Non-sparking soft face hammers have no exposed metal. Soft face does not mar surfaces or equipment. A – Brass or lead hammer. B – Plastic-faced hammer. C – Rubber mallet.

Saws are commonly used cutting devices. The **hacksaw** and **coping saw** are two very useful types. Hacksaws, Figure 2.29, are used to cut through metal. Coping saws, Figure 2.30, are used to cut through wood and other softer materials. Coping saws can be used to do finely detailed cutting. Larger projects can require electric saws. **Circular saws, reciprocating saws**, and **band saws** are popular types.

Drills are used for boring holes. Figure 2.31 shows a cordless electric drill. Figure 2.32 shows a selection of drill bits. The choice of bits is

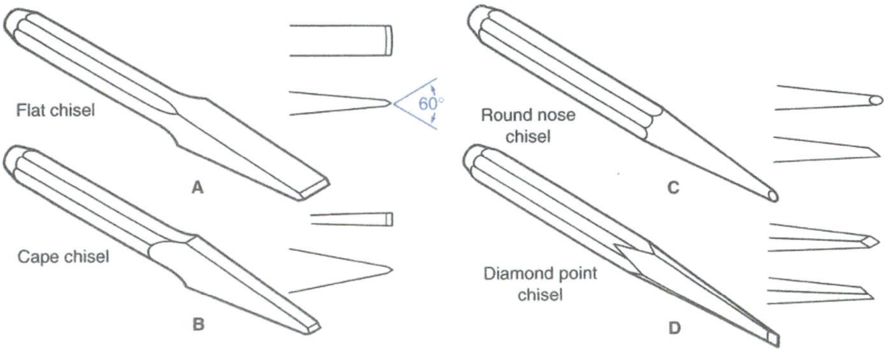

Figure 2.28 Shown are four types of cold chisels.

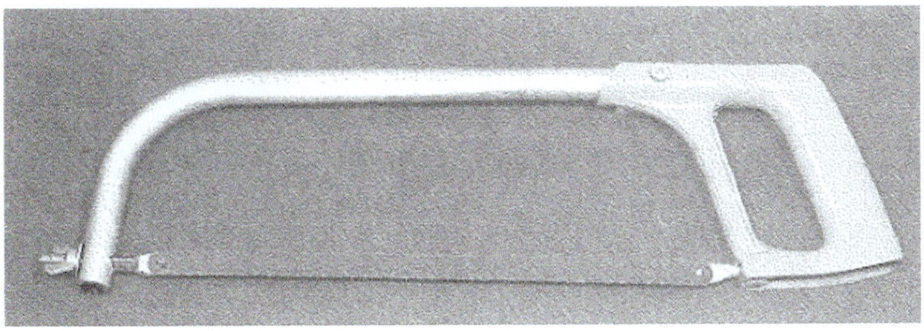

Figure 2.29 This hacksaw can be adjusted to hold blades of many different lengths.

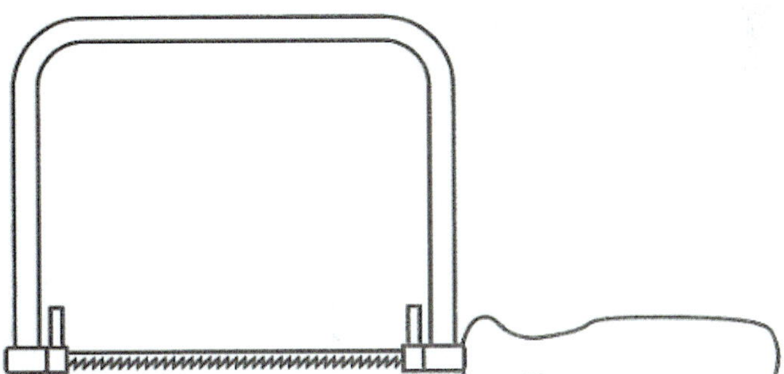

Figure 2.30 The blade of the coping saw is very narrow. Be careful not to twist the blade too far when cutting. The blades break easily.

determined by many factors. These factors include the size of the hole, the depth of the hole, and the type of material being drilled.

Running electrical wires requires special tools. These tools include the **pipe cutter, conduit bender**, and **fish tape**. **Pipe cutters**, Figure 2.33, are used by electricians to cut electrical conduit. The pipe cutter cuts more slowly

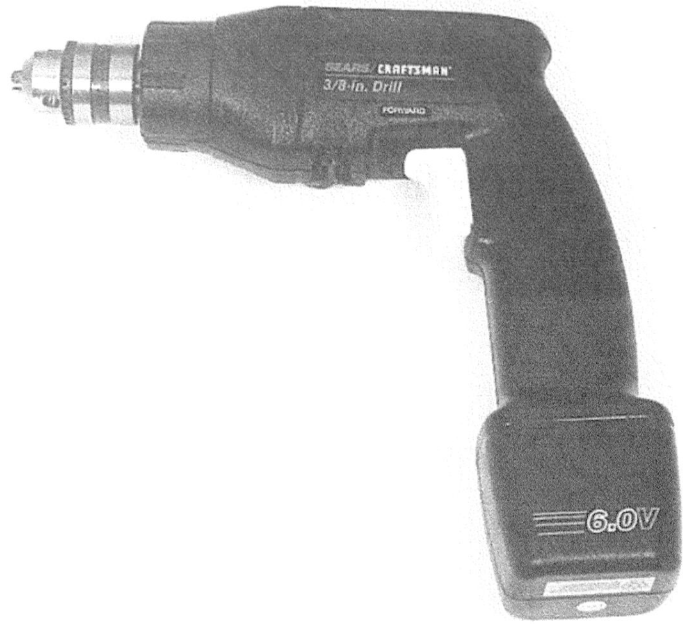

Figure 2.31 Cordless drills are made in a variety of sizes.

than a hacksaw. However, the pipe cutter cuts conduit cleanly, and it can be used on conduit with wires already inside. **Conduit benders**, Figure 2.34, are used to bend electrical conduit around corners. **Fish tape**, Figure 2.35, is used to pull wire through an electrical conduit.

Measuring devices are very important. Figures 2.36–2.39 show some common measuring tools.

Safety should never be ignored. There are a number of safety items that should be used when working with electricity and when working in an industrial setting. Safety glasses and hard hats, Figure 2.40, should always be worn in industrial settings. When working with electrical equipment that has been powered down, lock-out tags and padlocks (Figure 2.41) are important safety tools. They will prevent anybody from turning on the power to the equipment you are working on. Figure 2.42 shows three different types of voltage testers. On the left is the simplest tester. A bulb glows if a voltage is present. The tester in the center is a high-voltage tester. The tester on the right is a receptacle analyzer. It plugs into standard electrical outlets.

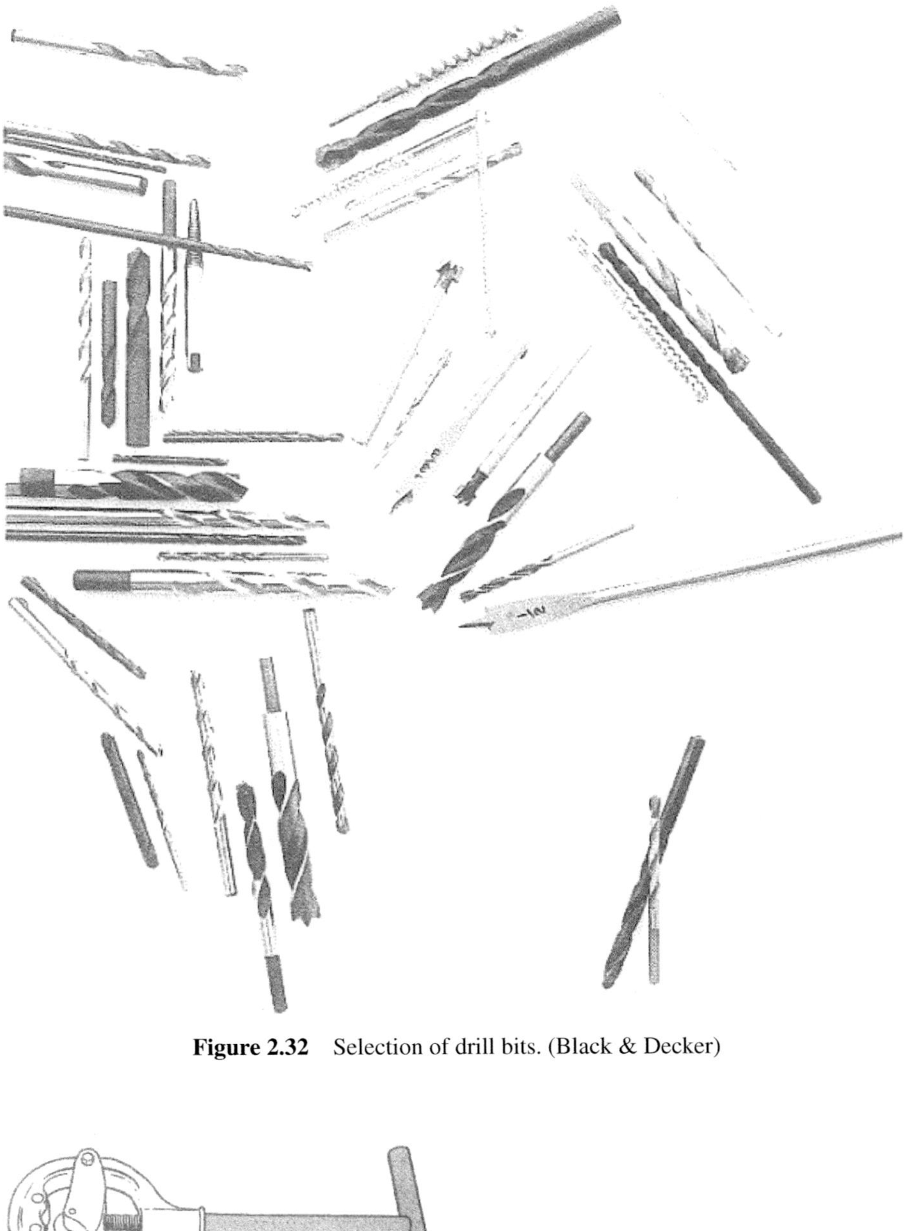

Figure 2.32 Selection of drill bits. (Black & Decker)

Figure 2.33 Pipe cutters safely cut electrical conduit.

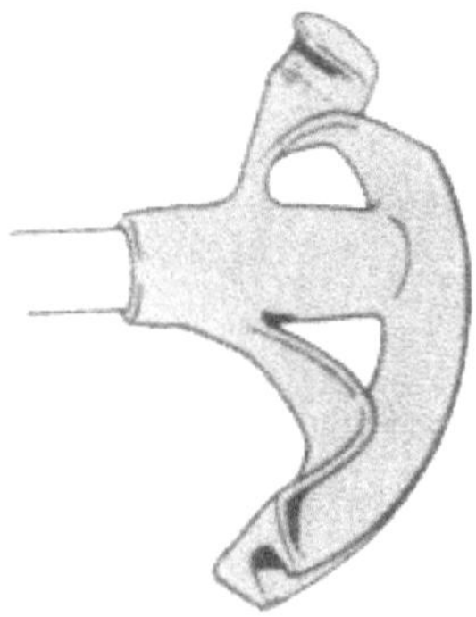

Figure 2.34 Conduit bender. Fitted with a long pipe handle for leverage, it turns metal conduit around a corner.

Figure 2.35 Fish tape. The flat, springy metal band is used to pull, or "fish," cables and wires through enclosed spaces.

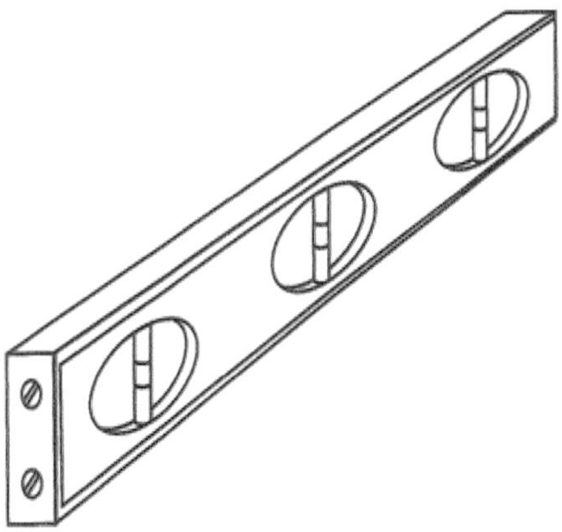

Figure 2.36 Level. Useful in leveling long conduit runs and bus bar installations.

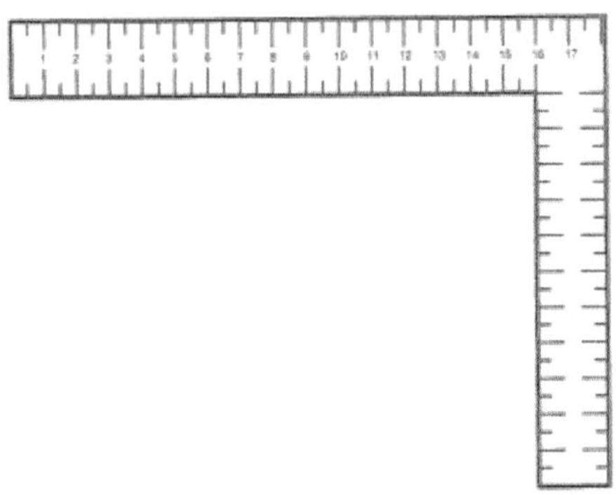

Figure 2.37 Carpenter square.

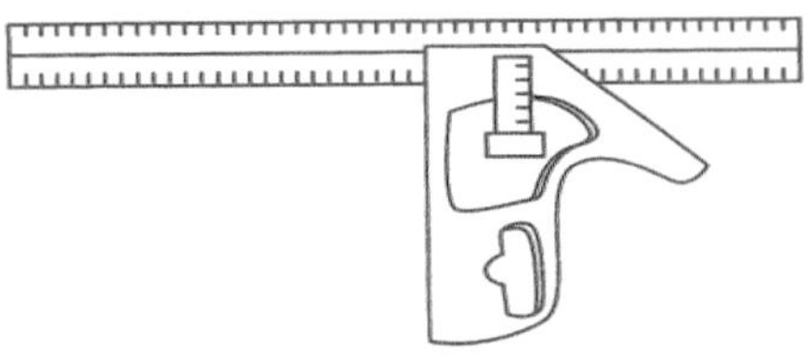

Figure 2.38 Combination square.

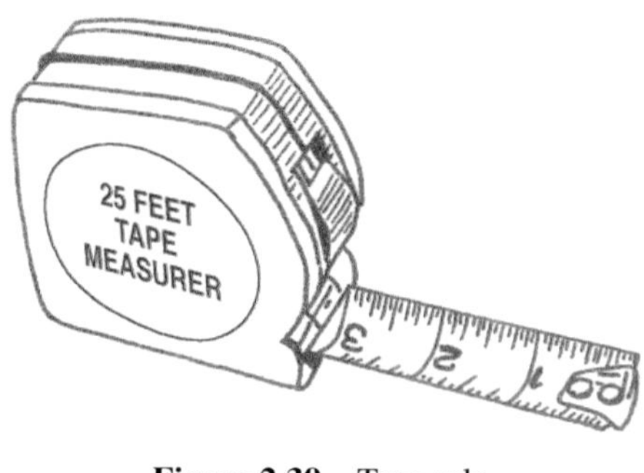

Figure 2.39 Tape rule.

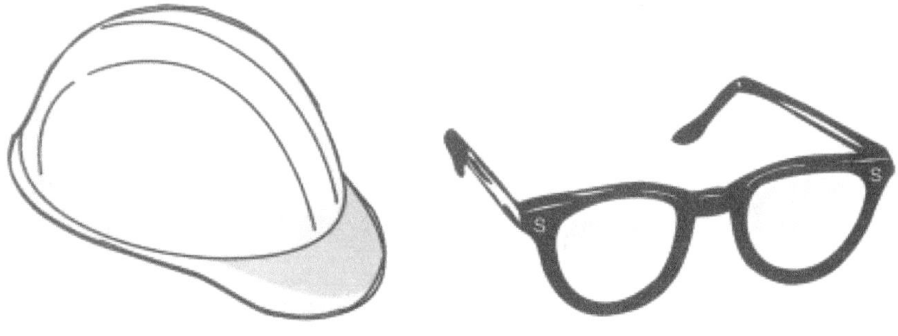

Figure 2.40 A-Safety glasses provide eye protection from flying objects. They are required on all construction sites. B-Safety hard hats are usually made of high-density polyethylene to protect against impact, chemicals, and high voltage. They are also required on all construction sites.

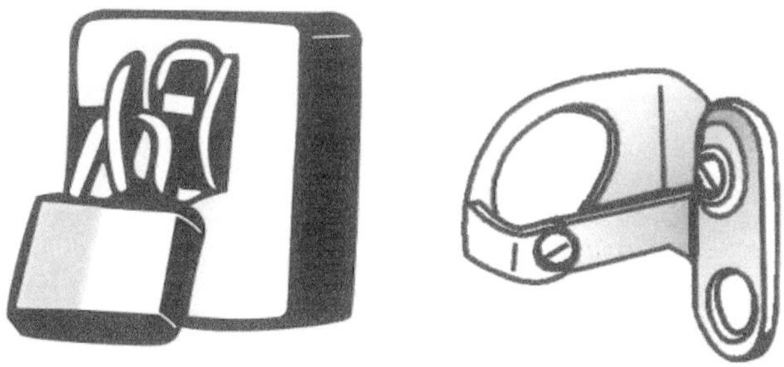

Figure 2.41 A-Padlocks are used to provide lock-out protection for electrical equipment. B-Lock-out tags are used in conjunction with padlocks to warn others that someone is working on the equipment.

2.6 Organizing Your Tools

It is important to use a method to organize your tools so that you can find them when needed. Tools should be marked in some way to show where they belong and to whom they belong. An organized tool storage system provides easy location and a way to protect your tools. Figures 2.43–2.48 show several different ways to organize your tools. These organization helpers include:

- Electrician's belt (Figure 2.43).
- Tool roll storage (Figure 2.44).
- Electronic tool kit (Figure 2.45).
- Mobile storage boxes (Figure 2.46).
- Tool chest (Figure 2.47).

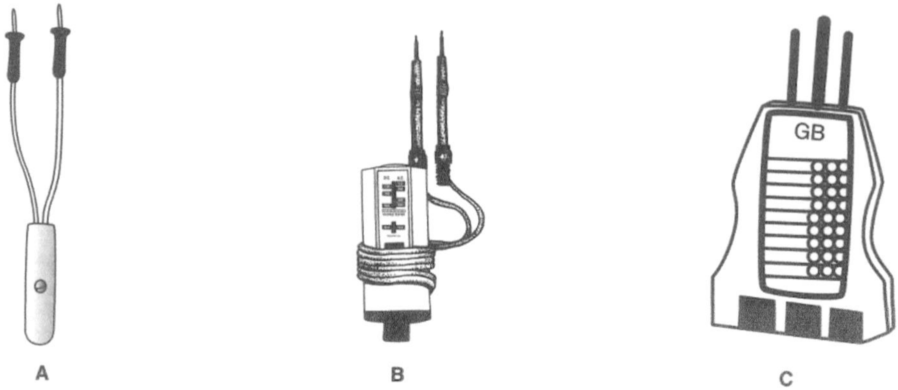

Figure 2.42 A – This **voltage tester** has a neon bulb that glows when the two probes are touched to a live circuit. It is used to check that the current is off. B – This high-voltage tester has insulated probes. It can be used to check voltages up to 600 volts. C – When plugged into a receptacle, a receptacle analyzer tests for electrical faults in the circuit. Three lights at the bottom of the tester glow in different combinations to diagnose problems.

Figure 2.43 Organizing your tools. This electrician's belt allows easy access to tools. (Jensen Tools, Inc.)

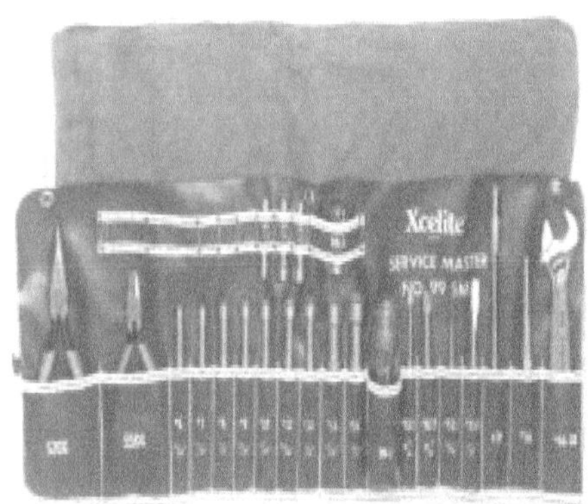

Figure 2.44 Roll of tools for servicing. A good way to organize tools. (The Cooper Group)

- Roller storage cabinet (Figure 2.48).

In addition to the organization methods shown in Figures 2.43 through 2.48, another helpful method is to use a pegboard. It is available in foot sheets for mounting on walls. Tool outlines can be painted or otherwise marked on the pegboard for proper storage.

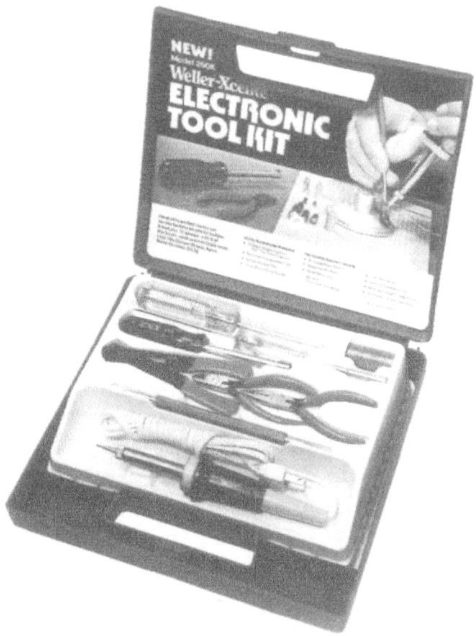

Figure 2.45 Electronic tool kit. Another way to organize tools. (The Cooper Group)

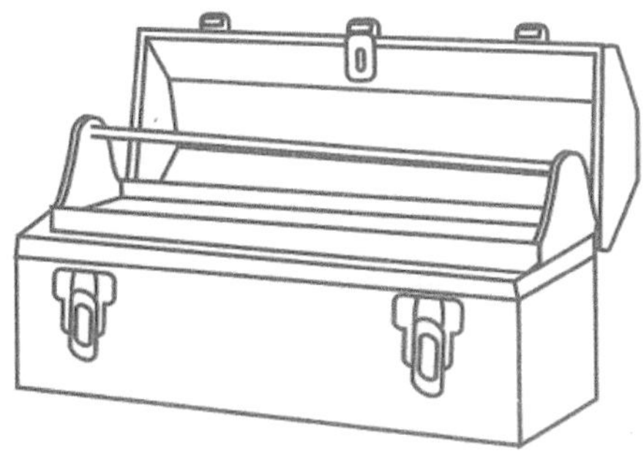

Figure 2.46 Mobile storage box.

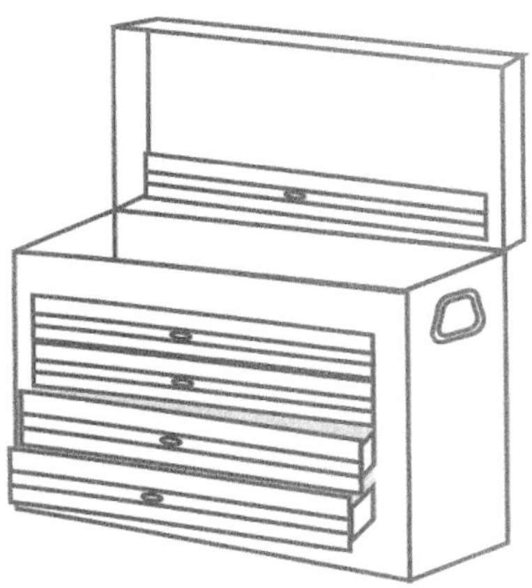

Figure 2.47 Tool chest.

Figure 2.48 Roller cabinet. (Snap-on Tools Corp.)

2.7 Soldering Tools and Techniques

Soldering is an important skill for electrical technicians. Good soldering is important for the proper operation of equipment. **Solder** is an alloy of tin and lead. The solder that is most used is 60 /40 solder. This means that it is

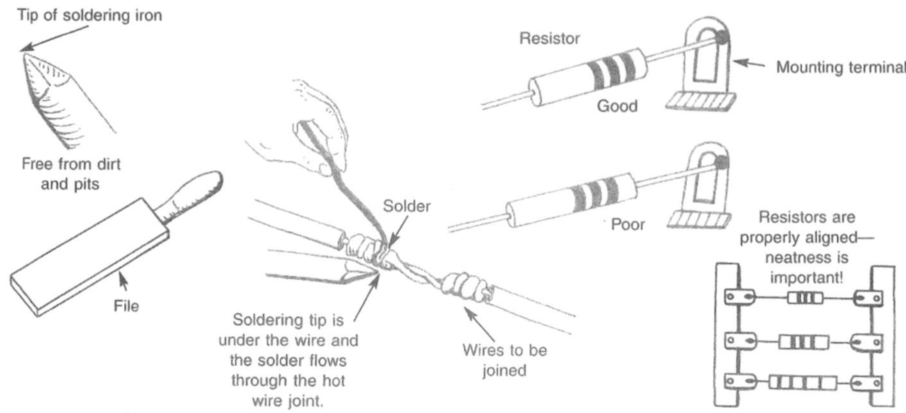

Figure 2.49 Proper soldering technique

made from 60 % tin and 40 % lead. Solder melts at a temperature of about 400 degrees F. (205 degrees C.). For solder to adhere to a joint, the parts must be hot enough to melt the solder. The parts must be kept clean to allow the solder to flow evenly. Rosin flux is contained inside the solder. It is called rosin-core solder.

A good mechanical joint must be made when soldering. Heat is then applied to the joint until the materials are hot. When they are hot, solder is applied to the joint. The heat of the metal parts (not the soldering tool) is used to melt the solder. Only a small amount of heat should be used. Solder should be used sparingly. The joint should appear smooth and shiny. If the joint does not appear smooth and shiny, it could be a cold solder joint. The term **cold solder joint** describes a bad solder connection. Be careful not to move the parts when the joint is cooling. This could also cause a cold solder joint. Some proper soldering techniques are shown in Figure 2.49.

When parts that could be damaged by heat are soldered, be very careful not to overheat them. Semiconductor components, such as diodes and transistors, are very heat-sensitive. One way to prevent heat damage is to use a heat sink, such as a pair of pliers. A **heat sink** is clamped to a wire between the joint and the device being soldered. A heat sink absorbs heat and protects delicate devices. **Printed circuit boards**, such as the one shown in Figure 2.50 are also very sensitive to heat. Care should be taken not to damage PC boards when soldering parts onto them.

A soldering iron and a soldering gun are shown in Figures 2.51 and 2.52. Small, low-wattage irons should be used with PC boards and semiconductor devices.

Figure 2.50 Printed circuit (PC) board soldering. (TRW/UTC Transformers)

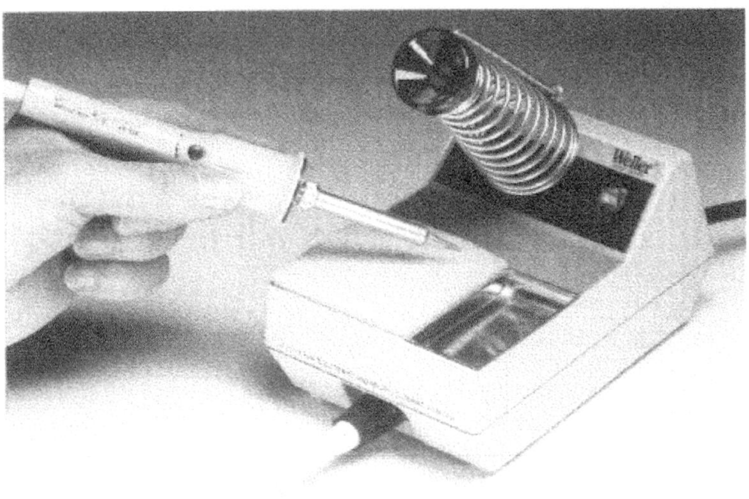

Figure 2.51 Soldering iron with temperature-control station and cleaning pad. (The Cooper Group)

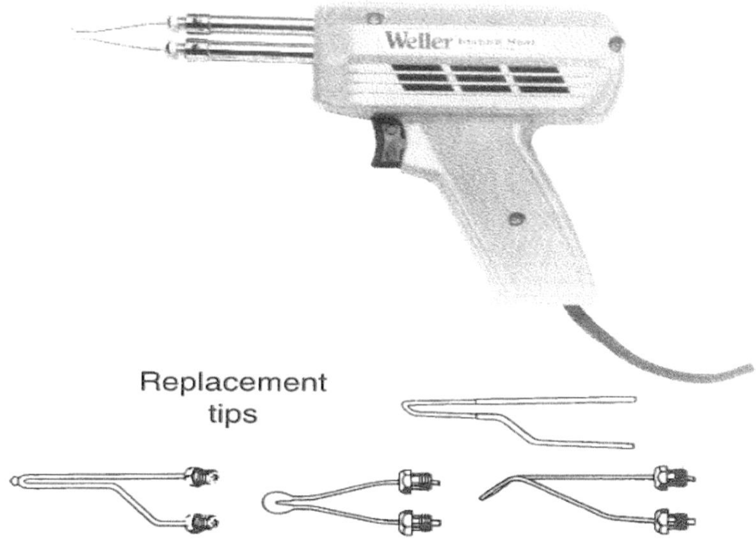

Figure 2.52 High-wattage soldering gun with spade tip. (The Cooper Group)

The following are some rules for good soldering:

1. Use the right kind and size of solder and soldering tools.
2. Be sure that the tip of the soldering iron is clean and tinned.
3. Be sure that all the parts to be soldered are heated. Place the tip of the soldering iron so that the wires and the soldering terminal are heated evenly.
4. Be sure not to overheat the parts.
5. Do not melt the solder onto the joint. Let the solder flow onto the joint.
6. Use the right amount of solder to do the job, but not enough to leave a "blob."
7. Be sure not to allow parts to move before the solder joint cools.

Desoldering tools are shown in Figure 2.53. The removal of electrical components from a printed circuit board is done with a desoldering tool. They can be similar to some of the tools shown in Figure 2.53, with a vacuum bulb to remove solder when heated. They can also be a more elaborate desoldering station with a controlled heat level.

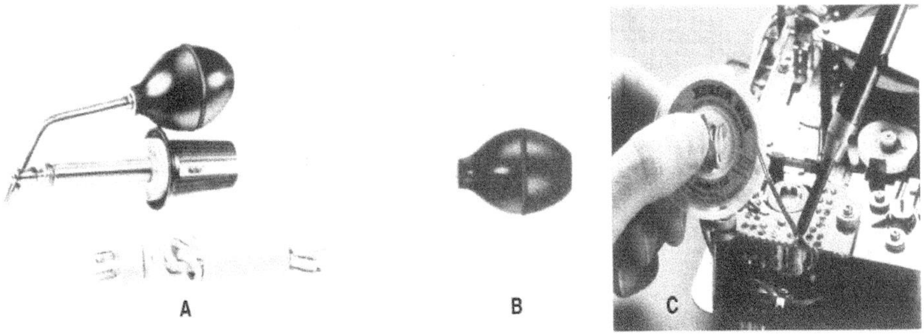

Figure 2.53 Desoldering equipment. A – Desoldering tool. B – Desoldering bulb. C – Desoldering wick.

2.8 Grounding

All power tools and equipment should be properly grounded. Grounding techniques are discussed in detail in **Unit II-Electrical Distribution Basics**. There are a few practical, safety-related items that you should always keep in mind when using power tools. A three-wire conductor should be used with power tools and electrical equipment. A three-prong plug connects the equipment to the ground of the electrical outlet to ensure safety. Equipment grounds reduce the possibility of shock due to electrical shorts or leakage. It is dangerous to use a three-prong to two-prong adapter to plug into an outlet unless a separate ground is connected.

2.9 Safety Rules

Any work area should have strictly enforced safety rules. These rules apply to the use of tools and electrical equipment. Many of them are common sense, but we need to constantly be "safety aware." Here are some safety rules you should remember:

- Always be "safety aware." Be careful and use common sense when using any type of tool or equipment.
- Clean tools after using them to keep them in good condition.
- Check lubrication on tools and equipment and lubricate as needed.

1. Store tools in a safe, protected area.

OSHA SAFETY COLOR CODES	
Red	Fire protection equipment and apparatus Portable containers of flammable liquids Emergency stop buttons and switches
Yellow	Caution and for marking physical hazards Waste containers for explosive or combustible materials Caution against starting, using, or moving equipment under repair Identification of the starting point or power source of machinery
Orange	Dangerous parts of machines Safety starter buttons The exposed parts (edges) of pulleys, gears, rollers, cutting devices, power jaws
Purple	Radiation hazards
Green	Safety Location of first-aid equipment (other than fire fighting equipment)

Figure 2.54 OSHA safety color codes.

2. Wear safety glasses and a hard hat when the work requires them.
3. Pay close attention to your work and do not become distracted.
4. Inspect electrical cords to be sure they are in good condition.
5. Remember the Occupational Safety and Health Administration (OSHA) safety color codes shown in Figure 2.54.
6. Always use the right tools for the job.
7. Learn how to use tools safely and properly.

8. Repair or replace damaged tools immediately-do not use them.
9. Keep sharp-edged tools away from the end of a machine or work area so that they won't fall off.
10. When repairing electrical equipment, be sure that the power is turned off and no one else can turn it on while you are working.
11. Wear proper clothing and protective equipment for the job. This may include safety shoes, hard hat, safety glasses, and tight sleeves and trouser legs. This can also include the absence of neckties, rings, and wristwatches.
12. Be sure that rags containing oil, gasoline, alcohol, or paints are kept in an approved type of container. The chances of a fire are decreased with good, neat housekeeping.
13. Respect all electrical voltages and maintain an awareness of safe practices for electrical equipment.
14. If possible, become trained in first aid and safety.
15. Remember that if you are not safe, you can make it unsafe for others.

Review Questions

Answer the following questions.

1. List five safety practices for an electrical lab.

2. List two electrical fire hazards.

3. Why is it important to use GFCIs?

4. What is the difference in operation between a fuse and a circuit breaker?

5. Describe the following tools and explain how each is used:
   a. Screwdriver.
   b. Wrench.
   c. Socket set.
   d. Allen wrench.
   e. Pliers.
   f. Wire strippers.
   g. Hammer.
   h. Saw.
   i. Chisel.
   j. Pipe cutters.
   k. Conduit bender.
   l. Soldering iron.

6. List six tool storage systems.

7. What is a cold solder joint?

8. List at least five rules for soldering.

9. Why is equipment grounding important?

10. List six safety rules.

3

Electrical Components, Symbols, and Measurement

Objectives

Upon completion of this chapter, you will be able to:

- Diagram a simple electrical circuit.
- Identify schematic electrical symbols.
- Identify different types of resistors.
- Identify resistor value by color code and size.
- Explain the operation of potentiometers (variable resistors).
- Demonstrate how the voltmeter, ammeter, and ohmmeter are connected to a circuit.
- Measure current, voltage, and resistance of basic electrical circuits.
- Explain the operation of potentiometers (variable resistors).
- Demonstrate safety while making electrical measurements.

Important Terms

analog meter	digital meter
block diagram	digital multimeter (DMM)
component	function-select switch
continuity check	multimeter
diagram	ohms adjust

Introduction

Most electrical equipment is made of several parts, or components, that work together. It would be almost impossible to explain how equipment operates without using symbols and diagrams. Electrical diagrams show how the

DOI: 10.1201/9788743810056-4

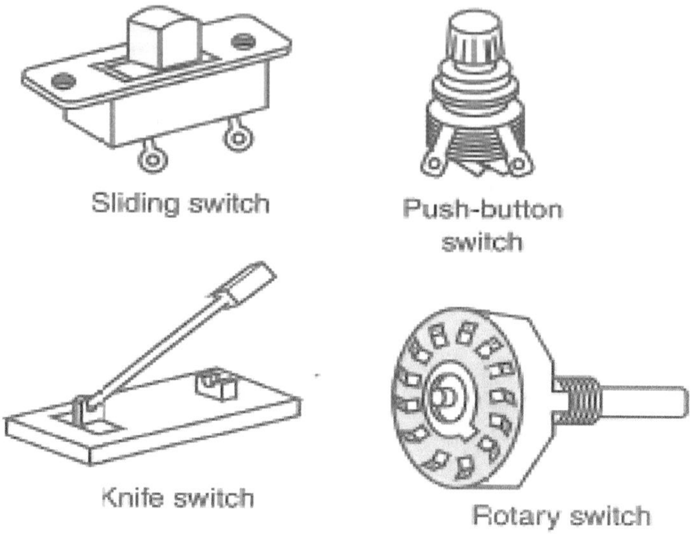

Figure 3.1 Types of switches.

components of equipment fit together. Common electrical components are easy to identify from their symbols. It is easy to learn these symbols. Any technician must be able to identify the components used in electrical circuits. The components of electrical equipment work together to form a system.

Another important activity for electrical technicians is measurement using electrical equipment. Measurements are made on many types of electrical control systems. The proper ways of measuring resistance, voltage, and current are basic for all electrical technicians. These are the three most common electrical measurements.

3.1 Electrical Components, Symbols, and Diagrams

Electrical **components** are the pieces that make up a circuit. Components can be represented by symbols. **Symbols** represent the components in diagrams. A **diagram** shows how the components are connected in a circuit. For example, it is easier to show symbols for a battery connected to a lamp than to draw a pictorial diagram of the battery and the lamp connected together.

3.1.1 Components

A common electrical component is a switch. Several different types of switches are shown in Figure 3.1. The simplest switch is a single-pole,

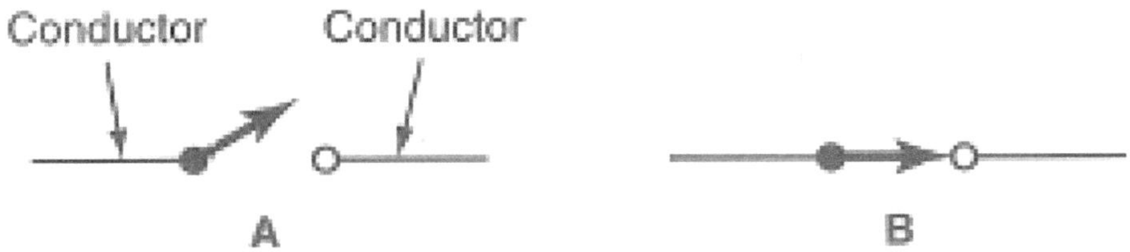

Figure 3.2 Symbol for a single-pole, single-throw (SPST) switch. A – Open or off condition. B – Closed or on condition.

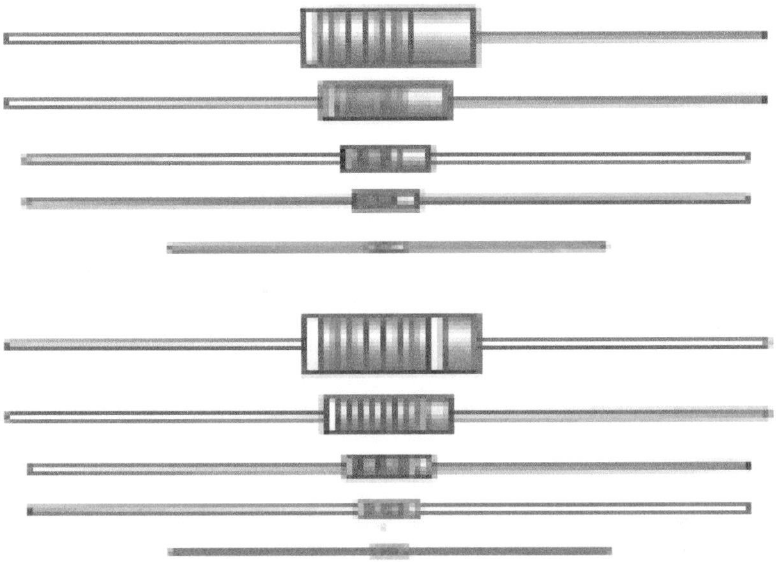

Figure 3.3 These are color-coded carbon resistors; larger sizes are capable of handling higher currents and have higher power ratings.

single-throw (SPST) switch. This switch turns a circuit on or off. In part A of Figure 3.2, the symbol for a switch in the off, or open, position is shown. With the switch in this position, there is no path for current to flow. Part B of Figure 3.2 shows a switch in the on, or closed, position. This switch position completes the circuit and allows current to flow.

In many electrical circuits, a component called a resistor is used. Resistors are usually small, cylinder-shaped components, such as those shown in Figure 3.3. Resistors are used to control the flow of electrical current. A typical color-coded resistor and its symbol are shown in Figure 3.4, The most common type of resistor uses color coding to mark its value. Each color on the resistor represents a specific number. Resistor color-code values are easy

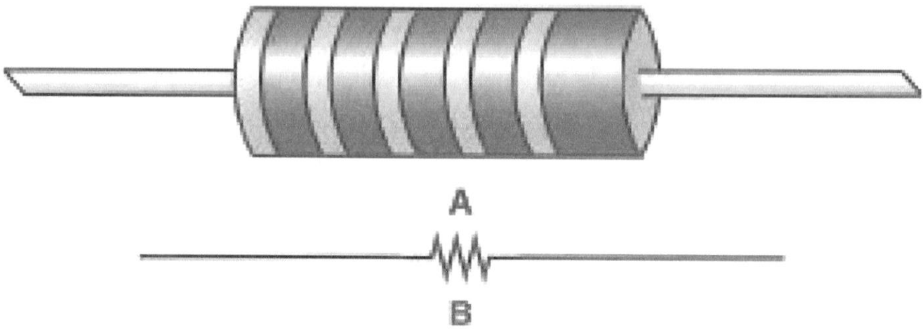

Figure 3.4 A – Color-coded resistor. B – Symbol for a resistor.

to leam. The color code will be covered in Section 3.3. Resistor values are always given in ohms. For instance, a resistor might have a value of 100 Ohms.

Another type of resistor is called a potentiometer, or "pot." A **potentiometer** is a variable resistor whose value can be changed, by adjusting a rotary shaft. For example, a 1000-Ohm pot can be adjusted to any resistance value from zero to 1000 Ohm. by rotating the shaft.

The pictorial representation and symbol of this component are shown in parts A and B of Figure 3.5. In the example shown in part C of Figure 3.5, potentiometer 1 is adjusted so the resistance between points A and B is zero. The resistance between points B and C is 1000 Ohms. By turning the shaft as far in the opposite direction as it will go, the resistance between points B and C becomes zero (see potentiometer 2). Between points A and B, the resistance is now 1000 Ohms. By rotating the shaft to the center of its movement, as shown by potentiometer 3, the resistance is split in half. Now the resistance from point A to point B is about 500 Ohms, and the resistance from point B to point C is also about 500 Ohms.

3.1.2 Symbols

There are several symbols that you should learn to recognize. These symbols are used in many electronic diagrams. These diagrams are used for installing, troubleshooting, and repairing equipment. The use of symbols makes it easy to draw diagrams and to understand the purpose of each circuit. Common symbols are listed in Appendix C.

Most equipment uses wires (conductors) to connect its components together. The symbol for a conductor is a narrow line. If two conductors cross one another on a diagram, it can be seen in the diagram. Part A of Figure 3.6 shows

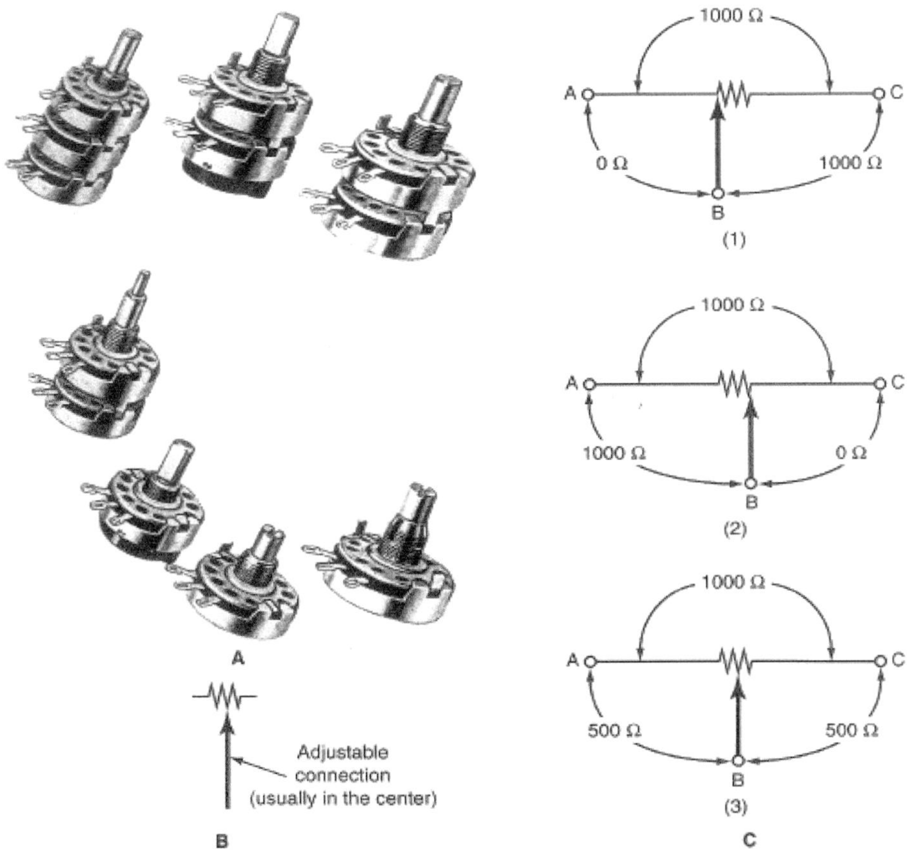

Figure 3.5 Potentiometers. A – Pictorials. B – Symbol. C – Examples.

two conductors crossing one another that do not connect. If two conductors are connected, the diagram will appear as shown in part B of Figure 3.6. Note that the dot indicates the connection. If there is no dot, the wires do not connect.

The symbols for a cell and for a battery are shown in Figure 3.7. The symbol for a single cell is indicated by one set of lines. A battery, which is made up of two or more cells, is shown with two sets of lines. The voltage of a cell or battery is marked near its symbol.

The long line in the symbol is always the **positive (+)** side, and the short line is the **negative (–)** side of the battery. Some common types of batteries are shown in Figure 3.8.

The symbols for two lamps connected across a battery are shown in Figure 3.9 using symbols for the battery and lamps. Notice the part of the diagram where the conductors are connected together.

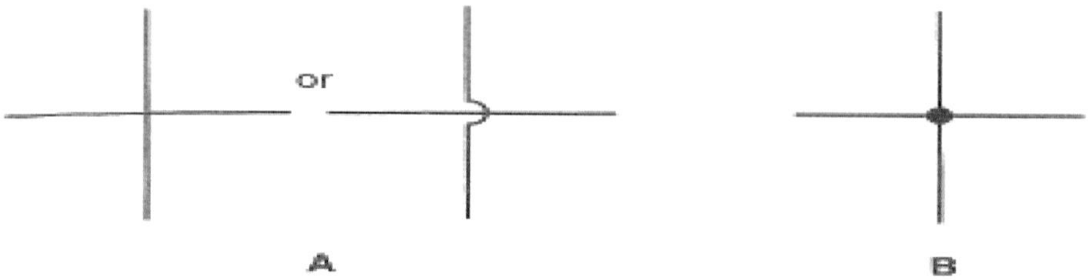

Figure 3.6 Symbols for electrical conductors. A) Conductors crossing. B) Conductors connected.

Figure 3.7 Batteries. A – Single cell. B) Battery.

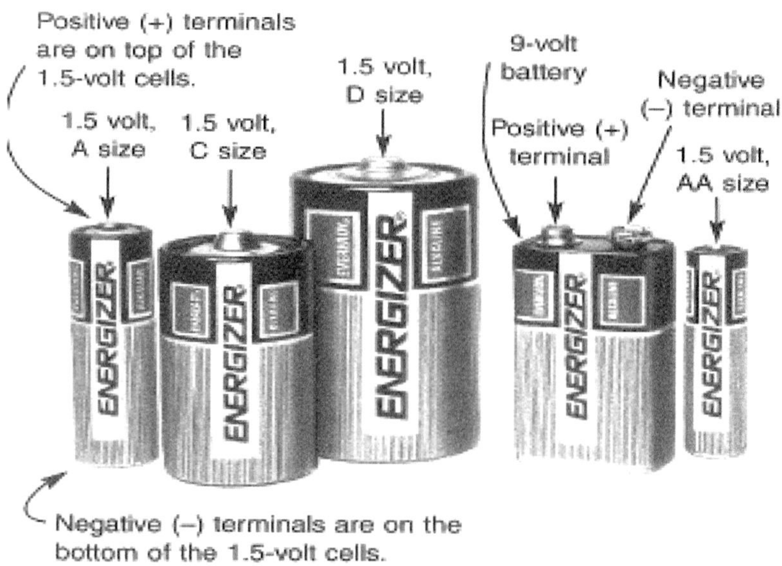

Figure 3.8 Batteries of different sizes and voltage values.

3.1.3 Diagrams

A simple circuit diagram using symbols is shown in Figure 3.10. This is a schematic diagram. There are three common types of diagrams you should be familiar with. They are the schematic diagram, the block diagram, and the wiring diagram.

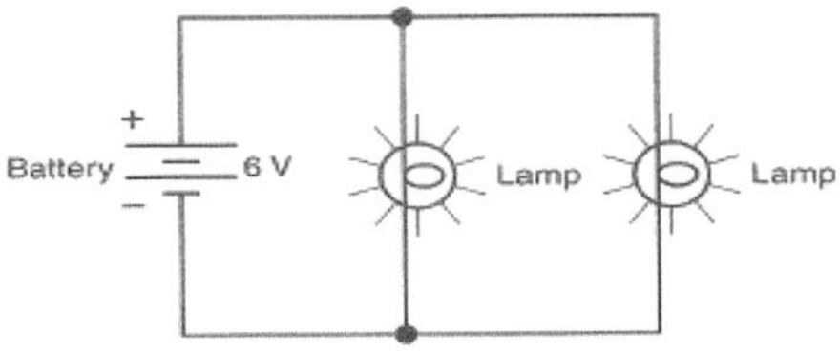

Figure 3.9 Symbols for a battery connected across two lamps.

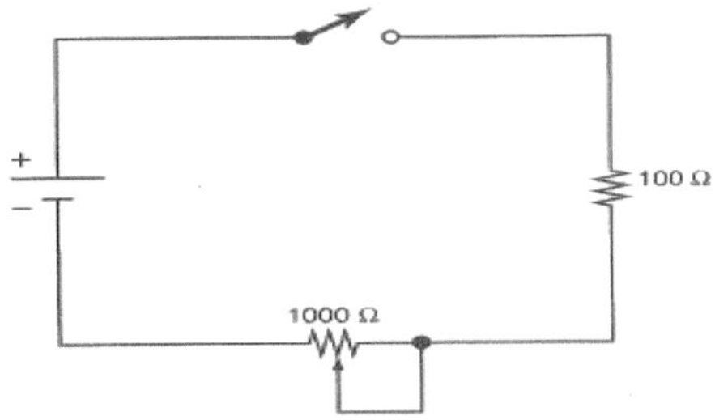

Figure 3.10 Simple circuit diagram.

3.1.3.1 Schematic diagrams

Schematic diagrams, or schematics, are used to represent the parts of equipment or circuits. They show how the components in each circuit fit together. Schematic diagrams are used to show the details of the electrical connections of any type of circuit or system. Schematics are used by manufacturers of electronic equipment showing operation and as an aid in servicing the equipment. A typical schematic diagram is shown in Figure 3.11. Notice the symbols that are used. Symbols are used to represent components in schematic diagrams. Standard symbols are used by all equipment manufacturers. Some common basic electrical symbols are shown in Appendix C. These symbols should be studied for future reference.

3.1.3.2 Block diagrams

Another way to show how electrical equipment operates is to use block diagrams. **Block diagrams** show the functions of the subparts of any electrical system. A block diagram of an **electrical system** was shown in Figure 1.1. The same type of diagram is used to show the parts of a radio in Figure 3.12. Inside the blocks, symbols or words are used to describe the function of the block. Block diagrams usually show the operation of the whole system. They provide an idea of how a system operates. However, they do not show detail like a **schematic diagram**. It is easy to see the major subparts of a system by looking at a block diagram.

3.1.3.3 Wiring diagram

Another type of diagram is called a **wiring diagram** (sometimes called a **cabling diagram**).

Wiring diagrams show the actual location of parts and wires on equipment. The details of each connection are shown on a wiring diagram. Schematic and block diagrams show only how parts fit together electrically. Wiring diagrams show the details of actual connections. A simple wiring diagram is shown in Figure 3.13.

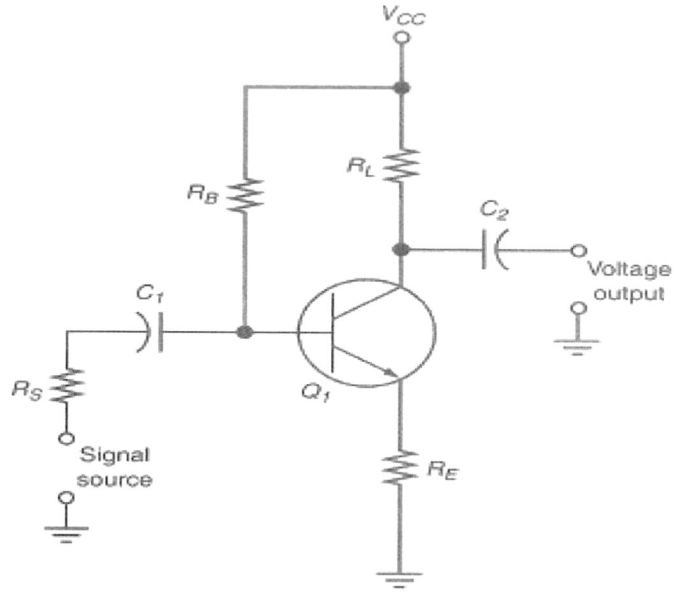

Figure 3.11 Schematic diagram of a transistor amplifier circuit.

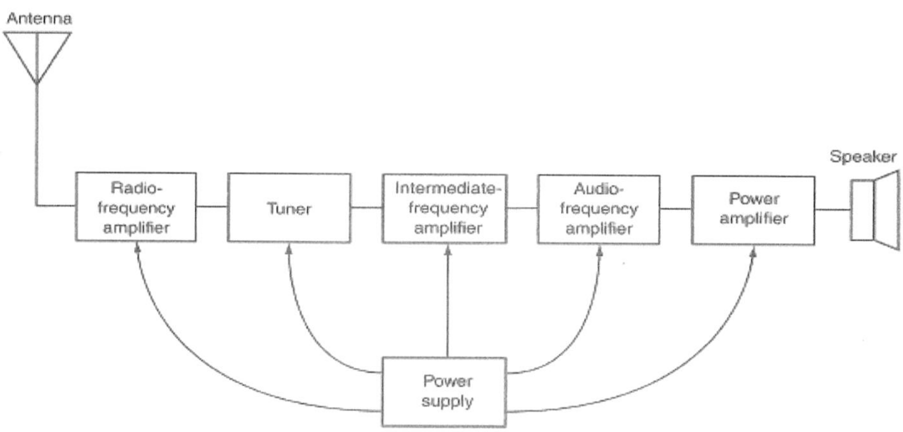

Figure 3.12 Block diagram of the parts of a radio.

3.2 Resistors

A wide variety of **resistors** are used. Most resistors are made of either a special carbon material or metal film. Some types of resistors are shown in Figure 3.14. **Carbon resistors** are constructed of a small cylinder of compressed material. Wires are attached to each end of the cylinder. The cylinder is then covered with an insulating coating. Carbon resistors are used to control small currents.

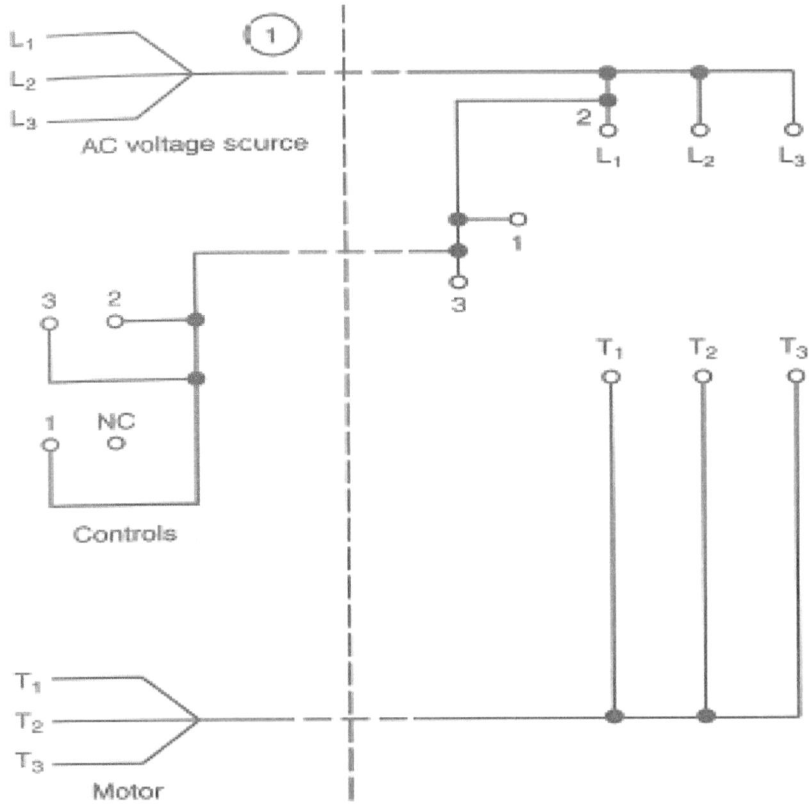

Figure 3.13 Simple wiring diagram.

Wire-wound resistors are ordinarily used to control large currents. Wire-wound resistors are constructed by winding resistance wire on an insulating material. The wire ends are attached to metal terminals. An enamel coating is used to protect the wire and to conduct heat away from it. Wire-wound resistors can have fixed taps, which can be used to change the resistance value in steps. They can also have sliders, which can be adjusted to change the resistance to any fraction of their total resistance. **Precision-wound resistors** are used where the resistance value needs to be very accurate, such as in measuring instruments.

Variable resistors are used to change resistance while equipment is in operation. They are ordinarily called **potentiometers or rheostats**. Both carbon and wire-wound variable resistors are made on a circular form, Figure 3.15. In the wire-wound resistor, part A, a contact arm is attached to the form to make contact with the wire. The contact arm can be adjusted to any position on the circular form by a rotating shaft. A wire connected to

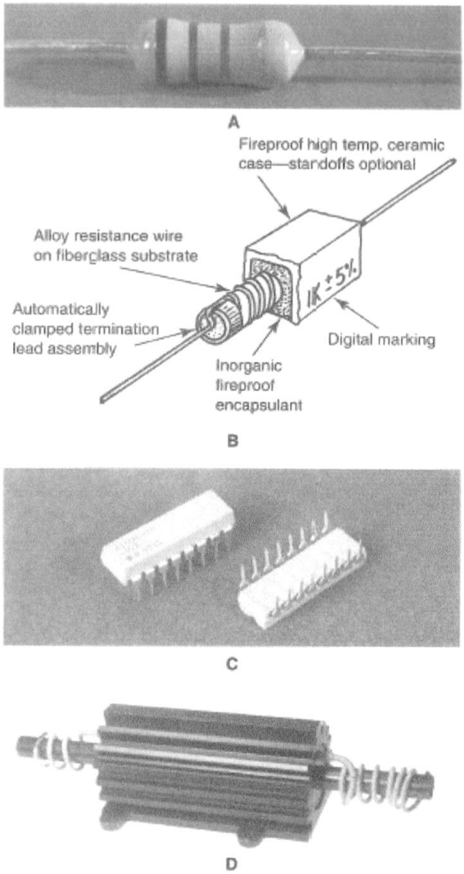

Figure 3.14 Resistor types. A – Carbon color-coded. B – Molded wire-wound resistor. C – Sixteen-pin dual-in-line package (DIP). D) Water-cooled, high-voltage resistor. (Dale Electronics, Inc.)

the movable contact is used to vary the resistance from the contact arm to either of the two outer wires of the variable resistor. For controlling smaller currents, carbon variable resistors are made by using a carbon compound mounted on a fiber disk, part B of Figure 3.15. A contact on a movable arm varies the resistance as the arm is turned by rotating a metal shaft.

3.3 Resistor Color Code

It is easy to find the value of a resistor using its **color code** or marked value. Most wire-wound resistors have resistance values (in ohms) printed on the

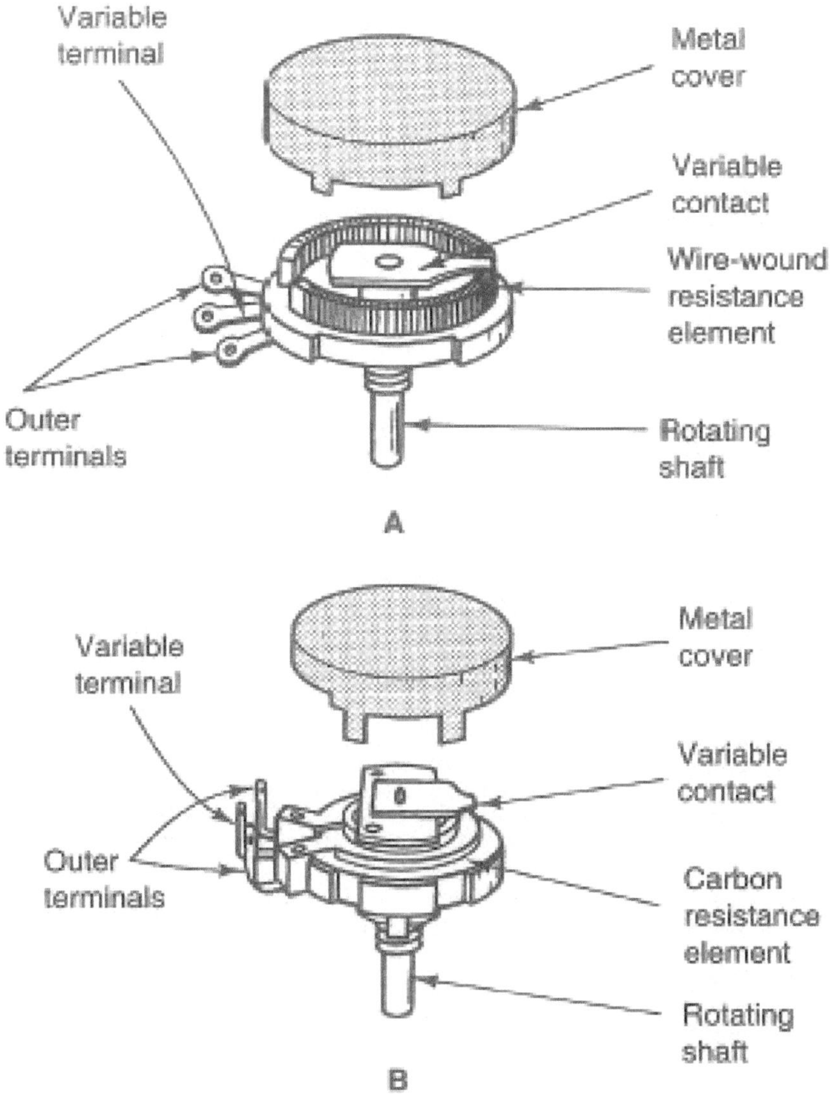

Figure 3.15 Variable resistor construction. A) Wire-wound variable resistor. B) Carbon variable resistor.

resistor. If they are not marked in this way, an ohmmeter must be used to measure the value. Most carbon resistors use colored bands to identify their value. Resistors are color-coded by an **end-to-center color band system** of marking. The first three colored bands are used to indicate the resistance value in ohms. A fourth color is used to indicate the tolerance of the resistor.

Color	1st digit	2nd digit	Number of zeros or multiplier	Tolerance (%)
Black	0	0	1	
Brown	1	1	10	
Red	2	2	100	
Orange	3	3	1,000	
Yellow	4	4	10,000	
Green	5	5	100,000	
Blue	6	6	1,000,000	
Violet	7	7		
Gray	8	8		
White	9	9		
Gold*	—	—	0.1	5
Silver*	—	—	0.01	10
No color	—	—		20

*When resistors have a value of less than 10 Ω, the third color band is a decimal multiplier. The two colors used are: gold = × 0.1 and silver = × 0.01.

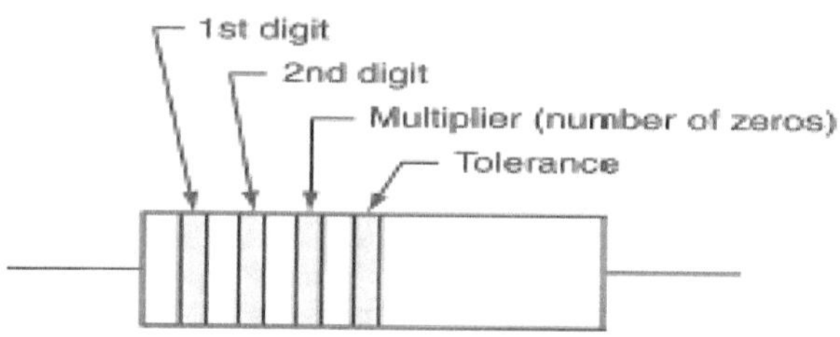

Figure 3.16 Resistor color code.

The colors are read in the correct order from the end of a resistor. Numbers from the resistor color code, shown in Figure 3.16, are substituted for the colors. Through practice using the resistor color code, the value of a resistor can be determined at a glance.

It is difficult to manufacture a resistor to the exact value required. For many uses, the actual resistance value can be as much as 20% higher or lower than the value marked on the resistor without causing any problem. In most uses, the actual resistance does not need to be any closer than 10% higher or lower than the marked value. This percentage of variation between the marked color-code value and the actual value of a resistor is called **tolerance**. A resistor with a 5% tolerance should be no more than 5% higher or lower

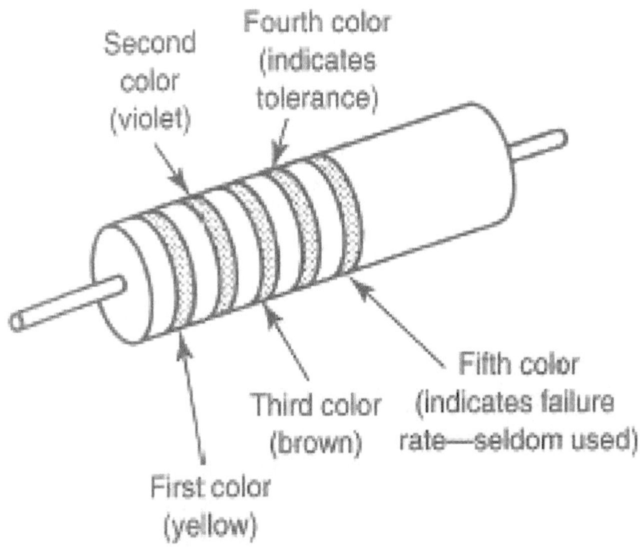

Figure 3.17 End-to-center system, resistor value = 470 Ohms.

than the marked value. Resistors with tolerances of lower than 5% are called **precision resistors**.

Resistors are marked with color bands starting at one end and leading to the center of the resistor. The body color is not used as part of the resistor value. For example, a resistor that has three color bands (yellow, violet, and brown) at one end has the color bands read from the end toward the center. See Figure 3.17. The resistance value is 470 Ohms. A resistor with a green band, a red band, and a black band has a value of 52 Ohms. Remember that black, as the third color, means that no zeros are to be added to the digits.

Another example of a resistor with the band color-code system is shown in Figure 3.18. Read the colors from left to right from the end of the resistor where the bands begin. Use the resistor color chart to determine the value of the resistor in ohms and its tolerance. This resistor also has a band indicating its **failure rate**.

The first color is orange, so the first digit in the value of the resistor is 3. The second color is black, so the second digit in the value of the resistor is 0. The first two digits of the resistor value are 30. The third color is yellow, which indicates the number by which the first two digits are to be multiplied. Sometimes it is easier to think of the third color as the number of zeros to add to the first two digits. The color yellow is a multiplier of 10,000.

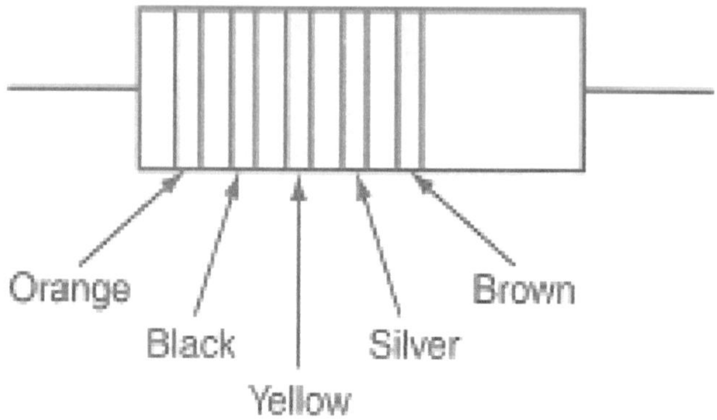

Figure 3.18 Resistor color code example.

Multiply 30 × 10,000 to obtain the value of the resistor (30 × 10,000 = 300,000 Ohms). A yellow color means that four zeros are added to the first two digits to find the value of the resistor (30 with four zeros added equals 300,000 Ohms).

The fourth color is silver. The ***tolerance*** of the resistor is given with this band. The tolerance shows how close the actual value of the resistor should be to the color-code value. Tolerance is a percentage of the actual value. In this example, silver shows a tolerance of ±10%. This means that the resistor should be within 10% of 300,000 Ohms in either direction. The value of the resistor could be as low as 270,000 Ohms (300,000 × 0.10 - 30,000 and 300,000 – 30,000 = 270,000). The value could be as high as 330,000 (300,000 × 0.10 = 30,000 and 300,000 + 30,000 – 330,000).

The fifth color is brown. The fifth band shows the percentage of failure per 1000 hours of use. A brown fifth color band shows a **failure rate** of 1% per 1000 hours of resistor use.

Sometimes there is no fourth color on the body of the resistor. This means that the tolerance is 20%. Frequently, no fifth color appears on the resistor. Using the color code, it is easy to list the ohms value, tolerance, and failure rate of resistors.

An easy way to remember the resistor color-code system is to remember the following:

Big Brown Rabbits Often Yield Great Big Vocal Groans When Gingerly Slapped.

The first letter of each word in the quote is the same as the first letter in each of the colors used in the color code. The words of the quote are counted (beginning with zero) to find the word corresponding to each digit or the number of zeros to be added (refer to Figure 3.19).

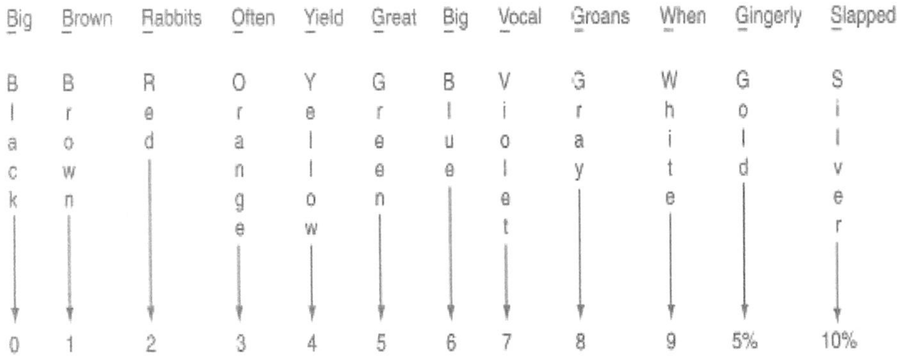

Figure 3.19 Quote used to memorize resistor color

3.4 Power Rating of Resistors

The size of a resistor helps to determine its **power rating**. Larger resistors are used in circuits that have high power ratings. Small resistors will be damaged if they are put in high-power circuits. The power rating of a resistor indicates its ability to give off or dissipate heat. Common power (wattage) ratings of color-coded resistors are 1/4, 1/2, 1, and 2 Watts. Notice the difference in the sizes of the resistors in Figure 3.3. Resistors that are larger in physical size will produce more heat and have higher power ratings.

When choosing a resistor, always use one that is safely **above** the power rating for the circuit. Always err on the side of safety. A resistor with a power rating that is too low could burn and possibly start a fire.

3.5 Measuring Resistance

Another important activity in the study of electrical control is measurement. Measurements are made in many types of circuits. The proper ways of measuring resistance, voltage, and current should be learned. These are the three most common electrical measurements.

Many important electrical tests may be made by measuring resistance. Resistance is opposition to the flow of current in an electrical circuit. The current that flows in a circuit depends upon the amount of resistance in that circuit. You should learn to measure resistance in an electrical circuit by using a meter.

Resistance tests are sometimes called **continuity checks**. A continuity check is made to see if a circuit is open or closed (a continuous path). A meter

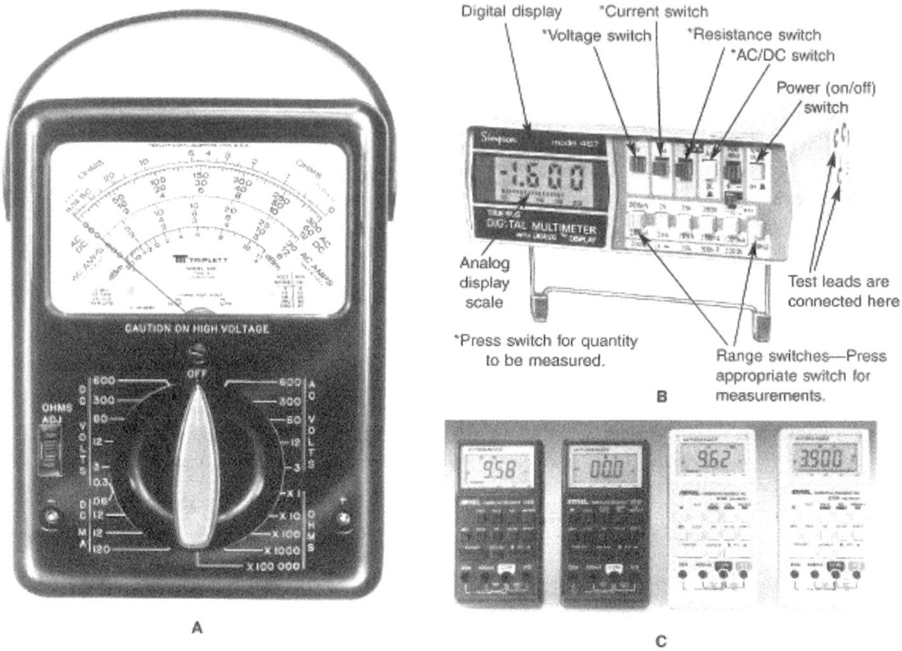

Figure 3.20 Meters used for electrical work. A – Analog multimeter: Analog meters have a needle that moves over a scale for making measurements. (Triplett Co.) B – Digital multimeter: This meter has a digital display for measurements. (Simpson Electric Co.) C – Programmable digital autoranging meters: Functions are set with the push buttons. (American Reliance Inc.)

is also used to measure exact values of resistance. Resistance must always be measured with no voltage applied to the component being measured.

Resistance can be measured with an **ohmmeter** or with a multimeter. An ohmmeter is a meter that just measures resistance. A **multimeter** is used to measure resistance, voltage, or current. This type of meter is often used by electrical technicians. Figure 3.20 shows some types of meters for doing electrical work. The meter in part A of Figure 3.20 is an **analog multimeter**. The meters shown in parts B and C of Figure 3.20 are **digital multimeters**.

Digital meters give their readings in a digital format. No calculations are necessary. When using an **analog meter**, you must first read the needle on the face of the meter. Next, you may need to multiply or divide that reading by a factor of 10 to get the correct result. Reading analog meters is covered in the following sections. The methods for connecting both digital and analog meters for taking readings are the same. **Information on analog scales is provided so you will know how to use either type of meter. Digital meters are much easier to use for most applications.**

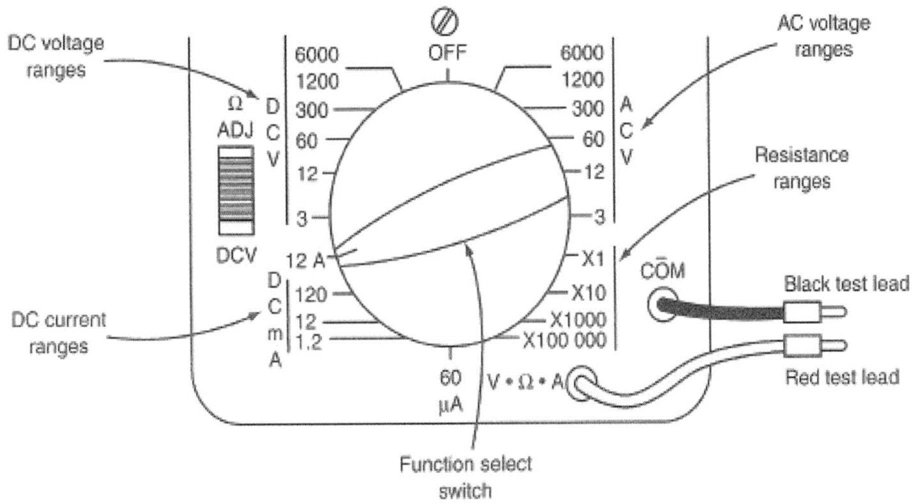

Figure 3.21 Controls of a typical multimeter.

The type of measurement is changed by adjusting the **function-select switch** to the desired measurement. By adjusting the function-select switch, the meter can be set to measure either resistance, voltage, or current. The meter switch shown in Figure 3.21 has settings for:

- Direct current (DC) voltage.
- Direct current (DC) amps and milliamps.
- Alternating current (AC) voltage.
- Resistance (ohms).

Figure 3.21 shows the controls of an **analog meter**. The function-select switch is in the center of this meter. Notice that the lower right section of the ranges is used for measuring ohms, or resistance. The ohms measurement area is divided into four ranges: × 1, × 10, × 1000, and × 100,000. These ranges are called **multipliers**. The meter may be adjusted to any of the four ranges for measuring resistance, depending on the estimated value to be measured.

The **test leads** used are ordinarily black and red. These colors are used to help identify which lead is the positive or negative side of the meter. This is important when measuring DC values. Red indicates **positive polarity (+)** and black indicates **negative (–) polarity**.

Refer again to the diagram of the meter controls shown in Figure 3.21. The red test lead is put in the hole, or jack, marked with **V-Ω-A**, or

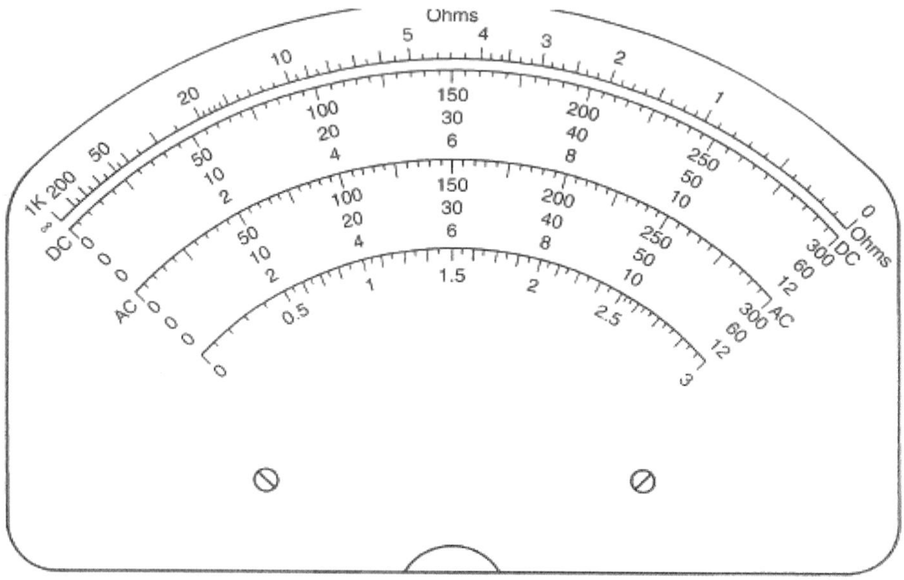

Figure 3.22 Multimeter scale.

volts-ohms-amperes. The black test lead is put in the hole or jack labelled COM, or negative common. The function-select switch should be placed on one of the resistance ranges. When the test leads are touched together or shorted, the meter needle moves from the left side of the meter scale to the right side. This test shows that the meter is operational.

The meter's **scale** should now be studied. Reading an analog meter requires some practice. Figure 3.22 shows the scale of a common type of multimeter. Notice that the top scale, from 0 to infinity, is labelled Ohms. Infinity indicates that a very high resistance is being measured. Zero indicates that a very low resistance is being measured. This scale is used for measuring ohms only. On most analog multimeters, the top scale is the resistance or ohms scale.

To measure any resistance, first select the proper meter range. On the meter range shown in Figure 3.21, the choices are × 1, × 10, × 1000, and × 100,000. Next, the ohmmeter must be properly **zeroed** before attempting to measure resistance accurately. When the test leads are touched together, the needle on the scale of the meter should move from infinity on the left to zero (0) on the right side of the scale. This indicates zero resistance. The needle of the meter is adjusted so that it is exactly over the zero mark. This is called **zeroing** the meter. If the needle does not reach zero or goes past zero when the test leads are shorted, the control marked **ohms adjust** is used.

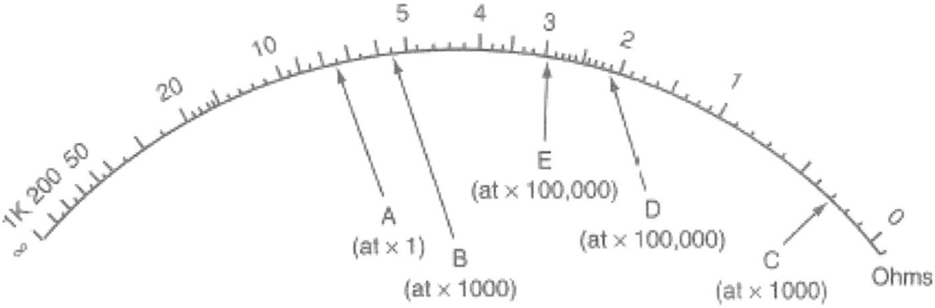

Figure 3.23 Ohm's scale of a multimeter.

The ohms-adjust control is indicated by Ω ADJ on the diagram. The ohmmeter should always be zeroed prior to each resistance measurement and after changing ranges. If the meter is not zeroed, the measurement will be incorrect.

The more accurate measurement of resistance is made when the meter's needle stops somewhere between the **center** of the ohms scale and **zero**. This is because of the greater distance between the numbers on the right of the scale. Choosing the proper range adjustments controls how far the needle moves. If the range selected is x 1, this means that the number the needle points to is multiplied by 1. If the function-select switch is adjusted to the × 100,000 range, this means that the number to which the needle points is multiplied by 100,000. Always zero the meter when changing ranges, and always multiply the number indicated on the scale by the multiplier of the range.

To take a measurement, the test leads are placed across a resistance. Assume that the needle of the meter moves to point A on the scale of Figure 3.23. This reading was taken using the × 1 range. Thus, the resistance equals 7.5 × 1 = 7.5 Ohms. For reading B, the meter range is × 1000. The reading at point B equals 5.5 × 1000 = 5500 Ohms. At point C, the reading is 0.3 × 1000 = 300 Ω.

The reading at point D uses the × 100,000 range. The needle shows 2.2 on the scale, so the reading is equal to 2.2 × 100,000, or 220,000 Ohms. For point E, the meter range is also set on × 100,000. The needle shows 2.9 on the scale. The reading is 2.9 × 100,000, or 290,000 Ohms.

It is very important not to measure the resistance of a component until it has been disconnected, or the reading may be wrong. Also, voltage should never be applied to a component when measuring resistance. This can damage the meter.

To learn to measure resistance, it is easy to use color-coded resistors. These resistors are small and easy to handle. Practice in the use of the

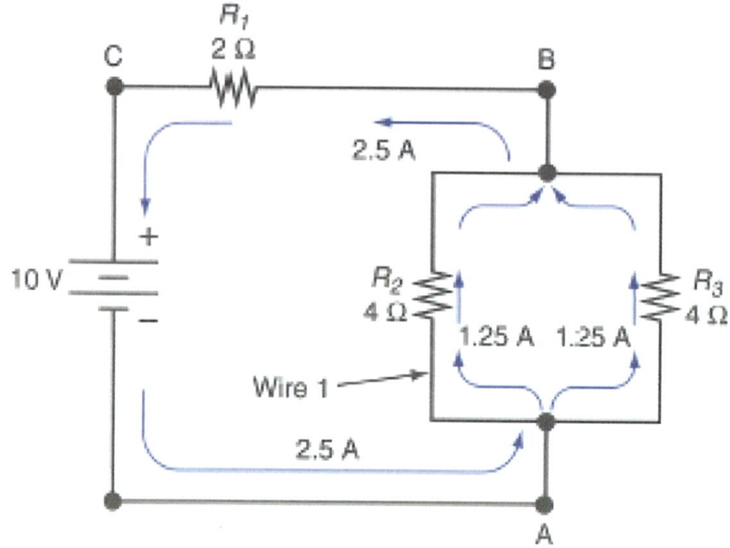

Figure 3.24 Combination circuit.

meter to measure several values of resistors makes reading the meter much easier.

To measure the total resistance of a parallel circuit or combination (series–parallel) circuit, examine Figure 3.24. First, remove the voltage source from the circuit. Prepare the meter to measure resistance. Be sure to zero the meter. Connect the meter across the points where the circuit was connected to the voltage source (points A and C in the figure). Adjust the meter range if necessary to get an accurate resistance reading. Once the proper range is selected, read the measured resistance on the meter scale.

A **multimeter** can also be used to measure the resistance of a potentiometer, as shown in Figure 3.25. If the shaft of the pot is adjusted while the ohmmeter is connected to points A and C, no resistance change will take place. The resistance of the potentiometer is measured in this way. Connecting to points B and C or to points B and A allows changes in resistance as the shaft is turned. The potentiometer shaft can be adjusted both clockwise and counterclockwise. The resistance varies from zero to maximum and from maximum back to zero as the shaft is adjusted.

3.6 Measuring Voltage

Voltage is the electrical pressure that causes current to flow in a circuit. A common voltage source is a battery. Batteries come in many sizes and voltage

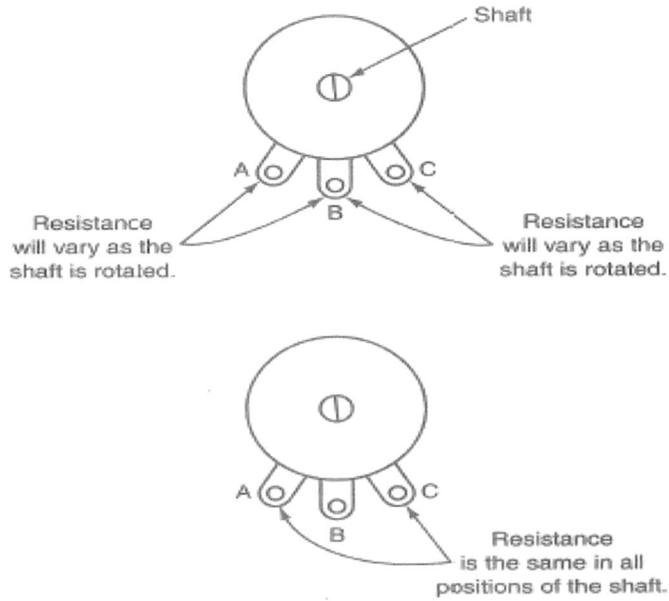

Figure 3.25 Measuring the resistance of a potentiometer.

values. The voltage applied to a component determines how much current will flow through it.

It is important to be able to measure voltage to check the operation of equipment. Many electrical problems develop due to either too much or too little voltage being applied to the equipment. A voltmeter is used to measure voltage in an electrical circuit. A multimeter is also used to measure voltage. **Refer to the controls of the multimeter shown in** Figure 3.21. The upper left part of the function-select switch is used for measuring DC voltage. The voltage ranges shown are 3,12, 60, 300, 1200, and 6000 V. When the function-select switch is adjusted to 3 V on the DC volts range, the meter measures up to 3 V. The same is true for the other ranges of DC voltage. The voltage value of each range is the maximum value of voltage that may be measured with the multimeter set on that range. The meter needle rests on the zero until some voltage is measured.

If the maximum voltage level is known, the proper DC voltage range can be chosen. The value of the range being used is the maximum value of voltage that can be measured on that range. For example, when the range selected is 12 V, the maximum voltage the meter can measure is 12 V. Any voltage above 12 V could damage the meter.

To measure a voltage that is unknown, start by using the **highest range** on the meter. Then slowly adjust the range downward until a voltage reading

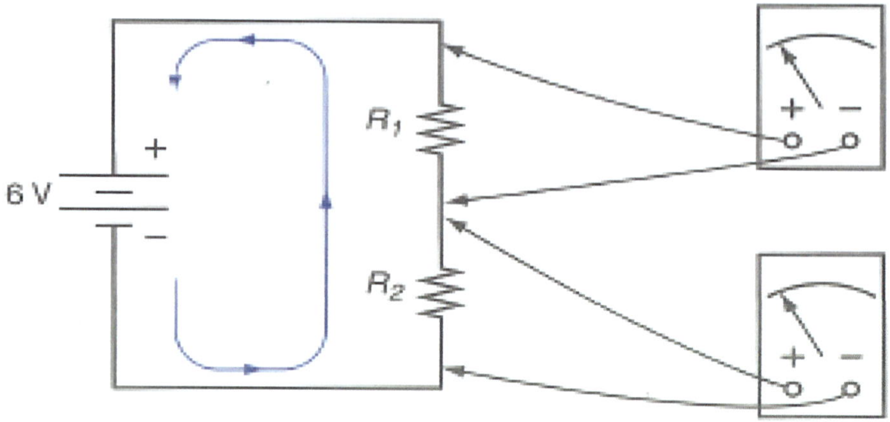

Figure 3.26 Measuring voltage drop in a DC circuit.

is indicated on the right side of the meter scale. This will prevent overloading and possibly damaging the meter.

Connect the red and black test leads to the meter by putting them into the proper jacks. The red test lead should be put into the jack labeled ***V*-Ω-*A***. The black test lead should be put into the jack labeled **-COM.** Matching the meter polarity to the voltage polarity is very important when measuring DC voltage. The meter needle moves backward, possibly causing damage to the meter, if polarities are not connected properly. Meter polarity is simple to determine. The positive (+) red test lead is connected to the positive side of the DC voltage being measured. The negative (–) black test lead is connected to the negative side of the DC voltage being measured. The meter is always connected across (in parallel with) the DC voltage being measured.

A certain amount of voltage is needed to cause an electrical current to flow through a resistance in a circuit. This voltage is called the **voltage drop**. Voltage drop is measured across any component through which current flows. The polarity of a voltage drop depends on the direction of the current. Current flows from the negative polarity of a battery to the positive polarity. In Figure 3.26, the bottom of each resistor is negative. The top of each resistor is positive. The negative test lead of the meter is connected to the bottom of the resistor. The positive test lead is connected to the top. The meters are connected as shown to measure each of the voltage drops in the circuit. If the meter polarity is reversed, the meter needle would move in the wrong direction.

It is easy to become familiar with the part of the meter's scale that is used to measure DC voltage. **Refer to the multimeter scale back in** Figure 3.22**.** Notice that the part of the scale below the ohms scale is the DC voltage scale.

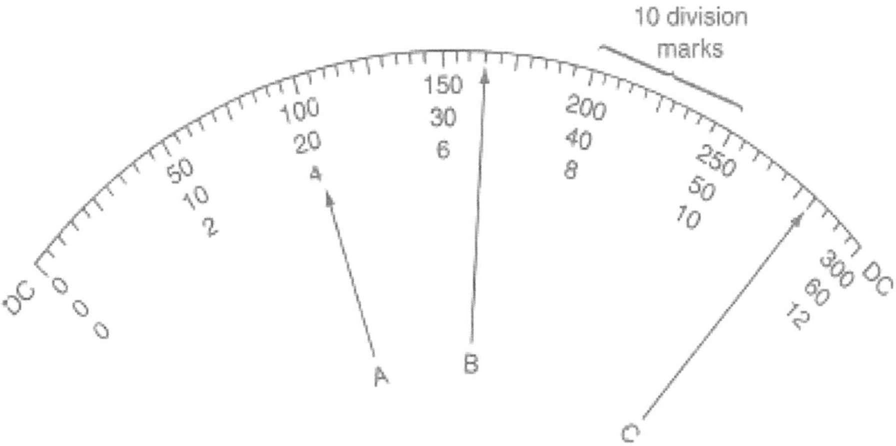

Figure 3.27 DC voltage scale of a multimeter.

This scale is usually black. Notice that there are three DC voltage scales: 0 to 12 V, 0 to 60 V, and 0 to 300 V. All DC voltages are measured using one of these scales. Notice that each of the DC voltage ranges on the function-select switch corresponds exactly to a number on the right side of the meter scale, or it corresponds to a number that can be easily multiplied or divided by a factor of 10 to equal the number on the **function-select switch.**

Notice that when the 12-, 60-, or 300-V **range** is used, the scale is read directly. On these ranges, the number to which the needle points is the actual value of the voltage being measured. When the 3-, 1200-, or 6000-V range is used, the number to which the needle points must be multiplied or divided by some factor of 10. If the meter's needle points to the number 50 while the meter is adjusted to the 60-V range, the measured voltage is 50 V. If the meter's needle points to the number 250 while the meter is adjusted to the 3-V range, the measured voltage is 2.5 V (250 100 = 2.5). When the 1200-V range is used, the numbers on the 0- to 12-V scale are read and then multiplied by 100.

You may have some difficulty in reading the voltage divisions on the scales. Look at the division marks from 200 to 250 in Figure 3.27. The difference between 200 and 250 is 50 units (250 – 200 = 50). There are 10 division marks between 200 and 250. The voltage per division mark is 50 10, or 5 volts per division. Each division mark between 200 and 250 is equal to 5 V. This procedure is like reading a ruler or other types of scales.

Some examples of DC voltage measurements with the meter set on the 3-V range are given next. If the test leads of the meter are placed across a voltage source and the meter's needle moves to point A on the **scale** of Figure 3.27, the DC voltage is equal to 100 - 100, or 1 V. The reading at point B is 165 100, or 1.65 V. At point C, the reading is 280 100, or 2.8 V.

If the ***range switch*** is changed to the 12-V position, the voltage is read directly from the meter scale. For example, if the range is set on 12 V and the meter needle moves to point A in Figure 3.27, the voltage equals 4 V. The reading at point B equals 6.6 V. At point C, the reading is 11.2 V. The same procedure is used for all other ranges.

3.7 Measuring Current

Current flows through a complete electrical circuit when voltage is applied. Many important tests are made by measuring current in electrical circuits. An ammeter is a meter designed to measure current. Current, like voltage and resistance, can also be measured with a **multimeter**. Learning to measure current properly is important.

Refer to the controls of the multimeter shown in Figure 3.21. There are a total of five current ranges. The function-select switch can be adjusted to any of five ranges of direct current: 12 A, 120 mA, 12 mA, 1.2 μA, and 60 μA. For example, when the function-select switch is placed in the 120 mA range, the meter is capable of measuring up to 120 mA of current. When measuring current, make it a habit to start with the meter set on its highest range. Move the range setting to a lower value if the meter needle moves only a small amount. The most accurate reading is when the meter needle is between the center of the scale and the right side. The same scales on the multimeter are often used for measuring direct current and DC voltage.

Current is measured by connecting the meter **into** a circuit, as shown in Figure 3.28. This is referred to as connecting the meter in *series* with the circuit. As the voltage applied to a circuit increases, the current also increases. If 12 V is applied to the lamp in Figure 3.28, a larger current will flow through the lamp. If 24 V is applied to the same lamp, an even larger current will flow. As resistance gets smaller, the current increases.

If the meter range is set on the 12 A range, the scale is read directly. The bottom DC scale, which has the number 12 on the right side, is used. Some examples are shown in Figure 3.29. Assume the meter is set on the 12-A range. At point A on the scale, the reading is 4.6 A. The reading at point B is 8.8 A.

The 60-uA range on the meter is for measuring very small currents. Measurements in this range are also read directly from the meter scale.

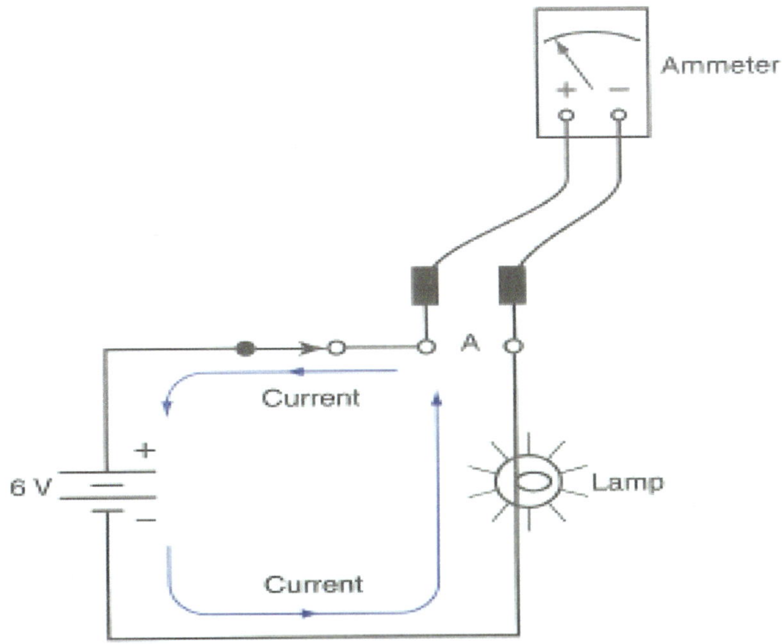

Figure 3.28 Meter connection for measuring DC current.

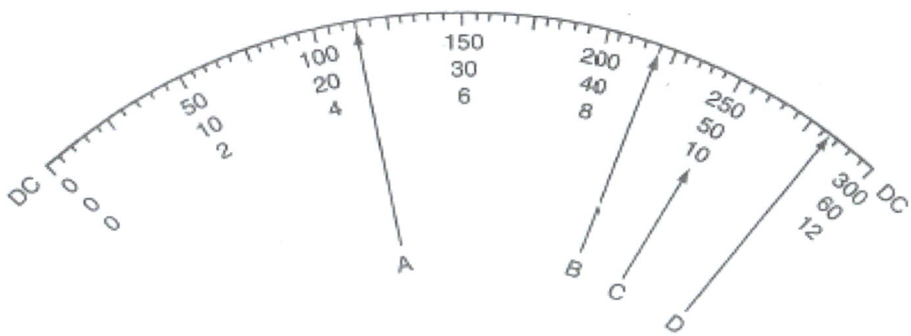

Figure 3.29 DC current scale of a multimeter.

Notice that the number 60 is the middle number on the right of the DC scale.

When the meter is set on the 120-mA range, the meter will measure up to 120 mA of direct current. The readings on the bottom DC scale are multiplied by 10 on this range setting. Examine Figure 3.29 again. This time assume that the function-select switch is set for the 120-mA range. Point A – 46 mA (4.6 × 10) Point B = 88 mA (8.8 × 10) Point C = 100 mA (10 × 10)

Point D = 113 mA (11.3 × 10) Notice that the reading at point D is halfway between the 11.2 and 11.4 divisions on the scale. The reading is 11.3 × 10, or 113 mA.

The test lead **polarity** of the meter is important when measuring direct current. The meter is connected to allow current to flow through the meter in the correct direction. The **negative test lead** is connected nearest to the negative side of the voltage source, and the **positive test lead** nearest to the positive side. The meter is then connected to the circuit. To measure current, a wire is removed to place the meter into the circuit. No voltage should be applied to the circuit when connecting the meter. The meter is placed in series with the circuit. Series circuits have one path for current to flow.

The following is a step-by-step procedure for properly measuring **current** through point A in the circuit of Figure 3.28.

- Turn off the circuit's voltage source by opening the switch.
- Set the meter to the highest current range (12 A).
- Remove the wire at point A.
- Connect the negative test lead of the meter to the negative side of the voltage source.
- Connect the positive test lead to the end of the wire that was removed from point A.
- Turn on the switch to apply voltage to the circuit.
- Look at the meter needle to see how far it has moved up the scale.
- Adjust the meter range until the needle moves to between the center of the scale and the right side.

A similar procedure can be used to measure the current in one branch of a **parallel circuit**. To measure current through path 1 in the parallel circuit of Figure 3.30, use the following:

- Make sure that no voltage is applied to the circuit by opening the switch.

1. Remove wires 1 and 2 from point A.
2. Set the meter to the highest current range.
3. Connect wires 1 and 2 to the positive test lead of the meter.

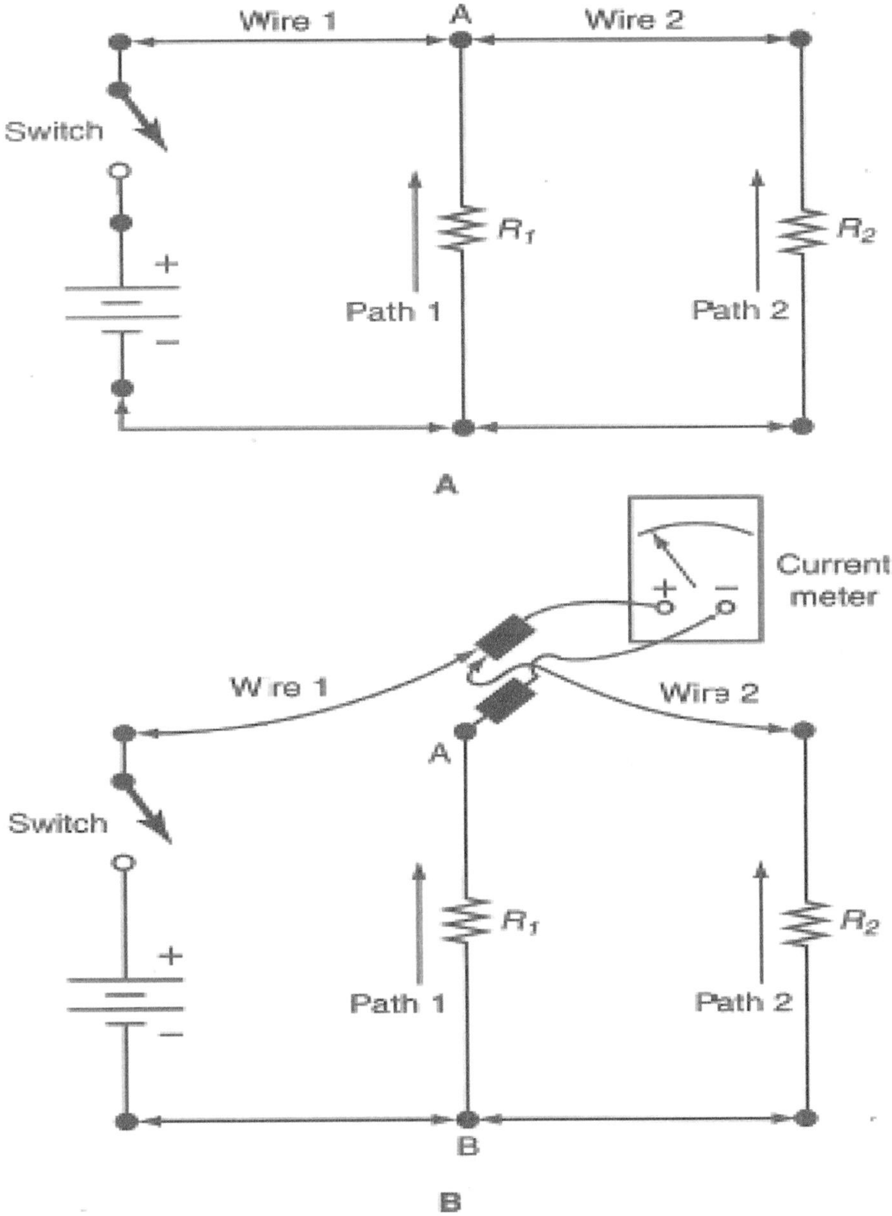

Figure 3.30 Making measurements in a parallel circuit. A – Original circuit. B – Circuit set up to measure current through path 1.

4. Connect the negative test lead of the meter to point A.
5. Turn on the switch to apply voltage to the circuit.
6. Adjust the meter, if necessary, to a lower range to get an accurate reading.
7. Read the current value on the scale of the meter.

Always remember the following **safety tips** when measuring current with a **multimeter**:

1. Turn off the voltage before connecting the meter in order not to get an electrical shock. This is an important habit to develop.
2. Set the meter to its highest current range. This assures that the meter needle will not move too far to the right of the scale and possibly cause damage to the meter.
3. Always remember to disconnect a wire and reconnect the wire to one of the meter test leads. If a wire is not removed to put the meter into the circuit, the meter will not be connected properly.
4. Use the proper meter polarity. The negative test lead is connected so that it is nearest the negative side of the voltage source. Similarly, the positive test lead is connected so that it is nearest the positive side of the voltage source.

3.8 Digital Meters

Many different **digital meters** are used. They employ numerical readouts to simplify the measurement process and to make more accurate measurements. Digital meters rely on the operation of digital circuitry to produce a numerical readout of the measured quantity.

The readout of a digital meter is designed to transform electrical signals into numerical data. Both letter and number readouts are available, as are seven-segment, discrete-number, and bar- matrix displays. Each method has a device designed to change electrical energy into light energy on the display.

Digital multimeters (DMMs) have numerical readouts which are called seven-segment displays. Part A of Figure 3.31 shows the seven-segment display and part B of the figure indicates the parts of the display that are illuminated when each number is displayed. Many meters have 3K digits in the display. A 3M-digit multimeter has three digit positions that can indicate from 0 to 9 and one digit position that can indicate only a value of 1. This latter digit is called the half-digit and is always the digit on the left of the display.

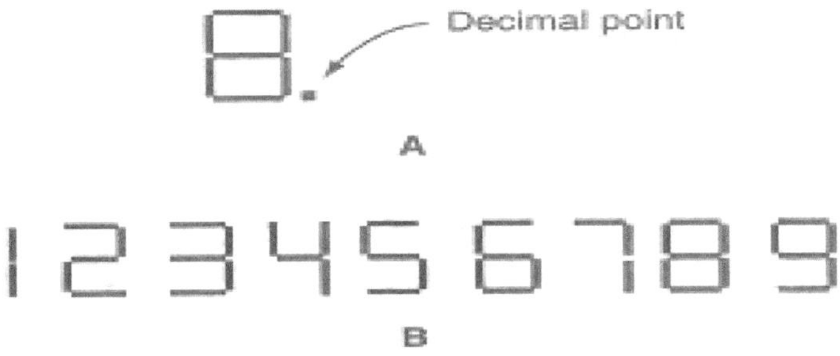

Figure 3.31 Seven-segment display used with digital multimeters (DMMs).

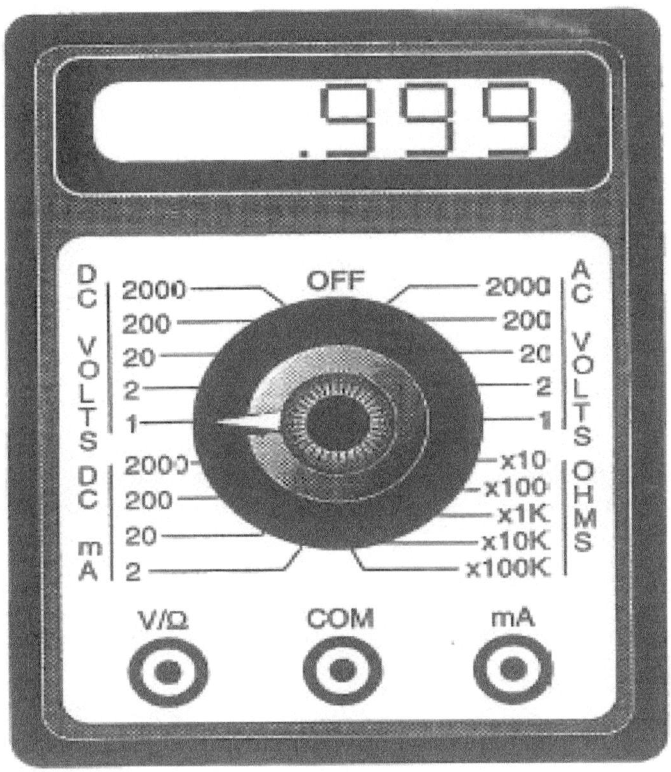

Figure 3.32 Three-and-a-half-digit digital multimeter.

The **resolution** of a digital multimeter is the smallest change of a quantity that the meter can measure. The smaller the change, the better the meter's resolution. The resolution of a meter is determined by the number of digits in the display. For example, if the reading of 0.999 V, as shown in Figure 3.22, increased by 0.001 volt to 1 volt, the display shows 1.000 volt.

A change of 0.001 volt is the resolution of the 3 ½ digit multimeter shown in Figure 3.32.

Review Questions

Answer the following questions.

1. Why are symbols used to represent electrical components?

2. Draw the symbols for the components listed here.
   a. Resistor.
   b. Ground connection.
   c. SPST switch.
   d. Potentiometer.
   e. Lamp.
   f. Fuse.

3. Draw a diagram of a circuit that has one battery, two resistors, a single-throw switch, and a current meter connected to form one path.

4. What values are shown by the first and second color bands of a color-coded resistor?

5. What value is shown by the third color band of a color-coded resistor?

6. What value is shown by the fourth color band of a color-coded resistor?

7. If a fourth color band is not indicated on a resistor, what is its tolerance value?

8. On which part of the ohms scale of an analog meter are the most accurate resistance measurements made?

9. What is meant by the x 1000 range of the ohmmeter?

10. What is meant by zeroing a meter?

11. If the range of an analog ohmmeter is adjusted to x 100,000 and the needle points to 10.6 on the ohms scale, what is the value of the resistance being measured?

12. When measuring an unknown voltage, what multimeter range should be used?

13. What determines how the meter is connected to measure the voltage drop across a resistor?

14. How is the proper voltage range of a multimeter selected?

15. What determines the resolution of a digital multimeter?

4

Review of Electrical Circuits

Objectives

Upon completion of this chapter, you will be able to:

- Solve basic DC and AC circuit problems.
- Explain Ohm's law.
- Explain the characteristics of a series and parallel AC circuits.
- Define energy, work, and power.
- Calculate power using the proper power formulas.
- Explain the relationship between AC voltages and current in resistive circuits.
- Define capacitive reactance and inductive reactance.
- Solve problems using the capacitive reactance formula and inductive reactance formula.
- Draw diagrams illustrating the phase relationship between current and voltage in a capacitive circuit or an inductive circuit.
- Define impedance.
- Calculate impedance of series and parallel AC circuits.
- Solve problems involving true power, apparent power, power factor, and reactive power in AC circuits.
- Describe voltage, current, and power relationships in three-phase AC circuits for wye and delta connections.

Important Terms

active power
admittance triangle
apparent power
average voltage
capacitance (C)
capacitive reactance (X_C)
capacitor
delta configuration

DOI: 10.1201/9788743810056-5

dielectric
effective value
energy
farad (F)
henry (H)
impedance (Z)
impedance triangle
inductance (L)
inductive reactance (X_L)
inductor
in-phase
kilovolt ampere (kVA)
kilowatt-hour (kWh)
kinetic energy
negative power
Ohm's law
peak-to-peak
phase angle (θ)
phasor diagram
positive power
potential energy
power
power factor (PF)
reactive power
rms value
total reactance (X_T)
true power
unity power factor
vector
vector diagram
voltampere (VA)
voltampere reactive (VAR)
watt
watt-second
work
wye configuration

To understand electrical circuits, it is necessary to know how to apply basic electrical theory. Electrical circuits are easy to understand by applying proper mathematical operations. The basic theory that is used is Ohm's law. Ohm's law should be learned as it applies to electrical circuits.

This chapter provides a review of the three major types of electrical circuits:

1. direct current (DC)
2. single-phase alternating current (AC)
3. three-phase alternating current (AC) circuits.

4.1 Review of Direct Current Circuits

Many important operations depend on **direct current (DC)** power. Industries use direct current power for many operations. Electroplating and variable-speed motor drives are two examples of direct current use. Direct-current energy is used to start automobiles and for many types of portable equipment used in homes. Three-phase or single-phase AC power is easily converted to direct current. Direct current is available from primary and secondary chemical cells. Direct current generators are also used to supply DC power for specialized applications.

To understand electrical systems, it is necessary to know how to apply basic electrical theory. Electrical control is a somewhat mathematical subject. The math is easy to understand since it has practical applications that are easy to see. One very important theory used is called **Ohm's law**. Ohm's law should be learned, as it applies to the basic theory of electrical circuits. The examples that follow are de circuits. AC circuits are more complex.

Ohm's law is the most basic and most used of all electrical theories. *Ohm's law* explains the relationship of voltage (the force that causes current to flow), current (the movement of electrons), and resistance (the opposition to current). Ohm's law is stated as follows: an increase in voltage increases current if resistance remains the same. Ohm's law stated in another way, is: an increase in resistance causes a decrease in current if voltage remains the same. The electrical values used with Ohm's law are usually represented with capital letters. For example, voltage is represented by the letter *V*, current, by the letter *I*, and resistance, by the letter *R*. The mathematical relationship of the three units is shown in the following formulas.

$$V = I \times R$$

$$I = \frac{V}{R}$$

$$R = \frac{V}{I}$$

Voltage (*V*) is measured in volts. Current (*I*) is measured in amperes. Resistance (*R*) is measured in ohms. Memorize these formulas. The Ohm's law circle of Figure 4.1 is helpful in remembering the formulas.

4.1.1 Series DC circuits

- There are several important characteristics of series DC circuits:
- The same current flows through each part of a series circuit. The total resistance of a series circuit is equal to the sum of the individual resistances. It can be found using the formula:

  $$R_T = R_1 + R_2 + R_3 + \cdots + R_N$$

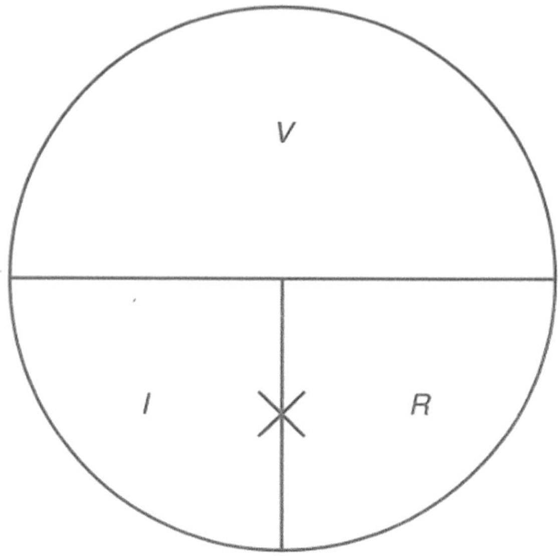

Figure 4.1 Ohm's law circle. To use the circle, cover the value you want to find and read the other values as they appear in the formula $V = I \times R$, $I = V/R$, $R = V/I$.

- The voltage applied to a series circuit is equal to the sum of the individual voltage drops.
- The voltage drop across a resistor in a series circuit is directly proportional to the size of the resistor.
- If the circuit is broken at any point, no current will flow.

Sample series circuit problems are shown in Figures 4.2 and 4.3. Helpful calculator procedures of Appendix A should be reviewed at this time.

4.1.2 Parallel DC circuits

There are several important basic rules for parallel circuits.

- There are two or more paths for current to flow.
- Voltage is the same across each component of the circuit.
- The sum of the currents through each path is equal to the total current that flows from the source.

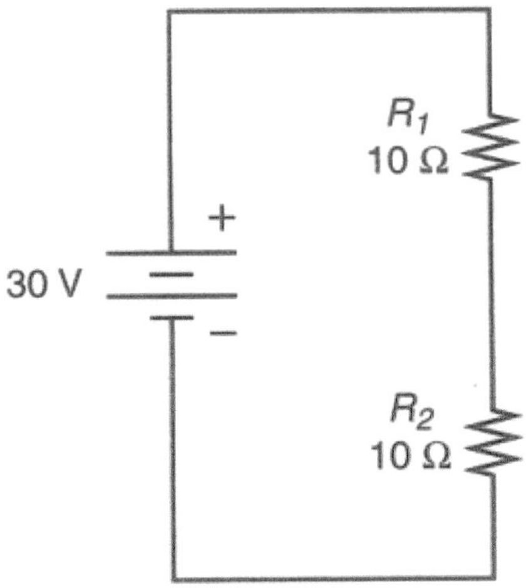

$$R_T = R_1 + R_2 = 10\ \Omega + 10\ \Omega = 20\ \Omega$$

$$I_T = \frac{V}{R_T} = \frac{30\ \text{V}}{20\ \Omega} = 1.5\ \text{A}$$

$$V_1 = I \times R_1 = 1.5\ \text{A} \times 10\ \Omega = 15\ \text{V}$$

$$V_2 = I \times R_2 = 1.5\ \text{A} \times 10\ \Omega = 15\ \text{V}$$

$$V_T = V_1 + V_2$$

$$30\ \text{V} = 15\ \text{V} + 15\ \text{V}$$

Figure 4.2 Series circuit example.

- Total resistance is found by using the formula:

$$\frac{1}{R_T} = \frac{1}{R_1} + \frac{1}{R_2} + \frac{1}{R_3} + \cdots + \frac{1}{R_N}$$

- If one of the parallel paths is broken, current will continue to flow in all the other paths.

Sample parallel circuit problems are shown in Figures 4.4 and 4.5.

4.1.3 Combination DC circuits

Problems for combination circuits are solved by combining rules for series and parallel circuits. Combination circuit problems can be solved by using the following step-by-step procedure:

1. Combine series and parallel parts to find the total resistance of the circuit.
2. Find the total current that flows through the circuit.
3. Find the voltage across each part of the circuit.
4. Find the current through each resistance of this circuit.

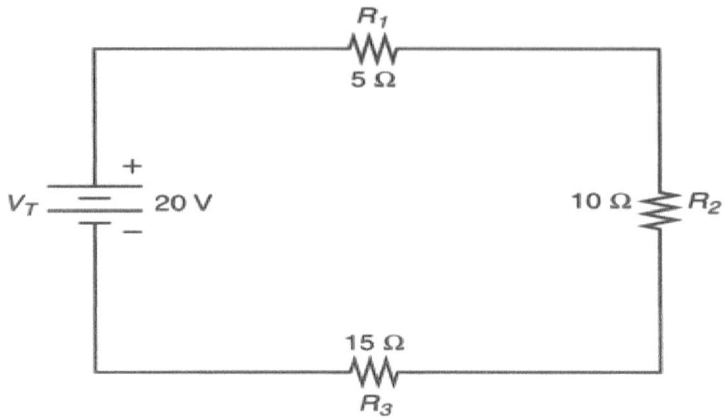

1. Find total resistance (R_T).
   $R_T = R_1 + R_2 + R_3$
   $= 5\ \Omega + 10\ \Omega + 15\ \Omega$
   $R_T = 30\ \Omega$
2. Find total current (I_T).
   $$I_T = \frac{V_T}{R_T} = \frac{20\ \text{V}}{30\ \Omega} = .667 \text{ amperes}$$
3. Find voltage drop across R_1 (V_1).
   $V_1 = I \times R_1$
   $= .667\ \text{A} \times 5\ \Omega$
   $V_1 = 3.33$ volts
4. Find voltage drop across R_2 (V_2).
   $V_2 = I \times R_2$
   $= .667\ \text{A} \times 10\ \Omega$
   $V_2 = 6.67$ volts
5. Find voltage drop across R_3 (V_3).
   $V_3 = I \times R_3$
   $= .667\ \text{A} \times 15\ \Omega$
   $V_3 = 10$ volts
6. Verify that the sum of the voltage drops is equal to source voltage (V_T).
   $V_T = V_1 + V_2 + V_3$
   $20\ \text{V} = 3.33\ \text{V} + 6.67\ \text{V} + 10\ \text{V}$

Figure 4.3 A series DC circuit problem with solving procedure.

Steps 3 and 4 must often be done in combination with each other, rather than doing one and then the other.

Look at the sample circuit of Figure 4.6. The value that should first be calculated is the resistance of R_2 and R_3 in parallel. When this quantity is found, it can be added to the value of the series resistor R_1 to find the total resistance of the circuit.

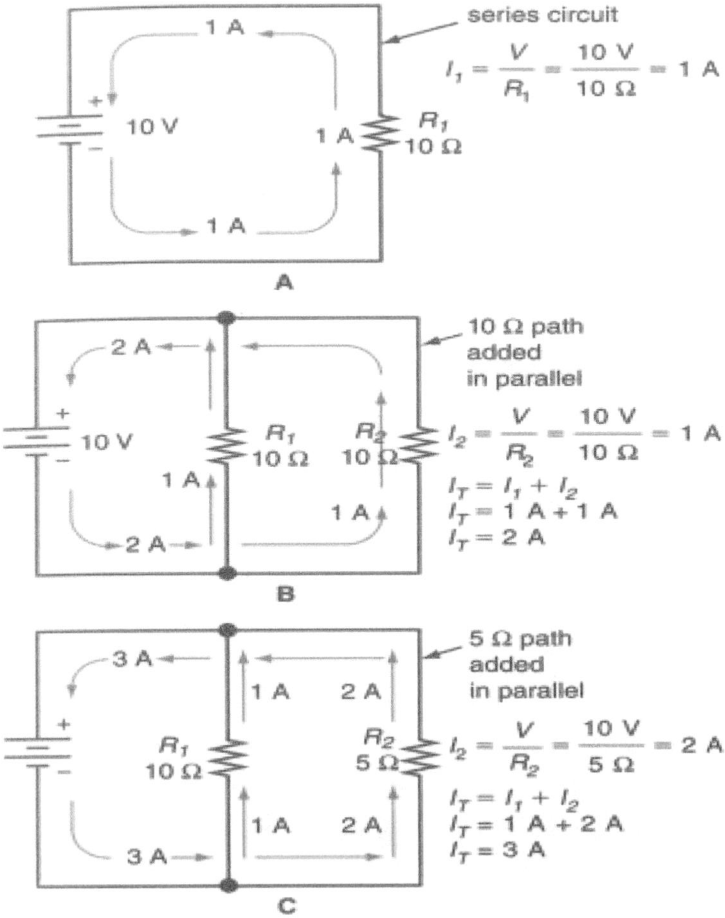

Figure 4.4 Current in a parallel circuit: A-A series circuit with only one path. B-A parallel circuit with two paths for current. $C - R$ changed to 5Ω.

$$R_T = R_1 + R_2 \| R_3$$

$$R_T = 30\Omega + \frac{(10\Omega \times 20\Omega)}{(10\Omega + 20\Omega)}$$

$$R_T = 30\Omega + 6.67\Omega$$

$$R_T = 36.67\Omega$$

(The symbol || means R_2 is in parallel with R_3.) When the total resistance (R_T) is found, the total current (I_T) can be found.

$$I_T = \frac{V_T}{R_T} = \frac{40\text{ V}}{36.67\Omega} = 1.09\text{ A}$$

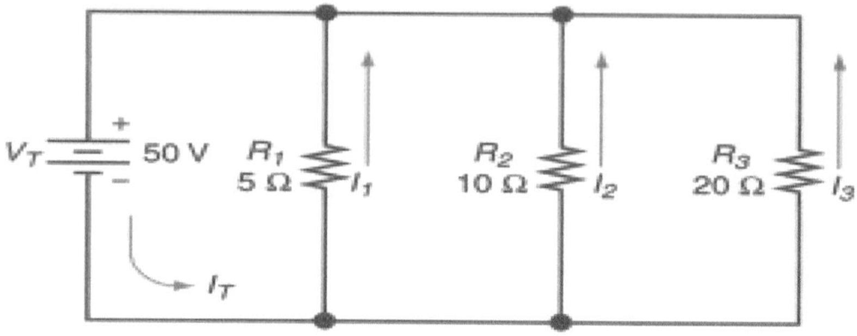

1. Find total resistance (R_T).

$$\frac{1}{R_T} = \frac{1}{R_1} + \frac{1}{R_2} + \frac{1}{R_3}$$

$$\frac{1}{R_T} = \frac{1}{5} + \frac{1}{10} + \frac{1}{20}$$

$$\frac{1}{R_T} = .2 + .1 + .05$$

$$\frac{1}{R_T} = .35$$

$$R_T = \frac{1}{.35} = 2.857\ \Omega$$

(This is less than any branch resistance.)

2. Find total current (I_T).

$$I_T = \frac{V_T}{R_T} = \frac{50\text{ V}}{2.85\ \Omega} = 17.5 \text{ amperes}$$

3. Find current through R_1 (I_1).

$$I_1 = \frac{V_1}{R_1} = \frac{50\text{ V}}{5\ \Omega} = 10 \text{ amperes}$$

4. Find current through R_2 (I_2).

$$I_2 = \frac{V_2}{R_2} = \frac{50\text{ V}}{10\ \Omega} = 5 \text{ amperes}$$

5. Find current through R_3 (I_3).

$$I_3 = \frac{V_3}{R_3} = \frac{50\text{ V}}{20\ \Omega} = 2.5 \text{ amperes}$$

6. Verify that the sum of the branch currents is equal to the total current (I_T).

$$I_T = I_1 + I_2 + I_3$$
$$= 10\text{ A} + 5\text{ A} + 2.5\text{ A}$$
$$I_T = 17.5 \text{ amperes}$$

Figure 4.5 A parallel DC circuit problem-solving procedure.

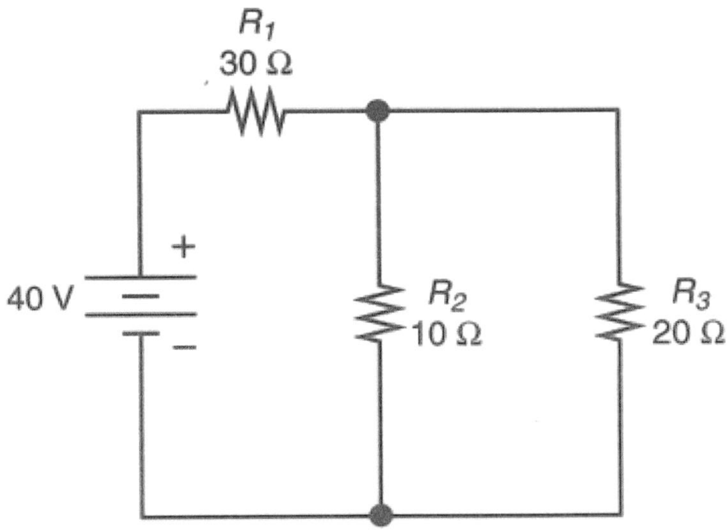

Figure 4.6 Combination circuit example.

Notice that the total current flows through resistor R_1 since it is in series with the voltage source. The voltage drop across resistor R_1 is:

$$V_1 = I_T \times R_1 = 1.09\text{A} \times 30\Omega = 32.7 \text{ V}$$

The applied voltage is 40 V, and 32.7 V is dropped across resistor R_1. The remaining voltage is dropped across the two parallel resistors R_2 and R_3:40 V – 32.7 V = 7.3 V across R_2 and R_3. The currents through R_2 and R_3 are:

$$I_2 = \frac{V_2}{R_2} = \frac{7.3 \text{ V}}{10\Omega} = 0.73 \text{ A}$$

$$I_3 = \frac{V_3}{R_3} = \frac{7.3 \text{ V}}{20\Omega} = 0.365 \text{ A}$$

Another type of combination circuit is shown in Figure 4.7. Resistors R_2 and R_3 are in series. When they are combined by adding their values, the circuit becomes a two branch parallel circuit. Total resistance (R_T) is found by using the product-over-sum method.

$$R_T = \frac{10\Omega \times 50\Omega}{10\Omega + 50\Omega} = \frac{500}{60} = 8.33\Omega$$

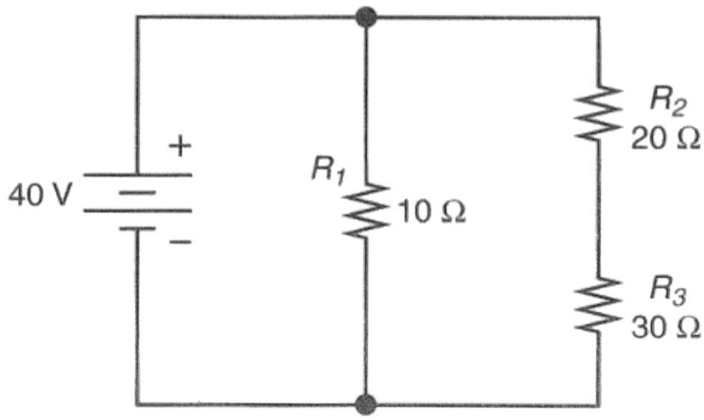

Figure 4.7 Combination circuit example.

Total current is:

$$I_T = \frac{V_T}{R_T} = \frac{40\ \text{V}}{8.33\Omega} = 4.8\ \text{A}$$

The voltage across R_1 is 40 V, since it is in parallel with the voltage source. Voltage drops across R_2 and R_2 are:

$$I_2 = I_3 = \frac{V_T}{R\ \text{of branch}} = \frac{40\ \text{V}}{50\Omega} = 0.8\ \text{A}$$
$$V_2 = I_2 \times R_2$$
$$V_2 = 0.8\,\text{A} \times 20\Omega$$
$$V_2 = 16\,\text{V}$$
$$V_3 = I_3 \times R_3$$
$$V_3 = 0.8\ \text{A} \times 30\Omega$$
$$V_3 = 24\ \text{V}$$

Notice that the sum of V_2 and V_3(16 V + 24 V) is equal to the source voltage.

Some additional combination circuits are shown in Figure 4.8, while Figure 4.9 shows another example of a combination de circuit problem. Work through the problem in Figure 4.9.

Some simple electrical circuit examples have been discussed. They become easy to understand after practice with each type of circuit. It is very important to understand the characteristics of series, parallel, and combination circuits.

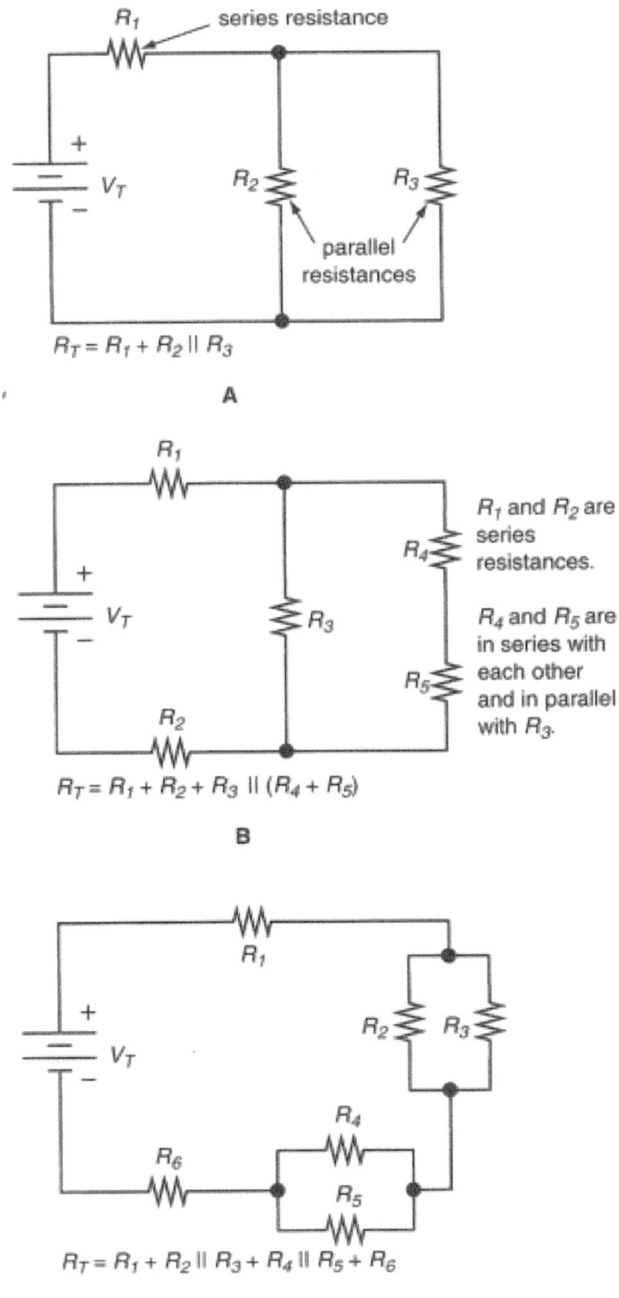

Figure 4.8 Three types of combination circuits.

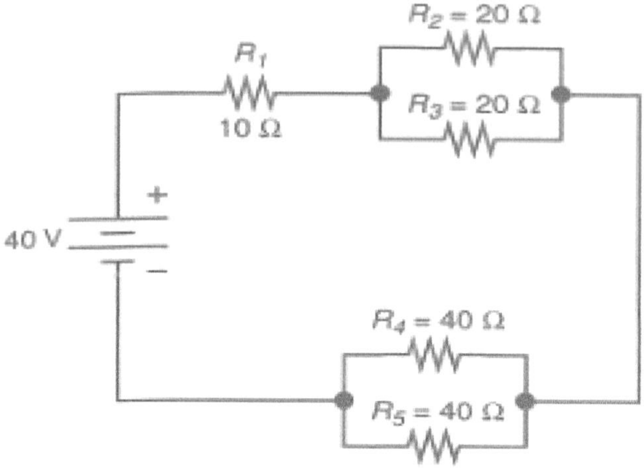

1. Find total resistance (R_T).

$$R_T = R_1 + R_2 \parallel R_3 + R_4 \parallel R_5$$
$$= 10\ \Omega + 10\ \Omega + 20\ \Omega$$
$$R_T = 40\ \Omega$$

2. Find total current (I_T).

$$I_T = \frac{V_T}{R_T} = \frac{40\text{ V}}{40\ \Omega} = 1\text{ ampere}$$

3. Find voltage across R_1 (V_1).

$$V_1 = I_1 \times R_1$$
$$V_1 = 1\text{ A} \times 10\ \Omega$$
$$V_1 = 10\text{ volts}$$

4. Find $V_2 = V_3$.

$$V_2 = I_T \times R_2 \parallel R_3$$
$$V_2 = 1\text{ A} \times 10\ \Omega$$
$$V_2 = 10\text{ volts}$$

5. Find $V_4 = V_5$.

$$V_4 = I_T \times R_4 \parallel R_5$$
$$V_4 = 1\text{ A} \times 20\ \Omega$$
$$V_4 = 20\text{ volts}$$

6. Find $I_2 = I_3$.

$$I_2 = \frac{V_2}{R_2} = \frac{10\text{ V}}{20\ \Omega} = 0.5\text{ amperes}$$

7. Find $I_4 = I_5$.

$$I_4 = \frac{V_4}{R_4} = \frac{20\text{ V}}{40\ \Omega} = 0.5\text{ amperes}$$

Figure 4.9 A combination dc circuit problem-solving procedures.

4.2 Energy, Work, and Power

An understanding of the terms energy, work, and power is necessary in the study of electrical systems. The first term, **energy**, means the capacity to do work. For example, the capacity to light a light bulb, to heat a home, or to move something requires energy. Energy exists in many forms, such as electrical, mechanical, chemical, and heat. If energy exists because of the movement of some item, such as a ball rolling down a hill, it is called **kinetic energy**. If energy exists because of the position of something, such as a ball that is at the top of the hill but not yet rolling, it is called **potential energy**. Energy has become one of the most important factors in our society.

A second important term is work. Work is the transferring or transforming of energy. Work is done when a force is exerted to move something over a distance against opposition, such as when a chair is moved from one side of a room to the other. An electrical motor used to drive a machine performs work. Work is performed when motion is accomplished against a force that opposes the motion. Work is also done each time energy changes from one form into another. Work can be expressed with the equation:

$$W = F \times d$$

where W is work in joules, F is force in newtons, and d is the distance the force moves in meters.

Sample Problem: Work
Given: An object with a mass of 22 kilograms is lifted 55 meters.

Find: The amount of work done when the object is moved.
Solution:
The force of gravity is the force opposing the work. The force of gravity acting on the object is equal to 9.8 m/s^2 (a constant that applies to objects on earth) multiplied by the mass of the object or:

$$F = 9.8\,\text{m/s}^2 \times 22\text{kg} = 215.6\,\text{newtons}$$
$$W = F \times d$$
$$W = 215.6\,\text{N} \times 55\text{ m}$$
$$W = 11{,}858\text{ joules}$$

A third important term is power. Power is the rate at which work is done. It considers not only the work that is performed but the amount of time in which the work is done. For instance, electrical power is the rate at which work is done as electrical current flows through a wire. Mechanical power is the rate at which work is done as an object is moved against opposition over a certain distance. Power is either the rate of production or the rate of use of energy and is measured in watts. Power can be expressed with the equation:

$$P = \frac{W}{t}$$

where P is power in watts, W is work done in joules, t is time taken to do the work in seconds.

Power can also be measured in horsepower. To convert watts to horsepower, divide the power in watts by 746 . The equation can be written:

$$\text{Power (in watts)} = \text{Power (in horsepower)} \times 746$$

Sample Problem: Power
Given: An electric motor is used to move an object along a conveyor line. The object has a mass of 150 kilograms and is moved 28 meters in 8 seconds.

Find: The power developed by the motor in watts and horsepower units.
Solution:

$$\text{Force } (F) = 9.8\,\text{m/s}^2 \times \text{mass}$$
$$F = 9.8\,\text{m/s}^2 \times 150\,\text{kg}$$
$$F = 1470\,\text{newtons}$$
$$\text{Work}(W) = F \times d$$
$$W = 1470\,\text{N} \times 28\,\text{m}$$
$$W = 41{,}160\,\text{joules}$$
$$\text{Power}\,(\text{P}) = \frac{W}{t}$$
$$P = \frac{41{,}160}{8}$$
$$P = 5145\,\text{watts}$$
$$\text{Power}\,(\text{in horsepower}) = \frac{\text{Power (in watts)}}{746} P = \frac{5145}{746} = 6.9\text{hp}$$

4.3 Power in DC Electrical Circuits

In terms of voltage and current, power (P) in watts (W) is equal to voltage (in volts) multiplied by current (in amperes). The formula is:

$$P = V \times I$$

For example, a 120 – V electrical outlet with 4 A of current flowing from it has a power value of:

$$P = V \times I = 120\,\text{V} \times 4\,\text{A} = 480\ \text{W}$$

The unit of electrical power is the watt. In the example, 480 W of power is converted by the load portion of the circuit. Another way to find power is:

$$P = \frac{V^2}{R}$$

This formula is used when voltage and resistance are known, but current is not known. Another formula that can be used is:

$$P = R^2 \times R$$

This formula is used when current and resistance are known.

It is easy to find the amount of power converted by each of the resistors in a series DC circuit, such as the one shown in Figure 4.10. In the circuit shown, the amount of power converted by each of the resistors and the total power is found as follows:

1. Power converted by resistor R_1:

   $$P_1 = I^2 \times R_1 = (2\,\text{A})^2 \times 20\Omega = 80\ \text{W}$$

2. Power converted by resistor R_2:

   $$P_2 = R^2 \times R_2 = (2\,\text{A})^2 \times 30\Omega = 120\ \text{W}$$

3. Power converted by resistor R_3:

   $$P_3 = R^2 \times R_3 = (2\,\text{A})^2 \times 50\Omega = 200\ \text{W}$$

4. Power converted by the circuit:

   $$\begin{aligned} P_T &= P_1 + P_2 + P_3 \\ &= 80\,\text{W} + 120\,\text{W} + 200\,\text{W} \\ &= 400\ \text{W} \end{aligned}$$

or $$P_T = V_T \times I = 200\,\text{V} \times 2\,\text{A} = 400\ \text{W}$$

When working with electrical circuits, it is possible to check your results by using other formulas.

Power in parallel circuits is found in the same way as it is for series circuits. In the example shown in Figure 4.11, the power converted by

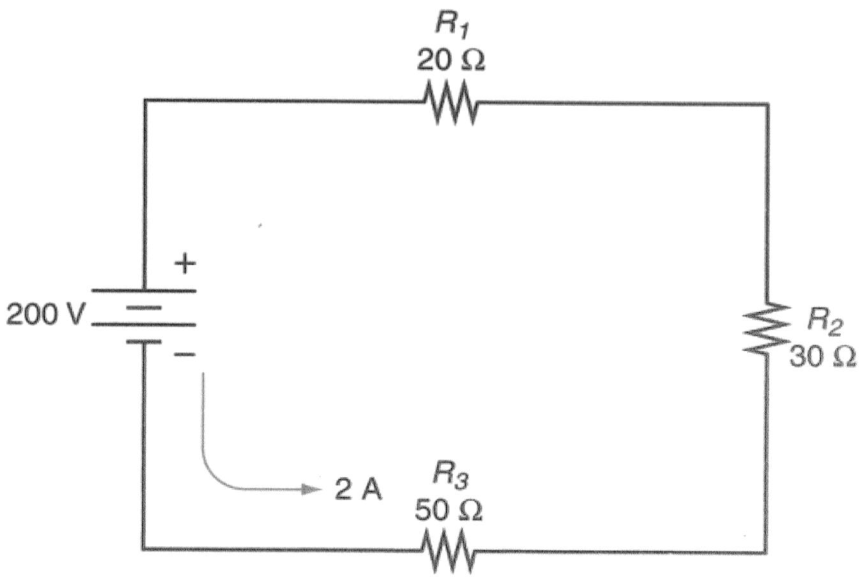

Figure 4.10 Power in a series DC circuit.

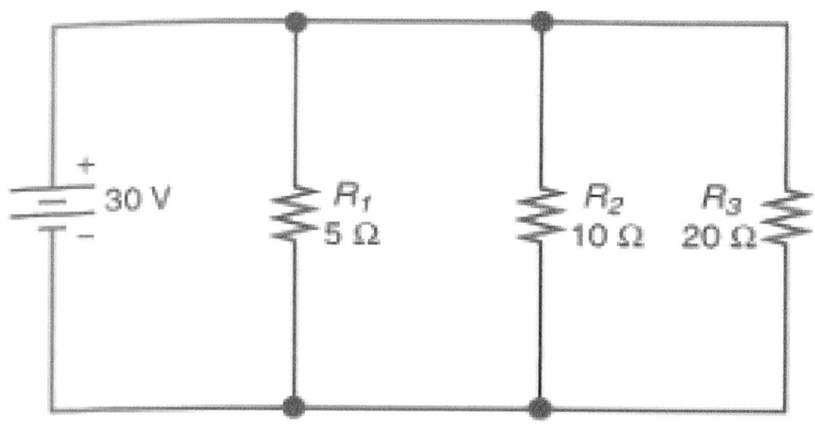

Figure 4.11 Power in a parallel DC circuit.

each of the resistors and the total power of the parallel circuit is found as follows:

1. Power converted by resistor R_1:

$$P_1 = \frac{V^2}{R_1} = \frac{(30\,\text{V})^2}{5\Omega} = \frac{900}{5} = 180\ \text{W}$$

2. Power converted by resistor R_2:

$$P_2 = \frac{V^2}{R_2} = \frac{(30\,\text{V})^2}{10\Omega} = \frac{900}{10} = 90\text{ W}$$

3. Power converted by resistor R_3:

$$P_3 = \frac{V^2}{R_3} = \frac{(30\,\text{V})^2}{20\Omega} = \frac{900}{20} = 45\text{ W}$$

4. Total power converted by the circuit:

$$P_T = P_1 + P_2 + P_3 = 180\,\text{W} + 90\,\text{W} + 45\,\text{W} = 315\text{ W}$$

Refer also to the calculator procedures for electrical power in Appendix A (Figure A.16).

The **watt** is the basic unit of electrical power. To determine the quantity of electrical energy, a factor that indicates power over a period of time must be used. Such a unit of electrical energy is called a watt-second. It is the product of power (in watts) and time (in seconds). It is more common to measure electrical energy in **kilowatt hours** (kWh). It is the kWh quantity of electrical energy that is used to determine the amount of electrical utility bills. A kilowatt-hour is 1000 watts in one hour of time.

As an example, if an electrical heater operates on 120 V, and has a resistance of 20Ω, what is the cost to use the heater for 200 hours at a cost of 5 cents per kWh?

1. $P = \frac{V^2}{R} = \frac{(120\text{ V})^2}{20\Omega} = \frac{14{,}400}{20} = 720\text{ W}$

2. There are 1000 W in a kilowatt $(1000\text{ W} = 1\text{ kW})$, so $720\text{ W} = 0.72\text{ kW}$.

3. Multiply the kW that the heater has been used by the hours of use.

   $\text{kW} \times 200\,\text{h} = \text{ kilowatt-hours (kWh)}$
   $0.72\,\text{kW} \times 200\,\text{h} = 144\text{kWh}$

4. Multiply the kWh by the cost.

   $\text{kWh} \times \text{cost} = \ 1.44\,\text{kWh} \times \$0.05\text{h} = \$7.20$

4.4 Review of Alternating Current Circuits

The following discussion provides a review of **alternating current (AC)** circuits. The basic characteristics of these AC circuits should be reviewed to gain a fundamental understanding of electrical systems.

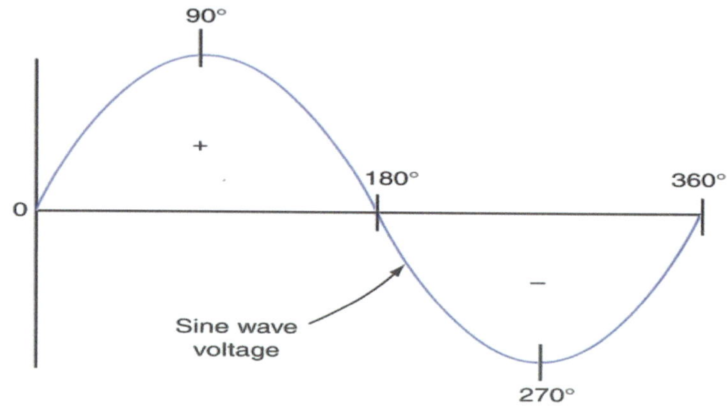

Figure 4.12 One complete cycle of an AC waveform.

4.4.1 Alternating-current basics

When an AC source is connected to some type of load, the current direction changes several times in a given unit of time. Remember that direct current (DC) flows in one direction only. A diagram of one **cycle** of alternating current is compared with a DC waveform in Figure 4.12. This waveform is called an AC sine wave. One complete revolution (360°) of an **AC sine wave** is illustrated in Figure 4.12. Note that the AC sine wave has a positive peak at 90°, then decreases to zero at 180°. It then increases to a peak negative voltage at 270°, then decreases to zero at 360°. The cycle then repeats itself. Current flows in one direction during the positive part and in the opposite direction during the negative half-cycle.

Figure 4.13 shows five cycles of **alternating current**. If the time required for an AC generator to produce five cycles were one second, the frequency of the AC would be five cycles per second. AC generators at power plants in the United States operate at a frequency of 60 cycles per second, or 60 Hz. Hertz is the international unit for frequency measurement. If 60ac sine waves are produced every second, a speed of 60 revolutions per second is needed. This produces a frequency of 60 cycles per second.

Shown in Figure 4.14 are several voltage values associated with alternating current. Among these are peak positive, peak negative, and peak-to-peak AC values. Peak positive is the maximum positive voltage reached during a cycle of AC. Peak negative is the maximum negative voltage reached. **Peak-to-peak** is the voltage value from peak positive to peak negative. A common AC voltage is 120 V, which is used in homes, is an effective value voltage. Its peak value is about 170 V. The **effective value** of AC is defined as the AC

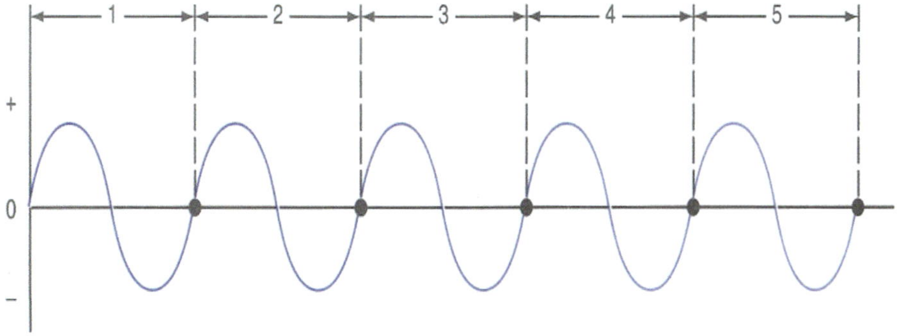

Figure 4.13 Five cycles of alternating current.

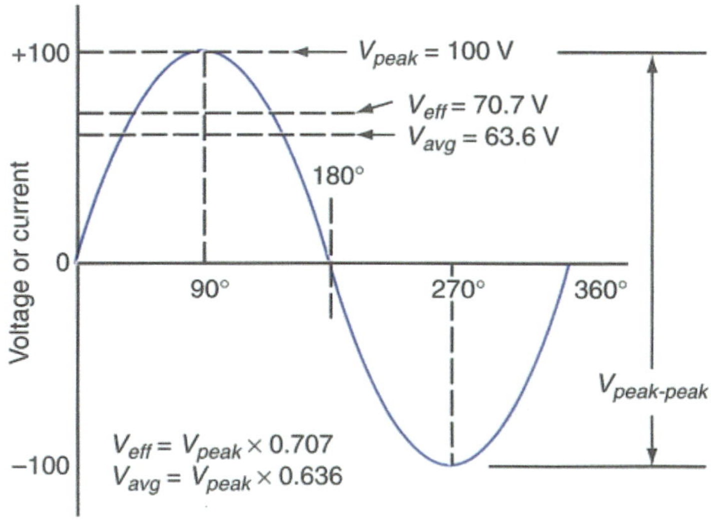

Figure 4.14 Voltage values of an AC waveform.

voltage that will do the same amount of work as a DC voltage of the same value. For instance, in the circuit of Figure 4.15, if the switch is placed in position 2, a 10-volt AC effective value is applied to the lamp. The lamp should produce the same amount of brightness with 10 volts of DC applied. When AC voltage is measured with a meter, the reading indicated is the effective value.

In some cases, it is important to convert one AC value to another. For instance, the voltage rating of electronic devices must be greater than the peak AC voltage applied to them. If 120 volts AC is the measured voltage applied to a device, the peak voltage is about 170 V. So, the device must be rated over 170 V rather than 120 V.

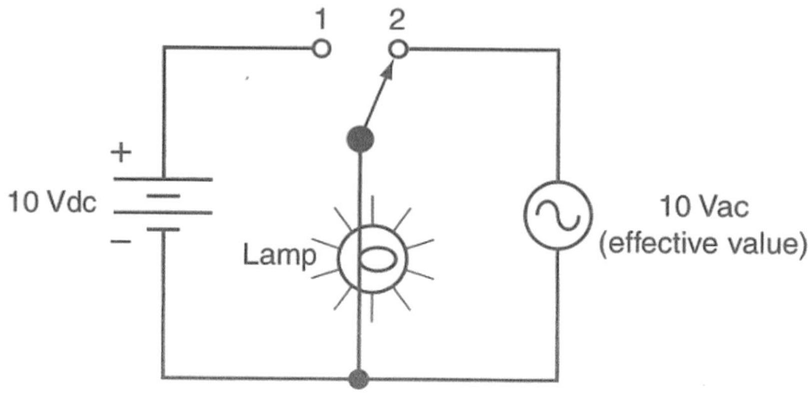

Figure 4.15 Comparison of effective AC voltage and DC voltage.

To determine peak AC, when the measured or effective value is known, use the formula:

$$\text{peak} = 1.41 \times \text{effective value}$$

When 120 V is multiplied by the 1.41 conversion factor, the peak voltage is found to be approximately 170 V.

Two other terms that should be mentioned are **rms value** and **average value**. The letters rms stand for root mean square, and the rms value is equal to 0.707 × peak value. **Root mean square** refers to the mathematical method used to determine effective voltage. Average voltage is the mathematical average of all **instantaneous voltages** that occur at each period of time throughout an alternation. The average value is equal to 0.636 times the peak value for one-half cycle (alternation) AC voltage.

4.4.2 Resistive AC circuits

The simplest type of AC electrical circuit is a resistive circuit, such as the one shown in part *A* of Figure 4.16. The purely resistive circuit offers the same type of opposition to alternating current power sources as it does to pure direct current power sources. In purely resistive AC circuits:

$$\text{voltage}\,(V) = \text{current}\,(I) \times \text{resistence}\,(R)$$

$$\text{current}\,(I) = \frac{\text{voltage}\,(V)}{\text{resistance}\,(R)}$$

$$\text{resistance}\,(R) = \frac{\text{voltage}\,(V)}{\text{current}\,(I)}$$

$$\text{power}\,(P) = \text{voltage}\,(V) \times \text{current}\,(I)$$

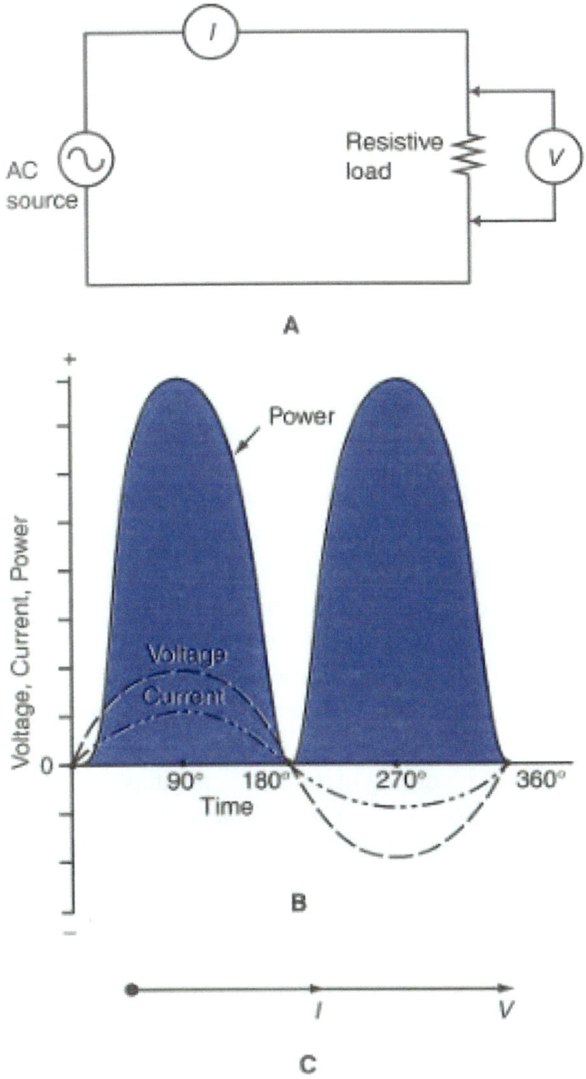

Figure 4.16 Resistive AC circuit. A – Circuit diagram. B – Waveforms. C – Vector representation.

These basic electrical relationships show that when voltage is increased, the current in the circuit increases proportionally. Also, as resistance is increased, the current in the circuit decreases. By looking at the waveforms of part B Figure 4.16, we can see that the voltage and current in a purely resistive circuit, with AC applied, are in phase. An in-phase relationship exists when the minimum and maximum values of both voltage and current occur at the same time interval.

The power curve is also shown in part B of Figure 4.16. The curve shows that the power converted by the circuit is a product of voltage times current ($P = V \times I$). Thus, when an AC circuit contains only resistance, its behavior is very similar to a DC circuit. Purely resistive circuits are seldom encountered in the design of electrical systems, although some devices are primarily resistive in nature.

4.4.3 Inductive AC circuits

Inductance is very commonly encountered in electrical power systems. Any inductive circuit exhibits the property of **inductance (*L*)**, which is the opposition to a change in current in a circuit. This property is evident due to a **counterelectromotive force (cemf)** that is produced when a magnetic field is developed around a coil of wire. The magnetic flux produced around the coils affects circuit action. Thus, the inductive property (cemf) produced by a magnetic field offers an *opposition* to change in the current flow in a circuit.

This property is found in coils of wire (which are called **inductors**) and in rotating machinery and transformer windings. Inductance is also present in electrical power transmission and distribution lines to some extent. The unit of measurement for inductance is the **henry (*H*)**. A circuit has a one-henry inductance if a current changing at a rate of one ampere per second produces an induced counterelectromotive force of one volt.

Inductors can be connected in series or parallel. When inductors are connected to prevent the magnetic field from affecting the other, these formulas are used to find total inductance (L_T).

Inductance in series is found with the formula:

$$L_T = L_1 + L_2 + L_3 + \cdots + L_N$$

where L_1, L_2, and L_3, are individual inductance values and N is the total number of inductors.

Inductance in parallel is found with the formula:

$$\frac{1}{L_T} = \frac{1}{L_1} + \frac{1}{L_2} + \frac{1}{L_3} + \cdots + \frac{1}{L_N}$$

where L_1, L_2, and L_3, are individual inductance values and N is the total number of inductors.

When inductors are connected so that the magnetic field of one affects the other, mutual inductance increases or decreases the total inductance. The effect of mutual inductance depends on the physical positioning of the inductors. Their distance apart and the direction in which they are wound affect mutual inductance (*M*). The formulas used to find total inductance are:

1. *Series aiding*: $L_T = L_1 + L_2 + 2M$
2. *Series opposing*: $L_T = L_1 + L_2 - 2M$
3. *Parallel aiding*: $\frac{1}{L_T} = \frac{1}{L_1 + M} + \frac{1}{L_2 + M}$
4. *Parallel opposing*: $\frac{1}{L_T} = \frac{1}{L_1 - M} + \frac{1}{L_2 - M}$

In an inductive circuit with AC applied, an opposition to current is created by the inductance. This type of opposition is known as inductive reactance (X_L). It is measured in ohms. The inductive reactance of an AC circuit depends upon the inductance of the circuit and the rate of change of current. The frequency of the applied alternating current establishes the rate of change of the current.

Inductive reactance (X_L) can be expressed as:

$$X_L = 2\pi fL$$

where, X_L is the inductive reactance in ohms, 2π (6.28) is the mathematical expression of one sine wave (0° to 360°), *f* is the frequency of the AC source in hertz, and *L* is the inductance of the circuit in henrys.

Sample Problem:
Given: A frequency of 60 HZ and an inductance of 20H.
Find: Inductive reactance.

Solution:

$$X_L = 2\pi \times 60\,\text{Hz} \times 20\text{H}$$
$$X_L = 7536\Omega$$

Inductance, shown in part A of Figure 4.17, adds more complexity to the relationship between voltage and current in an AC circuit. Examine the figure and work through the following problem.

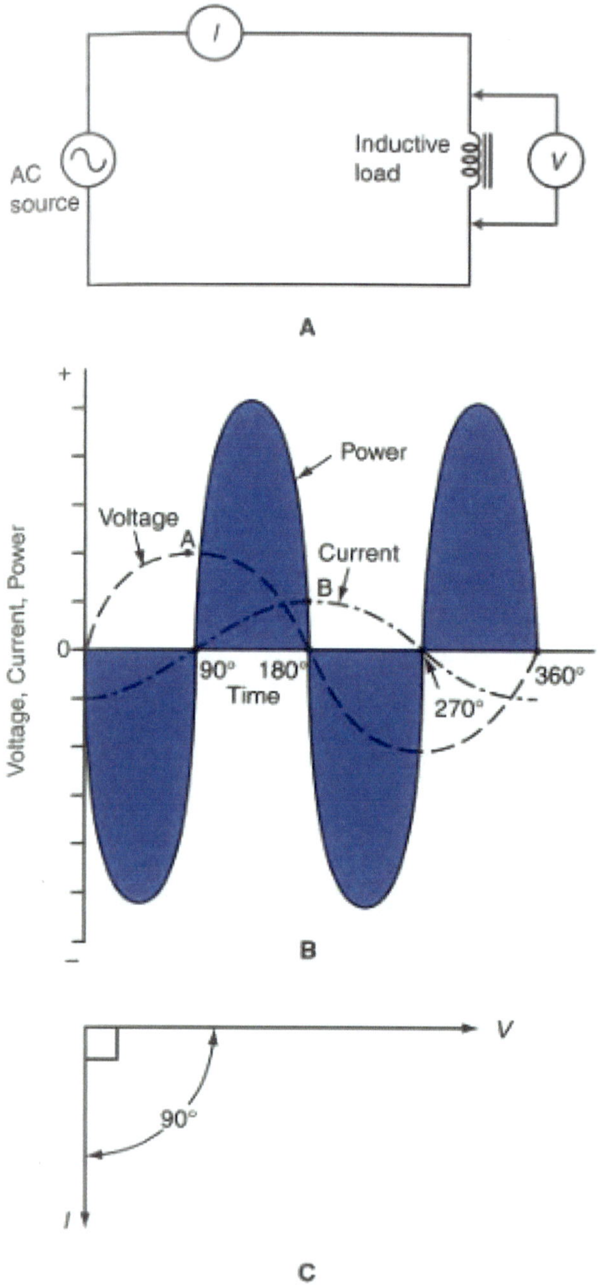

Figure 4.17 Inductive AC circuit. A – Circuit diagram. B – Waveforms. C – Vector representation.

Sample Problem: Energy stored in an inductor.

Given: A 10 henry coil has 5.8 amperes of current flowing through it.
Find: The amount of energy stored in the coil.
Solution:

A coil stores energy in its magnetic field due to current flowing through it. The amount of energy is defined by the equation:

$$W = \frac{1}{2} \times L \times R^2$$

where *W* is the energy stored in joules, *L* is the coil inductance in henrys, and *I* is the current through the coil in amperes.

$$W = \frac{1}{2} \times L \times I^2$$
$$W = \frac{1}{2} \times 10\text{H} \times (5.8\,\text{A})^2$$
$$W = 168.2\,\text{joules}$$

The opposition to change of current is evident in the diagram of part B of Figure 4.18. In an inductive circuit, we can say that voltage leads current or that current lags voltage. If the circuit were purely inductive (contained no resistance), the voltage would lead the current by 90° Figure 4.17B, and no actual power would be converted in the circuit. However, since all actual circuits have some resistance, the inductive characteristic of a circuit might typically cause the condition shown in Figure 4.18B to exist. Here, the voltage is leading the current by 30°. The angular separation between voltage and current is called the **phase angle** (θ). The phase angle increases as the inductance of the circuit increases. This type of circuit is called a *resistive-inductive* (RL) circuit.

In terms of power conversion, a purely inductive circuit would not convert any actual power in a circuit. All AC power would be delivered back to the power source. Refer back to part B Figure 4.17 and note points *A* and *B* on the waveforms. These points show that at the peak of each waveform the power is zero. The power curves shown are equal and opposite in value and will cancel each other out. Where both voltage and current are positive, the power is also positive since the product of two positive values is positive. When one value is positive and the other is negative, the product of the two values is negative. Therefore, the power converted is negative. Negative power means that electrical energy is being returned from the load device to the power source without being converted to another form of energy.

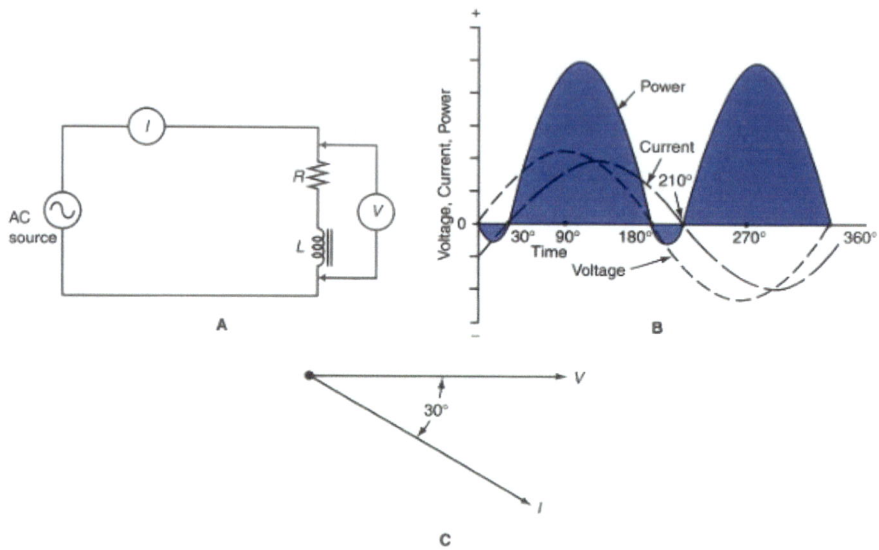

Figure 4.18 Series RL circuit. A – Circuit diagram. B – Waveforms. C – Vector representation.

Therefore, the power converted in a purely inductive circuit (90° phase angle) would be equal to zero.

Compare the purely inductive circuit waveforms to those of Figure 4.18. In the practical resistive-inductive (RL) circuit, part of the power supplied from the source is converted in the circuit. Only during the intervals from 0° to 30° and from 180° to 210° does negative power result. The remainder of the cycle produces positive power. Therefore, most of the electrical energy supplied by the source is converted to another form of energy.

4.4.4 Capacitive AC circuits

Figure 4.19 shows a capacitive device connected to an AC source. We know that whenever two conductive materials (plates) are separated by an insulating material, called a dielectric, the property of capacitance is exhibited. **Capacitance (C)** is a property that allows the storage of an electric charge. Capacitors are devices that have the capability of storing an electrical charge. They have many applications in electrical systems. Some capacitors are shown in Figure 4.20.

Always be careful when working with capacitors. A capacitor can hold a charge for a long period of time. Some capacitors hold dangerous levels of charge. Be sure to test and discharge all capacitors before you handle them.

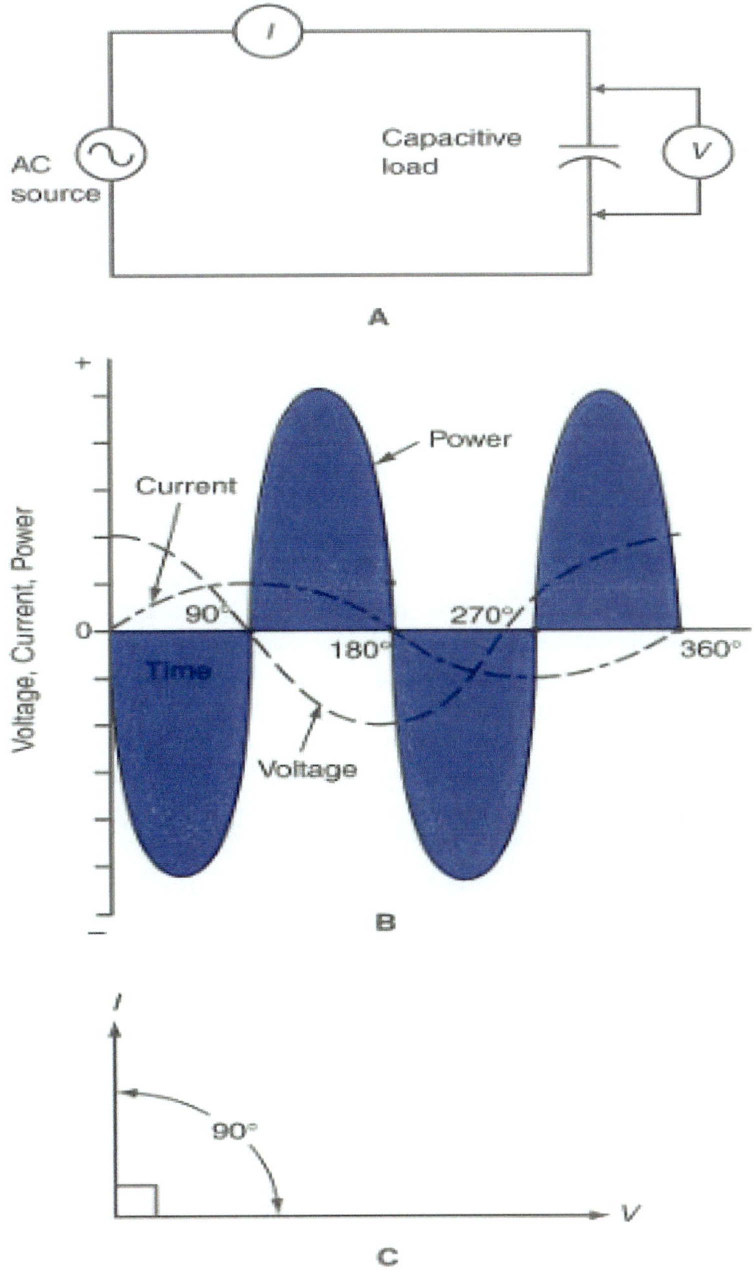

Figure 4.19 Capacitive AC circuit. A – Circuit diagram. B – Waveforms. C – Vector representation.

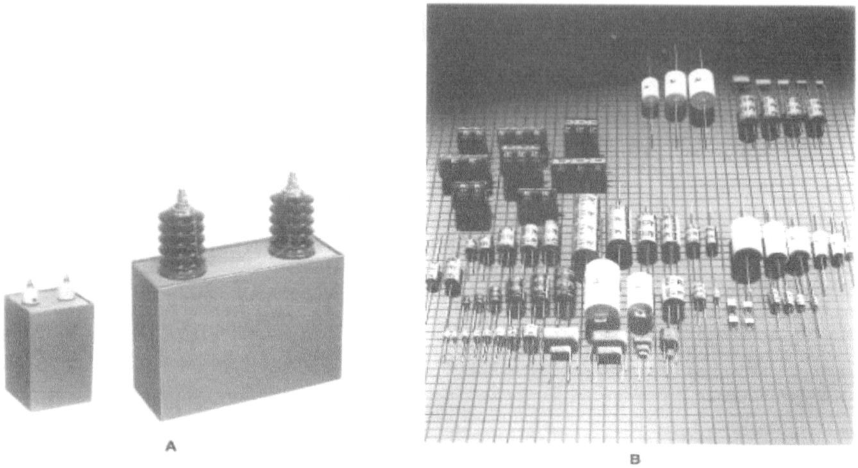

Figure 4.20 Some types of capacitors. A – High voltage capacitors. (Aerovox, Inc.) B – Various designs of small capacitors. (Illinois Capacitors Co.)

The operation of a capacitor in a circuit is dependent upon its ability to charge and discharge. When a capacitor **charges**, an excess of electrons (negative charge) is accumulated on one plate and a deficiency of electrons (positive charge) is created on the other plate. Capacitance is determined by the size of the conductive material (plates), by their distance apart, and by the type of dielectric used as the insulating material. Capacitance is **directly proportional** to the plate size and **inversely proportional** to the distance between the plates.

The unit of capacitance is the **farad (F).** A capacitance of one farad results when a potential of one volt causes an electrical charge of one coulomb to accumulate on a capacitor. Since the farad is a very large unit, **microfarad (μF)** values are ordinarily assigned to capacitors.

Sample Problem: Energy stored in a capacitor.

Given: A 100μF capacitor has 120 volts applied.
Find: The amount of energy stored in the capacitor.
Solution:
Energy is stored by a capacitor in its electrostatic field when voltage is applied to it. The amount of energy is defined by the equation:

$$W = \frac{1}{2} \times C \times V^2$$

where W is the energy stored in a capacitor in joules, C is the capacitance of the capacitor in farads, and V is the applied voltage in volts.

$$W = \frac{1}{2} \times C \times V^2$$

$$W = \frac{1}{2} \times 100^{-6} \times 120^2 = 0.72 \text{ joules}$$

Adding capacitors in series has the same effect as increasing the distance between plates, which reduces total capacitance. The total capacitance of capacitors connected in series is less than any individual capacitance. Total capacitance (C_T) is found in the same way as parallel resistance. The reciprocal formula is used:

$$\frac{1}{C_T} = \frac{1}{C_1} + \frac{1}{C_2} + \frac{1}{C_3} + \cdots + \frac{1}{C_N}$$

where C_1, C_2, and C_3, are individual capacitance values and N is the total number of capacitors.

Capacitors in parallel are similar to one capacitor with its plate area increased. Doubling the plate area would double the capacitance. Capacitance in parallel is found by adding individual values, just as with series resistors, or:

$$C_T = C_1 + C_2 + C_3 + \cdots + C_N$$

where C_1, C_2, and C_3, are individual capacitance values and N is the total number of capacitors.

If a direct current is applied to a capacitor, the capacitor will charge to the value of that de voltage. After the capacitor is **fully charged**, it will block the flow of direct current. However, if AC is applied to a capacitor, the changing value of current will cause the capacitor to alternately charge and discharge. In a purely capacitive circuit, the situation shown in Figure 4.19 would exist.

The greatest amount of current flows in a capacitive circuit when the voltage changes most rapidly. The most rapid change in voltage occurs at the 0° and 180° positions where the polarity changes. At these positions, maximum current is developed in the circuit. When the rate of change of the voltage value is slow, such as near the 90° and 270° positions, a small amount of current flows. In examining Figure 4.19, we can observe that, in a purely capacitive circuit, the current leads voltage by 90° (or the voltage lags the current by 90°). Since a 90° phase angle exists, no power would be converted

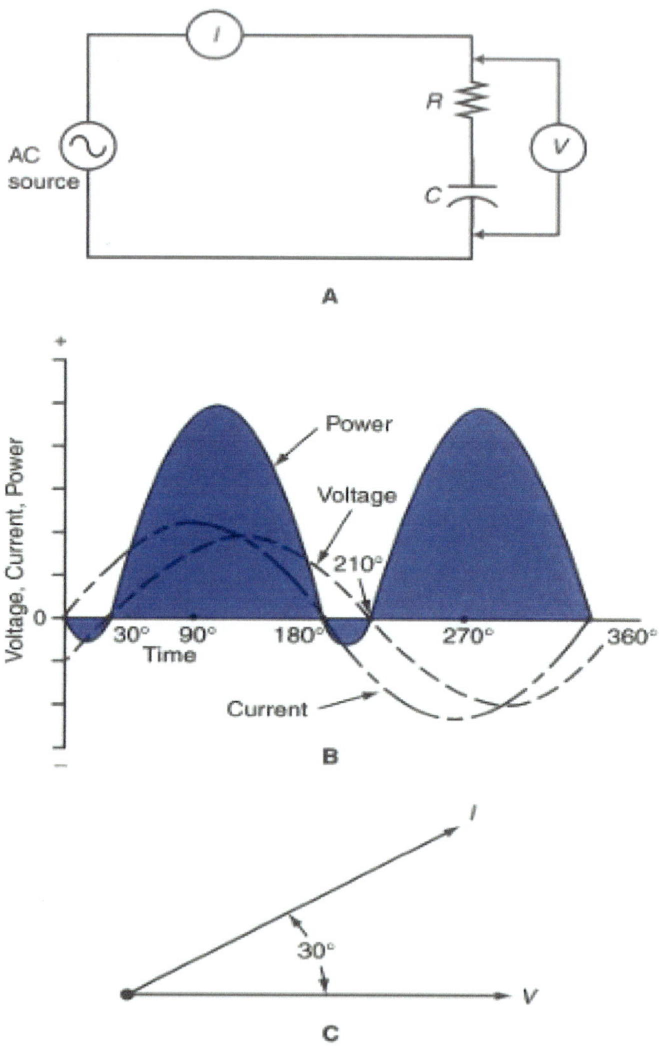

Figure 4.21 Series RC circuit. A – Circuit diagram. B – Waveforms. C – Vector representation.

in this circuit, just as no power was developed in the purely inductive circuit. As shown in part B of Figure 4.19, the positive and negative power waveforms will cancel one another out.

Since all circuits contain some resistance, a more practical circuit is the resistive-capacitive (RC) circuit shown in part A of Figure 4.21. In an RC circuit, the current leads the voltage by some phase angle between 0° and 90°. As capacitance increases, with no corresponding increase in resistance, the phase angle becomes greater. The waveforms of part B of Figure 4.21 show an RC circuit in which current leads voltage by 30°. This circuit is similar to the circuit of Figure 4.18. Power is converted in the circuit except during the 0° to 30° interval and the 180° to 210° interval. In the RC circuit shown, most of the electrical energy supplied by the source is converted to another form of energy in the load.

Due to the electrostatic field that is developed around a capacitor, an opposition to the flow of alternating current exists. This opposition is known as **capacitive reactance** (X_C). It is measured in ohms. Capacitive reactance is expressed as:

$$X_C = \frac{1}{2\pi fc}$$

where X_C is the capacitive reactance in ohms, $2\pi(6.28)$ is the mathematical expression of one sine wave (0° to 360°), f is the frequency of the source in hertz, and C is the capacitance in farads.

Sample Problem:

Given: A frequency of 50 Hz and a capacitance of 200μF.
Find: Capacitive reactance.
Solution:

$$X_c = \frac{1}{2\pi fC}$$

$$X_C = \frac{1}{2\pi \times 50\,\text{Hz} \times 200^{-6}\ \text{F}}$$

$$X_C = 15.92\Omega$$

4.5 Vector and Phasor Diagrams for AC Circuits

In part *C* of Figures 4.16–4.19 , and 4.21, a **vector diagram** was shown for each circuit condition that was illustrated. **Vectors** are straight lines that have a specific direction and length. They are used to represent voltage or current values. An understanding of vector diagrams (sometimes called **phasor diagrams**) is important when dealing with alternating current. Rather than using waveforms to show **phase relationships**, it is possible to use a vector or phasor representation.

Ordinarily, when beginning a vector diagram, a horizontal line is drawn with its left end as the reference point. Rotation in a counterclockwise direction from the reference point is then considered to be a positive direction. Note that in the preceding diagrams, the voltage vector was the reference. For

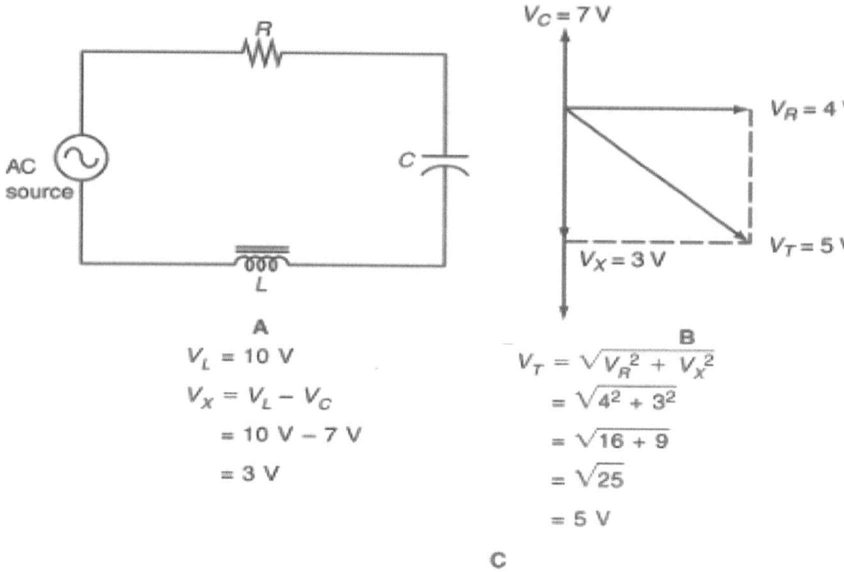

Figure 4.22 Series RCL circuit. A – Circuit diagram. B – Vector diagram. C – Problem solution.

the inductive circuits, the current vector was drawn in a clockwise direction, indicating a lagging condition. A leading condition is shown for the capacitive circuits by the use of a current vector drawn in a counterclockwise direction from the voltage vector. Examine the phasor diagram in Figure 4.22. This is a phasor diagram for circuits with resistance, capacitance, and inductance. It is called an RCL circuit.

Series RCL circuits have resistance (R), capacitance (C), and inductance (L). The total reactance (X_T) is found by subtracting the smaller reactance (X_L or X_C) from the larger one. Reactive voltage (V_X) is found by obtaining the difference between V_L and V_C. The effects of the capacitance and inductance are 180° out of phase with each other. Right triangles are used to show the simplified relationships of the circuit values.

4.6 Impedance in AC Circuits

An application in the use of vectors is for determining the total opposition of an AC circuit to the flow of current. This total opposition is called **impedance** (**Z**) and is measured in ohms.

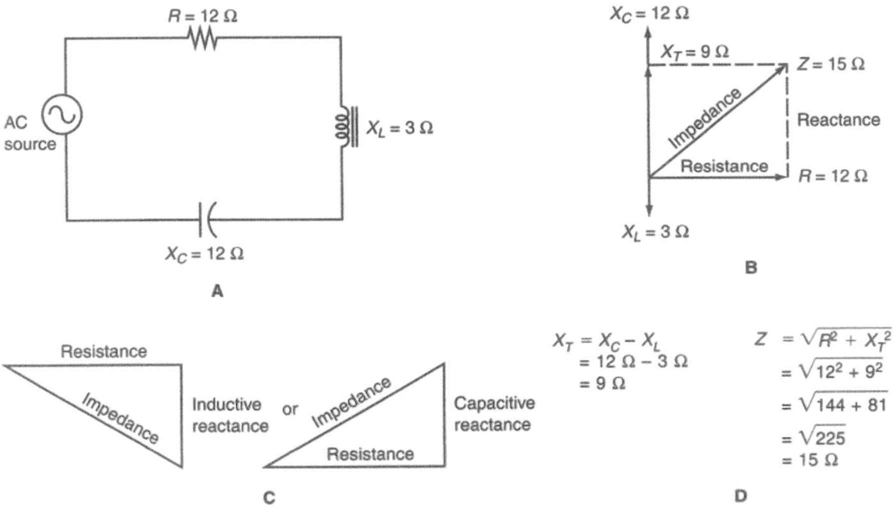

Figure 4.23 Finding total impedance (Z) in a series RCL circuit. A – Circuit diagram. B – Vector diagram. C – Impedance triangles. D – Problem solution.

4.6.1 Impedance in series AC circuits

Both resistances and **reactances** in AC circuits affect the opposition to current. The impedance of an AC circuit can be expressed as:

$$Z = \frac{V}{I}$$

or

$$Z = \sqrt{R + (X_L - X_C)^2}$$

This formula can be clarified by using the **vector diagram** shown in Figure 4.23. The total reactance (X_T) of an AC circuit can be found by subtracting the smallest reactance (X_L or X_C) from the largest reactance. The impedance of a series AC circuit is determined by using the preceding formula since a **right triangle** (called **an impedance triangle**) is formed by the three quantities that oppose the flow of alternating current. A sample problem for finding the total impedance of a series as circuit is shown in Figure 4.23.

Sample problem:
Given: A series AC circuit has the following component values:
$R = 100\Omega$, $L = 1\text{H}$, $C = 10\mu\text{F}$, $f = 60$ Hz.

Find: Total impedance of the circuit.
Solution:
First, find the inductive reactance.

$$X_L = 2\pi \times f \times L$$
$$X_L = 6.28 \times 60 \text{ Hz} \times 1\text{H}$$
$$X_L = 376.8\Omega$$

Then find the capacitive reactance.

$$X_C = \frac{1}{2\pi fC}$$
$$X_C = \frac{1}{6.28 \times 60\text{Hz} \times 10^{-6} \text{ F}}$$
$$X_C = 265.4\Omega$$

Then calculate the impedance.

$$Z = \sqrt{R + (X_L - X_C)^2}$$
$$Z = \sqrt{100^2 + (376.8 - 265.4)^2}$$
$$Z = \sqrt{10,000 + 12,410}$$
$$Z = 149.7\Omega$$

4.6.2 Impedance in parallel AC circuits

When components are connected in parallel, the calculation of impedance becomes more complex. Figure 4.24 shows a simple parallel RCL circuit. Since the total impedance in the circuit is smaller than the resistance or reactance, an impedance triangle such as the one shown in Figure 4.23 cannot be developed. A simple method used to find impedances in parallel circuits is the admittance triangle shown in Figure 4.24. The following quantities can be plotted on the triangle:

$$\text{Admittance} = \frac{1}{Z}$$

$$\text{Conductance} = \frac{1}{R}$$

$$\text{Inductive susceptance} = \frac{1}{X_L}$$

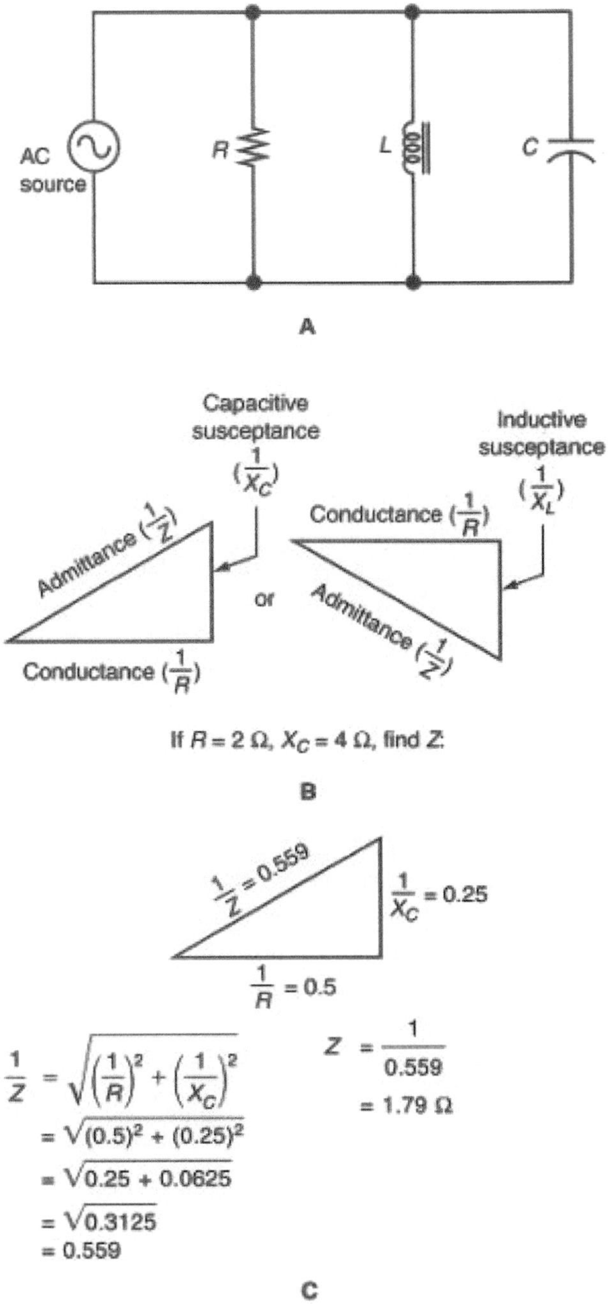

Figure 4.24 Finding total impedance in a parallel RCL circuit. A – Circuit diagram. B – Admittance triangles. C – Problem solution.

$$\text{Capacitive susceptance} = \frac{1}{X_C}$$

Notice that these quantities are the reciprocals of each type of opposition to alternating current. Therefore, since total impedance (Z) is the smallest quantity in a parallel AC circuit, it becomes the largest value on the admittance triangle. The sample problem of Figure 4.24 shows the procedure used to find the total impedance of a parallel RCL circuit.

4.7 Power Relationships in Single-phase AC Circuits

An understanding of basic power relationships in AC circuits is very important when studying electrical systems. In the previous sections, resistive, inductive, and capacitive circuits were discussed. Power converted in these circuits was discussed in terms of power waveforms that were determined by the phase angle between voltage and current. In a DC circuit, power is equal to the product of voltage and current ($P = V \times I$). This formula is also true for purely resistive circuits. However, when a **reactance** (either inductive or capacitive) is present in an AC circuit, power is no longer a product of voltage and current.

Since reactive circuits cause changes in the method used to compute power, the following techniques express the basic power relationships in AC circuits. The product of voltage and current is expressed in **voltamperes (VA)** or **kilovoltamperes (kVA)** and is known as **apparent power**. When using meters to measure power in an AC circuit, apparent power is the voltage reading multiplied by the current reading. The actual power that is converted to another form of energy by the circuit is measured with a wattmeter. This actual power is referred to as true power. It is useful to know the ratio of true power converted in a circuit to apparent power. This ratio is called the **power factor (PF)** and is expressed as:

$$\text{PF} = \frac{\text{P}}{\text{VA}}$$

or

$$\%\text{PF} = \frac{\text{P}}{\text{VA}} \times 100$$

where PF is the power factor of the circuit, P is the true power in watts, and VA is the apparent power in voltamperes. Power factor is a ratio and has no units.

The maximum value of power factor is 1.0 , or 100%, which would be obtained in a purely resistive circuit. This is referred to as a **unity power factor**.

The phase angle between voltage and current in an AC circuit determines the power factor. If a purely inductive or capacitive circuit existed, the 90° phase angle would cause a power factor of zero to result. In practical circuits, the power factor varies according to the relative values of resistance and reactance.

Sample Problem:

Given: A 240-volt, 60-hertz, 30-amp electric motor is rated at 6000 watts.

Find: Power factor at which the motor operates.

Solution:

$$\%PF = \frac{P}{VA} \times 100$$

$$\%PF = \frac{(6000\ W \times 100)}{(240\ V \times 30\ A)}$$

$$\%PF = 83.3\%$$

The power relationships we have discussed can be simplified by looking at the power triangle shown in Figure 4.25. There are two components that affect the power relationship in an AC circuit. The in-phase (resistive) component that results in power conversion in the circuit is called **active power**. Active power is the true power of the circuit and is measured in watts. The second component is that which results from an inductive or capacitive reactance and is 90° out of phase with the active power. This component, called reactive power, does not produce an energy conversion in the circuit. The unit of measure of reactive power is the **voltampere reactive (VAR)**.

The power triangle of Figure 4.25 shows **true power (watts)** on the horizontal axis, **reactive power (VAR)** at a 90° angle from the true power, and **voltamperes (VA)** as the longest side (hypotenuse) of the right triangle. This right triangle has a horizontal axis that corresponds to the resistive component of the circuit, while the vertical axis corresponds to the reactive component. The hypotenuse represents the resultant, which is based on the relative values of resistance and reactance in the circuit. We can now see how important vector representation and an understanding of the right triangle are in analyzing AC circuits.

We can further examine the power relationships of the **power triangle** by expressing each value mathematically, based on the value of apparent

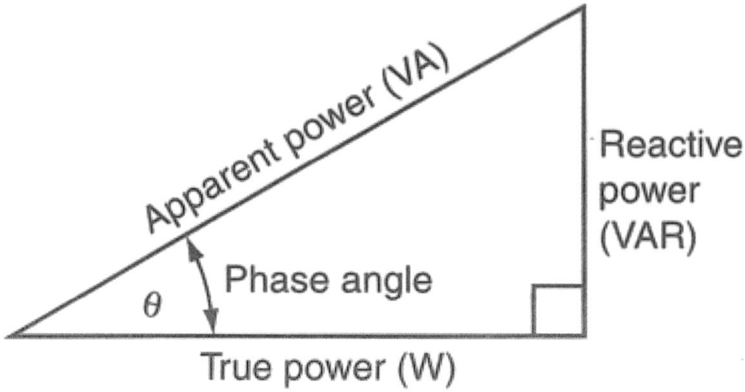

Figure 4.25 Power triangle.

power (VA) and the **phase angle (θ)**. Remember that the phase angle is the amount of phase shift, in degrees, between voltage and current in the circuit. **Trigonometric ratios**, which are discussed in **Appendix B**, show that the sine of an angle of a right triangle is expressed as:

$$\text{sine } \theta = \frac{\text{opposite side}}{\text{hypotenuse}}$$

Since this is true, the phase angle can be expressed as:

$$\text{sine } \theta = \frac{\text{reactive power (VA)}}{\text{apparent power (VA)}}$$

Therefore:

$$\text{VAR} = \text{VA} \times \text{sine } \theta$$

We can determine either the phase angle or the VAR value by using trigonometric functions. The cosine of an angle of a right triangle is expressed as:

$$\text{cosine } \theta = \frac{\text{adjacent side}}{\text{hypotenuse}}$$

Thus, in terms of the power triangle:

$$\text{cosine } \theta = \frac{\text{true power (W)}}{\text{apparent power (VA)}}$$

Therefore, true power can be expressed as:

$$\text{W} = \text{VA} \times \text{cosine } \theta$$

Sample Problem:

Given: A circuit has the following values: applied voltage = 240 V, current = 12 A; power factor = 0.83.

Find: True power of the circuit.

Solution:

$$W = VA \times \text{cosine } \theta$$
$$W = 240V \times 12\,A \times 0.83$$
$$W = 2390\,W$$

Note that the expression

$$\frac{\text{true power}}{\text{apparent power}}$$

is the **power factor** of a circuit. Therefore, the power factor is equal to the cosine of the **phase angle** ($PF = \text{cosine } \theta$).

Right triangle relationships can also be expressed to determine the value of any of the sides of the power triangle when the other two values are known. These expressions are as follows:

$$VA = \sqrt{W^2 + VAR^2}$$
$$W = \sqrt{V^2 - VAR^2}$$
$$VAR = \sqrt{V^2 - W^2}$$

Sample Problem:

Given: Total reactive power is 54 VAR, applied voltage is 120 V, and current is 0.5 A.

Find: True power of the circuit.

Solution:

$$W = \sqrt{VA^2 - VAR^2}$$
$$W = \sqrt{(120 \times 0.5)^2 - 54^2}$$
$$W = \sqrt{3600 - 2916}$$
$$W = 26.15W$$

Appendix B should be reviewed in order to gain a better understanding of the use of **right triangles** and **trigonometric ratios** for solving AC circuit problems.

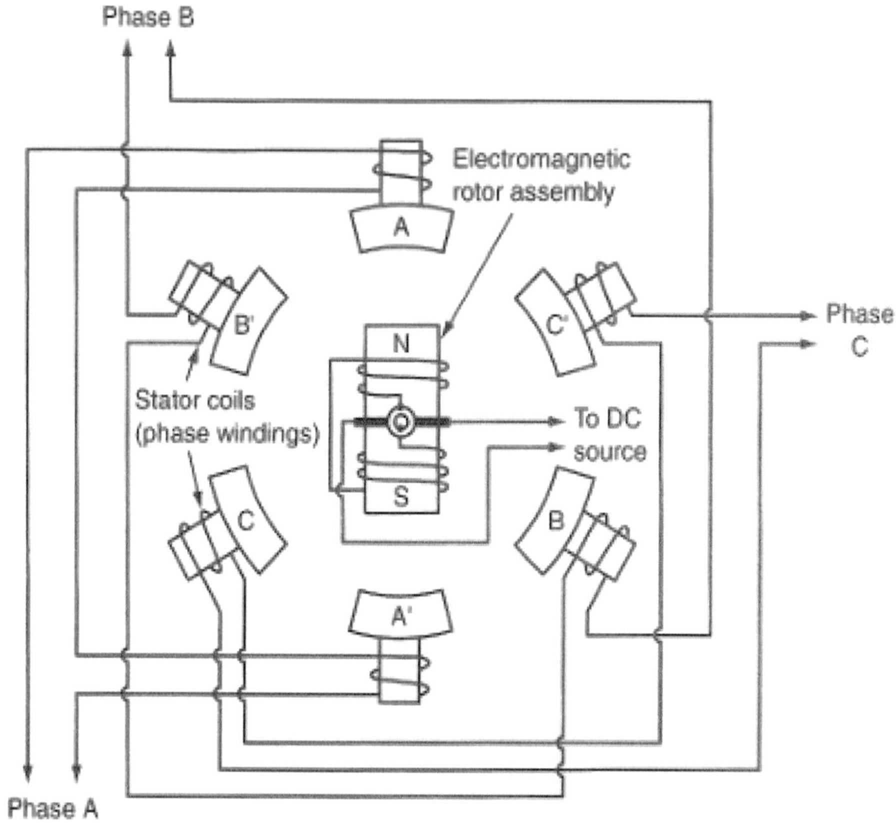

Figure 4.26 Three-phase AC generator pictorial diagram.

4.8 Power Relationships in Three-phase AC Circuits

To illustrate the basic concepts of **three-phase power systems**, we will use the example of a three-phase AC generator. A simplified pictorial diagram of a three-phase generator is shown in Figure 4.26. A three-phase voltage diagram is shown in Figure 4.27.

Single-phase AC voltage is produced by single-phase AC generators, or it can be obtained across two power lines of a three-phase system. A single-phase AC source has a hot wire and a neutral wire to carry electrical current. The neutral is grounded to help prevent electrical shocks. Single-phase power is the type of power distributed to our homes. A three-phase AC source has three power lines that carry electrical current. Three-phase voltage is produced by three-phase generators at power plants. Three-phase voltage is a combination of three single-phase voltages that are electrically connected.

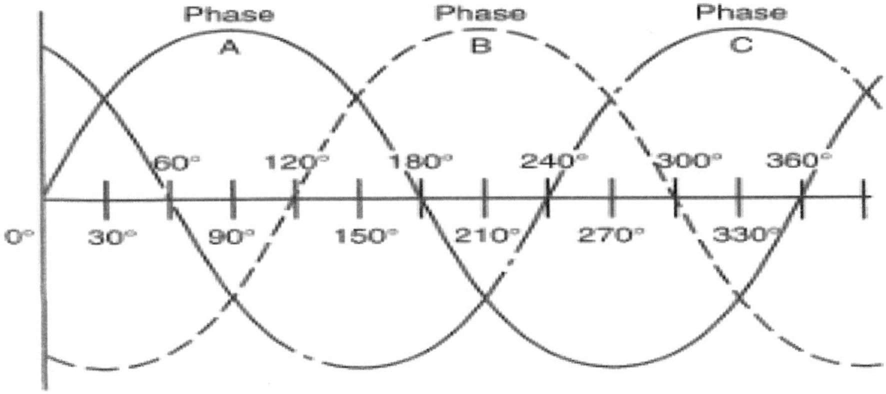

Figure 4.27 Three-phase AC voltage diagram.

This voltage is similar to three **single-phase AC sine waves** separated in phase by 120°, see Figure 4.27 . **Three-phase AC** is used to power large equipment in industry and commercial buildings. It is not distributed to homes. There are three current-carrying power lines on a three-phase system. A three-phase wye system has a neutral connection, and a three-phase delta system does not.

Remember that single-phase AC voltage is in the form of a sine wave. Single-phase AC voltage is used for low-power applications. Almost all electrical power is generated and transmitted over long distances as three-phase ac. Three coils are placed 120° apart in a generator to produce **three-phase AC voltage**. Most AC motors over one horsepower in size operate with three-phase AC power applied. Most industries and commercial buildings have three-phase equipment.

Three-phase AC systems have several advantages over single-phase systems. In a single-phase system, the power is said to be "pulsating." The peak values along a single-phase AC sine wave are separated by 360°. This is similar to a one-cylinder gas engine. A three-phase system is somewhat like a multicylinder gas engine. The power is more steady. One cylinder is compressing when the others are not, which is similar to the voltages in three-phase AC systems. The power of one separate phase is pulsating, but the total power is more constant. The peak values of three-phase AC are separated by 120°. This makes three-phase AC power more desirable to use.

The **power ratings** of motors and generators are greater when three-phase AC power is used. For a certain frame size, the rating of a three-phase AC motor is almost 50 percent larger than a similar single-phase AC motor.

In Figure 4.26, poles A′, B′, and C′ represent the beginnings of each of the phase windings, while poles A, B, and C represent the ends of each of the windings. These windings can be connected in either of two ways. These methods of connection, called the wye configuration and the delta configuration, are the basic types of three-phase power systems. These three-phase connections are shown schematically in Figure 4.28. Keep in mind that these methods apply not only to three-phase AC generators, but also to three-phase transformer windings and three-phase motor windings.

In the **wye connection** in part A of Figure 4.28, the beginnings or the ends of each winding are connected together. The other sides of the windings become the AC lines from the generator. The voltage across the AC lines (V_L) is equal to the square root of 3 (1.73) multiplied by the voltage across the phase windings (V_p), or:

$$V_L = V_p \times 1.73$$

Sample Problem:
Given: Phase voltage of 120 volts for a three-phase wye system.
Find: Line voltage.
Solution:

$$V_L = V_p \times 1.73$$
$$V_L = 120\text{ V} \times 1.73$$
$$V_L = 208\text{ V}$$

The **line currents** (I_L) are equal to the phase currents or (I_p), or:

$$I_L = I_p$$

In the **delta connection** in part B of Figure 4.28, the end of phase winding is connected to the beginning of the adjacent phase winding. The line voltages (V_L) are equal to the **phase voltages** (V_p) .The **line currents** (I_L) are equal to **the phase currents** (I_p) multiplied by 1.73 .

The power developed in each phase (P_p) for either a wye or a delta circuit is expressed as:

$$P_p = V_p \times I_p \times \text{PF}$$

Sample Problem:

Given: A three-phase delta system has a phase voltage of 240 volts, phase current of 20 amperes, and a power factor of 0.75 .

Find: Power per phase.

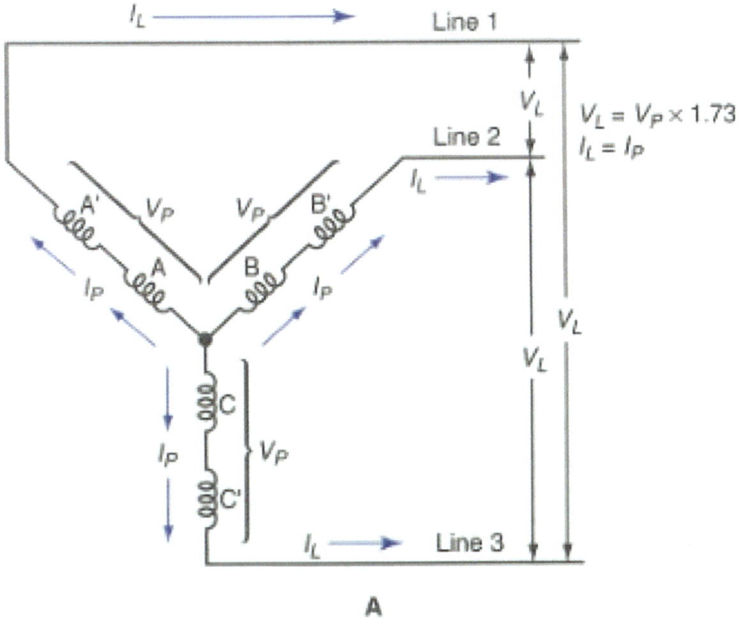

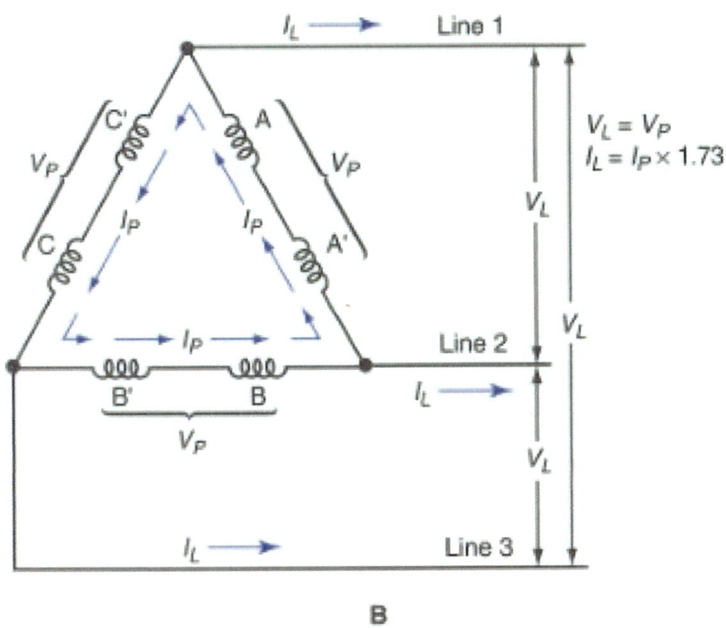

Figure 4.28 Three-phase connections. A – The wye (or star) connection. B – The delta connection.

Solution:

$$P_p = V_p \times I_p \times \text{PF}$$
$$P_p = 240\,\text{V} \times 20\,\text{A} \times 0.75$$
$$P_p = 3600\,\text{W}$$

where PF is the **power factor** of the load.

The **total power** (**P_T**) developed by all three phases of a three-phase system is expressed as:

$$P_T = 3 \times P_P$$
$$P_T = 3 \times V_p \times I_p \times \text{PF}$$
$$P_T = 1.73 \times V_L \times I_L \times \text{PF}$$

Sample Problem:

Given: A three-phase wye system has a phase voltage of 277 volts, phase current of 10 amperes, and a power factor of 0.85 .

Find: Total three-phase power.

Solution:

$$P_T = 3 \times V_p \times I_p \times \text{PF}$$
$$P_T = 3 \times 277\,\text{V} \times 10\,\text{A} \times 0.85$$
$$P_T = 7063.5\,\text{W}$$

We can summarize three-phase power relationships as follows:

Voltamperes per phase = $V_p \times I_p$
Total voltamperes = $3 \times V_p \times I_p$
Total voltamperes = $1.73 \times V_L \times I_L$

$$\text{Power factor (PF)} = \frac{\text{true power}}{1.73 \times V_L \times I_L}$$

$$\text{Power factor (PF)} = \frac{\text{true power}}{3 \times V_p \times I_p}$$

$$\text{Power per phase } (P_p) = V_p \times I_p \times \text{PF}$$

$$\text{Total power } (P_T) = 3 \times V_p \times I_p \times \text{PF}$$

$$\text{Total power } (P_T) = 1.73 \times V_L \times I_L \times \text{PF}$$

where V_p is the **phase voltage** in volts, I_p is the phase current in amperes, V_L is the **line voltage** in volts, and I_L is the **line current** in amperes.

Sample Problem:
Given: A three-phase delta system has a line voltage of 208 volts and a line current of 4.86 amps.
Find: Total three-phase voltamperes.

Solution:
Total voltamperes = $1.73 \times V_L \times I_L$
Total voltamperes = $1.73 \times 208\ \text{V} \times 4.86\ \text{A}$
Total voltamperes = 1748.8VA

Sample Problem:
Given: A three-phase wye system has the following values: phase voltage = 120 V, phase current = 18.5 A, and power factor = 0.95.
Find: Total three-phase power of the circuit.
Solution:

$$P_T = 3 \times V_p \times I_p \times \text{PF}$$
$$P_T = 3 \times 120\ \text{V} \times 18.5\ \text{A} \times 0.95$$
$$P_T = 6327\ \text{W} = 6.327\ \text{kW}$$

Calculations involving three-phase power are somewhat more complex than single-phase power calculations. Keep in mind the difference between phase values and line values to avoid making mistakes.

Review Questions

Answer the following questions.

1. State Ohm's law.

2. What are the symbols used for voltage, current, and resistance?

3. What is the relationship between voltage and current in a circuit?

4. What is the relationship between resistance and current in a circuit?

5. Define a series circuit.

6. Define a parallel circuit.

7. Define a combination circuit.

8. What is meant by kilowatt-hour?

9. What type of alternating current is used in homes?

10. What type of alternating current is used in industries?

11. Draw a sine wave.

12. What are the formulas used to find the following AC values when the effective value of AC is known?

    a. Peak.

    b. Peak-to-peak.

13. What is inductance?

14. What is inductive reactance?

15. What is impedance?

16. What is capacitance?

17. What is capacitive reactance?

18. What is the phase relationship between voltage and current in an inductive circuit?

19. What is the phase relationship between voltage and current in a capacitive circuit?

20. Why are vector diagrams used?

21. Draw a voltage triangle and label each side with $V_R = 10$ V, $V_L = 20$ V, and $V_C = 15$ V.

22. Draw a current triangle for a parallel AC circuit with $I_C = 8$ A, $I_L = 4$ A, and $I_R = 9$ A.

23. What is the power factor of a circuit?

24. What is the phase angle of a circuit?

5

Magnetism and Electromagnetism

Objectives

Upon completion of this chapter, you will be able to:

- Define the various terms related to magnets and electromagnetism
- Apply the left-hand rule for magnetic flux.
- Apply the left-hand rule for polarity.
- List three factors that affect the strength of electromagnets.
- Define the terms residual magnetism, permeability, retentivity, magnetic saturation, and magnetizing force.
- Describe the domain theory of magnetism.
- State Faraday's law for electromagnetic induction.
- Describe the construction of a relay and solenoid.
- Explain the operation of various electromagnetic devices.

Important Terms

alloy	flux density
ampere-turn	induced pole
armature	keeper
B-H curve	left-hand rule for polarity
compass	left-hand rule of magnetic flux
contact current	magnet
domain theory	magnetic field
dropout current	magnetic flux
electromagnet	magnetic motor contactor
electromagnetic induction	magnetic saturation
electromagnetism	magnetization
Faraday's law	magnetizing force

DOI: 10.1201/9788743810056-6

magnetomotive force
normally closed
normally open
oersted
permeability
pickup current
pole
relative permeability
relay
reluctance
residual magnetism
retentivity
solenoid

Magnetism is important since it is used in electric motors, generators, transformers, relays, and many other electrical devices. The Earth itself has a large magnetic field. Electromagnetism is magnetism that is produced due to the flow of an electrical current. There are many electrical machines that operate because of electromagnetism. This chapter reviews magnetism, electromagnetism, and some important applications for electrical control.

5.1 Permanent Magnets

Magnets are made of iron, cobalt, or nickel materials, usually in the form of an alloy. An alloy is a mixture of these materials. Each end of the magnet is called a **pole**. If a magnet were broken, each part would become a complete magnet. Each magnet would have two poles. **Magnetic poles** are always in pairs. When a magnet is suspended in air so that it can turn freely, one pole will point to the **North Pole**. The Earth acts like a large permanent magnet. This is why compasses can be used to determine direction.

The **north pole** of a magnet attracts the south pole of another magnet. A north pole ***repels*** another north pole, and a **south pole** repels another south pole. The two laws of magnetism are: (1) like poles repel, and (2) unlike poles attract.

The magnetic field patterns when two permanent magnets are placed end to end are shown in Figure 5.1. This type of permanent magnet is called a bar magnet. When the magnets are placed farther apart, a smaller force of attraction or repulsion exists.

Some materials retain magnetism longer than others. Hard steel holds its magnetism much longer than soft steel. A **magnetic field** is set up around any magnetic material. The field is made up of lines of force, or **magnetic flux** (Φ). These magnetic flux lines are invisible. They never cross one another. They always form individual closed loops around a magnetic material. They have a definite direction from the north to south pole along the outside of a magnet. When magnetic flux lines are close together, the magnetic field is strong. When magnetic flux lines are farther apart, the field is weaker. The magnetic field is strongest near the poles. Lines of force pass through all

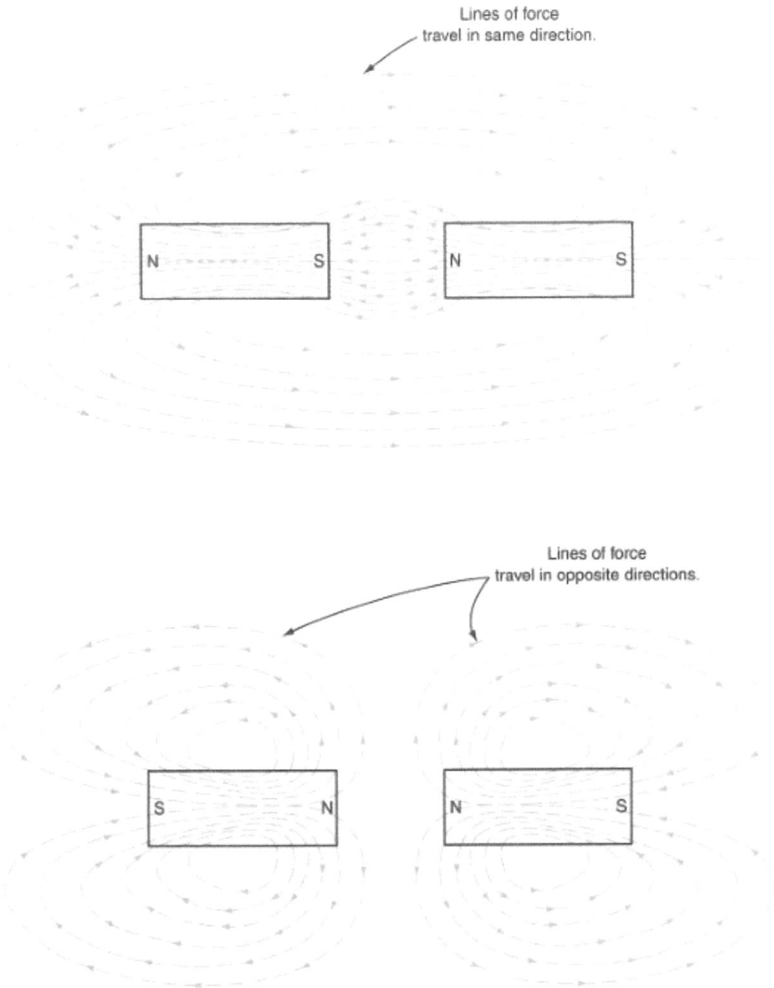

Figure 5.1 Magnetic field patterns when magnets are placed end to end.

materials. It is easy for lines of force to pass through iron and steel. Magnetic flux passes through a piece of iron as shown in Figure 5.2.

When **magnetic flux** passes through a piece of iron, the iron acts like a magnet. Magnetic poles are formed due to the influence of the flux lines. These poles are called induced poles. The induced poles and the magnet's poles attract and repel each other. Magnets attract pieces of soft iron in this way. It is possible to magnetize pieces of metal temporarily by using a bar magnet. If a magnet is passed over the top of a piece of iron several times in

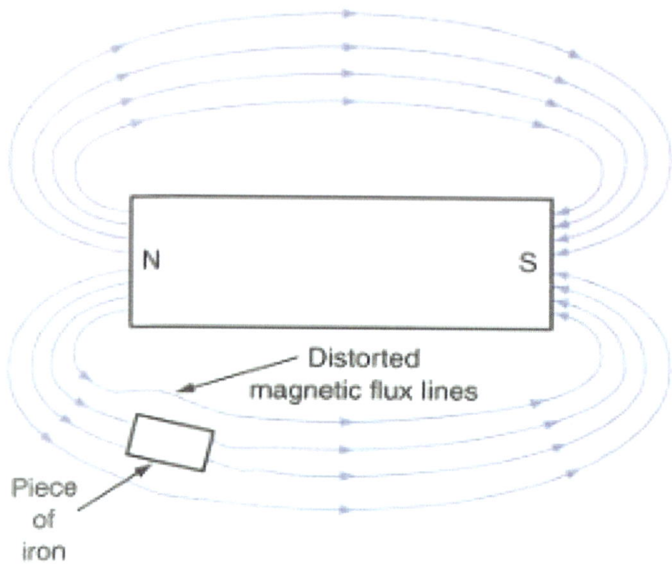

Figure 5.2 Magnetic flux lines distorted while passing through a piece of iron.

the same direction, the soft iron becomes magnetized. It will stay magnetized for a short period of time.

Compasses detect the presence of magnetic fields. When a compass is brought near the north pole of a magnet, the north-seeking pole of the compass is attracted to it. The polarities of the magnet can be determined by observing a compass brought near each pole.

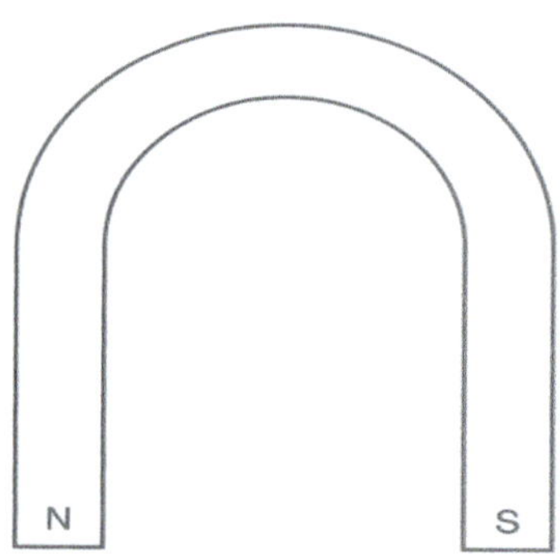

Figure 5.3 Horseshoe magnet.

Horseshoe magnets are similar to bar magnets. They are bent in the shape of a horseshoe, as shown in Figure 5.3. This shape gives the horseshoe magnet more magnetic field strength than a similar bar magnet. Since the magnetic poles are closer, the magnetic field strength is concentrated in one area. Many electrical devices use horseshoe magnets.

A **magnetic material** can lose some of its magnetism if it is jarred or heated. People must be careful when handling equipment that contains permanent magnets. A magnet also becomes weakened by loss of magnetic flux.

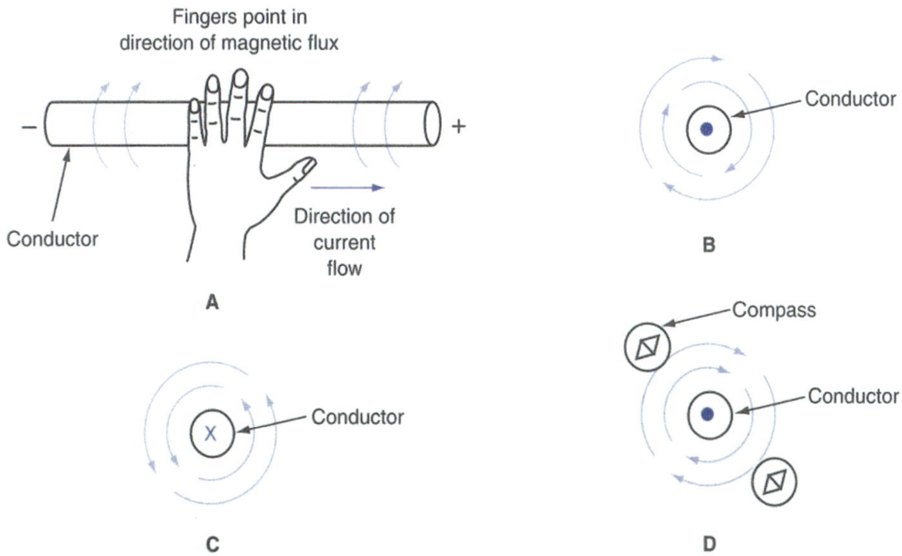

Figure 5.4 A – Left-hand rule of magnetic flux. B – Cross section of conductor with current flowing toward the observer. C – Cross section of conductor with current flowing away from the observer. D – Compass aligns tangent to magnetic lines of force.

Magnets should always be stored with a keeper, which is a soft-iron piece used to join magnetic poles. The keeper provides the **magnetic flux** with an easy path between poles. The magnet will retain its greatest strength for a longer period of time if keepers are used. **Bar magnets** should always be stored in pairs with a north pole and a south pole placed together. A complete path for magnetic flux is made in this way.

5.2 Magnetic Field around Conductors

Current-carrying conductors produce a magnetic field. It is possible to show the presence of a magnetic field around a current-carrying conductor. A compass is used to show that the magnetic flux lines are circular in shape. The conductor is in the center of the circular shape. The direction of the current and the magnetic flux lines can be found by using the left-hand rule of magnetic flux. A conductor is held in the left hand, as shown in part A of Figure 5.4. The thumb points in the direction of the current from negative to positive. The fingers then encircle the conductor in the direction of the magnetic flux lines.

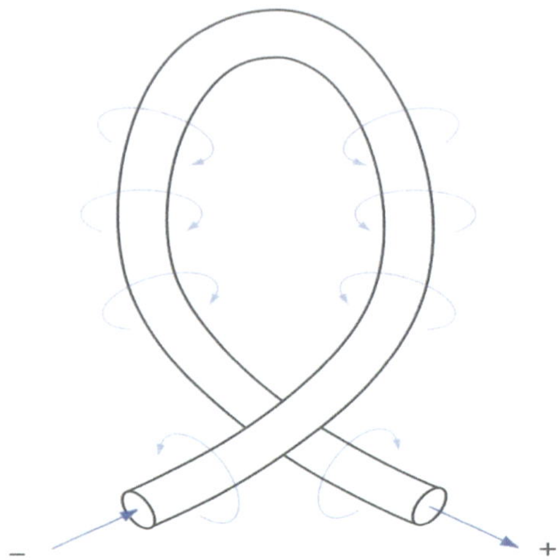

Figure 5.5 Magnetic field around a loop of wire.

A circular magnetic field is produced around a conductor. The field is stronger near the conductor and becomes weaker farther away from the conductor. A cross-sectional end view of a conductor with current flowing toward the observer is shown in part B of Figure 5.4. Current flow toward the observer is shown by a circle with a dot in the center. Notice that the direction of the magnetic flux lines is clockwise. This can be verified by using the left-hand rule.

When the direction of the current through a conductor is reversed, the direction of the ***magnetic lines of force*** is also reversed. The cross-sectional end view of a conductor, part C of Figure 5.4, shows current flowing in a direction away from the observer. Notice that the direction of the magnetic lines of force is now counterclockwise.

The presence of **magnetic lines of force** around a current-carrying conductor can be observed by using a compass. When a compass is moved around the outside of a conductor, the needle aligns itself tangent to the lines of force, as shown in part D of Figure 5.4. The needle does not point toward the conductor. When the current flows in the opposite direction, the compass polarities reverse. The compass needle aligns itself tangent to the conductor.

5.3 Magnetic Field around a Coil

The magnetic field around one loop of wire is shown in Figure 5.5. Magnetic flux lines extend around the conductor as shown. Inside the loop, the magnetic

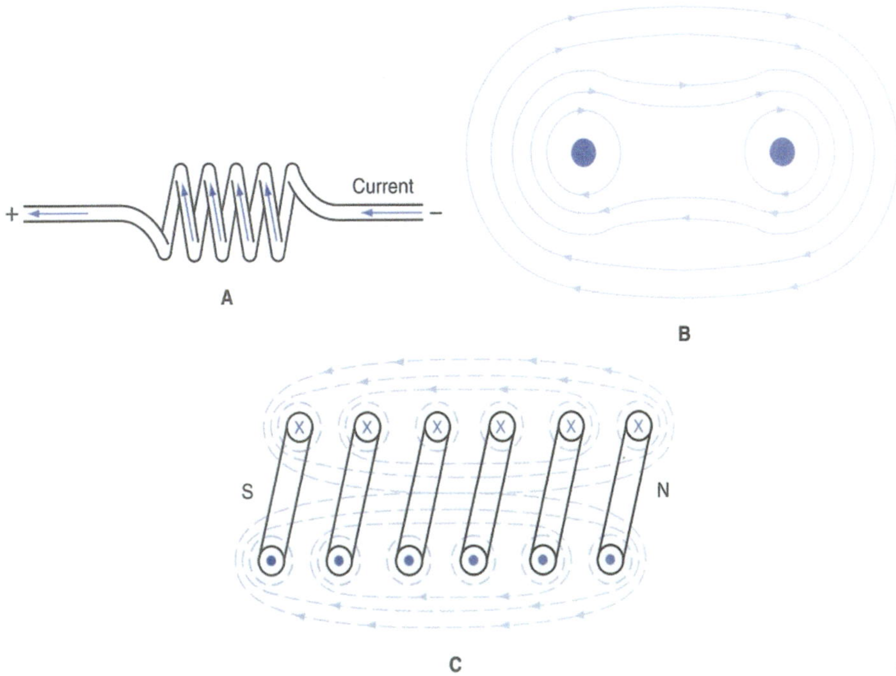

Figure 5.6 Magnetic field around a coil. A – Coil of wire showing the current. B – Lines of force around two loops that are parallel. C – Cross section of coil showing lines of force.

flux is in one direction. When many loops are joined together to form a coil, the magnetic flux lines surround the coil as shown in Figure 5.6. The field around a coil is much stronger than the field of one loop of wire. The field around the coil is the same shape as the field around a bar magnet. A coil that has an iron or steel core inside it is called an electromagnet. A core increases the magnetic flux density of a coil.

5.4 Electromagnets

Electromagnets have several turns of wire around a soft-iron core. An electrical power source is then connected to the ends of the turns of wire. When current flows through the wire, magnetic polarities are produced at the ends of the soft iron core. The three basic parts of an electromagnet are:

1. An iron core.
2. Wire windings.

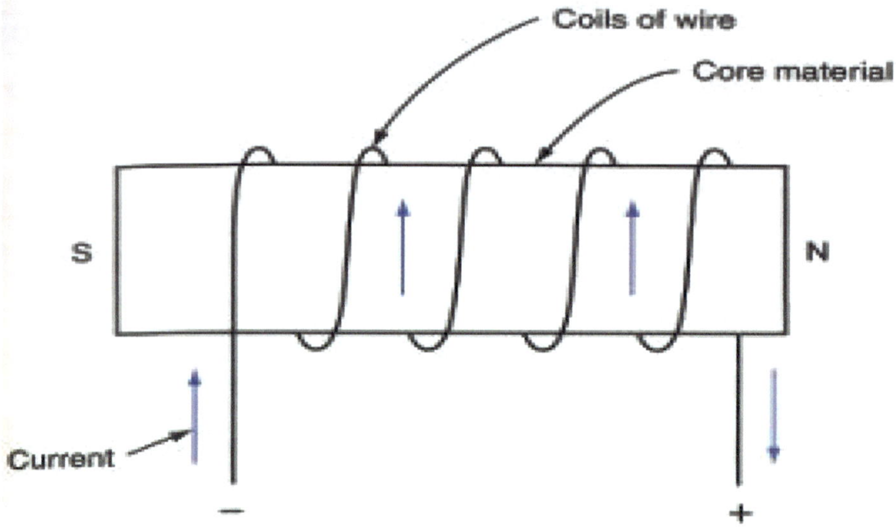

Figure 5.7 Selection of electromagnets. A-Pictorial diagram of an electromagnet. B-An electromagnet in operation. C-Battery-operated lift magnet. This magnet can lift up to 11,500 pounds of metal. (Magnetool, Inc.)

3. An electrical power source.

Electromagnetism is made possible by the electrical current, which produces a magnetic field. When electrical current flows through the coil, the properties of magnetic materials are developed. A simple electromagnet is shown in Figure 5.7.

The north pole of a coil of wire is the end where the lines of force come out. The south pole is the end where the lines of force enter the coil. This is like the field of a bar magnet. To find the north pole of a coil, use the **left-hand rule for polarity**, as shown in Figure 5.8. Grasp the coil with the left hand. Point the fingers in the direction of current flow through the coil. The thumb points to the north pole of the coil.

When the ***polarity*** of the voltage source is reversed, the magnetic poles of the coil also reverse. The poles of an electromagnet can be checked with a compass. The compass is placed near a pole of the electromagnet. If the north seeking pole of the compass points to the coil, that side is the north pole.

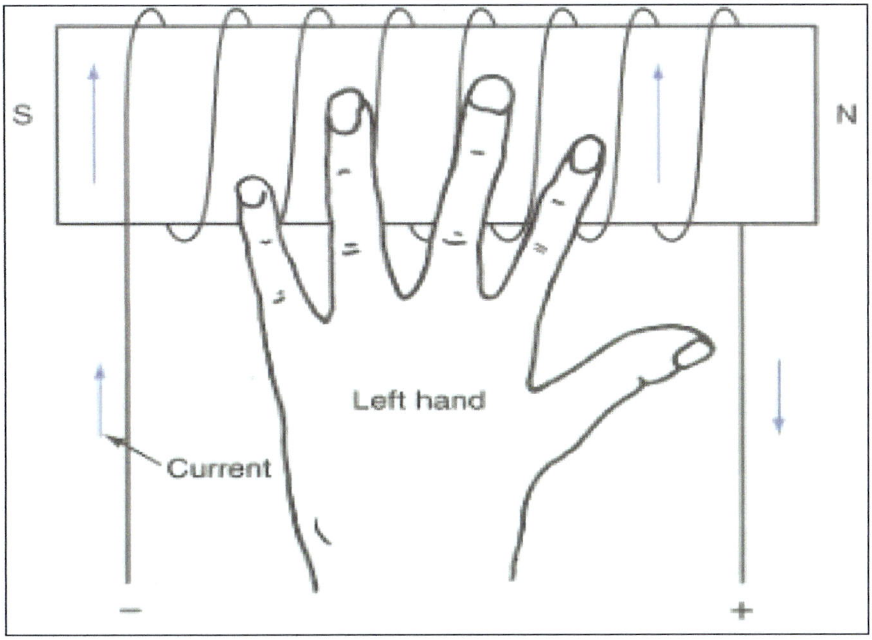

Figure 5.8 Left-hand rule for finding the polarities of an electromagnet.

5.5 Magnetic Strength of Electromagnets

In many ways, electromagnetism is similar to magnetism produced by natural magnets such as bar magnets. The main advantage of electromagnetism is that it is easily controlled. The magnetic strength of an electromagnet depends primarily on three factors:

- The amount of current passing through the coil.
- The number of turns of wire.
- The type of core material.

The strength of an electromagnet can be easily increased by increasing the current through the coil. This is done by increasing the voltage applied to the coil. A second way to increase the strength of an electromagnet is to have more turns of wire around the core. A greater number of turns produces more magnetic lines of force around the electromagnet. The strength of an electromagnet is also affected by the type of core material used. Different alloys of iron are used to make the cores of electromagnets. These materials transfer a magnetic field better than air or other nonmagnetic materials. Iron cores increase the flux density of an electromagnet. Figure 5.9 shows that an iron core causes the magnetic flux to be more dense. Some materials aid in the development of magnetic lines of force to a greater extent. Other types of core materials offer greater resistance to the development of magnetic flux around an electromagnet.

The **magnetic strength** of electromagnets is determined by the ampere-turns of each coil. The number of **ampere-turns** is equal to the current in amperes multiplied by the number of turns of wire. For example, 200 ampere-turns is produced by current through a 100-turn coil. One ampere of current through a 200-turn coil produces the same magnetic field strength. Figure 5.10 shows how the magnetic field strength of an electromagnet changes with the number of ampere-turns.

An electromagnet loses its field strength when the current stops flowing. However, an electromagnet's core retains a small amount of magnetic strength after the current stops flowing. This is called **residual magnetism**, or "leftover" magnetism. Residual magnetism can be reduced by using soft iron cores or increased by using hard-steel core material. Residual magnetism is very important in the operation of some types of electrical generators.

5.6 Ohm's Law for Magnetic Circuits

Ohm's law for electrical circuits was discussed in **Chapter 4**. A similar relationship exists in magnetic circuits. Magnetic circuits have magnetomotive force, magnetic flux, and reluctance. **Magnetomotive force (MMF)** is the force that causes **magnetic flux** to be developed. Magnetic flux describes the lines of force around a magnetic material. **Reluctance** is the opposition to the development of magnetic flux. These terms can be compared to voltage, current, and resistance in electrical circuits, as shown in Figure 5.11. When MMF increases, magnetic flux increases. Remember that in an electrical circuit, when voltage increases, current increases. When resistance in an electrical circuit increases, the current decreases. When the reluctance of a magnetic circuit increases, magnetic

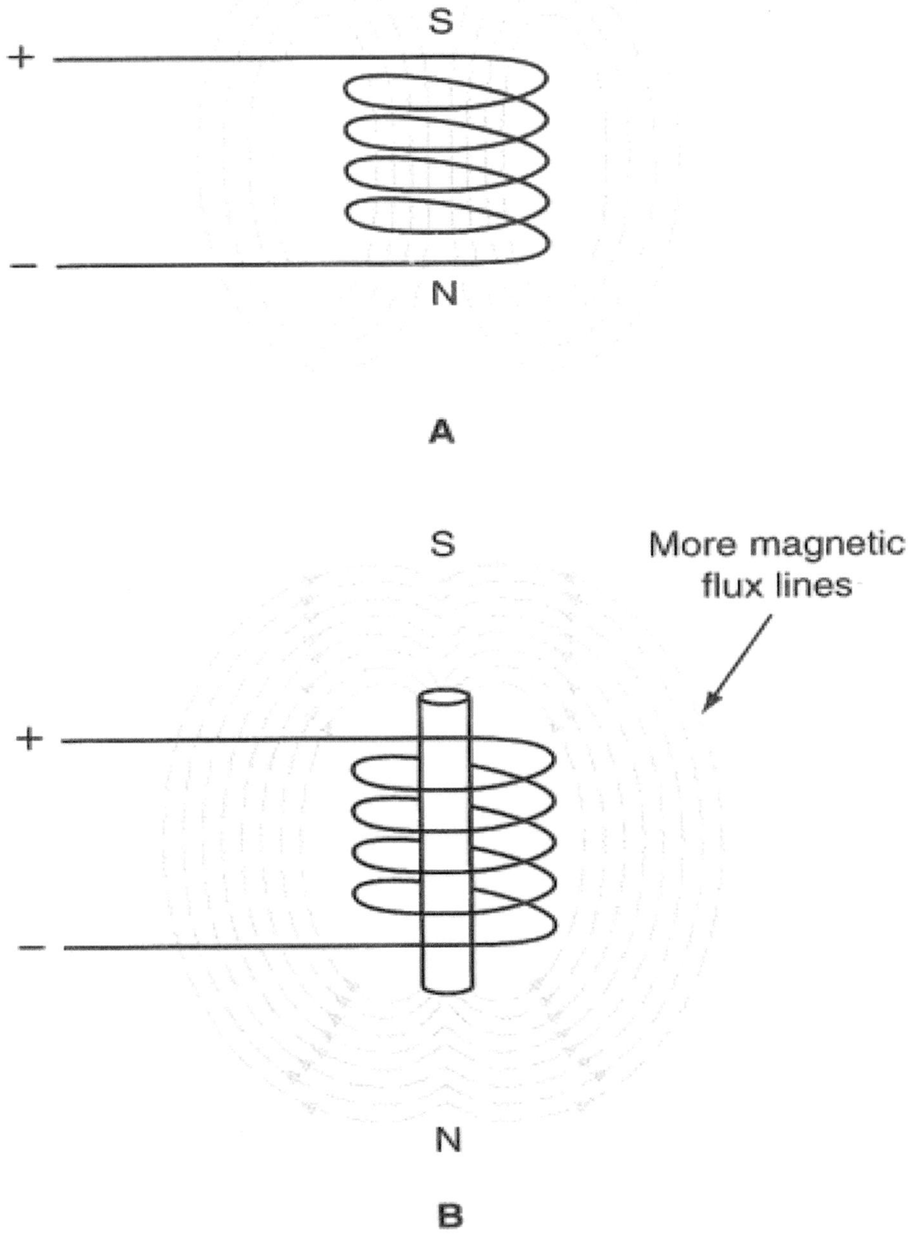

Figure 5.9 Effect of iron core on magnetic strength. A – Coil without core. B – Coil with core.

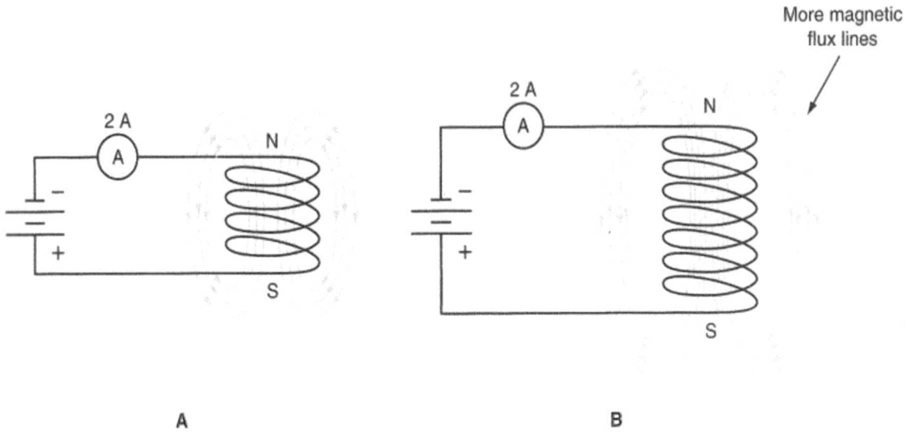

Figure 5.10 Effect of ampere-turns on magnetic field strength. A – Five turns with two amperes; B – Eight turns with two amperes.

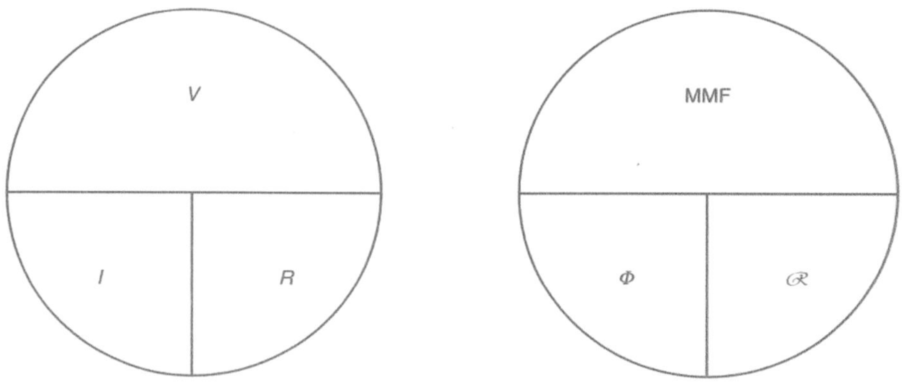

Figure 5.11 Relationship of magnetic and electrical terms.

flux decreases. The relationship between magnetic and electrical terms in Figure 5.11 should be studied.

5.7 Domain Theory of Magnetism

A theory of magnetism was presented in the nineteenth century by a German scientist named Wilhelm Weber. **Weber's theory** of magnetism was called the molecular theory. It dealt with the alignment of molecules in magnetic materials. Weber felt that molecules were aligned in an **orderly** arrangement

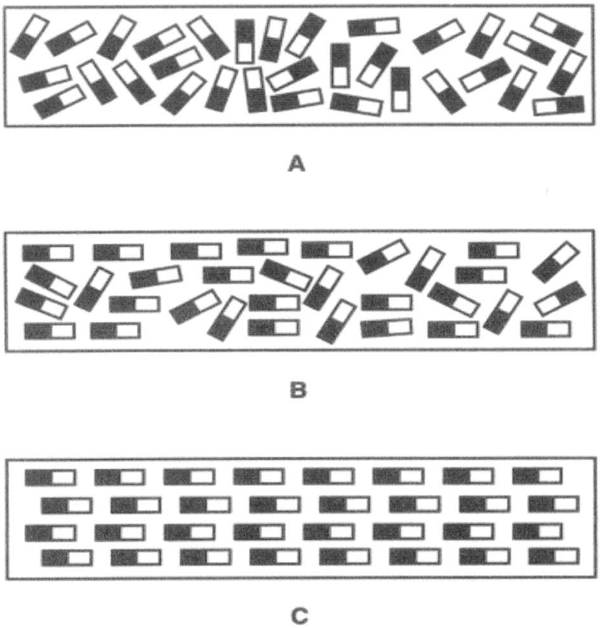

Figure 5.12 Domain theory of magnetism. A – Unmagnetized. B – Slightly magnetized. C – Fully magnetized (saturation).

in magnetic materials. In nonmagnetic materials, he thought that molecules were arranged in a **random** pattern.

Weber's theory has now been modified somewhat to become the **domain theory** of magnetism. This theory deals with the alignment of domains in materials rather than molecules. A domain is a group of atoms that acts like a tiny magnet. The rotation of electrons around the nucleus of these atoms is important. **Electrons** have a negative charge. As they orbit around the **nucleus** of atoms, their electrical charge moves. This moving electrical field produces a magnetic field. The **polarity** of the magnetic field is determined by the direction of electron rotation.

The domains of magnetic materials are atoms grouped together. Their electrons are believed to spin in the same direction. This produces a magnetic field due to electrical charge movement. Figure 5.12 shows the arrangement of domains in magnetic, nonmagnetic, and partially magnetized materials. In nonmagnetic materials, half the electrons spin in one direction and half spin in the opposite direction. Their charges cancel each other out. There is no magnetic field produced, since the charges are cancelled. Electron rotation in magnetic materials is in the same direction. This

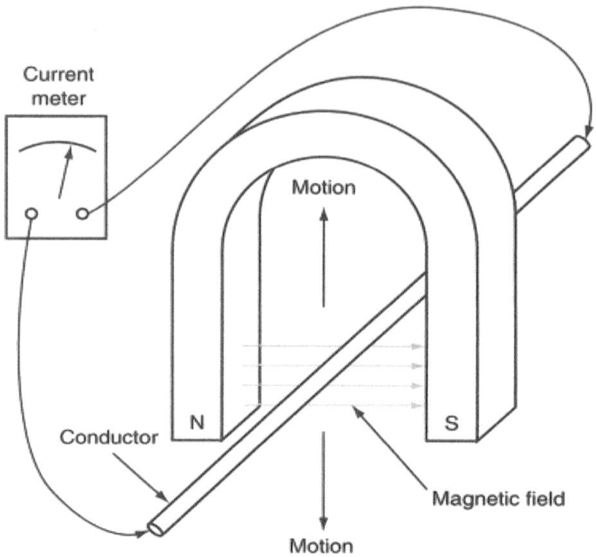

Figure 5.13 Faraday's law – electrical current is produced when there is relative motion between a conductor and a magnetic field.

causes the domains to act like tiny magnets that align to produce a magnetic field.

5.8 Electricity Produced by Magnetism

A scientist named Michael Faraday discovered in the early 1830s that electricity is produced from magnetism. He found that if a magnet is placed inside a coil of wire, electrical current is produced when the magnet is moved. Faraday found that electrical current is caused by magnetism and motion.

Faraday's law is stated as follows: When a coil of wire moves across the lines of force of a magnetic field, electrons flow through the wire in one direction. When the coil of wire moves across the magnetic lines of force in the opposite direction, electrons flow through the wire in the opposite direction. This law is the principle of electrical power generation produced by magnetism. Figure 5.13 shows Faraday's law as it relates to electrical power generation.

Current flows in a conductor placed inside a magnetic field only when there is motion between the conductor and the magnetic field. If a conductor is stopped while moving across the magnetic lines of force, the current stops flowing. The operation of electrical generators depends on conductors moving across a magnetic field. This principle is called **electromagnetic induction**.

5.9 Relays

Relays are electrical devices that rely on magnetism to operate. They control other equipment, such as motors, lights, or heating elements. Relays are very important devices. They are used to start the operation of other equipment. They use a small amount of electrical current to control a larger current, such as the current through a motor.

The basic construction and symbols of a relay are shown in Figure 5.14. Relays have an electromagnetic coil with electrical power applied to its two external leads. When the power is turned on, the electromagnet is energized. The electromagnet part of the relay controls a set of contacts. The contacts are called **normally open (NO)** or **normally closed (NC),** depending on their condition when the electromagnet is not energized. There is also a **common** contact.

If a lamp and its power source are connected in series with the common and NO contact as shown in Figure 5.14, the lamp will be off when the relay is not energized. If the relay is energized by applying power to it, the common contact is attracted to the NO contact by magnetic energy. The common contact is built onto an armature that moves when the electromagnet is energized. When the relay is energized, the light connected to the NO contact turns on. Notice that the lamp, or any load connected to the relay contacts, requires a separate power source.

In a similar way, the **NC contacts** are used (refer to Figure 5.15). When the relay is off, the circuit from the **common** terminal through the power source and lamp 1 is closed. This causes lamp 1, which is in series with the NC contacts, to be turned on. Note that the power source for the lamps is in the common line, since it is common to both of the other contacts. When the relay is turned on, the lamps change states. Lamp 2, connected to the **NO contact**, will be turned on, whereas lamp 1, connected to the NC contact, will be turned off. Notice in part B of Figure 5.14 how the common contact moves from the NC contact so that it touches the NO contact. This shows the basic operation of a relay with NO and NC contacts. Such a relay is very common. It is called a **double-pole, single-throw (DPST) relay**. There are many other types used.

The **coil resistance** of a relay is determined by the size of the wire used to wind the coil and the number of windings. A coil with only a few turns of large-diameter wire has a low resistance. This causes a high current to flow through the relay coil. If a relay coil has many turns of small-diameter wire,

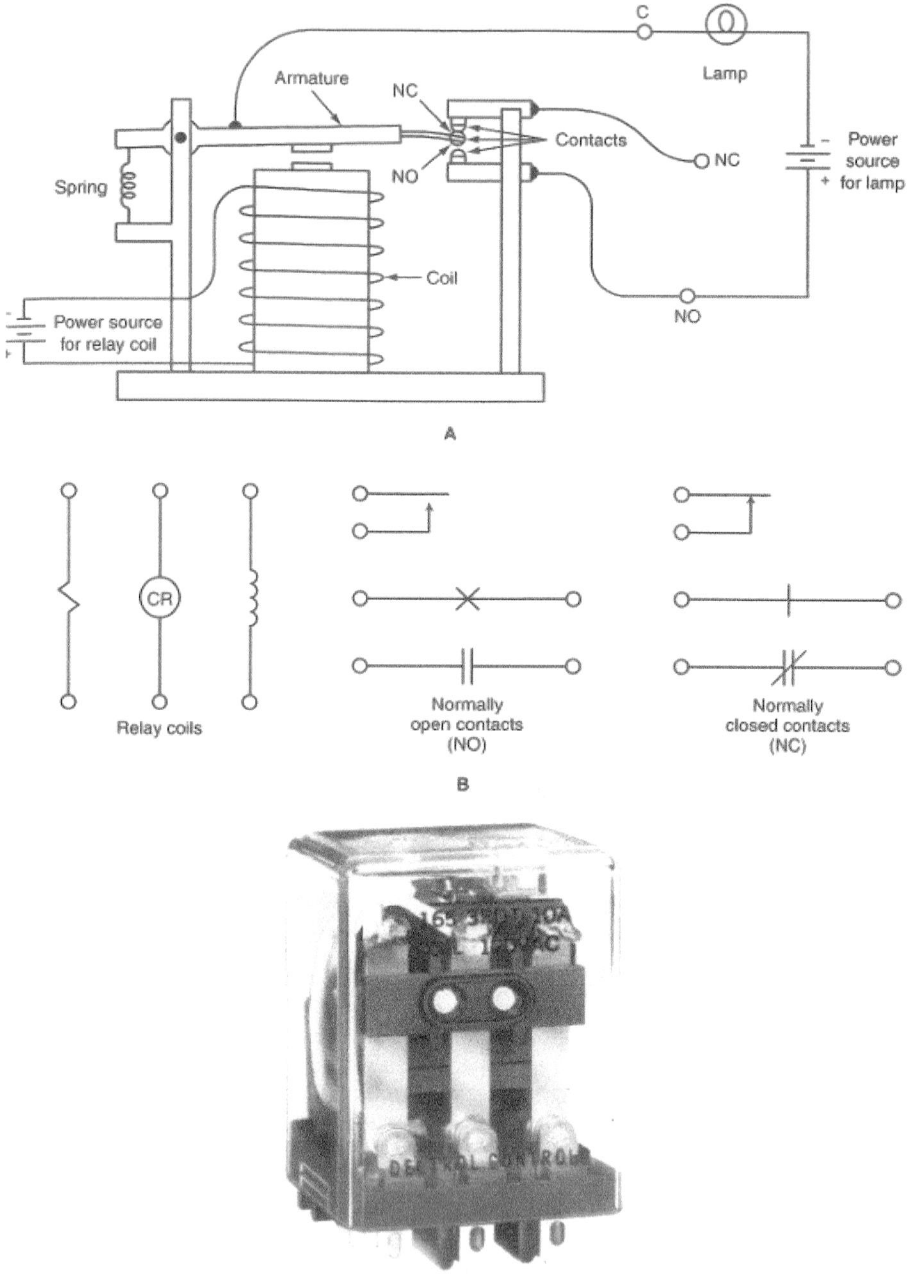

Figure 5.14 A – Diagram of a relay with circuit. B – Symbols used for relays. Plug-in type relay for industrial applications. (Deltrol Controls)

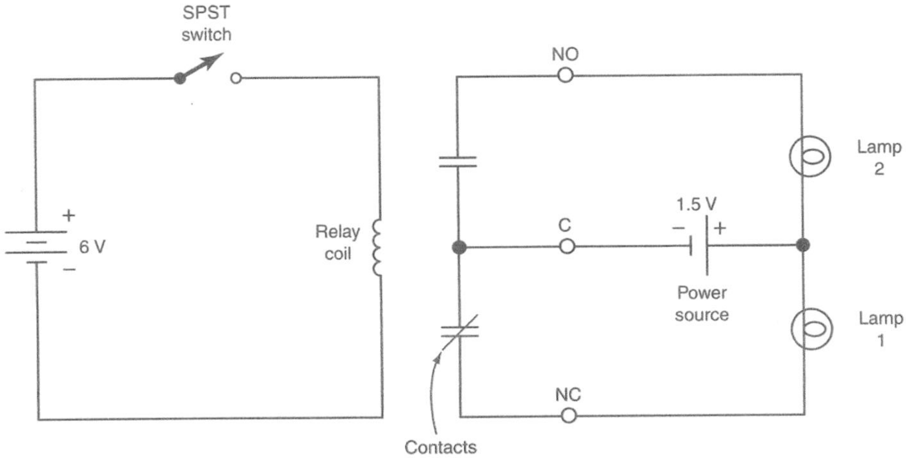

Figure 5.15 Schematic diagram of a relay circuit.

it has a high resistance. Remember that small wires have high resistance, and large wires offer lower resistance to current. Coil resistance is usually marked on a relay. It may also be measured with a meter.

There are some important **current ratings** for relays. Two of these ratings are called pickup current and dropout current. These ratings are usually specified on the relay. They can also be found by using a variable power source and a current meter to measure the actual values. When the voltage applied to the relay coil is increased to a point where the relay turns on, the current indicated on the meter is the pickup current. Pickup current is the minimum current required to energize the relay. If the applied voltage is decreased until the relay de-energizes, the meter indicates dropout current. **Dropout current** is the minimum current that will keep the relay energized. The contact current rating is also very important. A large current usually flows through the relay contacts and the load connected to the relay. The **contact current** is the maximum current that can safely flow through the contact circuit.

5.10 Solenoids

Solenoids are similar to relays, since they also use electromagnetic coils. They are sometimes called actuators. Instead of having contacts, a solenoid has a **plunger** that moves when the coil is energized. The back of the plunger is attached to a spring. It causes the plunger to return to its original position when the solenoid de-energizes. The movement of the plunger of the solenoid is used to activate some type of load connected to it. A solenoid is shown in Figure 5.16. When the solenoid is energized, the plunger moves to the

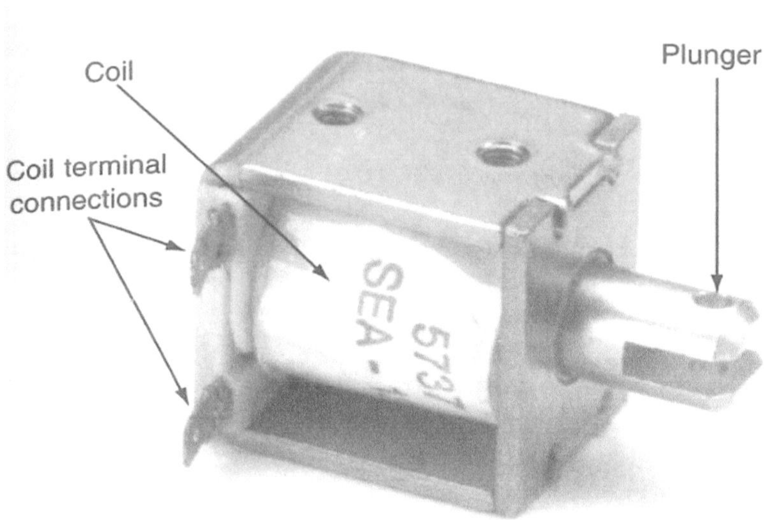

Figure 5.16 Solenoid. (Magnet-Schultz of America, Inc)

center. A solenoid could be used to open a control valve to allow liquid from a tank to flow into a container. When the solenoid is de-energized, the control valve closes and stops the flow of liquid.

5.11 Magnetic Motor Contactors

A very important type of relay is a magnetic motor contactor. **A magnetic motor contactor** is shown in Figure 5.17. The motor contactor is a control element that starts and stops motors. It operates due to electromagnetic relay action. A start **push-button switch** is pressed to close a contactor. This completes a low current path through the contactor coil. The contactor coil produces a magnetic field that causes a set of contact points to close. The movement of a part called its **armature** completes an electrical path between the power line and a motor. When this action takes place, the motor starts.

Releasing the start button does not de-energize the contactor coil. A path to the coil's voltage source is completed through the stop **push-button switch.** The motor continues to run as long as electric power is applied.

Stopping a contactor-controlled motor is achieved by pushing the stop button. This opens the contactor coil's voltage source. The coil deenergizes and causes the armature to move. The contact points then break contact. This removes the power from the motor. The path becomes open, and the motor stops. Motor contactors are designed to latch in place, which holds the contactor

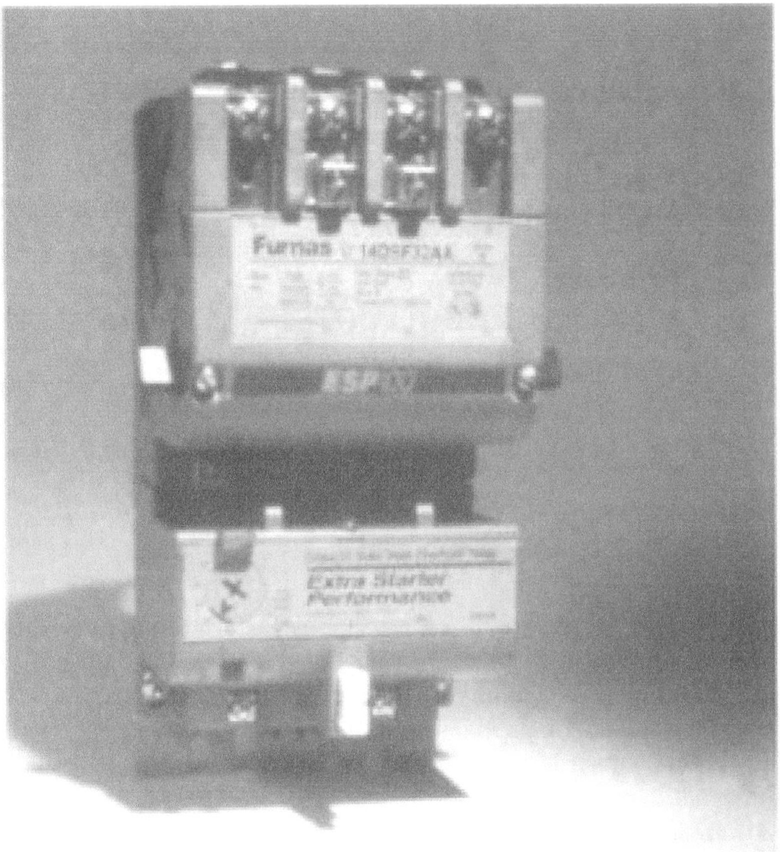

Figure 5.17 Magnetic motor contactor. (Furnas Electric Co.)

in place once the coil is energized. This characteristic is important for motor control.

5.12 Magnetic Effects

There are several magnetic effects that are important. Among these effects are residual magnetism, permeability, retentivity, and magnetic saturation. **Residual magnetism**, discussed in Section 5.5, is an effect that is important in the operation of some types of electrical generators. Residual magnetism is the ability of electromagnets to retain a small magnetic field after the electrical current is turned off. A small magnetic field remains around an electromagnet after it is demagnetized. This magnetic field is very weak.

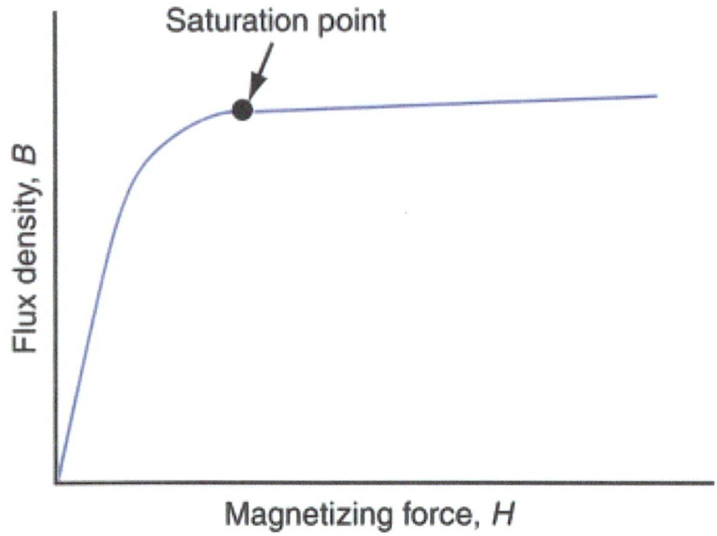

Figure 5.18 Magnetization or B–H curve.

Permeability is the ability of a magnetic material to transfer magnetic flux. It is the ability of a material to magnetize and demagnetize. Soft iron material has a high permeability. It transfers magnetic flux very easily. Soft iron magnetizes and demagnetizes rapidly. This makes soft iron a good material to use in the construction of generators, motors, transformers, and other electromagnetic machines.

A related term is **relative permeability**. This is a comparison of the permeability of a material to the permeability of air (1.0). Suppose that a material has a relative permeability of 1000. This means that the material will transfer 1000 times more magnetic flux than an equal amount of air. The highest relative permeability of materials is over 5000.

Another magnetic effect is called **retentivity**. The retentivity of a material is its ability to retain a magnetic flux after a magnetizing force is removed. Some materials retain magnetic flux for a long period of time. Other materials lose their magnetic flux almost immediately after the magnetizing force is removed.

Magnetic saturation is important in the operation of machines that have electromagnets, especially generators. Saturation is best explained by the curve shown in Figure 5.18. This is called **magnetization** or **B–H curve**. The curve shows the relationship between a magnetizing force and flux density. Notice that as the **magnetizing force** increases, so does the flux density. Magnetizing force is measured in **oersteds**. The basic unit is the number of ampere-turns per meter of length. **Flux density** is the number of lines of force

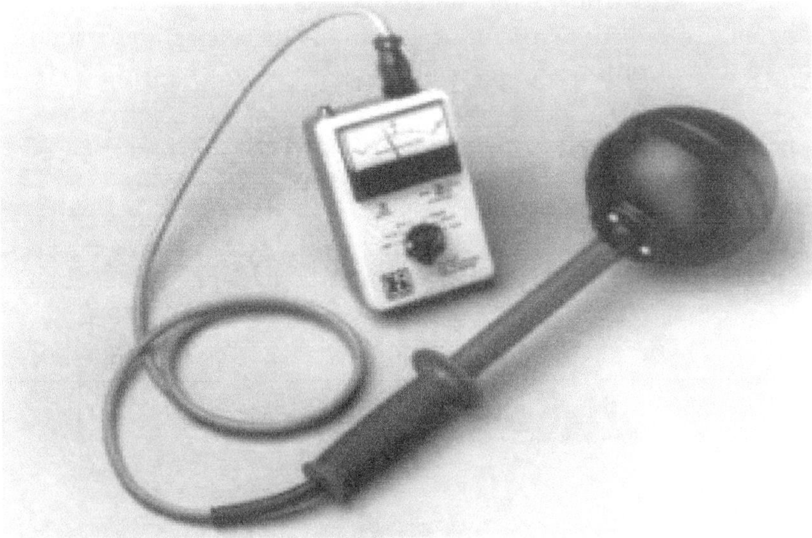

Figure 5.19 Magnetic field meter. (Holaday Industries, Inc.)

per unit area of a material. The unit of flux density is the gauss per square centimeter of area. An increase in flux density occurs until magnetic saturation is reached. This **saturation point** depends on the type of material. At the saturation point, the maximum alignment of domains takes place in the material.

Figure 5.19 shows a meter used to detect the presence of magnetic fields. This meter measures the strength of magnetic fields (in **gauss**).

In addition to being used to start motors turning, magnetism can be used to slow and stop them as performed with this hysteresis brake and clutch. (Gelia Wells & Mohr)

Review Questions

Answer the following questions.

1. List the three materials used in the construction of permanent magnets.

2. What are the two laws of magnetism?

3. How can a piece of iron be temporarily magnetized?

4. Why should magnets be stored in a keeper?

5. Define electromagnetism.

6. State the left-hand rule of magnetic flux around a conductor.

7. State the left-hand rule for determining the polarity of an electromagnet.

8. List the three basic parts of an electromagnet.

9. What are three ways to increase the strength of an electromagnet?

10. What is residual magnetism?

11. State Faraday's law.

12. What is a relay?

13. What are the pickup current and dropout current ratings of relays?

14. What is a solenoid?

15. What is a magnetic contactor?

UNIT TWO

Electrical Distribution Basics

Once electrical energy has been produced, it must be distributed to the location where it is used. Individuals who work with electrical control systems must be familiar with distribution systems. Unit II deals with electrical distribution basics.

The fundamentals of distribution systems are discussed in **Chapter 6**. This chapter provides an overview of distribution systems, transformer operation, and conductor characteristics.

Chapter 7 examines specialized power distribution equipment such as protective devices, high-voltage equipment, and switchgear. Then, **Chapter 8** moves into the types of distribution systems used today. Ground-fault circuit interrupters are also discussed extensively in this chapter, since their use is widespread in electrical distribution systems.

DOI: 10.1201/9788743810056-7

6

Distribution Systems

Objectives

Upon completion of this chapter, you will be able to:

- Describe the transmission and distribution of electrical power.
- Compare underground and overhead transmission systems.
- List several purposes of transformers.
- Describe the construction of a transformer.
- Explain transformer action.
- Calculate turns ratio, voltage ratio, current ratio, power, and efficiency of transformers.
- Explain the purpose of multiple secondary transformers, autotransformers, and current transformers.
- Explain factors that cause losses in transformer efficiency.
- Explain circular mil sizing of copper and aluminum electrical conductors.
- Calculate cross-sectional area, resistance or ampacity of round, square or rectangular copper or aluminum conductors.
- Describe the types of insulation used for conductors in electrical distribution systems.

Important Terms

American Wire Gage (AWG)
Ampacity
Autotransformer
circular mil (cmil)
control center
copper loss
cross-sectional area
current transformer
eddy current
efficiency
iron loss
mutual induction
polarity markings
primary windings
rectification system
resistivity (ρ)

DOI: 10.1201/9788743810056-8

- secondary windings
- square mil
- step-down transformer
- step-up transformer
- substation
- thermoplastic
- thermosetting
- transformer
- transmission system
- distribution system
- variable autotransformer

Introduction

Power distribution systems are very important electrical systems. In order to transfer electrical energy from a source to the place where it will be used, some type of distribution network must be utilized. The method used to distribute power from where it is produced to where it is used can be quite simple. For example, a battery could be connected directly to a motor, with only a set of wires to connect them together.

More complex distribution systems are used to transfer electrical power from the power plant to industries, homes, and commercial buildings. Distribution systems usually employ such equipment as transformers, circuit breakers, and protective devices. Technicians who work with electrical control systems must be familiar with distribution systems. These systems supply the proper type of power to machines to cause them to do their intended work.

6.1 Overview of Electrical Power Distribution

The distribution of electrical energy in the United States is normally in the form of three-phase 60 Hz alternating current. This energy, of course, can be manipulated or changed in many ways by the use of electrical circuitry. For instance, a **rectification system** is capable of converting the alternating current into a form of direct current. Also, single-phase power is generally suitable for lighting and small appliances, such as those used in the home or residential environment. However, where a large amount of electrical power is required, three-phase power is more economical.

The distribution of electrical power involves a very complex system of interconnected power transmission lines. These transmission lines originate at the electrical power-generating stations located throughout the United States. The ultimate purpose of these power transmission and distribution systems is to supply the electrical power necessary for industrial, residential, and commercial use. From the point of view of the systems, we can say that the overall electrical power system delivers power from the source to the load that is connected to it. A typical electrical distribution system is shown in Figure 6.1 in pictorial form.

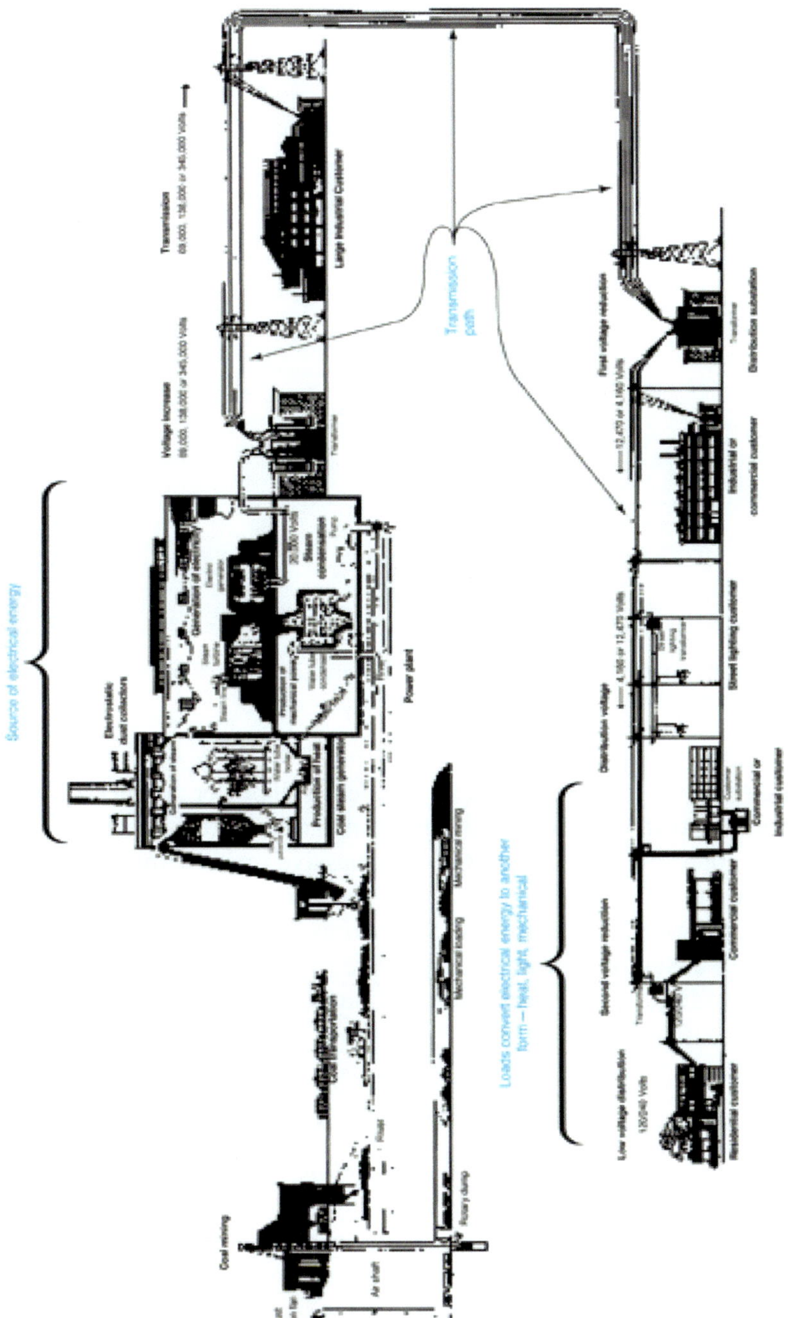

Figure 6.1 A typical electrical power distribution system. (Kentucky Utilities Co.)

Figure 6.2 An outdoor substation showing the construction of high-voltage circuit breakers and disconnect switches. (Kentucky Utilities Co.)

Industries use almost 50% of all the electrical power produced, so three-phase power is distributed directly to most industries. **Electrical substations**, such as the one shown in Figure 6.2, use massive **transformers** and associated equipment (such as oil-filled circuit breakers, high-voltage conductors, and huge strings of insulators) in distributing electrical power to industry. From these substations, power is distributed to industrial sites to energize industrial machinery, to residential homes, and to commercial users of electrical power.

6.2 Transmission and Distribution Systems

Transmission and distribution systems are used to interconnect electrical power production systems and to provide a means of delivering electrical power from a generating station to its point of utilization. Most electrical power systems east of the Rocky Mountains are interconnected with one another in a parallel circuit arrangement. These interconnections of power production systems are monitored and controlled, in most cases, by a computerized control center. Such **control centers** provide a means of data

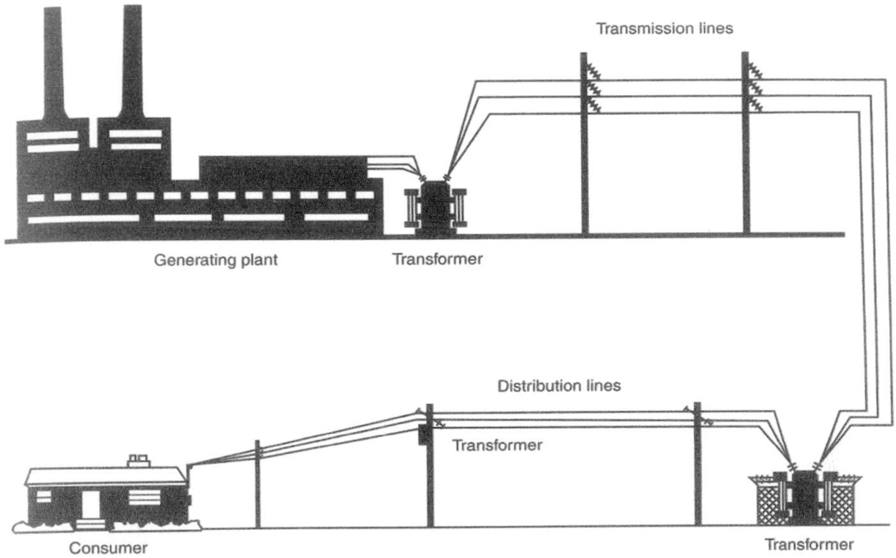

Figure 6.3 The generation and transmission of electricity.

collection and recording, system monitoring, frequency control, and signaling. Computers have become an important means of ensuring the efficient operation of electrical systems.

The **transmission** of electrical power requires many long interconnected power lines to carry the electrical current from where it is produced to where it is used (see Figure 6.3). **Power transmission lines** ordinarily operate at voltage levels from 15 kilovolts to 500 kilovolts of alternating current. Common transmission line voltages are in the range of 50 to 150 kilovolts of alternating current. High-voltage direct current overhead transmission lines may become economical, although they are not being used extensively at the present time.

6.2.1 Overhead distribution

Overhead power transmission lines require much planning to ensure the best use of our land. The location of overhead transmission lines is limited by zoning laws, populated areas, highways, railroads, and waterways, as well as other topographical and environmental factors. Today, an increased importance is being placed upon environmental and aesthetic factors.

Another option is **ultra-high-voltage** transmission lines that use higher AC transmission voltages. Alternating-current overhead transmission voltages have increased to levels in the range of 765 kV to 1000 kV.

6.2.2 Underground distribution

The use of **underground cable** is ordinarily confined to the short lengths required in congested urban areas. The cost of underground cable is much more than for **aerial cable**. Also, underground cabling does not dissipate heat as well as overhead cabling. To improve underground cable power-handling capability, research is being done in forced-cooling techniques, such as circulating oil, and with compressed-gas insulation. Another possible method is the use of cryogenic cables or **superconductors** that operate at extremely low temperatures and have a large power-handling capability.

6.3 Use of Transformers for Power Distribution

The heart of an electrical distribution system is a control device known as a **transformer**. A transformer is an electrical device that uses magnetism to link one coil of wire to another. This device is capable of controlling massive amounts of power for efficient distribution. Transformers are also used for many other applications. A knowledge of transformer operation is essential for understanding electrical systems. The distribution of alternating-current power is dependent upon the use of transformers at many points along the line of the power distribution system.

It is economically feasible to transmit electrical power over long distances at high voltages since less current is required at high voltages. Less current means that the **line loss (I^2R loss)** is reduced significantly. A typical high-voltage transmission line can extend a distance of 50 to 100 miles from the generating station to the first substation. These high-voltage power transmission lines typically operate at 100,000 to 500,000 volts. The voltage is raised by using step-up transformers to increase the voltage produced by the AC generators at the power station. Various **substations** are encountered along the power distribution system. At these substations, transformers are used to reduce the high transmission voltages to a voltage level such as 480 volts, which is suitable for industrial motor loads, or to 120/240 volts for residential use. Notice the transformer location in Figure 6.2.

Transformers provide a means of converting an alternating-current voltage from one value to another. The basic construction of a transformer is illustrated in Figure 6.4. Notice that the transformer shown consists of two sets of windings that are not physically connected. The **primary windings** make up the coil that receives the input energy from an ACsource. **Secondary windings** make up the coil that receives the energy from the primary windings and delivers it to the load. The only connection between the primary

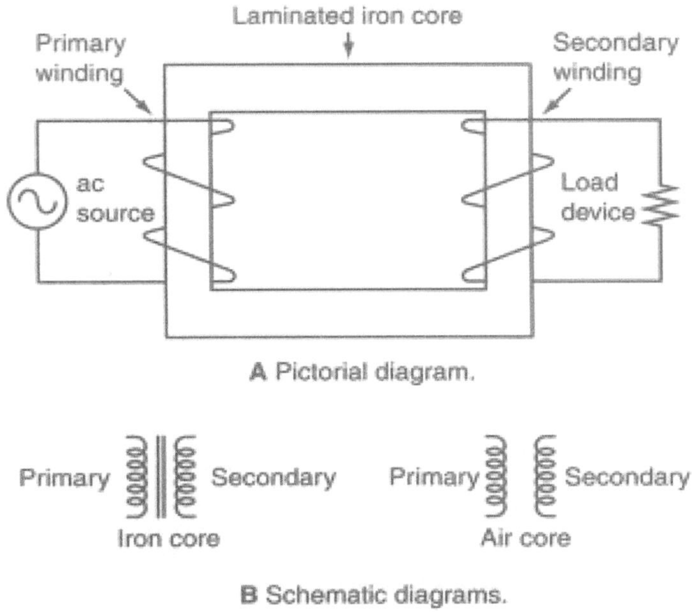

Figure 6.4 Basic transformer construction.

windings and secondary windings is the magnetic coupling effect known as **mutual induction**, which takes place when the circuit is energized by an alternating current voltage. The laminated iron core plays an important role in transferring magnetic flux from the primary winding to the secondary winding.

6.4 Transformer Operation

If an AC current (which is constantly changing in value) flows in the primary winding, the magnetic field produced around the **primary winding** will be transferred to the **secondary winding**. Thus, an induced voltage is developed across the secondary winding. In this way, electrical energy can be transferred from the source (primary-winding circuit) to a load (secondary-winding circuit).

The efficient transfer of energy from the primary to the secondary windings of a transformer depends on the coupling of the magnetic field between these two windings. Ideally, all **magnetic lines of force** developed around the primary winding would be transferred by magnetic coupling to the secondary winding. However, a certain amount of magnetic loss takes place as some lines of force escape to the surrounding air.

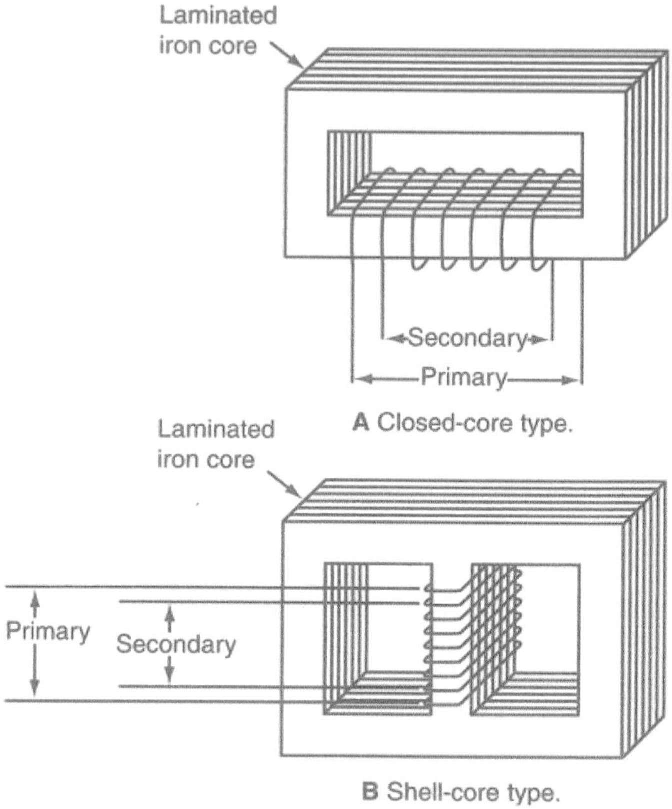

Figure 6.5 Types of iron-core transformers.

6.4.1 Transformer core construction

In order to decrease the amount of magnetic losses, transformer windings are wound around iron cores. Iron cores concentrate the magnetic lines of force so that better coupling between the primary and secondary windings is accomplished. Two types of transformer cores are illustrated in Figure 6.5. These cores are made of laminated iron to reduce undesirable eddy currents. **Eddy currents** are small currents induced into the core material. These eddy currents cause power losses. The diagram of part A of Figure 6.5 shows a closed-core transformer construction. The transformer windings of the closed-core type are placed along the outside of the metal core. Part B of Figure 6.5 shows the shell-core type of construction. The shell-core construction method produces better magnetic coupling since the transformer windings are surrounded by metal on both sides. Note that the primary and secondary windings of both types are placed adjacent to one another for better magnetic coupling.

6.4.2 Transformer efficiency and losses

Transformers are very efficient electrical devices. A typical efficiency rating for a transformer would be around 98%. **Efficiency** of electrical equipment is expressed as:

$$\text{efficiency}\,(\%) = \frac{P_{\text{out}}}{P_{\text{in}}} \times 100$$

where P_{out} is the power output in watts and P_{in} is the power input in watts.

Sample Problem:

Given: A transformer circuit has a power output of 2.5 kW and a power input of 2550 W.

Find: The efficiency of the transformer.

Solution:

$$\%\ \text{eff} = \frac{2500\ \text{W}}{2550\ \text{W}} \times 100$$
$$\%\ \text{eff} = 98\%$$

The losses that reduce efficiency, in addition to flux leakage, are copper losses and iron losses. **Copper loss** is the loss of the windings. **Iron losses** are those caused by the eddy currents in the metallic core material. The insulated laminations of the iron core help to reduce iron losses.

6.5 Step-up and Step-down Transformers

Transformers are functionally classified as step-up or step-down types. These types are illustrated in Figure 6.6. The **step-up transformer** in part A of Figure 6.6 has fewer turns of wire on the primary than on the secondary. If the primary winding has 50 turns of wire and the secondary has 500 turns, a turns ratio of is developed. If 12 volts AC is applied to the primary from the source, 10 times that voltage, or 120 volts AC, will be transferred to the secondary load (assuming no losses).

The example of part B of Figure 6.6 is a step-down transformer. The **step-down transformer** has more turns of wire on the primary than on the secondary. The primary winding of the example has 200 turns, while the secondary winding has 100 turns. This is a 2:1 ratio. If 120 volts AC is applied to the primary from the source, then one-half that amount, or 60 volts AC, will be transferred to the secondary load.

6.6 Transformer Voltage and Current Relationships

The following examples (Figures 6.6A and 6.6 B) show that there is a direct relationship between the primary and secondary turns and the voltages across each winding. This voltage relationship can be expressed:

$$\frac{V_P}{V_S} = \frac{N_P}{N_S}$$

where V_p is the voltage across the primary winding, V_s is the voltage across the secondary winding, N_p is the number of turns in the primary winding, and N_s is the number of turns in the secondary winding.

Sample Problem:

Given: A transformer circuit has the following values: V_p = 240 volts, N_P = 1000 turns, and V_S = 120 volts.

Find: The number of secondary turns of wire required to accomplish this step-down of voltage.

Solution:

$$\frac{V_P}{V_S} = \frac{N_P}{N_S}$$

$$\frac{240\ \text{V}}{120\ \text{V}} = \frac{1000}{N_S}$$

$$N_S = 500\ \text{turns}$$

6.6.1 Transformer power

The transformer is a power-control device. With an ideal transformer, the power in would be equal to the output power. Since there are no ideal transformers, the following relationship can be expressed for power:

$$P_P = P_S + \text{losses}$$

where P_p is the primary power and P_s is the secondary power.

Sample Problem:

Given: The power output (secondary power) of a transformer is 15 kW and its losses are as follows: iron losses = 200 W, copper loss = 350 W.

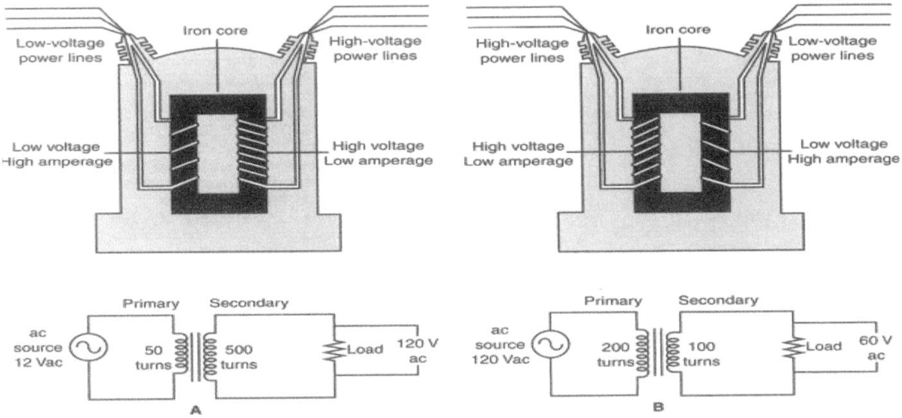

Figure 6.6 Step-up and step-down transformers. A – Step-up type. B – Step-down type.

Find: The power input (primary power) required.
Solution:

$$P_P = P_S + \text{ losses}$$
$$P_P = 15{,}000\text{ W} + (200\text{ W} + 350\text{ W})$$
$$P_P = 15{,}550\text{ Watts}$$

The losses are those that ordinarily occur in a transformer. In transformer theory, an ideal device is usually assumed, and losses are not considered. Thus, since $P_P = P_s$ and $P = V \times I$, then:

$$V_P \times I_P = V_S \times I_S$$

where I_P is the primary current and I_S is the secondary current.

If the voltage across the secondary is stepped up to twice the voltage across the primary, the secondary current will be stepped down to one-half the primary current. The current relationship of a transformer is thus expressed as:

$$\frac{I_P}{I_S} = \frac{N_S}{N_P}$$

Note that whereas the voltage-turns ratio is a direct relationship, the current ratio is an inverse relationship.

Sample Problem:

Given: An ideal transformer circuit has the following values: $V_P = 600$ V, $V_S = 2400$ V, and $I_S = 80$ A.

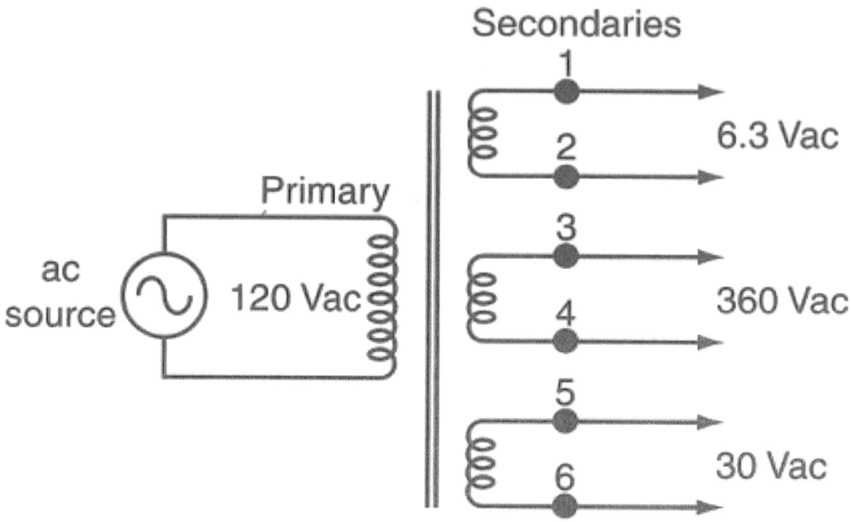

Figure 6.7 Transformer with multiple secondary windings.

Find: The primary current drawn by the step-up transformer.
Solution:

$$\frac{I_P}{I_S} = \frac{N_S}{N_P}$$

$$\frac{I_P}{80\text{ A}} = \frac{2400\text{ V}}{600\text{ V}}$$

$$I_P = 3200\text{ A}$$

6.7 Multiple Secondary Transformers

It is also possible to construct a transformer that has **multiple secondary windings**, as shown in Figure 6.7. This transformer is connected to a 120-volt AC source that produces the primary magnetic flux. The secondary has two step-down windings and one step-up winding. Between points 1 and 2, a voltage of 6.3 volts AC could be supplied. Between point 5 and point 6, 30 volts AC can be obtained, and between points 3 and 4 , a voltage of 360 volts AC can be supplied to a load. This type of transformer is used for the power supply of various types of electronic equipment and instruments.

6.8 Autotransformers

Another specialized type of transformer is the **autotransformer**, shown in Figure 6.8. The autotransformer has only one winding, with a common

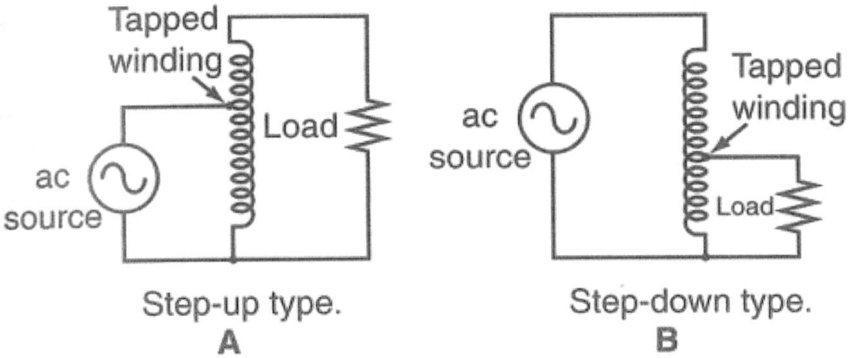

Figure 6.8 Autotransformers.

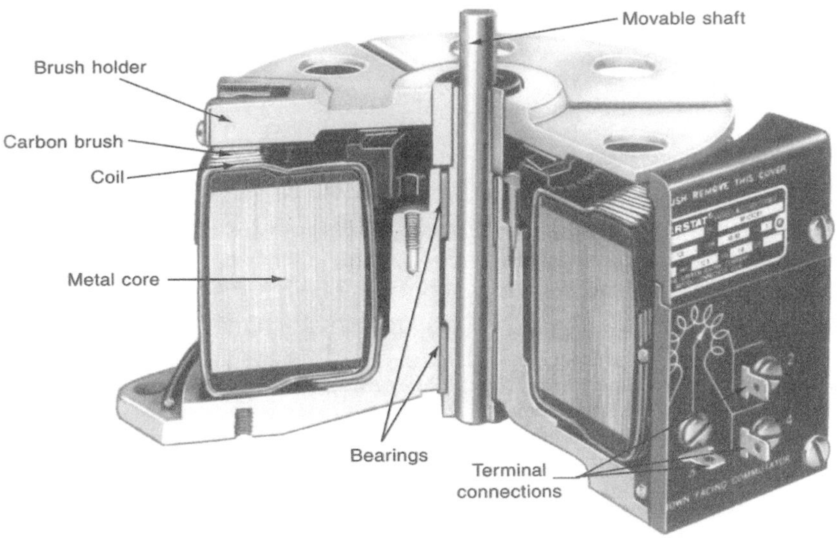

Figure 6.9 Cutaway view of a variable autotransformer. (Superior Electric Co.)

connection between the primary and secondary. The principle of operation of the autotransformer is similar to other transformers. Both the step-up and step-down types are shown in Figure 6.8. The control device shown in Figure 6.9 is a **variable autotransformer** in which the winding tap may be adjusted along the entire length of the winding to provide a variable AC voltage to a load.

6.9 Current Transformers

Current transformers are often used to reduce the large value of line current to a smaller value for measurement or control purposes. These transformers

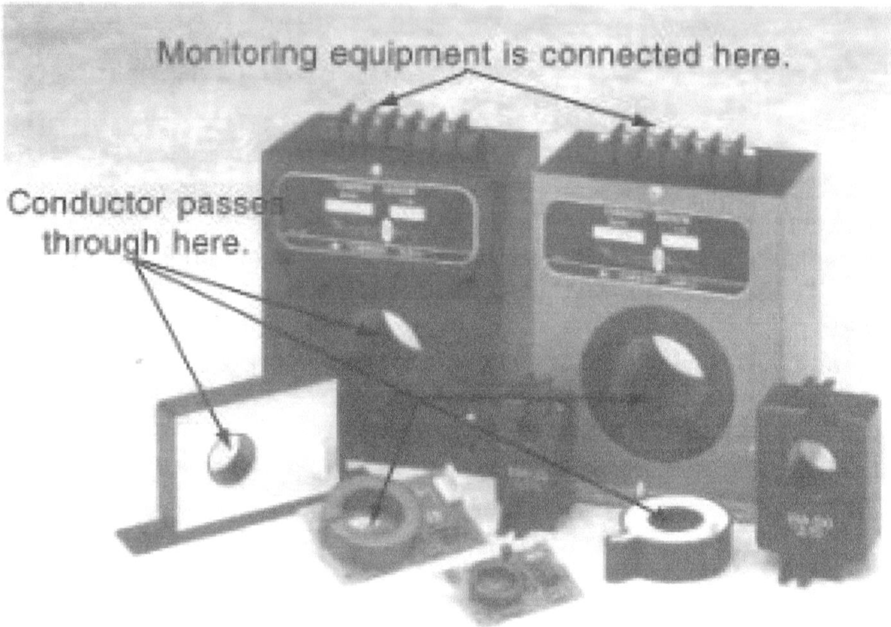

Figure 6.10 Single-phase current transformers. (F.W. Bell)

are used to measure the current magnitude of high-current systems. Since most metering systems respond linearly to current changes, the current transformer principle could also be used to measure quantities other than current in high-power systems. Single-phase current transformers are shown in Figure 6.10.

6.10 Transformer Polarity and Ratings

Power distribution transformers usually have **polarity markings** so that their windings can be connected in parallel to increase their current capacity. The standard markings are H_1, H_2, H_3, etc., for the high-voltage windings, and X_1, X_2, X_3, etc., for the low-voltage windings. Many power transformers have two similar primary windings and two similar secondary windings to make them adaptable to different voltage requirements simply by changing from a series to a parallel connection. The voltage combinations available from this type of transformer are shown in Figure 6.11.

The ratings of power transformers are very important. Usually, transformers are rated in **kilovolt amperes (kVA)**. A kilowatt rating is not used since it would be misleading due to the various power-factor ratings of

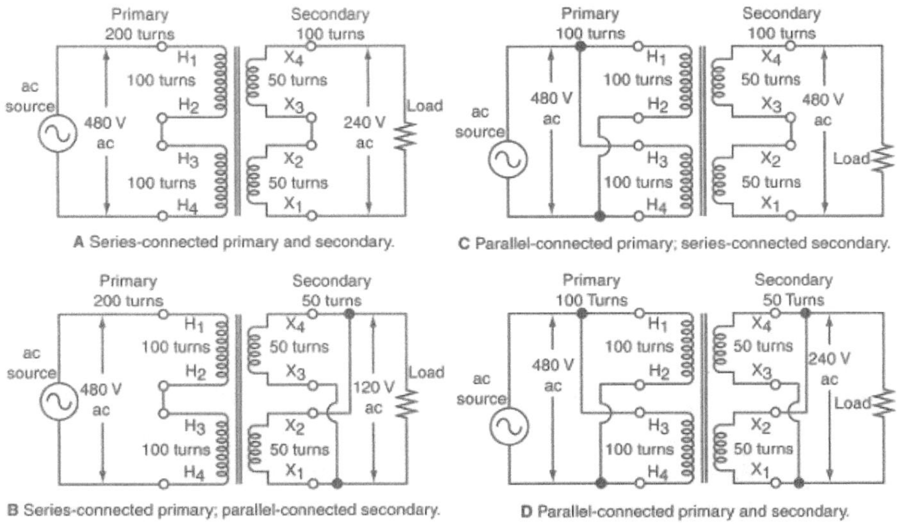

Figure 6.11 Some transformer connection methods for various voltage combinations.

industrial loads. Other power-transformer ratings usually include frequency, rated voltage of each winding, and an impedance rating.

Power transformers located along a power distribution system operate at very high temperatures. The three-phase transformers shown in Figure 6.12 have attached cooling equipment. This equipment is necessary for large transformers. The purpose of the cooling equipment is to conduct heat away from the transformer windings. Several power transformers are of the **liquid-immersed type**. The windings and core of the transformer are immersed in an insulating liquid that is contained in the transformer enclosure. The liquid insulates the windings, and it conducts heat away from them. One insulating liquid that is used extensively is called askarel. Some transformers, called dry types, use forced air or inert gas as coolants. Some locations, particularly indoors, are considered hazardous for the use of liquid-immersed transformers. However, most transformers rated at over 500 kVA are liquid-filled.

6.11 Conductors in Power Distribution Systems

The portions of the electrical distribution system that carry current are known as conductors. Conductors may be in the form of solid or stranded wires, cable assemblies, or large metallic bus-bar systems. A conductor can have insulation or, in some cases, be bare metal. Figure 6.13 shows some common types of conductors.

Figure 6.12 Three-phase transformers used in an electrical substation.

6.11.1 Conductor characteristics

Round conductors are measured by using the **American Wire Gage (AWG)** system. See Figure 6.14. Wire sizes in the AWG system range from No. 40 (smallest) to No. 0000 (largest). The cross-sectional area of a conductor doubles with each increase of three sizes. The diameter doubles with every increase of six sizes. The area of conductors is measured in **circular mils (cmil)**. Sizes larger than No. 0000 AWG are expressed in thousand circular mil (MCM) units.

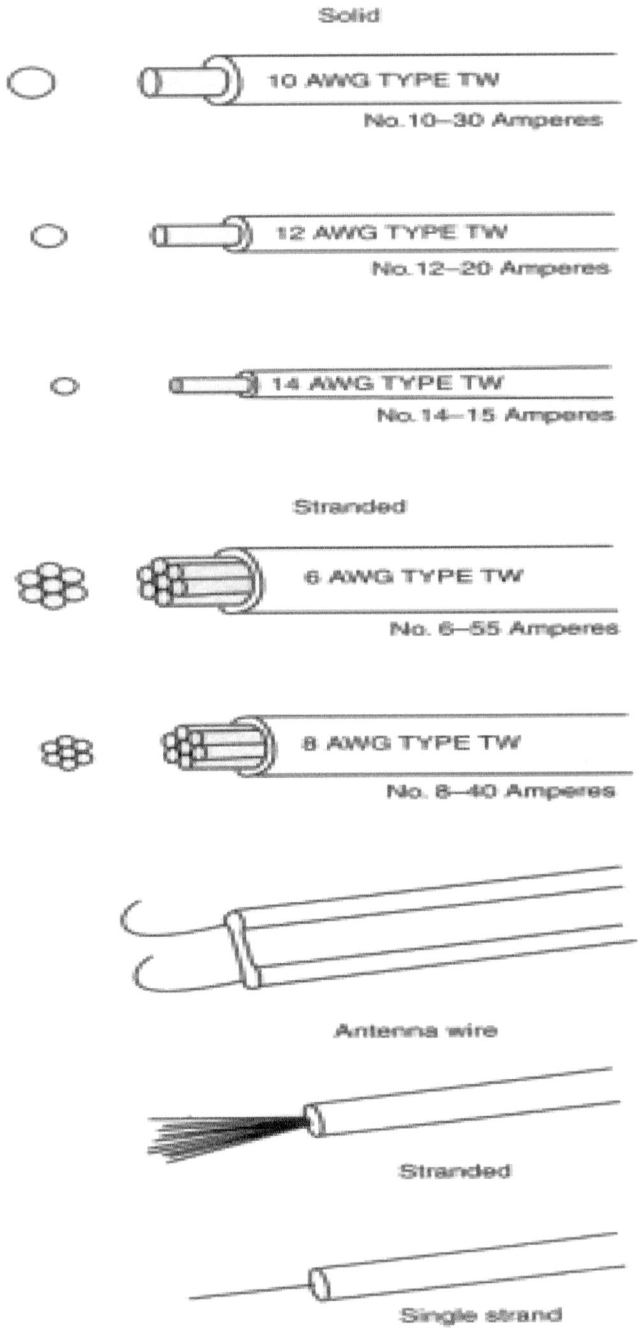

Figure 6.13 Different types of conductors.

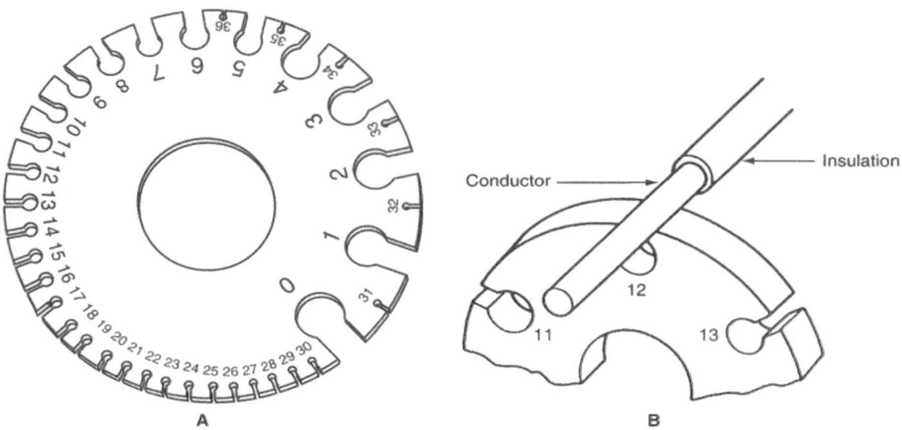

Figure 6.14 A – American Wire Gage (AWG) used to measure conductor size. B – Using the AWG.

Almost all **conductors** are made of either copper or aluminum. Both of these metals possess the necessary flexibility, current-carrying ability, and economical cost to act as efficient conductors. Copper is a better conductor. It has a lower resistance than aluminum. However, aluminum is 30% lighter in weight. Therefore, aluminum conductors are used when weight is a factor for consideration in conductor selection. One specialized overhead power-line conductor is the **aluminum conductor steel-reinforced (ACSR)** type. This type of conductor is used for long distance power transmission. It has stranded aluminum wires.

6.11.2 Conductor types

Copper is still the most widely used conductor material, both for solid and for stranded electrical wire. The availability of a variety of thermosetting and thermoplastic insulating materials offers great flexibility in meeting the requirements for most conductor applications. The operating temperature ranges for various types of insulation are given in Figure 6.15.

Copper has a good combination of various properties. These properties include malleability, good strength, and high electrical and thermal conductivity. Copper also has the capability of being alloyed or coated with other metals. Copper can be plated with silver to produce better solderability and, also, a conductor that has better high-frequency characteristics. This quality is due to the high **conductivity** of the silver and the **skin effect** present at higher frequencies.

Type of Insulation	Temperature Range (°C)
Neoprene	−30° to 90°
Teflon	−70° to 200°
Polyethylene	−60° to 80°
Rubber	−40° to 75°
Vinyl	−20° to 80°
Polypropylene	−20° to 105°

Figure 6.15 Operating temperature ranges of various types of insulation.

Where little vibration and no flexing are required of a wire or cable, single-strand conductors can be used. The advantage of a single-strand conductor is its lower cost compared to that of equivalent types of stranded wire. Wire and cable with solid conductors can be used as interconnection wires for electrical instruments and similar equipment.

Stranded conductors are used to provide more flexibility. They also have a longer usage life than do solid conductors. If a solid conductor were cut during installation by wire strippers, it would probably break after being bent a few times. However, a stranded wire would not break in this situation. Wires having from 26 to 41 strands are used where much flexibility is needed. Wires with from 65 to 105 strands are used for special purposes.

Flat or round **braided conductors** are sometimes used for situations where they are better suited than solid or stranded cables. These applications include use on tools, cranes, hoists, and other heavy industrial and construction applications. These conductors are seldom insulated since this would hinder their flexibility.

6.11.3 Conductor area

The unit of measurement for conductors is the **circular mil (cmil)** since most conductors are round. One mil is equal to 0.001 inch. Thus, one circular mil is equal to a circle whose diameter is 0.001 inch. The **cross-sectional area** of a conductor (in circular mils) is equal to its diameter (in mils) squared, or cmil = D^2. For example, if a conductor is a 1/4 inch in diameter, its circular mil area can be found as follows. The decimal equivalent of 1/4 inch is 0.250 inch. This equals 250 mils. Inserting this value into the formula for the cross-sectional area of a conductor gives you:

$$\begin{aligned} \text{area (in cmils)} &= D^2 \text{ (in mils)} \\ \text{area} &= (250)^2 \\ \text{area} &= 62{,}500 \text{ cmils} \end{aligned}$$

If the conductor is square or rectangular in shape, its area can be found by applying the following formula:

$$\text{area}(\text{in cmil}) = \frac{\text{area}(\text{in square mils})}{0.7854}$$

This formula allows us to convert the dimensions of a conductor to **square mils** from **circular mils**. For example, if a conductor is 1⁄2 inch × 1⁄4 inch, the circular mil area can be found using the following method. Again, convert the fractional inches into decimal values and then into an equivalent mil value. Thus, 1⁄2 inch equals 0.5 inch, which equals 500 mils, and inch is 0.75 inch, which equals 750 mils. Using the formula for square mils and substituting:

$$\begin{aligned} \text{area}(\text{in cmil}) &= \frac{\text{area (in square mils)}}{0.7854} \\ \text{area} &= \frac{500\text{mil} \times 750\text{mils}}{0.7854} \\ \text{area} &= \frac{375{,}000\text{mils}^2}{0.7854} \\ \text{area} &= 477{,}463.7\text{cmils} \end{aligned}$$

We can also use the circular mil area of a conductor, which can be found in any standard conductor table, to find the diameter of a conductor. If a table shows that the circular mil area of a conductor is equal to 16,510 cmils (a No. 8 AWG conductor), its diameter is found by the following method.

$$\begin{aligned} D^2 &= \sqrt{\text{cmil}} \\ D &= \sqrt{\text{cmil}} \\ D &= \sqrt{16{,}510\text{cmils}} \end{aligned}$$

$$\begin{aligned} D &= 128.5 \text{ mils} \\ D &= 0.1285 \text{ inch} \end{aligned}$$

Conductor	Resistivity in ohms per cmil-ft
Silver	9.8
Copper	10.4
Aluminum	17.0
Tungsten	33.0
Nickel	50.0
Iron	60.0

Figure 6.16 Resistivity of common conductors.

6.11.4 Resistance of conductors

The **resistance** of a conductor expresses the amount of opposition it will offer to the flow of electrical current. The unit of measurement for resistance is the ohm (Ω). The **resistivity** (ρ) of a conductor is the resistance for a specified cross-sectional area and length. This measurement is given in **ohms per circular mil-feet (cmil-ft)**. The resistivity of a conductor changes with temperature, so resistivity is usually specified at a temperature of 20° Celsius. The resistivities for some common types of conductors are listed in Figure 6.16.

We can use Figure 6.16 to calculate the resistance of any size conductor. We know that resistance increases as the length increases and decreases as the cross-sectional area increases. The following method can be used to find the resistance of 500 feet of aluminum conductor that is 1⁄4 inch in diameter. According to Figure 6.16, aluminum has a resistivity of 17ohms. The diameter (D) equals 1⁄4 inch (0.250 inch), which is the equivalent of 250 mils. Using the formula and substituting, we have:

$$\text{resistance} = \frac{\text{resistivity} \times \text{length}\left(\text{in feet}\right)}{\text{diameter}^2\left(\text{in mils}\right)}$$

$$\text{resistance} = \frac{17\Omega \text{ per cmil-ft} \times 500 \text{ ft}}{(250 \text{ mils})^2}$$

$$\text{resistance} = \frac{8500}{62{,}500}$$

$$\text{resistance} = 0.136\Omega$$

Size (AWG or MCM)	Area (cmil)	Number of Wires	Diameter of Each Wire (in.)	DC Resistance (Ω/1000 ft) 25°C	
				Copper	Aluminum
18	1,620	1	0.0403	6.51	10.7
16	2,580	1	0.0508	4.10	6.72
14	4,110	1	0.0641	2.57	4.22
12	6,530	1	0.0808	1.62	2.66
10	10,380	1	0.1019	1.018	1.67
8	16,510	1	0.1285	0.6404	1.05
6	26,240	7	0.0612	0.410	0.674
4	41,740	7	0.0772	0.259	0.424
3	52,620	7	0.0867	0.205	0.336
2	66,360	7	0.0974	0.162	0.266
1	83,690	19	0.0664	0.129	0.211
0	105,600	19	0.0745	0.102	0.168
00	133,100	19	0.0837	0.0811	0.133
000	167,800	19	0.0940	0.0642	0.105
0000	211,600	19	0.1055	0.0509	0.0836
250	250,000	37	0.0822	0.0431	0.0708
300	300,000	37	0.0900	0.0360	0.0590
350	350,000	37	0.0973	0.0308	0.0505
400	400,000	37	0.1040	0.0270	0.0442
500	500,000	37	0.1162	0.0216	0.0354
600	600,000	61	0.0992	0.0180	0.0295
700	700,000	61	0.1071	0.0154	0.0253
750	750,000	61	0.1109	0.0144	0.0236
800	800,000	61	0.1145	0.0135	0.0221
900	900,000	61	0.1215	0.0120	0.0197
1000	1,000,000	61	0.1280	0.0108	0.0177

Figure 6.17 Sizes of copper and aluminum conductors.

6.11.5 Conductor sizes and types

Figure 6.17 lists the sizes of copper and aluminum electrical conductors. As mentioned earlier, the **American Wire Gage (AWG)** is the standard used to measure the diameter of conductors. The sizes range from No. 40 AWG, which is the smallest, to No. 0000 AWG.

Note, in Figure 6.17, that as the AWG size number becomes smaller, the conductor is larger. Sizes up to No. 8 AWG are solid conductors, while larger wires have from 7 to 61 strands. Figure 6.17 also lists the DC resistance (in ohms per 1000 feet) of the copper and aluminum conductors. These values are used to determine conductor voltage drop in distribution systems.

Some stranded special-purpose conductors that are used in distribution wiring for industrial and commercial facilities are shown in Figure 6.18. Also notice in the figure the special connectors used to permanently connect the

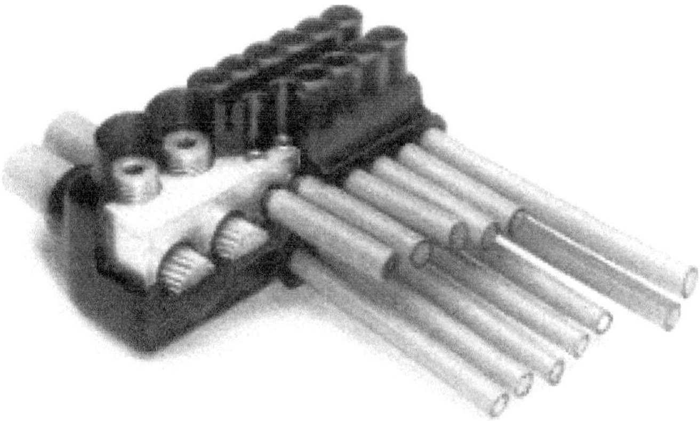

Figure 6.18 Stranded conductors showing special connector. (Polaris Co.)

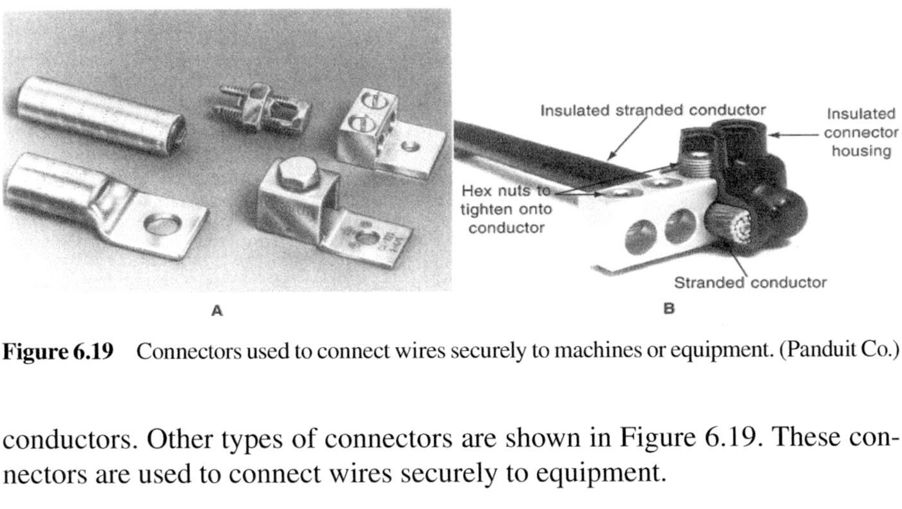

A B

Figure 6.19 Connectors used to connect wires securely to machines or equipment. (Panduit Co.)

conductors. Other types of connectors are shown in Figure 6.19. These connectors are used to connect wires securely to equipment.

6.11.6 Ampacity of conductors

A measure of the ability of a conductor to carry electrical current is called **ampacity**. The ampacity of conductors depends on several factors such as the type of material, cross-sectional area, and the type of area in which they are installed. Conductors in the open or in **free air** dissipate heat much more rapidly than they do if they are enclosed in a metal **raceway** or plastic cable. When several conductors are contained within the same enclosure, **heat dissipation** is a greater problem.

All metal materials will conduct electrical current to some extent. Copper and aluminum are the two most desirable types used. Copper is the

Wire Size	Copper: With R, T, TW Insulation	Copper: With RH, RHW, TH, THW Insulation	Aluminum: With R, T, TW Insulation	Aluminum: With RH, RHW, TH, THW Insulation
14	20	20		
12	25	25	20	20
10	40	40	30	30
8	55	65	45	55
6	80	95	60	75
4	105	125	80	100
3	120	145	95	115
2	140	170	110	135
1	165	195	130	155
0	195	230	150	180
00	225	265	175	210
000	260	310	200	240
0000	300	360	230	280
250	340	405	265	315
300	375	445	290	350
350	420	505	330	395
400	455	545	355	425
500	515	620	405	485
600	575	690	455	545
700	630	755	500	595
750	655	785	515	620
800	680	815	535	645
900	730	870	580	700
1000	780	935	625	750

Figure 6.20 Allowable ampacities of single conductors in free air.

most used metal since it is the better conductor of the two and is physically stronger. However, aluminum is usually used where weight is a factor. The weight of copper is almost three times that of a similar volume of aluminum. However, the resistance of aluminum is over 150% that of copper. The **ampacity** of an aluminum conductor is, therefore, less than a similar size copper conductor. In a wiring design, aluminum conductors that are one size larger than their copper counterparts are ordinarily used to allow for this difference. See Figure 6.20.

6.12 The National Electrical Code (NEC)

The **National Electrical Code (NEC)**^ is a very important document to understand. All industrial equipment and wiring in the United States must conform to the NEC standards. The NEC is not difficult to use. You should become familiar with the comprehensive index contained in the NEC and the organization of the various sections. For instance, if you wish to review the standards related to **system grounding**, you should look at the index and locate this term. The index will refer you to the appropriate sections in the NEC that discuss system grounding. A table of contents of the current NEC is shown below. This listing provides an overview of the organization of the NEC. It is important for electrical technicians to learn to use the NEC.

6.12.1 National Electrical Code (NEC) Table of Contents

Chapter 1 – General

Definitions
Requirements for Electrical Installations

Chapter 2 – Wiring Design and Protection

Use and Identification of Grounded Conductors
Branch Circuits
Feeders
Branch-Circuit, Feeder, and Service Calculations
Outside Branch Circuits and Feeders
Services
Overcurrent Protection
Grounding
Surge Arresters

Chapter 3 – Wiring Methods and Materials

Wiring Methods
Temporary Wiring
Conductors for General Wiring
Cable Trays
Open Wiring on Insulators
Messenger Supported Wiring
Concealed Knob-and-Tube Wiring
Integrated Gas Spacer Cable
Medium Voltage Cable
Flat Conductor Cable

Mineral-Insulated Metal-Sheathed Cable
Electrical Nonmetallic Tubing
Armored Cable
Metal-Clad Cable
Nonmetallic-Sheathed Cable
Service-Entrance Cable
Underground Feeder and Branch-Circuit Cable
Power and Control Tray Cable
Nonmetallic Extensions
Nonmetallic Underground Conduit with Connectors
Intermediate Metal Conduit
Rigid Metal Conduit
Rigid Nonmetallic Conduit
Electrical Metallic Tubing
Flexible Metallic Tubing
Flexible Metal Conduit
Liquid-Tight Flexible Metal Conduit and Liquid-Tight Flexible Nonmetallic Conduit
Surface Metal Raceways and Surface Non-metallic Raceways
Multioutlet Assembly
Underfloor Raceways
Cellular Metal Floor Raceways
Cellular Concrete Floor Raceways
Metal Wireways and Nonmetallic Wireways
Flat Cable Assemblies
Busways
Cable bus
Outlet, Device, Pull and Junction Boxes, Conduit Bodies and Fittings
Cabinets, Cutout Boxes, and Meter Socket Enclosures
Auxiliary Gutters
Switches
Switchboards and Panelboards

Chapter 4 – Equipment for General Use

Flexible Cords and Cables
Fixture Wires
Lighting Fixtures, Lamp holders, Lamps, and Receptacles
Lighting Systems Operating at 30 Volts or Less
Appliances

Fixed Electric Space-Heating Equipment
Fixed Outdoor Electric Deicing and Snow-
Melting Equipment
Fixed Electric Heating Equipment for
Pipelines and Vessels
Motors, Motor Circuits, and Controllers
Air-Conditioning and Refrigerating Equipment
Generators
Transformers and Transformer Vaults
Phase Converters
Capacitors
Resistors and Reactors
Storage Batteries
Equipment, Over 600 Volts

Chapter 5 – Special Occupancies

Hazardous (Classified) Locations, Classes I, II, and III, Divisions 1 and 2
Class I Locations
Class II Locations
Class III Locations
Intrinsically Safe Systems
Class I, Zone 0, 1, and 2 Locations
Hazardous (Classified) Locations-Specific
Commercial Garages, Repair and Storage
Aircraft Hangers
Gasoline Dispensing and Service Stations Bulk Storage Plants
Spray Application, Dipping, and Coating Processes
Health Care Facilities
Places of Assembly
Theaters, Audience Areas of Motion Picture and Television Studios, and Similar Locations
Carnivals, Circuses, Fairs, and Similar Events
Motion Picture and Television Studios and
Similar Locations
Motion Picture Projectors
Manufactured Building
Agricultural Buildings
Mobile Homes, Manufactured Homes, and Mobile Home Parks
Recreational Vehicles and Recreational

Vehicle Parks
Park Trailers
Floating Buildings
Marinas and Boatyards

Chapter 6 – Special Equipment

Electric Signs and Outline Lighting
Manufactured Wiring Systems
Office Furnishings
Cranes and Hoists
Elevators, Dumbwaiters, Escalators, Moving Walks, Wheelchair Lifts, and Stairway Chairlifts
Electric Vehicle Charging System
Equipment
Electric Welders
Audio Signal Processing, Amplification, and Reproduction Equipment Information Technology Equipment Pipe Organs
X-Ray Equipment
Induction and Dielectric Heating
Equipment
Electrolytic Cells
Electroplating
Industrial Machinery
Electrically Driven or Controlled Irrigation Machines
Swimming Pools, Fountains, and Similar Installations
Integrated Electrical Systems
Solar Photovoltaic Systems
Fire Pumps

Chapter 7 Special Conditions

Emergency Systems
Legally Required Standby Systems
Optional Standby Systems
Interconnected Electrical Power Production Sources
Circuits and Equipment Operating at Less than 50 Volts
Class 1, Class 2, and Class 3 Remote-
Control, Signaling, and Power-Limited
Circuits
Instrumentation Tray Cable: Type ITC
Fire Alarm Systems

Wire Size	Copper: With R, T, TW Insulation	Copper: With RH, RHW, TH, THW Insulation	Aluminum: With R, T, TW Insulation	Aluminum: With RH, RHW, TH, THW Insulation
14	15	15		
12	20	20	15	15
10	30	30	25	25
8	40	45	30	40
6	55	65	40	50
4	70	85	55	65
3	80	100	65	75
2	95	115	75	90
1	110	130	85	100
0	125	150	100	120
00	145	175	115	135
000	165	200	130	155
0000	195	230	155	180
250	215	255	170	205
300	240	285	190	230
350	260	310	210	250
400	280	335	225	270
500	320	380	260	310
600	355	420	285	340
700	385	460	310	375
750	400	475	320	385
800	410	490	330	395
900	435	520	355	425
1000	455	545	375	445

Figure 6-21 Ampacities of conductors in a raceway or cable (3 or less).

Optical Fiber Cables and Raceways
Closed-Loop and Programmed Power
Distribution

Chapter 8 Communication systems

Communication Circuits
Radio and Television Equipment
Community Antenna Television and Radio Distribution Systems
Network-Powered Broadband Communications Systems

6.13 Ampacity Tables

Figures 6.20 through 6.22 are used for conductor ampacity calculations for electrical wiring design. These tables are simplified versions of those given in the National Electrical Code. Figure 6.20 is used to determine conductor ampacity when a single conductor is mounted in free air. Figure 6.21 is used to find the ampacity of conductors when not more than three are mounted in a raceway or cable. These two tables are based on ambient temperatures of

Ambient Temperature		Conductor Correction Factor	
C°	F°	R, T, TW	RH, RHW, TH, THW
40	104	0.82	0.88
45	113	0.71	0.82
50	122	0.58	0.75
55	131	0.41	0.67
60	140	—	0.58
70	158	—	0.35

Figure 6.22 Correction factors for temperatures above .

30° Celsius (86° Fahrenheit). Figure 6.22 lists the correction factors that are used for temperatures over 30° C.

As an **example**, we will find the ampacity of three No. 10 copper conductors with RHW insulation that are mounted in a raceway. They will be located in a foundry area where temperatures reach 50° C. The ampacity for No. 10 RHW copper wire is 30 amperes (from Figure 6.21). The correction factor for an RHW-insulated conductor at 50° C ambient temperature is 0.75 (Figure 6.22). Therefore, 30 amperes × 0.75 = 22.5 amperes.

6.14 Insulation in Power Distribution Systems

Synthetic insulation for wire and cable is classified into two broad categories-thermosetting and thermoplastic. The mixtures of materials within each of these categories are so varied as to make the available number of insulations almost unlimited. Most insulation is composed of compounds made of synthetic rubber polymers ***(thermosetting)*** and from synthetic materials ***(thermoplastics)***. These synthetic materials are combined to provide specific physical and electrical properties.

Thermosetting materials are characterized by their ability to be stretched, compressed, or deformed (within reasonable limits) under mechanical strain and, then, return to their original shape when the stress is removed.

Thermoplastic insulation materials are best known for their excellent electrical characteristics and relatively low cost. These materials are popular since much thinner insulation thicknesses can be used to obtain good electrical properties, particularly at the higher voltages. Some insulators used with electrical power distribution systems are shown in Figure 6.23.

There are many types of insulation used today for electrical conductors. Some new materials have been developed that will last for exceptionally long periods of time and will withstand very high operating temperatures.

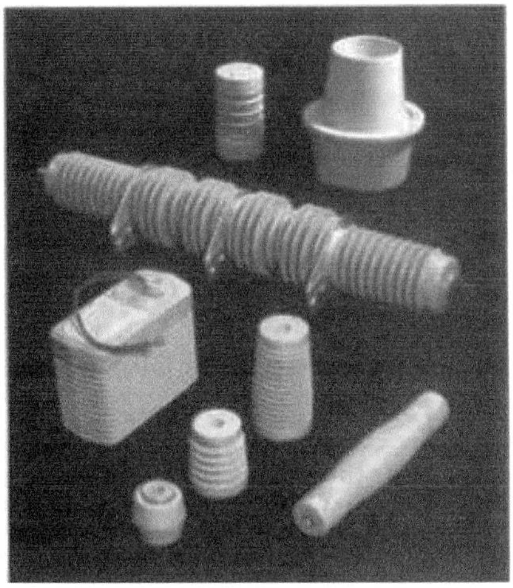

Figure 6.23 Insulators used with electrical power distribution systems. (S & C Electric Co.)

Abbreviation	Type of Insulation
R	Rubber—140° F
RH	Heat-Resistant Rubber—167° F
RHH	Heat-Resistant Rubber—194° F
RHW	Moisture and Heat-Resistant Rubber—167° F
T	Thermoplastic—140° F
THW	Moisture and Heat-Resistant Thermoplastic—167° F
THWN	Moisture and Heat-Resistant Thermoplastic with Nylon—194° F

Figure 6.24 Common abbreviations for types of electrical insulation.

The operating conditions where the conductors are used mainly determine the type of insulation required. For instance, system voltage, heat, and moisture affect the type of insulation required. Insulation must be used that will withstand both the heat of the surrounding atmosphere and the heat developed by the current flowing through the conductor. Exceptionally large currents cause excessive heat to be developed in a conductor. Such heat could cause insulation to melt or burn. This is why overcurrent protection is required as a safety factor to prevent fires. The ampacity or current-carrying capacity of a conductor depends upon the type of insulation used. The NEC has developed a system of abbreviations for identifying various types of insulation. Some of the abbreviations are shown in Figure 6.24.

Review Questions

Answer the following questions.

1. Why are current transformers used?

2. List several types of transformers.

3. How is transformer efficiency determined?

4. A 1500 -turn primary winding has 120 volts applied. The secondary voltage is 20 volts. How many turns of wire does the secondary winding have?

5. When a 200 -ohm resistance is placed across a 240 -volt secondary winding, the primary current is 30 amperes. What is the value of the primary voltage of the transformer?

6. A transformer has a 20:1 turns ratio. The primary voltage is 4800 volts. What is the secondary voltage?

7. A transformer has a 2400 -volt primary winding and a 120 -volt secondary. The primary winding has 1000 turns. How many turns of wire does the secondary winding have?

8. A 240 -volt to 4800 -volt transformer has a primary current of 95 amperes and a secondary current of 4 amperes. What is its efficiency?

9. A transformer has a 20-kVA power rating. Its primary voltage is 120 volts and its secondary voltage is 480 volts. What are the values of its maximum primary current and secondary current?

10. Convert the following diameters of electrical conductors into mils:

    a. 1/8 inch

    b. 1/4 inch.

    c. 3/8 inch.

    d. 1/2 inch.

    e. 0.2 inch.

    f. 0.28 inch.

    g. 0.15 inch.

    h. 0.66 inch.

11. Calculate the cross-sectional area of the following electrical conductors (in circular mils):

    a. 3/16 inch diameter.

    b. 0.5 inch square conductor.

    c. 0.38 inch diameter.

    d. 1/2 inch × 3/8 inch rectangular conductor.

12. Calculate the resistance of 480 feet of copper conductor that is 0.28 inch in diameter.

13. Calculate the resistances of the following electrical conductors:

    a. 380 feet of No. 10 AWG copper.

b. 2500 feet of No. 000 AWG copper.

c. 230 feet of 500MCM copper.

d. 1900 feet of No. 8 AWG aluminum.

e. 220 feet of No. 10 AWG aluminum.

14. Calculate the resistances of the following electrical conductors:

 a. 600 feet of No. 0000 AWG copper.

 b. 500 feet of No. 1 AWG aluminum.

 c. 300 feet of No. 12 AWG copper.

 d. 200 feet of 250 MCM aluminum.

 e. 1800 feet of No. 10 AWG copper.

15. In order to carry the following currents, what are the minimum sizes of conductors needed? No more than three conductors will be used in the same raceway.

 a. 380 amperes using THW copper.

 b. 95 amperes using THW aluminum.

 c. 180 amperes using RH copper.

 d. 230 amperes using TW copper.

 e. 230 amperes using *R* copper.

 f. 650 amperes using RHW copper.

 g. 5–80 amperes using T copper.

 h. 82 amperes using TH aluminum.

7

Electrical Distribution Equipment

Objectives

Upon completion of this chapter, you will be able to:

- Describe the following equipment used at substations for electrical power distribution: high- voltage fuses, high-voltage circuit breakers, high-voltage disconnect switches, lightning arresters, high-voltage insulators, high-voltage conductors, and voltage regulators.
- Describe the following types of protective equipment used in electrical distribution systems: plug fuses, cartridge fuses, time-lag fuses, low-voltage circuit breakers, protective relays, motor fault current protection, overheating protection, and undervoltage protection.
- Describe the following in relation to electrical power distribution systems: feeder lines, branch circuits, safety switches, distribution panel boards, low-voltage switchgear, and service entrances.

Important Terms

air-break disconnect switch
branch circuit
compressed-air circuit breaker
conduit
distribution substation
duct
expulsion-type fuse
feeder line
flashover
insulator
lightning arrester
liquid fuse
magnetic air breaker
oil-filled circuit breaker
oilless circuit breaker
panel board
raceway
safety switch
service entrance
solid-material fuse
switchboard
switchgear
tap changer
voltage regulator

DOI: 10.1201/9788743810056-9

Figure 7.1 Electrical substation. (Kentucky Utilities Co.)

Introduction

In order to distribute electrical power, it is necessary to use many types of specialized equipment. The electrical distribution system consists of such specialized equipment as power transformers, high-voltage fuses and circuit breakers, lightning arresters, power-factor-correcting capacitors, and power metering systems. Some types of electrical distribution equipment which technicians should understand will be discussed in this chapter. This equipment is often used with electrical control systems.

7.1 Equipment Used at Substations

Substations are very important parts of the electrical distribution systems. The link between the high-voltage transmission lines and the low- voltage power distribution systems is the substation. The function of a *distribution substation*, such as the one shown in Figure 7.1, is to receive electrical power from a high-voltage system and convert it to voltage levels suitable for industrial, commercial, or residential use. The major functional component of a substation is the transformer, whose basic characteristics were discussed in

Chapter 6. However, there are many other types of specialized equipment required for the operation of a substation.

Since power lines may be ***short circuited***, various protective equipment is used to prevent damage to both the power lines and the equipment. This protective equipment must be designed to handle high voltages and currents. Either fuses or circuit breakers can be used to protect high-voltage power lines. ***High-voltage fuses*** (those used for over 600 volts) are made in several ways. An *expulsion-type fuse* has an element that will melt and vaporize when it is overloaded. This causes the power line it is connected in series with to open. *Liquid fuses* have a liquid filled metal enclosure that contains the fuse element. The liquid acts as an arc suppressing medium. When the fuse element melts due to an excessive current in a power line, the element is immersed in the liquid to extinguish the arc. This type of fuse reduces the problem of high-voltage arcing. A *solid-material fuse* is similar to a liquid fuse except that the arc is extinguished in a chamber filled with a solid material. An enclosure used for high-voltage fuses in power distribution systems is shown in Figure 7.2.

Ordinarily, high-voltage fuses at substations are mounted adjacent to air-break disconnect switches. ***Air-break disconnect switches*** provide a means of switching power lines and disconnecting them for repair. The fuse and switch ***enclosure*** is usually mounted near the overhead power lines at a substation.

Circuit breakers that control high voltages are also located at electrical substations. It should be pointed out that large arcs are present whenever a high-voltage circuit is interrupted. This problem is not encountered to any great extent in low- voltage protective equipment. There are two major types of high-voltage circuit breakers—oil filled and oilless. These circuit breakers are designed to operate on voltages of 1000 volts to over 500,000 volts.

Oil-filled circuit breakers, Figure 7.3, are used primarily for outdoor substations, except for very high voltages in the range of 500,000 volts and higher. In this type of circuit breaker, the contacts are immersed in an insulating oil contained in a metal enclosure. ***Oilless circuit breakers*** are ordinarily used for indoor operation.

Another type of high-voltage circuit breaker is the ***magnetic air breaker*** in which the contacts separate, in the air, when the power line is overloaded. Magnetic blowout coils are used to develop a magnetic field, which causes the arc (produced when the contacts break) to be concentrated into arc chutes where it is extinguished. A modification of this type is the ***compressed-air circuit breaker***. In this breaker, a stream of compressed air is concentrated on the contacts when the power line is opened. The compressed air aids in extinguishing the arc that is developed when the contacts open.

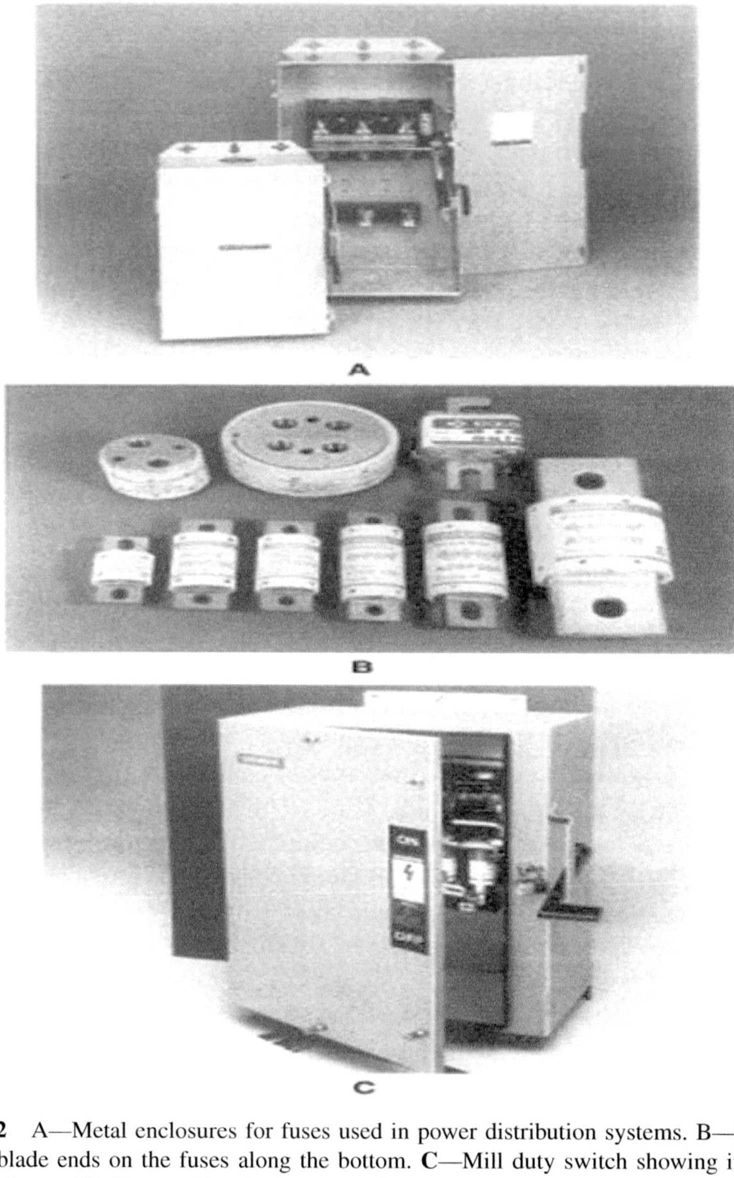

Figure 7.2 A—Metal enclosures for fuses used in power distribution systems. B—Notice the knife-blade ends on the fuses along the bottom. **C**—Mill duty switch showing internal view with fuses. (Challenger Electrical Equipment Corp., Gould, Inc., Siemens Energy and Automation)

High voltage disconnect switches, mentioned in **Section 7.2**, are used to disconnect electrical equipment from the power lines that supply the equipment. Ordinarily, disconnect switches are not operated when current is flowing through them. A high voltage arcing problem would occur if

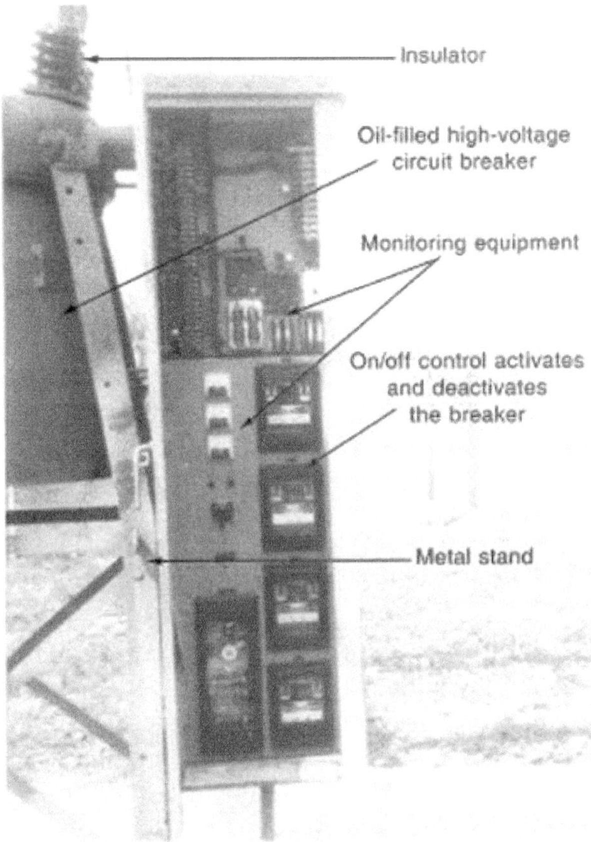

Figure 7.3 Circuit breaker—The circuit breaker is on the top left; its control panel is located next to it on the right.

disconnect switches were opened while current was flowing through them. They are opened mainly to isolate equipment from power lines for safety purposes. Most disconnect switches are the "air-break" type which is similar in construction to knife switches. These switches are available for indoor or outdoor use in both manual and motor-operated designs. Air-break high-voltage disconnect switches are shown in the substation of Figure 7.1.

The purpose of using **lightning arresters** on power lines is to cause the conduction to ground of excessively high voltages caused by lightning strikes or other system problems. Without lightning arresters, power lines and associated equipment could become inoperable when struck by lightning.

Arresters are designed to operate rapidly and repeatedly if necessary. Their response time must be more rapid than the other protective equipment

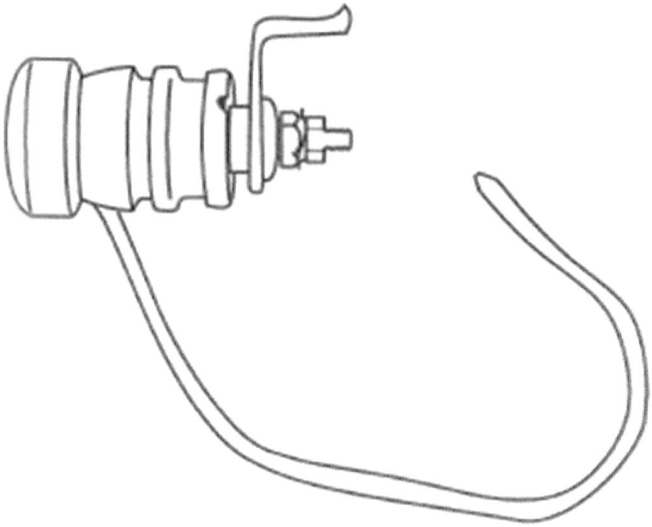

Figure 7.4 Drawing of a lightning arrester.

used on power lines. Lightning arresters must have a rigid connection to ground on one side. The other side of the arrester is connected to a power line. Sometimes, they are connected to transformers or the insides of switchgear. Lightning is a major cause of power-system failures and equipment damage, so lightning arresters have a very important function.

Lightning arresters are also used at outdoor substations. The lightning arrester, such as the one drawn in Figure 7.4, provides a path for the lightning to eliminate the flashover between power lines. This **flashover** causes short circuits.

Valve-type lightning arresters are used frequently. They are two-terminal devices in which one terminal is connected to the power line and the other is connected to ground. The path from line to ground is of such high resistance that it is normally open. However, when lightning, which is a very high voltage, strikes a power line, it causes conduction from line to ground. Thus, voltage surges are conducted to ground before flashover between the lines occurs. After the lightning surge has been conducted to ground, the valve assembly then causes the lightning arrester to become nonconductive once more.

All power transmission lines must be isolated so as not to become safety hazards. Large strings of **insulators** are used at substations and at other points along the power distribution system to isolate the current-carrying conductors

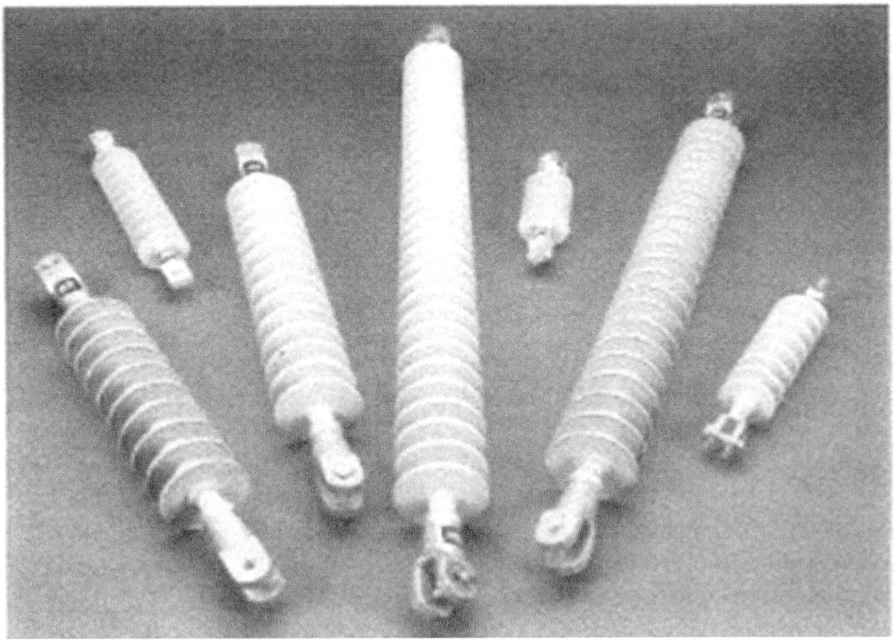

Figure 7.5 High-voltage insulators. (Porcelain Products, Talley, Indiana)

from their steel supports or any other ground-mounted equipment. Insulators can be made of porcelain, rubber, or a thermoplastic material. High-voltage insulators are shown in Figure 7.5.

Power transmission lines require many insulators to electrically isolate the power lines from the steel towers and wooden poles that support the lines. Insulators must have enough mechanical strength to support power lines under all weather conditions. They must also have sufficient insulating properties to prevent any arcing between the power lines and their support structures. High-voltage insulators are usually made of porcelain. ***Insulators*** are constructed in "strings" that are suspended from steel or wooden towers. The design of these insulators is very important since design affects their capacitance and their ability to withstand weather conditions.

The conductors used for long-distance power distribution are, ordinarily, uninsulated aluminum wires or ***aluminum-conductor steel reinforced (ACSR)*** wires. For shorter distances, insulated copper wires are used.

Voltage regulators are an important part of the power distribution system. **Voltage regulators** are used to maintain the voltage levels at the proper value. A constant voltage must be maintained in order for the electrical

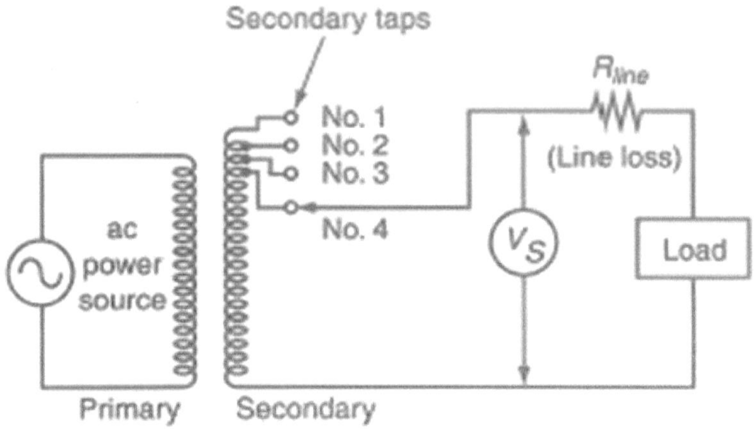

Figure 7.6 Transformer tap changer voltage regulator.

equipment to function properly. For instance, motors do not operate properly when a reduced or an excessive voltage is applied to them.

Transformer ***tap changers***, illustrated in Figure 7.6, can be used as voltage regulators. By either a manual or an automatic changing of the secondary tap, the voltage output can be changed to compensate for changes in the load voltage. As load current increases, line loss (I × R) also increases. Increased line loss causes the secondary voltage (Vg) to decrease. If the secondary tap is initially connected to tap No. 4, the secondary voltage can be boosted by reconnecting to either tap No. 3, No. 2, or No. 1. This can be done automatically with a motor-controlled tapchanger. There are various other types of automatic voltage regulators that can be used with electrical power distribution systems.

7.2 Power-System Protective Equipment

There are many devices used to protect electrical systems from damages due to abnormal conditions. Switches, fuses, circuit breakers, lightning arresters, and protective relays are all used for this purpose. Some of these devices automatically disconnect the equipment from power lines before any damage can occur. Other devices sense changes from the normal operation of the system and make the changes necessary to compensate for abnormal circuit conditions. The most common electrical problem which requires protection is ***short circuits***.

Other problems include overvoltage, under- voltage, and changes in frequency. Generally, more than one method of protection is used to protect

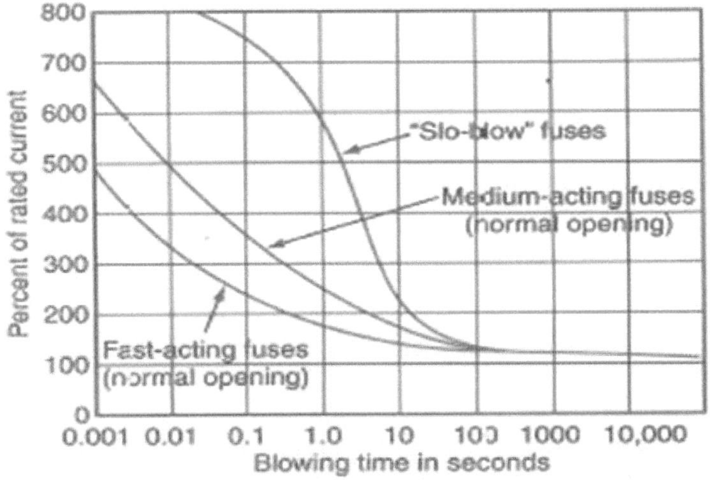

Figure 7.7 Typical operating characteristic curves for three types of fuses.

electrical circuits from faulty conditions. The purpose of any type of protective device is to cause a current-carrying conductor to become inoperative when an excessive amount of current flows through it.

All protective devices, including fuses, have an operating-characteristic time curve, such as the one shown in Figure 7.7. To prevent any possible damage to equipment, circuit protection should be planned utilizing these curves. They show the response time required when an overload occurs for a protective device to interrupt a circuit.

The simplest type of protective device is a ***fuse***. Fuses are low-cost items and have a fast operating speed. However, in three-phase systems, since each hot line must be fused, two lines are still operative if only one fuse burns out. Three-phase motors *will* continue to run with one phase removed. This condition is undesirable since motor torque is greatly reduced and overheating can result. Another obvious disadvantage of fuses is that replacements are required.

Plug fuses

Fuses could be used in ***safety switches*** and power distribution panels. The **plug fuse**, shown in Figure 7.8, is a common type of fuse. Standard sizes for this fuse are 10, 15, 20, 25, and 30 amperes at voltages of 125 volts or below. These fuses have a zinc or metallic-alloy fusible element enclosed in a case made of an insulating material. Their application could be in ***safety switches*** and fuse ***panel boards***.

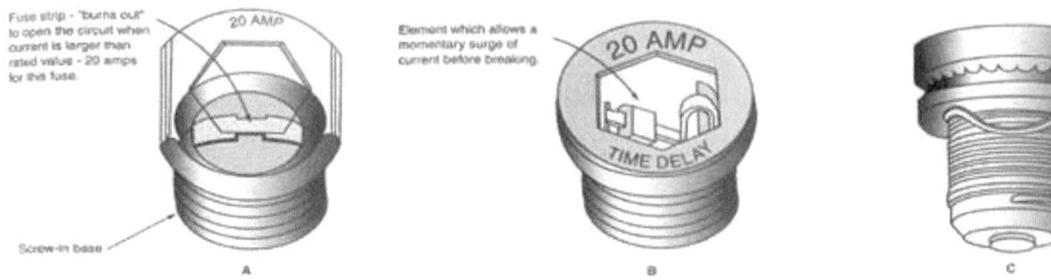

Figure 7.8 A—Typical plug fuse. B—Time delay fuse. C—Side view of a plug fuse.

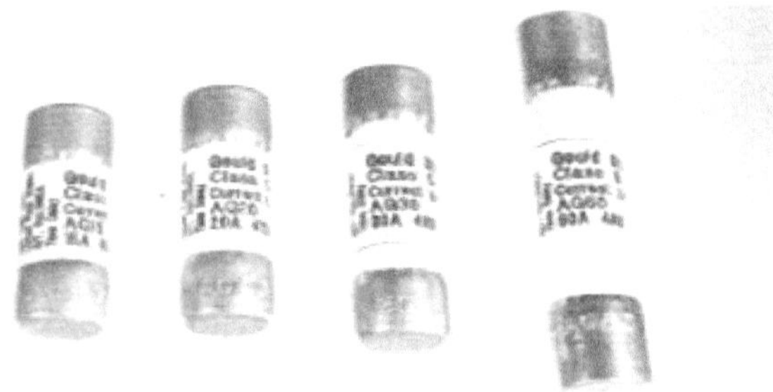

Figure 7.9 Cartridge fuses. Notice that the larger fuses have a higher current rating. (Gould, Inc.)

Cartridge fuses, such as the ones shown in Figure 7.9, are commonly used in power distribution systems for voltages up to 600 volts. They have a zinc or alloy fusible element that is housed in a round fiber enclosure. Some types have non-renewable elements. Other types have renewable elements. Cartridge fuses can be used to protect high-current circuits since they come in sizes of 60, 100, 200, 400, 600, and 1000 amperes.

A modified version of the plug or cartridge fuse is called a time delay fuse. This type of fuse is used to delay the circuit-interrupting action. It is useful where momentary high currents exist periodically, such as in motor-starting currents. The fuse element melts only when an excessive current is sustained over the time-lag period. Thus, sufficient circuit protection is still provided.

Time delay fuses are used to limit current on systems including electric motors that draw higher currents during their start cycle than during normal operation. These devices allow the system to start up at a higher current and then protect the system during normal operation without disrupting the distribution system.

The type of metal used in fuses is ordinarily an ***alloy*** material or, possibly, aluminum. All metals have resistance. When current flows through metal,

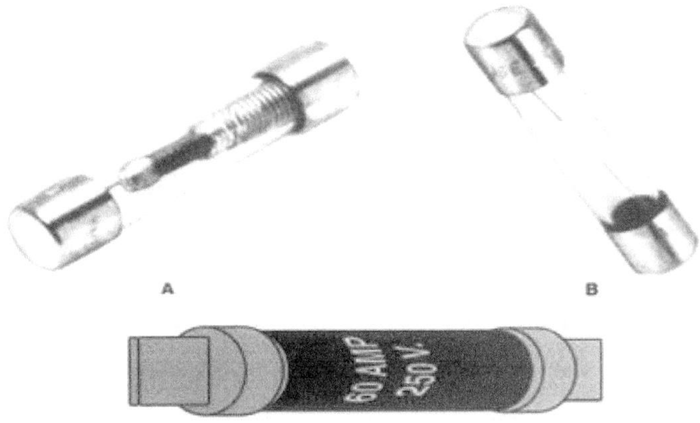

Figure 7.10 Types of fuses. A—Slow-blow fuse element (miniature). B—Regular fuse element (miniature). C—Knife-blade fuse. (Littlefuse, Inc.)

heat energy is produced. As the current increases, more heat is produced. This causes the temperature of the metal to increase. When the melting point of the fuse metal is reached, the fuse will open, causing the circuit to which it is connected to open.

Metals that decompose rapidly are used rather than metals that produce small metallic globules when they melt. This reduces the likelihood of any arcing occurring after the fuse metal has melted. The current rating of fuses depends upon the melting temperature of the fuse metal, as well as its shape, size, and the type of enclosure used. Some additional types of fuses are shown in Figure 7.10.

Circuit breakers, Figure 7.11, are somewhat more sophisticated overload devices than fuses. Although their function is the same as that of fuses, circuit breakers are much more versatile. In three-phase systems, circuit breakers can open all three hot lines when an overload occurs. They can also be activated by remote control relays. Relay systems can cause circuit breakers to open due to changes in frequency, voltage, current, or other circuit variables.

Circuit breakers that are used in industrial plants are usually the ***low-voltage*** variety (less than 600 volts). They are not nearly as complex as their ***high-voltage*** counterparts (which were discussed previously). Most low-voltage circuit breakers are housed in molded-plastic cases that mount in metal power distribution panels.

Circuit breakers are designed so that they will automatically open when a current occurs that exceeds the rating of the breaker. Ordinarily, the circuit

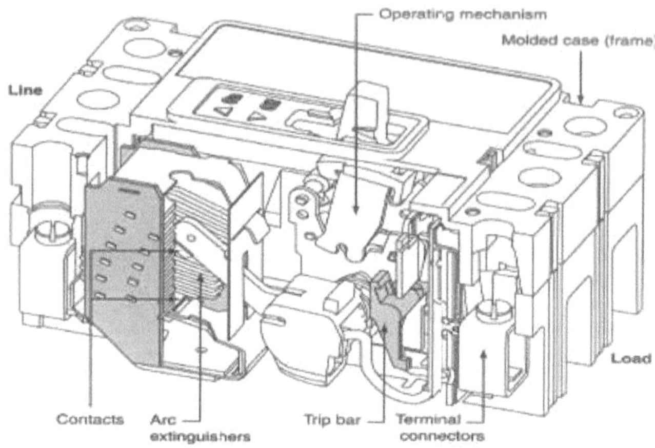

Figure 7.11 Circuit breaker construction.

breakers must be reset manually. Most circuit breakers employ either a thermal gripping element or a magnetic trip element. Ratings of circuit breakers extend into current ranges as high as 800 to 2000 amperes.

Protective relays provide an accurate and sensitive method of protecting electrical equipment from short circuits and other abnormal conditions. ***Overcurrent relays*** are used to cause the rapid opening of electrical power lines when the current exceeds a predetermined value. The response time of the relays is very important in protecting the equipment from damage. Some common types of faults that can be protected by relays are line-to- ground short circuits, line-to-line short circuits, double line-to-ground short circuits, and three- phase line short circuits. Each of these conditions is caused by faulty circuit conditions that draw an abnormally high current from the power lines.

7.3 Motor Protective Devices

The distribution of electrical power to motors is a particularly important function of industrial and commercial power distribution equipment. Basic functions that motors are expected to perform are starting, stopping, reversing, and speed variation. These functions can be manually or automatically controlled. Various types of protective devices are used to provide for the efficient distribution of power to electric motors.

Overload protection is the most important motor protection function. Such protection should serve the motor, its branch circuit, and associated control equipment. The major cause of motor overload is an excessive

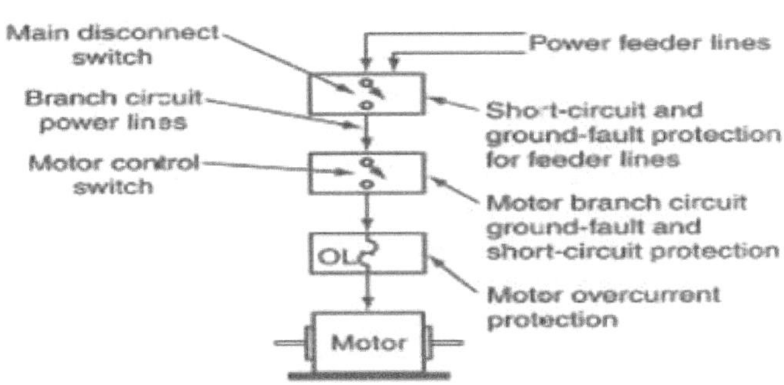

Figure 7.12 Block diagram of a motor-protection circuit.

mechanical load on the motor, which causes it to draw too much current from the power source. A block diagram of a motor protection system is shown in Figure 7.12.

Thermal overload relays are often used as protective devices. Thermal relays can be reset either manually or automatically. One type of thermal overload relay uses a bimetallic heater element. The bimetallic element bends as it is heated by the current flowing through it. When the current reaches the rating of the element, the relay opens the branch circuit. Another type of element is the melting-alloy type. This device has contacts held closed by a ratchet wheel. At the current capacity of the device, the fusible alloy melts causing the ratchet wheel to turn. A spring then causes the device to open the circuit. Thermal motor overload devices are shown in Figure 7.13.

Motor-fault currents are excessive currents that occur in motors due to some unnatural malfunction. Since motor-fault currents cannot be withstood for any duration of time, some type of protection must be provided to disconnect the motor from the power distribution system when a fault condition occurs. Such protection can be provided by motor starter, circuit breakers, or fuses. The type of protection used is dependent upon several characteristics of the power distribution system and the motor.

Motors must be protected from excessive overheating. This protection is provided by magnetic or thermal protective devices that are ordinarily within the motor-starter enclosure. Protective relays or circuit breakers can also perform this function. When an operational problem causes the motor to overheat, the protective device is used to automatically disconnect the motor from its power supply.

Motors do not operate efficiently when less than their rated voltage is applied. Some types of motors can be destroyed if they are operated

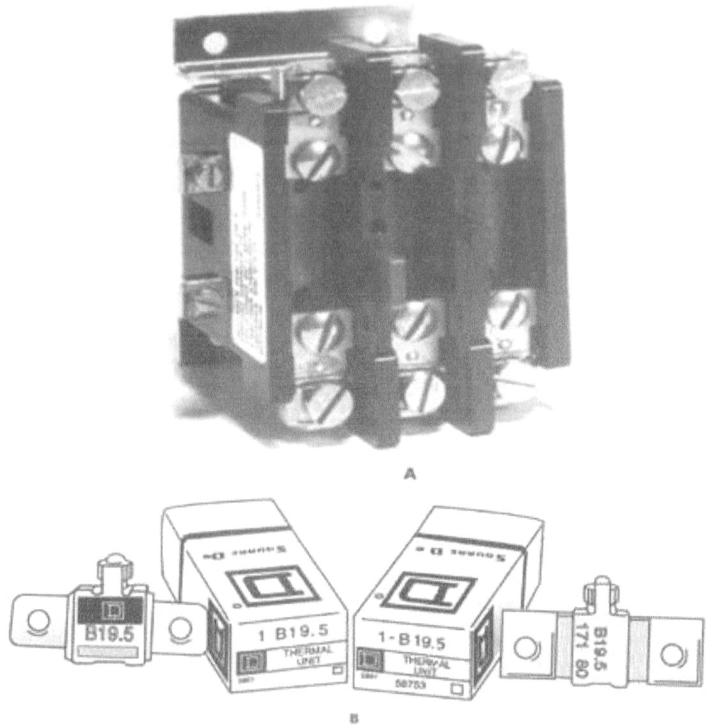

Figure 7.13 A—Plastic mounting for three-phase motor thermal overload devices. B—Thermal motor overload devices. (Furnas Electric Co.)

continuously at reduced voltages. ***Magnetic contactors*** can be used effectively to protect against under voltages. A specific level of voltage is required to cause magnetic contactors to operate. If the voltage is reduced below a specified level, the magnetic contactor will open, disconnecting the circuit between the power source and the motor and stopping the motor before any damage can be done. Motors can be protected by electronic sensors that provide overcurrent control and monitoring.

7.4 Interior Electrical Distribution

Electrical power is delivered to the location where it is to be used and then distributed within a building by the power distribution system. Various types of circuit breakers and switchgear are employed for interior power distribution. One factor involved in power distribution is the distribution of electrical energy to the many types of loads that are connected to the system. This part

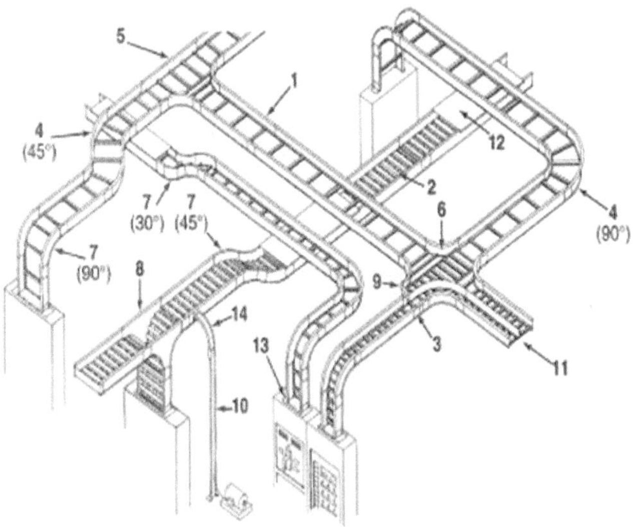

1. Ladder tray straight section.
2. Ventilated trough tray.
3. Splice plate-bolted connector.
4. Horizontal bend (30° to 90°).
5. Tee.
6. Cross.
7. Vertical bends (30° to 90°). INSIDE VERTICAL = transition from level to upward direction. OUTSIDE VERTICAL = transition from level to downward direction.
8. Vertical tee.
9. Reducer.
10. Channel tray.
11. Barrier.
12. Cover.
13. Tray-to-box connector.
14. Channel-outside bend.

Figure 7.14 Industrial cable tray raceway design. (Chalfant Manufacturing Co.)

of the distribution system is concerned with the conductors, feeder systems, branch circuits, grounding methods, and protective and control equipment that is used.

Most electrical distribution to industrial and commercial loads is through wires and cables contained in *raceways*. These ***raceways*** carry the conductors that carry the power to the various equipment throughout a building. Copper conductors are ordinarily used for indoor power distribution. The physical size of each conductor is dependent upon the current rating of the branch circuit. Raceways can be large metal *ducts* or rigid metal conduit. Raceways provide a compact and efficient method of routing cables and wires throughout an industrial or commercial building. A cable tray raceway design for industrial applications is shown in Figure 7.14.

The conductors that carry current to the electrical load devices in a building are called feeders and branch circuits. ***Feeder lines*** supply power to branches that are connected to them. Primary feeder lines can be either overhead or underground. Usually, overhead lines are preferred because they permit flexibility for future expansion. ***Underground systems*** cost more, but they are much more attractive in appearance.

Secondary feeders are connected to the primary feeder lines and supply power to individual sections within the building. Either aluminium or copper feeder lines can be used, depending on the specific power requirements. The distribution is from the feeder lines, through individual protective equipment, to branch circuits that supply the various loads. Each ***branch circuit*** has various protective devices according to the needs of that particular branch. The overall feeder-branch system can be a very complex network of switching Equipment, transformers, conductors, and protective equipment.

In addition to circuit protection, power distribution systems must have equipment that can be used to connect or disconnect the entire system or parts of the system. Various types of switching devices are used to perform this function. A simple type of switch is the *safety switch.* This type of switch is mounted in a metal enclosure and operated by means of an external handle. ***Safety switches*** are used only to turn a circuit off or on. However, fuses are often mounted in the same **enclosure** with the safety **switch**.

Another type of switch is the kind used in conjunction with a circuit-breaker panel board. **Panelboards** are metal cabinets that enclose the main disconnect switch and the branch-circuit protective equipment. Distribution panel boards are usually located between the power feed lines within a building and the branch circuits that are connected to it.

Metal-enclosed ***switchgear*** or metal-clad **switchgear** is a type of equipment that houses all the necessary control devices for the electrical circuits that are connected to them. The control devices contained inside the switchgear include circuit breakers, disconnect switches, interconnecting cables and buses, transformers, and the necessary measuring instruments. Switchgear is used for indoor and outdoor applications at industrial plants, commercial buildings, and substations. The voltage ratings of switchgear are usually from 13.8 to 138 kilovolts with 1 mega- voltampere to 10 megavolt ampere power ratings.

Metal-enclosed low-voltage switchgear are used in many industrial and commercial buildings. They are used as a ***distribution control center*** to house the circuit breakers, bus bars, and terminal connections that are part of the power distribution system. Ordinarily, a combination of switchgear and distribution transformers is placed in adjacent metal enclosures. This combination is referred to as a load-center unit substation since it is the central control for several loads. The rating of these load centers is usually 15,000 volts or lower for the high-voltage section and 600 volts or less for the low-voltage section. Load centers provide flexibility in the electrical power distribution design of industrial plants and commercial buildings.

7.5 Electrical Service Entrance

Electrical power is brought from the overhead power lines or from the underground cable into a building by what is called a **service entrance**. A good working knowledge of the National Electrical Code (NEC) specifications and definitions is necessary to understand service-entrance equipment. The NEC sets the minimum standards that are necessary for wiring design inside a building.

The type of equipment used for an electrical service entrance of a building can include high-current conductors and insulators, disconnect switches, protective equipment for each load circuit that will be connected to the main power system, and meters needed to measure power, voltage, current, or frequency. It is also necessary to ground the power system at the service-entrance location. This is done by a "grounding electrode/' which is a metal rod driven deep into the ground. The grounding conductor is attached securely to this grounding electrode. Then, the grounding conductor is used to make contact with all neutral conductors and safety grounds of the system. A service entrance is shown in Figure 7.15.

There are several terms associated with service- entrance equipment. The service-entrance conductors are a set of conductors that are brought to a building by the local electrical utility company. These conductors must be capable of carrying all the electrical current that is to be delivered to the various loads inside the building that are to be supplied by the power system. Conductors that extend from the service entrance to a power distribution panel or other type of overcurrent protective equipment are called feeders. Feeders are power lines that supply branch circuits. A **branch *circuit*** is defined as conductors that extend beyond the last overcurrent protective equipment of the power system. Usually, each branch circuit delivers electrical power to a small percentage of the total load of the main power system.

In commercial and industrial installations, switchboards and panel boards are used to supply power to various loads throughout the power system. A *switchboard* is a large enclosure that has several overcurrent protective devices (fuses or circuit breakers). Each feeder is connected to the proper type of overcurrent device. Often, switchboards contain metering equipment for the power system. Panelboards, are smaller than switchboards but are used for a similar purpose. They are enclosures for overcurrent devices for either branch circuits or feeder circuits. A common example of a panel board is the main power-distribution panel which houses the circuit breakers used for the branch circuits of a home. For more specific definitions of terms, you should refer to the most recent edition of the National Electrical Code.

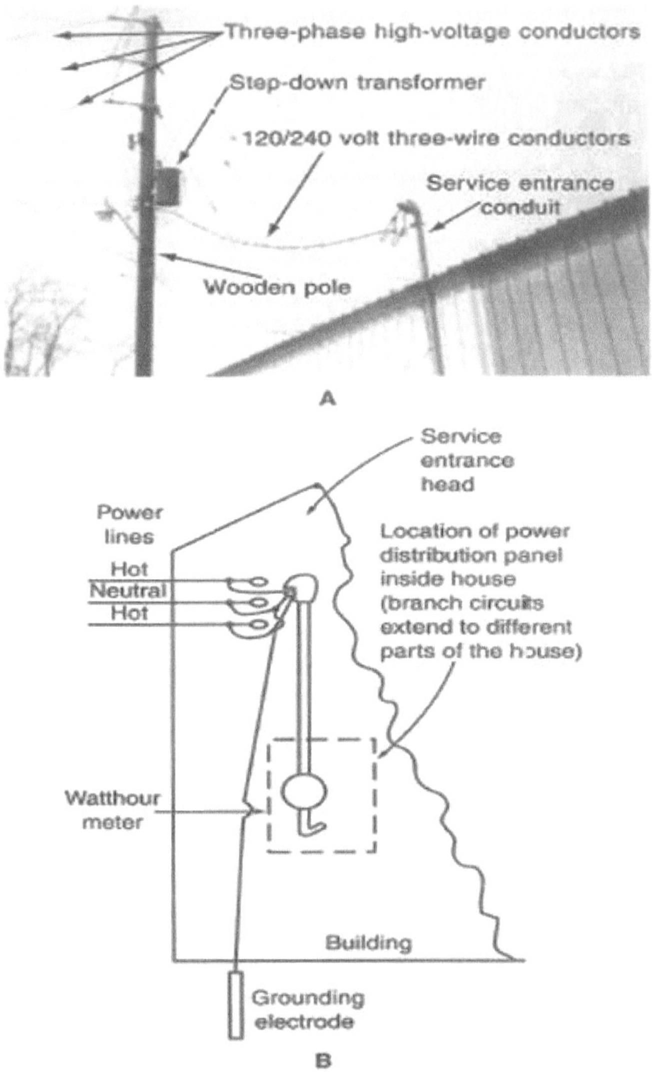

Figure 7.15 A—Service entrance showing three-phase high-voltage conductors, step-down transformer, 120/240-volt conductors and service entrance conduit. B—Single-phase electrical service entrance.

Review Questions

Answer the following questions.

1. What is the function of an electrical substation?

2. What is the purpose of an air-break disconnect switch?

3. What is the purpose of a lightning arrester?

4. What is the purpose of a high-voltage insulator?

5. What is the purpose of a voltage regulator?

6. List three types of fuses.

7. What is the purpose of a protective relay?

8. What is the purpose of a thermal overload relay?

9. What is the purpose of a raceway?

10. Define feeder circuit.

11. Define branch circuit.

12. What is the purpose of a distribution panel-board?

13. What is a service entrance?

14. What is a switchboard?

8

Single-phase and Three-phase Distribution Systems

Objectives

Upon completion of this chapter, you will be able to:

- Explain the operation of single-phase, two-wire and single-phase, three-wire power distribution systems.
- Describe and sketch diagrams of delta-delta, delta-wye, wye-wye, wye-delta, and open delta three-phase transformer connection methods.
- Explain the operation of three-phase, three-wire power distribution systems.
- Explain the operation of three-phase, four-wire power distribution systems.
- Explain the operation of three-phase, three-wire with neutral power distribution systems.
- Describe the following in relation to grounding of electrical power systems: system ground, equipment ground, and ground fault circuit interrupters.
- Explain the use of the National Electrical Code as a standard for electrical power distribution design in the United States.
- Calculate voltage drop in an electric circuit using a conductor table or circular mil formula.
- Calculate voltage drop in a branch circuit or feeder circuit for a single-phase or three-phase distribution system.
- Determine grounding conductor size by using the proper tables.
- Describe the following in relation to electrical power systems: hot conductor, neutral conductor, safety ground, AWG and MCM conductor sizing, non-metallic sheathed cable, metal-clad cable, rigid conduit, electrical metallic tubing (EMT).

DOI: 10.1201/9788743810056-10

Important Terms

- BX cable
- delta-delta connection
- delta-wye connection
- electrical metallic tubing (EMT)
- equipment ground
- GFCI circuit breaker
- GFCI receptacle
- ground-current relaying
- ground-fault circuit interrupter (GFCI)
- grounding electrode, metal-clad cable
- nonmetallic-sheathed cable (NMC)
- open-delta connection
- open-ground condition
- plug-in GFCI receptacle
- rigid conduit
- Romex cable
- single-phase, three-wire system
- single-phase, two-wire system
- system grounding
- thin wall conduit
- three-phase, four-wire system
- three-phase, three-wire system
- three-phase, three-wire system with neutral
- wye-delta connection
- wye-wye connection

Introduction

When electrical power is distributed to its point of utilization, it is normally either in the form of single-phase or three-phase ac voltage. Single-phase ac voltage is distributed into residences and other smaller commercial buildings. Normally, three-phase ac voltage is distributed to industries and larger commercial buildings. Thus, the main types of power distribution systems are ***residential (single-phase)*** and **industrial or commercial (three-phase).**

An important aspect of both single-phase and three-phase distribution systems is grounding. Two grounding methods, ***system grounding*** and ***equipment grounding***, are discussed along with ground-fault protective equipment.

8.1 Single-Phase Systems

Most electrical power when produced at the power plants is produced as three-phase ac voltage. Electrical power is also transmitted in the form of three-phase voltage over long-distance power transmission lines. At its destination, three phase voltage can be changed into three separate single-phase voltages for distribution into residential areas.

Although single-phase systems are used mainly for residential power distribution systems, there are some industrial and commercial applications of single-phase systems. Single-phase power distribution usually originates from three-phase power lines so electrical power systems are capable of supplying both three-phase and single-phase loads from the same power lines.

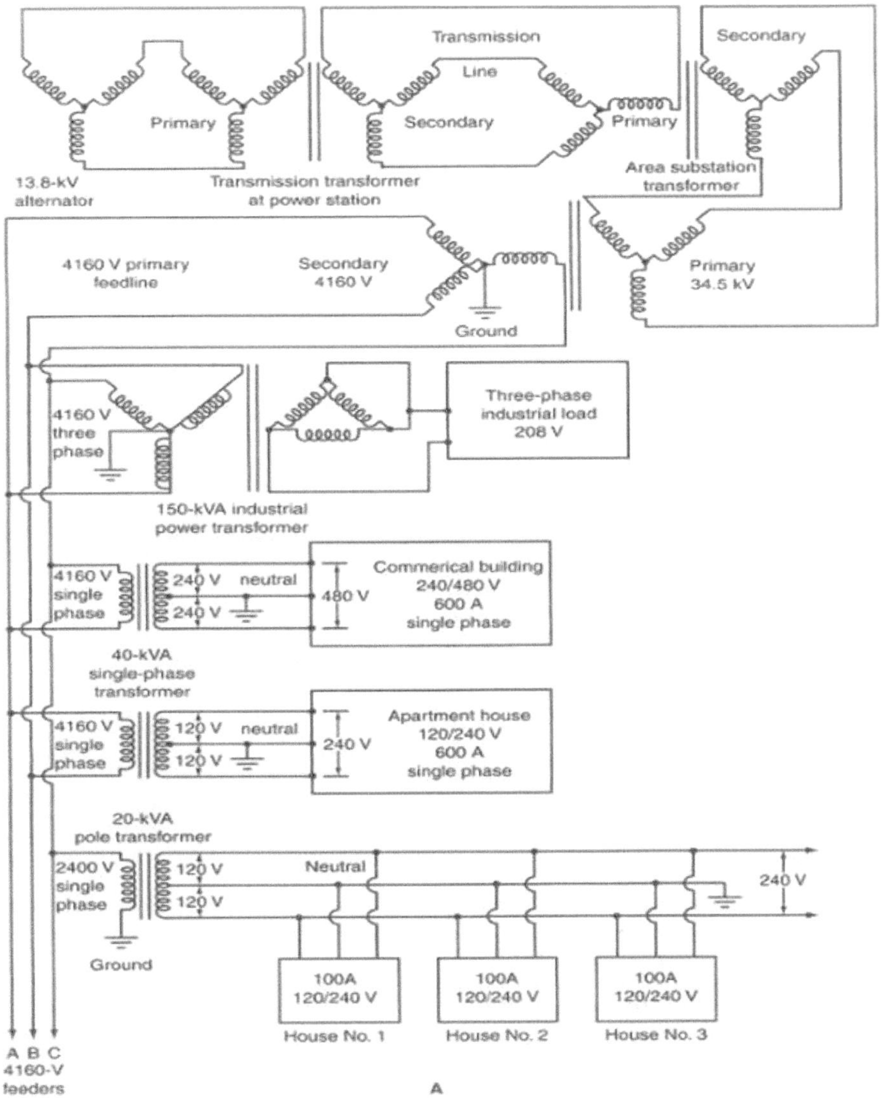

Figure 8.1 A-Typical power distribution system. B-Relationship of feeders and branch circuits.

Figure 8.1 shows a typical power distribution system-from the power station (source) to the various single-phase and three-phase loads that are connected to the system.

Single-phase systems can be of two major types-single-phase, two-wire systems and single-phase, three-wire systems. A ***single-phase, two-wire***

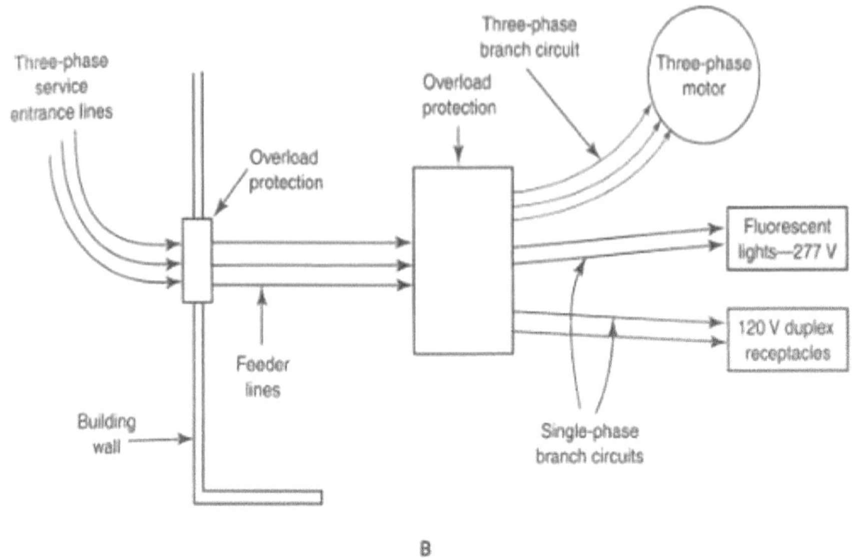

Figure 8.1 Continued.

system is shown in part of Figure 8.2 (the top circuit). This system uses a transformer that has a secondary that produces one single-phase voltage, such as 120 or 240 volts. This system has one hot line and one neutral line. In residential distribution systems, in the past, this was the most used type to provide 120 -volt service. However, as appliance power requirements increased, the need for a dual-voltage system was evident. To meet the demand for more residential power, the ***single-phase, three-wire system*** is used. A home service entrance can be supplied with -volt energy by the methods shown in parts B and C of Figure 8.2 (center and bottom circuits). Each of these systems is derived from a three-phase power line.

The ***single-phase, three-wire system*** has two hot lines and a neutral line. The hot lines, which usually have black and red insulation, are connected to the outer terminals of the transformer secondary windings. The neutral (white insulated wire) is connected to the center tap of the distribution transformer. Thus, from neutral to either hot line, 120 volts for lighting and small-power requirements can be obtained. Across the two hot lines, 240 volts is supplied for higher-power requirements. Therefore, the current requirement for large power-consuming equipment is cut in half, since 240 volts rather than 120 volts is used. Either the single-phase, two-wire or the single-phase, three-wire system can be used to supply single-phase power for industrial or commercial use. However, these single-phase systems are mainly for residential power distribution.

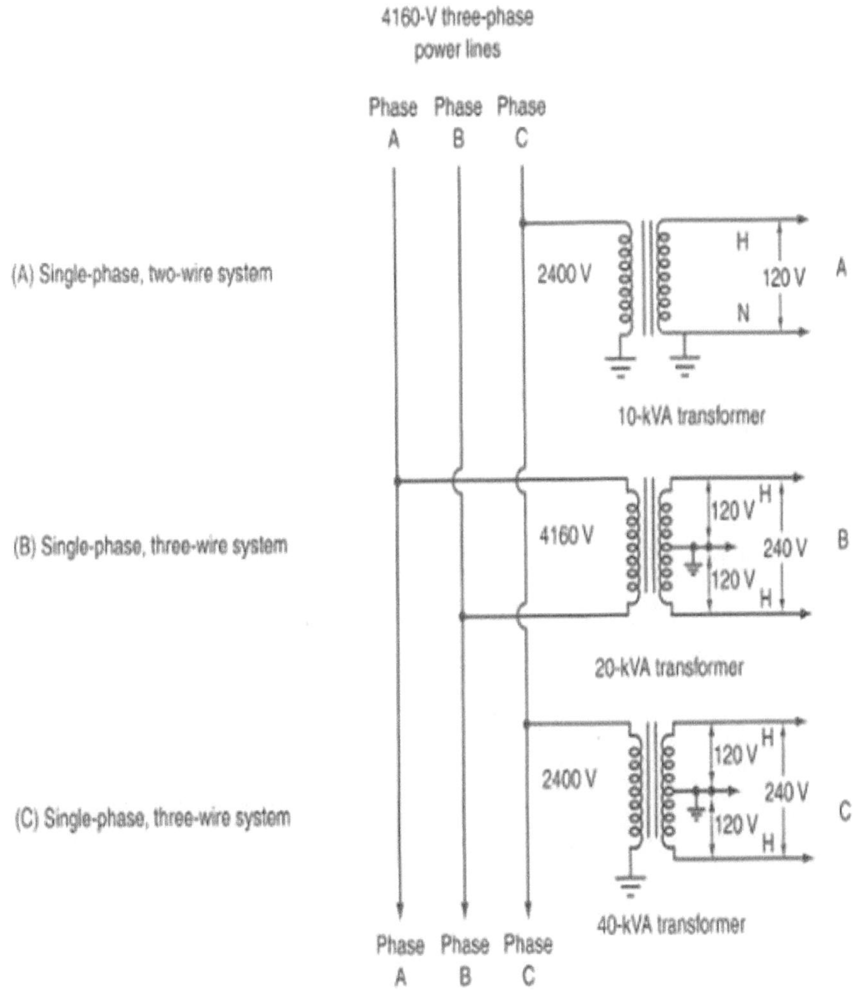

Figure 8.2 Single-phase power distribution systems.

8.2 Three-phase Systems

Since industries and commercial buildings use three-phase power predominantly, they rely upon three-phase distribution systems to supply this power. Large three-phase distribution transformers are usually located at substations adjoining the industrial plants or commercial buildings. Their purpose is to supply the proper ac voltages to meet the necessary load requirements. The ac voltages that are transmitted to the distribution substations are high voltages that must be stepped down by three-phase transformers.

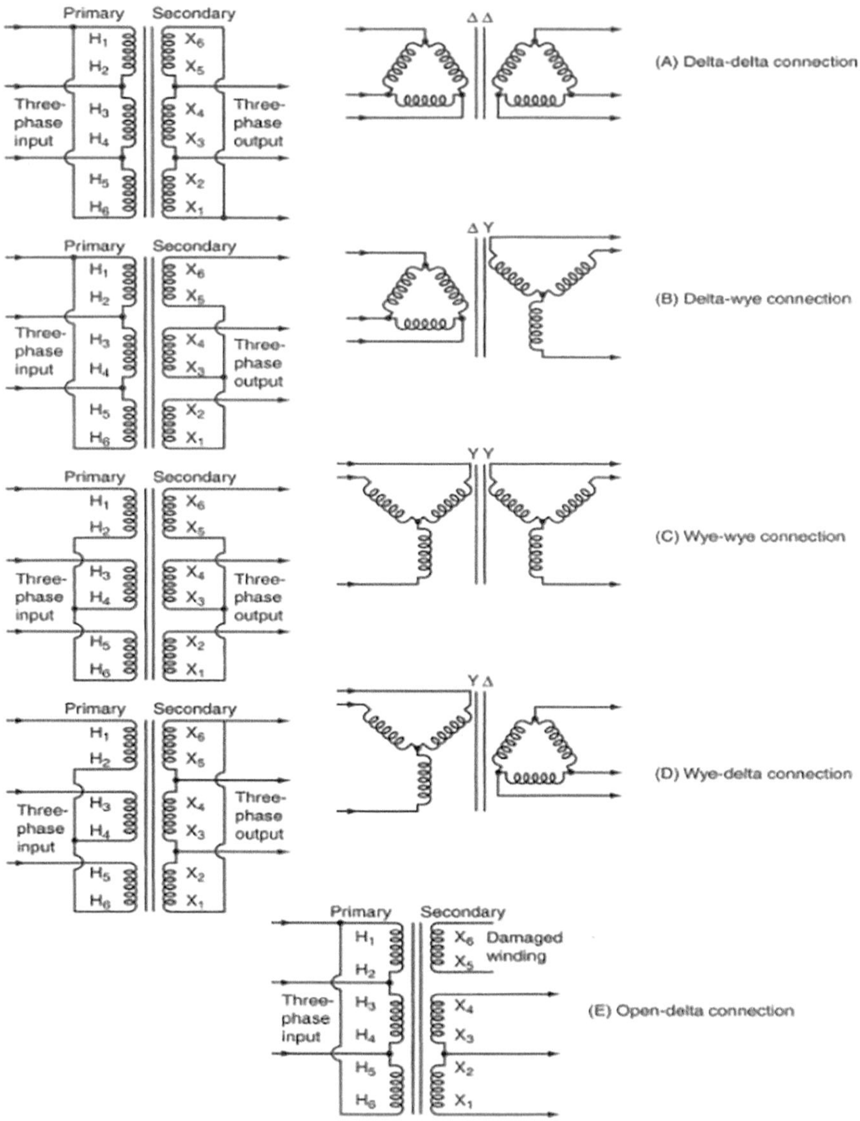

Figure 8.3 Basic three-phase transformer connection methods.

8.2.1 Three-phase transformer connections

There are five ways in which the primary and secondary windings of three-phase transformers can be connected. These are the delta-delta, the delta-wye, the wye-wye, the wye-delta, and the open-delta connections. These basic methods are illustrated in Figure 8.3.

The ***delta-delta*** connection (Figure 8.3, part A) is used for some lower voltage applications. The delta-wye connection (Figure 8.3, part B) is commonly used for stepping up voltage. An inherent step-up factor of 1.73 times is obtaned due to the voltage characteristic of the wye-connected secondary. The wye-wye connection in part C of Figure is ordinarily not used, while the wye-delta connection (Figure 8.3, part D) can be used advantageously to step voltages down. The open-delta connection (Figure 8-3, part E) is used if one transformer winding becomes damaged or is taken out of service. The transformer still delivers three-phase power, but at a lower current and power capacity. This connection can also be desirable when the full capacity of three transformers is not needed until a later time. Two identical single-phase transformers can be used to supply power to the load until, at a later time, the third transformer is needed to meet increased load requirements.

8.2.2 Types of three-phase distribution systems

Three-phase power distribution systems that supply industrial and commercial buildings are classified according to the number of phases and number of wires required. These systems, shown in Figure 8.4 are the three-phase, three-wire system, the three-phase, three-wire system with neutral, and the three-phase, four-wire system. The primary winding connection is not considered here.

The ***three-phase, three-wire system***, shown in part A of Figure 8.4, can be used to supply motor loads of 240 volts or 480 volts. Its major disadvantage is that it only supplies one voltage as only three hot lines are supplied to the load. The usual insulation color code for these hot lines is black, red, or blue, as specified in the National Electrical Code.

The disadvantage of the three-phase, three wire system may be partially overcome by adding one center-tapped winding as shown in the ***three-phase, three-wire system*** with neutral in part B of Figure 8.4. This system can be used as a supply for 120/240 volts or 240/480 volts. Assume that it is used to supply 120/240 volts. The voltage from the hot line at point 1 and the hot line at point 2 to neutral would be 120 volts, due to the center-tapped winding. However, 240 volts would still be available across the other two hot lines. The ***neutral wire*** is color coded with a white or gray insulation.

The disadvantage of this system is that, when making wiring changes, it would be possible to connect a 120-volt load between the neutral and point 3 (sometimes called the "wild" phase). The voltage present here would be the combination of three-phase voltages between points 1 and 4 and point 1 and 3. This would be a voltage in excess of 300 volts! Although the wild-phase

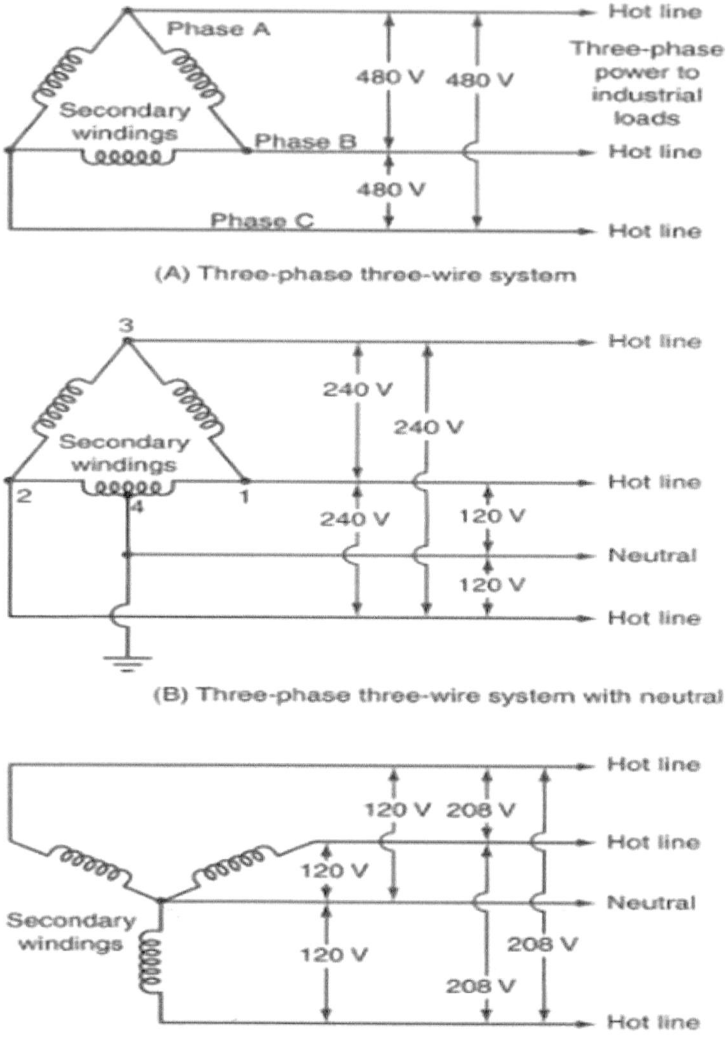

Figure 8.4 Industrial power distribution systems

situation exists, this system is capable of supplying both high-power loads and low-voltage loads, such as for lighting and small equipment.

The most widely used three-phase power distribution system is the ***three-phase, four-wire system***. This system, shown in part C of Figure 8.4, commonly supplies 120/208 volts and 277/480 volts for industrial or commercial load requirements. The 120/208-volt system is illustrated here. From neutral to any hot line, 120 volts for lighting and low-power loads can be obtained. Across any two hot lines, 208 volts is present for supplying motors or other high power loads. The 120/208 -volt system is often used for underground distribution in urban areas.

The most popular system for industrial and commercial power distribution is the 277/408-volt system, which is capable of supplying both three-phase and single-phase loads. A 240/416-volt system is sometimes used for industrial loads. Note that this system is based on the voltage characteristics of the three-phase wye connection and that the relationship $V_L = V_P \times 1.73$ exists for each application of this system.

8.3 Grounding of Distribution Systems

The concept of ***grounding*** in an electrical power distribution system is very important. Distribution systems must have continuous uninterrupted grounds. If a grounded conductor is opened, the ground is no longer functional. An ***open-ground condition*** can present severe safety problems and cause abnormal system operation.

Distribution systems must be grounded at substations and at the end of the power lines before the power is delivered to the load. Grounding is necessary at substations for the safety of the public and the power company maintenance personnel. Grounding also provides points for transformer neutral connections for equipment grounds. Safety and equipment grounds will be discussed in more detail later.

At substations, all external metal parts must be grounded and all transformer, circuit breaker, and switch housings must be grounded. Also, metal fences and any other metal that is part of the substation construction must be grounded. Grounding ensures that any person who touches any of the metal parts will not receive a high voltage shock. If a high-voltage line comes in contact with any of the grounded parts, the system is opened by protective equipment. Thus, the danger of high voltages at substations is substantially reduced by grounding.

The actual ground connection is made by welding, brazing, or bolting a conductor to a metal rod or bar. This rod is then physically placed in the earth. The rod device is called a ***grounding electrode***. Proper grounding techniques are required for safety as well as circuit performance. There are two types of grounding: system grounding and equipment grounding.

8.3.1 System grounding

System grounding involves the actual grounding of a current-carrying conductor (usually called the neutral) of a power distribution system. Three phase systems can be either the wye or delta type. The wye system has an obvious advantage over the delta system in that one side of each phase winding is connected to ground. We will define a ground as a reference point of zero-volt potential, which is usually an actual connection to earth ground. The common terminals of the wye system, when connected to ground, become the neutral conductor of the three-phase, four-wire system.

The delta system does not readily lend itself to grounding since it does not have a common neutral. The problem of ground faults (line-to-ground shorts) occurring in ungrounded delta systems is much greater than in wye systems. A common method of grounding a delta system is to use a wye-delta transformer connection and ground the common terminals of the wye-connected primary. However, the wye system is now used more often for industrial and commercial distribution since the secondary is easily grounded and it provides overvoltage protection from lightning or line-to ground shorts.

Single-phase 120/240 - or 240/480 -volt systems are grounded in a manner similar to a threephase ground. The neutral of the single-phase, three-wire system is grounded by a metal rod (grounding electrode) driven into the earth at the transformer location. System grounding conductors are insulated with white or gray material for easy identification.

8.3.2 Equipment grounding

The second type of ground is the ***equipment ground***. The equipment ground, as the name implies, places operating equipment at ground potential. The conductor that is used for this purpose is either bare wire or a green insulated wire.

The ***National Electrical Code*** describes conditions that require fixed electrical equipment to be grounded. Usually all fixed electrical equipment located in industrial plants or commercial buildings should be grounded. Types of equipment that should be grounded include enclosures for switching and protective equipment for load control, transformer enclosures, electric motor frames, and fixed electronic test equipment. Industrial plants should use 120 -volt single-phase duplex receptacles of the grounded type for all portable tools. The grounding of these receptacles can be checked by using a plug-in tester.

8.4 Ground-Fault Protection

Ground-fault circuit interrupters (GFCIs), such as the one shown in Figure 8.5, are now used extensively in industrial, commercial, and residential

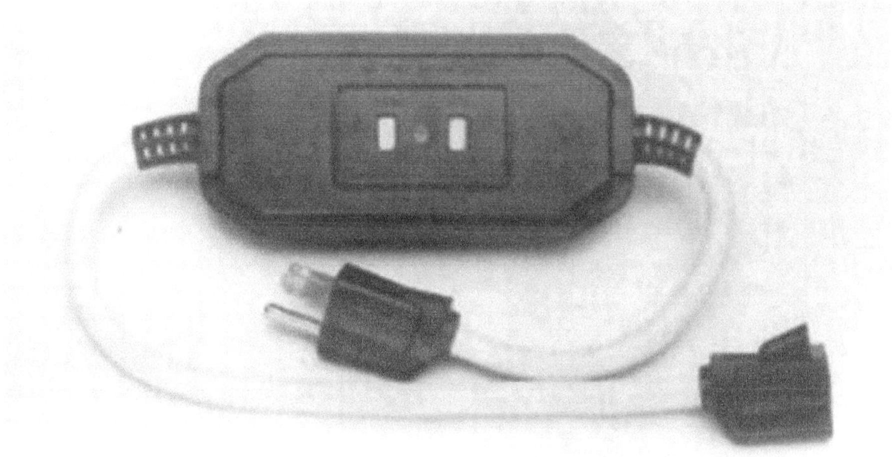

Figure 8.5 In-line type of ground-fault circuit interrupter (GFCI). (Leviton Co.)

power distribution systems. It is required by the National Electrical Code that all 120 -volt single-phase 15 - or 20 -ampere receptacle outlets that are installed outdoors or in bathrooms have ground-fault circuit interrupters connected to them. These devices are also called ground fault interrupters (GFIs).

8.4.1 GFCI operation

GFCI devices are designed to eliminate electrical shock hazard resulting from individuals coming in contact with a hot ac line (line-to ground short). The circuit interrupter is designed to sense any change in circuit conditions, such as would occur when a line-to-ground short exists.

One type of GFCI has control wires that extend through a magnetic toroidal loop, see Figure 8.6. Ordinarily, the ac current flowing through the conductors inside the loop is equal in magnitude and opposite in direction. Any change in this equal and opposite condition is sensed by the magnetic toroidal loop. When a line-to-ground short occurs, an instantaneous change in circuit conditions occurs. The change causes a magnetic field to be induced into the toroidal loop. The induced current is amplified to a level sufficient to cause the circuit-breaker mechanism to open. Thus, any line to-ground short causes the ground-fault circuit interrupter to open. The operating speed of the GFCI is so fast (since only a minute current opens the circuit) that the shock hazard to individuals is greatly reduced.

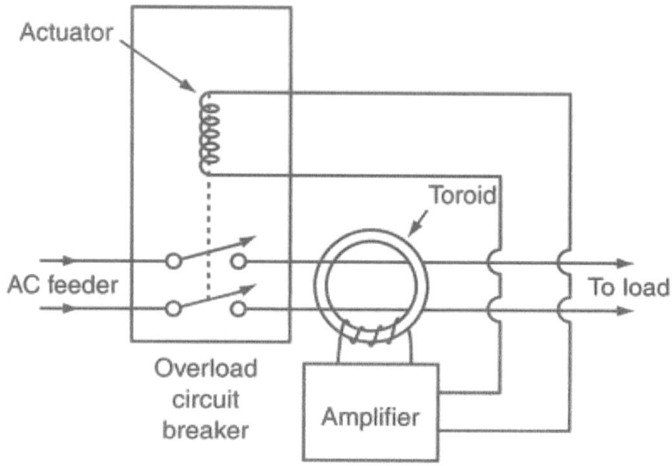

Figure 8.6 Simplified schematic of a ground-fault interrupter.

8.4.2 Types of ground-fault protection systems

There are four basic types of ground-fault protection systems in use today. They are hospital applications, residential applications, motor protection applications, and specific electrical power distribution system applications. These ground fault systems can be classified either by what they are to protect or the type of protection they are to provide. Hospital applications and residential applications are designed to protect people from excessive shock. Motor and electrical power applications are for protecting electrical equipment.

Another classification method is according to the amount of current required before an alarm system sounds or the disconnect of an electrical circuit occurs. Typical current values that will cause alarms or disconnects to activate are 0.002 amperes (2 mA) for hospital applications, 0.005 amperes for residential applications, 5 to 100 amperes for motor-protective circuit applications, and 200 to 1200 amperes for electrical power distribution equipment applications.

8.4.3 Need for ground-fault protection

In order to understand the need for a ground-fault circuit interrupter (for the protection of people), certain basic facts must first be understood. These facts relate to people as well as to ground faults.

One important fact is that a person's body resistance varies with the amount of moisture present on the skin, the muscular structure of the body,

Amount of Current	Effect on Body
1 mA or less	No sensation (not felt).
More than 5 mA	Painful shock.
More than 10 mA	Muscle contractions; could cause a "freezing" to the electrical circuit for *some* people.
More than 15 mA	Muscle contractions; could cause "freezing" to the electrical circuit for *most* people.
More than 30 mA	Breathing difficult; could cause unconsciousness.
50 to 100 mA	Ventricular fibrillation of the heart is possible.
100 to 200 mA	Ventricular fibrillation of the heart is certain.
Over 200 mA	Severe burns and muscular contractions; the heart is more apt to stop beating than to fibrillate.
1 ampere and above	Permanent damage to body tissues.

Figure 8.7 Body reaction to alternating current.

and the voltage to which the body is subjected. Experiments have shown that the body resistance from one hand to the other hand is somewhere between 1000 and 4000 ohms. These estimates are based upon several assumptions concerning moisture and muscular structure. We also know that resistance of the body (from hand to hand) is lower for higher voltages. This is because higher voltages have the capability of breaking down the outer layers of the skin. Thus, higher voltages are more dangerous.

We can use Ohm's law to estimate that the typical current resulting from the average body resistance (from hand to hand) is about 115 mA at 240 volts ac. It is about 40 mA at 120 volts ac. The effects of a 60- Hz alternating current on the human body are generally accepted to be as given in Figure 8.7.

Ventricular fibrillation is a contraction of the heart. Once ventricular fibrillation occurs, it will continue and death will occur within a few minutes. Resuscitation techniques, when applied immediately, can save a victim. Deaths caused by electrical shock account for a high percentage of the deaths that occur in the home and in industry. Many of these deaths are due to contact with low voltage circuits (600 volts and under), mainly 120 - and 240 -volt systems.

8.4.4 GFCI applications

There are many GFCI applications. Construction sites, where temporary wiring is set up, are required to use GFCIs for the protection of workers using electrical equipment. Ground-fault protection of individuals and commercial equipment must be provided for wye-connected systems of 150 to 600 volts for each distribution panelboard rated at over 800 amperes. In this situation, the GFCI will open all ungrounded conductors at the panelboard when a line-to-ground short occurs. GFCIs are used for all types of residential, commercial, and industrial applications.

8.4.5 Ground-fault protection for the home

All ground-fault protection devices are constructed according to standards developed by the Underwriter's Laboratories. Ground-fault circuit interrupters for homes consist of three types.

- Circuit breaker.
- Receptacle.
- Plug-in types.

The **GFCI circuit breakers** combine ground fault protection and circuit interruption in the same overcurrent and short-circuit protective equipment as does a standard circuit breaker. A GFCI circuit breaker fits the same space required by a standard circuit breaker. It provides the same branch-circuit wiring protection as the standard circuit breaker, as well as ground-fault protection. The GFCI sensing system continuously monitors the current balance in the ungrounded (hot) conductor and the grounded (neutral) conductor.

The current in the neutral wire becomes less than the current in the hot wire when a ground fault develops. This means that a portion of the circuit current is returning to ground by some means other than the neutral wire. This produces an imbalance. When an imbalance in current occurs, the sensor (a differential current transformer) sends a signal to the solid-state circuitry, which activates a trip mechanism. This action opens the hot line. A differential current as low as 5 mA in residential protection will cause the sensor to send a fault signal and cause the circuit breaker to interrupt the circuit.

Ordinarily, ***GFCI receptacles*** provide ground fault protection on 120-, 208-, or 240 -volt ac systems. The GFCI receptacles come in 15- and 20-ampere designs. The 15-ampere unit has a receptacle configuration for use with 15 -ampere plugs only. The 20 -ampere device has a receptacle configuration for use with either 15- or 20-ampere plugs. These GFCI receptacles have connections for hot, neutral, and ground wires. All GFCI receptacles have a two-pole tripping mechanism that breaks both the hot and the neutral load connections at the time a fault occurs. The GFCI shown in Figure 8.5 is this type.

The ***plug-in GFCI*** receptacles provide protection by plugging into a standard wall receptacle. Some manufacturers provide units that will fit either two-wire or three-wire receptacles. The major advantage of this type of unit is that it can be moved from one location to another.

8.5 Ground-fault Protection for Distribution Equipment

Ground faults can destroy electrical equipment if allowed to continue. Phase-to-phase short circuits and some types of ground faults are usually high current. Normally, they are adequately handled by conventional overcurrent protective equipment. However, some ground faults produce an arcing effect from relatively low currents that are not large enough to trip conventional protective devices. Arcs can severely burn electrical equipment. A 480- or 600-volt system is more susceptible to arcing damage than a 120-, 208-, or 240-volt system because of the higher voltages that sustain the arcing effect. High-current faults are quickly detected by conventional overcurrent devices. Low-current values must be detected by GFCIs.

Ground faults that cause an arcing in the equipment are probably the most frequent faults. They can result from damaged or deteriorated insulation, dirt, moisture, or improper connections. They usually occur between one hot conductor and the grounded equipment enclosure, conduit, or metal housing. The line-to-neutral voltage of the source will cause current to flow in the hot conductor, through the arc path, and back through the ground path. The impedance of the conductor and the ground-return path (enclosure, conduit, or housing) depends on many factors. As a result, the fault-current value cannot be predicted. The ***fault current*** can also increase or decrease as the fault condition continues.

It is apparent that many factors influence the magnitude, duration, and the effect of an arcing ground fault. Some conditions produce a large amount of fault current, while others limit the fault current to a relatively small amount. Damage to equipment that is caused by arc-current magnitude and the time that the arc persists can be very great. Probably the more important factor is the time period of the arcing voltage. The larger the arcing time, the more chance that the arcs will spread to different areas within the equipment.

Ground-current relaying is one method used to protect equipment from ground faults. Current flows through a load or fault along the hot and neutral conductors and returns to the source on these conductors and, to some extent, along the ground path.

The normal ground path current is very small. Therefore, essentially all the current flowing from the source is also returning on the same hot line and neutral conductors. However, if a ground fault occurs, the ground current will increase to the point where some current will escape through the fault and return via the ground path. As a result, the current returning on the hot and neutral conductors is less than the amount going out. The difference is an indication of the amount of current in the ground path. A relay, which senses this difference in currents, can act as a ground fault protective device.

Motor protective systems offer protection in the 5- to 100-ampere range. This type of ground fault protective system offers protection against ground faults in both the single-phase and the three-phase systems. Many insulation system failures begin with a small leakage current that builds up with time until damage results. These ground fault systems detect ground leakage currents while they are still small and, thus, prevent any extensive damage happening to the motors.

8.5.1 National Electrical Code (NEC) Use

The National Electrical Code sets forth the minimum standards for electrical wiring in the United States. The standards contained in the NEC are enforced by being incorporated into the different city and community ordinances that deal with electrical wiring in residences, industrial plants, and commercial buildings. These local ordinances conform to the standards set forth in the NEC.

In most areas of the United States, a license must be obtained by any individual who does electrical wiring. Usually, a test administered by the city, county, or state must be passed in order to obtain this license. These tests are based on local ordinances and the NEC. The rules for electrical wiring that are established by the local electrical power company are also sometimes incorporated into the license test.

8.5.2 Electrical inspections

When new buildings are constructed, they must be inspected to see if the electrical wiring meets the standards of the local ordinances, the NEC, and the local

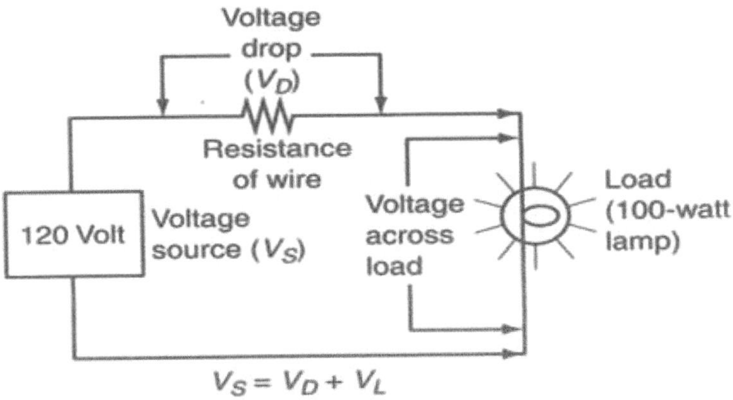

Figure 8.8 Voltage drop in an electrical circuit.

power company. The organization that supplies the electrical inspectors varies from one locality to another. Ordinarily, the local power company can advise individuals about whom to contact for information about electrical inspections.

8.6 Distribution System Wiring Design

The wiring design of electrical power distribution systems can be very complex. There are many factors that must be considered in the wiring design of a distribution system installed in a building. Wiring design standards are specified in the National Electrical Code (NEC), which is published by the National Fire Protection Association (NFPA). The NEC, local wiring standards, and electrical inspection policies should be considered in an electrical wiring design.

There are several distribution system wiring design considerations that are pointed out specifically in the NEC. The following sections deal with voltage-drop calculations, branch circuit design, feeder-circuit design, and the design for grounding systems.

8.6.1 Voltage Drop in Electrical Conductors

Although the resistance of electrical conductors is very low, a long length of wire could cause a substantial voltage drop. This is illustrated in Figure 8.8. Remember that a voltage drop is current times resistance ($I \times R$). Therefore, whenever current flows through a system, a voltage drop is created. Ideally, the voltage drop caused by the resistance of a conductor will be very small.

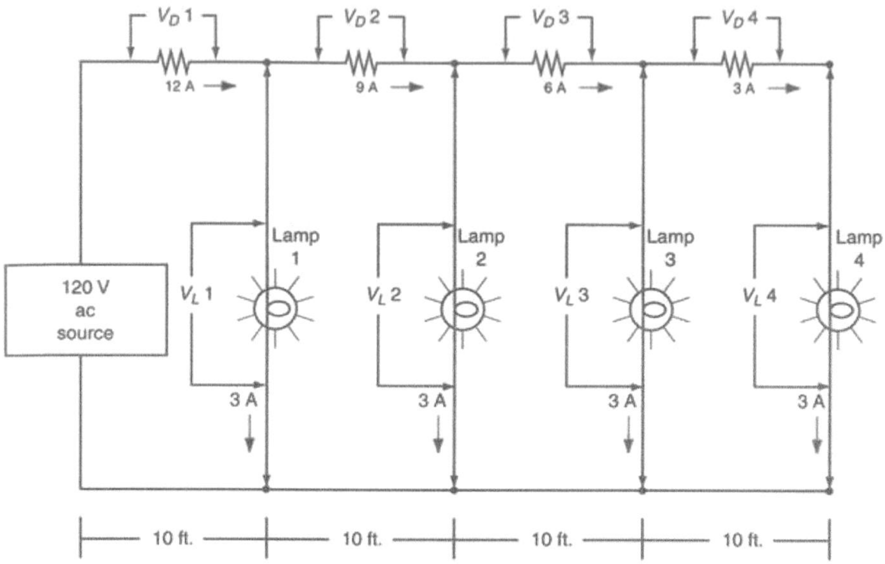

Figure 8.9 Circuit for calculating the voltage drop in a branch circuit.

However, a longer section of electrical conductor has a higher resistance. Therefore, it is sometimes necessary to limit the distance a conductor can extend from the power source to the load that it supplies. Many types of loads do not operate properly when a value less than the full source voltage is available.

You can also see from Figure 8.8 that as the voltage drop (V_D) increases, the voltage applied to the load (V_L) decreases. As current in the system increases V_D, increases causing V_L to decrease since the source voltage stays the same.

8.6.2 Calculations using conductor table

It is important when dealing with electrical wiring design to be able to determine the amount of voltage drop caused by conductor resistance. Figure 8.9 is used to make these calculations. The National Electrical Code limits the amount of voltage drop that a system can have. This means that long runs of conductors must ordinarily be avoided. Remember that a conductor with a large cross-sectional area will cause a smaller voltage drop since its resistance is smaller per unit length.

To better understand how to determine the size of conductor required to limit the voltage drop in a system, examine a sample problem.

Sample Problem:

Given: A 200-ampere load located 400 feet (121.92 meters) from a 240 -volt single-phase source. Limit the voltage drop to 2% of the source.

Find: The size of an RH copper conductor needed to limit the voltage drop of the system.

Solution:

1. The allowable voltage drop equals 240 volts times 0.02(2%). This equals 4.8 volts.

2. Next, determine the maximum resistance for 800 feet (243.84 meters). This is the equivalent of 400 feet (121.92 meters) × 2, since there are two current-carrying conductors for a single-phase system.

$$R = \frac{V_D}{I}$$

$$R = \frac{4.8\text{ V}}{200\text{ A}}$$

$$R = 0.024\Omega\text{(resistance for 800 feet)}$$

3. Determine the maximum resistance for 1000 feet (304.8 meters) of conductor.

$$\frac{800\text{ feet}}{1000\text{ feet}} = \frac{0.024\Omega}{R}$$

$$800R = (1000)(0.024)$$

$$\text{R} = 0.030\Omega$$

4. Use Figure 8.10 to find the size of copper conductor that has the nearest dc resistance (ohms per 1000 feet) value that is equal to or less than the value calculated in Step 3. The conductor chosen is conductor size 350 MCM, RH copper.

5. Check this conductor with the proper ampacity table to ensure that it is large enough to carry 200 amperes. Figure 8.11 shows that a 350MCM, RH copper conductor will handle 310 amperes of current. Therefore, it is safe to use 350MCM conductors. (Always remember to use the largest conductor if steps 4 and 5 produce conflicting values.)

6. If the current is larger than listed on the tables, use more than one conductor of the same size for design calculations.

Size (AWG or MCM)	Area (cmil)	Number of wires	Diameter of each Wire (in.)	DC Resistance (Ω/1000 ft.) 25°C Copper	Aluminum
18	1,620	1	0.0403	6.51	10.7
16	2,580	1	0.0508	4.10	6.72
14	4,110	1	0.0641	2.57	4.22
12	6,530	1	0.0808	1.62	2.66
10	10,380	1	0.1019	1.018	1.67
8	16,510	1	0.1285	0.6404	1.05
6	26,240	7	0.0612	0.410	0.674
4	41,740	7	0.0772	0.259	0.424
3	52,620	7	0.0867	0.205	0.336
2	66,360	7	0.0974	0.162	0.266
1	83,690	19	0.0664	0.129	0.211
0	105,600	19	0.0745	0.102	0.168
00	133,100	19	0.0837	0.0811	0.133
000	167,800	19	0.0940	0.0642	0.105
0000	211,600	19	0.1055	0.0509	0.0836
250	250,000	37	0.0822	0.0431	0.0708
300	300,000	37	0.0900	0.0360	0.0590
350	350,000	37	0.0973	0.0308	0.0505
400	400,000	37	0.1040	0.0270	0.0442
500	500,000	37	0.1162	0.0216	0.0354
600	600,000	61	0.0992	0.0180	0.0295
700	700,000	61	0.1071	0.0154	0.0253
750	750,000	61	0.1109	0.0144	0.0236
800	800,000	61	0.1145	0.0135	0.0221
900	900,000	61	0.1215	0.0120	0.0197
1000	1,000,000	61	0.1280	0.0108	0.0177

Figure 8.10 Sizes of copper and aluminum conductors.

8.6.3 Alternative method

In some cases, an easier method to determine the conductor size for limiting the voltage drop is to use one of the following formulas to find the cross-sectional (cmil) area of the conductor.

$$\text{cmil} = \frac{\rho \times I \times 2d}{V_D} \text{(single-phase systems)}$$

Wire Size	Copper: With R, T, TW Insulation	Copper: With RH, RHW, TH, THW Insulation	Aluminum: With R, T, TW Insulation	Aluminum: With RH, RHW, TH, THW Insulation
14	15	15		
12	20	20	15	15
10	30	30	25	25
8	40	45	30	40
6	55	65	40	50
4	70	85	55	65
3	80	100	65	75
2	95	115	75	90
1	110	130	85	100
0	125	150	100	120
00	145	175	115	135
000	165	200	130	155
0000	195	230	155	180
250	215	255	170	205
300	240	285	190	230
350	260	310	210	250
400	280	335	225	270
500	320	380	260	310
600	355	420	285	340
700	385	460	310	375
750	400	475	320	385
800	410	490	330	395
900	435	520	355	425
1000	455	545	375	445

Figure 8.11 Ampacities of conductors in a raceway or cable (three or less).

or

$$\text{cmil} = \frac{\rho \times I \times 1.73d}{V_D} \text{(three-phase systems)}$$

where ρ is the resistivity from Figure 8.12, I is the load current in amperes, V_D is the allowable voltage drop, and d is the distance from source to load in feet.

Conductor	Resistivity in ohms per cmil-ft.
Silver	9.8
Copper	10.4
Aluminum	17.0
Tungsten	33.0
Nickel	50.0
Iron	60.0

Figure 8.12 Resistivity of common conductors.

The **sample problem** for a single-phase system that you just solved using a conductor table can be set up as follows.

$$\text{cmil} = \frac{\rho \times I \times 2d}{V_D}$$
$$\text{cmil} = \frac{10.4 \times 200 \times 2 \times 400}{240 \times 0.02}$$
$$\text{cmil} = \frac{1{,}664{,}000}{4.8}$$
$$\text{cmil} = 346{,}666$$
$$\text{cmil} = 347\text{MCM}$$

The next largest size is a conductor.

8.10 Branch-circuit Design Considerations

A branch circuit is defined as a circuit which extends from the last overcurrent protective device of the power system. Branch circuits, according to the National Electrical Code, are either 15, 20, 30, 40 , or 50 amperes in capacity. Loads larger than 50 amperes would not be connected to a branch circuit. Lighting circuits are one of the most common types of branch circuits. They are usually either 15-ampere or 20-ampere circuits.

There are many rules in the National Electrical Code (NEC) that apply to branch-circuit design. The following information is based on the NEC. First, each circuit must be designed so that accidental short circuits or grounds do not cause damage to any part of the system. Then, fuses or circuit breakers are to be used as branch-circuit overcurrent protective devices. Should a short circuit or ground condition occur, the protective device should open and interrupt the flow of current in the branch circuit.

One important NEC rule is that No. 16 or No. 18 (extension cord) wire can be tapped from No. 12 or No. 14 conductors-but not larger than No. 12. This means that an extension cord of No. 16 wire should not be plugged into a receptacle that uses No. 10 wire. Damage to smaller wires (due to the heating effect) before the overcurrent device can open is eliminated by applying this rule.

The maximum rating of an individual load (such as a portable appliance connected to a branch circuit) is 80% of the branch-circuit current rating. So, a 20 -ampere circuit could not have a single load that draws more than 16 amperes. If the load is a permanently connected appliance, its current rating cannot be more than 50% of the branch circuit capacity if portable appliances or lights are connected to the same circuit.

Branch circuits must be designed so that sufficient voltage is supplied to all parts of the circuit. The distance that branch circuit can extend from the voltage source or power distribution panel is, therefore, limited. A voltage drop of 3% is specified by the National Electrical Code as the maximum allowed for branch circuits in electrical wiring design.

The method for calculating the voltage drop in a branch circuit is a step-by-step process that is illustrated by the following problem. Refer to the circuit diagram given in Figure 8.9.

Sample Problem:

Given: A 120-volt, 15-ampere branch circuit supplies a load consisting of four lamps. Each lamp draws 3 amperes of current from the source. The lamps are located at 10-foot (3.05-meter) intervals from the power distribution panel.

Find: The voltage across lamp No. 4.

Solution:

1. Find the resistance for 20 feet (6.1 meters) of conductor (same as 10-foot conductor × 2). A No. 14 copper wire is used for 15 -ampere branch circuits. From Figure 8.10, we find that the resistance of 1000 feet (304.8 meters) of No. 14 copper wire is 2.57 ohms. Therefore, the resistance of 20 feet of wire is:

$$\frac{20\text{ feet}}{1000\text{ feet}} = \frac{R}{2.57\Omega}$$

$$1000R = (20) \times (2.57)$$

$$R = 0.0514\Omega$$

2. Calculate voltage drop No. 1($V_D 1$). (R equals the resistance of 20 feet of wire.)

$$V_D 1 = I \times R$$
$$V_D 1 = 12\,\text{A} \times 0.0514\Omega$$
$$V_D 1 = 0.6168\text{Vac}$$

3. Calculate the voltage at load No. 1($V_L 1$). (The source voltage minus $V_D 1$.)

$$V_L 1 = 120\,\text{V} - 0.6168\,\text{V}$$
$$V_L 1 = 119.383 \text{ volts}$$

4. Calculate voltage drop No. 2(V_{D^2})[2].

$$V_{D^2} = I \times R$$
$$V_{D^2} = 9\,\text{A} \times 0.0514\Omega$$
$$V_{D^2} = 0.4626\text{Vac}$$

5. Calculate the voltage at load No. 2(V_{L^2}).

$$V_{L^2} = 119.383\,\text{V} - 0.4626\,\text{V}$$
$$V_{L^2} = 118.920 \text{ volts}$$

6. Calculate voltage drop No. 3(V_{D^3}).

$$V_{D^3} = I \times R$$
$$V_{D^3} = 6\,\text{A} \times 0.0514\Omega$$
$$V_{D^3} = 0.384\text{Vac}$$

7. Calculate the voltage at load No. 3(V_{L^3}).

$$V_{L^3} = 118.920\,\text{V} - 0.3084\,\text{V}$$
$$V_{L^3} = 118.612 \text{ volts}$$

8. Calculate voltage drop No. 4(V_{D^4}).

$$V_{D^4} = I \times R$$
$$V_{D^4} = 3A \times 0.0514\Omega$$
$$V_{D^4} = 01.1542\text{Vac}$$

9. Calculate the voltage at load No. 4(V_{L4}).

$$V_{L^4} = 118.612\,\text{V} - 0.1542\,\text{V}$$
$$V_{L^4} = 118.458 \text{ volts}$$

Notice that the voltage across lamp No. 4 is substantially reduced from the 120 -volt source value due to the voltage drop of the conductors. Also, notice that the resistances used to calculate the voltage drops represented both wires (hot and neutral) of the branch circuit. Ordinarily, 120-volt branch circuits do not extend more than 100 feet (30.48 meters) from the power distribution panel. The preferred distance is 75 feet (22.86 meters). The voltage drop in branch-circuit conductors can be reduced by making the circuit shorter in length or by using larger conductors.

In residential electrical wiring design, the voltage drop in many branch circuits is difficult to calculate since the lighting and portable appliance receptacles are placed on the same branch circuits. Since portable appliances and plug-in lights are not used all of the time, the voltage drop will vary according to the number of lights and appliances in use. This problem is usually not encountered in an industrial or commercial wiring design for lights, since the lighting units are usually larger and are permanently installed on the branch circuits.

A branch circuit usually consists of a nonmetallic sheathed cable that is connected into a power distribution panel. Each branch circuit that is wired from the power distribution panel is protected by a fuse or circuit breaker. The power panel also has a main switch that controls all of the branch circuits that are connected to it.

A diagram of a ***single-phase, three-wire*** (120/240-volt) power distribution panel is shown in Figure 8.13. Notice that eight 120 -volt branch circuits and one 240 -volt circuit are available from the power panel. This type of system is used in most homes where several 120 -volt branch circuits and, typically, three or four 240-volt branch circuits are required. Notice in Figure 8.13 that each hot line has a circuit breaker while the neutral line connects directly to the branch circuits. Neutrals should never be opened (fused). This is a safety precaution in electrical wiring design.

A diagram for a ***three-phase, four-wire*** (120/208-volt) power distribution panel is shown in Figure 8.14. There are three single-phase 120volt branch circuits and two 208-volt three-phase branch circuits shown. The single-phase branches are balanced (one hot line from each branch). Each hot line has an individual circuit breaker. Three phase lines should be connected so that an overload in the branch circuit will cause all three lines to open.

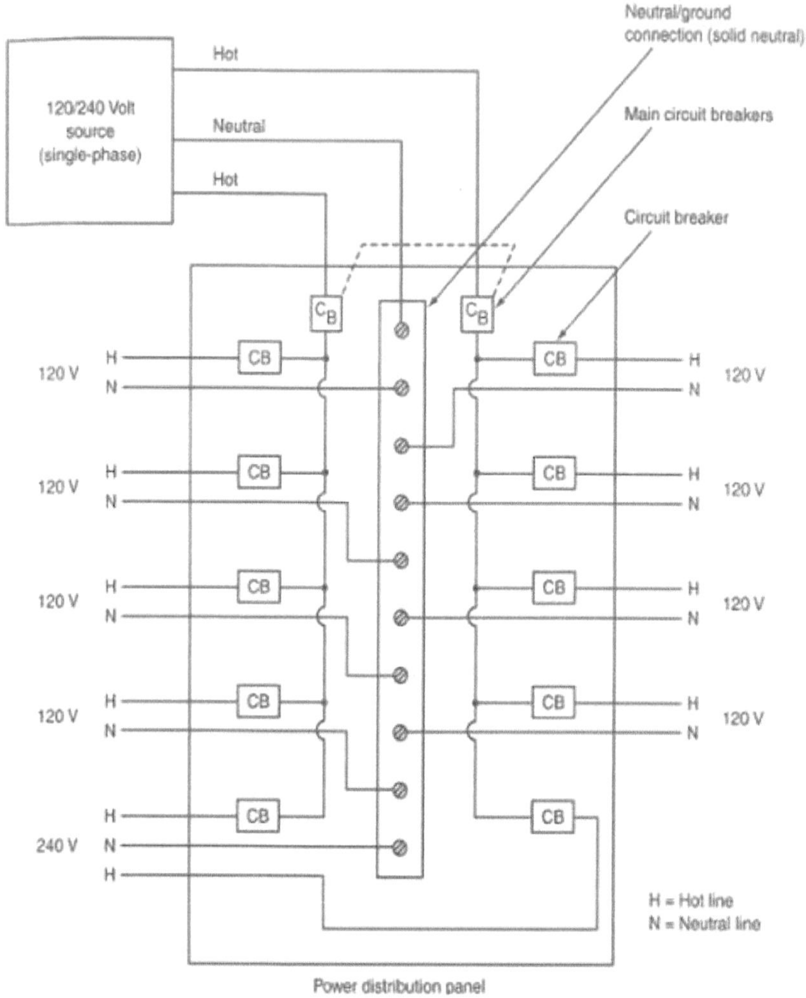

Figure 8.13 Diagram of a power distribution panel for a single-phase, three-wire branch circuit.

This is accomplished by using a three phase circuit breaker. This is arranged internally as shown in Figure 8.15.

8.11 Feeder-circuit Design Considerations

Feeder circuits are used to distribute electrical power to power distribution panels. Many feeder circuits extend for very long distances. Therefore, ***voltage drop*** must be considered in feeder-circuit design. In higher voltage feeder

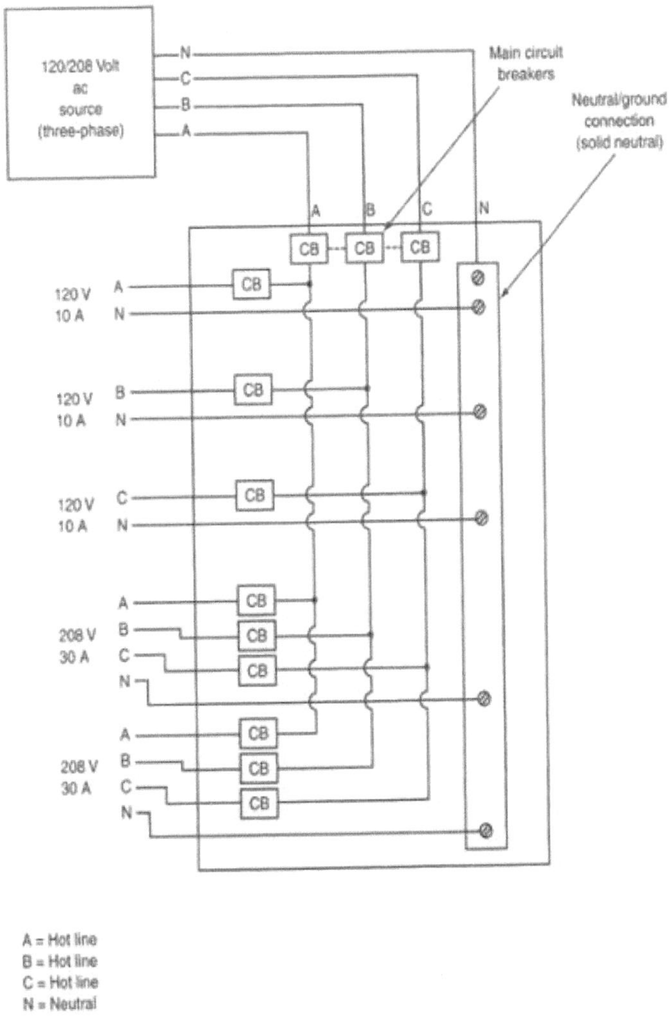

Figure 8.14 Diagram of a power distribution panel for a three-phase, four-wire branch circuit.

circuits, the voltage drop is reduced. However, many lower voltage feeder circuits require large-diameter conductors to provide a tolerable level of voltage drop. High current feeder circuits also present a problem in terms of the massive overload protection that is sometimes required. This protection is usually provided by system ***switchgear*** or ***load centers*** where the feeder circuits originate.

The amount of current that a feeder circuit must be designed to carry depends upon the actual load demanded by the branch-circuit power

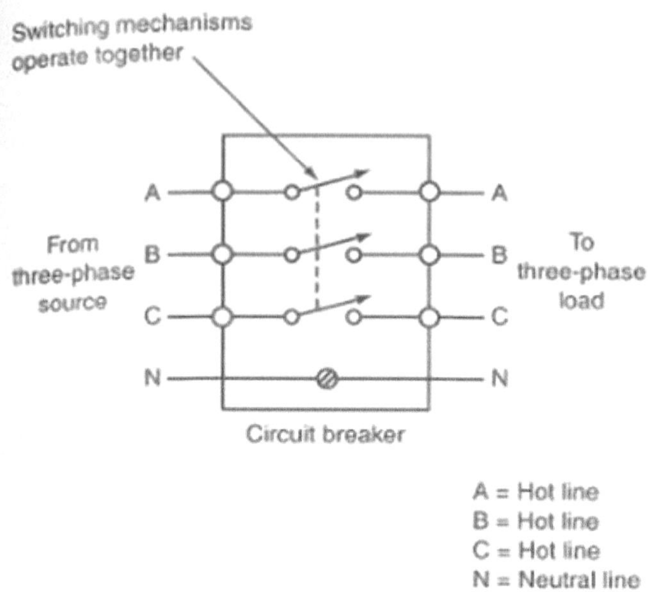

Figure 8.15 Diagram of a three-phase circuit breaker.

distribution panels that it supplies. Each power distribution panel will have a separate feeder circuit. Also, each feeder circuit must have its own overload protection.

The following problem is an example of size calculation for a feeder circuit.

Sample Problem:

Given: Three 15-kW fluorescent lighting units are connected to a three-phase, four-wire (277/480-volt) system. The lighting units have a power factor of 0.8 .

Find: The size of THW aluminum feeder conductors required to supply this load.

Solution:

1. Find the line current:

$$I_L = \frac{P_T}{1.73 \times V_L \times \text{PF}}$$

$$I_L = \frac{45{,}000\ \text{W}}{1.73 \times 480\ \text{V} \times 0.8}$$

$$I_L = 67.74\text{amps}$$

2. From Figure 8.11, we find that the conductor size that will carry 67.74 amperes of current is a No. 3 AWG THW aluminum conductor.

8.11.1 Voltage drop calculation

Feeder-circuit design must take the conductor voltage drop into consideration. The voltage drop in a feeder circuit must be kept as low as possible so that maximum power can be delivered to the loads connected to the feeder system. The NEC allows a maximum 5% voltage drop in the combination of a branch and a feeder circuit. However, a voltage 5% reduction represents a significant power loss in a circuit. We can calculate power loss due to voltage drop as V^2/R, where V^2 is the voltage drop of the circuit and R is the resistance of the conductors of the circuit.

The calculation of feeder conductor size is similar to that for a branch-circuit voltage drop. The size of the conductors must be large enough to: (1) have the required ampacity, and (2) keep the voltage drop below a specified level. If the second requirement is not met, possibly due to a long feeder circuit, the conductors chosen must be larger than the ampacity rating requires.

The following problem illustrates the calculation of feeder conductor size based upon the voltage drop in a single-phase circuit.

Sample Problem:

Given: A single-phase 240 -volt load in a factory is rated at 85 kilowatts. The feeders (two hot lines) will be 260 -foot (79.25 meter) lengths of RHW copper conductor. The maximum conductor voltage drop allowed is 2%.

Find: The feeder conductor size required.

Solution:

1. Find the maximum voltage drop of the circuit.

$$V_D = \% \times \text{load}$$
$$V_D = 0.02 \times 240$$
$$V_D = 4.8 \text{ V}$$

2. Find the current drawn by the load.

$$I = \frac{\text{power}}{\text{voltage}}$$
$$I = \frac{85{,}000 \text{ W}}{240 \text{ V}}$$
$$I = 354.2 \text{ A}$$

3. Find the minimum circular-mil conductor required. Use the formula given for finding the cross-sectional area of a conductor in singlephase systems that was previously given in the Alternative method portion of Section 8.9.

$$\text{cmil} = \frac{\rho \times I \times 1.73d}{V_D} \text{(single-phase systems)}$$

$$\text{cmil} = \frac{10.4 \times 354.2 \times 2 \times 260}{4.8}$$

$$\text{cmil} = 399{,}065.33\text{cmil}$$

4. Determine the feeder conductor size. The next larger size conductor in Figure 8-10 is also 400 MCM. Check Figure 8-11 and you will see that a 400 MCM RHW copper conductor will carry 335 amperes. This is less than the required 354.2 amperes, so use the next larger size, which is a 500 MCM conductor.

The conductor size for a three-phase feeder circuit is determined in a similar way. In the following problem, the feeder size will be determined as based upon the circuit voltage drop.

Sample Problem:

Given: A 480-volt three-phase, three-wire delta) feeder circuit supplies a 45-kilowatt balanced load to a commercial building. The load operates at a 0.75 power factor. The feeder circuit three hot lines) will be a 300 -foot (91.44-meter) ength of RH copper conductor. The maximum roltage drop is 1%.

Find: The feeder size required (based on the roltage drop of the circuit).

Solution:

1. Find the maximum voltage drop of the circuit.

$$V_D = 0.01 \times 480\text{V}$$

$$V_D = 4.8 \text{ V}$$

2. Find the line current drawn by the load.

$$I_L = \frac{P}{(1.73 \times V \times \text{PF})}$$

$$I_L = \frac{45{,}000 \text{ W}}{1.73 \times 480 \text{ V} \times 0.75}$$

$$I_L = 72.25 \text{ A}$$

3. Find the minimum circular-mil conductor area required. Use the formula for finding cmil in three-phase systems that was given in Section 8.9 .

$$\text{cmil} = \frac{\rho \times I \times 1.73d}{V_D}$$

$$\text{cmil} = \frac{10.4 \times 72.25 \times 1.73 \times 300}{4.8}$$

$$\text{cmil} = 81{,}245\text{cmil}$$

4. Determine the feeder conductor size. The closest and next larger conductor size in Figure 8.10 is No. 1 AWG. Check Figure 8.11 and you will see that a No. 1 AWG RH copper conductor will carry 130 amperes. This is much more than the required 72.25 amperes. Therefore, use No. 1 AWG RH copper conductors for the feeder circuit.

8.12 Determining Grounding Conductor Size

Grounding considerations in electrical wiring design were discussed previously. Another aspect in wiring design is to determine the size of the grounding conductor required in a circuit. All circuits that operate at 150 volts or less must be grounded. Therefore, all residential electrical systems must be grounded. Higher voltage systems used in industrial and commercial buildings have grounding requirements that are specified by the NEC and by local codes. A ***ground*** at the service entrance of a building is usually a metal water pipe that extends uninterrupted underground or a grounding electrode that is driven into the ground near the service entrance.

The size of the ***grounding conductor*** is determined by the current rating of the system. Figure 8.16 lists equipment grounding-conductor sizes for interior wiring. The sizes of grounding conductors listed in Figure 8.16 are for equipment grounds that connect to raceways, enclosures, and metal frames for safety purposes. Note that a No. 12 or a No. 14 wiring cable, such as 12-2 WG NMC can have a No. 18 equipment ground. The ground is contained in the same cable sheathing as the hot conductors.

Figure 8.17 lists the minimum grounding conductor sizes for system grounding of service entrances. It is used to find the minimum size of grounding conductors needed for service entrances, based upon the size of the hot line conductors used with the system.

8.13 Parts of Interior Electrical Systems

Some parts of interior electrical distribution systems have been discussed previously. Such types of equipment as transformers, switchgear, conductors,

Ampere Rating of Distribution Panel	Grounding Conductor Size (AWG)*	
	Copper	Aluminum
15	14	12
20	12	10
30	10	8
40	10	8
60	10	8
100	8	6
200	6	4
400	3	1
600	1	00
800	0	000
1000	00	0000

Figure 8.16 Equipment grounding-conductor sizes for interior wiring.

Conductor Size (Hot Line)		Grounding Conductor Size	
Copper	Aluminum	Copper (AWG)	Aluminum (AWG)
No. 2 AWG or smaller	No. 0 AWG or smaller	8	6
No.1 or 0 AWG	No. 00 or 000 AWG	6	4
No. 00 or 000 AWG	No. 0000 AWG or 250 MCM	4	2
No. 000 AWG to 350 MCM	250 MCM to 500 MCM	2	0
350 MCM to 600 MCM	500 MCM to 900 MCM	0	000
600 MCM to 1100 MCM	900 MCM to 1750 MCM	00	0000

Figure 8.17 System grounding-conductor sizes for service entrances.

insulators, and protective equipment are parts of interior wiring systems. There are, however, certain parts of interior electrical distribution systems that are unique to the wiring system itself. These parts include the nonmetallic-sheathed cables (NMC), the metal-clad cables, the rigid conduit, and the electrical metallic tubing (EMT).

Nonmetallic-sheathed cable (NMC) is a common type of electrical cable used for interior wiring. NMC, sometimes referred to as ***Romex cable***,

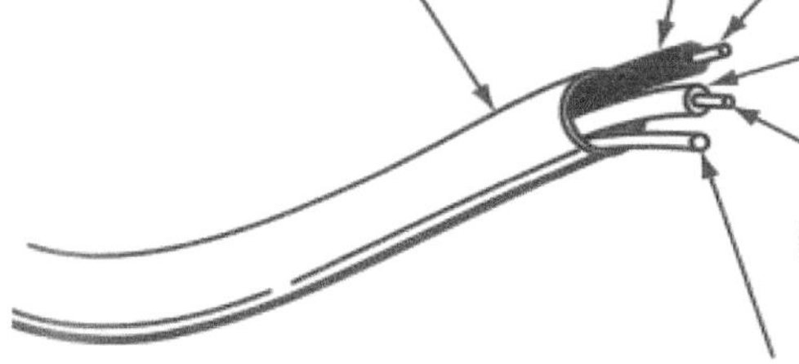

Figure 8.18 Nonmetallic-sheathed cable.

is used in residential wiring systems almost exclusively. The most common type used is No. 12-2 WG, which is illustrated in Figure 8.18. This type of NMC comes in 250-foot rolls for interior wiring. The cable has a thin plastic outer covering with three conductors inside.

The conductors have colored insulation that designates whether the conductor should be used as a hot, neutral, or equipment ground wire. For instance, the conductor connected to the hot side of the system has black or red insulation. The neutral conductor has white or gray insulation. The equipment grounding conductor has either a green insulation or no insulation (bare conductor). There are several different sizes of bushings and connectors used for the installation of NMC in buildings.

The designation No. 12-2 WG means that:

- The copper conductors used are No. 12 AWG as measured by an American Wire Gage (AWG).
- There are two current-carrying conductors.
- The cable comes with a ground (WG) wire. A No. 14-3 WG cable, in comparison, would have three No. 14 conductors and a grounding conductor. NMC ranges in size from No. 14 to No. 1 AWG copper conductors and from No. 12 to No. 2 AWG aluminum conductors.

Metal-clad cable is similar to NMC except that it has a flexible spiral metal covering rather than a plastic covering. A common type of metal clad cable is called ***BX cable***. Like NMC, ***BX cables*** contain two or three conductors. There are also several sizes of connectors and bushings used in the installation of BX cable. The primary advantage of this type of metal-clad cable is that it is contained in a metal enclosure that is flexible so that it can be bent easily. Other metal enclosures are usually more difficult to bend.

The exterior of ***rigid conduit*** looks like water pipe. It is used in special locations for enclosing electrical conductors. Rigid conduit comes in 10 -foot lengths that must be threaded for joining the pieces together. The conduit is secured to metal wiring boxes by locknuts and bushings. It is bulky to handle and takes a long time to install.

Electrical metallic tubing (EMT) or thin wall conduit is somewhat like rigid conduit except that it can be bent with a special conduit-bending tool. EMT is easier to install than rigid conduit since no threading is required. It also comes in 10-foot lengths. EMT is installed by using compression couplings to connect the conduit to metal wiring boxes. Interior electrical wiring systems use EMT extensively since it can be easily bent, can be connected together, and can be connected to metal wiring boxes.

Review Questions

Answer the following questions.

1. List the two types of single-phase distribution systems.

2. List the five types of three-phase transformer connections.

3. List three types of three-phase distribution systems.

4. What is the purpose of distribution system grounding?

5. What is the difference between a system ground and an equipment ground?

6. What is the purpose of a ground fault circuit interrupter (GFCI)?

7. List three types of GFCIs.

8. What is the purpose of the National Electrical Code?

9. What is the purpose of calculating voltage drop in electrical conductors?

10. What is a branch circuit?

11. What is the purpose of a feeder circuit?

12. How is grounding conductor size determined?

13. What is nonmetallic sheathed cable?

14. What is the purpose of metal-clad cable?

15. Describe the difference between rigid conduit and thin wall conduit.

UNIT THREE

Electric Machinery Basics

Motors are defined as machines that convert electrical power into mechanical power for many different applications. In order to understand electric machinery control, you must be aware of the types and characteristics of electric motors. **Unit III** provides thorough coverage of many types of electric motors.

Chapter 9 introduces the types of direct current (DC) motors and their applications. An introduction to basic motor operation is also included in this chapter. Single-phase alternating current (AC) motors are the topic of **Chapter 10**. These motors have many uses in the home and for industrial/commercial applications.

Chapter 11 discusses three-phase motors, which are used for automated equipment and commercial applications. Servo systems and motion control for automation and robotic applications are covered in **Chapter 12**. Modern control systems for electric motors use the basics of motion control for automated processes.

These four chapters provide comprehensive explanations of many common types of electric motors that are used for automated control systems.

DOI: 10.1201/9788743810056-11

9

Direct-Current Motors

Objectives

Upon completion of this chapter, you will be able to:

- Discuss the basic construction and principles of electric motors.
- Describe torque development in DC motors.
- List characteristics of DC motors.
- Discuss the basic construction and operating principles of the following five motors: permanent magnet de motor, series-wound de motor, shunt-wound DC motor, compound-wound DC motor, and brushless DC motor.
- Describe the method used to reverse the rotation of DC motors.
- Compare the characteristics of DC motors.
- Describe the construction and operation of a dynamotor.
- Discuss the operation of a brushless DC motor and its advantages in automated control systems.
- Explain the need for DC motor starting circuits.
- List and describe five types of losses in DC motors.

Important Terms

armature current (I_A)
armature reaction
brush
brushless DC motor
commutator assembly
compound-wound DC motor
counter electromotive force (CEMF)
cumulative compound DC motor
differential compound DC motor
dynamotor
generator action
lamination
load
long-shunt
motor action

DOI: 10.1201/9788743810056-12

permanent-magnet DC motor
rotor runaway
speed
series-wound DC motor
short shunt
shunt-wound de motor
speed regulation
stator
torque
Generator
Motor
Stator
Yoke (frame)
Field Poles
Field Windings
Concentrated Windings
Distributed Windings
Wound Rotor
Squirrel Cage Rotor
Slip Ring Commutator
Split Ring Commutator
Brushes
Rotor Shaft
Bearings
Rotor Core
Winding Polarity
Multiple Field Windings
Armature
Series Connection
Parallel Connection
Armature Rating
Average Induced Voltage
Full-Pitch Coil
Fractional-Pitch Coil
Pitch Factor
Lap Windings
Wave Windings
Simplex Windings
Multiplex Windings
Chorded Windings
Multiplicity
Triplex Windings
Generated Voltage

Introduction

Machines that convert electrical energy into mechanical energy represent a broad category of electrical loads used today. The motor load is the major electrical power-consuming load used in the United States. Motors of various sizes are used for processes ranging from precise machinery control to the movement of massive pieces of equipment in industry. They are used in home appliances, commercial building heating and cooling systems, and for many other applications.

9.1 Basic Motor Principles

The function of a motor is to convert electrical energy into mechanical energy in the form of a rotary motion. To produce a rotary motion, a motor must have an electrical power input. Generator action is brought about due to a magnetic field, a set of conductors within the magnetic field, and relative motion between the two. Motion is similarly produced in a motor due to the interaction of a magnetic field and a set of conductors. This interaction is referred to as motor action.

All motors, regardless of whether they operate from an AC or a DC power line, have several basic characteristics in common. Their basic parts include:

- A stator-the frame and other stationary components.
- A rotor- the rotating shaft and its associated parts
- Auxiliary equipment-such as a **brush** and **commutator assembly** for DC motors and a starting circuit for single-phase AC motors. The basic parts of a DC motor can be seen in Figure 9.1. A cutaway diagram of a DC motor is shown in Figure 9.2 with the basic parts labeled.

The motor principle (motor action) is illustrated in Figure 9.3. In part A of Figure 9.3, no current is flowing through the conductors due to the position of the brushes in relation to the commutator. During this condition, no motion is produced. When current flows through the conductor, a circular magnetic field is developed around the conductor. The direction of the current determines the direction of the circular magnetic fields, as shown in the cross-sectional diagram in part B of Figure 9.3.

When current flows through the conductors within the main magnetic field, it produces a field that interacts with the main field. The interaction of these two magnetic fields results in motion being produced. The circular magnetic field around the conductors causes a compression of the main magnetic flux at points and in part of Figure 9.3. This compression causes the magnetic field to produce a reaction in the opposite direction of the compression. Therefore, motion is produced away from points A and B.

In actual motor operation, a rotary motion in a clockwise direction would be produced. The **right-hand rule** shown in part D of Figure 9.3 is used to predict the direction of motion. To change the direction of rotation, the direction of the current through the conductors would have to be reversed.

The rotating effect produced by the interaction of two magnetic fields is called **torque**. The torque produced by a motor depends on the strength of the main magnetic field and the amount of current flowing through the conductors. As the magnetic field strength (or the current through the conductors) increases, the amount of torque or rotary motion will also increase.

9.2 Torque Development in DC Motors

Motor action, or the development of torque, is a significant electromechanical principle. There are some basic factors of torque development that should

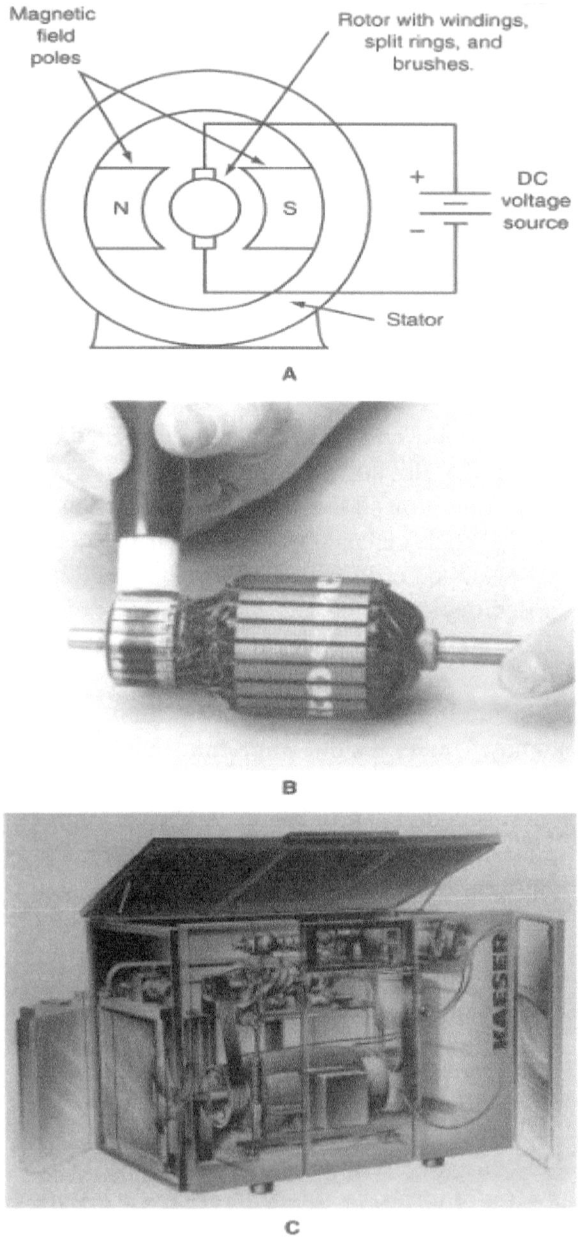

Figure 9.1 A-Basic parts of a DC motor. B – Wound armature with split-ring commutator; notice the stick brush that is used to clean the commutator. C – Belt-driven AC motor used with an air compressor. (The Eraser Company, Inc., Kaeser Compressors, Inc.)

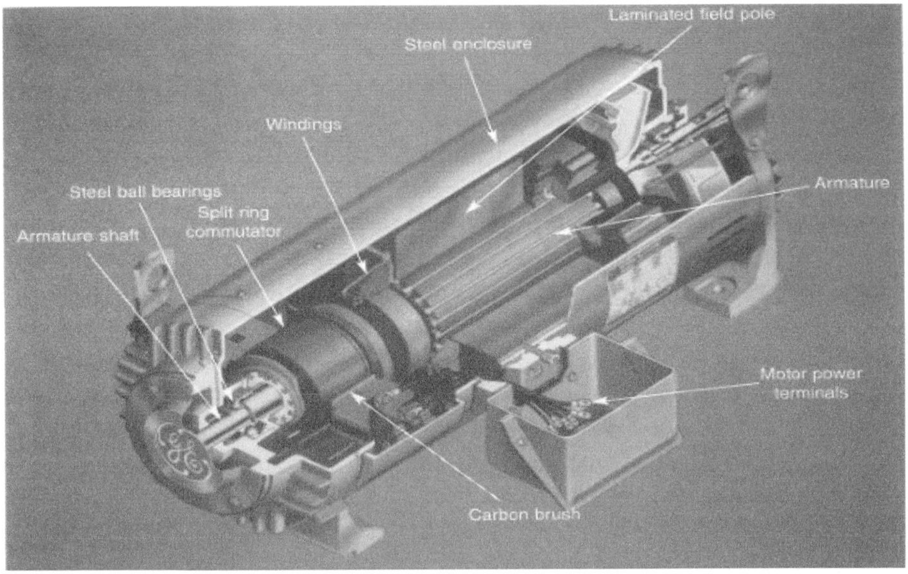

Figure 9.2 Cutaway diagram of a DC motor. (General Electric Co, Fort Wayne, Indiana)

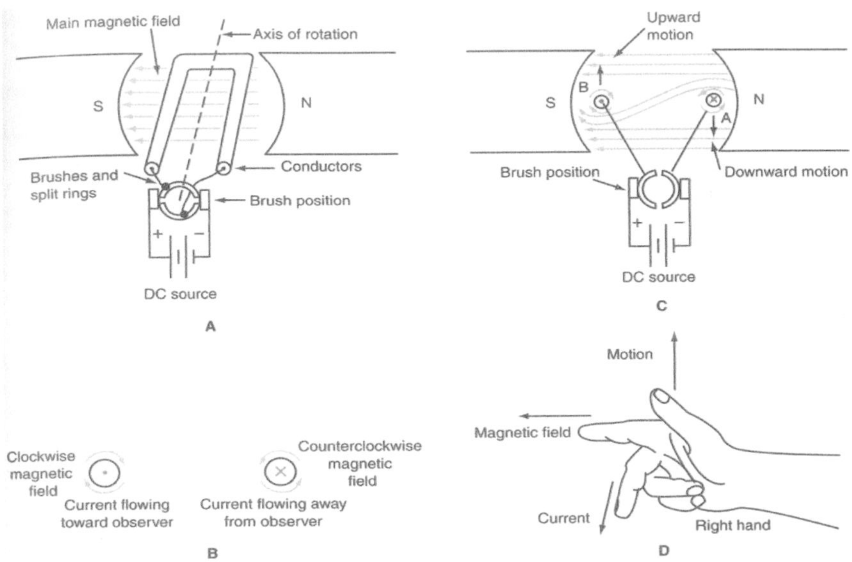

Figure 9.3 Motor principle (motor action). A – Condition with no current flowing through conductors. B – The direction of current through the conductors determines the direction of the magnetic field around the conductors. C – Condition with current flowing through conductors. D – Right-hand rule.

be kept in mind. Motor action allows electrical current to produce rotation (torque). During this process, a generated voltage or **counterelectromotive force (CEMF)** is produced. This phenomenon is called the generator action of a motor. The generated voltage due to CEMF (V_C) is always less than the voltage applied to the motor , and its direction is such that it opposes the **armature current** (I_A).

A comparison should be made between motor action and generator action. When a rotating electrical machine is operated as a DC generator, a torque is developed in the armature conductors (counter torque) that opposes the torque of the prime mover. The armature current of a DC generator flows in the same direction as the polarity of the generated voltage (V_R) and V_R exceeds the voltage applied to the load circuit. When a rotating DC machine is operated as a motor, the generated CEMF (V_C) opposes the armature current and is less than the voltage applied to the machine (V_A). These relationships can be summarized as follows:

$$\text{For DC motors: } V_A = V_C + I_A \times R_A$$

$$\text{For DC generators: } V_R = V_A + I_A \times R_A$$

where V_A is the applied voltage (also called terminal voltage), V_C is the CEMF developed in a motor armature circuit, V_R is the voltage generated or induced into a generator armature, and $I_A \times R_A$ is the voltage drop across the armature circuit.

The CEMF (V_C) is expressed as:

$$V_C = V_A - I_A \times R_A$$

The torque developed by a motor is dependent on armature current. The value of armature current can be calculated by the formula:

$$I_A = \frac{V_A - V_C}{R_A}$$

The force (F) per conductor developed due to motor action is expressed as:

$$F = \frac{B \times I \times l}{1.13} \times 10^{-7}\,\text{lb}$$

where B is the flux density in lines of force per square inch, I is the current in amperes, l is the length of the conductor, and 1.13 is a constant.

For example, if a conductor is 10in long, the current flowing through it is 15 A, and the magnetic field strength is 30,000 lines/in.², the force (*F*) developed by the conductor is:

$$F = \frac{B \times I \times l}{1.13} \times 10^{-7}\,\text{lb.}$$

$$F = \frac{30{,}000\ \text{lines / in.}^2 \times 15\ \text{A} \times 10\text{in.}}{1.13} \times 10^{-7}\,\text{lb.}$$

$$F = \frac{4{,}500{,}000 \times 10^{-7}\,\text{lb.}}{1.13} = 0.398\text{lb.}$$

Torque developed in motors is a force that produces mechanical energy in the form of rotation. It is the product of force and distance and is commonly expressed in units of foot-pounds (ft.-lb.), ounce-inches (oz.-in.), or kilogram-meters (kg – m).

The calculation of **torque** of a DC motor can be accomplished by considering the percentage of armature conductors that lie directly adjacent to a field pole. This calculation is based on the assumption that the armature conductors that are positioned directly adjacent to a field pole contribute in an equal manner to the production of torque. The average force (F_{avg}) is found by the equation:

$$F_{\text{avg}} = F_C \times C_A$$

where F_C is the average force per armature conductor adjacent to a field pole in pounds and C_A is the number of armature conductors.

The average torque(T_{avg}) of the machine (in ft-lb) can be computed as:

$$T_{\text{avg}} = F_{\text{avg}} \times r = F_C \times C_A \times r$$

where *r* is the radial distance to the axis of rotation in feet.

As an **example**, the **armature** of a DC motor has 200 conductors, a diameter of 6 in, and a conductor length of 12in. The flux density of the field is 30,000 lines/in² and 75% of the armature conductors are adjacent to a field winding. The armature current flow is 10 A. The force per conductor (F_C) is found as follows:

$$F_C = \frac{B \times I \times l}{1.13} \times 10^{-7}\,\text{lb.}$$

$$F_C = \frac{30{,}000\ \text{lines / in.}^2 \times 10\ \text{A} \times 12\text{in.}}{1.13} \times 10^{-7}\,\text{lb.}$$

$$F_C = \frac{3,600,000 \times 10^{-7}}{1.13} = 0.319\text{lb.}$$

The average force (F_{avg}) tending to rotate the armature can then be calculated as:

$$F_{avg} = F_C \times C_A$$
$$F_{avg} = 3.18\text{lb.} \times (200 \times 0.75)$$
$$F_{avg} = 477\text{lb.}$$

The average torque (T_{avg}) is then found as:

$$T_{\text{avg}} = F_{\text{avg}} \times r$$
$$T_{\text{avg}} = 477\text{lb.} \times 0.25\text{ft.} (3\text{in.} = 1/2 \text{ of} 6\text{in.} = 0.25\text{ft.})$$
$$T_{\text{avg}} = 119.25\text{ft.} - \text{lb.}$$

The torque of a DC motor field can also be determined by the formula:

$$T = \text{K} \times \Phi \times I_A \text{ ft-lb.}$$

where K is a constant for a specific machine, **Φ** is the magnetic flux per pole in lines/in.², and I_A is armature current in amperes. The formula shows that as either the magnetic flux or the armature current values increase, torque increases in direct proportion.

9.2.1 Horsepower rating of motors

The **horsepower** rating of a motor is based on the amount of torque produced at the rated full load values. Horsepower, which is the usual method of rating motors, can be expressed mathematically as:

$$\text{hp} = \frac{2\pi \times N \times T}{33,000}$$
$$\text{hp} = \frac{N \times T}{5252}$$

where hp is the horsepower rating, 2π is a constant, *N* is the speed of the motor in revolutions per minute, and *T* is the torque developed by the motor in ft.-lb.

Figure 9.4 Typical DC motors. A – Notice the field coils and the cooling fins on the rotor shaft. B – Notice the construction of the commutator and brushes. (General Electric Co, Fort Wayne, Indiana)

9.3 DC Motor Characteristics

Motors that operate from DC power sources have many applications in industry when speed control is desirable. DC motors are classified as series, shunt, or compound machines, depending on the method of connecting the armature and field windings. Also, permanent-magnet DC motors are used for certain applications. Typical DC motors are shown in Figure 9.4.

The operational characteristics of all de motors can be generalized by referring to Figure 9.5. Most electric motors exhibit characteristics similar to those shown in the block diagram. When discussing DC motor characteristics, it is necessary to be familiar with the following terms: load, speed, counterelectromotive force (CEMF), armature current, and torque.

The amount of mechanical load applied to the shaft of a motor determines its operational characteristics. As the mechanical load is increased, the speed of a motor tends to decrease. As the speed decreases, the voltage induced into the conductors of the motor due to generator action (CEMF) decreases. The generated voltage, or counterelectromotive force, depends on the number of rotating conductors and the speed of rotation. Therefore, as the speed of rotation decreases, so does the CEMF.

The counterelectromotive force generated by a motor is in opposition to the supply voltage. Since the CEMF is in opposition to the supply voltage, the actual working voltage of a motor will increase as the CEMF

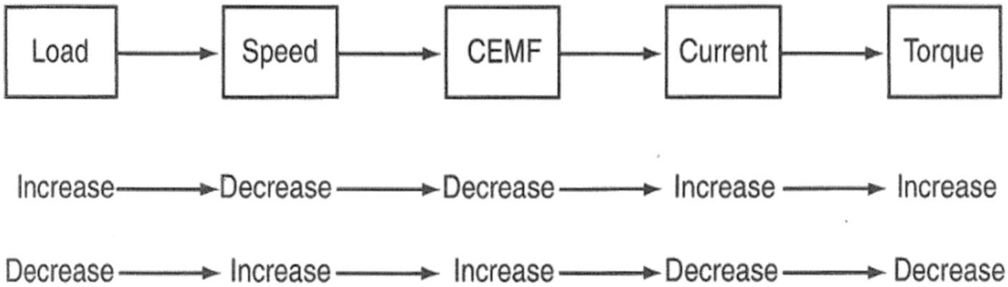

Figure 9.5 Operational characteristics of DC motors.

decreases. When the working voltage increases, more current flows through the armature windings. Since the torque of a motor is directly proportional to the armature current, the torque will increase as the armature current increases.

To discuss briefly the opposite situation, if the mechanical load connected to the motor decreases, the speed tends to increase. An increase in speed causes an increase in the CEMF. Since the CEMF is in opposition to the supply voltage, as the CEMF increases, the armature current decreases. A decrease in armature current causes a decrease in torque.

Torque varies directly with changes in load. As the load on a motor is increased, its torque also increases to try to meet the increased load requirement. However, the current drawn by a motor also increases when the load is increased.

The presence of a CEMF to oppose the armature current is very important in motor operation. The lack of any CEMF when a motor is being started explains why motors draw a very large initial starting current compared to their running current when full speed is reached. Maximum armature current flows when there is no CEMF. As the CEMF increases, the armature current decreases. Thus, resistances in series with the armature circuit are often used to compensate for the lack of CEMF and to reduce the starting current of a motor. After the motor has reached full speed, these resistances are bypassed by automatic or manual switching systems in order to allow the motor to produce maximum torque. Keep in mind that the armature current, which directly affects torque, can be expressed as:

$$I_A = \frac{V_T - V_C}{R_A}$$

where I_A is the armature current in amperes, V_T is the terminal voltage of the motor in volts, V_C is the CEMF generated by the motor in volts, and R_A is the armature resistance in ohms.

Another DC motor characteristic is called armature reaction. **Armature reaction** is an effect that distorts the main magnetic field in de motors. The brushes of a DC motor can be shifted to counteract the effect of armature reaction. However, devices called interpoles or compensating windings are ordinarily used to control armature reaction in DC motors.

9.4 Speed Characteristics of DC Motors

The most desirable characteristic of DC motors is their speed-control capability. By varying the applied DC voltage with a rheostat, speed can be varied from zero to the maximum revolutions per minute (r/min) of the motor. Some types of DC motors have more desirable speed characteristics than others. **Speed regulation** is a measure of how much motor speed varies with varying loads. Comparative speed regulation for different types of motors can be calculated.

Speed regulation is expressed as:

$$\%R = \frac{S_{NL} - S_{FL}}{S_{FL}} \times 100$$

where $\%R$ is the percentage of speed regulation, S_{NL} is the no-load speed in r/min, and S_{FL} is the rated full-load speed in r/min.

Good speed regulation (low %R) results when a motor has nearly constant speeds under varying load situations.

9.5 Types of DC Motors

The manner in which the armature and field circuits of a DC motor are connected determines its basic characteristics. The types of commercially available DC motors basically fall into four categories:

- Permanent-magnet DC motors.
- Series-wound DC motors.
- Shunt-wound DC motors.
- Compound-wound DC motors.

Each of these motors has different characteristics due to its basic circuit arrangement and physical properties.

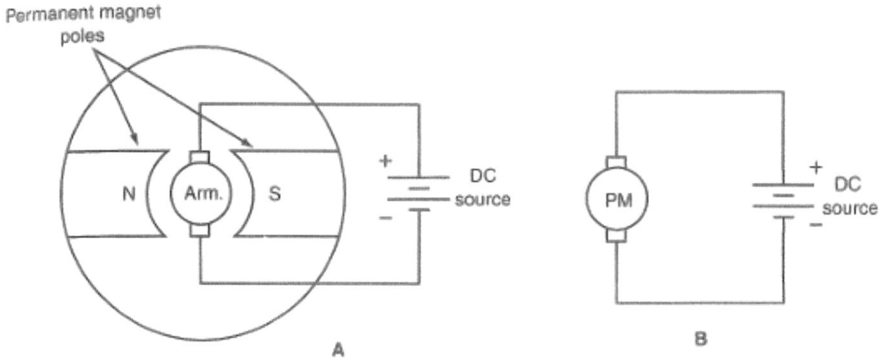

Figure 9.6 Permanent-magnet DC motor. A – Pictorial diagram. B – Schematic diagram.

9.5.1 Permanent-magnet DC motors

The permanent-magnet DC motor, shown in Figure 9.6, is used for low-torque applications. When this type of motor is used, the DC power supply is connected directly to the armature conductors through the brush/commutator assembly. The magnetic field is produced by permanent magnets mounted on the stator. The rotor of permanent magnet motors is a wound armature.

This type of motor ordinarily uses either alnico or ceramic permanent magnets rather than field coils. The alnico magnets are used with higher horsepower applications. **Ceramic magnets** are ordinarily used for lower-horsepower, slow-speed motors. Ceramic magnets have the advantage of being highly resistant to demagnetization. However, they are relatively low in magnetic flux level. The magnets are usually mounted in the motor frame and then magnetized prior to the insertion of the armature. A permanent-magnet DC motor is shown in Figure 9.7.

The permanent-magnet motor has several advantages over conventional types of DC motors. One advantage is reduced operational cost. The speed characteristics of the permanent-magnet motor are similar to those of the shunt-wound DC motor. The direction of rotation of a permanent magnet motor can be reversed by reversing the two power lines.

9.5.2 Series-wound DC motor

The series-wound DC motor, shown in Figure 9.8, has the armature and field circuits connected in a series arrangement. There is only one path for current to flow from the DC voltage source. Because of this, the field is wound of relatively few turns of large-diameter wire, giving the field a low resistance.

Figure 9.7 Permanent-magnet DC motor. (Bodine Electric Co.)

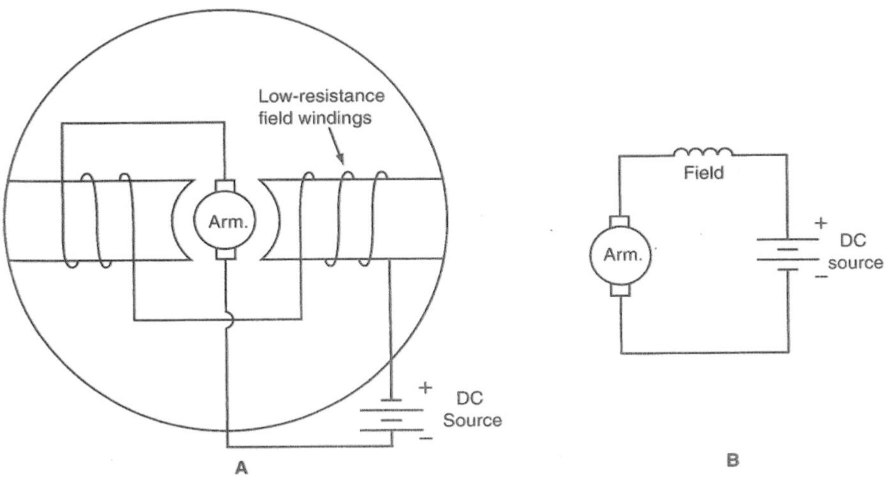

Figure 9.8 Series-wound DC motor. A – Pictorial diagram. B – Schematic diagram.

Changes in load applied to the motor shaft cause changes in the current through the field. If the mechanical load increases, the current also increases. The increased current creates a stronger magnetic field.

The speed of a series motor varies from very fast at no load to very slow at heavy loads. Since large currents can flow through the low-resistance field,

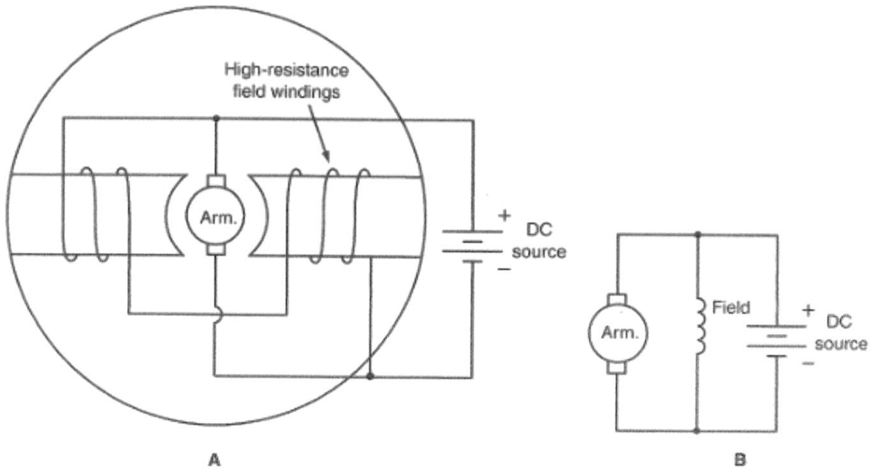

Figure 9.9 Shunt-wound DC motor. A – Pictorial diagram. B – Schematic diagram.

the series motor produces a high-torque output. Series motors are used where heavy loads must be moved and speed regulation is not important. A typical application is for automobile starter motors.

9.5.3 Shunt-wound DC motors

Shunt-wound DC motors are more commonly used than any other type of DC motor. As shown in Figure 9.9, the **shunt-wound DC motor** has field coils connected in parallel with its armature. This type of DC motor has **field coils** that are wound of many turns of small-diameter wire and have relatively high resistance. Since the field is a high-resistance parallel path of the circuit of the shunt motor, a small amount of current flows through the field. A strong electromagnetic field is produced due to the many turns of wire that form the field windings.

A large majority (about 95%) of the current drawn by the shunt motor flows in the armature circuit. Since the field current has little effect on the strength of the field, motor speed is not affected appreciably by variations in load current. The relationship of the currents that flow through a DC shunt motor is as follows:

$$I_T = I_A + I_F$$

where I_T is the total current drawn from the power source, I_A is the armature current, and I_F is the field current.

The field current can be varied by placing a variable resistance in series with the **field windings**. Since the current in the field circuit is low, a low-wattage rheostat can be used to vary the speed of the motor using the variation in field resistance. As field resistance increases, field current decreases. A decrease in field current reduces the strength of the electromagnetic field. When the field flux is decreased, the armature rotates faster due to reduced magnetic field interaction. Thus, the speed of a DC shunt motor can be easily varied by using a field rheostat.

The shunt-wound DC motor has very good speed regulation. The speed does decrease slightly when the load increases due to the increase in voltage drop across the armature. Because of its good speed regulation characteristic and its ease of speed control, the shunt-wound DC motor is commonly used for industrial applications. Many types of variable-speed machine tools are driven by DC shunt motors.

9.5.4 Compound-wound DC motors

The **compound-wound DC motor**, shown in Figure 9.10, has two sets of field windings. One set is in series with the armature and one is in parallel. This motor combines the desirable characteristics of the series- and shunt-wound motors.

There are two methods of connecting compound-wound DC motors: cumulative and differential. A **cumulative compound DC motor** has series and shunt fields that aid each other. **Differential compound DC motors** have series and shunt fields that oppose each other. There are also two ways in which the series windings are placed in the circuit. One method is called a short shunt. With this method the shunt field is placed across the armature. The schematic diagram for a short shunt can be seen on the lower left side of Figure 9.10. The **long-shunt** method has the shunt field winding placed across both the armature and the series field (see the lower right side of Figure 9.10).

Compound motors have high torque (similar to a series-wound motor) together with good speed regulation (similar to a shunt motor). Therefore, when good torque and good speed regulation are needed, the compound-wound DC motor can be used. A major disadvantage of a compound-wound DC motor is its expense.

9.6 DC Motor Rotation Reversal

A **DC motor** is designed so that its shaft will rotate in either direction. It is a very simple process to reverse the direction or rotation of any DC motor. By reversing

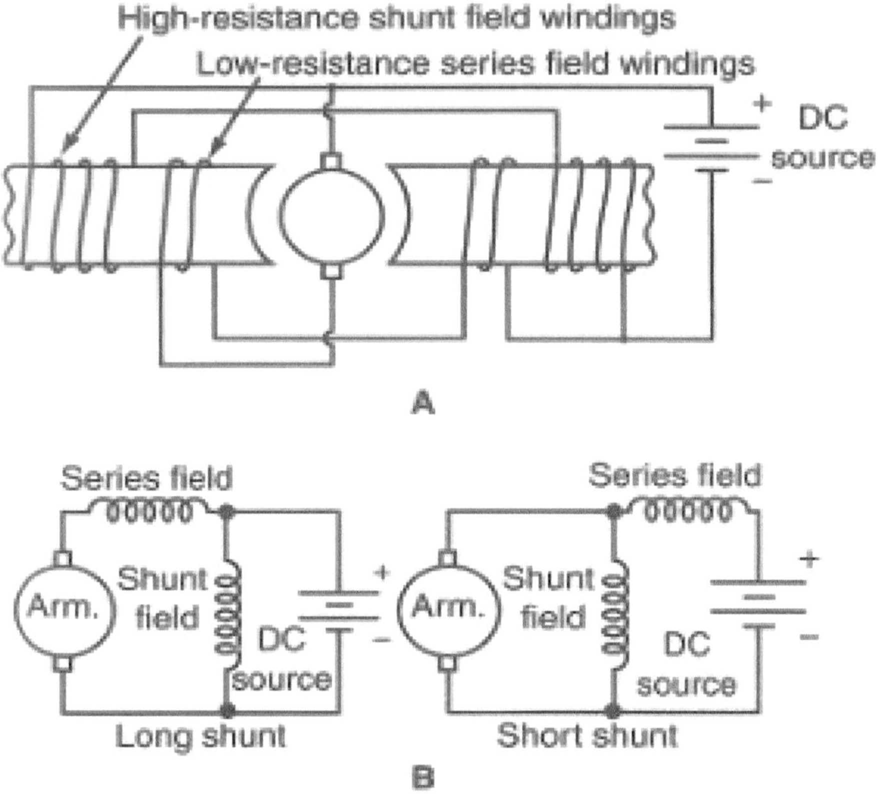

Figure 9.10 Compound-wound DC motor. A – Pictorial diagram. B – Schematic diagram.

the relationship between the connections of the **armature windings** and **field windings**, reversal of rotation is achieved. Usually, this is done by changing the terminal connections where the power source is connected to the motor.

Four terminals are ordinarily used for interconnection purposes. They can be labeled A1 and A2 for the armature connections and F1 and F2 for the field connections. If either the armature connections or the field connections are reversed, the rotation of the motor will reverse. However, if both are reversed, the motor shaft will rotate in its original direction since the relationship between the armature and field windings would still be the same.

9.7 Comparison of DC Motor Characteristics

The characteristics of DC motors should be considered when selecting motors for particular applications. Figures 9.11 and 9.12 show comparative graphs that illustrate the relative torque and speed characteristics of DC motors.

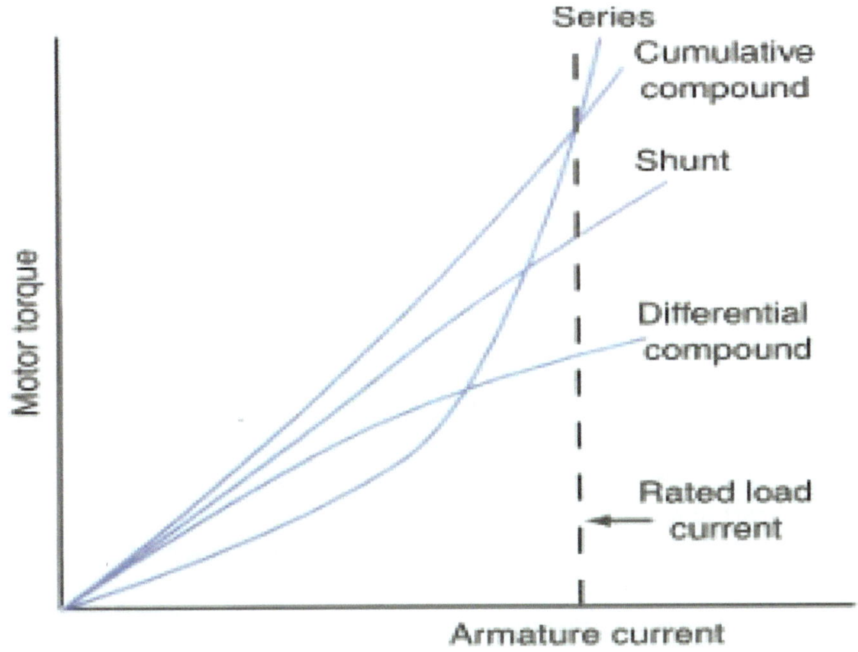

Figure 9.11 Torque versus armature current curves of DC motors.

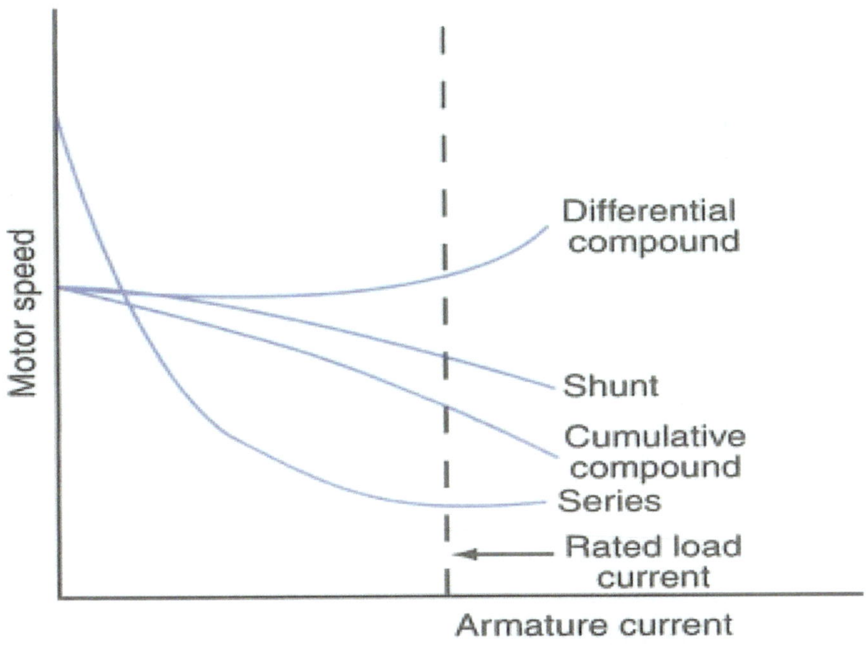

Figure 9.12 Speed versus armature current curves of DC motors.

9.7.1 Torque relationships

A comparative set of torque versus armature current curves for DC motors is shown in Figure 9.11. The effect of increased mechanical load on the shaft of each type of motor can be predicted.

Series-wound DC motors have equal armature current, field current, and load current ($I_A = I_F = I_L$). The magnetic flux (Φ) produced by the field windings is proportional to the armature current (I_A). The torque produced by a series-wound motor with low values of I_A is less than other motors due to a lack of field flux development. However, at rated full-load armature current, the torque is greater than other types of DC motors.

Shunt-wound DC motors have a fairly constant magnetic field flux due to the high-resistance field circuit. An almost linear torque versus armature current curve is a characteristic of shunt-wound DC motors. Since torque is directly dependent on armature current, as I_A increases, torque increases in direct proportion.

Cumulative compound DC motors, as noted in Section 9.5, have series and shunt field windings that aid each other in the production of an overall magnetic field. In this type of motor circuit, the general torque equation is:

$$T = K\left(\Phi_S + \Phi_P\right)I_A$$

where Φ_S is the series field flux and Φ_P is the parallel (shunt) field flux. The series field flux increases as I_A increases, and the shunt field flux remains fairly constant. Therefore, the torque curve (Figure 9.11) for a cumulative compound DC motor is always higher than that of a similar shunt-wound DC motor.

The **differential compound DC motor**, as noted in Section 9.5, has series and shunt fields that oppose each other in the production of an overall magnetic field. The general torque equation for this type of motor is:

$$T = K\left(\Phi_S - \Phi_R\right)I_A$$

The value of Φ_S depends primarily on armature current and Φ_P is fairly constant. Since these magnetic flux values oppose each other, the torque curve of a differential compound motor is always less than that of a shunt-wound DC motor.

9.7.2 Speed relationships

A comparative set of speed versus armature current curves for DC motors is shown in Figure 9.12. The effect of increased armature current on motor speed for each type of motor can be predicted.

Series-wound DC motors have speed characteristics that conform to the basic speed equation:

$$S = \frac{V_A - I_A \times (R_A + R_F)}{K\Phi}$$

where V_A is the rated armature voltage, R_A is the armature current, R_F is the armature resistance, K is the resistance of the series field winding,is a constant, and **Φ** is the magnetic flux of the field windings.

Since the magnetic flux produced by the series field is proportional to armature current flow, the speed equation can be written as:

$$S = K\frac{V_A - I_A \times (R_A + R_F)}{I_A}$$

With a small mechanical **load** applied to the shaft of a **series-wound DC motor**, the armature current is very low. A low value of I_A causes the numerator of the speed equation to be large and the denominator small. The result is an extremely high speed. At the no-load condition the I_A is small, field flux (**Φ**) is small, and speed is excessively high. Series motors have a runaway speed characteristic at no load. Thus, a load should always be connected to a series motor shaft to prevent the motor speed from becoming so high as to damage the motor. When the load on a series-wound DC motor increases, the numerator of the speed equation decreases faster than the denominator increases, and the speed drops appreciably.

Shunt-wound DC motors have speed characteristics based on the speed equation:

$$S = \frac{V_A - (I_A \times R_A)}{K\Phi}$$

The magnetic field flux (**Φ**) of a shunt-wound DC motor is practically constant. As the mechanical load applied to the shaft of a shunt-wound DC motor increases, the CEMF decreases. The speed also decreases slightly due to the difference between and the product of $I_A \times R_A$. The speed of a shunt-wound DC motor is considered to be almost constant.

Cumulative compound-wound DC motors have the following speed equation:

$$S = \frac{K(V_A - I_A \times (R_A + R_F))}{\Phi_S + \Phi_P}$$

As the load applied to the motor increases, I_A increases, the flux produced by the series field (**Φ**$_S$) increases, the CEMF decreases, and the numerator decreases at an even faster rate. The speed thus decreases with increases in load.

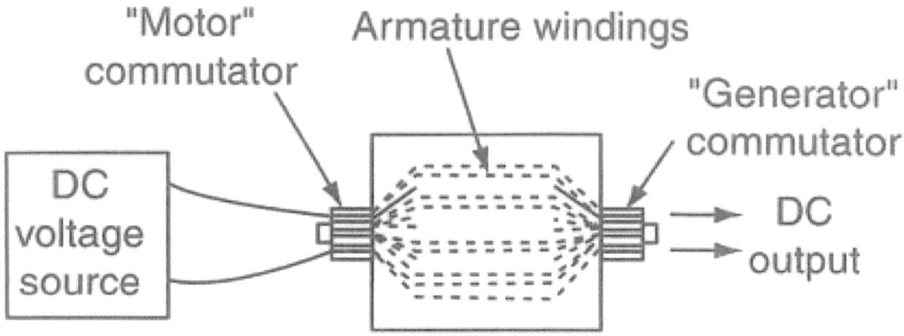

Figure 9.13 Dynamotor construction.

The **differential compound DC motor** has a speed equation of:

$$S = \frac{K\left(V_A - I_A \times \left(RA + RF\right)\right)}{\Phi_P - \Phi_S}$$

The parallel and series field windings develop an overall magnetic flux in which the two (Φ_P and Φ_S) are in opposition. As load increases, the numerator of the speed equation decreases, but the denominator decreases more rapidly. Thus, speed increases with increases in load. For this reason, differential compound motors are seldom used.

9.8 Dynamotors

Another type of DC motor is called a dynamotor. A **dynamotor**, depicted in Figure 9.13, converts DC voltage of one value to DC voltage of another value. It is actually a motor/generator housed in one unit.

The **armature** has two separate windings. One winding is connected to the **commutator** of the motor section and the other winding is connected to the commutator of a generator unit. A magnetic field, developed by either permanent magnets or electromagnetic windings, surrounds the armature assembly. Since the magnetic field remains relatively constant, the generator voltage output depends on the ratio of the number of motor windings to the number of generator windings. For instance, if there are twice as many generator windings as motor windings, the generated DC voltage output will be twice the value of the DC voltage that is input to the motor section of the dynamotor. In most cases, the dynamotor has been replaced by solid-state conversion systems. Typically, applications of the dynamotor have been to supply higher values of DC voltage from a low DC voltage source.

Figure 9.14 Brushless DC motor. (BEI Motion Systems)

9.9 Brushless DC Motors

The major problem with most DC motors is the low reliability of the **commutator/brush assembly**. The brushes have a limited life and cause the commutator to wear. This wearing produces brush dust that can cause other maintenance problems. The use of transistors has resulted in the development of brushless *dc* motors, Figure 9.14, that have neither brushes nor commutator assemblies. Instead, they make use of solid-state switching circuits.

Although some **brushless DC motors** use other methods, the transistor-switched motor is the most common. The motor itself is actually a single-phase AC permanent-capacitor induction motor with a center-tapped main winding. See Figure 9.15. Transistors, which are operated by an oscillator circuit, conduct alternately through the paths of the main winding. The oscillator circuit requires a feedback winding wound into the stator slots. This generates a control voltage to determine the frequency. A capacitor (C_2)

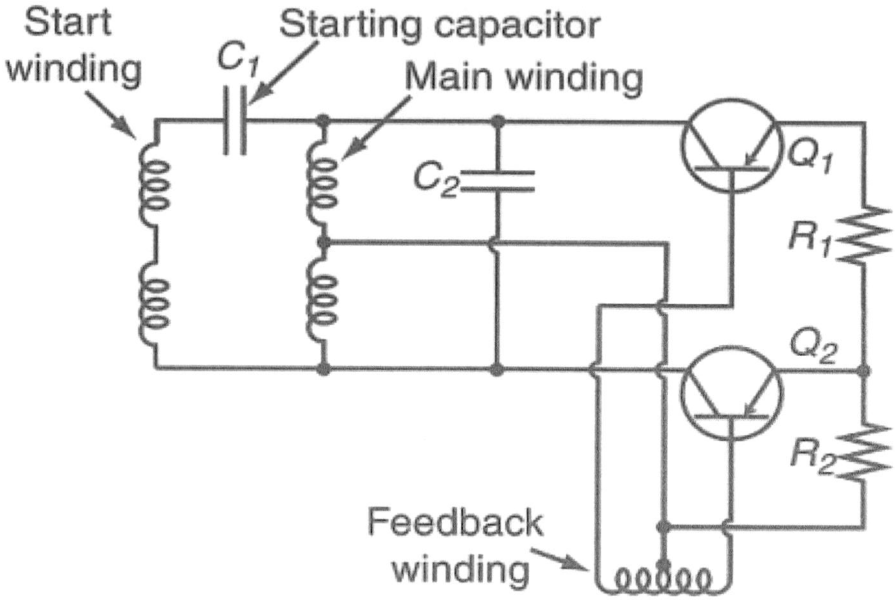

Figure 9.15 Brushless DC motor circuit.

is placed across the main winding to reduce voltage peaks and to keep the frequency of the circuit at a constant value.

The main disadvantage of this motor is its inability to develop a very high starting torque. As a result, it is suitable only for driving very low torque loads. When used in a low-voltage system, this motor is not very efficient. Also, since only half of the main winding is in use at any instant, copper losses are relatively high. However, the advantages outweigh the disadvantages for certain applications. These advantages include (since there are no brushes and a commutator) that motor life is limited mainly by the bearing. With proper lubrication, a brushless DC motor can be used for an indefinite period. Also, the motor frequency and thus the speed can be adjusted by varying the oscillator circuit.

9.10 DC Motor Starting Circuits

When DC motors are initially started, a problem exists caused by a lack of CEMF in the armature circuit. Excessive starting current flows through the low resistance of the armature conductors. Once CEMF builds up as speed increases, the problem of high starting current is less evident.

The **example** below illustrates the need for DC motor starting circuits that limit the current until the speed of rotation of the motor has increased.

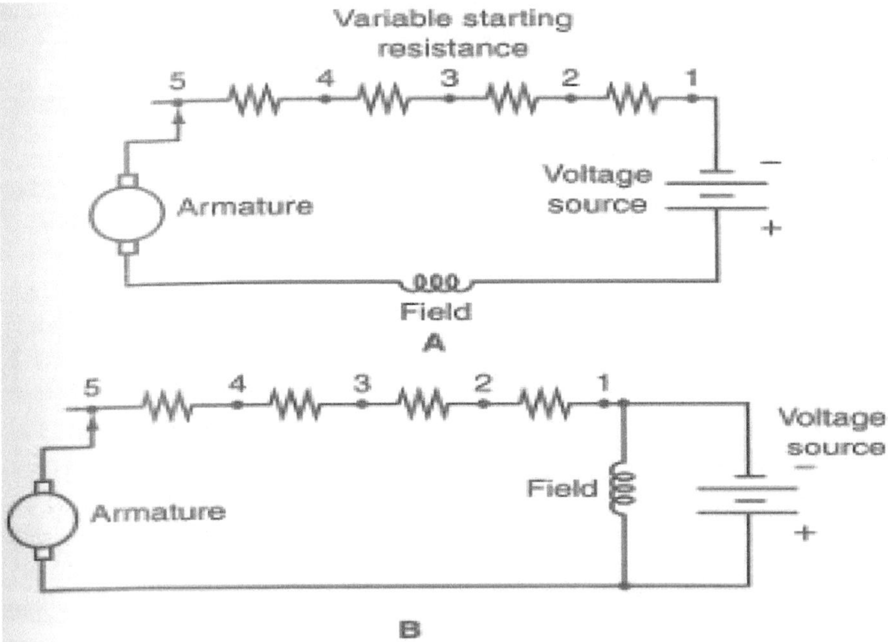

Figure 9.16 DC motor starting circuits. A – Series-wound motor control. B – Shunt-wound motor control.

Assume that a 240V DC shunt motor has an armature resistance of 0.2Ω and a rated full-load armature current of 30 A. The starting current (I_S) equals:

$$I_S = \frac{V_A}{R_A} = \frac{240\ V}{0.2\Omega} = 120\ A\ (\text{with CEMF} = 0\ V)$$

The percent of rated full-load armature current is:

$$\%\ \text{full-load} = \frac{120\ A}{30\ A} \times 100 = 400\%.$$

DC motor starting circuits employ resistances connected in series with the power lines to limit starting current. Since motor torque is dependent on the current, the resistances are used only during the starting cycle of the motor. They are either manually or automatically removed from the motor circuit once operating speed is reached. Starting resistances are often used in steps as shown in Figure 9.16. A decreasing value of starting resistance is needed as motor speed increases.

The value of **starting resistance** required for various values of rated full-load current can be calculated as follows. Assume that the 240-V DC motor of the previous example is used. The starting resistance (R_{ST}) required in series with the armature to limit starting current to 150% of rated full-load armature current can be calculated:

$$R_{\text{total}} = \frac{V_A}{I_A} = \frac{240\text{ V}}{150\%\text{ of}30\text{ A}} = \frac{240\text{ V}}{45\text{ A}} = 5.33\Omega$$
$$R_{ST} = R_{\text{total}} - R_A$$
$$R_{ST} = 5.33\Omega' - 0.2\Omega$$
$$R_{ST} = 5.13\Omega$$

Ordinarily, DC motors use starting circuits that limit armature current during initial starting to a value greater than the rated full-load current. This allows a motor to produce greater starting torque.

9.11 Losses in DC Motors

DC motors are designed to convert electrical power into mechanical power. However, all the input power is not converted into mechanical power due to the electrical resistance of the armature and other rotational losses. This causes heat to be produced within the motor. Motor losses can be divided into two areas: those that depend on the load and those that depend on speed.

9.11.1 Winding losses

Winding losses are caused by the electrical resistance of the motor windings and are equal to $R^2 \times R$ (where I is the armature current and R is the armature resistance). As the torque output of the motor is increased, I is increased, giving rise to additional losses. These losses are very important since heating of the armature winding will cause an increase in R that will result in further losses and more heating. This process can lead to self self-destruction of the motor if the maximum current is not limited.

9.11.1.1 Brush contact losses

Brush contact losses are fairly complex to analyze since they depend on a number of factors that will vary with motor operation. Brush contact resistance can represent a high proportion of the total resistance of the motor. This resistance is increased by heating due to $I^2 \times R$ losses in the brushes and contact area.

9.11.2 ron losses

Iron losses fall into two categories: (1) **Eddy current losses** are common in all machine parts experiencing a changing magnetic field. Eddy currents are induced into the motor armature as it undergoes changes in magnetization. These currents depend on speed and have a significant heating effect at high speeds. Eddy currents are reduced by making the armature core of thin, insulated sections or laminations, stacked to produce the required core length. (2) **Hysteresis losses** are brought about by the resistance of the core material to constant changes of the magnetic field. This causes additional heat generation with increases in speed. Iron losses are the major factor in determining the maximum speed that may be attained by an iron core motor.

9.11.3 Friction losses

Friction losses are associated with the mechanical characteristics of the motor. They are caused by brush friction, bearing friction, and resistance. These areas in themselves generate heat in addition to adding to armature losses.

9.11.4 Short circuit currents

As the brushes slide over the commutator, there is a period during which the brush is in contact with two commutator segments. During this period, the brush will short out the coil connected to those segments. The effect of this is to contribute a torque that opposes the main driving torque. It tends to increase with motor speed.

All of these losses will contribute heat to the motor. It is this heating that will limit the motor application and lifetime.

9.12 Construction of Electrical Machinery

Electrical machines accomplish electromechanical energy conversion. Generators convert mechanical energy into electrical energy, while motors convert electrical energy input into mechanical energy output. Generators and motors have basic construction characteristics that are common among many types of machines. The functions of various machines differ even though their construction is similar. Generators have rotary motion supplied by prime movers, which provide mechanical energy input. Relative motion between the conductors and a magnetic field of generators produces an electrical energy output. Motors have electrical energy supplied to their windings and a magnetic field that develops an electromagnetic interaction to produce mechanical energy or torque. This chapter discusses the construction and basic characteristics of electrical machines.

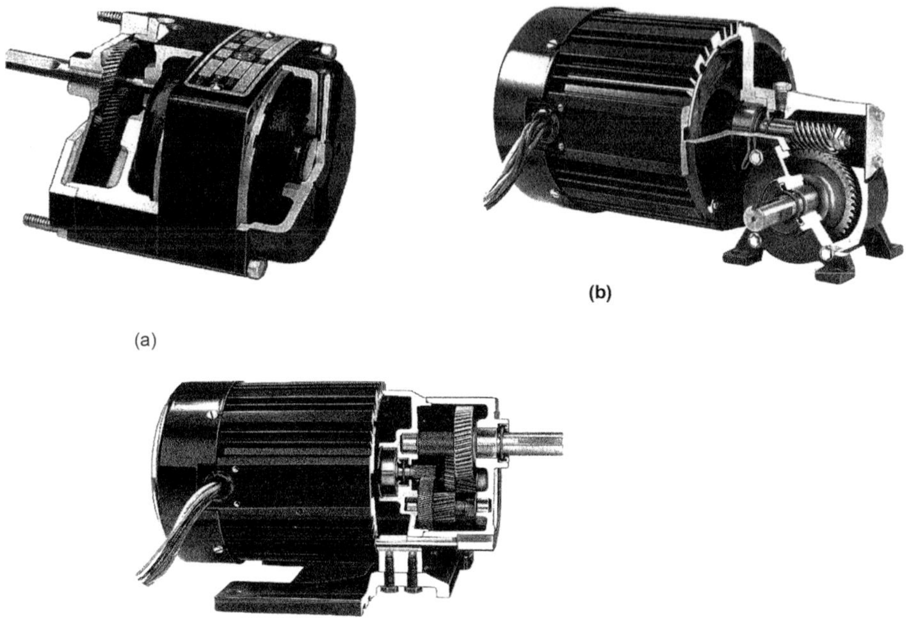

Figure 9.17 Machine stator assemblies. A, B – and (b) courtesy Bodine Co.; C – courtesy Louis-Allis Co.

9.12.1 Rotating Machinery Construction

The construction of most rotating electrical machines is somewhat similar. Most machines have a stationary part called the **stator** and a rotating set of conductors called the **rotor**. The stator consists of a **yoke** or **frame,** which serves as a support and a metallic path for the magnetic flux developed in a machine. Machine stator assemblies are shown in Figure 9.17. Figure 9.17 (a) shows a cutaway view of a small gear-reduction motor. Larger motor stator assemblies are shown in Figure 9.17 (b) and (c). Notice the design of the mounting brackets on the stator of each type.

9.12.1.1 Field poles and windings

Rotating machines also have *field* poles, which are part of the stator assembly. Field poles are constructed of laminated sheets of steel and secured to the machine frame. They are usually curved on the portion near the rotor to provide a low-reluctance path for magnetic flux. The **field windings** or **field coils** are placed around the poles as shown in the cutaway view of Figure 9.18. The field coils are electromagnets that develop an electromagnetic field interaction with the rotor to generate a voltage or to produce torque in a machine. There are two major types of electromagnetic field coils used

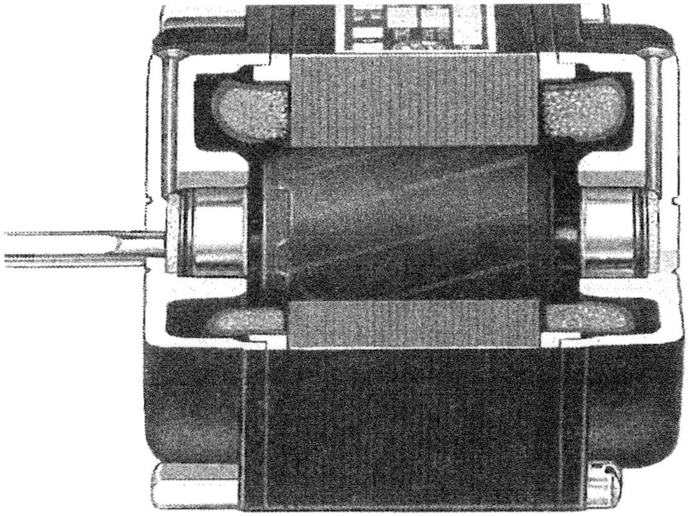

Figure 9.18 Field coils placed around laminated field poles (cutaway view). Courtesy Bodine Co.

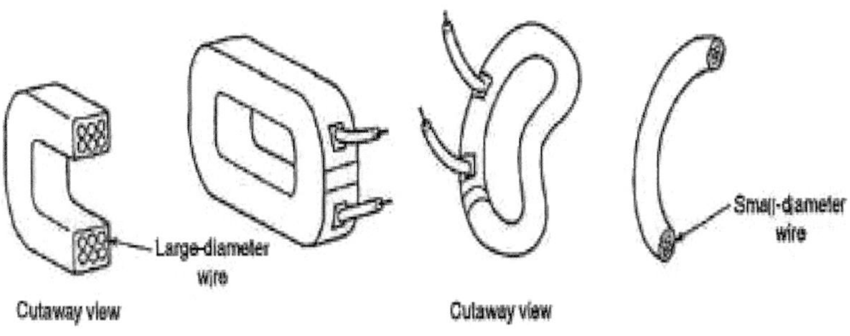

Figure 9.19 Coil windings: (a) concentrated or salient type – left side; (b) distributive type – left side.

in the construction of electrical machines. These are the (1) **concentrated** or salient coils and (2) **distributive** field coils. As a general rule, salient coils are used more frequently in DC motors and generators, while distributive coils are found in most AC applications.

Figure 9.19 shows examples of these two types of coil windings. Coil windings are insulated from each other and from the field poles. Figure 9.20 shows distributive windings of stator assemblies. Figure 9.20 (a) shows a shows a heavily insulated distributive wound stator of a small servomotor. A stator assembly of a machine with distributive windings is illustrated in Figure 9.20 (b).

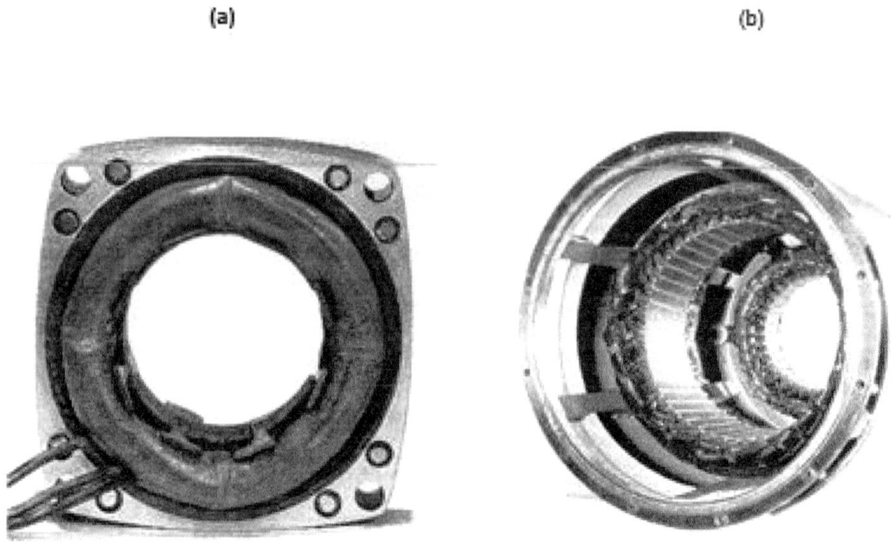

Figure 9.20 Distributive windings used on machine stator assemblies: (a) Heavily insulated distributive wound stator (courtesy Bodine Electric Co.). (b) Stator assembly with distributive windings (courtesy Lima Electric Co.).

9.12.2 Rotor construction

In the study of electrical machines, there is a need to understand the electromagnetic fields produced by the rotating section of a motor or a generator. This section is called the armature or rotor, and its coils are also of either salient or distributive type. Some types of machines use solid metal rotors called **squirrel-cage rotors**. Types of rotor assemblies are shown in Figure 9.21.

9.12.3 Slip rings, spilt rings, and brushes

In order for electrical energy to be supplied to a rotating device such as the armature, some sort of sliding brush contact must be established. Sliding brush contacts are either **slip rings** or **split rings**. **Slip rings** are constructed of a cylinder of insulating material with two separate solid metal rings glued to it. Sliding **brushes** made of carbon and graphite ride on the metal rings and permit application or extraction of electrical energy from the rings during rotation. A **brush** is shown in Figure 9.22**.** The **split ring commutator** is similar to the slip ring, except that a solid metal ring is cut into two or more separate sections. As a general rule, slip rings are used in AC motors and

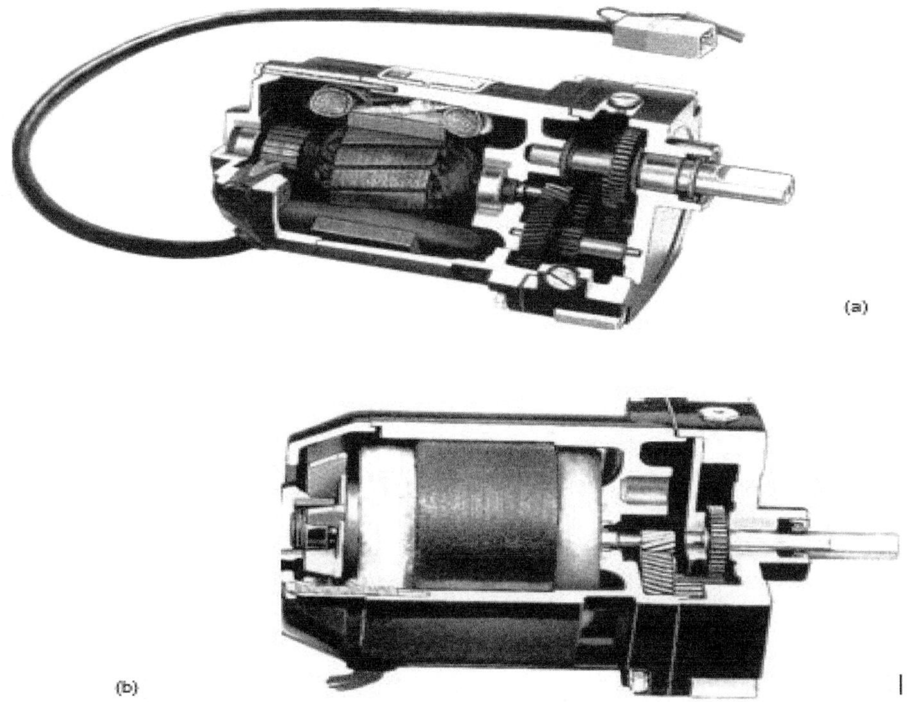

Figure 9.21 Machine rotor assemblies: A – with windings and split ring commutator/brush assembly (courtesy of Litton Industrial Products, Inc.-Louis-Allis Div. B – with laminated solid rotor (courtesy of Bodine Electric Co.}.

generators, while DC machines employ the split-ring commutating device. The gap or split in the commutator is kept at a minimum to reduce sparking of the brushes. Slip rings and split rings are shown in Figure 9.23.

9.12.4 Other machine parts

There are several other arts used in the construction of rotating machines. Among these are the rotor shaft, which rotates between a set of **bearings**.

Bearings may be either the ball, roller, or sleeve type. Bearing *seals,* often made of felt material, are used to keep lubricant around the bearing and keep dirt out.

A rotor **core** is usually constructed of laminated steel to provide a low-reluctance magnetic path between the field poles of a machine and to reduce eddy currents.

Figure 9.22 A carbon brush used with a rotating electrical machine. Courtesy Helwig Co.

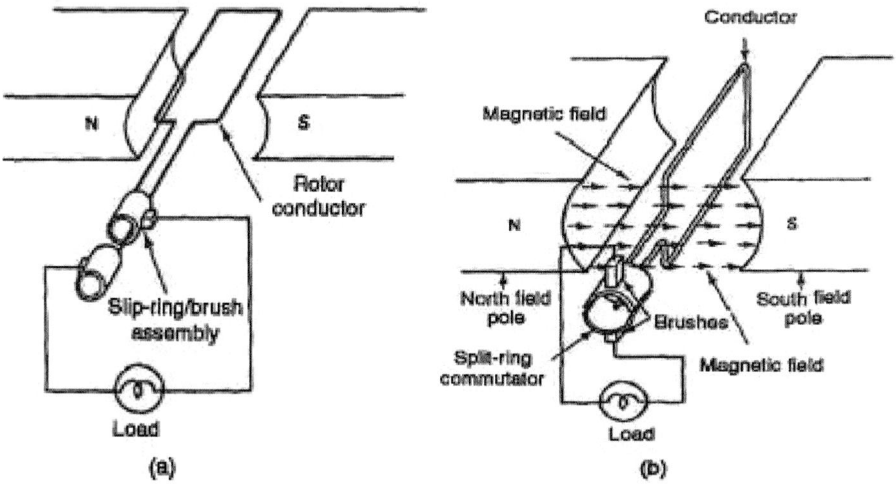

Figure 9.23 (a) – slip rings; (b) – split rings.

9.13 Stator Design of Electrical Machines

The electromagnetic coils used to produce a magnetic field in rotating electrical machines are called **field coils**. In a two-pole machine, two coils are used to develop a magnetic circuit as shown in Figure 9.25. The coils are wound around laminated metal cores called pole pieces. The strength of the magnetic field developed around the pole pieces is affected by their magnetic properties and the properties of the stator assembly. The primary purpose of the field coils is to develop a magnetic circuit for an electrical machine.

9.13.1 Winding polarities

Basic magnetic laws apply to the development of magnetic circuits in electrical machines. The coil shown in Figure 9.26 is energized by a direct-current (DC) voltage source. If a compass is placed along a central axis near the

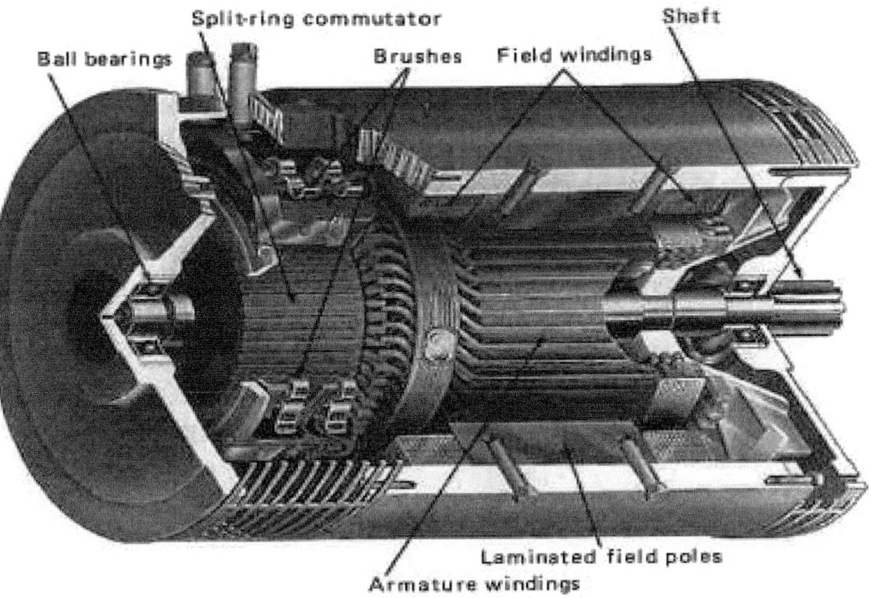

Figure 9.24 Cutaway view showing construction features of a machine. Courtesy General Electric Co.

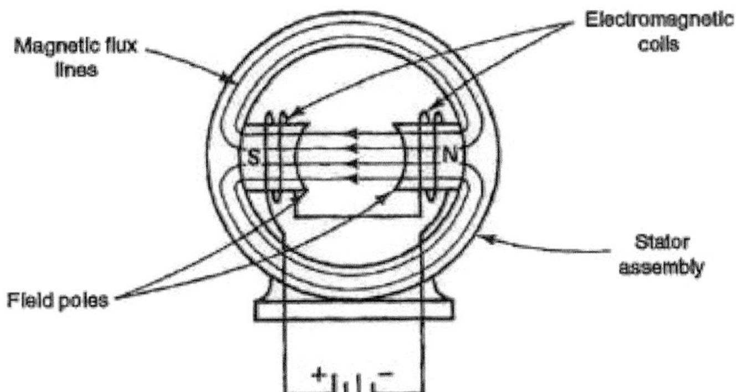

Figure 9.25 Two-pole electrical machine.

energized coil, a **polarity** is developed on either end. If the power source polarity is reversed, the magnetic polarities are also reversed. The same principle applies to electromagnetic coils used in electrical machines.

Figure 9.27 shows an electromagnetic coil energized by a DC power source. The dark band on the lower right section indicates the beginning of

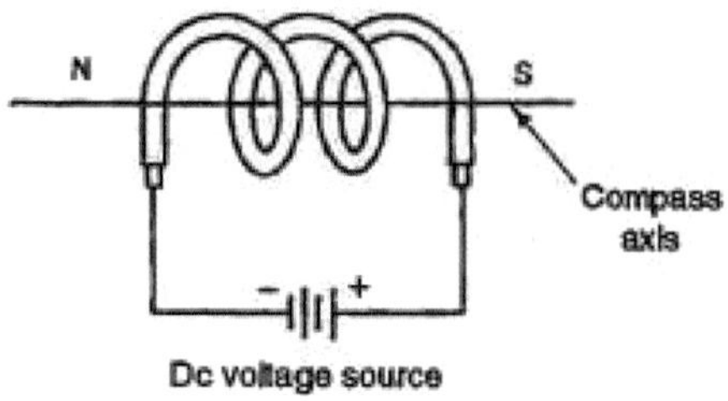

Figure 9.26 Coil energized by a DC source.

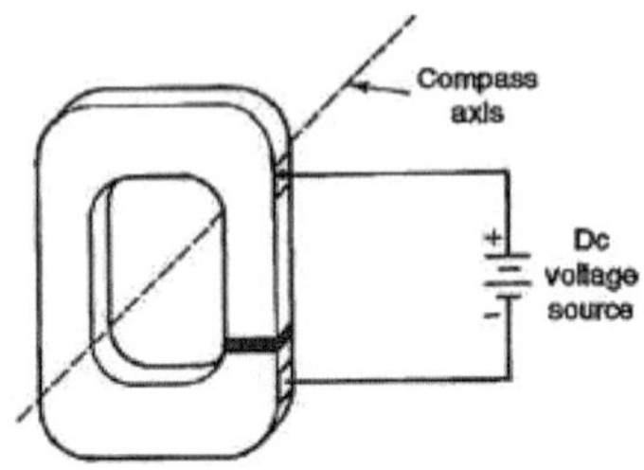

Figure 9.27 Electromagnetic coil energized by a DC source.

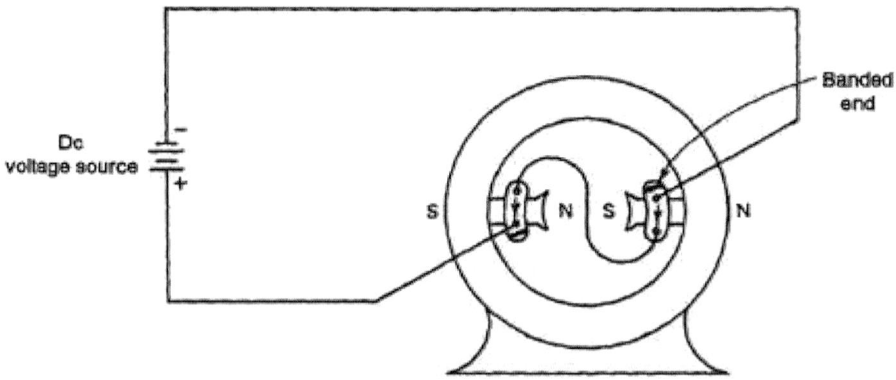

Figure 9.28 Proper method of connecting the windings of a two-pole machine.

the coil winding for reference purposes. The positive side of the power source connects to the end of the winding and the negative side to the beginning. If the power source polarity is reversed, the magnet polarity indicated along the compass axis of the figure is also reversed.

Polarities developed in an electrical machine must be a result of proper coil connections. Figure 9.28 shows a method of connecting the windings of a two-pole electrical machine. Dark bands are used to represent the beginnings of each of the windings. Notice that these coils are connected in series to form opposite polarities. The wire that connects the two coils extends from the end of one winding to the end of the next winding. The assumption here is that current flow from beginning to end through a coil produces an internal south polarity and current flow from end to beginning causes an internal north polarity. The current directions are shown on the diagram with arrows. If a compass is used to check polarities, the external and internal polarities would be as shown. Notice that in either case, the polarities are opposite.

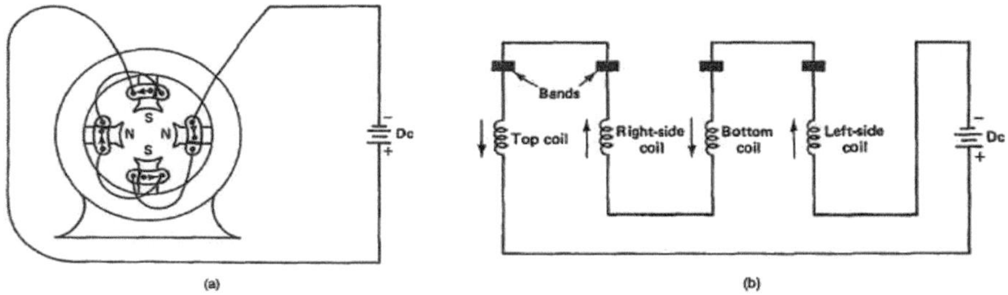

Figure 9.29 Series coil connections for a four-pole machine: A – pictorial; B – schematic.

Opposite polarities are used to develop the proper internal magnetic circuit for an electrical machine.

9.13.2 Multiple field windings

Many rotating electrical machines have more than two poles. There must always be an even number of **field windings**, such as a four-pole or six-pole machine. Additional poles in an electrical machine provide a better magnetic circuit by reducing the reluctance of the magnetic flux lines. Reduced reluctance allows a more efficient internal transfer of magnetic flux through a machine. The practical effect of adding more field windings is to increase generator output capability and motor torque characteristics.

A pictorial diagram of a four-pole electrical machine is shown in Figure 9.29 (a). The coils are connected in **series** to form the proper alternate **north and south polarities**. Notice the direction of current flow through each coil. The internal polarities produce an alternate north-south arrangement due to the direction of current flow through each coil. The schematic shown in Figure 9.29 (b) may be helpful in understanding the proper field winding connections for a four-pole electrical machine with a series circuit design.

Figure 9.30 (a) shows a parallel connection for a four-pole machine with a schematic representation in Figure 9.30 (b). Parallel connections supply full source voltage across each coil, while series connections distribute the source voltage in four equal parts across the coils. An eight-pole concentrated winding stator assembly is shown in Figure 9.31.

9.13.3 Rotor winding design

Several rotating electrical machines use wound-rotor assemblies. These assemblies are ordinarily called **armatures**. Armature coils are wound into

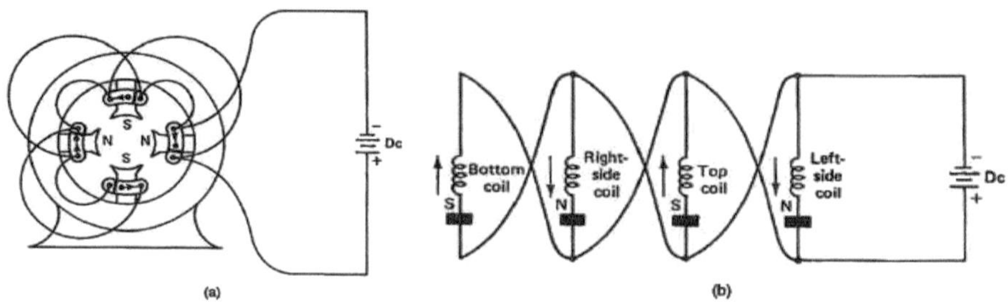

Figure 9.30 Parallel coil connections for a four-pole machine: A – pictorial diagram; B – schematic diagram.

Figure 9.31 An eight-pole stator assembly. Courtesy Bodine Electric Co.

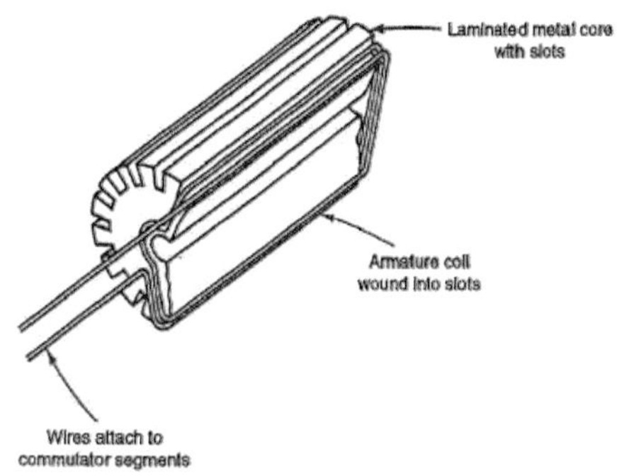

Figure 9.32 Armature coils wound onto a laminated metal core.

slots on laminated metal cores as shown in Figure 9.32. Figure 9.33 shows the wound-rotor assembly of a four-pole rotor with concentrated windings. The coils make contact with either a **split-ring commutator** or a **slip-ring/brush assembly**. The metallic core material provides a low-reluctance magnetic circuit for the armature.

The output characteristics of an electrical machine depend on the proper design of the armature circuit. There are two common methods used to connect armature windings to split-ring commutators of DC machines. These methods are called (1) lap windings and (2) wave windings. Pictorial representations of these winding methods are shown in Figure 9.34.

Lap windings (see Figure 9.34 (a)) are wound so that one coil overlaps the previous coil. The beginning and end of a lap-wound coil connect to adjacent split-ring commutator segments. Toe commutator segments make

Figure 9.33 Four-pole wound rotor assembly. Courtesy of Lima Electric Co., Inc.

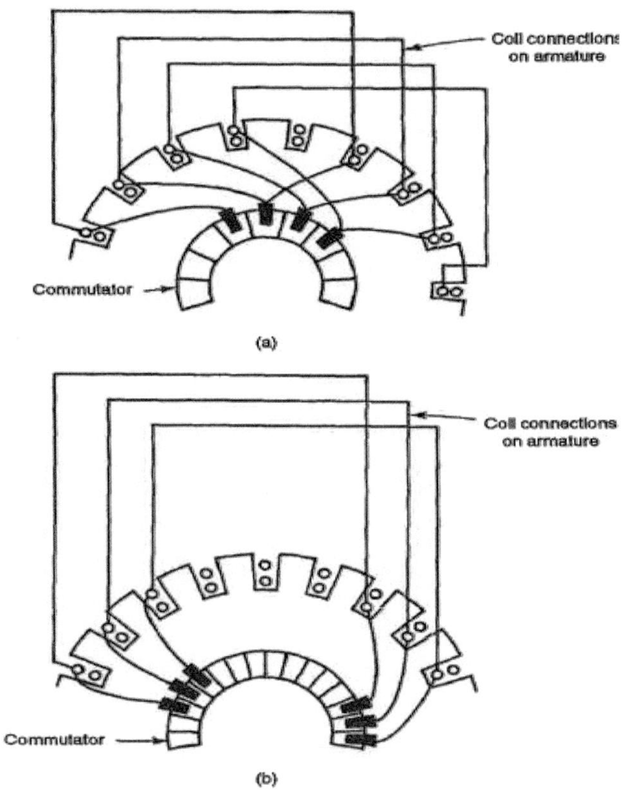

Figure 9.34 Methods of connecting windings of wound armatures: A – lap windings; B – wave windings.

connections to adjacent coils. The result is to place the coils in a **parallel** connection. In lap-wound armatures, there are the same number of parallel paths through the winding circuit as there are field poles. For each two parallel paths, there is one set of brushes. Lap windings are used where current capacity, rather than voltage, is the main consideration.

Wave windings (see Figure 9.34 (b)) have the beginning of one coil connected to a split-ring commutator segment a distance equal to two poles away from the segment where the end of the coil is connected. The result is to place the coils in a series connection. In wave-wound armatures, there are only **two** parallel paths for current flow, regardless of the number of field poles. Wave windings are used where high voltage, rather than current capacity, is the primary design factor.

9.14 Armature Ratings

Armature windings of electrical machines have two or more parallel paths in an even-numbered arrangement. Each path has a series-connected group of coils, in both lap-wound and wave-wound armatures. The **voltage rating** of the armature circuit is determined by the number of series-connected coils per path. The current rating of the armature circuit is dependent on the current-carrying capacity of the conductors in a series-connected group of coils. Increasing the number of parallel paths causes the current rating to increase. Since an armature circuit has a fixed number of coils. The **current rating** is increased at the expense of voltage rating. This relationship is analogous to a group of batteries connected together. Connecting batteries in series increases voltage, and connecting them in parallel increases current capacity (see Figure 9.35).

Notice in Figure 9.35 that the power rating of each circuit arrangement is the same. Since power equals the product of voltage and current (V × I), the power rating is the same regardless of circuit configuration. This same relationship is true for armature winding connections. The power rating is determined by the voltage and current ratings of individual coils in a path. The only method that can be used to increase the power rating of an armature winding is to design an armature that has more conductors and is larger in physical size. Thus, higher power ratings require armatures that are larger in physical size.

9.14.1 Example of average induced voltage

An **example** of average voltage induced in an armature circuit follows. The voltage induced into the armature circuit of a DC generator is produced by

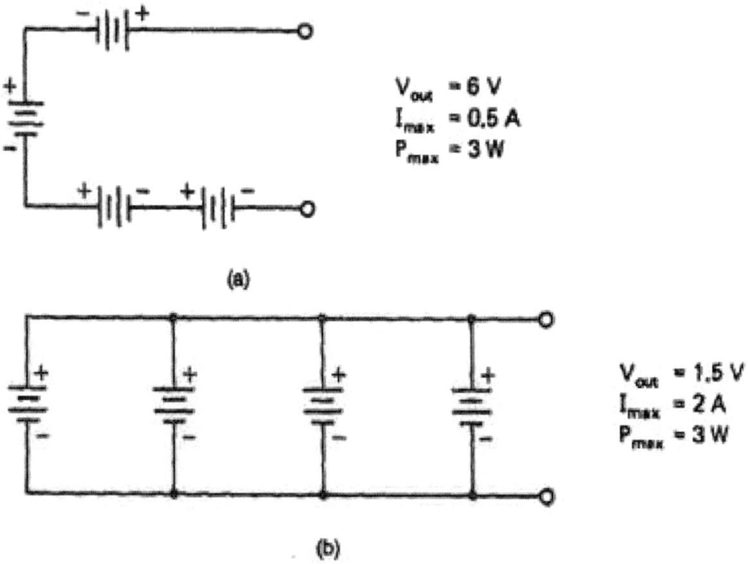

Figure 9.35 Analogy of armature windings and batteries in A – series and B – parallel connections. Each battery is rated at 1.5 V, 0.5 **A.**

many coils connected in series. To determine the voltage developed across the brushes of a DC generator, the average induced voltage per conductor in one-fourth revolution (90°) must be found. Rotation of a coil from 0 to 90° is from minimum magnetic flux linkage to maximum flux linkage. This average voltage for one-fourth revolution is found by the formula:

$$V_{avg} = \frac{\varnothing}{t} \times 10^{-8}\,\text{V}$$

where $\varnothing$ = number of magnetic flux lines per pole
t = time required for 1/4 revolution
10^{-8} = a constant

The average induced voltage per coil (V_{avg}/coil) for a complete revolution (four quarter-revolutions) may be found by applying the equation

$$V_{avg}/\text{coil} = 4\varnothing N_s \times 10^{-8}\,\text{V}$$

where N = number of turns per coil
s = relative speed in revolutions per second (r/s) between the coil and the magnetic field

For example, assume that the number of magnetic flux per pole is 5×10^6 lines, the number of turns per coil is two, and the speed is 100 r/min. The average voltage per coil is

$$
\begin{aligned}
V_{avg}/\text{coil} &= 4 \times (5 \times 10^6 \text{ lines/pole}) \times (2 \text{ turns}) \times (100 \text{ r/min} \times 1/60 \text{ min/s}) \times 10^{-6} \\
&= 4(5 \times 10^6)(2)(100 \times 0.16)(10^{-8}) \\
&= 64 \text{ V}
\end{aligned}
$$

9.14.2 Armature winding design

Most armature windings have pre-formed coils which are inserted into the slots of a laminated metal armature core. Each coil has many turns of small, insulated wire. Ordinarily, they are taped or dipped in lacquer to insulate them from the armature core material. Usually, a coil spans 180 electrical degrees and is either lap-wound or wave-wound. A coil that covers 180° is called a **full-pitch coil,** and one that spans less than 180° is called a **fractional-pitch coil.** Armatures wound with a fractional pitch are called **chorded windings**. Another term associated with fractional-pitch coils is **pitch factor (p).** A coil spanning 120 electrical degrees has a pitch factor of 120°/180° or $p = 0.667$ (66.7%).

Most armature windings are **two-layer** windings having the sides of two coils inserted into one armature core slot. Many series-connected coils having two or more current paths are used to form the armature winding circuit. The two methods used to form the end connections of the individual coils are **lap windings and wave windings**, as discussed previously. Both types of windings have one coil side adjacent to a north field pole, while the other is adjacent to a south pole. Lap windings have coil ends connected to adjacent commutator bars, while wave windings do not.

There are several methods of connecting either lap or wave windings to produce various voltage and current characteristics for DC machines. **Multiplex windings** have several sets of windings connected together in the armature core. In contrast, a **simplex winding** has only one set of coils to form a winding. If there are three winding sets connected together on an armature core, it is referred to as a **triplex winding**. Two sets of windings are called **duplex windings**. The number of winding sets is referred to as the **multiplicity** of the armature. Multiplicity affects the number of parallel paths in an armature winding circuit.

9.14.3 Generated voltage calculation example

To calculate the **generated voltage** of a DC generator, the multiplicity of the windings must be considered. The number of current paths for a lap winding is found as follows: *M x P,* where *M* is the multiplicity (2 for duplex, 3 for triplex, etc.), and *P* is the number of poles. For wave windings: paths = 2 × *M.* The generated voltage of a de generator is dependent on the number of current paths, as shown by the following formula:

$$V_g = \frac{\emptyset ZSP}{60(\text{number of paths})}$$

where Ø = magnetic flux per pole
Z = number of armature conductors (2 × number of turns)
S = speed of rotations in r/min
P = number of poles

For example, assume that a duplex, lap-wound amature is connected in an eight-pole machine, the magnetic flux per pole is 5×10^5 lines, the speed of rotation is 100 r/min, and there are 500 coils, each having two turns of wire. The following method is used to calculate the generated voltage (V_g)

1. Calulate the number of paths.

$$\begin{aligned} \text{paths} &= \text{M} \times \text{P} \\ &= 2 \times 8 \\ &= 16 \end{aligned}$$

2. Calulate the number of

$$\begin{aligned} \text{Z} &= 500 \text{ coils} \times 2 \text{ turns/coil} \times 2 \text{ conductors/turn} \\ &= 2000 \text{ conductors} \end{aligned}$$

3. Calculate V_g.

$$\begin{aligned} V_g &= \frac{\emptyset ZSP}{60(\text{number of paths})} \times 10^{-8} \\ &= \frac{(5 \times 10^5)(2000)(100)(8)}{60(16)} \times 10^{-8} \\ &= 83.33 \text{ V} \end{aligned}$$

9.14.4 Lap-winding design

For **lap windings**, the number of current paths in parallel is determined by the product of multiplicity and the number of poles *(M × P).* The current flow through each armature coil is equal to the total armature current divided by the number of paths. Lap windings require the same number of brushes as poles. The example shown in Figure 9.36 is an illustration of a four--pole lap-wound simplex winding. The circuit connections are simplified to show

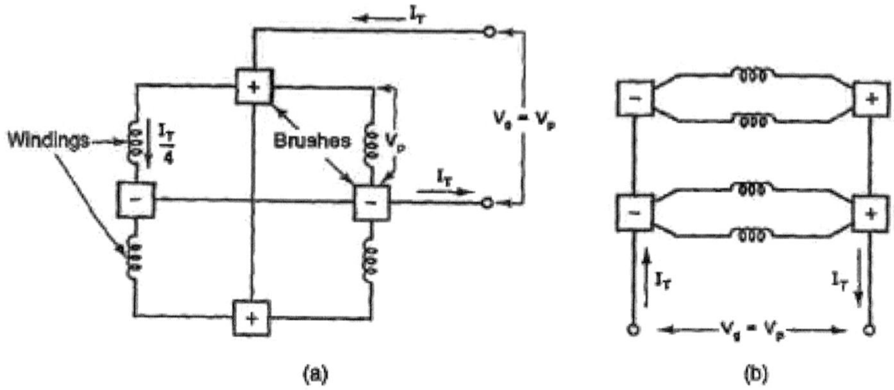

Figure 9.36 Lap-wound dc generator armature circuit: A – simplified armature circuit; B – equivalent circuit.

how current is distributed and the voltage output is developed. There are four poles and four paths. Each path carries one-fourth of the total armature current. A voltage *(Vp)* is generated by each path. The total power *(Pr)* rating of the armature is *Vp* × *Ir*.

9.14.5 Wave-winding design

For **wave windings**, the number of current paths in the armature is determined by the multiplicity of the winding only (2M). Current paths are independent of the number of poles. The current flow through each armature coil is equal to the total armature current divided by the number of paths (like lap windings). Wave windings require only two brushes, regardless of the number of poles. However, in large machines, more brushes may be used to reduce the current flow per brush. The example shown in Figure 9.27 is an illustration of a four-pole simplex wave winding. The circuit connections are simplified to show how current is distributed and voltage output is developed. There are four poles and two paths. The current flow in each path is equal to the total current divided by the number of poles. Since there are two paths connected in series, the generated voltage per path is 2 × *Vp*. The power rating is equal to 2 × *Vp* × *Ir*, as in lap-wound armatures.

9.16.1 Comparison of lap and wave windings.

Ordinarily, lap windings are used for **high-current and high-voltage** applications. It is relatively easy to determine whether an armature is lap-wound

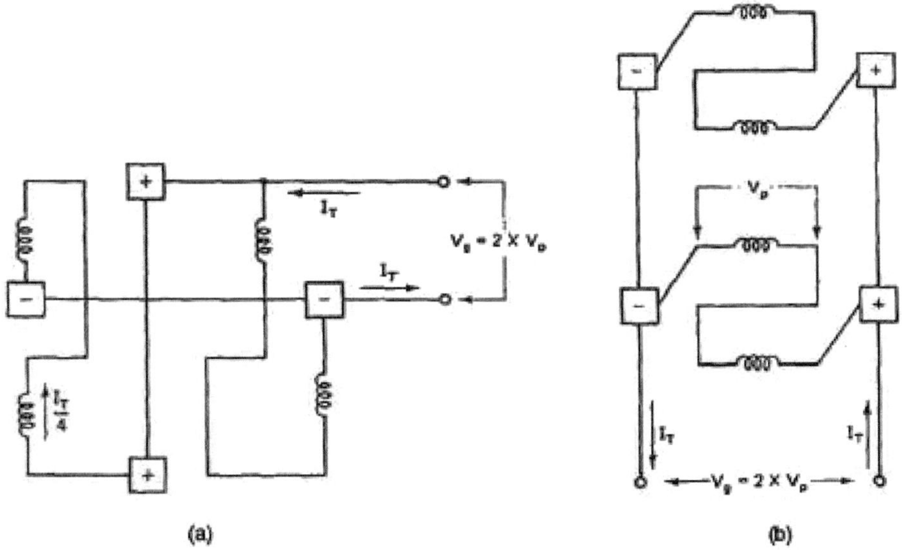

Figure 9.37 Wave-wound DC generator armature circuit: A – simplified armature circuit; B – equivalent circuit.

or wave-wound. The ends of lap-wound coils are connected to adjacent commutator bars, while wave-wound coils are not. Wave windings require only two brushes, regardless of the number of poles. This characteristic of wave windings is important in applications where brush maintenance and replacement are a serious problems.

Review Questions

Answer the following questions.

1. List several applications of electric motors.

2. State the right-hand motor rule.

3. What is the relationship between applied voltage and generated voltage in DC motors and generators?

4. How is the force developed by motor action calculated?

5. How is the average torque of a motor calculated?

6. How is the horsepower of a motor calculated?

7. What is the relationship between load, speed, CEMF, armature current, and torque of a DC motor?

8. How does armature reaction affect DC motor operation?

9. What is the speed regulation of a motor?

10. List the four types of DC motors.

11. How can a DC motor's rotation be reversed?

12. List the major operational characteristics of each major type of DC motors.

13. How is speed calculated for the following types of DC motors?

    a. Series-wound.

    b. Shunt-wound.

    c. Cumulative compound.

14. What method is ordinarily used to start DC motors?

15. What is the main problem of starting DC motors?

16. What is the basic difference between a generator and a motor?

17. Define the following parts of rotating machines: (a) stator, (b) rotor, (c) yoke, (d) field poles, and (e) field windings.

18. What is the difference between concentrated windings and distributive windings?

19. What two basic types of rotors are commonly used in rotating electrical machines?

20. Discuss the construction and uses of slip rings and split-ring commutators.

21. What is the purpose of brushes used with rotating electrical machines?

22. Discuss the construction of the electrical machines shown in Figure 9.23.

23. What is the purpose of the field coils of a rotating electrical machine?

24. Discuss the method used to connect four field coils in series to produce **N-S-N-S** polarities.

25. What is the purpose of using more than two poles in the construction of a rotating electrical machine?

26. Discuss lap-wound and wave-wound armatures.

27. How are the voltage, current, and power ratings of an armature determined?

28. Define the following terms used in winding design: (a) full-pitch coil, (b) fractional-pitch coil, (c) chorded windings, (d) simplex windings, (e) multiplicity, (t) pitch factor, and (g) triplex windings.

29. How is the generated voltage of a DC generator calculated?

30. How is the number of current paths for a lap-wound armature determined? For a wave-wound armature?

Additional Problems

Calculate the correct answers for the following.

1. A DC motor hasapplied, an armature resistance of 1.8 Ω, and an armature current of 32 A. What is the value of the CEMF developed?

2. A de motor has 90 V applied, a CEMF of 33 V, and an armature resistance of 3.2 Ω. What is the value of the armature current?

3. An armature conductor of a de motor is 9 inches long, draws 7.5 A of current, and has a surrounding magnetic field strength of 24,000 lines/in.^{2}. What is the force developed by the conductor?

4. The armature of Problem 3 has 120 conductors, a diameter of 6 inches, and 80% of the armature conductors adjacent to a field winding. Find:
   a. Average force.
   b. Average torque.

5. What is the horsepower of a motor running at 1750 revolutions per minute and having a torque of 25ft.-lb.?

6. If the no-load speed of a motor is 3550 revolutions per minute and the full-load speed is 3450 revolutions per minute, what is its speed regulation?

7. A de generator has 8×10^5 flux lines per pole, two turns per coil, and rotates at a speed of 1000 r/min. What is the average voltage induced per coil?

8. A six-pole DC generator has a triplex lap-wound armature. The magnetic flux per pole is 6 × 10 lines, the speed of rotation is 500 r/min, and there are 250 coils, each having two turns of wire. Calculate the generated voltage.

9. A DC generator has 6×10^4 magnetic flux lines per pole, two turns per coil, and a speed of 2000 r/min. What is the Vg per coil?

10. A duplex wave-wound armature is used in a four-pole de generator. The magnetic field flux per pole is 2.4×10^5 lines, the speed of rotation is 1000 r/min, and there are 200 coils, each having two turns. What is the generated voltage?

11. A DC generator has 5×10^6 flux lines per coil and rotates at a speed of 600 r/min. Calculate the average voltage induced per coil.

12. An eight-pole DC generator has a duplex lap-wound armature. The magnetic flux per pole is 5 × 10 lines, the speed of rotation is 1000 r/min, and there are 200 coils, each having two turns of wire. What is the generated voltage?

13. A DC generator has 5.5×10^5 flux lines per pole, two turns per coil, and rotates at a speed of 1800 r/min. Calculate the average voltage per coil.

14. A triplex lap-wound armature is used in an eight-pole de generator. The magnetic field flux per pole is 3.5×10^5 lines, the speed of rotation is 600 r/min, and there are 100 coils, each having two turns. Calculate the generated voltage.

10

Single-phase Alternating-current Motors

Objectives

Upon completion of this chapter, you will be able to:

- Discuss the operation of a universal motor on a direct current or alternating current power source.
- Describe the basic parts used to construct an induction motor.
- Discuss the principle of operation of an induction motor.
- Discuss the construction characteristics and operation of the following motors: split-phase induction motor, capacitor-start motor, shaded-pole motor, repulsion motor, and synchronous motor.
- Recognize the dimension system used with electric motors.
- Recognize and describe the nameplate data typically found on an electric motor.

Important Terms

capacitor-start single-phase induction motor
capacitor-start, capacitor-run motor
flange mounting
fractional-horsepower motor
integral horsepower motor
permanent capacitor motor
pole shading
repulsion motor
resilient mounting
rigid mounting
rotor frequency
run windings
shaded-pole motor
slip
split-phase AC induction motor
squirrel-cage rotor
start windings
synchronous motor
synchronous speed
two-value capacitor motor
universal motor

DOI: 10.1201/9788743810056-13

Figure 10.1 A – Cutaway of a typical single-phase motor. B – Motor-driven electric hoist. Can lift up to 1 ton. C – Machine tool driven by an electric motor. D – Compressor unit driven by an electric motor. (Gould, Inc., Electric Motors Division; Lift-Tech International, Inc.; Govro-Nelson Co.; Bauer Compressors, Inc.)

Introduction

Single-phase AC motors are common for industrial as well as commercial and residential usage. They operate from a single-phase AC power source. There are three basic types of single-phase AC motors. These types are universal motors, induction motors, and synchronous motors. Some applications of single-phase AC motors are shown in Figure 10.1.

10.1 Universal Motors

Universal motors can be powered by either AC or DC power sources. The universal motor, illustrated in Figure 10.2, is constructed in the same way as

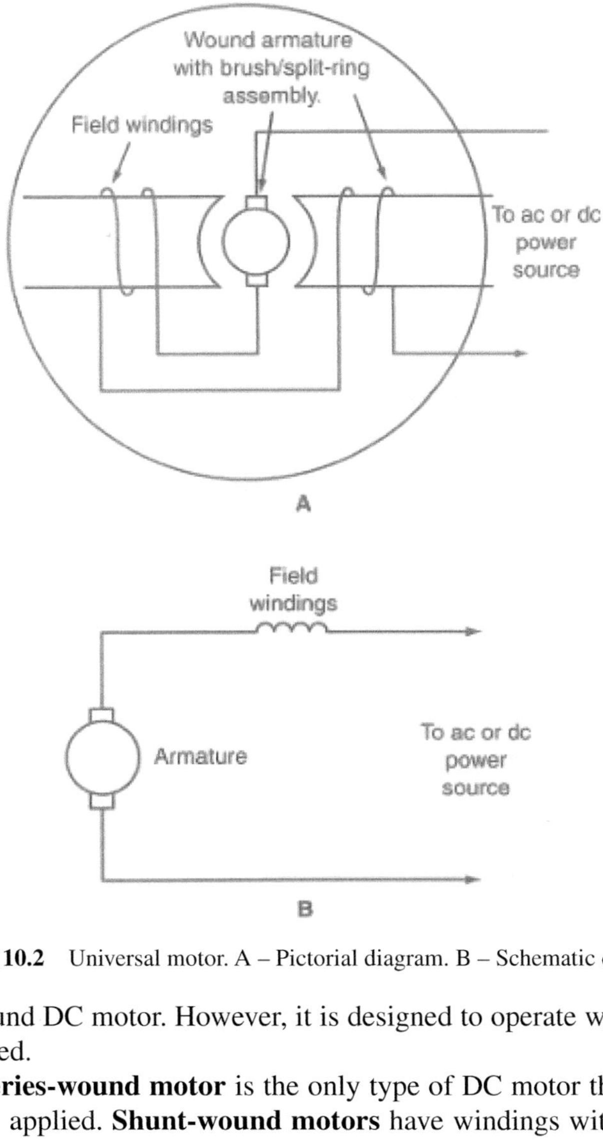

Figure 10.2 Universal motor. A – Pictorial diagram. B – Schematic diagram.

a series-wound DC motor. However, it is designed to operate with either AC or DC applied.

The **series-wound motor** is the only type of DC motor that will operate with AC applied. **Shunt-wound motors** have windings with inductance values that are too high to function with AC applied. Series motors have windings that have low inductance (a few turns of large-diameter wire) and therefore offer a low impedance to the flow of alternating current. The universal motor is one type of AC motor that has concentrated (or salient) field windings. These field windings are similar to those of all DC motors.

The operational principle of the **universal motor** with AC applied involves the instantaneous change of both field and armature polarities. Since the field windings have low inductance, the reversals of field polarity brought about by the changing nature of alternating current also create reversals of current direction through the armature conductors at the proper time intervals. The universal motor operates in the same manner as a **series-wound DC motor** with one exception. The field polarity and the direction of armature current change at a rate of 120 times per second when connected to a 60 Hz AC source.

The **speed** and **torque** characteristics of universal motors are similar to those of DC series-wound motors. Applications of universal motors in industry are mainly for portable tools and small motor-driven equipment. Home applications include mixers, blenders, and electric drills.

10.2 Induction Motors

A **single-phase AC induction motor** is illustrated and shown in Figure 10.3. The coil symbols along the stator represent the field coils of an induction motor. These coils are energized by an AC source. Therefore, their instantaneous polarity changes 120 times per second when 60 Hz AC is applied.

Induction motors have a solid rotor, referred to as a **squirrel-cage rotor**. This type of rotor, shown in Figure 10.4, has large-diameter copper conductors soldered at each end to a circular connecting plate. This plate actually short-circuits the individual conductors of the rotor. When current flows in the stator windings, a current is induced in the rotor. This current is developed due to **transformer action** between the stator and rotor. The stator, which has AC voltage applied, acts as a transformer primary. The rotor is similar to a transformer secondary since current is induced into it.

Since the **stator polarity** changes in step with the applied AC, it develops a rotating magnetic field. The rotor becomes instantaneously polarized due to the induced current flowing through the short-circuited copper conductors. The rotor will, therefore, tend to rotate in step with the revolving magnetic field of the stator. Because of inertia, a rotor must be initially put into motion by some auxiliary method. But if some method of initially starting the rotation is used, the rotor will continue to rotate.

10.2.1 Induction motor speed

The **speed** of an AC induction motor is based on the speed of the rotating magnetic field and the number of stator poles that the motor has. The speed of the rotor will never be as high as the speed of the rotating stator field. If the two speeds were equal, there would be no relative motion between the rotor and stator, and, therefore, no induced rotor current, and no **torque** would be developed.

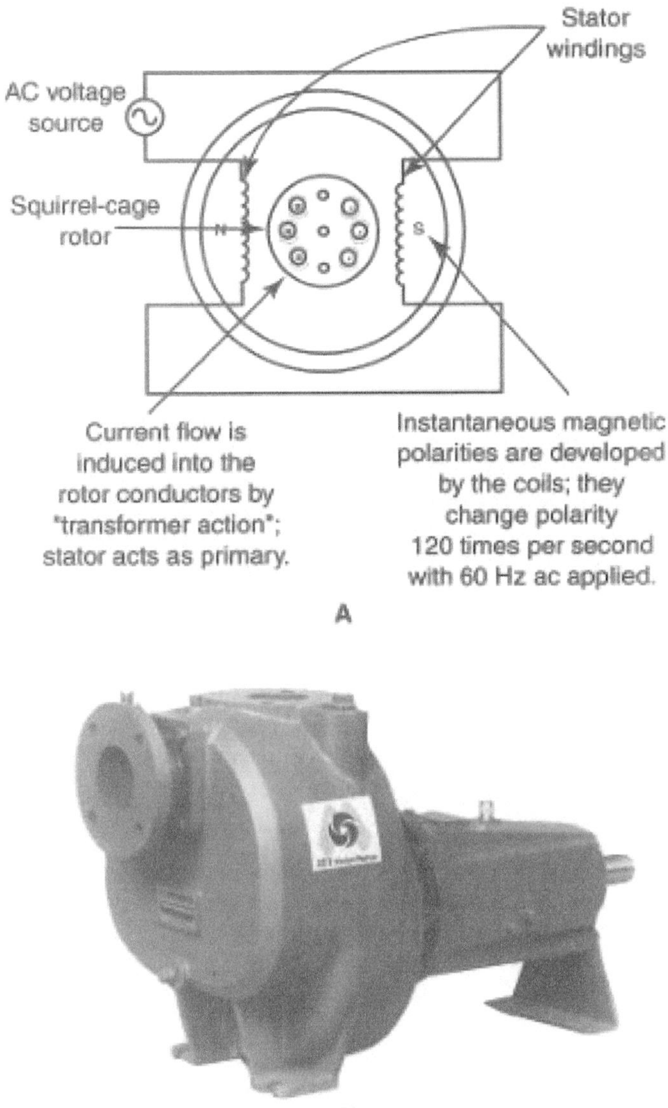

Figure 10.3 Single-phase AC induction motor. A – Schematic diagram. B – Application. Induction motor used to power a centrifugal pump. (ITT A-C Pump Co.)

The **speed** of the rotating stator field can be expressed as:

$$S = \frac{f \times 120}{N}$$

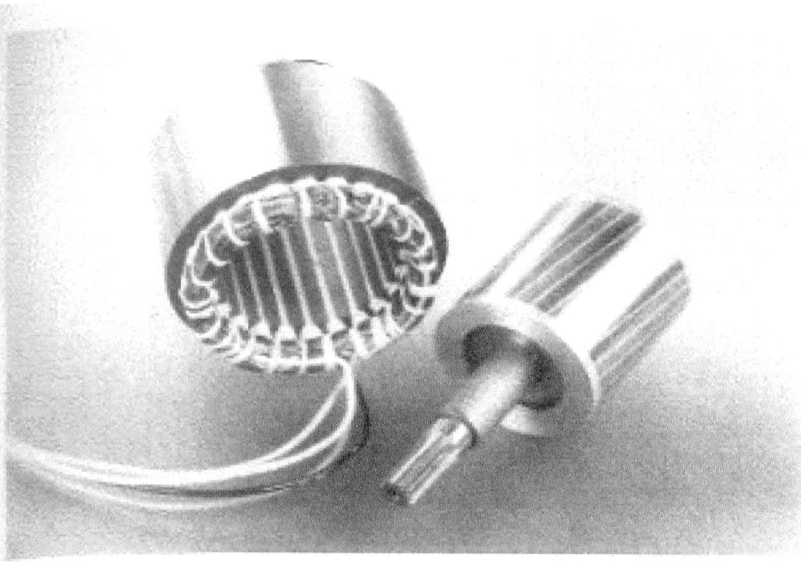

A

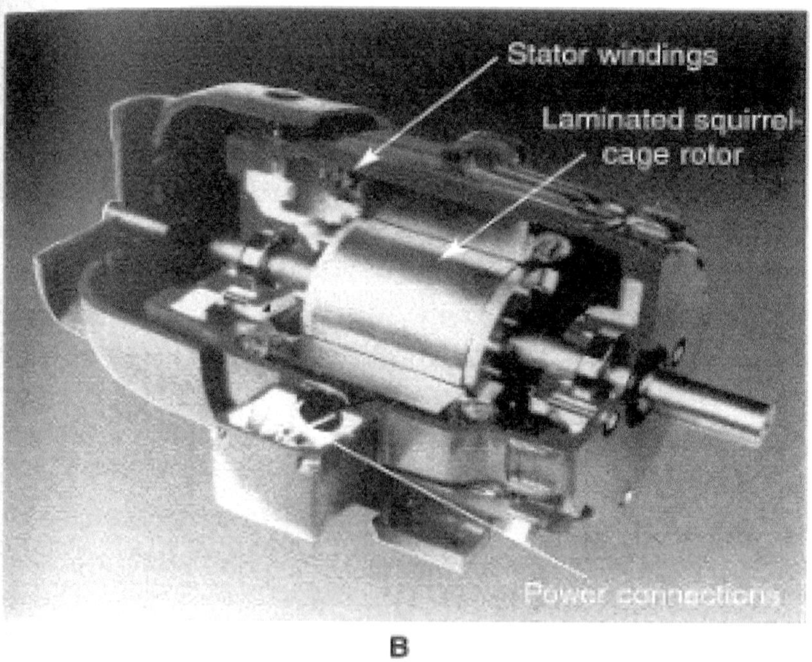

B

Figure 10.4 A – Squirrel-cage rotor and wound stator unit. B – Cutaway showing squirrel-cage rotor. (Eastern Air Devices, Inc.; General Electric Co.)

where S is the speed of the rotating stator field in r/min, f is the frequency of the applied AC voltage in hertz, N is the number of poles of the stator windings, and 120 is a conversion constant.

A two-pole motor operating from a 60 Hz source would have a stator speed of 3600r/min. The stator speed is also referred to as the **synchronous speed** of a motor.

Assume that an induction motor has four stator poles and operates on 60 Hz. Its synchronous (stator) speed is:

$$S = \frac{f \times 120}{N} = \frac{60 \times 120}{4} = 1800\text{r}/\text{min}$$

10.2.2 Slip

The difference between the revolving stator speed of an induction motor and the rotor speed is called **slip**. The rotor speed must lag behind the revolving stator speed in order to develop torque. The more the rotor speed lags behind, the more torque is developed. Slip is expressed mathematically as:

$$\%\ \text{slip} = \frac{S_S - S_R}{S_S} \times 100$$

where S_S is the **synchronous (stator) speed** in r/min. and S_R is the rotor speed in r/min. As the rotor speed approaches the stator speed, the percentage of slip becomes smaller.

Assume that an induction motor has a synchronous speed of 3600r/min and a rotor operating speed of 3400r/min. The **percentage of slip** can be found:

$$\%\ \text{slip} = \frac{S_S - S_R}{S_S} \times 100$$

$$\%\ \text{slip} = \frac{3600 - 3400}{3600} \times 100$$

$$= \frac{200}{3600} \times 100$$

$$= 0.055 \times 100 = 5.5\%$$

10.2.3 Rotor frequency

Another factor, referred to as **rotor frequency**, affects the operational characteristics of an induction motor under load. As the load on the shaft of the motor increases, the rotor speed tends to decrease. The stator speed, however, is unaffected. When a two-pole induction motor connected to a 60 Hz source operates at 10%, the rotor frequency would equal 360r/min (3600r/min × 10%).

Class	Starting characteristics
A	Moderate torque, moderate starting current
B	Moderate torque, low starting current
C	High torque, low starting current
D	High slip
E	Low starting torque, moderate starting current
F	Low starting torque, low starting current

Figure 10.5 Classes of AC induction motors.

Functionally, this means that a revolving stator field passes across a rotor conductor 360 times per minute. Current is induced into a rotor conductor each time the stator field revolves past the conductor. As the slip is increased, more current is induced into the rotor. This causes more torque to be developed.

The **rotor frequency** depends on the amount of slip. It can be expressed as:

$$f_r = f_s \times \text{slip}$$

where f_r is the frequency of the rotor current in r/min. and f_S is the frequency of the stator current in r/min. **Slip** is expressed as a decimal (10% would be 0.1). **Rotor frequency** affects the operational characteristics of induction motors.

10.3 Classes of Single-phase AC Induction Motor

Single-phase AC induction motors are classified according to the method used for starting. Common types of single-phase AC induction motors include:

- Split-phase motors.
- Capacitor motors.
- Shaded-pole motors.
- Repulsion motors.

Figure 10.5 shows the classes of induction motors according to starting characteristics.

10.3.1 Split-phase AC motors

The **split-phase AC induction motor**, illustrated in Figure 10.6, has two sets of stator windings. One set, called the **run windings**, is connected directly across the AC line. The other set, called the **start windings**, is also connected across the AC line. However, the start windings are connected in series with a centrifugal switch mounted on the shaft of the motor. The **centrifugal switch** is in the closed position when the motor is not rotating.

Before discussing the functional principle of the split-phase AC motor, an understanding of how rotation is developed by an AC motor is important. Refer to Figure 10.7. In part A of Figure 10.7, a two-pole stator with single-phase AC applied is shown. For purposes of this discussion, a permanent magnet is placed inside the stator to represent the squirrel-cage rotor of an induction motor.

At time t_0 of the AC sine wave, no stator field is developed. Time interval t_1 will cause a stator field to be produced. Assume a north polarity on the right pole of the stator and a south polarity on the left pole. These polarities will cause the rotor to align itself horizontally based on the laws of magnetic attraction. At time t_2, the stator poles become demagnetized and then begin to magnetize in the opposite direction. At the time interval t_y the stator poles are magnetized in the opposite direction. The rotor will now align itself horizontally in the opposite direction.

This effect will continue at a rate of 120 polarity changes per second if 60 Hz AC is applied to the stator. However, the rotor would not start unless it was positioned initially to be drawn toward a pole piece. Therefore, some starting method must be used for single-phase AC motors, since they are not self-starting.

Assume that we have a **two-phase** situation as shown in part B of Figure 10.7. We now have two sets of stator windings with one phase connected to each set. (Two-phase voltage is, of course, not produced by power companies in the United States. But this example will show the operational principle of a split-phase motor.)

In the two-phase voltage diagram, when one phase is at minimum value, the other is at maximum value. At the time interval t_1, phase 1 is maximum while phase 2 is minimum. Assume that the right stator pole becomes a north polarity and the left stator pole becomes a south polarity. The rotor will align itself horizontally since no polarity is developed in the vertical poles at this time. Now, as we progress to time t_2, phase 1 is minimum and phase 2 is maximum. Assume that the upper stator pole becomes a south polarity and the bottom stator pole becomes a north polarity. The rotor will now align itself vertically by moving 90°. At time t_3, phase 1 becomes maximum in the opposite direction and phase 2 is minimum. This time interval results in a north pole on the left and a south

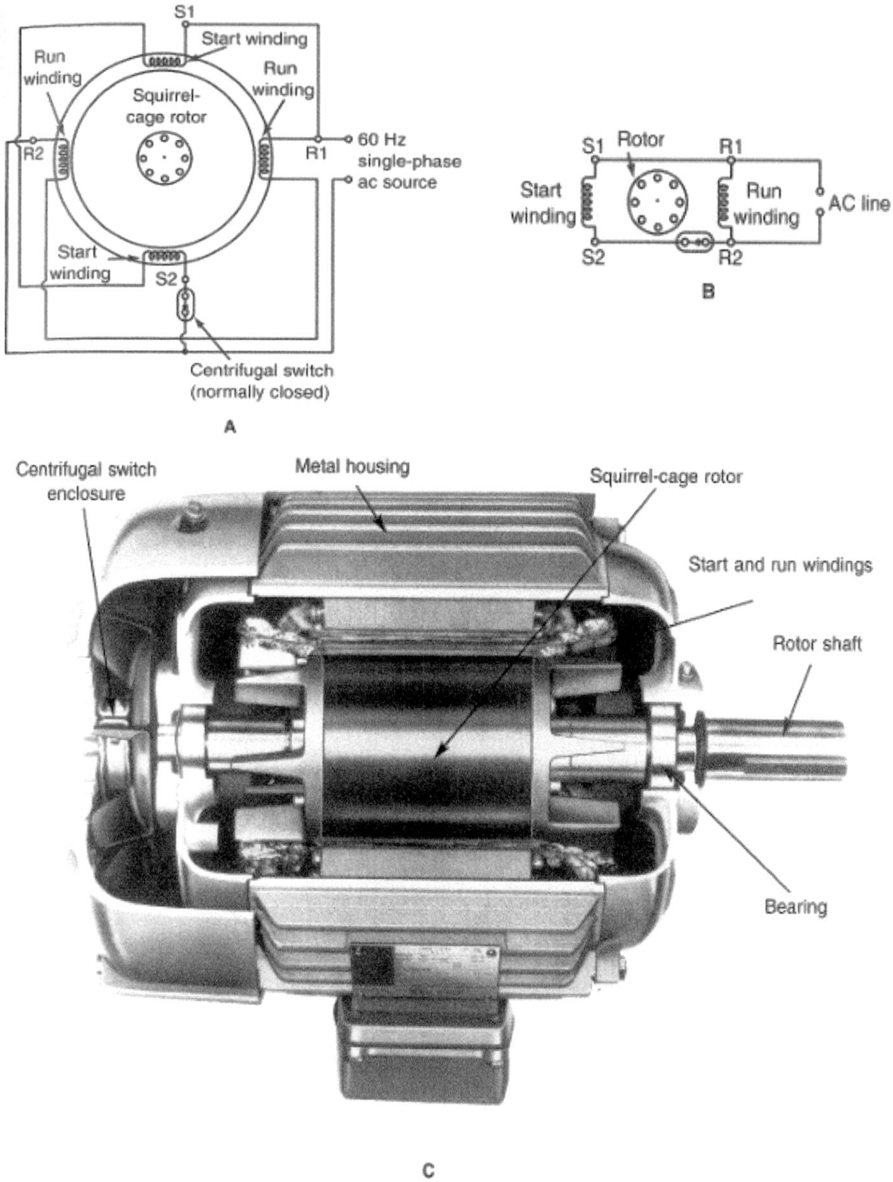

Figure 10.6 Split-phase AC induction motor. A – Pictorial diagram. B – Schematic diagram. C – Cutaway view. (Westinghouse Electric)

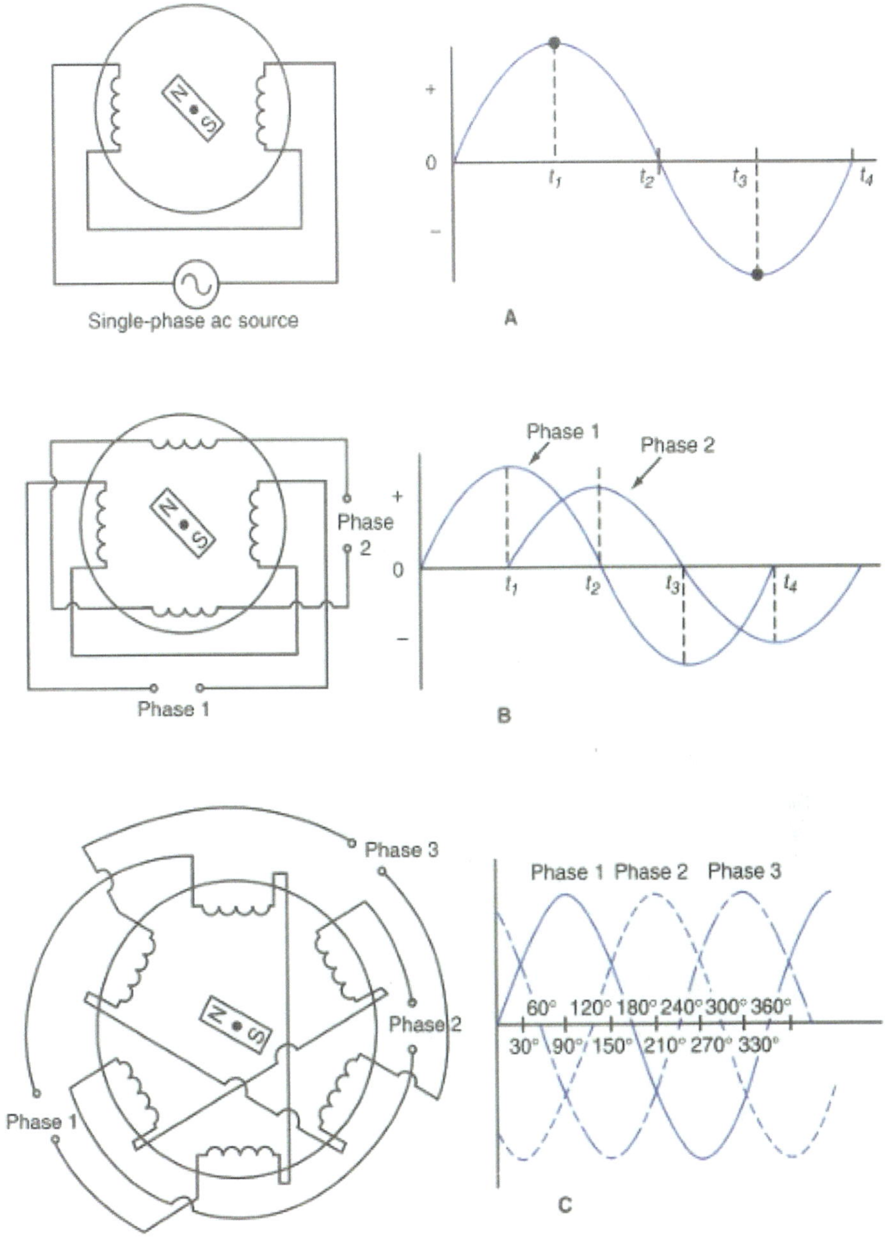

Figure 10.7 Method of rotation development in AC motors. A – Single-phase situation. B – Two-phase situation. C – Three-phase situation.

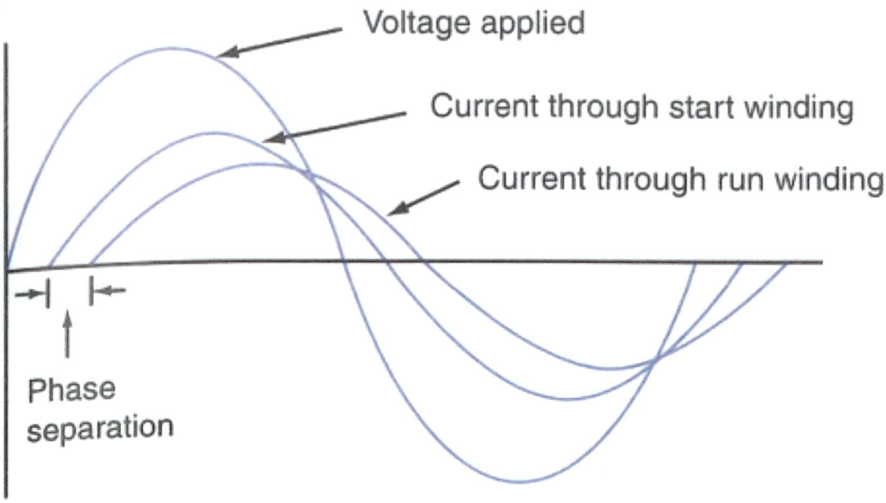

Figure 10.8 Relationship of voltage and current in a split-phase AC induction motor.

pole on the right. Thus, the rotor moves 90° farther in a clockwise direction. This effect will continue as two-phase voltage is applied to the stator poles.

We can see from the two-phase situation that a direction of rotation is established by the relationship of phase 1 and phase 2, and that an AC motor with two-phase voltage applied would be self-starting. The same is true for a three-phase situation, which is illustrated in part C of Figure 10.7. Rotation of the rotor would result due to the 120° phase separation of the three-phase voltage applied to the stator poles. **Three-phase induction motors** are therefore self-starting with no auxiliary starting method required.

Referring back to the split-phase motor of Figure 10.6, it can be seen that the purpose of the two sets of windings is to establish a simulated two-phase condition to start the motor. The single-phase voltage applied to this motor is said to be split into a two-phase current. A rotating magnetic field is created by phase splitting.

The **start winding** of the split-phase motor is made of relatively few turns of small-diameter wire, giving it a high resistance and a low inductance. The **run winding** is wound with many turns of large-diameter wire, causing it to have a lower resistance and a higher inductance. We know that inductance in an AC circuit causes the current to lag the applied AC voltage. The more inductance present, the greater is the lag in current.

When single-phase AC is applied to the stator of a **split-phase induction motor**, the situation illustrated in Figure 10.8 results. Notice that the current

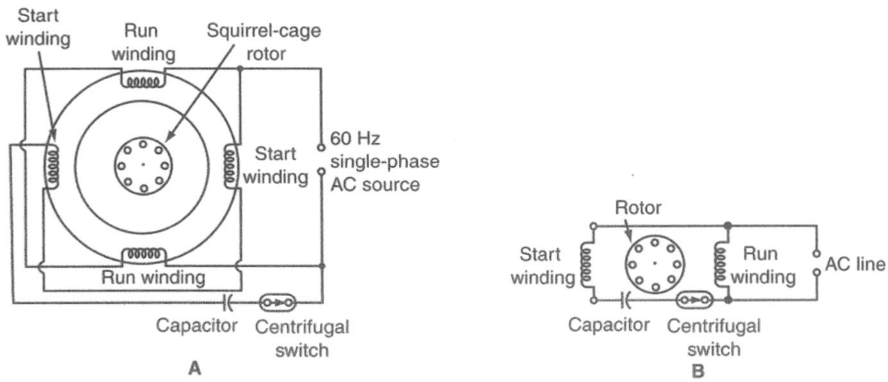

Figure 10.9 Capacitor-start single-phase AC induction motor. A – Pictorial diagram. B – Schematic diagram.

in the start winding lags the applied voltage due to its inductance. However, the current in the run winding lags by a greater amount due to its higher inductance. The phase separation of the currents in the start and run windings creates a two-phase situation. The phase displacement, however, is usually around 30° or less. This gives the motor a low starting torque as this phase separation does not nearly approach the 90° separation of two-phase voltage.

When the **split-phase AC induction motor** reaches about 80% of its normal operating speed, the centrifugal switch will open the start winding circuit, since it is no longer needed. The removal of the start winding minimizes energy losses in the machine and prevents the winding from overheating. When the motor is turned off and its speed reduces, the centrifugal switch closes to connect the start winding back into the circuit. Split-phase motors are fairly inexpensive compared to other types of single-phase motors. They are used where low torque is required to drive mechanical loads such as small machinery.

10.3.2 Capacitor motors

Capacitor motors are an improvement over the split-phase AC motor. **Capacitor-start single-phase induction motors** are diagrammed in Figure 10.9. Notice that this diagram is the same as the split-phase motor except that a capacitor is placed in series with the start winding. The purpose of the capacitor is to cause the current in the start winding to lead (rather than lag) the applied voltage. Figure 10.10 shows several typical capacitors used in capacitor-start motors.

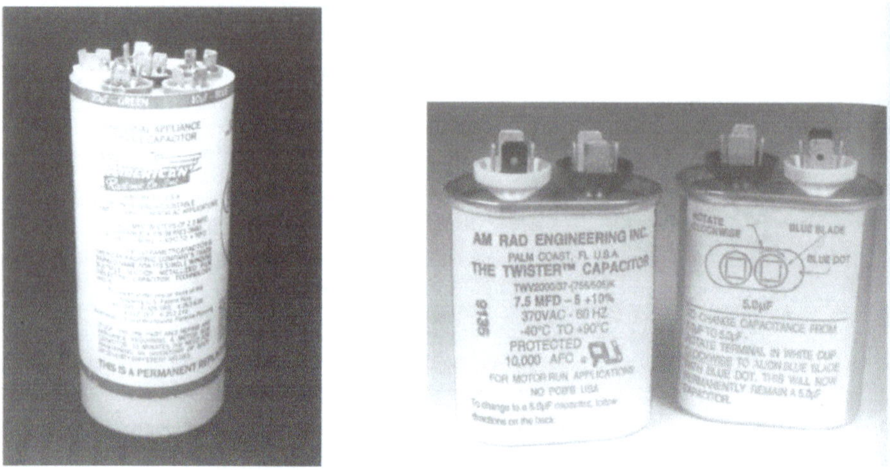

A B

Figure 10.10 Capacitors used with capacitor-start single-phase AC induction motors. (American Radionic Co.)

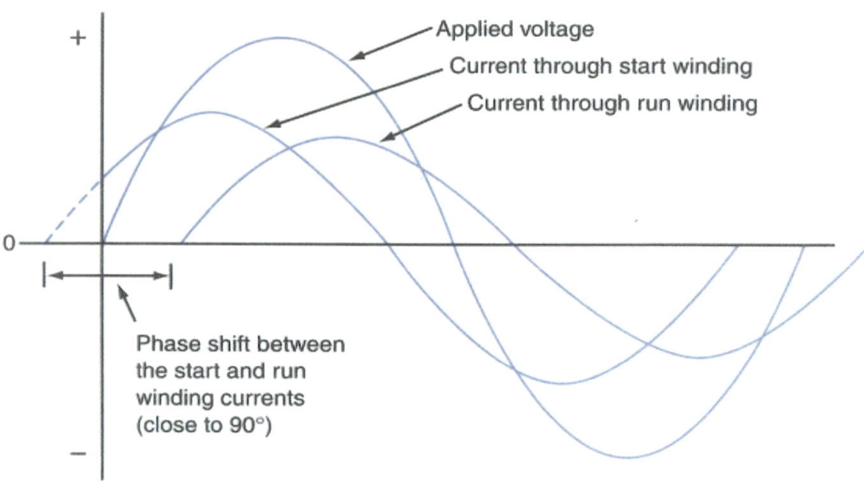

Figure 10.11 Relationship of voltage and current in a capacitor-start single-phase AC induction motor.

This situation is illustrated in Figure 10.11. The current in the **start winding** now **leads** the applied voltage due to the high value of capacitance in the circuit. Since the **run winding** is highly inductive, the current through it **lags** the applied voltage. Notice that the amount of phase separation now

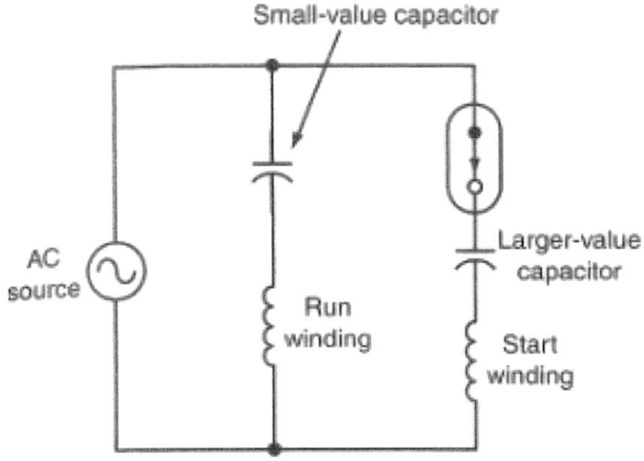

Figure 10.12 Schematic of a capacitor-start, capacitor-run AC induction motor.

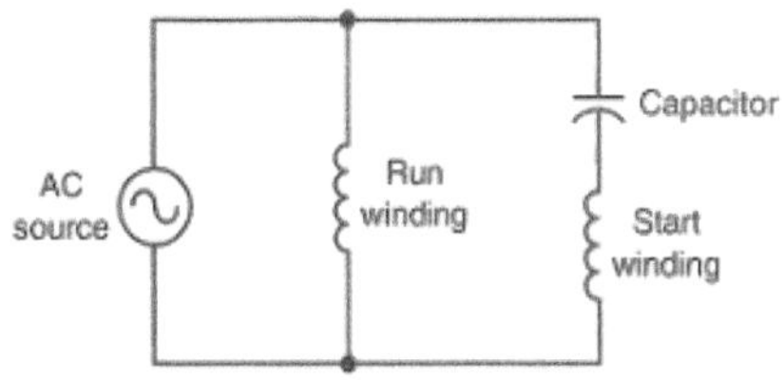

Figure 10.13 Permanent capacitor AC induction motor schematic diagram.

approaches 90°, or an actual twophase situation. The starting torque produced by a capacitor-start induction motor is much greater than that for the split-phase motor. Thus, this type of motor can be used for applications requiring greater initial torque. However, these motors are somewhat more expensive than split-phase AC motors. Most capacitor motors, as well as split-phase motors, are used in **fractional-horsepower** sizes (less than one hp).

Another type of capacitor motor is called a **capacitor-start**, capacitor-run, or **two-value capacitor motor**. Its circuit is shown in Figure 10.12. This motor employs two capacitors. One of higher value is in series with the start winding and centrifugal switch. The other, of lower value, is in series with the **start winding** and remains in the circuit during operation. The larger capacitor is used only to increase starting torque. It is removed from the circuit by the centrifugal switch during normal operation. The smaller capacitor and the entire start winding are part of the operational circuit of the motor. The smaller capacitor helps to produce a more constant running torque as well as quieter operation and improved power factor.

Yet another type of capacitor motor is called a **permanent capacitor motor**, diagrammed in Figure 10.13. This motor has no centrifugal switch, so its capacitor is permanently connected into the circuit. These motors

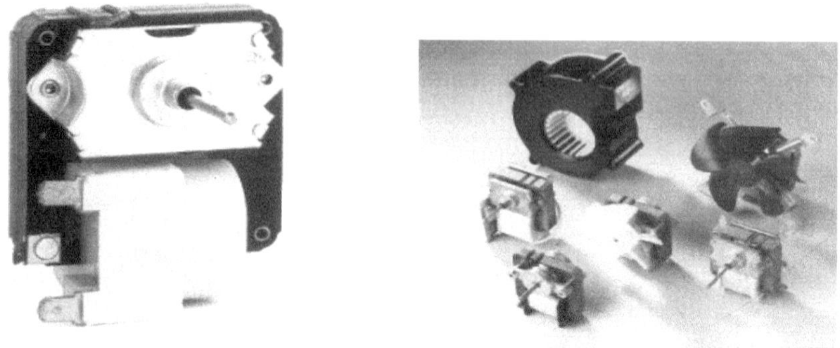

A B

Figure 10.14 Shaded-pole motors. A – The construction is very simple. B – Applications include fans and blowers. (Olsen Machine and Tool Co.)

are used only for very low torque requirements and are made in small, fractional-horsepower sizes.

Both split-phase motors and capacitor motors can have their direction of **rotation** reversed easily. Simply change the relationship of the start winding and run winding. By reversing either the start winding connections or the run winding connections (but not both), the rotational direction will be reversed.

10.3.3 Shaded-pole induction motors

Another method of producing torque by a simulated two-phase method is called **pole shading**. **Shaded-pole motors** are used for very low torque applications (see Figure 10.14). They are low-cost, rugged, and reliable motors that are made only in low-horsepower ratings, usually from 1/20 to 1/3hp.

The operational principle of a shaded-pole motor is shown in Figure 10.15. The single-phase alternation shown is for discussion purposes only. Note the shading coils in the upper right and lower left of the two poles. The shaded pole is encircled by a heavy copper conductor and is actually a part of the main field pole. This closed-loop conductor will have current induced into it when AC is applied to the field.

When AC voltage is applied to the **stator windings**, the magnetic flux in the main poles induces a voltage into the shaded sections of the poles. Since the shaded section acts like a transformer secondary, its voltage is out of phase with the main field voltage, as shown on the waveform diagram. The dashed lines represent induced voltage into the shaded section of the field poles. Note the four time intervals shown in sequence at the bottom of Figure 10.15. The voltage induced in the shaded pole from the main pole

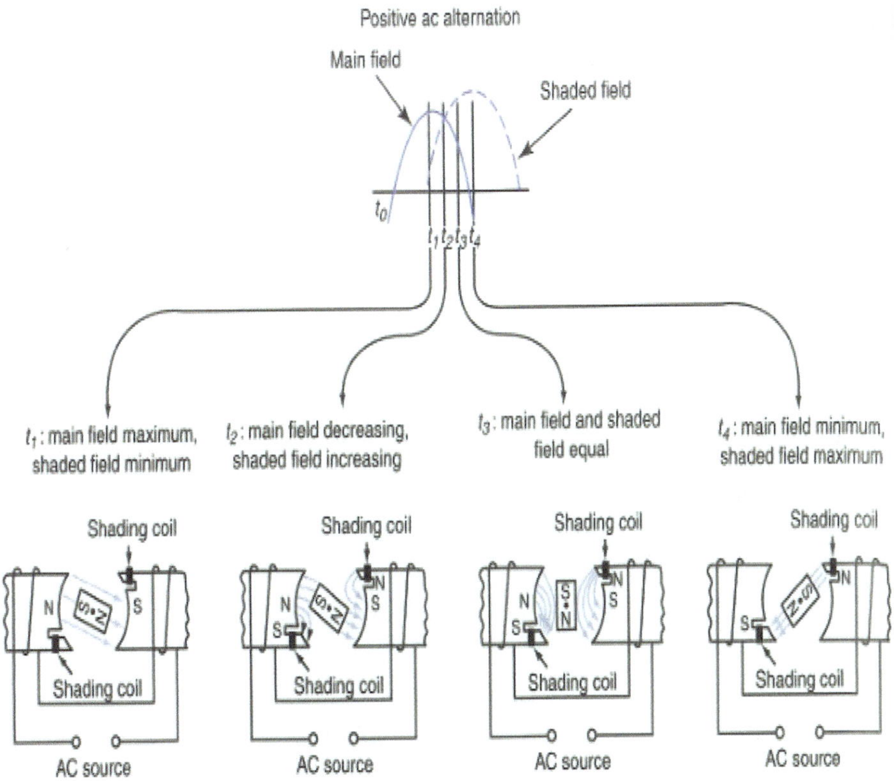

Figure 10.15 Operational principle of a shaded-pole AC induction motor.

field causes the movement of the rotor to continue. Study this illustration to understand the operational principle of the shaded-pole AC induction motor.

The **shaded-pole motor** is inexpensive since it uses a squirrel-cage rotor and has no auxiliary starting winding or centrifugal mechanism. Applications are limited mainly to small fans and blowers.

10.3.4 Repulsion motors

Another type of AC induction motor is the **repulsion motor**. This motor was used for many applications, but it is now being replaced by other types of single-phase motors. The principle of operation of the repulsion motor is an interesting contrast to other induction motors. Motors of this type should be studied in relation to their operational principles.

Figure 10.16 shows the operational principle of a **repulsion motor**. This motor has a wound rotor that functions similarly to a squirrel-cage rotor.

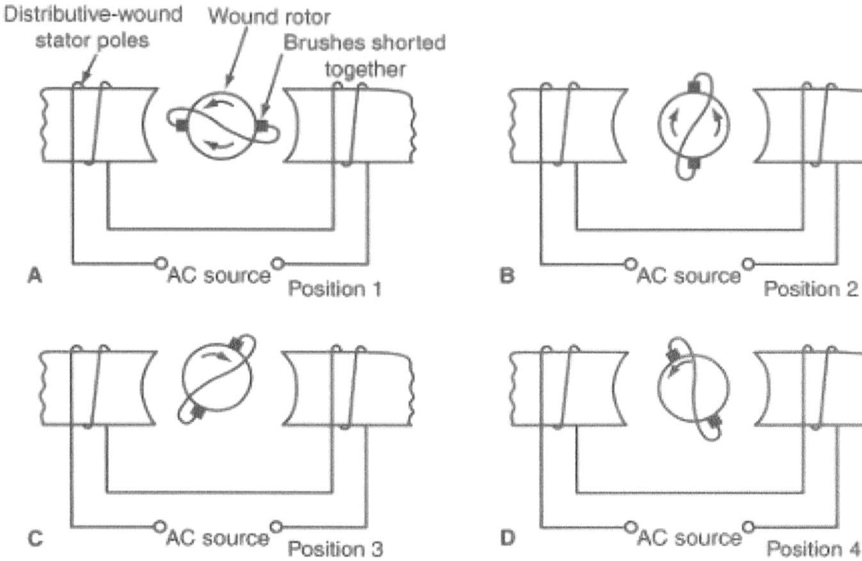

Figure 10.16 Operational principle of a repulsion motor.

It also has a commutator/brush assembly. The brushes are shorted together to produce an effect similar to the shorted conductors of a squirrel-cage rotor. The position of the brush axis determines the amount of torque developed and the direction of rotation of the repulsion motor.

In position 1 of Figure 10.16, the brush axis is aligned horizontally with the stator poles. Equal and opposite currents are now induced into both halves of the rotor. Thus, no torque is developed with the brushes in this position.

In position 2, the brushes are placed at a 90° angle to the stator field poles. The voltages induced into the rotor again counteract one another and no torque is developed.

In position 3, the brush axis is shifted about 60° from the stator poles. The current in the armature now causes a magnetic field to exist around the rotor. The rotor field will now follow the revolving stator field in a clockwise direction.

As might be expected, when we shift the brush axis in the opposite direction, as shown in position 4, rotation reversal will result. Thus, magnetic repulsion between the stator field and the induced rotor field causes the rotor to turn in the direction of the brush shift.

Figure 10.17 Single-phase AC synchronous motors. (Burgess SAIA, Inc.)

Repulsion-start induction motors and some similar types of modified repulsion motors have very high starting torque. Their speed can be varied by varying the position of the brush axis. However, the mechanical problems inherent with this type of motor have caused it to become obsolete.

10.4 Single-phase Synchronous Motors

It is often desirable, in timing or clock applications, to use a constant-speed drive motor. Such a motor, which operates from a single-phase AC line, is called a **synchronous motor**. Single-phase synchronous motors, such as those shown in Figure 10.17, have stator windings that are connected across the AC line. The rotor, however, is made of permanent-magnetic material. Some external method of starting must be used in order to initiate rotation. Once the rotor is started, it will rotate in synchronism with the revolving stator field since it does not rely on the induction principle. The speeds of synchronous motors are based on the speed formula. For 60-HZ operation, the following synchronous speeds would be obtained:

Two-pole	3600r/min.
Four-pole	1800r/min.
Six-pole	1200r/min.
Eight-pole	900r/min.
Ten-pole	720r/min.
Twelve-pole	600r/min.

Single-phase synchronous motors are used only for very low torque applications. Such applications include clocks and timing devices requiring constant speed.

10.5 Motor Dimensions

There are many different sizes of electric motors. **Fractional-horsepower motors** are those less than one hp in size, while **integral-horsepower motors** are larger than one hp. When specifying types of electric motors, one must know the dimensions of a motor to be used for a specific application. Dimensions are usually listed in a manufacturer's catalog. The dimensions of a typical motor are shown in Figure 10.18. For several applications, the motor method is also important.

The major types of motor mountings are resilient, rigid, and flange methods. **Resilient mountings provide** a vibration-reducing insulated contact between the motor and the mounting surface. **Rigid** and **flange mounting methods** are used to attach a motor directly to a machine. These methods are illustrated in Figure 10.19.

10.6 Motor Nameplate Data

It is important to be able to interpret the data on a **motor nameplate**. The information contained on a typical nameplate is summarized as follows:

Manufacturing company*:* the company that built the motor.

Motor type*:* A specific type of motor (e.g., split-phase ac, universal, three-phase induction, etc.).

Identification number*:* Number assigned by the manufacturer.

Model number*:* Number assigned by the manufacturer.

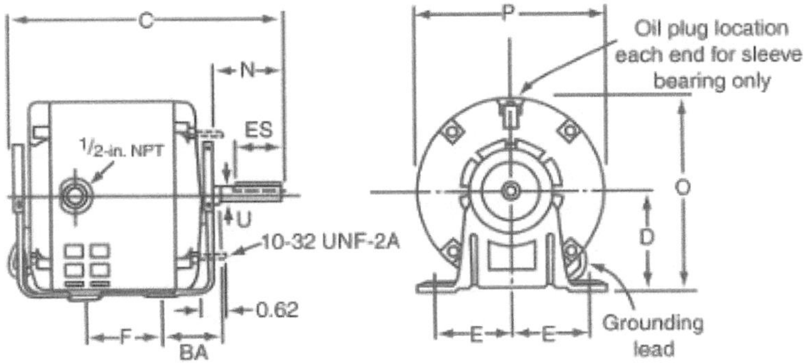

A

Various motor types available

Dimensions[a] (in.)									
C	D	E	F	N	O	P	U	BA	ES
9.09	3.00	2.12	2.75	1.59	5.84	5.66	0.50	2.50	1.25 flat
9.34	3.00	2.12	2.75	1.59	5.84	5.66	0.50	2.50	1.25 flat
10.47	3.50	2.44	3.00	1.97	6.33	5.66	0.62	2.75	0.19 x 0.19 x 1.38 key
11.22	3.50	2.44	3.00	1.97	6.33	5.66	0.62	2.75	0.19 x 0.19 x 1.38 key
10.09	3.00	2.12	2.75	1.59	5.83	5.66	0.50	2.50	1.25 flat
10.47	3.50	2.44	3.00	1.97	6.62	6.25	0.62	2.75	0.19 x 0.19 x 1.38 key
9.47	3.50	2.44	3.00	1.97	6.33	5.66	0.50[b]	2.75	0.19 x 0.19 x 1.38 key
9.72	3.50	2.44	3.00	1.97	6.33	5.66	0.50[b]	2.75	0.19 x 0.19 x 1.38 key
10.47	3.50	2.44	3.00	1.97	6.33	5.66	0.50[b]	2.75	0.19 x 0.19 x 1.38 key
9.47	3.00	2.12	2.75	1.97	5.83	5.66	0.50	2.50	1.62 flat
9.72	3.00	2.12	2.75	1.97	5.83	5.66	0.50	2.50	1.62 flat
10.47	3.00	2.12	2.75	1.97	5.83	5.66	0.50	2.50	1.62 flat

[a] Dimensions from side and end views.
[b] Motor is supplied with 0.62 adapter sleeve.

B

Figure 10.18 Typical AC motor dimensions. A – Resilient mounting: split-phase AC motor side and end views with dimensions lettered. B – Sample dimensions for the motor. (Marathon Electric)

Frame type*:* Frame size defined by NEMA.

Number of phases (AC): single phase or three phase.

Horsepower*:* The amount produced at rated speed.

Cycles (AC): Frequency the motor should be used with (usually 60 Hz).

Speed (r/min)*:* The amount at rated horsepower, voltage, and frequency.

Voltage rating*:* Operating voltage of motor.

Current rating (amperes): Current drawn at rated load, voltage, and frequency.

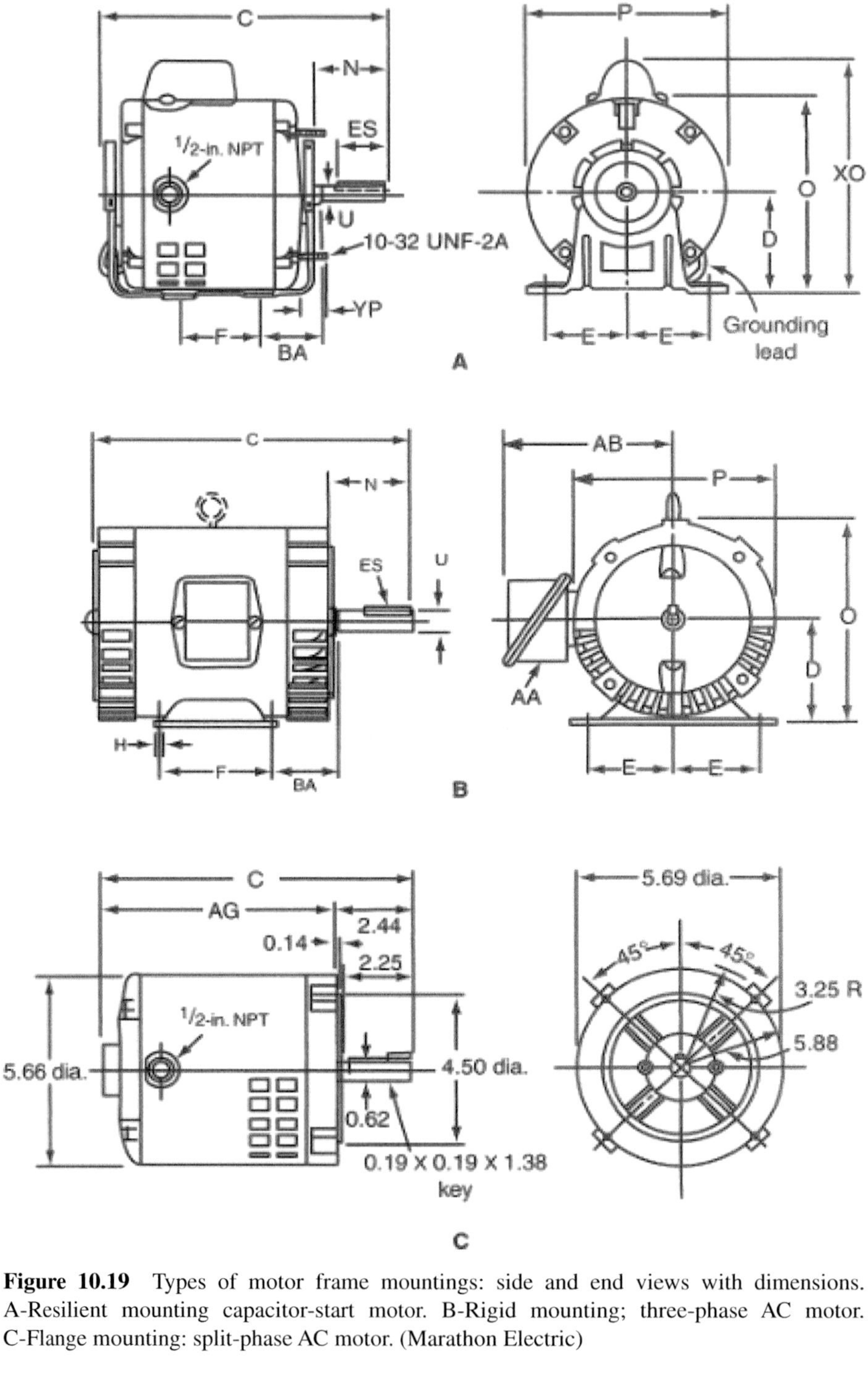

Figure 10.19 Types of motor frame mountings: side and end views with dimensions. A-Resilient mounting capacitor-start motor. B-Rigid mounting; three-phase AC motor. C-Flange mounting: split-phase AC motor. (Marathon Electric)

Thermal protection: Indicates the type of overload protection used.

Temperature rating (°C): Amount of temperature that the motor will rise over ambient temperature when operated.

Time rating: Time the motor can be operated without overheating (usually continuous).

Amperage: Current drawn at rated load, voltage, and frequency.

Review Questions

Answer the following questions.

1. What are the three basic types of single-phase AC motors?

2. What is a universal motor?

3. List four common applications of universal motors.

4. How is the speed of a single-phase AC induction motor determined?

5. What is meant by the term slip?

6. How is the rotor frequency of a single-phase AC induction motor determined?

7. What are some common types of single-phase AC induction motors?

8. What is the purpose of a centrifugal switch in a single-phase AC induction motor?

9. What are the classes of induction motors according to starting characteristics?

10. Describe a capacitor-start, capacitor-run motor.

11. Describe a permanent capacitor motor.

12. How is the direction of rotation of each of the following single-phase motors reversed?

    a. Split-phase motor.

    b. Capacitor-start motor.

    c. Shaded-pole motor.

    d. Repulsion motor.

    e. Synchronous motor.

13. What is a fractional-horsepower motor?

14. What is an integral-horsepower motor?

15. What are the three major types of motor mountings?

Additional Problems

Answer the following questions.

1. Calculate the synchronous speeds of the following single-phase AC motors.
   a. Two-pole, 50 Hz.
   b. Four-pole, 60 Hz.
   c. Eight-pole, 60 Hz.
   d. Six-pole, 50 Hz.

2. What is the percent slip of a single-phase AC induction motor with a synchronous speed of 3600 r/min and a rotor speed of 3450 r/min ?

3. What is the rotor frequency of the motor in the previous problem?

4. What is the percent slip of a four-pole AC induction motor operating at a speed of 1725 r/min on a 60 Hz power line?

5. What is the rotor frequency of the motor in the previous problem?

6. What is the efficiency of a 240 V single-phase AC induction motor that is rated at 3hp and converts 2650 W of true power?

7. What is the efficiency of a 1/2-hp 120 V single-phase AC induction motor that converts 500 W of true power?

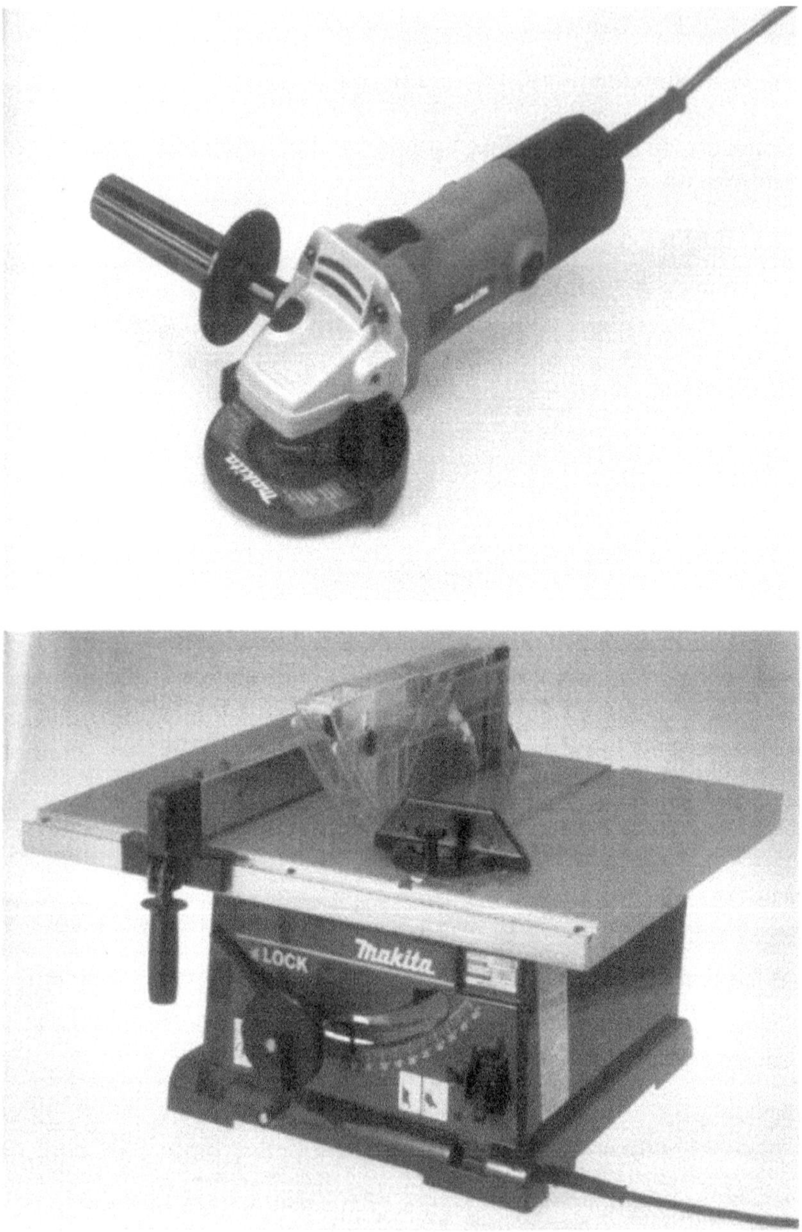

Most household power tools are powered with single-phase AC motors. (Makita)

11

Three-phase Alternating-current Motors

Objectives

Upon completion of this chapter, you will be able to:

- Discuss the construction and operation of a three-phase induction motor.
- Describe the special construction features of a three-phase synchronous motor.
- Compare the characteristics of three-phase induction motors and three-phase synchronous motors.
- Describe two methods of starting a three-phase synchronous motor.
- Explain power factor in AC systems.
- List and explain two power factor correction methods for AC systems.
- Describe the construction and characteristics of wound-rotor induction motors.

Important Terms

damper windings	three-phase induction motor
normal excitation	three-phase synchronous motor
overexcitation	under excitation
power factor-corrective capacitors	V-curve
synchronous capacitor	wound-rotor induction motor

DOI: 10.1201/9788743810056-14

Introduction

Three-phase AC motors are often called the "workhorses of industry." Most motors used in industry are operated from three-phase power sources. There are three basic types of three-phase motors:

- Three-phase induction motors.
- Three-phase synchronous motors.
- Three-phase wound-rotor motors.

11.1 Three-phase Induction Motors

The **three-phase induction motor** is illustrated in part A of Figure 11.1. A three-phase AC waveform is shown at the top of part B of Figure 11.1. These AC waveforms represent the input voltage to phases A, B, and C of the stator of a three-phase induction motor. At reference point t_2 on the diagram, phase A has a voltage value of zero, phase B is approaching maximum negative value, and phase C is approaching maximum positive value. The resulting polarities of the stator coils are shown in the diagram below point t_v. Phase B has a north polarity at area B2 and a south polarity at Bl. Phase C has a north polarity at area CI, and a south polarity at C2. Since the magnetic flux associated with each phase travels from north to south through the area of least reluctance, the resulting magnetic flux develops as shown in the diagram. The arrow represents the position of a rotor placed within the machine. The arrowhead represents the north polarity of the rotor.

At reference point t_2 of the illustration, phase C has a voltage value of zero, so minimum magnetic field flux will develop around the coil of phase C. Phase A is approaching maximum positive value and phase B is approaching maximum negative value. Phase A has a north polarity at Al and a south polarity at A2. Phase B has a south polarity at B, and a north polarity at B2. The resulting magnetic flux of the stator causes the rotor to shift a total of 60°, which is the amount of physical spacing between coils.

At reference point t_3, phase B is at the minimum value, phase A is approaching the maximum positive, and phase C is approaching the maximum negative. Phase B has a minimum magnetic flux. Phase A has a north polarity at Al and a south polarity at A2. Phase C has a north polarity at C2 and a south polarity at CI. The resulting magnetic flux causes the rotor to shift another 60°.

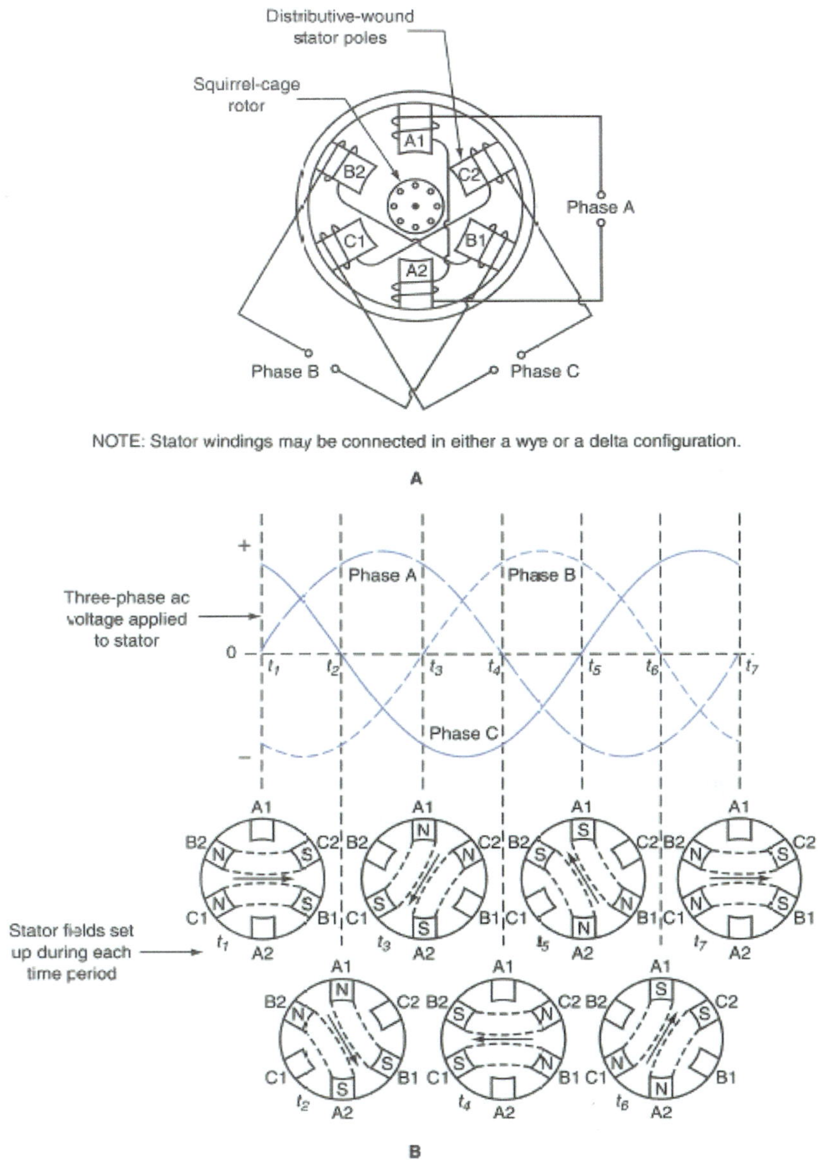

Figure 11.1 Three-phase AC induction motor operational principle. A – Pictorial diagram. B – Three-phase waveforms and the resulting stator fields.

At reference point t_4, the sequence begins with opposite polarities developing around the stator coils. Notice that the polarities of Bl, B2, CI, and C2 are opposite from those of reference point t_1. The resulting polarities cause the rotor to position itself 180° from the position of reference point t_v.

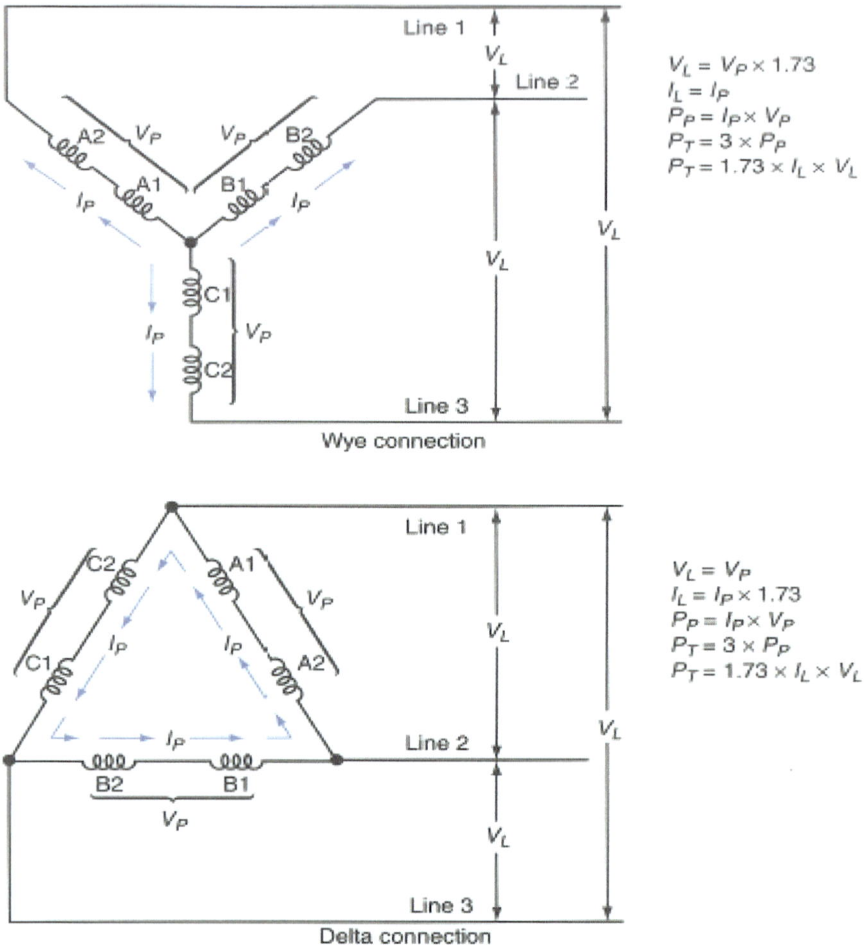

Figure 11.2 Diagrams of the windings connected in wye and delta arrangements.

The sequence continues as the three-phase induction motor continues to operate. The actual rotor of this type of motor is a squirrel-cage type that relies on the individual principle to sustain operation. The application of three-phase voltage to the motor provides phase separation that allows this type of motor to be self-starting.

Figure 11.2 shows the voltage, current, and power relationships of the stator coils of a three-phase induction motor. These are the same for any three-phase system. Notice that the voltage across each phase in the wye connection is less than in the delta connection since in the wye connection $V_p - V_L \div 1.73$. This is important for motor starting applications.

The construction of this motor is very simple. It has only a distributive-wound stator, which is connected in either a wye or a delta configuration, and a squirrel-cage rotor. See Figure 11.3. Since three-phase voltage is applied to the stator, phase separation is already established. No external starting mechanisms are needed. Three-phase induction motors are made in a variety of integral-horsepower sizes and have good starting and running torque characteristics.

The direction of rotation of a three-phase motor of any type can be changed very easily. If any two power lines coming into the stator windings are reversed, the direction of rotation of the shaft will change. **Three-phase induction motors** are used for many industrial applications. These applications include motor use as mechanical energy sources for machine tools, pumps, elevators, hoists, and conveyors.

11.2 Three-phase Synchronous Motors

Three-phase synchronous motors are unique and very specialized motors. They are considered a constant-speed motor, and they can be used to correct the power factor of three-phase systems. Figure 11.4 shows a pictorial diagram of a three-phase AC synchronous motor. Synchronous motors are usually employed in very large horsepower ratings.

The three-phase synchronous motor is physically constructed like a **three-phase alternator**. Direct current is applied to the rotor to produce a rotating electromagnetic field. The stator windings are connected in either a wye or a delta configuration. The only difference is that three-phase AC power is applied to the synchronous motor, while three-phase power is extracted from an alternator. Thus, the motor acts as an electrical load while the alternator functions as a source of three-phase power. This relationship should be kept in mind during the following discussion.

The **three-phase synchronous motor** differs from the three-phase induction motor in that the rotor is wound and is connected through a slip-ring brush assembly to a DC power source. Three-phase synchronous motors, in their pure form, have no starting torque. Some external means must be used to start the motor initially. Synchronous motors are constructed so that they will rotate at the same speed as the revolving stator field. At synchronous speed, rotor speed equals slip. Thus, the **speed** of a synchronous motor can be determined by using the following formula:

$$s = \frac{f \times 120}{N/3}$$

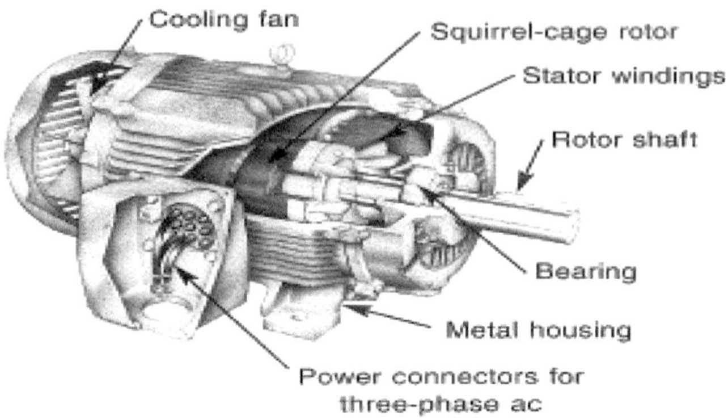

A

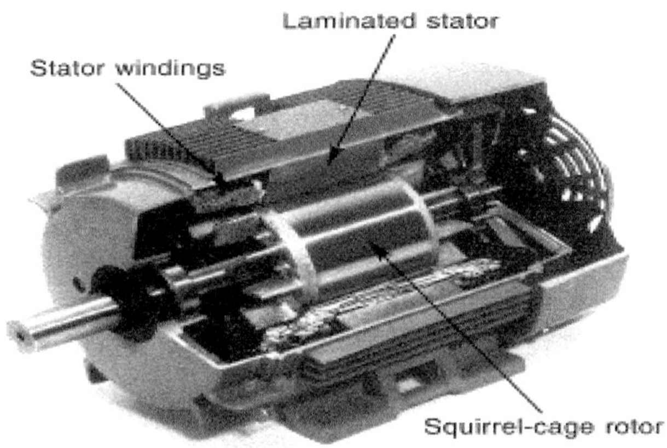

B

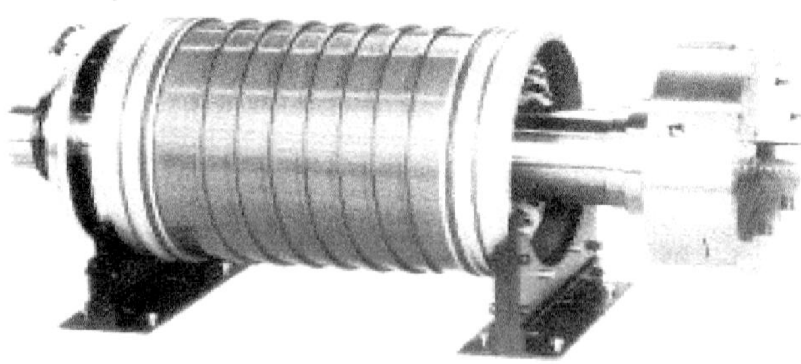

Figure 11.3 Three-phase AC induction motors. A and B – Cutaway views of typical three-phase motors. C – Frameless AC induction motor with a unique design used for lightweight applications where a frame is not required. (Siemens Energy and Automation, Inc.; General Electric Co., Small AC Motor Dept.; Rexroth Corp., Indramat Division)

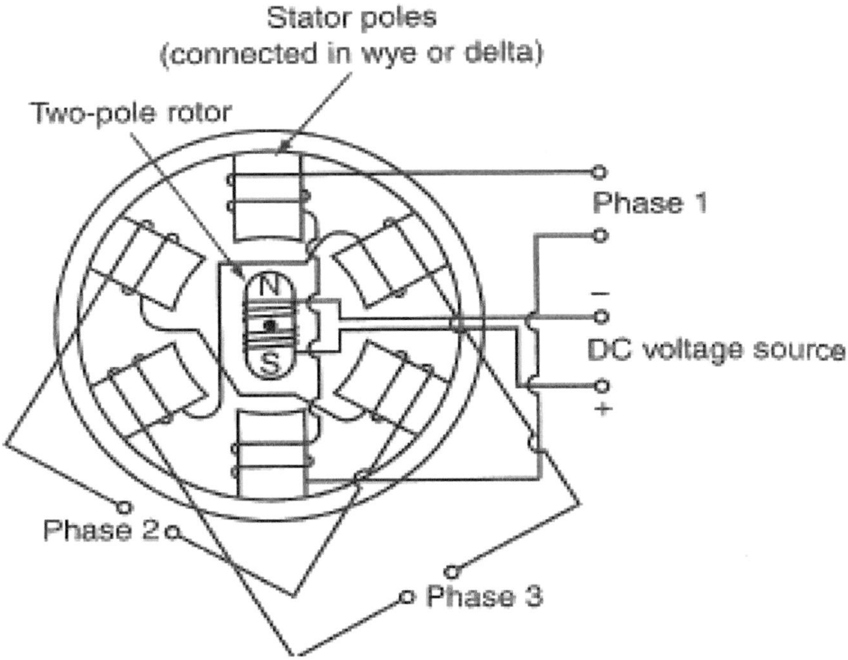

Figure 11.4 Pictorial diagram of a three-phase AC synchronous motor.

where S is the speed of a synchronous motor in r/min. *f*, is the frequency of the applied AC voltage in hertz, N/3 is the number of stator poles per phase, and 120 is a conversion constant.

Note this is the same as the formula used to determine the stator speed of a single-phase motor, except the number of poles must be divided by three (the number of phases). A three-phase motor with 12 actual poles would have four poles per phase. Therefore, its stator speed would be 1800 r/min. Synchronous motors have **operating speeds** that are based on the number of stator poles.

One method of starting a large synchronous motor is to use a smaller auxiliary DC machine connected to the shaft of the synchronous motor, as illustrated in Figure 11.5. The mcthod of starting would be as follows.

1. DC power is applied to the auxiliary motor, causing it to increase in speed. Three-phase AC power is applied to the stator.

2. When the speed of rotation reaches a value near the synchronous speed of the motor, the DC power circuit is opened. At the same time, the terminals of the auxiliary machine are connected across the commutator/brush assembly of the rotor.

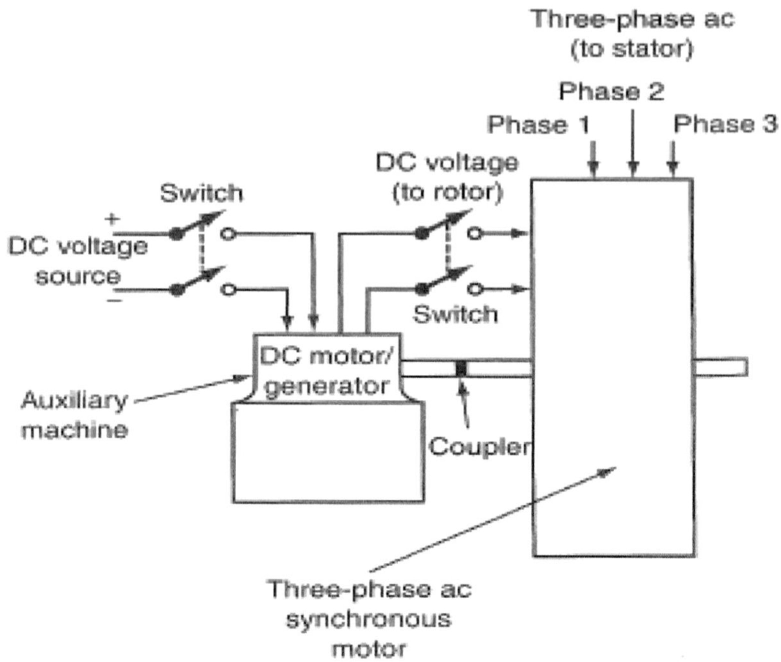

Figure 11.5 Auxiliary motor used to start a three-phase synchronous motor.

3. The auxiliary machine now converts to generator operation. It supplies exciter current to the rotor of the synchronous motor, using the motor as its prime mover.
4. Once the rotor is magnetized, it will lock in step, or *synchronize*, with the revolving stator field.
5. The speed of rotation remains constant under changes in load conditions.

Another starting method is shown in Figure 11.6. This method utilizes damper windings, which are similar to the conductors of a squirrel-cage rotor. These windings are placed within the laminated iron of the rotor assembly. No auxiliary machine is required when damper windings are used. The starting method used is as follows.

1. Three-phase AC power is applied to the stator windings.
2. The motor operates as an induction motor due to the **transformer action** of the damper windings.
3. The motor speed builds up so that the rotor speed is somewhat less than the speed of the revolving stator field.

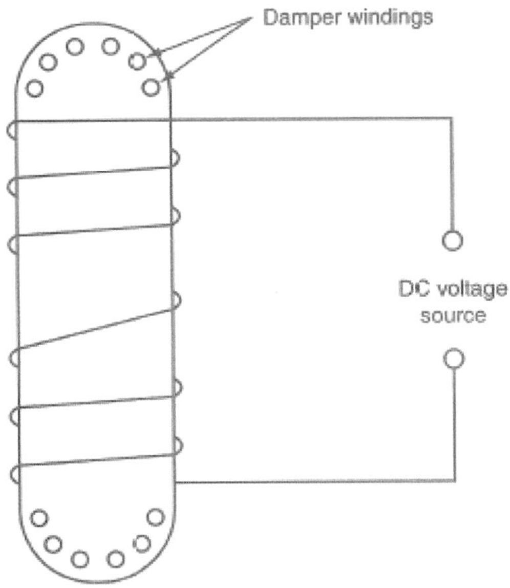

Figure 11.6 Damper windings used in the rotor of a three-phase synchronous motor.

4. DC power from a rotating DC machine or, more commonly, a rectification system is applied to the commutator/brush assembly of the rotor.
5. The rotor becomes magnetized and builds up speed until the rotor speed is equal to the stator speed.
6. The speed of rotation remains constant regardless of the load placed on the shaft of the motor.

An outstanding advantage of the **three-phase synchronous motor** is that it can be connected to a three-phase power system to increase the overall power factor of the system. The power factor correcting capabilities of the three-phase synchronous motor will be discussed in Section 11.4.

11.3 Power Factor in AC Systems

In DC circuits, **power** is equal to the product of voltage and current ($P = V \times I$). This formula is also true for purely resistive circuits. However, when reactance (either inductive or capacitive) is present in an AC circuit, power can no longer be calculated by that simple formula. Reactive circuits cause changes in the method used to compute power in AC circuits.

Reviewing from **Chapter 4**, the product of voltage and current in AC circuits is expressed in voltamperes (VA) or kilovolt amperes (kVA) and is known as apparent power. When using meters to measure power in an AC circuit, apparent power is the voltage reading multiplied by the current reading. The actual power that is converted to another form of energy by the circuit is measured with a wattmeter. This actual power is referred to as **true power**. The ratio of true power converted in a circuit to apparent power is called the **power factor** and is expressed as:

$$PF = \frac{P}{VA}$$

Or

$$\% PF = \frac{P}{VA} \times 100$$

where PF is the power factor of the circuit, P is the true power in watts, and VA is the apparent power in voltamperes. The maximum value of power factor is 1.0 or 100%, which would be obtained in a purely resistive circuit. This is referred to as the unity power factor.

The **phase angle** between voltage and current in an AC circuit determines the power factor. If a purely inductive or capacitive circuit existed, the 90° phase angle would cause a power factor of zero to result. In practical AC circuits, the power factor varies according to the relative values of resistance and reactance.

The power relationships are simplified by using the power triangle discussed in **Chapter 4.** There are two components that affect the power relationship in an AC circuit. The in-phase (resistive) component, which results in power conversion in the circuit, is called active power. Active power is the **true power** of the circuit and is measured in watts. The second component is that which results from an inductive or capacitive reactance and is 90° out of phase with the active power. This component, called **reactive power**, does not produce an energy conversion in the circuit. Reactive power is measured in volt-amperes reactive (VAR).

The **cosine** of the phase angle of the **power triangle** is expressed as:

$$\text{cosine } \theta = \frac{\text{adjacent side}}{\text{hypotenuse}}$$

or

$$\text{cosine}\theta = \frac{\text{true power (W)}}{\text{apparent power (VA)}}$$

Therefore, true power can be expressed as W = VA × cosine 6. Note that the expression true power apparent power is the power factor of a circuit. Therefore, the **power factor** is equal to the cosine of the phase angle (PF = cosine 6).

11.4 Power-factor Correction

Since most industries use many electric motors, the industrial plant represents a highly inductive load. This means that industrial power systems operate at a **power factor** of less than unity (1.0). It is undesirable for an industry to operate at a low power factor, since the utility company would have to supply more electrical power to the industry than is actually used.

A given value of voltamperes (voltage × current) is supplied to an industry. If the power factor of the industry is low, the current must be higher since the power converted by the total industrial load equals VA × PR The value of power factor decreases as the reactive power drawn by the industry increases. This is illustrated in Figure 11.7.

Assume a constant value of true power to see the effect of increases in reactive power drawn by an inductive load. The smallest reactive power shown (VAR_1) results in the voltampere value of VAj. As reactive power is increased, as shown by the VAR_2 and VAR_3 values, more voltamperes (VA_2 and VA_3) must be drawn from the source. This is true since the voltage component of the supplied voltamperes remains constant. This example represents the same effect as a decrease in power factor, since PF = W/VA, and as VA increases, PF will decrease if W remains constant.

Utility companies will charge industries for operating at power factors below a specified level. It is desirable for industries to correct their power factor to avoid such charges and to make more economical use of electrical energy. This correction is done by introducing capacitive reactance into the system.

Since the effect of capacitive reactance is opposite to that of inductive reactance, their reactive effects will counteract one another. Two methods can be used to increase the power factor.

- Power factor-corrective capacitors.
- Three-phase synchronous motors.

It can be beneficial for industries to invest in either power factor-corrective **capacitors or three-phase synchronous motors** to correct their power factor.

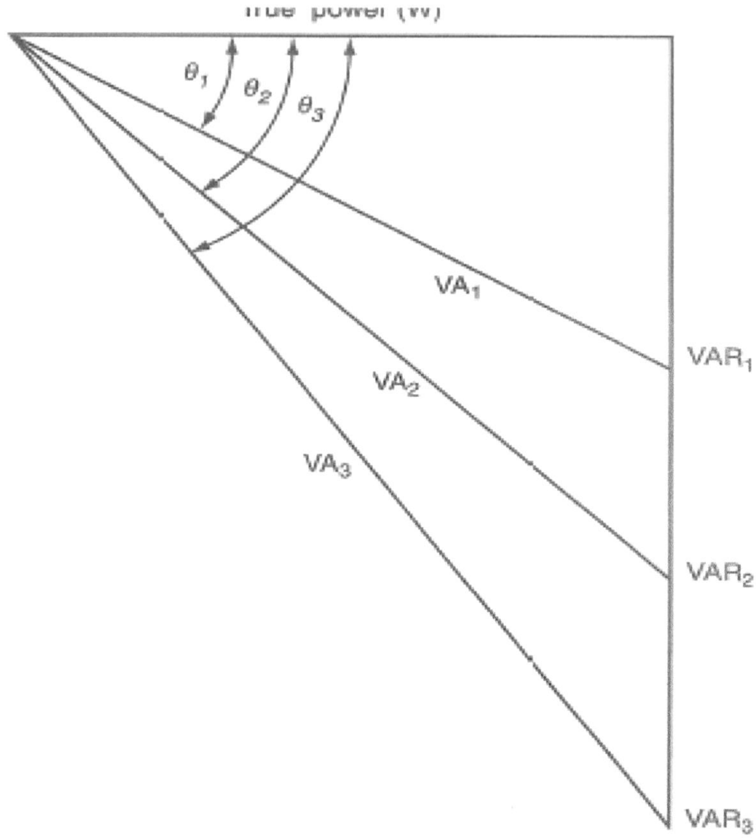

Figure 11.7 Effect of increases in reactive power on apparent power.

In the **example** shown in Figure 11.8, assume that both **true power** and **inductive reactive power** remain constant at values of 10 kW and 10 kVAR. In part A of Figure 11.6, the power factor equals 70%. If 5-kVAR capacitive reactive power is introduced into the AC power line, the net reactive power becomes 5 kVAR (10-kVAR inductive minus 5-kVAR capacitive), as shown in part B. With the addition of 5-kVAR capacitive, the power factor is increased to 89%. Now, in part C of the figure, if a 10-kVAR capacitive is added to the AC power line, the total reactive power becomes zero. The true power is now equal to the apparent power; therefore, the power factor is 1.0, or 100%, which is characteristic of a purely resistive circuit.

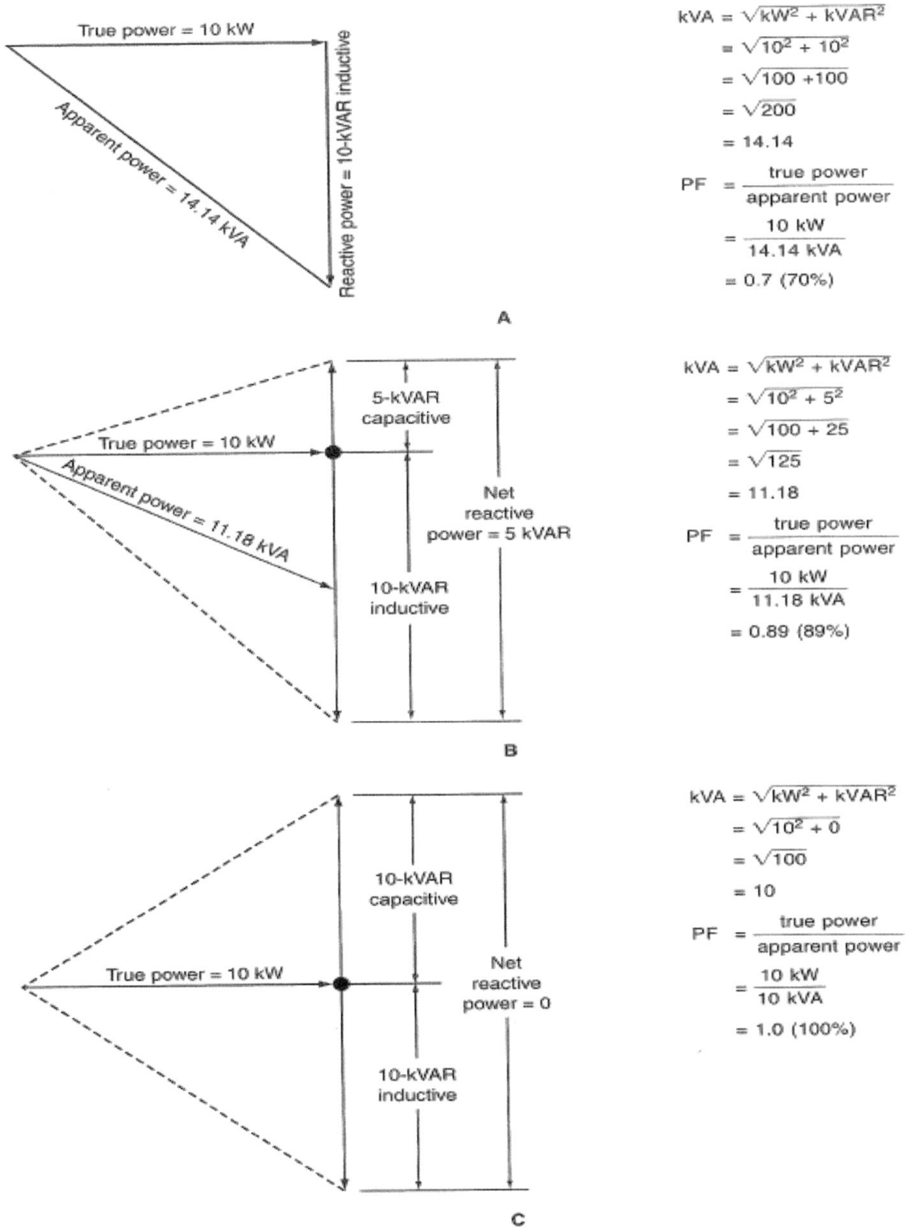

Figure 11.8 Effect of capacitive reactance on an inductive load. A – Reactive power equals 10-kVAR inductive. B – Reactive power equals 10-kVAR inductive plus 5-kVAR capacitive. C – Reactive power equals 10-kVAR inductive plus 10-kVAR capacitive.

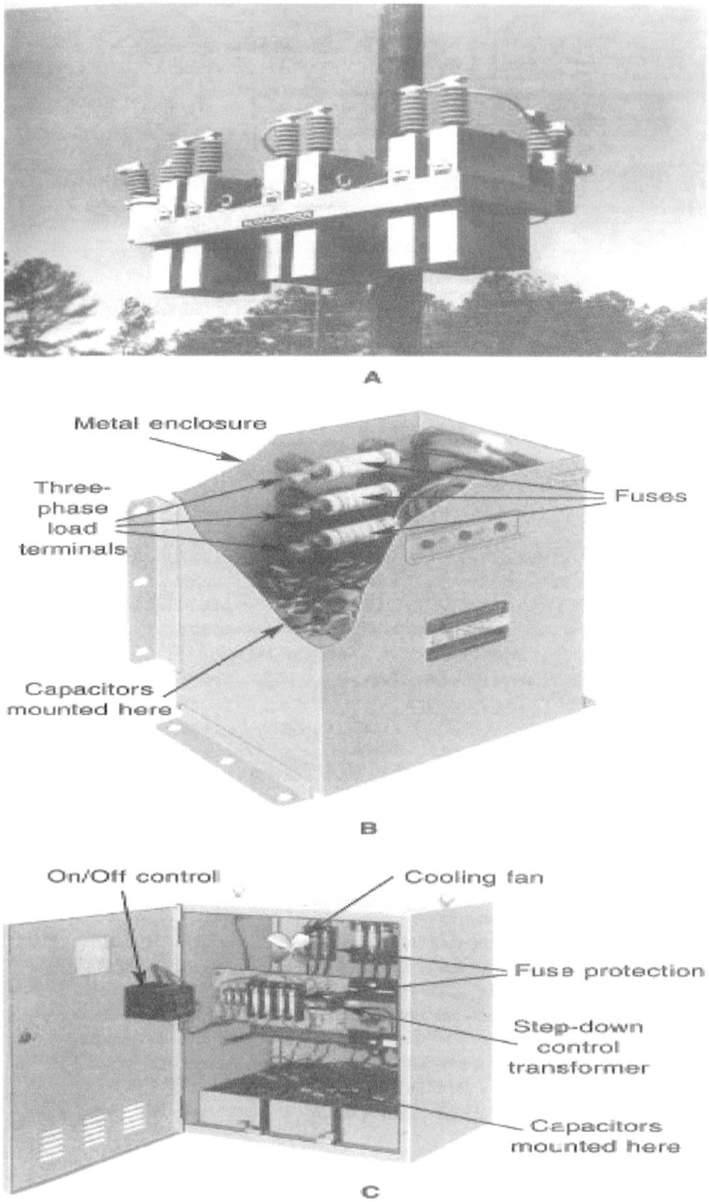

Figure 11.9 Power factor correction capacitors. A – Pole-mounted power factor correction capacitors. (Cooper Power Systems) B – Static capacitor set. (Square-D Co.) C – Static capacitor unit. (Square-D Co.)

The effect of the increased **capacitive reactive power** is to increase or correct the power factor. This reduces the current drawn from the AC power lines that supply the industrial loads.

Utility companies also attempt to correct the **power factor** of their systems. A certain quantity of inductance is present in most of the power distribution systems. This inductance can come from generator windings, transformer windings, and power lines. To counteract the inductive effects, utility companies use **power factor-corrective capacitors**, such as the pole-mounted units shown in Figure 11.9. This type of capacitor unit, plus larger banks of capacitors located at substations, can be used to correct the power factor of AC power systems.

We know from previous discussion that low power factors cannot be tolerated by industries. Thus, the expense of installing three-phase synchronous machines could be justified in order to appreciably increase system power factor. If no load is intended to be connected to the shaft of a three-phase synchronous motor, it is called a **synchronous capacitor**. It is designed to act only as a power factor corrective machine. Of course, it might be beneficial to use this motor as a constant-speed drive connected to a load as well as for power-factor correction.

To understand how a three-phase synchronous machine operates as a power factor-corrective machine, refer to the curves of Figure 11.10. Synchronous motors operate at a constant speed. Thus, variation in rotor DC excitation current has no effect on speed. However, the excitation level will change the power factor at which the machine operates.

Three operational conditions can exist, depending on the amount of DC excitation applied to the rotor. These conditions are:

- *Normal excitation:* operates at a power factor of 1.0.
- *Underexcitation:* operates at a lagging power factor (inductive effect).
- *Overexcitation:* operates at a leading power factor (capacitive effect).

Note the variation of stator current drawn by the synchronous motor as rotor current varies. Observe also that the stator current is minimum when the power factor equals 1.0 or 100%. The situations shown on the graph indicate no-load, half-load, and full-load conditions with power factors equal to 1.0, 0.8 leading, and 0.8 lagging. These curves are sometimes referred to as **V-curves** for a synchronous machine.

The second graph shows the variation of power factor with changes in rotor current under three different load conditions. Thus, a three-phase synchronous motor, when overexcited, can improve the overall power factor of a three-phase system.

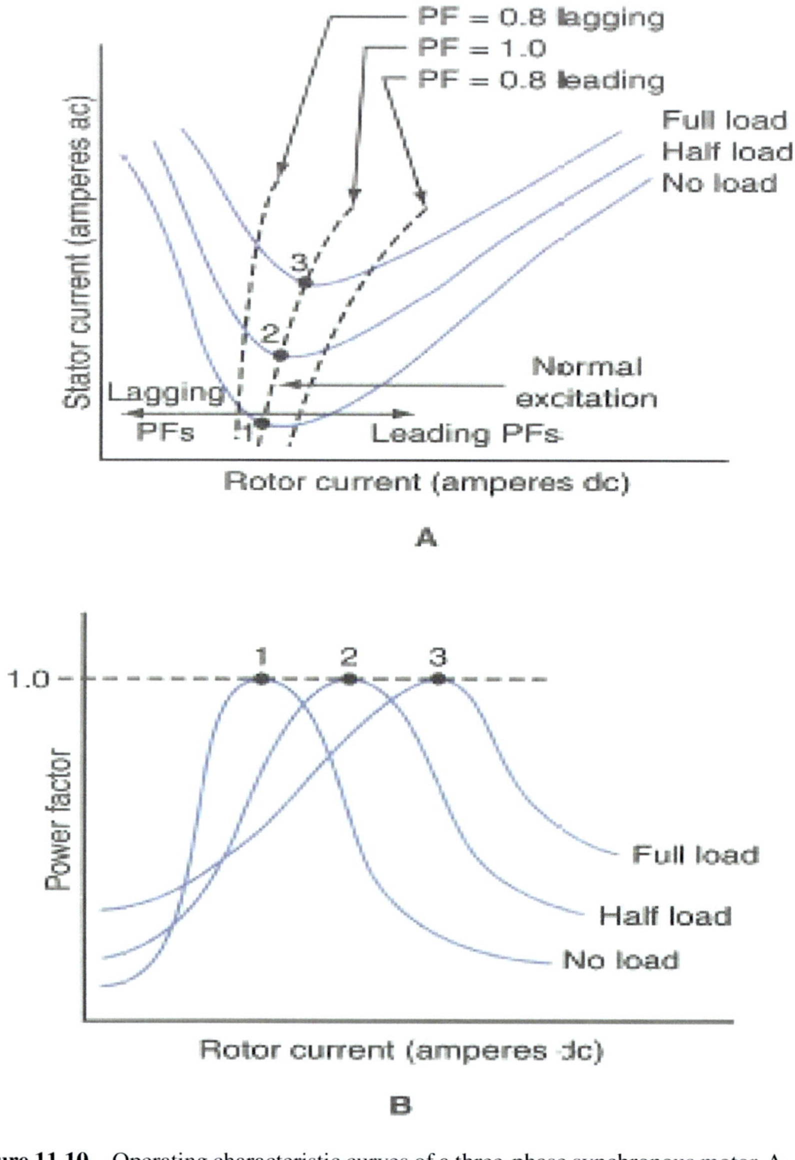

Figure 11.10 Operating characteristic curves of a three-phase synchronous motor. A – Rotor current versus stator current. B – Rotor current versus power factor.

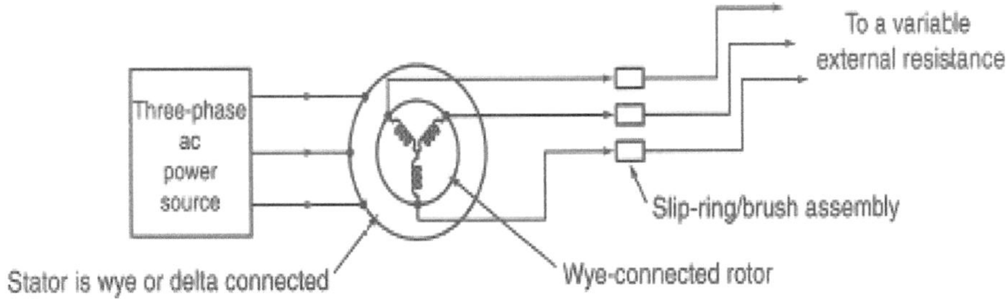

Figure 11.11 Three-phase wound-rotor induction motor.

As the **load** increases, the angle between the stator pole and the corresponding rotor pole on the synchronous machine increases. The **stator current** will also increase. However, the motor will remain synchronized unless the load causes "pull-out" to take place. The motor would then stop rotating due to the excessive **torque** required to rotate the load. Most synchronous motors are rated larger than 100 hp and are used for many industrial applications requiring constant-speed drives.

11.5 Wound-rotor Induction Motors

The **wound-rotor induction motor** shown in Figure 11.11 is a specialized type of three-phase motor. This motor can be controlled externally by placing resistances in series with its rotor circuit. The starting torque of a wound-rotor induction motor can be varied by the value of external resistance. The advantages of this type of motor are a lower starting current, a high starting torque, smooth acceleration, and ease of control.

The major disadvantage of this type of motor is that it costs a great deal more than an equivalent three-phase induction motor using a squirrel-cage rotor. Thus, they are not used as extensively as other three-phase motors.

Review Questions

Answer the following questions

1. What are the three basic types of three-phase motors?

2. How can the direction of rotation of a three-phase induction motor be reversed?

3. List the two methods of starting three-phase synchronous motors.

4. Define power-factor correction.

5. Why is power-factor correction important for industries?

6. Define the following terms:

   a. Normal excitation.

   b. Under excitation.

   c. Over excitation.

7. What is the relationship between rotor current and stator current in a three-phase synchronous motor at unity power factor?

8. What is the relationship between rotor current and power factor of a three-phase synchronous motor at full load?

9. What is a V-curve?

10. What are four advantages of wound-rotor induction motors?

Additional Problems

Answer the following questions.

1. Calculate the synchronous speeds of the following three-phase motors.

   a. 6-pole, 60 Hz.

   b. 12-pole, 60 Hz.

   c. 18-pole, 60 Hz.

   d. 24-pole, 60 Hz.

2. A 2000 kVA power system operates at a power factor of 0.6 lagging. Calculate the following:

   a. True power of the system.

   b. Total reactive power.

   c. kVAR required to obtain unity power factor.

   d. kVAR required to obtain 0.8 power factor.

   e. Capacitance required to obtain unity power factor.

3. An industry has an average load of 4500 kVA at a power factor of 0.6 lagging. An 800-hp three-phase synchronous motor with an efficiency of 80% is added to improve the system power factor to 0.9 lagging. Calculate:

   a. Power factor of the synchronous motor.

   b. kVA rating of the synchronous motor.

4. A three-phase synchronous motor rated at a 0.8 power factor (leading) is connected across a 240-V three-phase power line. The power factor of the system prior to connection of the synchronous machine was 0.7 (lagging). What effect (increase or decrease) will the addition of the

synchronous capacitor have on the following factors of the three-phase system?

a. Power factor.

b. Line current.

c. Apparent power.

d. Reactive power.

5. When the mechanical load applied to the shaft of a three-phase induction motor increases, what effect on the following operational factors can be expected?

   a. Speed.

   b. Line current.

   c. Power factor.

   d. Efficiency.

   e. Torque.

   f. Horsepower.

6. Discuss why each factor of Problem #5 changes as it does from a mathematical point of view.

12

Special Purpose Motors and Motion Control Systems

Objectives

Upon completion of this chapter, you will be able to:

- Describe, in general terms, synchro and servo systems.
- Discuss the basic types of servomotors used to drive automated equipment.
- Describe the construction features of servomotors and their operational differences.
- Define the terms bifilar, step angle, step accuracy, and step response time as related to DC step motors.
- Describe types of linear actuators used with automated equipment.
- Explain the operation of frequency conversion systems.
- Describe the operation of gear-reduction and adjustable speed motors.
- Recognize the parts of motion control systems.
- Identify the types of motors used with motion control systems.
- Describe the types of motor drive units used with motion control system.
- Discuss the use of encoders and other feedback devices used with motion control systems.

Important Terms

adjustable speed motor
bifilar
DC stepper motor
detent torque
forcer
frequency converter
gear-reduction motor
half stepping

DOI: 10.1201/9788743810056-15

holding torque
hybrid motor
linear actuator
magnetomotive force (MMF)
microstepping
microstepping drive
optical encoder
overshoot, platen
quadrature
ramping
recirculating chopper
rectification
regenerative chopper
resolver, resonance
running torque
servo system
servomechanism
servomotor
stacks
static torque curve
step accuracy
step angle
step response time
steps per second
synchro system
tooth pitch
torque-speed curve
torque-to-inertia ratio
translator
unipolar drive
voltage limited

Introduction

In recent years, some new and unique types of electric motors have been developed. Particular emphasis is now being placed on the design of motors controlled by computer systems. These motors are used for many applications in automated manufacturing. The technology of applying these systems is referred to as motion control.

12.1 Synchro and Servo Systems

In industry, there is a need for machines that produce a type of rotary motion that is somewhat different from that produced by electric motors. Synchro systems and servo mechanisms can be used to achieve this operation. Through these devices, it becomes possible to transmit rotary motion between locations without direct mechanical linkage.

One type of machine, called a **synchro system**, employs rotary motion to control the angular position of one shaft by positioning the shaft of a second device. Synchro systems were developed long ago.

Synchro systems are classified as two or more **motor-generator** units connected together to permit the transmission of angular shaft positions by electromagnetic field changes. When an operator turns the generator shaft of the generator unit to a certain position, it automatically rotates the motor shaft to an equivalent position at a remote location. Through this type of

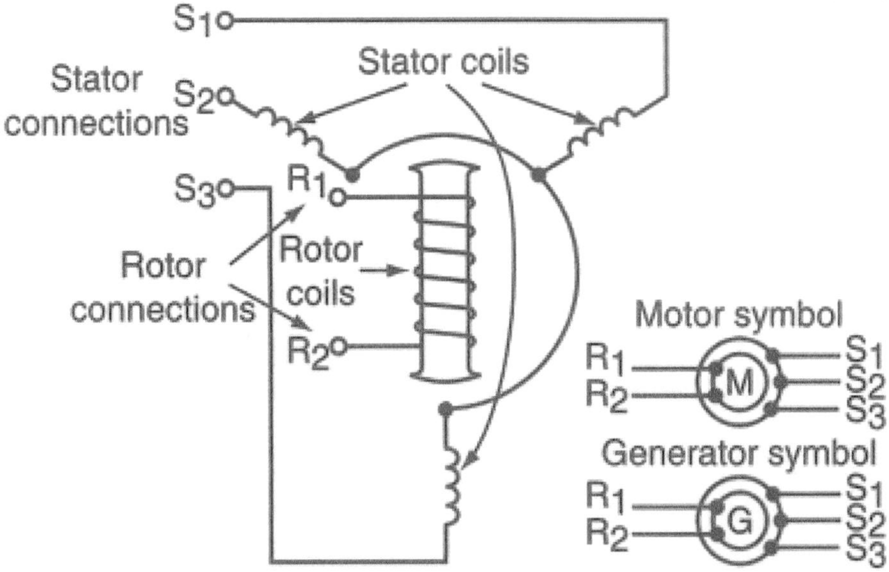

Figure 12.1 Schematic diagram and symbols of a synchro system.

system, it is possible to achieve accurate control over great distances. These systems show the concept of transferring position information.

In systems that require increased rotational torque or precise movements of a control device, servo systems are employed. A **servo system**, or **servomechanism**, is a special type of AC or DC motor that drives a precision piece of equipment in specific increments. Systems that include servo systems generally require amplifiers and error-detecting feedback devices to control angular displacement.

12.1.1 Synchro system operation

A **synchro system** contains two or more electromagnetic devices that are similar in appearance to small electric motors. These devices are connected together in such a way that the angular position of the generator shaft can easily be transmitted to the motor or receiver unit. Figure 12.1 shows a schematic diagram and symbols of a synchro system. As a general rule, the generator and motor units are identical electrically. Physically, the motor unit has a metal flywheel attached to its shaft to prevent shaft oscillations or vibrations when it is powered. The letters G and M inside the electrical symbol denote the generator or motor functions.

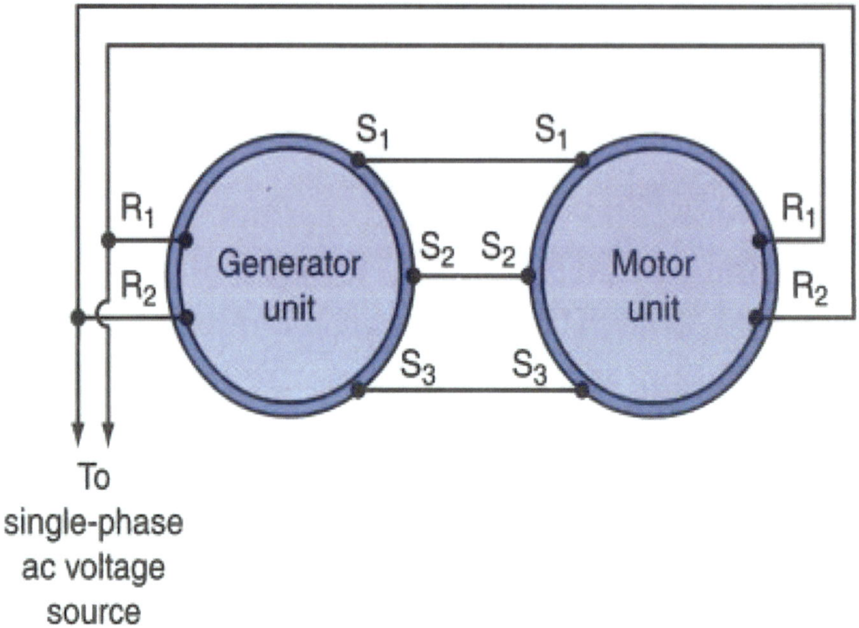

Figure 12.2 Circuit diagram of a synchro system.

Figure 12.2 shows the circuit diagram of a basic synchro system. Single-phase AC voltage is used to power this particular system. The line voltage is applied to the rotors of both the generator and motor. The stationary coils, or stator windings, are connected together as indicated.

When power is initially applied to the system, the motor positions itself according to the location of the generator shaft. No physical change takes place after the motor unit aligns itself with the generator position. Both units remain in a stationary condition until some further action takes place. Turning the generator shaft a certain number of degrees in a clockwise direction causes a corresponding change in the motor unit. If calibrated dials are attached to the shaft of each unit, they would show the same angular displacement.

The **stator coils** of a synchro system are wound inside a cylindrical laminated metal housing. The coils are uniformly placed in slots and connected to provide three poles spaced 120° apart. These coils serve functionally like the secondary windings of a transformer.

The **rotor coil** of the synchro system is also wound on a laminated core. This type of construction causes north and south magnetic poles to extend from the laminated area of the rotor. Insulated slip rings on the shaft are used to supply AC power to the rotor. The rotor coil responds functionally like the primary windings of a transformer.

When AC is applied to the rotor coil of the synchro system, it produces an alternating magnetic field. By transformer action, this field cuts across the stator coils and induces a voltage in each winding. The physical position of the rotor coil determines the amount of voltage induced in each stator coil. If the rotor coil is parallel with a stator coil, maximum voltage is induced. The induced voltage is minimum when the rotor coil is at right angles to a stator coil set.

The stator coils of the generator and motor of a synchro system are connected as indicated in the circuit diagram of Figure 12.2. Voltage induced in the stator coils of the generator, therefore, causes a resulting current to flow in the stator coils of the motor. This, in turn, causes a corresponding magnetic field to be established in the motor stator. Line voltage applied to the rotor of the motor unit wall causes it to align itself with the magnetic field of the stator coils.

Any change in rotor position of the generator unit is translated into an induced voltage and applied to the motor stator coils. Through this action, linear displacement changes can be effectively transmitted to the motor through three rather small stator coil wires.

12.1.2 Servo systems

Servo systems are rotating machines that accomplish such functions as changing the mechanical position or speed of a machine. Mechanical position applications include computer-controlled machinery and process control equipment. Speed applications are found in conveyor-belt control units, spindle speed control in machine tool óperations, and disk or magnetic-tape drives for computers. A servo system is usually a rather complex unit that follows the commands of a closed-loop control path. Figure 12.3 shows a block diagram for a typical servo system.

The input of a **servo system** serves as the reference element to which the controlled device responds. By changing the input, a command is applied to the error detector. This device receives data from both the input source and from the controlled output device. If a correction is needed with reference to the input command, it is amplified and applied to the actuator. The **actuator** is normally a **servomotor** that produces controlled shaft displacements. The controlled output device relays information back to the error detector for position comparison.

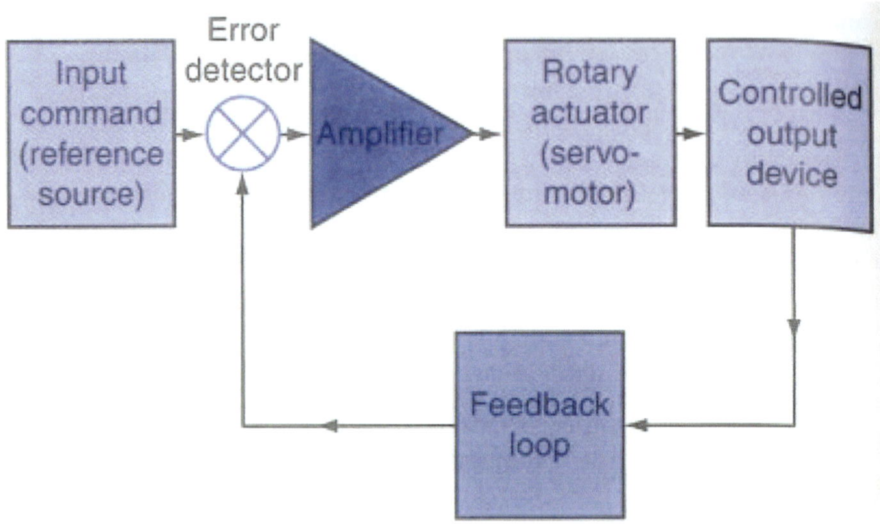

Figure 12.3 Block diagram of a servo system.

12.2 Servomotors

The **servomotor** is responsible for producing mechanical changes from an electromagnetic actuating device. A device of this type is normally coupled to the controlled device by a gear train or some mechanical linkage. Both AC and DC servomotors, such as those shown in Figure 12.4, are used to achieve this operation.

A **servomotor** is a very special type of device that is used to achieve a precise degree of rotary motion. Servomotors are unique when compared with other electric motors. Servomotors, for example, are designed to do something other than change electrical energy into rotating mechanical energy. Motors of this type must be able to respond accurately to signals developed by the system's amplifier. Second, they must be capable of reversing direction quickly when a specific signal polarity is applied. Also, the amount of torque developed by a servomotor must be quite high. As a general rule, torque is a function of the voltage and current levels of the supply source.

12.2.1 AC synchronous servomotors

The construction of an **AC synchronous servomotor** is quite simple. It contains no brushes, commutators, slip rings, or centrifugal switches. As shown in Figure 12.5, it is simply made up of a rotor and a stator assembly. There is no direct physical contact between the rotor and stator. A carefully maintained

A

B

Figure 12.4 A – Typical servomotors. B – Cutaway of a brushless servomotor. (Pacific Scientific Co.)

air gap is always present between these two components. As a result of this construction, the motor has a long operating life and is highly reliable.

The ***speed*** of a synchronous servomotor is directly proportional to the frequency of the applied AC and the number of stator poles. Since the number of stator poles cannot be effectively altered after the motor has been manufactured, frequency is the most significant speed factor. Speeds of 28, 72, and 200r/min are typical. The speed of 72r/min is a common industrial standard.

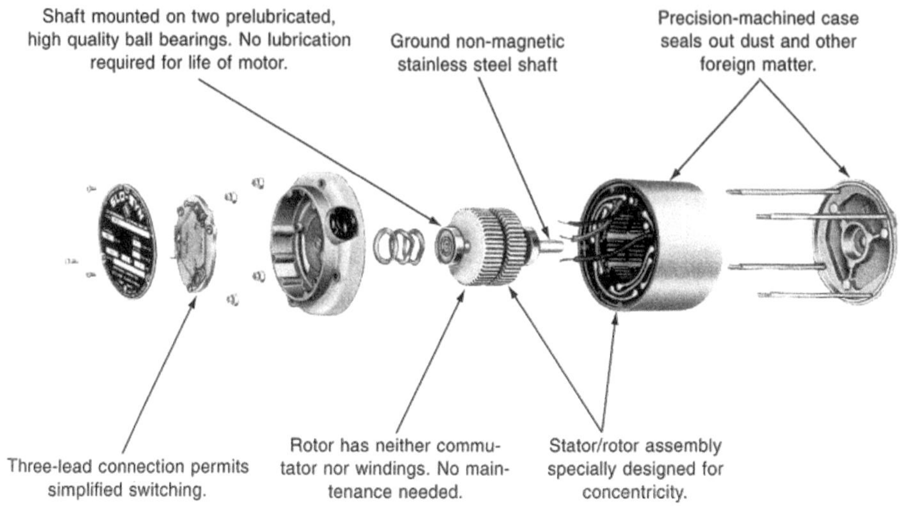

Figure 12.5 Exploded view of an AC synchronous motor. (Superior Electric Co.)

Figure 12.6 shows the stator layout of a **two-phase synchronous servomotor** with four poles per phase. Poles N1-S3 and N5-S7 represent one phase, while poles N2-S4 and N6-S8 represent the second phase. There are places for 48 teeth around the inside of the stator. One tooth per pole, however, has been eliminated to provide a space for the windings. Five teeth per pole, or a total of 40 teeth, are formed on the stator. The four coils of each phase are connected in series to achieve the correct polarity.

The **rotor** of the **synchronous servomotor** in **Figure** is an axially magnetized permanent magnet. There are 50 teeth cast into its form. The front section of the rotor has one polarity, while the back section has the opposite polarity. The physical difference in the number of stator teeth (40) and rotor teeth (50) means that only two teeth of each part can be properly aligned at the same time. With one section of the rotor being a north pole and the other section being a south pole, the rotor has the ability to stop very quickly. The rotor can also accomplish direction reversals very rapidly because of this gear-like construction.

A circuit diagram of a **single-phase synchronous servomotor** is shown in Figure 12.7. The resistor and capacitor of this circuit are used to produce a phase shift in one winding. As a result, the two windings are always out of phase regardless of whether the switch is in the clockwise (CW) or counterclockwise (CCW) position. When power is applied, the four coils of one phase produce an electromagnetic field. The rotor is attracted to and aligns itself with these stator coils. Then, 90 ° later, the four coils of the second phase produce a corresponding field. The stator is again attracted to this position.

Figure 12.6 Stator layout of a two-phase AC synchronous motor. (Superior Electric Co.)

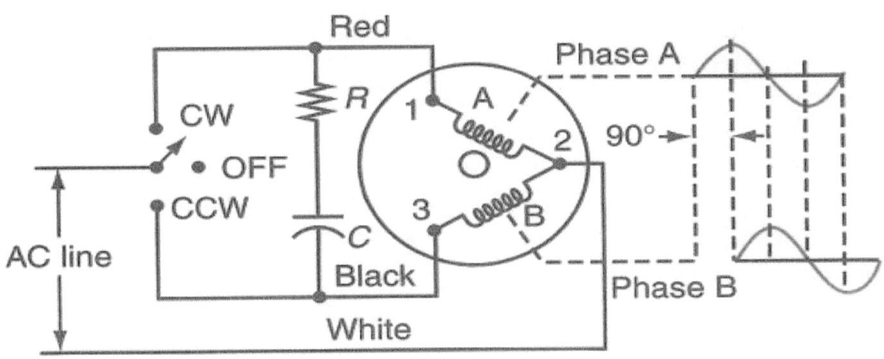

Figure 12.7 Circuit diagram of a single-phase synchronous motor. (Superior Electric Co.)

As a result of this action, the rotor sees a moving force across first one phase and then the other. This force gives the rotor the needed torque that causes it to start and continue rotation when power is applied.

The **synchronous servomotor** just described has the capability of starting in one and one-half cycles of the applied AC source frequency. In addition to this, it can be stopped in five mechanical degrees of rotation. These two characteristics are attributed primarily to the geared rotor and stator construction. Synchronous servomotors of this type have another important characteristic. They draw the same amount of line current when stalled as they do when operating. This characteristic is very important in automatic machine-tool applications where overloads occur frequently.

12.2.2 DC stepper motors

DC stepper motors are specialized electrical machines that are used to control automatic industrial equipment. DC motors of this type are used in many high-power servomechanisms. They are typically more efficient and develop significantly more torque than the synchronous servomotor. The DC stepper motor is used primarily to change electrical pulses into rotary motion that is used to produce mechanical movement.

The shaft of a **DC stepper motor** rotates a specific number of mechanical degrees with each pulse of electrical input voltage. The amount of rotary movement or angular displacement produced by each pulse can be repeated precisely with each succeeding pulse from the input source. The resulting output of this machine is used to accurately locate or position automatic industrial systems.

The velocity, distance, and direction of a specific machine can be moved or controlled by a DC stepper motor. The movement error of this device is generally less than 5% per step. This error factor is not cumulative, regardless of the distance moved or the number of steps taken. Motors of this type are energized by a DC drive amplifier that is controlled by computer logic circuits. The **drive amplifier** circuitry is a key factor in the overall performance of this motor.

The basic construction of a **DC stepper motor** is very similar to that of the AC synchronous motor of Figure 12.5. The stator construction and coil placement are also the same as that of the AC synchronous motor layout shown in Figure 12.6. Some manufacturers make servomotors that can be operated as either an AC synchronous motor or as a DC stepper motor. To achieve this, the rotor must be of permanent magnet construction.

12.2.3 DC stepper motor terminology

The rotation of a stepper motor shaft is in fixed repeatable increments (steps). Each time the stator winding polarity is changed by the input logic circuitry, the motor shaft rotates a specific amount. This degree of rotation is called the **step angle** (in degrees). The number of steps required for the shaft to rotate

Step angle (deg)	Steps per 360° rotation
0.72	500
1.8	200
2.0	180
2.5	144
5.0	72

Figure 12.8 Some step angles available.

360 ° or one complete revolution may be found by dividing the step angle into 360 °. Several step angles are available. See Figure 12.8.

Rather than using an r/min speed rating, as motors with free-running shafts do, the DC stepper motor is rated in steps per second. The steps per second rating is the number of steps the motor shaft rotates in one second. The term step accuracy is used to define a position accuracy tolerance. This percentage figure indicates the total possible error of the stepper motor in moving one step. The total error does not accumulate with each additional step. This condition is desirable in precision automated manufacturing systems in industry.

Step response time is another important rating of DC stepper motors. **Step response time** indicates the time required for the motor shaft to complete a single step after a DC input signal has been applied to the stator. This time rating is usually in milliseconds and relies on the torque-to-inertia ratio of the motor and the DC input source characteristics. The **torque-to-inertia ratio** is determined by the following method:

$$\text{torque-to-intertia ratio} = \frac{\text{holding torque (oz-in)}}{\text{rotor inertia}\left(\text{oz-in s}^2\right)}$$

The **holding torque** of a DC stepper motor is the amount of torque in ounce-inches (oz-in) required to move the motor shaft from a zero revolutions per minute condition.

12.2.4 DC stepper motor operation

An important principle that applies to the operation of DC stepper motors is the basic law of magnetism. Like magnetic poles repel and unlike magnetic poles attract. If a **permanent-magnet rotor** is placed between two series-connected stator coils, the situation of Figure 12.9 would result. With power applied to the stator, the rotor would be repelled in either direction. The direction of rotation in this case is unpredictable.

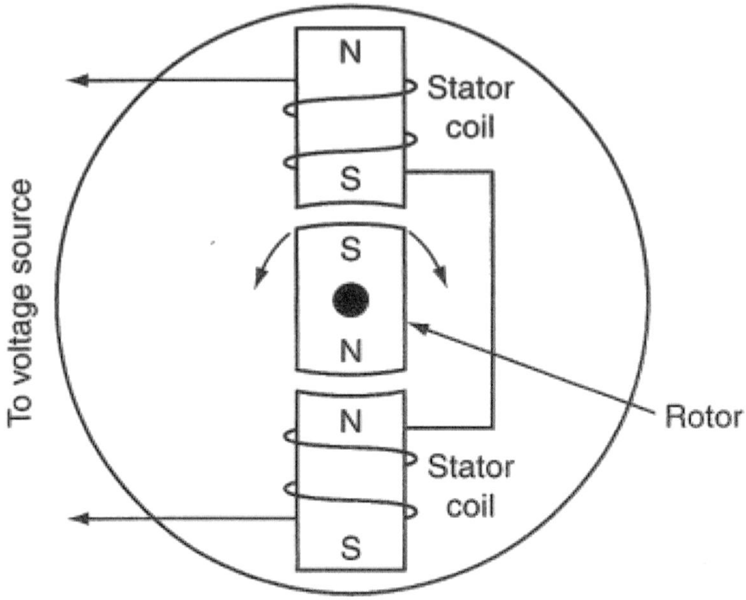

Figure 12.9 Stator coils with power applied.

Adding two more **stator coils** to this simple motor, as indicated in Figure 12.10, makes the direction of rotation predictable. With the stator polarities indicated, the rotor would align itself midway between the two pairs of stator coils. The direction of rotation can now be predicted and is determined by the polarities of the stator coil sets. Adding more stator coil pairs to a motor of this type improves its rotation and makes the stepping action very accurate. Figure 12.11 shows several types of DC stepper motors.

Figure 12.12 shows an electrical diagram and switching sequence of a DC stepper motor. The stator coils of this motor are wound by a special type of construction called ***bifilar***. With this type of construction, two separate wires are wound into the coil slots at the same time. The two wires are small in size, allowing more turns than could be achieved with a larger wire. Construction of this type simplifies the control circuitry and DC input source requirements to drive the motor.

Operation of the stepper motor of Figure 12.11 can be achieved in a four-step switching sequence. Any of the four combinations of switches 1 or 2 will produce a corresponding rotor position location. Each switching combination causes the motor to move one-fourth of a step. After the four switch combinations have been achieved, the switching cycle repeats itself.

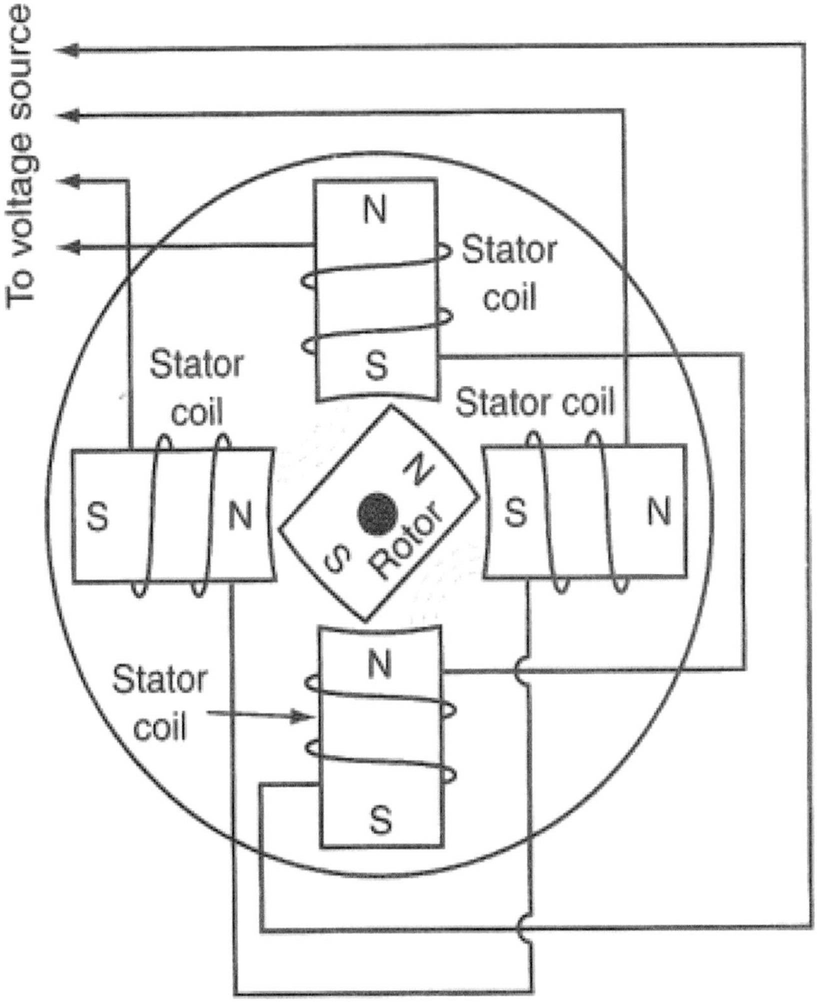

Figure 12.10 Two sets of stator coils to produce a constant rotor movement.

A rotor similar to the one shown in Figure 12.6 normally has 50 teeth. Using a 50-tooth rotor in the circuit of Figure 12.11 would permit four steps per tooth, or 200 steps per revolution. The teeth of a stepper motor's rotor are actually permanent magnet poles. They are alternate north–south–northsouth polarities that align with the opposite polarities of the stator poles to produce continuous motion. The amount of displacement, or step angle, of this motor is therefore determined by the number of teeth on the rotor and the switching sequence.

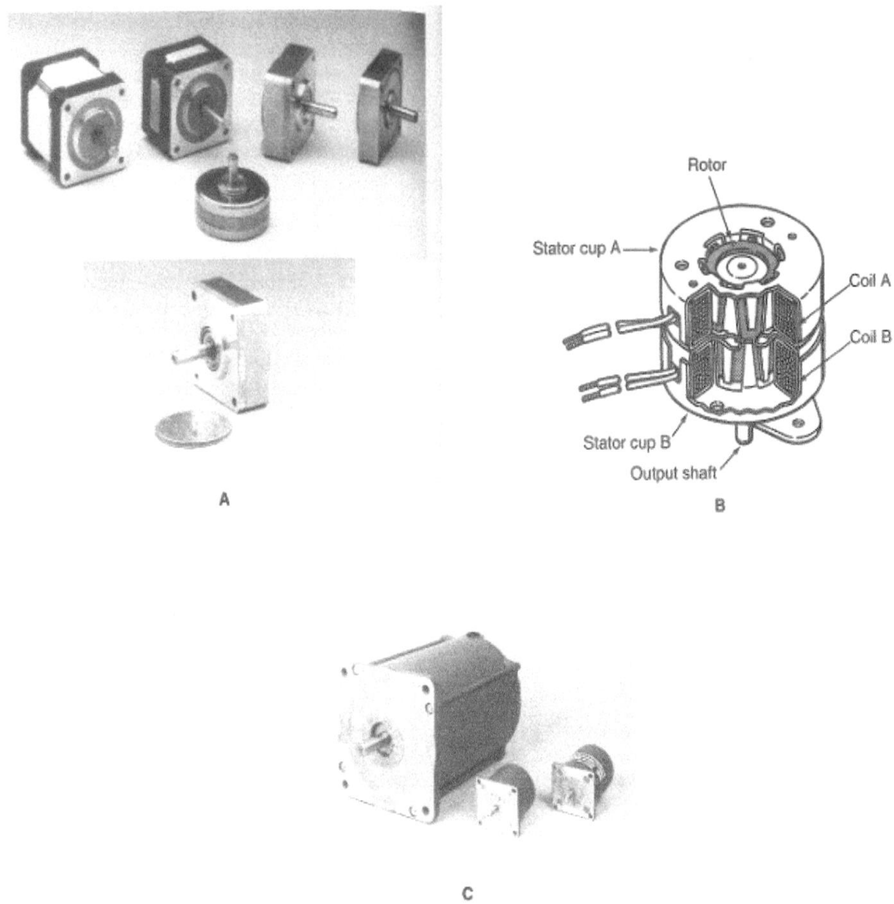

Figure 12.11 A – Miniature DC stepper motors. B – Cutaway diagram of a DC stepper motor. C – Large DC stepper motor showing size comparison. (Lin Engineering: Airpax Corp.; Superior Electric Co.)

A stepper motor that takes 200 steps to produce one revolution will move each 200 steps, or 1.8 ° per step. It is not unusual for stepper motors to use eight switching combinations to achieve one step. In this case, each switching combination could be used to produce 0.9 ° of displacement. Motors and switching circuits of this type permit a very precise controlled movement.

An **eight-step** switching sequence called half-stepping is also used. During this type of operation, the motor shaft moves half of its normal step angle for each input pulse applied to the stator. This allows more precise movement control and greater speed capability. Figure 12.13 shows the circuit diagram and switching arrangement of the eight-step switching mode.

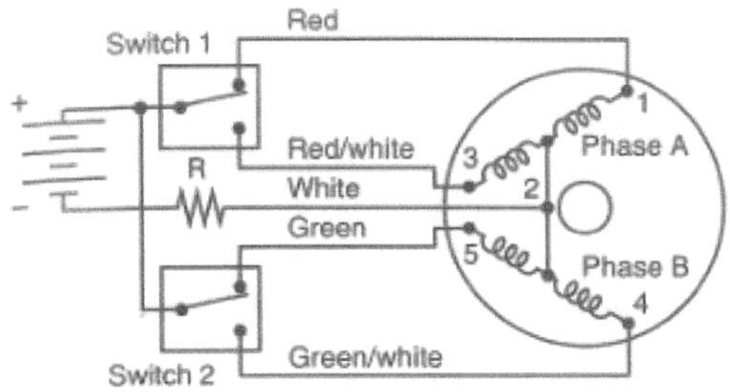

Switching Sequence*

Step	Switch 1	Switch 2
1	1	5
2	1	4
3	3	4
4	3	5
1	1	5

*To reverse direction, read chart up from bottom.

Figure 12.12 Circuit diagram and switching sequence of a bifilar-wound de stepper motor. (Superior Electric Co.)

Eight-step switching sequence*

Step	SW1	SW2	SW3	SW4
1	On	Off	On	Off
2	On	Off	Off	Off
3	On	Off	Off	On
4	Off	Off	Off	On
5	Off	On	Off	On
6	Off	On	Off	Off
7	Off	On	On	Off
8	Off	Off	On	Off
1	On	Off	On	Off

*To reverse direction, read chart from the bottom up.

A

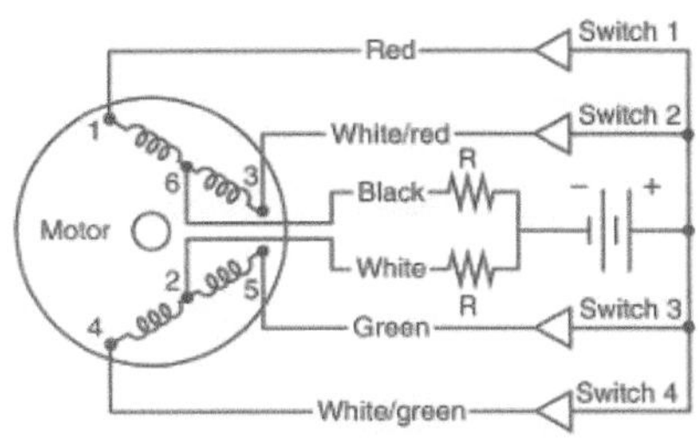

B

Figure 12.13 A – Circuit diagram and switching sequence of an eight-step DC stepper motor. (Superior Electric Co.)

12.3 Linear Actuators

A **linear actuator** is a device that converts some kind of power into linear motion. In industry, there is a need for linear actuators that provide movement changes that can be accurately controlled at different speeds, made to stop instantly, and hold a position without further demand for power. Driven linear actuators are designed to achieve these operating conditions. The term "driven," in this case, refers to electric-motor-operated mechanisms. A diagram and a selection of linear actuators are shown in Figure 12.14.

The operating principle of driven linear actuators is as follows. When a drive motor is activated, it rotates the threaded drive screw. A mating nut on the drive screw, which is attached to a drive rod, is thereby made to move axially. This action causes the drive rod to either extend or retract, depending on the direction of rotation of the drive motor. Examine again the diagram in part A of Figure 12.14.

When the drive motor of the actuator is set into motion, completion of the stroke or some type of overload will activate an overload cam. This action can be used to shut off the motor or reverse its rotation, depending on the design of the actuator. This mechanism can also be triggered by an outside control unit. The unit permits operation for certain time periods, or it can be triggered by a series of events. Feedback signals can also be generated to provide accurate positioning control in closed-loop applications. Driven actuators are used wherever linear motion is desired. This includes lever arms, cranks, slides, conveyor belt positioning, and diverter gate control.

The function of a **driven linear actuator** is similar to that of a hydraulic or pneumatic cylinder. However, electromagnetic actuators of this type have a number of unique advantages over their fluid counterparts. These advantages include such things as variable drive velocity, installation at remote locations, unlimited control capabilities, and overload protection. Notice the many different designs and possible applications that were shown in Figure 12.14.

12.4 Frequency Conversion Systems

The power system frequency used in the United States is 60 Hz (60 cycles per second). However, there are specific applications that require other frequencies in order to operate properly. Mechanical **frequency converters** are used to change an incoming frequency into some other frequency. Frequency converters are motor-generator sets that are connected together.

For example, a frequency of 60 Hz could be applied to a synchronous motor that rotates at a specific speed. A generator connected to its shaft could have the necessary number of poles to cause it to produce a frequency of 25 Hz. Recall that frequency is determined by the following relationship.

$$\text{frequency (Hz)} = \frac{\begin{matrix}\text{speed of rotation} \\ \text{(r / min.)}\end{matrix} \times \begin{matrix}\text{number} \\ \text{of poles}\end{matrix}}{120}$$

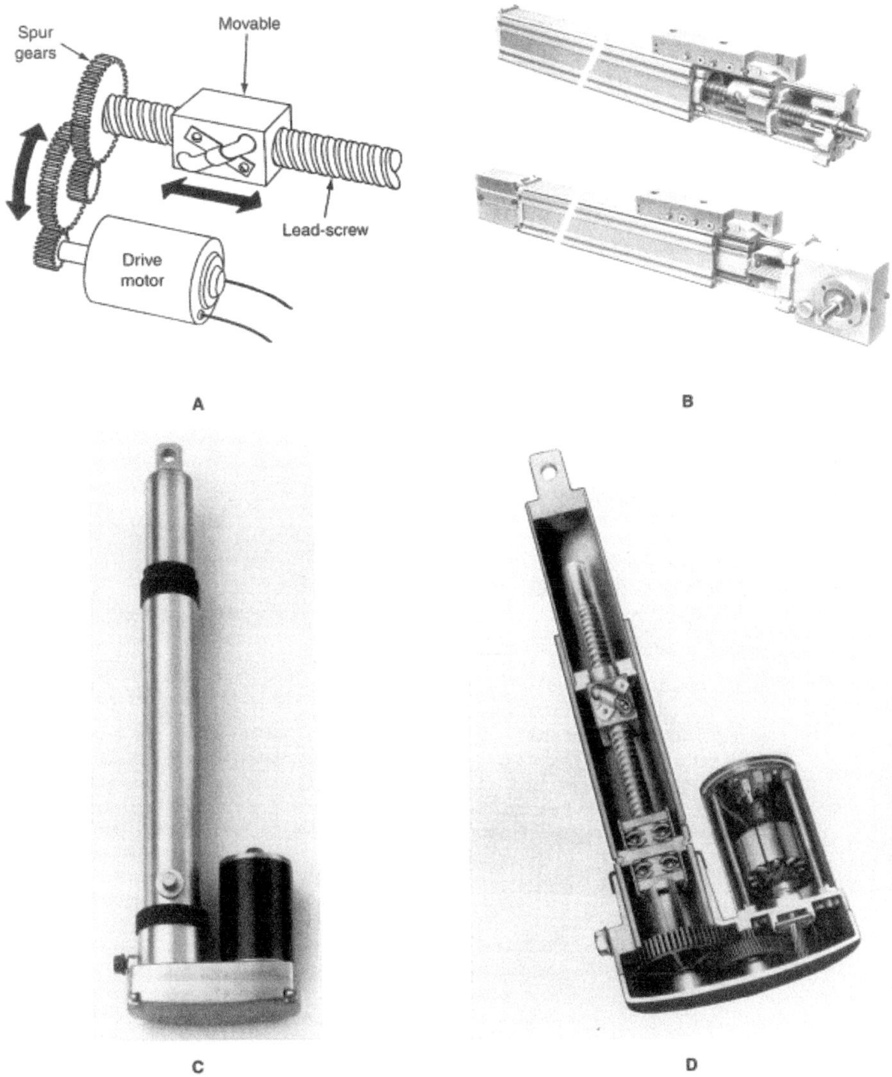

Figure 12.14 Linear actuators. A – Operating principle of B. B – Linear actuators showing detailed internal construction features. C – Exterior view of an electric motor-driven actuator. D – Interior view of C. E – Two sizes of actuators that use a lead-screw to control movement. F – Ball slide guide that is used with an actuator to provide accurate linear motion. G – Single-axis positioning table. A motor is connected to the lead screw to provide accurate linear motion. H – Intelligent actuator used to provide linear motion. I – High precision linear drive designed to produce a reciprocating, traverse motion. (Warner Electric, South Beloit, IL; Thompson Industries; Kollmorgen Motion Technologies, Radford, VA; Precision Industrial Components; IAI America, Inc., Torrence, CA; Amacoil, Inc.)

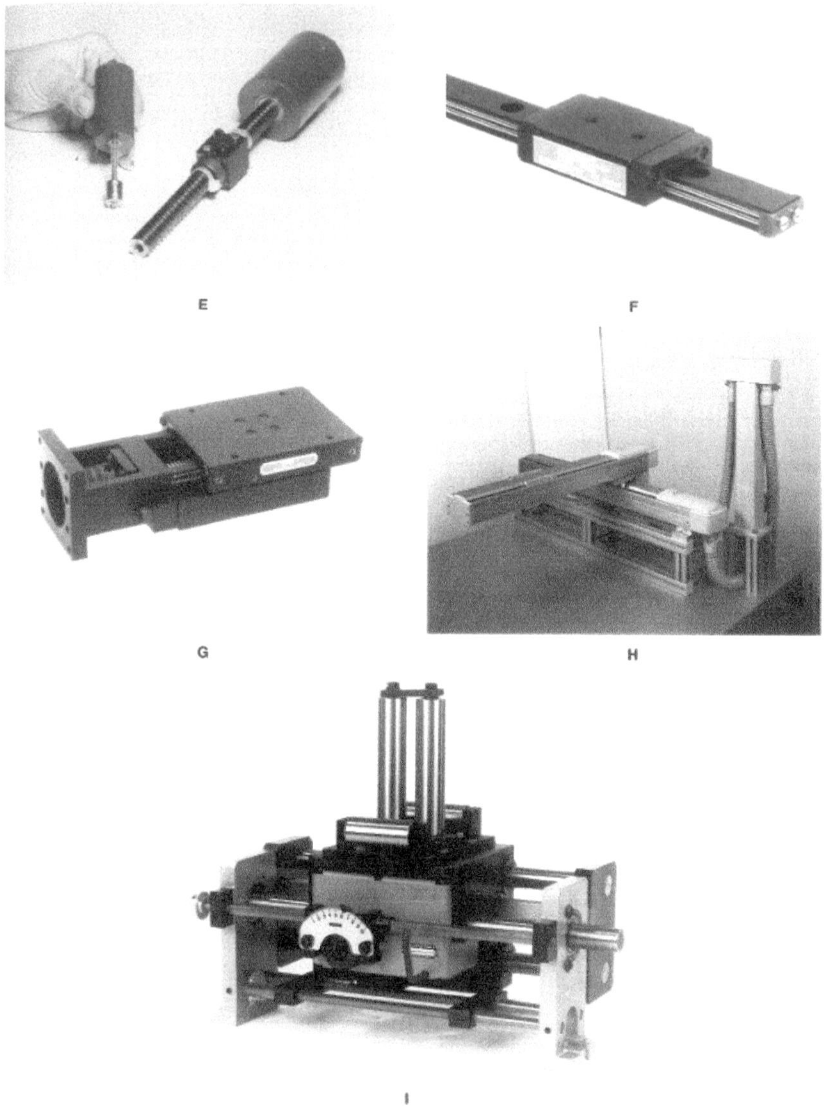

Figure 12.14 Continued

Examine Figure 12.15. Part A shows a block diagram of a frequency-conversion system. Parts B and C show two typical systems. Synchronous units such as the one shown are used wherever precise frequency control is required. It is also possible to design units that are driven by induction motors if some frequency variation can be tolerated.

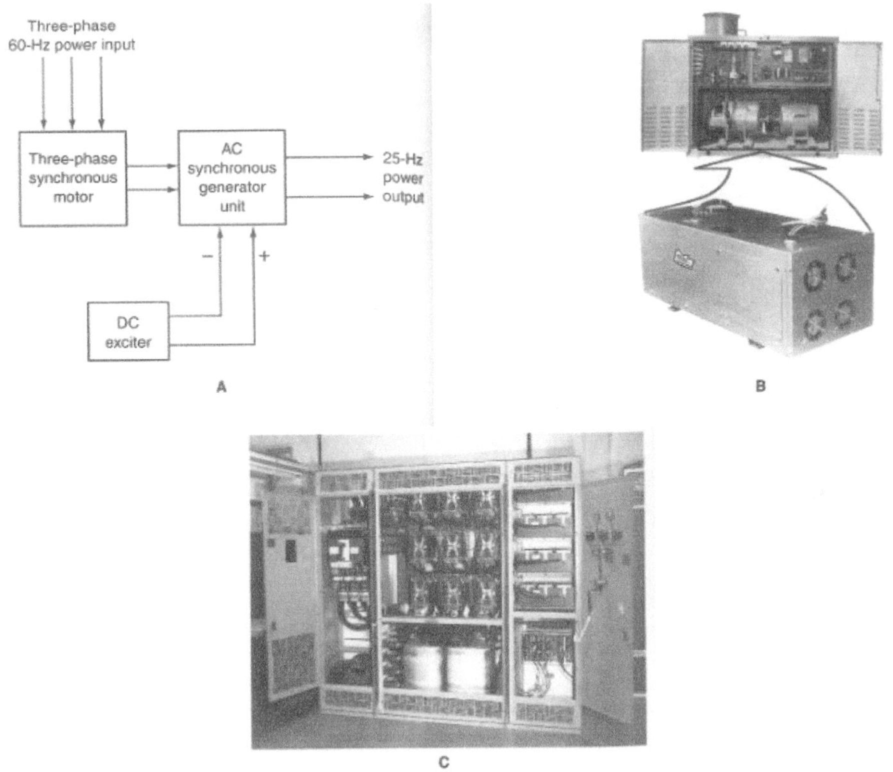

Figure 12.15 A – Block diagram of a motor-generator unit for frequency conversion. B – Typical motorgenerator unit. C-800 horsepower variable frequency drive. (Kinetics Industries; Robicon Corp.)

12.5 Single-phase to Three-phase Converters

Two methods are available for converting single-phase AC power to three-phase AC power. The static method involves the use of inductors and capacitors to accomplish phase separation near 120 electrical degrees. A static single-phase to three-phase conversion system is shown in Figure 12.16.

A rotary single-phase to three-phase conversion system is shown in Figure 12.17. This system is a motor-generator unit. A single-phase motor is used to drive a three-phase generator unit.

Either the static or rotary system can be used where three-phase power must be obtained from a single-phase system. Machines that use three-phase motors are more economical to operate. In some cases, it is more cost-effective

Figure 12.16 Static single-phase to three-phase conversion system. (Kay Industries, Inc.)

Figure 12.17 Rotary phase converter. Operates three-phase CNC machines on single-phase power. (Kay Industries, Inc.)

to install a single-phase to three-phase conversion system than to have three-phase power installed in a building.

12.6 Gear-reduction and Adjustable-speed Motors

Gear-reduction motors, such as the ones shown in Figure 12.18, are used where a reduced speed of operation is required for a motor. Most gear-reduction motors have either horizontal shafts or shafts at a right angle (90°) to the motor frame. The horsepower delivered at the output shaft of a gear-reduction motor is always less than that produced by the motor. This is due to gear reduction mechanical losses. Gear reduction motors should be selected by their output shaft speed and torque ratings rather than horsepower.

Adjustable speed motors are very desirable for industrial applications. To explain the operation of an adjustable speed system, look at part A of Figure 12.19. Adjustable speed is obtained by rotating the handwheel on top of the motor unit. The handwheel produces speed changes by changing the diameter of two variable-diameter pulleys (see part B) that are driven by a heavy-duty belt drive. Each variable-diameter pulley has two disks. One disk of each pulley is movable, allowing adjustable position due to the amount of

Figure 12.18 Gear-reduction motors. A – Helical bevel gear motor. B – Right angle gear motor. C – Harmonic drive differential transmission for more accurate speed and position control. D – Cutaway views showing internal detail of gear motors and speed reducers. (Nord Gear Corp.; Fairchild Industrial Products; Stober Drives, Inc.)

belt tension. Some additional types of adjustable speed motors are shown in parts C and D of Figure 12.19. The unit in part uses a traction-type adjustable speed drive. The drive system shown in part D uses a variable gear reduction system.

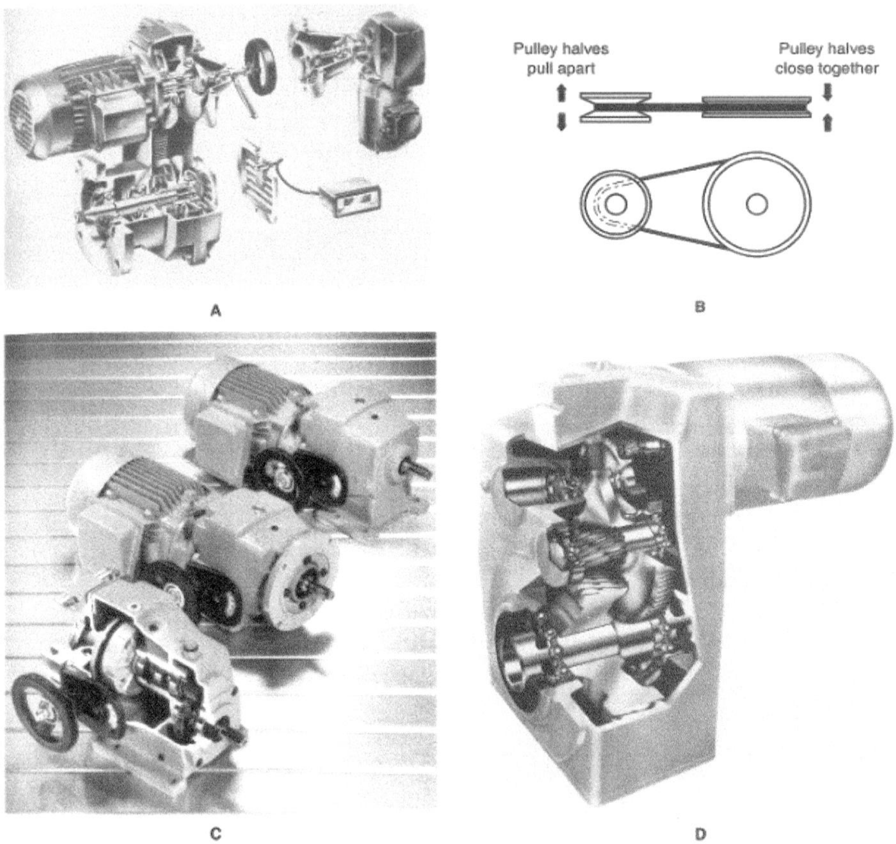

Figure 12.19 Adjustable speed motors. A – Belt-driven adjustable speed drive. B – Belt drive system used with A. C – Handwheel adjust. D – External control. (NORD Gear Corp.; Stober Drives, Inc.)

12.7 Specialized Machinery for Automated Manufacturing

The development of robotic systems and automated manufacturing in industry has brought about many new applications of specialized electrical machinery. The operation of automated manufacturing systems and industrial robots relies upon various types of electrical machinery. Electrical power systems are depended on to supply energy to manufacturing equipment so that industrial operations and processes can be performed.

A **robotic system** is a unique type of automated manufacturing system. These systems sometimes require several types of energy inputs and specialized machinery for proper operation. The operation of the electrical machinery used with robotic systems is dependent on the proper distribution and control of electrical power. Robotic systems use both AC and DC electrical energy inputs.

The AC electrical energy requirement for an automated manufacturing system can be either three-phase or single-phase. For larger machines, three-phase AC is ordinarily used. Several machines used for automated manufacturing require DC power supplies. Industries use DC sources for many specialized automated processes. Electroplating and DC variable-speed motor drives are two examples that show the need for DC energy for automated operations. Many robotic systems use DC servomotors and controls. Although most of the electrical power produced in the United States is three-phase alternating current, several methods are available to convert alternating current to direct current for automated manufacturing processes. Direct current is also made available through primary and secondary chemical cells, which are used extensively. The process of converting alternating current to direct current is called rectification. Rectification systems are usually the most convenient and inexpensive method to provide DC energy for automated industrial processes.

For automated manufacturing processes today, there is a need for devices that produce a type of rotary motion that is somewhat different from that produced by typical electric motors. This type of ***actuator*** employs rotary motion to control the angular position of a shaft. Servomechanisms of various types are used in industry to achieve this basic operation for automated manufacturing systems. Through these devices, it becomes possible to transmit rotary motion between locations without a direct mechanical linkage. Robotic systems ordinarily use rotary actuators such as DC stepper motors or AC synchronous motors to control their movement. Industrial robots are discussed in Chapter 22.

12.8 Application of Motors in Motion Control Systems

The tasks performed by the motor of a motion control system are determined by an indexer/controller, Figure 12.20, and other auxiliary systems. A motor's speed, distance of travel, direction, and acceleration rate can be controlled. The control function can be accomplished by a desktop computer or an attached control unit. One controller can operate several motor drives of a multi-axis system.

We will look at each of the system elements of motion control systems, Figure 12.21. Several new terms are highlighted in the following sections. To simplify your reading, a list of common stepper motor terms can be referenced in Figure 12.22.

A

B

Figure 12.20 Controllers for motion control systems. A – Multi-axis controller that can control up to four axes of motion. Such units have power supplies, drivers/amplifiers, and user I/O (input-output). B – Single-axis controller for a DC servomotor system. (IAI America, Inc.)

This section discusses the applications that are typically performed by each motor type in a motion control system. Certain applications are best accomplished by one motor type. However, there is a wide range of applications that can be equally well done by more than one motor type. A major consideration is the compatibility with existing equipment.

12.8.1 Applications of DC stepper motors

In applications where cost is a major factor, motion control can be done with a DC stepper motor. High torque, low speed, and continuous duty applications are also appropriate for a DC stepper motor. At low speeds, it is very

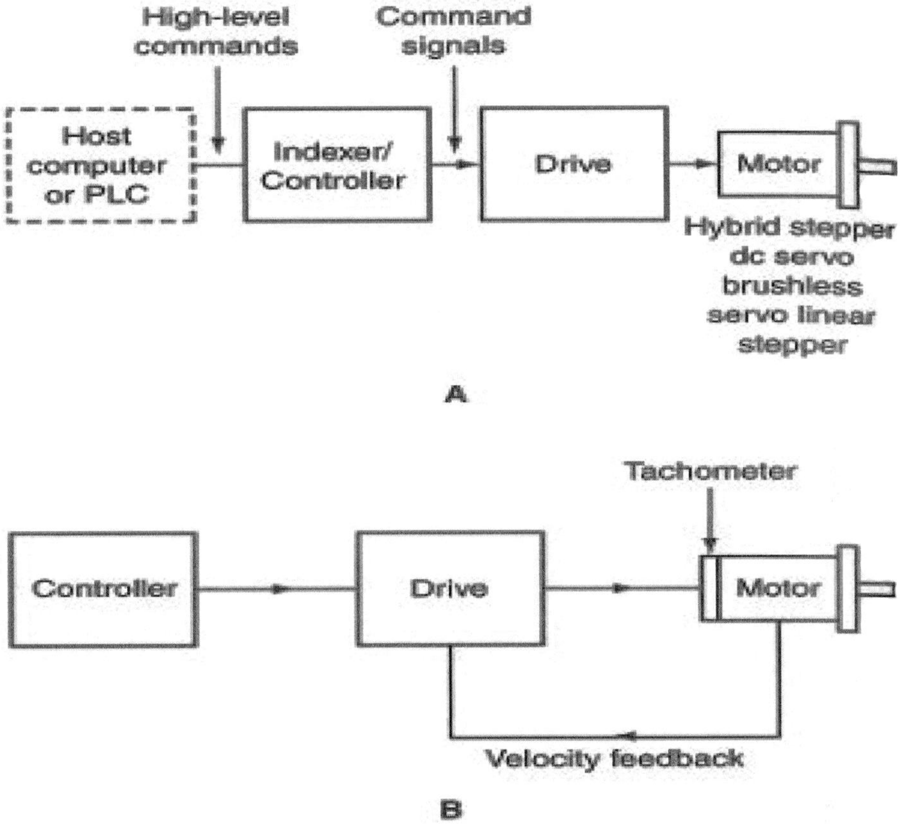

Figure 12.21 Motion control systems. A – Elements of motion control. B – Typical closed-loop servo system. (Parker-Hannefin Co.)

efficient in terms of torque output relative to both physical size and input power requirements. **Microstepping** can be used to improve low-speed application accuracy.

Short, rapid, repetitive moves are best done with a **DC stepper**. The lack of commutation problems is also important. The brushes of the DC stepper motor can cause problems with frequent starts, stops, and direction changes. Low-speed, high-smoothness applications are appropriate for microstepping or direct drive servos. Applications in hazardous environments or in a vacuum cannot use a brush motor. Either a stepper or a brushless motor should be used.

Stepper motors are simple to drive and control. They only require four input connections. They provide excellent torque at low speeds and up to five times the continuous torque of a brush motor of the same frame size. For some applications, this eliminates the need for a gear drive.

Stepper Motor Technology

As the use of stepper motor systems has increased in a broad base of applications, terminology has evolved which requires definition.

A. **Step Angle**—This is the specific angular increment the motor shaft will move each time the winding polarity is changed. It is specified in degrees.

B. **Steps Per Revolution**—This term describes the total number of steps required for the motor shaft to rotate 360°, or one complete revolution. The number is calculated by dividing the step angle into 360°.

C. **Steps Per Second**—This is the number of steps accomplished by the motor in one second of time. This figure replaces the rpm value of a standard drive motor.

D. **Step Accuracy**—Defined as positional accuracy tolerance. This value is generally expressed in percent and indicates the total error introduced by the stepper motor in a single step movement. The error is noncumulative, i.e., it does not increase as additional steps are taken. In a linear system with a resolution of 0.001 inch, a 3% accuracy motor would introduce a maximum of 0.00003 inch error into the system. This total error would not accumulate or increase with total distance moved or number of movements made.

E. **Holding Torque**—With the motor shaft at standstill (zero rpm condition), holding torque is the amount of torque, from an external source, required to break the shaft away from its holding position. It is measured with rated current and voltage applied to the motor. Holding torque is a basic characteristic of stepper motors and provides positioning integrity under standstill, or rest, conditions.

F. **Residual Torque**—This is the torque present at standstill under power off conditions and is a result of the permanent magnetic flux acting on the stator poles. Residual torque is present under power off conditions only with a motor of permanent magnet rotor design.

G. **Step Response**—When given a command to take a step, a stepper motor will respond within a specific time period. This time period, or "time for a single step," is a function of the torque-to-inertia ratio of the motor and of the characteristics of the electronic drive system. Ratings given are for no-load conditions and are generally expressed in milliseconds.

H. **Torque-To-Inertia Ratio**—This ratio is calculated by dividing the rated holding torque of the motor in ounce-inches by its rotor inertia in ounce-inch-seconds squared. The better the torque-to-inertia ratio, the better the step response. When dealing with step response problems, it is important to know the torque-to-inertia ratio of the motor.

I. **Resonance**—Stepper motors are a "spring response" system and, as such, have certain "natural" frequency characteristics. When a motor's natural frequency or "resonance" is reached, an increase in the audible level of the motor's operation can be detected. In cases of severe resonance, the motor may lose steps and/or oscillate about a point. The frequency at which this occurs depends on the motor and the load. In many applications, it may not occur to any perceptible degree. However, the designer should realize that this condition can exist and specific facts about the resonant characteristics of a particular motor should be obtained from the manufacturer.

J. **Drives**—This is a broad term used to describe the circuitry which controls the stepper motor. It usually consists of a power supply, sequencing logic and power output switching components. These drives are generally categorized as Translators, Translator/Oscillators, or Indexers.

K. **Translator**—An electronic control with circuitry to convert pulses into the switching sequence which will operate the motor one step for each pulse received. Translators usually have no counting capability.

L. **Translator/Oscillator**—Includes translator circuitry together with a built-in oscillator which can serve as a pulse source.

M. **Indexer**—An electronic control which includes the translator function plus additional circuitry to control the number of steps taken as well as direction and velocity.

N. **Pulse Rate**—The rate at which windings are switched. Where one pulse equals one motor step, the pulse rate is also the motor stepping rate.

O. **Ramping**—This is the process of controlling the pulse frequency to accelerate the motor from base speed to running speed as well as to decelerate the motor from running speed to base speed. Ramping increases the ability to drive the motor and load to higher speeds, particularly with large inertial loads.

P. **Slew Rate**—An area of high speed where the motor can run unidirectionally in synchronism. However, it cannot start, stop, or reverse at this rate. A stepper motor is brought to a slewing rate using acceleration and is then decelerated to a stop under conditions where no step loss can be tolerated.

Q. **Damping**—Damping is defined as the reduction or elimination of step overshoot. It is used where settling down time is important. Damping methods used include mechanical, electronic, and viscous means.

Figure 12.22 Stepper motor terminology. (Warner Electric/Superior Electric Co.)

There are three main types of stepper motors.

- Permanent magnet stepper motors.
- Variable reluctance stepper motors.
- Hybrid motors.

These types are discussed in the following sections.

12.8.2 Permanent magnet stepper motors

Permanent magnet stepper motors are one of the most widely used type of motor for nonindustrial applications. They are low-cost, low-torque, low-speed devices. It is their simplicity of construction that makes them a very low-cost item. The motor construction results in relatively large step angles. They are effectively used for applications requiring little inertia, such as positioning the print wheel in a printer for a computer system.

Disadvantages of permanent magnet stepper motors include resonance effects and relatively rough performance at low speeds, unless a microstepping drive is used. Microstepping drives subdivide the basic motor step by proportioning the current in the two adjacent windings. In this way, the step size is reduced.

Another disadvantage is that these motors consume current regardless of load conditions. They therefore tend to run hot. Losses at low speed are relatively high and can cause excessive heating. They are also frequently noisy, especially at high speeds.

12.8.3 Variable reluctance stepper motors

There is no permanent magnet in a variable reluctance stepper motor. Torque output for a given frame size is restricted, although the torque-to-inertia ratio is good. This type of motor is used in small sizes for applications such as micro-positioning tables.

Variable reluctance stepper motors are seldom used in industrial applications. Since they have no permanent magnet, they are not sensitive to current polarity. They require a complex drive system compared to other motor types.

12.8.4 Hybrid motors

The **hybrid motor** is a widely used stepper motor for industrial applications. The name, hybrid motor, derives from the fact that it combines the operating principles of the other two motor types (permanent magnet and variable reluctance). Most hybrid motors are considered two-phase motors.

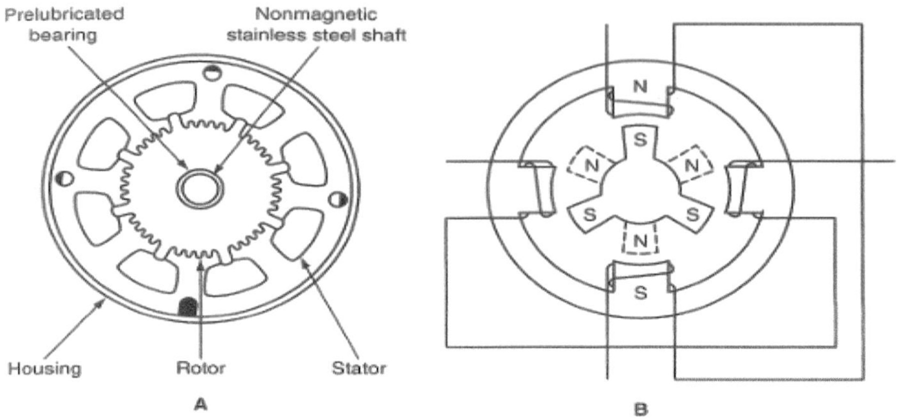

Figure 12.23 A – Hybrid stepper motor. B – Simple 12 step per revolution hybrid motor. (Parker-Hannefin Co.)

An enhanced hybrid stepper motor is a unique type that uses special magnets to improve performance.

The operation of the **hybrid motor** can be understood by looking at a simple model that will produce 12 steps per revolution, Figure 12.23. The rotor of this machine consists of two pole pieces with three teeth on each. Between the pole pieces is a permanent magnet that is magnetized along the axis of the rotor. This makes one end a north pole and the other a south pole. The teeth are offset at the north and south ends as shown in part B of Figure 12.23.

The stator consists of a shell that has four teeth that extend the full length of the rotor. Coils are wound on the stator teeth and are connected together in pairs.

With no current flowing in any of the motor windings, the rotor will tend to move to one of the positions shown in the diagrams of Figure 12.24. This is because the permanent magnet in the rotor is trying to minimize the magnetic resistance (reluctance) of the flux path from one end to the other. This minimization occurs when a pair of north and south pole rotor teeth are aligned with two of the stator poles.

The torque that holds the rotor in one of these positions is usually small. The torque is referred to as the **detent torque**. The motor shown has twelve possible detent positions. If current is passed through one pair of stator windings, as shown in part A of Figure 12.24, the resulting north and south stator poles will attract teeth of the opposite polarity on each end of the rotor. There are now only three stable positions for the rotor. This

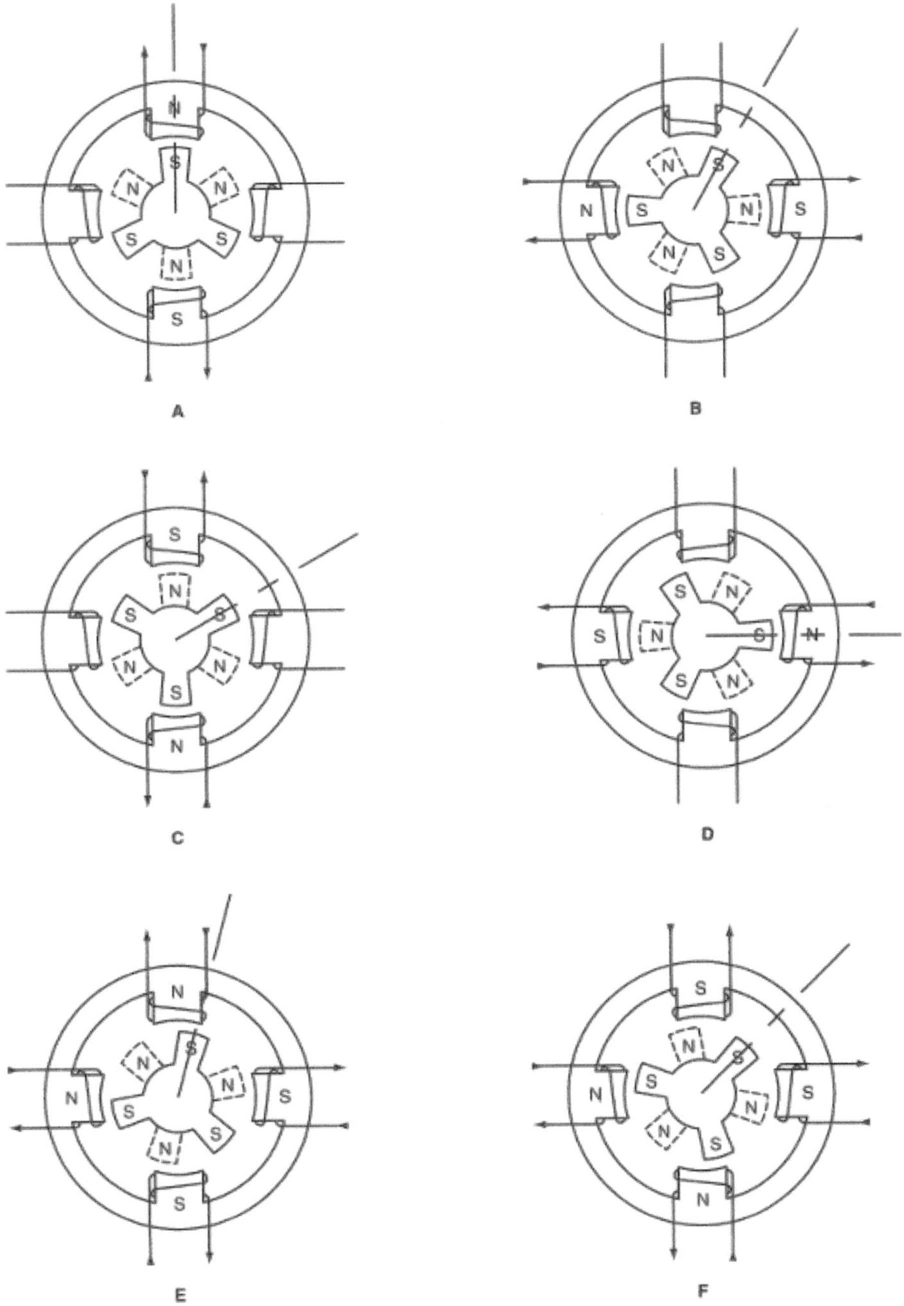

Figure 12.24 Full stepping. Parts A, B, C, and D: One phase on. Parts E and F: Two phase on. (Parker-Hannefin Co.)

Figure 12.25 Half stepping. (Parker-Hannefin Co.)

is the same as the number of rotor teeth. The torque required to move the rotor from its stable position is now much greater. This is referred to as the holding torque.

By changing the current flow from the first to the second set of stator windings (see part B of Figure 12.24), the stator field rotates through 90° and attracts a new pair of rotor poles. This results in the rotor turning through 30°, corresponding to one full step. Going back to the first set of stator windings and energizing them in the opposite direction will rotate the stator field through another 90°. The rotor then takes another 30° step, part *C* of Figure 12.24. Finally, the second set of windings is energized in the opposite direction, part D of Figure 12.24. This causes movement to a third step position. Now look back to position A. Notice that after four steps, the rotor will move through one tooth pitch. **Tooth pitch** is the angular separation between rotor teeth, in this case 30°. This motor, therefore, performs 12 steps per revolution ($12 \times 30° = 360°$).

If the coils are energized in the reverse sequence, the motor will rotate in the other direction. If two coils are energized at the same time, as shown in part E of Figure 12.24, the rotor takes the position shown. It is attracted to two different stator poles. Greater torque is produced under these conditions since more of the stator poles are influencing the rotor.

The motor can be made to take a full step simply by reversing the current in one set of windings. This causes a 90° rotation of the stator field. This would be the normal way of driving the motor. This method always keeps two windings energized and reverses the current in each winding alternately.

By alternately energizing one winding and then two windings, as shown in part A of Figure 12.25, the rotor moves through only 15°. The number of steps per revolution will be doubled. This is called half-stepping. Although

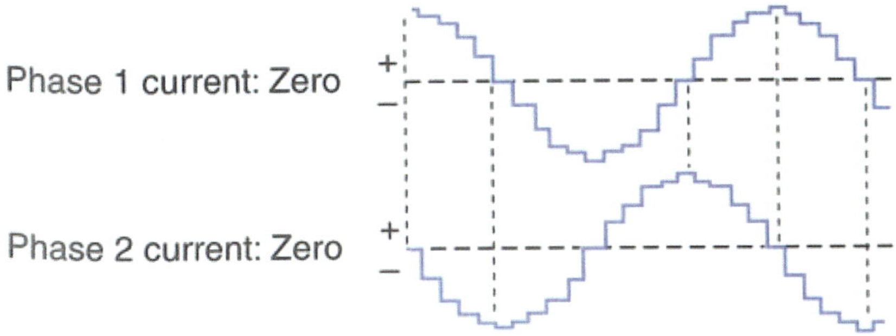

Figure 12.26 Microstepping. (Parker-Hannefin Co.)

there is sometimes a slight loss of torque, this mode results in much better smoothness of operation at low speeds.

As mentioned earlier, **microstepping** drives subdivide the basic motor step by proportioning the current in the two adjacent windings. In this way, the step size is reduced. The low-speed smoothness is also greatly improved. High-resolution microstep drives divide the full motor step into, for example, 500 microsteps, giving 100,000 steps per revolution. In this situation, the current pattern in the windings closely resembles two sine waves with a 90° phase shift between them. This is shown in Figure 12.26. The motor is now being driven in the same way as the AC synchronous motor. In fact, the stepper motor can be driven in this way from a 60 Hz sine wave source by including a capacitor in series with one phase. It will then rotate at 60 revolutions per minute.

A standard 200-step **hybrid motor** operates in the same way as the stepper just described. However, this motor type has a greater number of teeth on the rotor and stator, giving a smaller basic step size. The rotor is constructed in two sections, but each section has 50 teeth. The half tooth displacement between the two sections is also used. The stator has eight poles each with five teeth, making a total of 40 teeth. See Figure 12.27.

If a tooth were placed in each of the gaps between the stator poles, there would be a total of 48 teeth (two less than the number of rotor teeth). If rotor and stator teeth are aligned vertically, they will be aligned at the top and bottom. However, due to the displacement between the sets of rotor teeth, alignment of the rotor also occurs horizontally.

The windings are arranged in sets of four and wound such that opposite poles are the same. The north poles at the top and bottom attract the south pole teeth at the front of the rotor. The south poles at 3 and 9 o'clock attract the north pole teeth at the back. By switching the current to the second set of coils, the stator field pattern rotates through 45° to align with this new field.

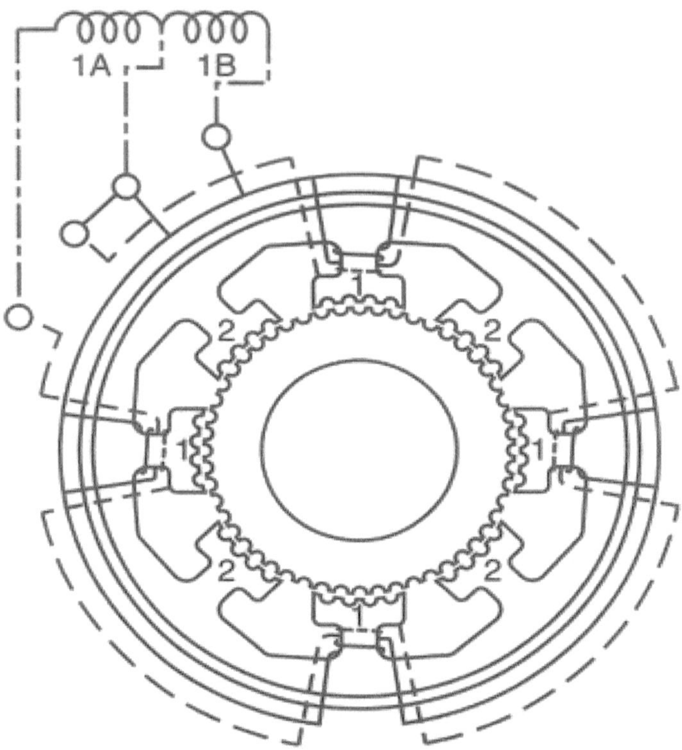

Figure 12.27 Two hundred-step hybrid motor. (Parker-Hannefin Co.)

The rotor only has to turn through 1.8°. This is equivalent to one quarter of a tooth pitch on the rotor, giving 200 full steps per revolution.

Note that there are as many **detent positions** as there are full steps per revolution. Normally, there are 200. The detent positions correspond with rotor teeth being fully aligned with stator teeth. When power is applied to a stepper drive, it is usual for it to energize in the "zero phase" state. An unloaded motor will always move at least one-half step when power is turned on.

Bifilar windings are used on these motors. This means there are two identical sets of windings on each pole. Two lengths of wire are wound together as though they were a single coil. This produces two windings that are almost identical. Rather than having to reverse the current in one winding, the field can be reversed by transferring current to a second coil wound

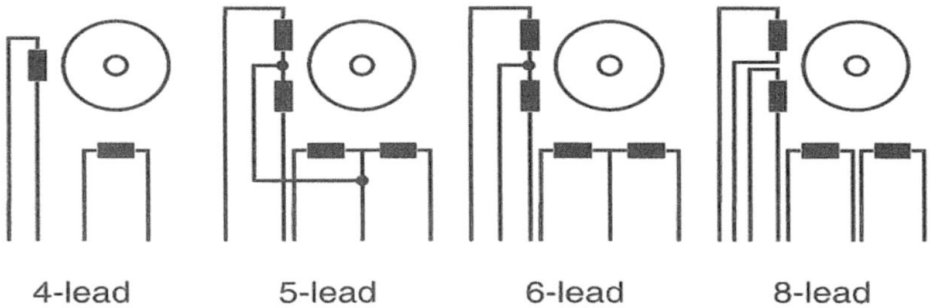

Figure 12.28 Motor lead configurations. (Parker-Hannefin Co.)

in the opposite direction. Although the two coils are wound the same way, interchanging the ends has the same effect on polarity. With a bifilar-wound motor, the drive can be kept simple. The two sets of windings give additional flexibility for different connection methods to give alternative torque–speed characteristics.

If all the coils of a **bifilar-wound motor** are brought out separately, there will be a total of eight leads as shown in Figure 12.28. This is a common configuration since it provides flexibility. Motors are still produced with only six leads. One lead serves as a common connection to each winding of a bifilar pair. This arrangement limits the range of applications of the motor. The windings cannot be connected in parallel. Some motors are made with only four leads. These are not bifilar-wound and cannot be used with a unipolar drive. There is only one connection method with a four-lead motor. For many applications, this is not a problem. Occasionally, a five-lead motor will be encountered. They cannot be used with bipolar drives.

12.8.5 Linear stepper motors

The **linear stepper motor** is similar to a rotary stepper motor that has been modified such that it operates in a straight line. The moving component is referred to as the forcer, and it travels along a fixed part or platen. The platen is equivalent to the rotor in a normal stepper. This system has no permanent magnet. The magnetic interaction is the force with the coils as shown in Figure 12.29.

The ***forcer*** shown is equipped with four pole pieces each having three teeth. The teeth are staggered in pitch with respect to those on the platen. This allows switching the current in the coils to bring the next set of teeth into alignment. A complete switching cycle (four full steps) is equivalent to one tooth pitch on the platen. Like the rotary stepper, the linear motor can be driven from a **microstep drive.** A typical linear resolution will be 12,500 steps per inch.

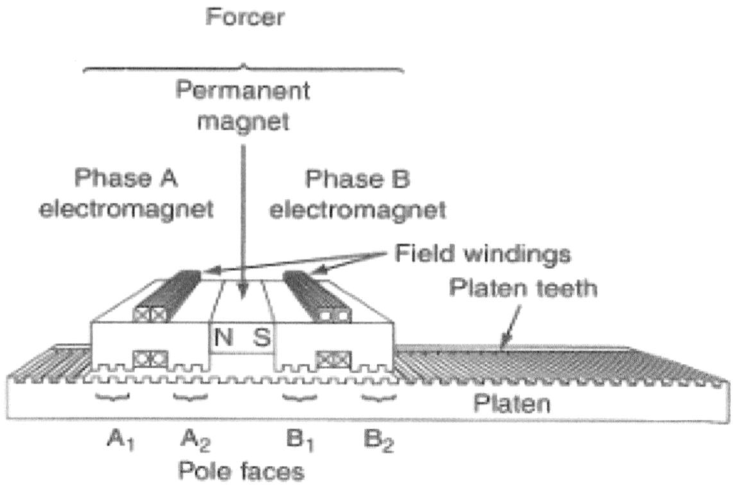

Figure 12.29 Linear stepper motor. (Parker-Hannefin Co.)

The **linear motor** is used for applications involving a low mass to be moved at high speed. In a leadscrew-driven system, most inertia is usually in the leadscrew rather than the load to be moved. Hence, most of the motor torque goes to accelerate the leadscrew. This problem becomes worse the longer the travel time required. Using a linear motor, the developed force is applied directly to the load, and the performance achieved is independent of the length of the move.

The **forcer** consists of two electromagnets (A and B), as shown in Figure 12.29, and a strong rare-earth permanent magnet. The two pole faces of each electromagnet are toothed to concentrate the magnetic flux. Four sets of teeth on the forcer are spaced in **quadrature** (separated by a quarter cycle) so that only one set at a time can be aligned with the **platen**. The magnetic flux passing between the forcer and the platen develops a very strong force of attraction between the two sections.

12.8.6 Stepper motor characteristics

There are a number of stepper motor performance characteristics that should be considered. Figure 12.30 shows a **static torque curve** for a hybrid step motor. This relates to a motor that is energized but stationary. It shows how the restoring torque varies with rotor position as it is moved from its stable point. Assume that there are no frictional or other static loads on the

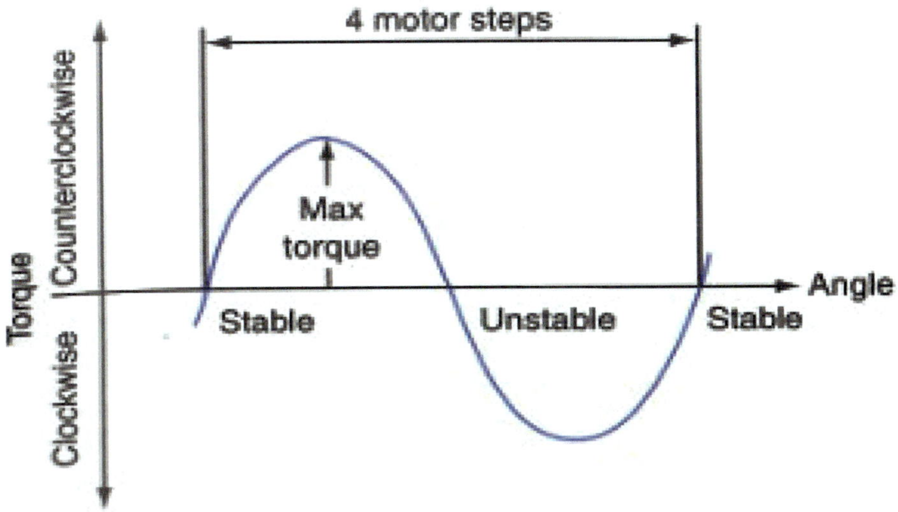

Figure 12.30 Static torque–displacement characteristic. (Parker-Hannefin Co.)

motor. As the rotor moves away from the stable position, the torque steadily increases until it reaches a maximum after one full step (1.8°). This maximum torque value, the holding torque, represents the largest static load that can be applied to the shaft without causing continuous rotation. The maximum **running torque** of the motor is always less than the holding torque (typically about 70%).

As the shaft is moved beyond one **full step**, the torque falls until it is again at zero after two full steps. This zero point is unstable, and the torque reverses just beyond it. The next stable point is found four full steps away from the first. This is equivalent to one tooth pitch on the rotor, or 1/50 of a revolution.

The **static torque** characteristic helps to explain some of the effects of the motor. For example, the stiffness of the system is determined by how the shaft position changes when a torque load is applied to a stationary motor. The shaft must move until the generated torque matches the applied load. If the load varies, the static position also varies. Position errors will result from effects such as friction or out-of-balance torque loads. It is important to remember that the static stiffness is not improved by using a microstepping drive. A given load on the shaft will produce the same angular deflection. While microstepping increases resolution and smoothness, it may not improve positioning accuracy. Under dynamic conditions with the

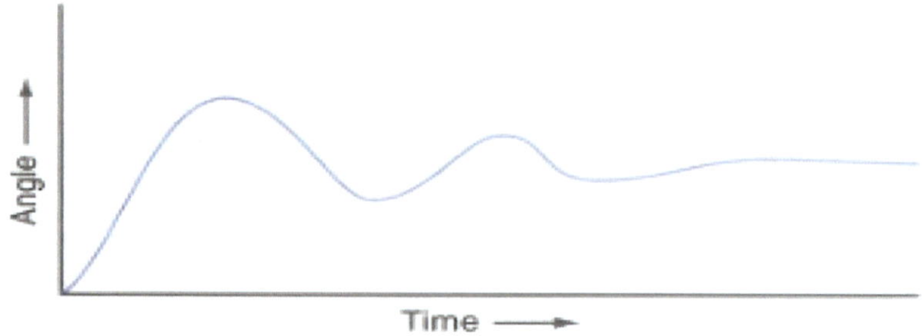

Figure 12.31 Single step response. (Parker-Hannefin Co.)

motor running, the rotor must be lagging behind the stator field to produce torque.

When the **stepper** performs a single step, the response is a form of oscillation, as shown in Figure 12.31. The system behaves like a "magnetic spring." When power is first applied, this accelerates the rotor until a new stable point is reached. When the momentum carries the rotor past the stable position, the torque now reverses. This slows the rotor down and brings it back in the opposite direction. The **amplitude** and **frequency** of this oscillation will depend on the friction and inertia in the system and the electrical characteristics of the motor and drive. This overshoot is also dependent on step amplitude. Thus, **half-stepping** produces less overshoot than **full stepping**. It is important to step the motor at its natural oscillation frequency. Higher rates can cause an exaggerated response known as resonance. This can lead to the motor desynchronizing or stalling. It is seldom a problem with half-step drives and even less with a microstepper.

To operate a **DC stepper motor** at faster speeds, it is necessary to start at a speed within the start-stop range as shown in Figure 12.32. When stopping the motor, it must be decelerated back into the start-stop range before the applied clock pulses end. Using acceleration and deceleration, called **ramping**, allows much higher speeds to be achieved.

Continuous operation at high speeds is not normally possible with a **DC stepper motor** due to rotor heating. High speeds can be used for many positioning applications. It is difficult to demagnetize a DC stepper motor. However, too much current will damage the motor. Excessive heating can melt the insulation or damage the rotor laminations. If this happens, the effects can be the same as demagnetizing the rotor.

To produce a motor with a higher **torque output**, it is necessary to increase the strength of both the permanent magnet in the rotor and the stator field. A stronger rotor magnet can be obtained by increasing the diameter. This

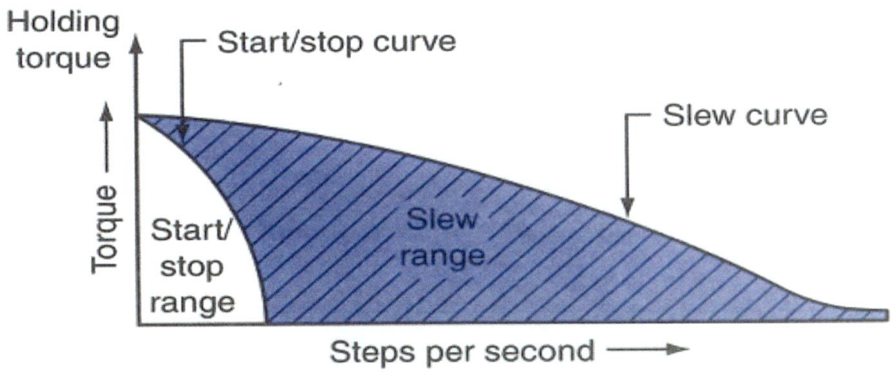

Figure 12.32 Start/stop and slew curves. (ParkerHannefin Co.)

reduces the acceleration performance of the motor because the **torque-to-inertia ratio** gets worse. To increase torque output without reducing acceleration performance, more magnet sections, or stacks, can be added to the same shaft. A second stack will enable twice the torque to be produced. It will also double the inertia, so the torque-to-inertia ratio remains the same. DC stepper motors are manufactured in single-, two-, and three-stack designs in each frame size.

12.8.7 Brushless DC motors

A major limiting factor in the performance of **DC brush motors** is internal heating. This heat can escape only through the shaft and bearings to the outer casing, or through the gap between the armature and field magnets and from there to the casing. Both of these routes are thermally inefficient. Cooling of the motor armature is very difficult.

In the **brushless DC motor**, the construction of the iron-cored motor is turned inside out, so the rotor becomes a permanent magnet, and the stator becomes a wound iron core. This design is shown in Figure 12.33. The current-carrying coils are located in the housing. This provides a short, thermal path to the outside. The ease of cooling the brushless motor allows it to produce a much higher power in relation to its conventional size.

Another major advantage is the lack of a **commutator and brush assembly.** These are a source of wear and potential maintenance problems. The brushless motor can be used in adverse environmental conditions.

A major disadvantage of the brushless DC motor is that in order to achieve high torque and low inertia, it becomes necessary to use **rare-earth**

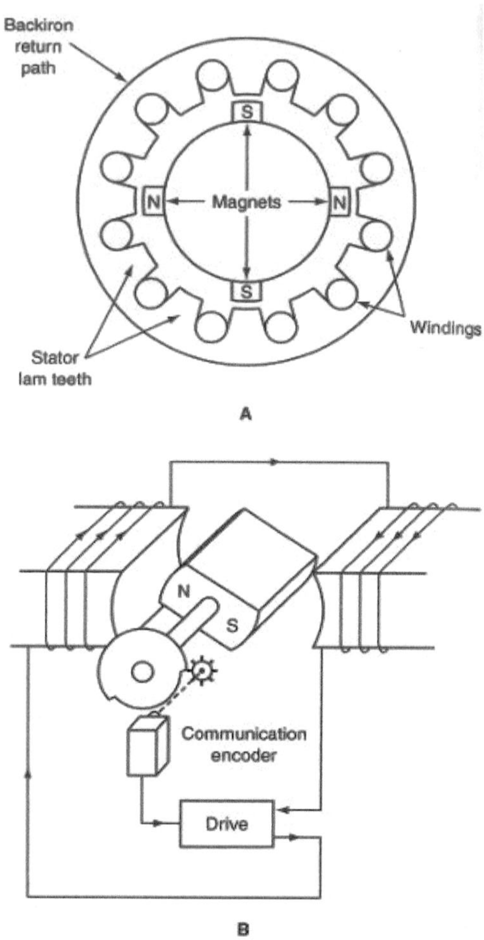

Figure 12.33 A – Brushless DC motor. B – Brushless motor. (Parker-Hannefin Co.)

magnets, which are much more expensive than **ceramic magnets**. The electronic circuits necessary to drive a brushless motor are also more complex than for a brush motor. A servoamplifier for a brushless DC motor is shown in Figure 12.34.

12.9 Stepper Motor Drives

The **stepper drive** delivers electrical power to a motor in response to signals from a control system. The drive must act as a source of current. The applied voltage acts as a means of controlling the current. The motor is a

Figure 12.34 Brushless DC motor servo-amplifier. (Automation, Inc.)

torque-producing device. This torque is generated by the interaction of magnetic fields. The driving force behind the stator field is the **magnetomotive force (MMF)**. MMF is proportional to current and to the number of turns in the windings. The product of the current and the number of turns is called the ampere-turns of the windings.

Input signals to the **stepper drive** consist of step pulses and a direction signal. One step pulse is required for every step the motor is to take. This is true regardless of the stepping mode. The drive can require between 200 and 50,000 pulses to produce one revolution of the shaft. A commonly used **stepping mode** in industrial applications is the half-step mode. The motor performs 400 steps per revolution. At a shaft speed of 1800 rpm, this corresponds to a step pulse frequency of 20 kHz. The same shaft speed at 25,000 steps per revolution requires a step frequency of

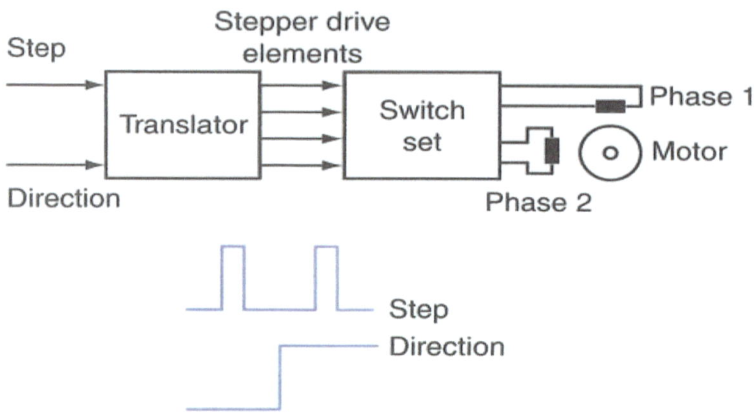

Figure 12.35 Stepper drive elements. (Parker-Hannefin Co.)

750 kHz. Motion controllers for microstep drives must be able to output a much higher step frequency.

The logic section of the stepper drive is referred to as the translator. Its function is to **translate** the step and direction signals into control waveforms for the switch set. Refer to Figure 12.35. The basic translator functions are common to most types of drives. The translator is more complex for a microstepping drive.

12.9.1 Unipolar drives

The simplest type of switch set is the **unipolar** arrangement shown in Figure 12.36. It is referred to as a unipolar drive because current can only flow in one direction through any particular motor terminal. A bifilar-wound motor must be used since reversal of the stator field is achieved by transferring current to the second coil. This type of drive functions well at low stepping rates. As **speed** is increased, the **torque** falls off rapidly due to the inductance of the windings.

A disadvantage of the **unipolar drive** is its inability to utilize all the coils on the motor. At any one time, there will only be current flowing in one half of each winding. If both sections at the same time could be used, a 40% increase in ampereturns for the same power dissipation would result. To achieve high performance combined with high efficiency, a bipolar drive is needed. This type can drive current in either direction through each motor coil.

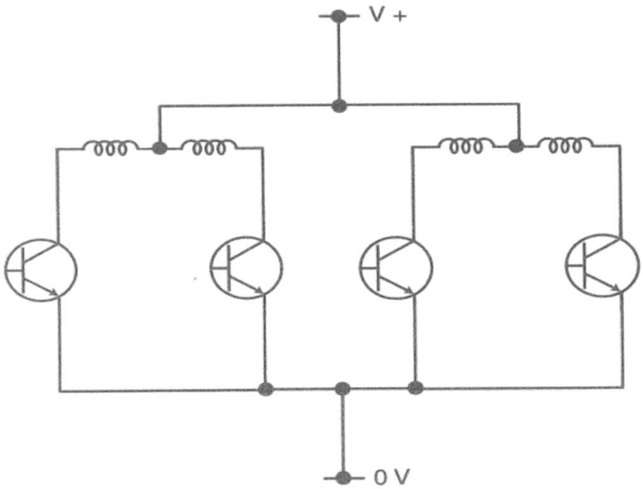

Figure 12.36 Basic unipolar drive. (Parker-Hannefin Co.)

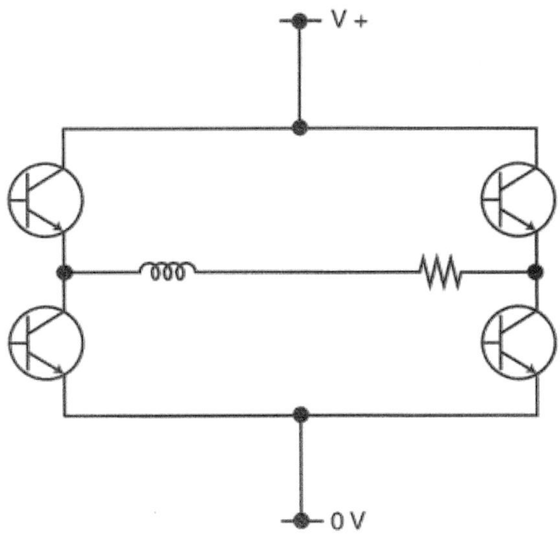

Figure 12.37 Bipolar RL drive. (Parker-Hannefin Co.)

12.9.2 RL drives

The **RL drive** principle is illustrated in the circuit of Figure 12.37. Using an applied voltage of, for example, 10 times the rated motor voltage, the current will reach its final value in one-tenth of the time. This produces an increase in motor speed.

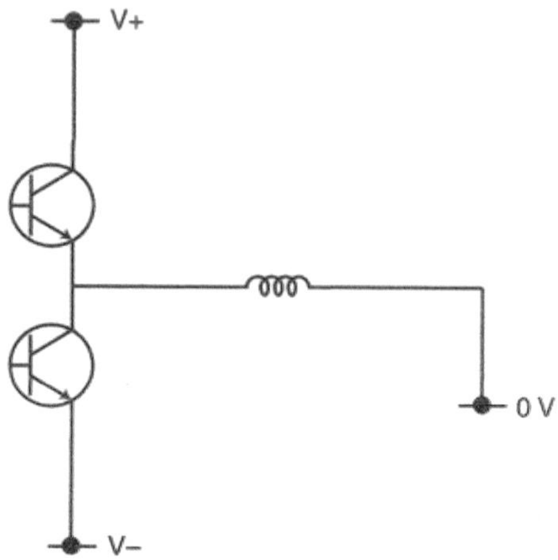

Figure 12.38 Simple bipolar drive. (Parker-Hannefin Co.)

Under normal conditions, there is much more powcr dissipated in the series resistor than in the motor itself. This produces a large amount of heat. The extra power must all come from the DC power supply. RL drives are only suited to low-power applications. They do have the benefits of simplicity, robustness, and low interference produced.

12.9.3 Bipolar drives

A **bipolar drive** is shown in Figure 12.38. Two power supplies are used with a pair of switching transistors. Current can be made to flow in either direction through the motor coil by turning on one transistor or the other.

However, there are problems with this scheme. Two power supplies capable of delivering the total current for both motor phases are needed. When all the current is coming from one supply, the other is not operating. The power supply utilization is poor. The transistors must be rated at double the voltage that can be applied across the motor. This requires the use of more costly components.

The standard arrangement used in **bipolar motor drives** is the bridge system shown in Figure 12.39. This circuit uses an extra pair of switching transistors. The problems of the previous circuit are overcome. Only one power supply is needed. Transistor voltage ratings are the same as those available for driving the motor.

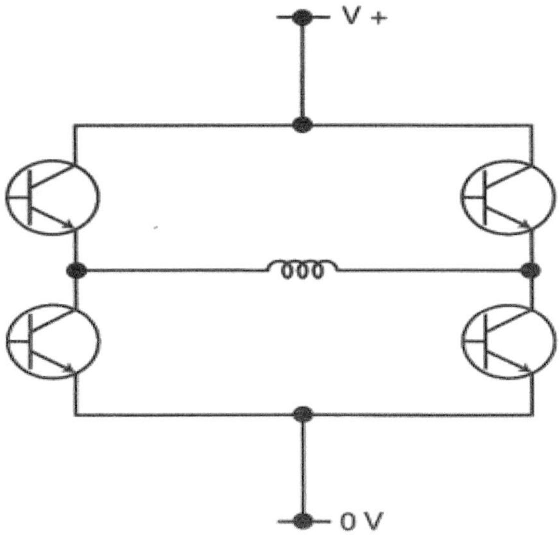

Figure 12.39 Bipolar bridge. (Parker-Hannefin Co.)

12.9.4 Chopper drives

The method of current control used in some stepper drives is called a **recirculating chopper**. This circuit utilizes a four-transistor bridge (see Figure 12.39) together with recirculation diodes. It provides a feedback voltage proportional to the current in the motor.

Current flows into the motor winding by turning on one top switch and one bottom switch. This applies the full supply voltage across the motor. Current will rise in an almost linear fashion. When the required current level has been reached, the top switch is turned off and the stored energy in the coil keeps the current circulating through the bottom switch and a parallel diode. Losses in the system cause this current to slowly decrease. When a preset value is reached, the top switch is turned back on and the cycle repeats. The current is therefore maintained at the correct value by switching or **chopping** the supply to the motor windings.

This method of current control is very efficient because very little power is dissipated in the **switching transistors**. Power drawn from the power supply is closely related to the mechanical power delivered by the motor shaft. This is not true of the RL drive, which draws maximum power from the supply at a standstill.

A variation of this circuit is the **regenerative chopper,** in which the supply voltage is applied across the motor winding in alternating directions. This causes the current to increase and decrease at approximately equal rates. This technique requires fewer components and is lower in cost. However, ripple currents in the motor cause an increase in motor heating.

12.9.5 Microstepping drives

Dividing the basic motor stop into smaller values is possible by proportioning the current in the two motor windings. This produces a series of intermediate step positions between the one-phase on points. These intermediate positions are equally spaced and produce approximately equal torque when the motor is running. Accurate microstepping places increased demands on the accuracy of the drive system. **Pulse-width modulation (PWM)** is frequently used to achieve higher accuracy than can be achieved using other systems.

In high-resolution **microstep drives** producing 10,000 steps per revolution or more, the best performance will only be obtained with a particular type of motor. This type has stator teeth that are on a 7.5-degree pitch, giving 48 equal pitches in 360°. In most hybrid steppers, the stator teeth have the same pitch as the rotor teeth. This gives equal increments of 7.2 degrees. This arrangement gives better torque output. It is less satisfactory as a microstepper since the magnetic poles are "harder." There is no progressive transfer of tooth alignment from one pole to the next.

12.9.6 Torque–speed curves

Motor inductance is the factor that opposes rapid changes of current. This makes it more difficult to drive a stepper at high speeds. A **torque–speed curve** is shown in Figure 12.40. At low speeds, the current has plenty of time to reach the required level, and so the average current in the motor is very close to the regulated value from the drive. Changing the regulated current setting or changing to another drive with a different current rating affects the available torque.

As speed is increased, the time taken for the current to rise becomes a large part of the interval between step pulses. This has the effect of reducing the average current level. Therefore, the torque starts to decrease. As speed is increased further, the interval between **step pulses** does not allow the current time to reach a level where the chopping action can begin. Under these conditions, the final value of current depends only on the supply voltage. If the

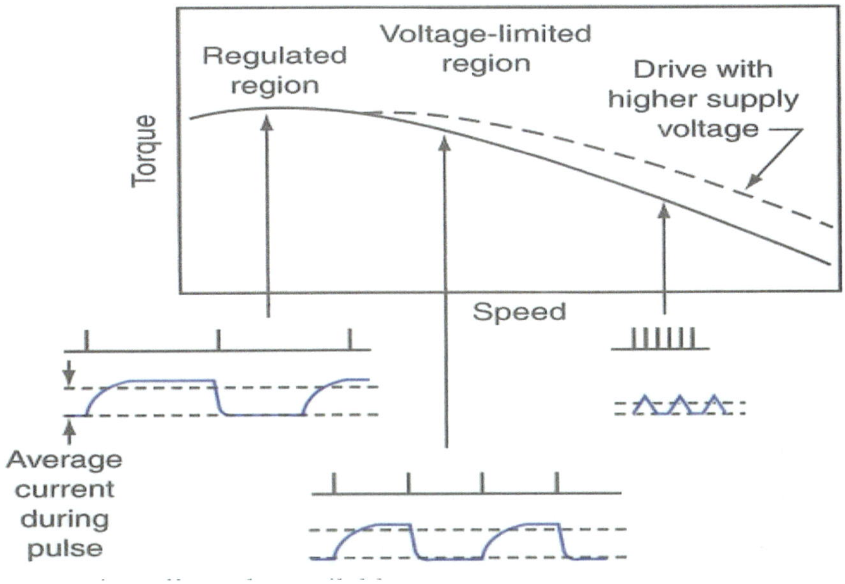

Figure 12.40 Regulated and voltage-limited regions of the torque–speed curve. (Parker-Hannefin Co.)

voltage is increased, the current will increase more rapidly. This will achieve a higher value in the available time. So this region of the curve is described as **voltage-limited**, as a change in the drive current setting would have no effect.

We can summarize by saying that at low speeds the torque depends on the drive current setting, whereas at high speeds it depends on the drive supply voltage. The high-speed performance is not affected by the drive current setting. Reducing the current simply "flattens out" the torque curve without restricting the ability to run at high speeds.

12.10 Feedback Devices

Several different types of **feedback devices** are used with motion control systems. These feedback devices are electronic devices that monitor the motion or position of a motor and transfer this information back to the controller or drive system. The feedback system is a very important part of a motion control system. Some common types of feedback devices are tachometers, optical encoders, incremental encoders, and resolvers.

12.10.1 Tachometers

A permanent magnet de motor can be used as a tachometer, since when driven mechanically it generates an output voltage that is proportional to shaft speed. The other main requirements for a tachometer are that the output voltage should be smooth over the operating range and that the output should be stabilized against temperature variations. Small permanent magnet DC motors are frequently used in servo systems as speed-sensing devices. In order to combine high performance and low cost, de servo motor designs often use a tachometer mounted on the motor shaft and enclosed within the motor housing.

12.10.2 Optical encoders

In **servo control systems**, where mechanical position is required to be controlled, some form of position-sensing device is needed. There are a number of types in use, but for accurate position control, the most used is the optical encoder. There are two forms of this encoder, absolute and incremental.

Optical encoders operate by means of a grating that moves between a light source and a detector. When light passes through the transparent areas of the grating, an output is seen from the detector. For increased resolution, the light intensity is increased, and a mask is placed between the grating and the detector. The grating and the mask produce a shuttering effect, so that only when their transparent sections are in alignment is light allowed to pass to the detector.

An **incremental encoder** generates a pulse for a given increment of shaft rotation (rotary encoder) or a pulse for a given linear distance travel (linear encoder). Total distance traveled or shaft angular rotation is determined by counting the encoder output pulses. An absolute encoder has a number of output channels, such that every shaft position can be described by its own unique code. Types of encoders are shown in Figure 12.41. Typical applications are shown in Figure 12.42.

12.10.3 Resolvers

A resolver, in principle, is similar to a rotating transformer. Figure 12.43 shows a brushless resolver. This type of feedback device provides accurate position and velocity feedback. They are also resistant to shock and vibration, which are often encountered in industrial settings.

Like other types of feedback devices, the resolver mounts directly on the motor shaft. No coupling devices are needed. The information output from the resolver will usually be converted from analog to digital form, for use in a digital positioning system, by means of a resolver-to-digital converter.

A B C

Figure 12.41 Encoders. A – Optical encoder that utilizes reflected light from a graduated disk attached directly to the rotor. It produces either a digital or analog output. B and C – Typical encoders. (Portescap Co.; BEI Sensors and Motion Systems; Dynamics Research Group)

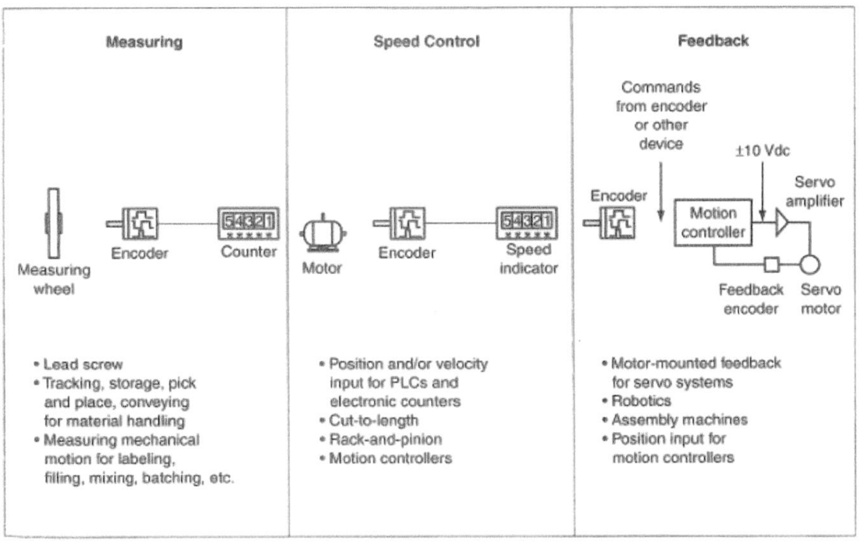

Figure 12.42 Applications of encoders. (Encoder Products Co.)

12.11 Special-purpose Systems

There are several types of special-purpose systems that are used for motion control applications. A few of these special types of systems are shown in Figures 12.44–52. The systems that are shown here illustrate some of the many changes that are taking place in motion control technology today.

Figure 12.43 Brushless resolver. (Clifton PrecisionNorth, Division of Litton Systems, Inc.)

Figure 12.44 Disk magnet stepper motor. (Portescap U.S., Inc.)

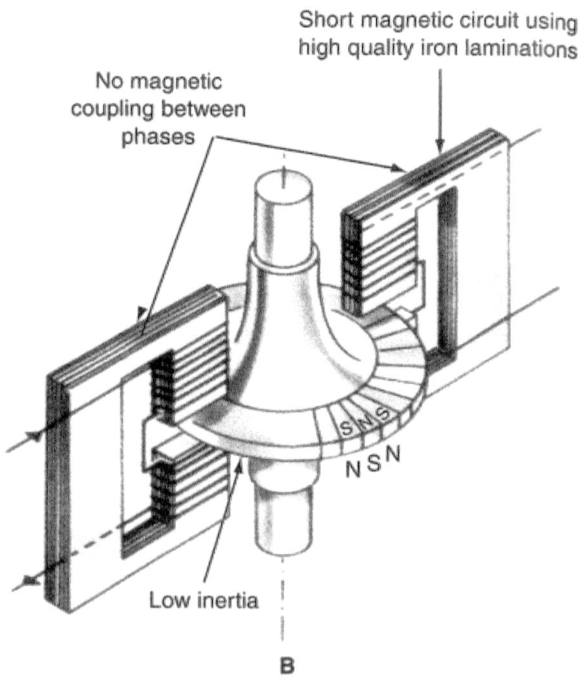

Figure 12.45 Operational principle for a disk magnet stepper motor. (Portescap U.S., Inc.)

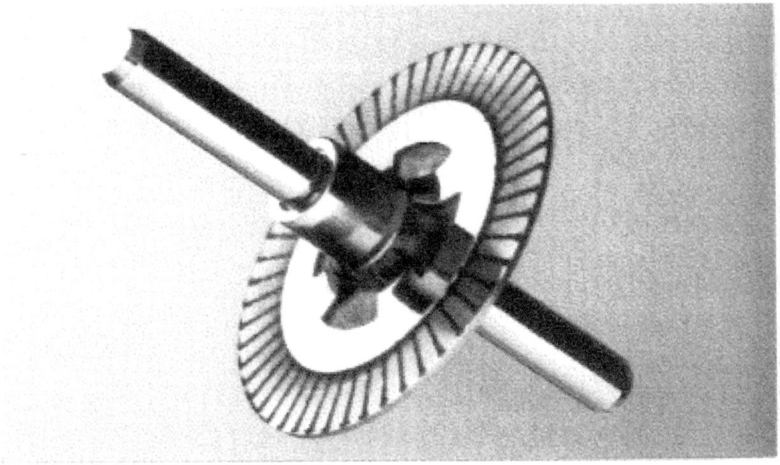

Figure 12.46 Rare-earth magnet low-inertia rotor for a disk magnet stepper motor. (Portescap U.S., Inc.)

Figure 12.47 DC motor with ironless rotor commutation system. Low low-inertia rotor allows rapid acceleration and deceleration. (Portescap U.S., Inc.)

Figure 12.48 Space-saver disk motor. (Panasonic Co.)

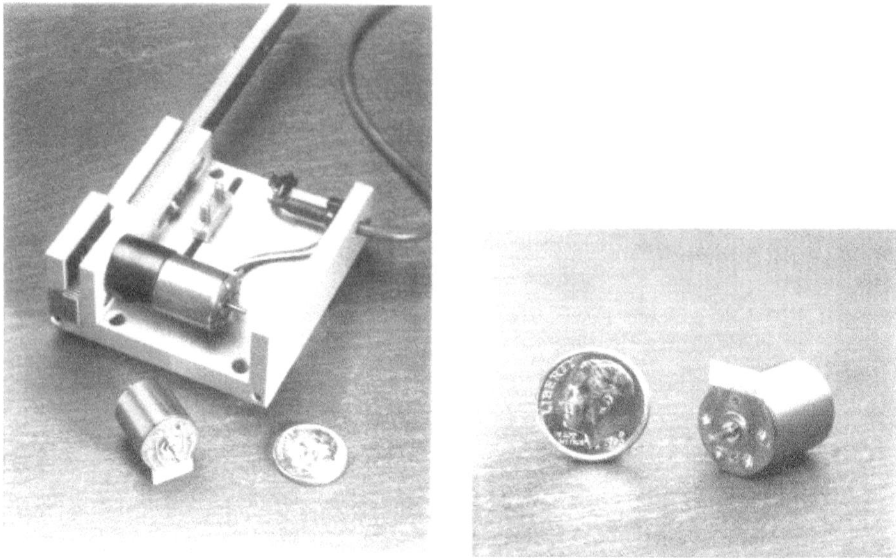

A B

Figure 12.49 Miniature stepper motors – available in 6, 12, and 24 volt DC. (Donovan Micro-Tek, Inc.)

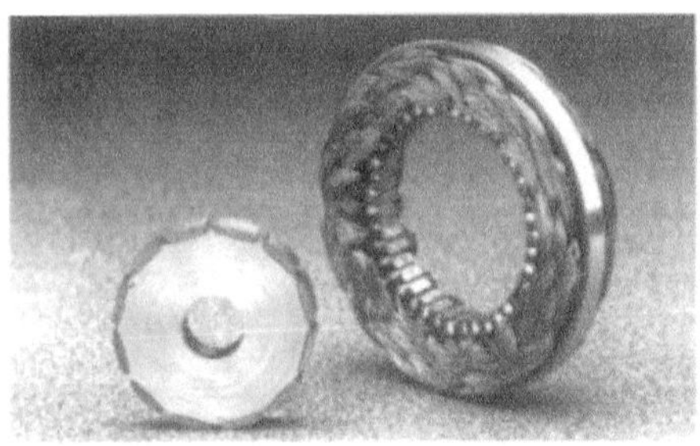

Figure 12.50 Frameless motor for applications that require light weight, low-friction, and in which a frame is not necessary. (Hathaway Corp.)

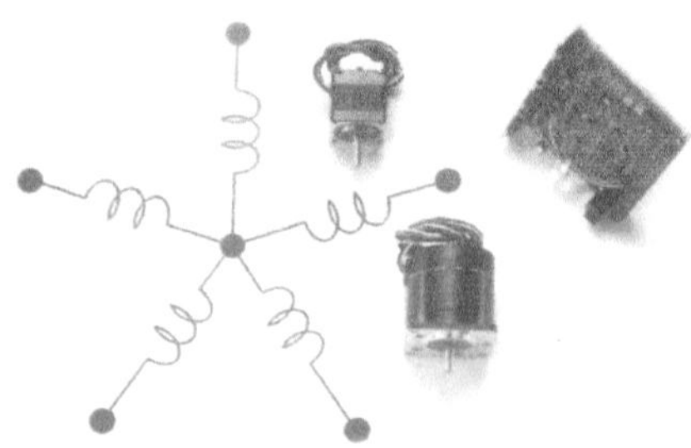

Figure 12.51 Five-phase stepper motor achieves smooth, vibration-free performance with high efficiency. (Nyden Corp.)

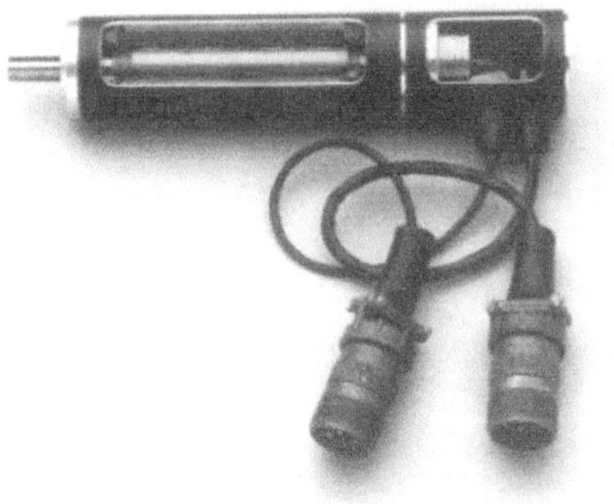

Figure 12.52 Round frame brushless DC motor showing internal detail, including low-inertia rotor and resolver used as a feedback device. (Pacific Scientific)

Review Questions

Answer the following questions.

1. What is a synchro system?

2. What is a servomechanism?

3. Describe the construction of a DC stepping motor.

4. What is meant by "half-stepping" of a DC stepping motor?

5. Define the following terms:
   a. Step angle.
   b. Steps per second.
   c. Step accuracy.
   d. Step response time.
   e. Torque-to-inertia ratio.

6. What is a linear actuator?

7. What is the purpose of a frequency conversion system?

8. List two types of single-phase to three-phase conversion systems.

9. What is the purpose of using a gear reduction motor?

10. What is a motion control system?

11. What is meant by microstepping?

12. List and briefly describe three types of stepper motors.

13. Define the following terms:
    a. Detent torque.
    b. Tooth pitch.
    c. Static torque.
    d. Running torque.

14. List and briefly describe four types of stepper motor drives.

15. What are three types of feedback devices?

UNIT FOUR

Electrical Motor Control

In the two chapters that make up **Unit IV**, basic electrical motor control will be studied. The fundamentals of electrical motor control are covered in **Chapter 13**. This chapter examines several of the fundamental devices and types of equipment that are used for electrical motor control. This equipment includes such electromechanical devices as contactors, relays, switches, and electronic equipment, such as power control devices.

Chapter 14 examines the mechanical and electronic power control systems that are used with electrical machinery. These motor control systems have been used extensively in day-to-day industrial, commercial, and residential applications.

Chapter 13 – Electrical Control Basics

Chapter 14 – Motor Control Systems

DOI: 10.1201/9788743810056-16

13

Electrical Control Basics

Objectives

Upon completion of this chapter, you will be able to:

- Describe the function of standards organizations, such as NEMA, ANSI, and IEEE, in electrical power control.
- Define the important terms used with electrical power control systems.
- Recognize and sketch symbols used for electrical power control systems.
- Describe the following types of switches used for electrical power control: toggle switch, rocker switch, push-button switch, rotary switch, limit switch, temperature switch, light level switch, pressure switch, foot switch, and drum controller switch.
- Describe pilot lights used with electrical power control.
- Recognize and describe various motor starting systems.
- List the functions of motor starters.
- Describe the method of assigning size ratings for motor contactors.
- Describe the classification system used for motor starters.
- Describe the method of assigning sizes to manual starters.
- Describe a combination starter.
- Describe bimetallic and melting alloy thermal overload devices and their current ratings.
- List the criteria used to select motor controllers.
- List types of motor controller enclosures used by NEMA.
- Explain the operation of a solenoid.
- Explain the operation of a relay.

DOI: 10.1201/9788743810056-17

Important Terms

- actuator
- American National Standards Institute (ANSI)
- control center
- drum controller switch
- float switch
- foot switch
- general-purpose relay
- heater
- Institute of Electrical and Electronics Engineers (IEEE)
- latching relay
- limit switch
- liquid level
- magnetic starter
- momentary contact switch
- motor-controller enclosure
- motor-starting device
- National Electrical Manufacturers' Association (NEMA)
- overtravel
- pilot light
- pneumatic timing relay
- pressure switch
- pretravel
- push-button switch
- rocker switch
- rotary switch
- solenoid
- solid-state relay
- sustained contact switch
- switch
- temperature switch
- timing relay
- toggle switch

Introduction

Control is a very important part of electrical system operation. Control is the most complex part of the electrical system. We will limit the discussion in this chapter primarily to some of the common types of electromechanical equipment and devices that are used for electrical motor control.

Control equipment and devices are used in conjunction with many types of electrical loads. Electric motors (mechanical loads) use about 50% of all electrical energy produced in the United States. In many cases, similar control equipment and devices are used to control electrical lighting and heating loads.

13.1 Control Standards, Symbols, and Definitions

A great amount of fundamental knowledge is needed in order to fully understand electric motor control. Individuals who are concerned with electrical control systems should have a good knowledge of the standards, symbols, and definitions that govern electrical motor control.

13.1.1 NEMA standards

The **National Electrical Manufacturers' Association (NEMA)** has developed standards for electrical control systems. NEMA standards are used extensively to obtain information about the construction and performance of various electrical power equipment. These standards provide information concerning the voltage, frequency, power, and current ratings for various equipment.

13.1.2 ANSI and IEEE standards

There are two well-known organizations that publish standards that are of importance to industry. These organizations are the **American National Standards Institute (ANSI)** and **the Institute of Electrical and Electronics Engineers (IEEE)**.

The **standards** published by these organizations are for the use of the manufacturing industries as well as consumers. These standards are subject to a periodic review. Therefore, industrial users should keep up to date by obtaining the revisions.

These standards were developed through input from manufacturers, consumers, government agencies, and scientific, technical, and professional organizations. Often, these are published standards by governmental agencies.

It should be noted that the **National Electrical Code**, discussed previously, is an American National Standards Institute (ANSI) standard.

13.1.3 Definitions

There are a number of basic definitions that are used when dealing with electrical motor control. These definitions are particularly important when interpreting control diagrams and standards. A listing of the important motor control definitions follows. You should study these definitions.

Across-the-line control – A method of motor starting in which a motor is connected directly across the power lines when it is started.

Automatic starter – A self-acting starter that is completely controlled by control switches or some other sensing mechanism.

Auxiliary contact – A contact that is part of a switching system. It is in addition to the main contacts and is operated by the main contacts.

Braking – A control method that is used to rapidly stop and hold a motor.

Circuit breaker – An automatic device that opens under abnormally high-current conditions and can be manually or automatically reset.

Contact – A current-conducting part of a control device that is used to open or close a circuit.

Contactor – A control device that is used to repeatedly open or close an electric circuit.

Controller – A device or group of devices that systematically control the delivery of electric power to the load or loads connected to it.

Disconnect – A control device or group of devices that will open so that electrical current in a circuit will be interrupted.

Drum controller – Sets of electrical contacts that are mounted on the surface of a rotating cylinder, usually for controlling the on/off or forward/reverse condition of a load.

Electronic control – Devices, usually solid state, that perform part of the control function of a system.

Full voltage control – A control system that connects equipment directly across the power lines when it is started.

Fuse – A circuit-protection device that disconnects a circuit when an overcurrent condition occurs. However, the fuse is destroyed in the process and must be replaced.

Horsepower – The power output or mechanical work rating of an electrical motor.

Jogging – Momentary operation that causes a small movement of the load being controlled.

Magnetic contactor – A contactor that is operated electromagnetically and usually controlled by push buttons activated by an operator.

Manual controller – An electric control device that functions when operated by mechanical means, usually a human operator.

Master switch – A switch that controls the power delivered to other parts of a system.

Motor – A device (mechanical load) for converting electrical power to mechanical power in the form of rotary motion.

Multispeed starter – An electrical power-control device that provides for varying the speed of a motor.

Overload relay – Overcurrent protection for a load. While it is in operation, it maintains the interruption of the load from the power supply until it is reset or replaced.

Overload relay reset – A push button that is used to reset a thermal overload relay after the relay has been overloaded.

Pilot device – A control device that directs the operation of another device or devices.

Plugging – A braking method that causes a motor to develop a retarding force in the reverse direction.

Push button – A button-type control switch that is manually operated for actuating or disconnecting some load device.

Push-button station - A housing for the push buttons that are used to control equipment.

Reduced-voltage starter – A control device that applies a reduced voltage to a motor when it is started.

Relay – A control device that is operated by one electrical circuit to control a load that is part of another electrical circuit.

Remote control – A system in which the control of an electrical load takes place from some distant location.

Safety switch – An enclosed manually operated disconnecting switch used to turn a load off when necessary.

Solenoid – An electromagnetically actuated control device that is used to produce linear motion for performing various control functions.

Starter – An electric controller that is used to start, stop, and protect the motor to which it is connected.

Timer – A control device that provides variable time periods so that a control function can be performed.

13.1.4 Symbols

One should have an understanding of the electrical symbols that are commonly used with electrical control systems. A few of these symbols are somewhat different from basic electrical symbols. Some typical **symbols** are shown in Figure 13.1. You should especially observe the symbols that are used for the various types of switches and push buttons.

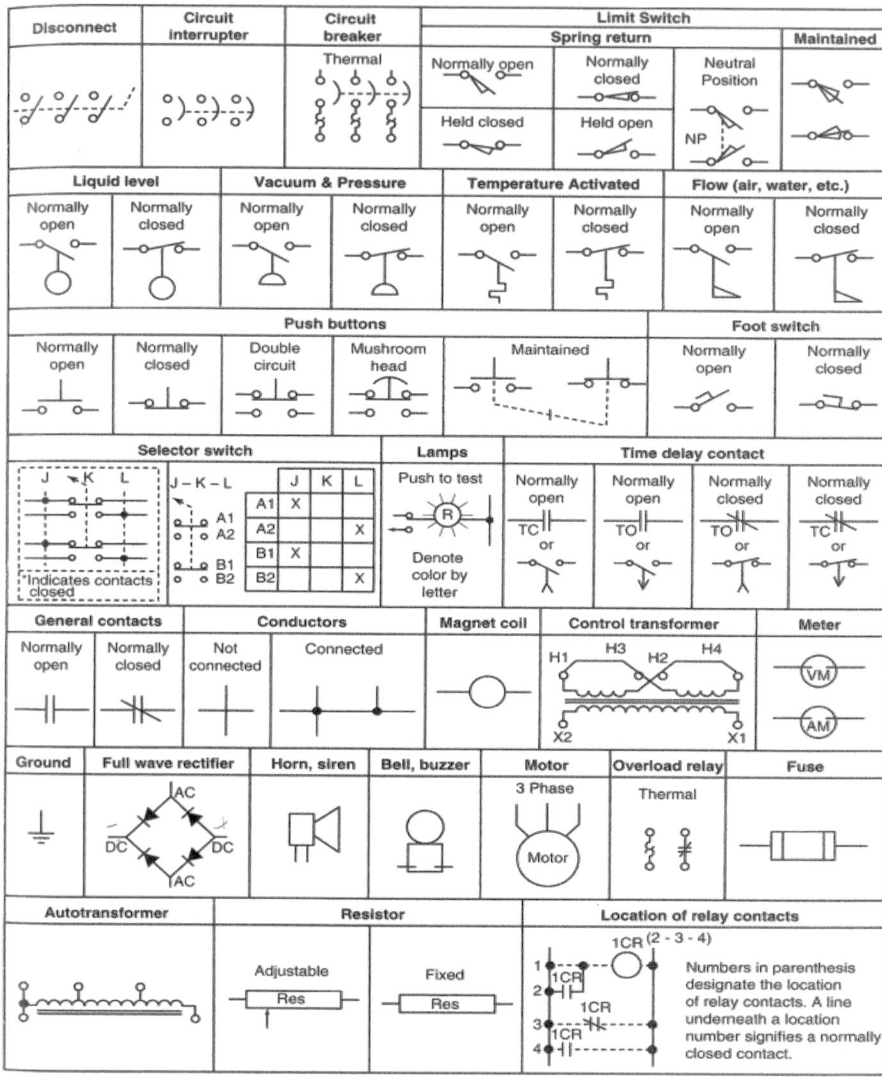

Figure 13.1 Common power control symbols. (Furnas Electric Co.)

13.2 Control Using Switches

An important, but often overlooked, part of electrical motor control is the various types of switches used. This section will examine the many types of switches that are used to control electrical systems. The primary function of a switch is to turn a circuit on or off. However, many more complex switching functions can be performed using switches. The emphasis in this section will be on switches that are used for motor control. Keep in mind that other load devices can also be controlled in a similar manner by switches.

13.2.1 Toggle switches

Among the simplest types of switches are **toggle switches**. The symbols for several kinds of toggle switches are shown in Figure 13.2. Figure 13.3 shows some typical types of toggle switches. You should become familiar with the symbols that are used for various types of toggle switches and with the control functions that they can accomplish such as turning electrical equipment on and off.

13.2.2 Rocker switches

Another type of switch is called a **rocker switch** and is shown in Figure 13.4. Rocker switches are used for on-off control. They can be either **momentary contact switches**, which accomplish temporary control, or sustained contact switches, which cause a load to remain in an on or off condition until the switch position is manually changed.

13.2.3 Push-button switches

Push-button switches, such as those shown in Figure 13.5, are commonly used. Many motor control applications use push-button switches as a means of starting, stopping, or reversing a motor. The push buttons are manually operated to close or open the control circuit of the motor. There are several types of push-button switches used in the control of motors. Figure 13.6 diagrams some push-button styles.

Ordinarily, **push-button switches** are either the normally closed or normally open type. Cutaway views and schematic symbols are shown in Figure 13.7. A **normally closed (NC) push button** is closed until it is depressed manually. It will open a circuit when it is depressed. The **normally open (NO) push button** is open until it is manually depressed. Once it is depressed, it will close a circuit. The start push button of a motor control station is a normally open type, while the stop switch is a normally closed type.

13.2.4 Rotary switches

Another common type of switch is the **rotary switch**. Many different switching combinations can be wired by using a rotary switch. The shaft of

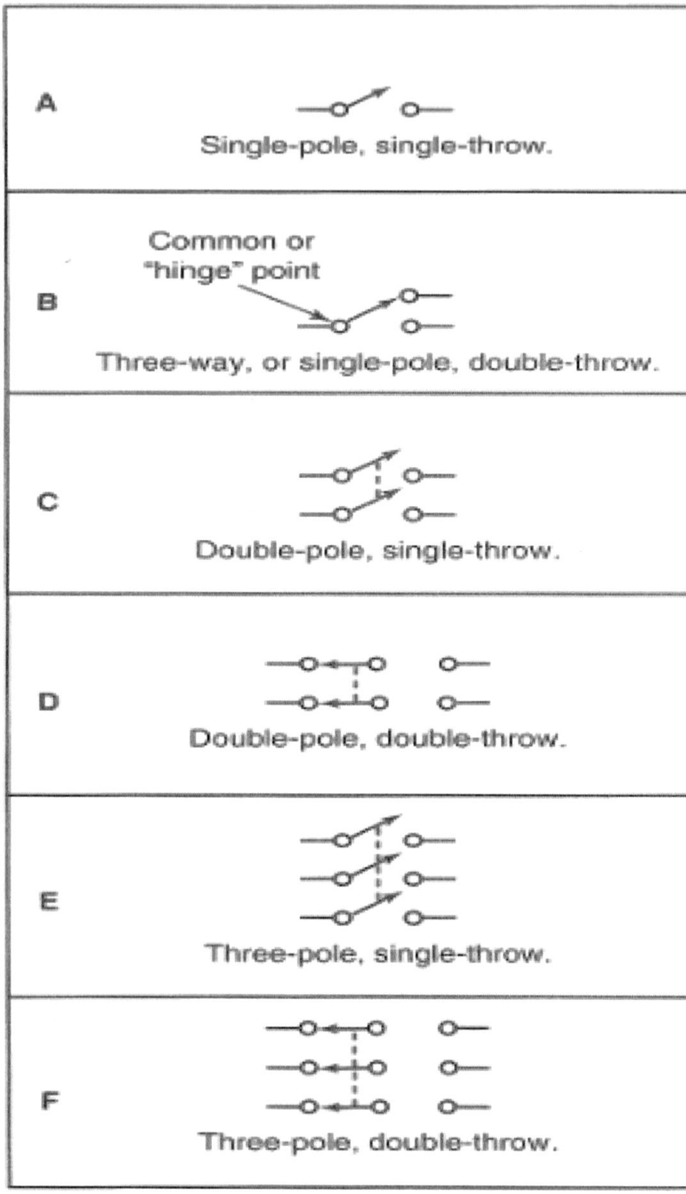

Figure 13.2 Types of toggle switches.

a rotary switch is attached to sets of moving contacts. These moving contacts touch different sets of stationary contacts that are mounted on ceramic segments when the rotary shaft is turned to different positions. The shaft can lock into place in any of several positions. A common type of rotary

Figure 13.3 Types of switches

Figure 13.4 Rocker switches.

switch is shown in Figure 13.8. Rotary switches are usually controlled by manually turning the rotor, shaft clockwise or counterclockwise. A knob is normally fastened to the end of the rotary shaft to permit easier turning of the shaft.

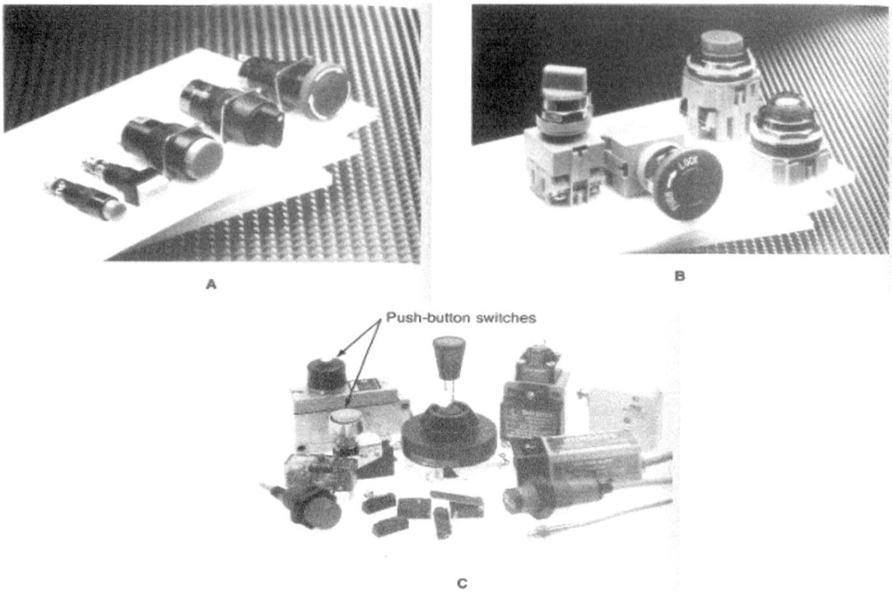

Figure 13.5 A and B – Push-button switches. C – Pushbutton switches with a variety of other switches. See if you can identify the following: limit switches, joystick switch, and proximity switches. (IDEC Corp; Burgess SAIA, Inc.)

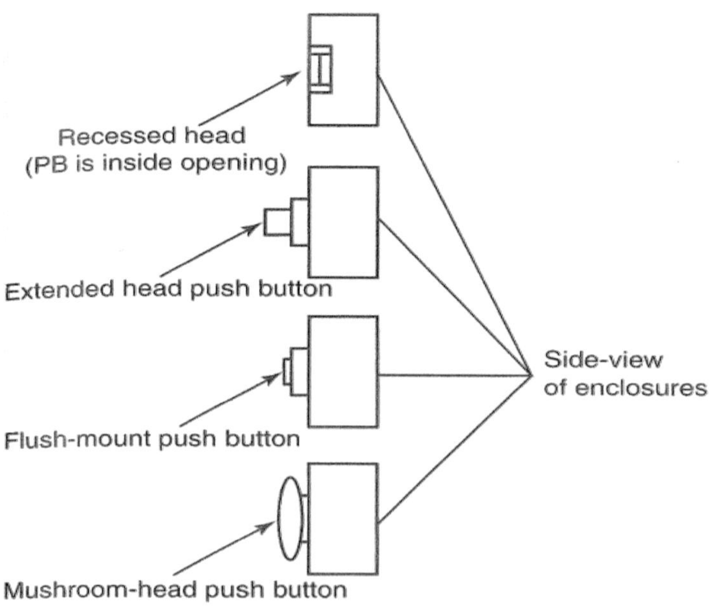

Figure 13.6 Push-button styles.

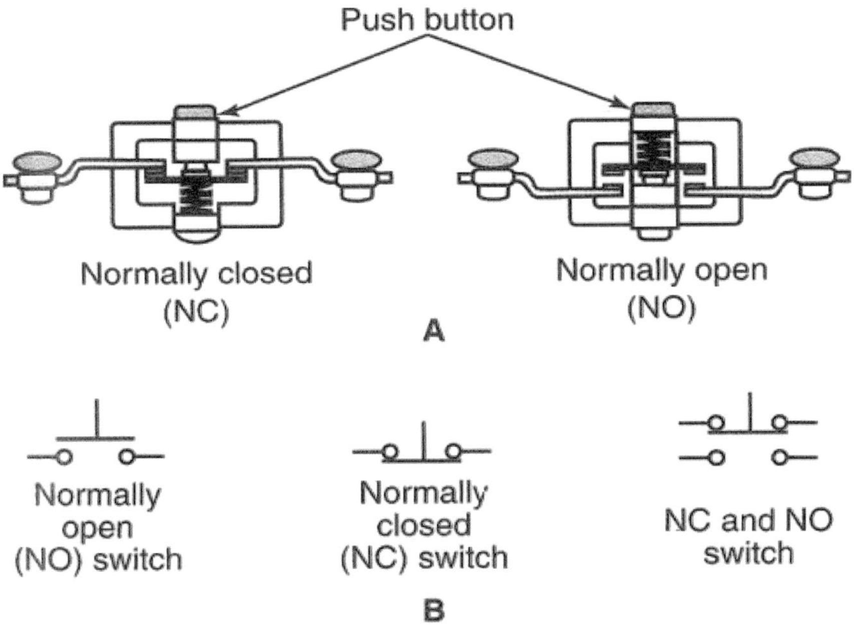

Figure 13.7 Push-button switches. A – Cutaway views. B – Symbols.

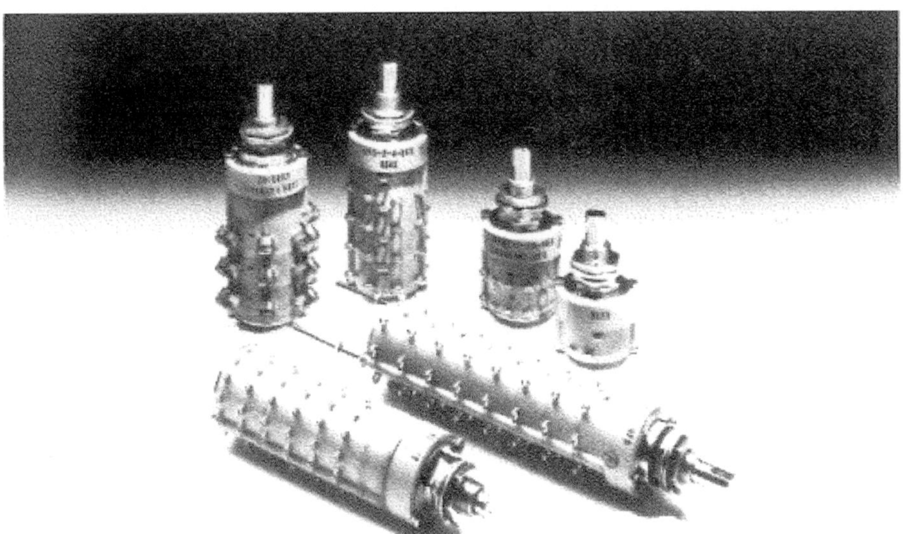

Figure 13.8 Rotary switches. (Janco Corp.)

13.2.5 Limit switches

Some typical limit switches are shown in Figure 13.9. They are made in a variety of sizes. **Limit switches** are merely on/off switches that use a mechanical movement to cause a change in the operation of the electrical control circuit of a motor or other load device. The electrical current developed as a result of the mechanical movement is used to limit the movement of the machine or to make some change in its operational sequence. Limit switches are often used in sequencing, routing, sorting, or counting operations in industry. Ordinarily, they are used in conjunction with hydraulic or pneumatic controls, electrical relays, or other motor-operated machinery. You can find limit switches in drill presses, lathes, and conveyor systems.

In its most basic form, a limit switch converts mechanical motion into an electrical control current. Some parts of the limit switch are labeled in part C of Figure 13.9. Notice that the cam is an external part that is usually mounted on a machine. The cam applies force to the actuator of the limit switch. The **actuator** is the part of the limit switch that causes the internal NO or NC contacts to change state. The actuator is operated by either a linear or rotary motion of the cam, applying force to the limit switch.

Two other terms associated with limit switches are **pretravel** and overtravel. Pretravel is the distance that the actuator must move to change the normally open or normally closed state of the limit-switch contacts. Overtravel is the distance the actuator moves beyond the point where the contacts change state. Both pretravel and overtravel settings are important in machine setups where limit switches are used.

13.2.6 Temperature switches

Temperature switches are among the most common types of control devices used in industry. A temperature switch is shown in Figure 13.10.

The control element of a temperature switch contains a specific amount of liquid. The liquid increases in volume when the temperature increases. Thus, changes in temperature can be used to change the position of a set of contacts within the temperature-switch enclosure. Temperature switches can be adjusted throughout a range of temperature settings.

13.2.7 Liquid level (float) switches

Liquid level or **float switches** are used when it is necessary to control the level of a liquid. Some designs have an operating lever connected to a rod and float

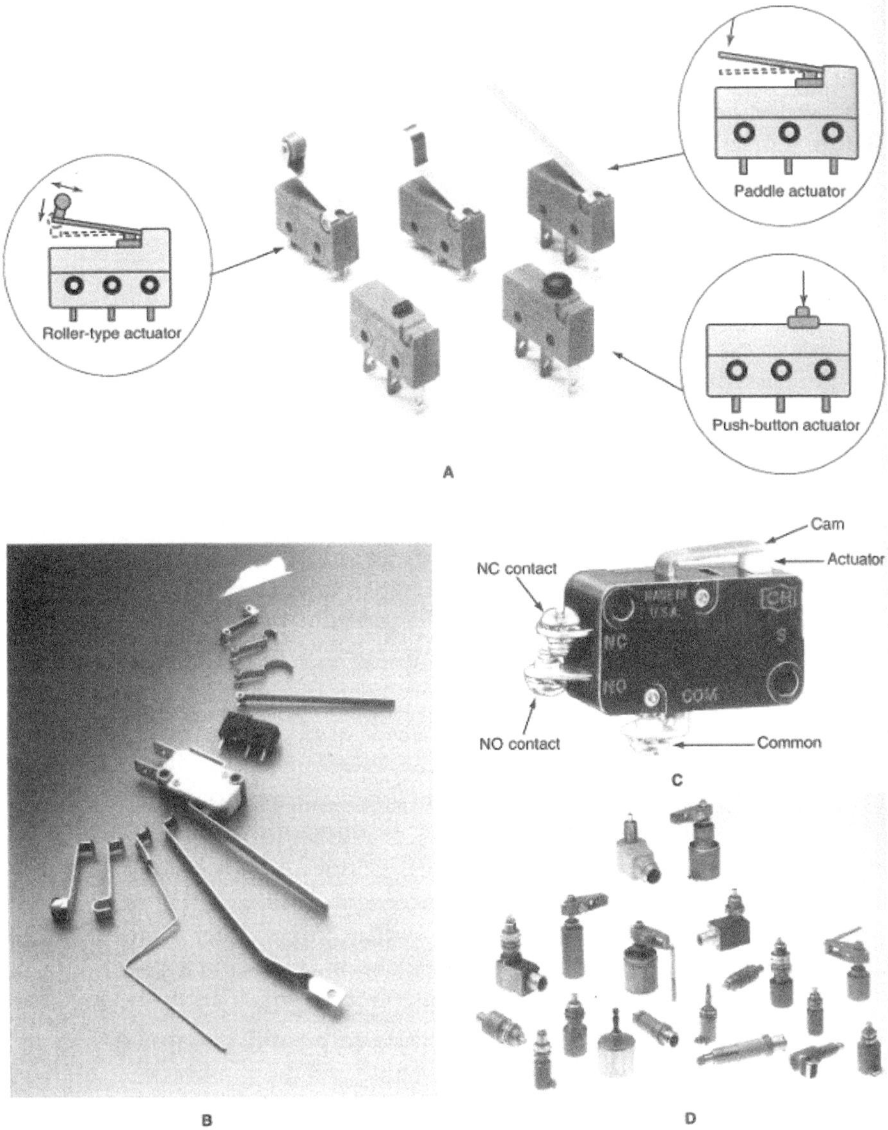

Figure 13.9 Limit switches. A – Snap-action switch with interchangeable levers: NC or NO. B – Notice the actuators of various shapes and sizes for these snap-action limit switches. C – Precision switch with NO and NC screw terminals. D – Limit switches. (Marquardt Switches, Inc.; Cherry Electrical Products; Eaton Corp., Cutler Hammer Products; C&K/Unimax, Inc.)

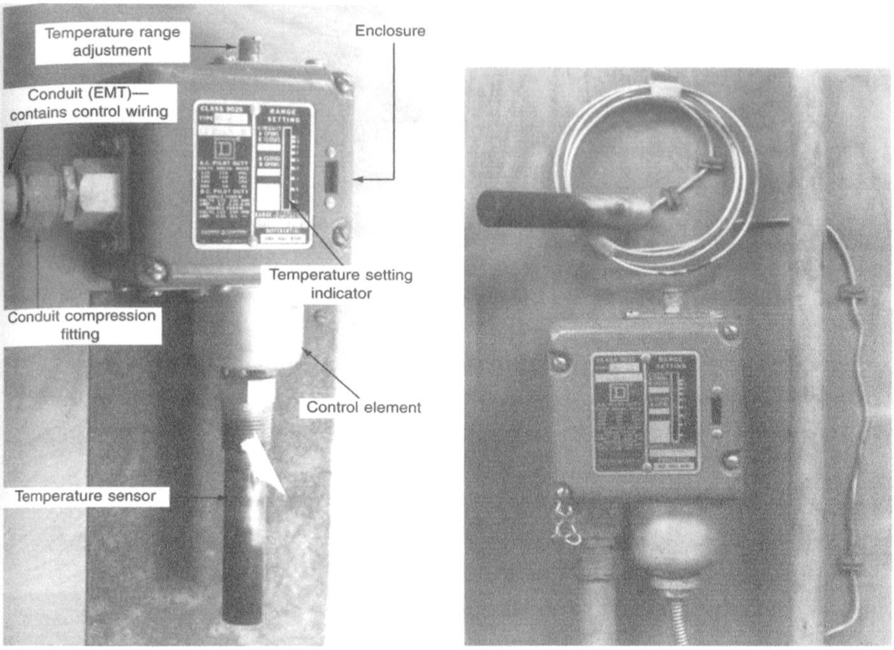

Figure 13.10 Temperature switches. A – Directly connected heat sensor. B – Remote heat sensor.

assembly. The float assembly is placed into a tank of liquid where the motion of the liquid controls the movement of the operating lever of the float switch.

The **float switch** usually has a set of **normally open** and **normally closed** contacts that are controlled by the position of the operating lever. The contacts are connected to a pump motor circuit. In operation, the normally open contacts are connected in series with a pump-motor control circuit. When the liquid level is reduced, the float switch is lowered to a point where the operating lever is moved far enough that the contacts are caused to change to a closed state. The closing of the contacts causes the pump motor to turn on. More liquid is then pumped into the tank until the liquid level rises high enough to cause the switch to turn the pump motor off. The designs shown in Figure 13.11 use sensors to sense changes in liquid level.

13.2.8 Pressure switches

Another type of electrical control device is called a **pressure switch**. A pressure switch has a set of electrical contacts that change states due to a variation in the pressure of air, hydraulic fluid, water, or some other medium.

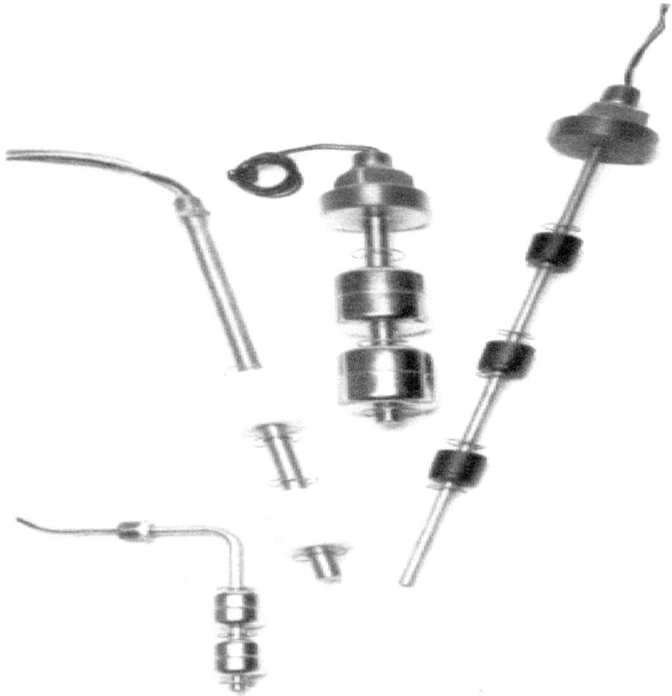

Figure 13.11 Liquid level switches. (Madison Co.)

Some pressure switches are diaphragm-operated. They rely upon the intake or expulsion of a medium, such as air, which takes place in a diaphragm assembly within the pressure-switch enclosure. Another type of pressure switch uses a piston mechanism to initiate the action of opening or closing the switch contacts. In this type of switch, the movement of the piston is controlled by the pressure of the medium (air, water, etc.). Pressure switches are shown in Figure 13.12.

13.2.9 Foot switches

A **foot switch** is a switch that is controlled by a foot pedal. This type of switch is used for applications where a machine operator has to use both hands during the operation of the machine. The foot switch provides an additional control position for the operation of a machine for such a time as when the hands cannot be used. A foot switch is shown in Figure 13.13.

Figure 13.12 Pressure switches. (Esko Industries)

Figure 13.13 Foot switch. (Furnas Electric Co.)

13.2.10 Drum controller switches

Drum controller switches are special-purpose switches that are ordinarily used to control large motors. They can be used with either single-phase or three-phase motors. The usual functions of a drum-controller switch are for start/stop control or for the forward/reverse/stop control of electrical motors. A drum controller switch is shown in Figure 13.14. Contacts are moved as the handle of the controller is turned to provide machine control.

13.2.11 Additional switches

There are numerous different types of switches. Large electronics parts manufacturers can have well over a hundred pages of their catalog dedicated just

Figure 13.14 Drum controller. (Furnas Electric Co.)

to pictures and descriptions of switches. To help familiarize yourself with other types of switches, it is good practice to flip through the pages of some of these catalogs. Figure 13.15 shows miscellaneous types of switches that you should be able to recognize.

13.3 Pilot Lights for Switches

Pilot lights usually operate in conjunction with switches. Motor-control devices often require that some visual indication of the operating condition of the motor be provided. Pilot lights of various types are used to provide such a visual indication. They usually indicate either an on or off condition. For instance, a pilot light could be wired in parallel with a motor to indicate when it is on. Some types of pilot lights are:

- Full-voltage across-the-line lights – These are less expensive pilot lights, but they do not last as long.
- Transformer-operated lights – These use a lower voltage to activate the lamp. They last longer, but they require the expense of an additional

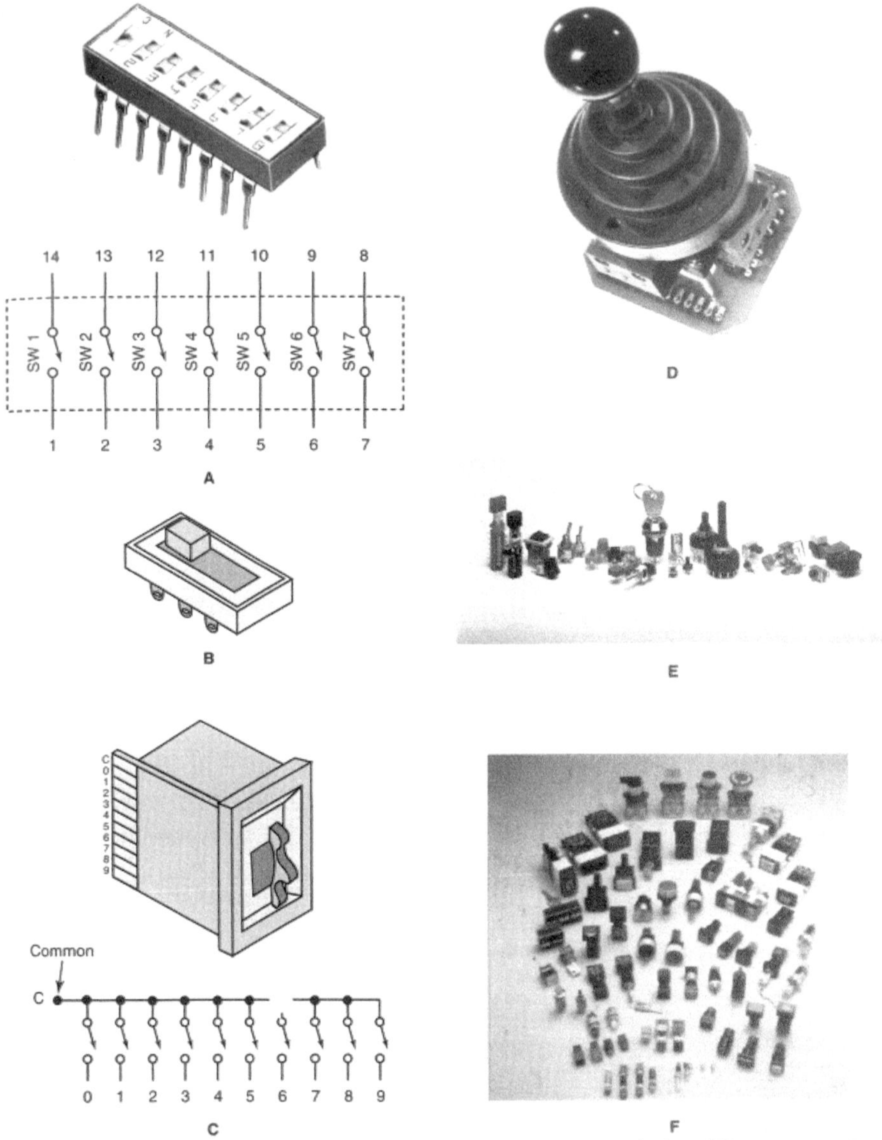

Figure 13.15 Miscellaneous switches. A – Dual-inline-package (DIP) 14-pin switch, shaped like an integrated circuit. B – Slide switch. C – Thumbwheel switch with schematic. D – Joystick switch with controller. E and F – Make a list of the types of switches you see in these diagrams. (Grayhill Co.; Euchner, USA, Inc.; E-Switch, A Division of Electronic Components Groupe, Inc.; EAO Switch Corp.)

Figure 13.16 Illuminated push-button switches. (Eaton Corp.)

transformer, which ordinarily reduces the operating voltage to six volts.

- Resistor-type lights – These lights use series resistors to reduce the voltage across the lamps.
- Illuminated push button – The functions of the push button and the pilot-light device are combined into one item of equipment. This reduces the mounting-space requirement. Figure 13.16 shows illuminated pushbutton switches.

13.4 Control Equipment for Electric Motors

There are several types of electromechanical equipment used for the control of electric motors. The selection of power control equipment affects the efficiency of the system operation and the performance of the machinery. It is very important to use the proper type of equipment for each power control application. This section concentrates on the types of equipment used for motor control.

13.4.1 Motor-starting control

A **motor-starting device** is a type of power control used to accelerate a motor from a stopped condition to its normal operating speed. There are many variations in motor-starter design. The simplest design is a manually operated on/off switch connected in series with one or more power lines. This type of starter is used only for smaller motors that do not draw an excessive amount of current.

Another type of motor starter is the **magnetic starter**, which relies upon an electromagnetic effect to open or close the power-source circuit of the motor. This type of starter is shown in Figure 13.17.

Often, **motor starters** are grouped together for the control of adjacent equipment in an industrial plant. Such groupings of motor starters and associated control equipment, shown in Figure 13.18, are called **control centers**. Control centers provide easier access to the power distribution system since they are more compact and the control equipment is not scattered throughout a large area.

13.4.2 Function of motor starters

Various types of **motor starters** are used for control of electric motors. The functions of a starter vary in complexity. Motor starters usually perform one or more of the following functions:

- On and off control.
- Acceleration.
- Overload protection.
- Reversing direction of rotation.

Some starters control a motor by being connected directly across the power input lines. Other starters reduce the level of input voltage that is applied to the motor when it is started so as to reduce the value of the starting current. Ordinarily, motor overload protection is contained in the same enclosure as the magnetic contactor.

13.4.3 Sizes of motor starter contactors

The **contactors** used with motor starters are rated according to their current capacity. The **National Electrical Manufacturers Association (NEMA)**

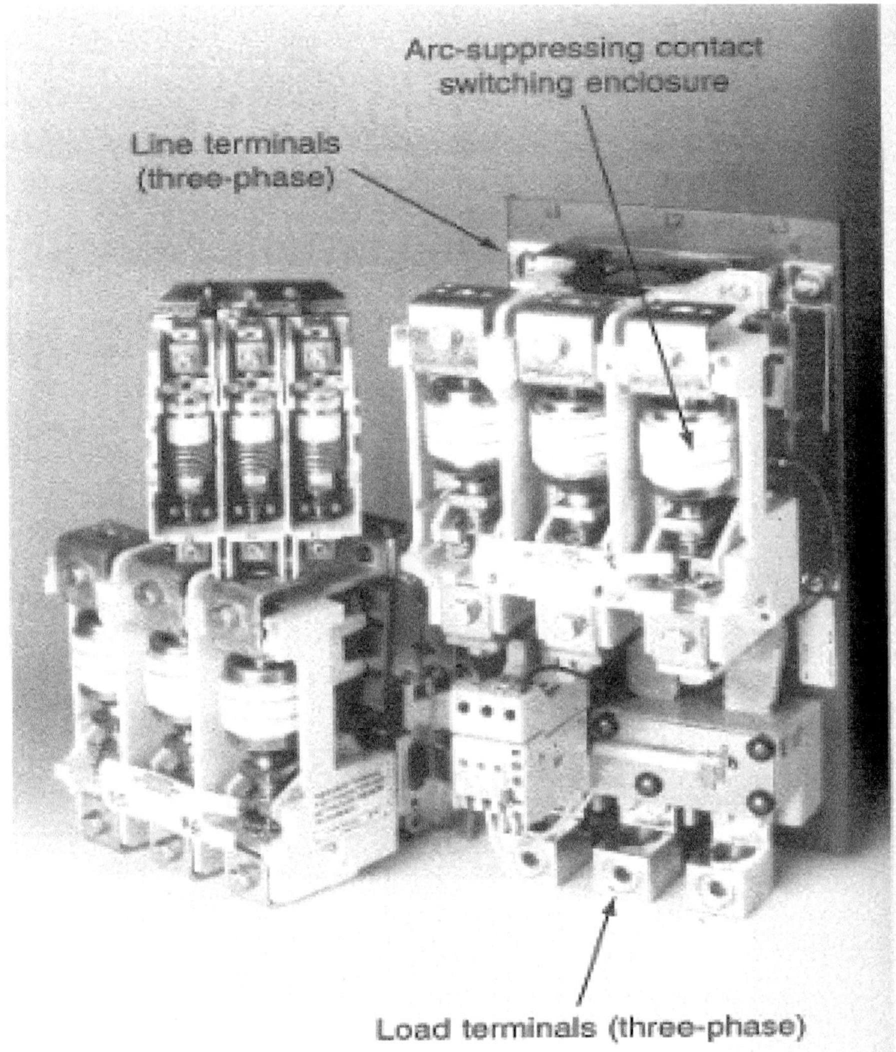

Figure 13.17 Magnetic contactors with vacuum-sealed contacts. (Eaton Corp., Cutler-Hammer Products)

has developed standard sizes for magnetic contactors according to the current capacity. Figure 13.19 lists the NEMA standard sizes for magnetic contactors. By looking at this chart, you can see that a NEMA size 1 contactor has a 30-ampere current capacity if it is open (not mounted in a metal enclosure) and a 27-ampere capacity if it is enclosed. The corresponding maximum horsepower ratings of loads for each of the NEMA contactor sizes are also shown in Figure 13.19.

Figure 13.18 Electrical machine control console (control center). (Siemens Energy and Automation, Inc.)

13.4.4 Classes of motor starters

The types of motor starters that are commercially available are divided into five classes. These classes, established by the National Electrical Manufacturers Association (NEMA), are:

- Class A – Alternating-current, manual or magnetic, air break or oil-immersed starters that operate on 600 volts or less.
- Class B – Direct-current, manual or magnetic, air-break starters that operate on 600 volts or less.
- Class C – Alternating-current, intermediate voltage starters.

	Ampere Rating		Maximum Horsepower Rating of Load					
			Single-Phase		Three-Phase			
NEMA Size	Open	Enclosed	115 V	230 V	115 V	200 V	230 V	460 V
00	10	9	0.33	1	0.75	1.5	1.5	2
0	20	18	1	2	2	3	3	5
1	30	27	2	3	3	7.5	7.5	10
2	50	45	3	7.5	7.5	10	15	25
3	100	90	7.5	15	15	25	30	50
4	150	135	—	—	25	40	50	100
5	300	270	—	—	—	75	100	200
6	600	540	—	—	—	150	200	400
7	900	810	—	—	—	—	300	600
8	1350	1215	—	—	—	—	450	900
9	2500	2250	—	—	—	—	800	1600

Figure 13.19 Sizes of magnetic contactors.

- Class D – Direct-current, intermediate voltage starters.
- Class E – Alternating-current, magnetic starters that operate on 2200 volts to 4600 volts. Class El uses contacts, and Class E2 uses fuses.

13.4.5 Manual motor starters

Some motors use **manual starters** to control their operation. Figure 13.20 shows manual starters that can be used to control motors.

A **manual starter** provides starting, stopping, and overload protection similar to that of a magnetic contactor. However, manual starters must be mounted near the motor that is being controlled. Remote-control operation is not possible as it would be with a magnetic contactor. Manual starters remain closed until they are manually turned off. They are usually limited to small sizes.

13.4.5.1 Sizes of manual motor starters

A uniform method has also been established for sizing manual motor starters. Some examples of the sizes of full-voltage manual starters are:

- Size M-0-For single-phase 112-volt motors up to 1 horsepower.
- Size M-1-For single-phase 112-volt motors up to 2 horsepower.
- Size M-IP-For single-phase 112-volt motors up to 3 horsepower.

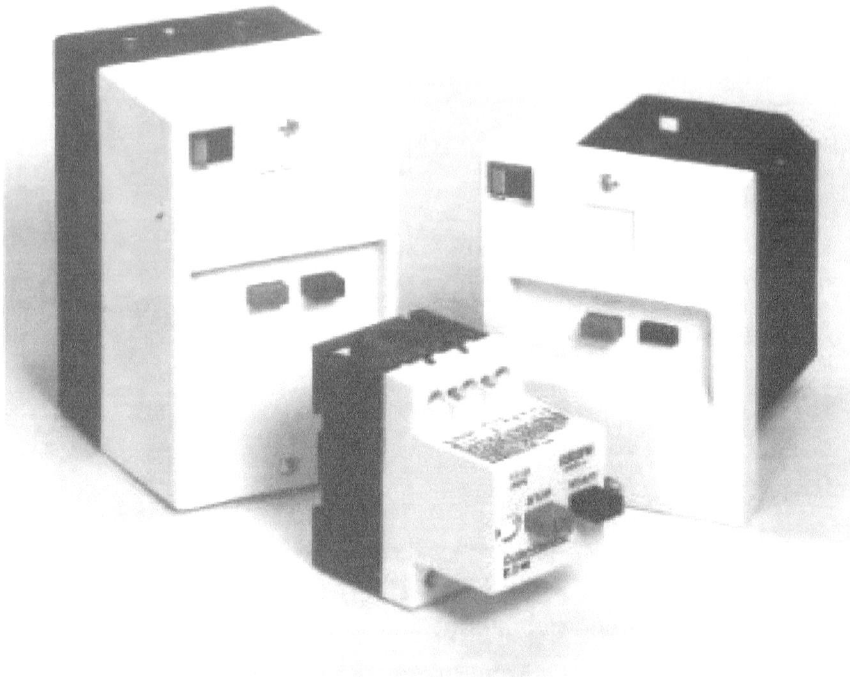

Figure 13.20 Manual starters, (Eaton Corp., CutlerHammer Products)

	Maximum Horsepower Rating of Load				
Starter Size	Single-Phase 115 V	Three-Phase 230 V	Ampere 200-230 V	460-575 V	Rating
M-0	1	2	3	5	20
M-1	2	3	7.5	20	30
M-1P	3	5	—	—	30

Figure 13.21 Sizes of manual starters.

These sizes are summarized in Figure 13.21.

13.4.6 Combination starters

A popular type of motor starter used in control applications is the **combination starter**. These starters are made with protective devices such as fused-disconnect switch air-type circuit breakers, or a system of fuses and circuit breakers mounted in a common enclosure. They are used on

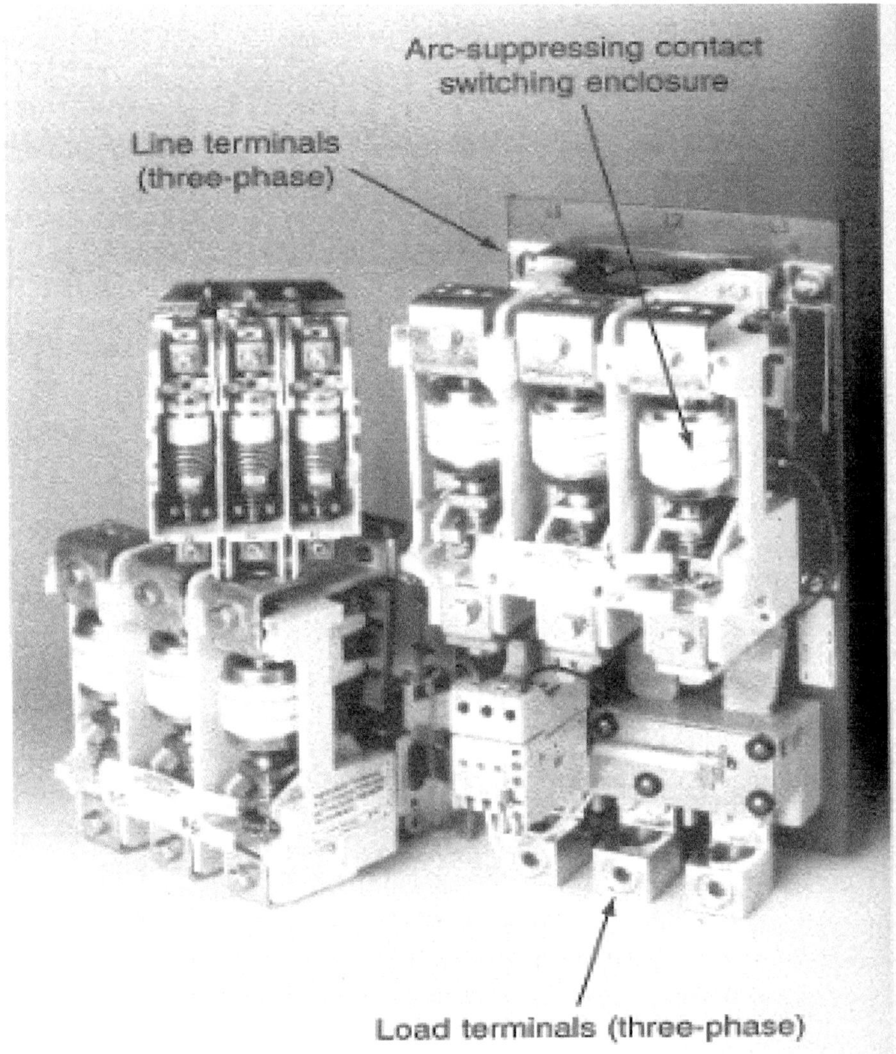

Figure 13.22 Combination starter. (Furnas Electric Co.)

systems of 600 volts or less. A combination magnetic starter is shown in Figure 13.22.

13.5 Motor Overload Protection

Both **manual and magnetic starters** can have overload protection contained in their enclosures to protect the motor. It is common practice to place

Table 26142					
Full Load Motor Amps	Heater Code	Full Load Motor Amps.	Heater Code	Full Load Motor Amps	Heater Code
12.6–14.5	H33	21.2–22.2	H38	34.6–38.9	H43
14.6–16.0	H34	22.3–23.6	H39	39.0–44.7	H44
16.1–16.9	H35	23.7–26.2	H40	44.8–48.8	H45
17.0–18.5	H36	26.3–30.1	H41	48.9–54.2	H46
18.6–21.1	H37	30.2–34.5	H42	54.3–60.0	H48

Figure 13.23 Melting alloy devices.

Table 62K					
Trip Amperes	Heater Code	Trip Amperes	Heater Code	Trip Amperes	Heater Code
1.95	K21	5.59	K36	17.0	K56
2.12	K22	6.39	K37	18.3	K57
2.31	K23	6.88	K39	19.8	K58
2.49	K24	7.78	K41	21.0	K60
2.75	K26	8.42	K42	22.5	K61
3.05	K27	9.54	K43	24.1	K62
3.29	K28	10.1	K49	25.7	K63
3.69	K29	11.5	K50	28.3	K64
4.01	K31	12.6	K52	31.1	K67
4.32	K32	13.9	K53	34.6	K68
4.77	K33	15.1	K54		
5.14	K34	16.0	K55		

Figure 13.24 Bimetallic devices.

either bimetallic or melting-alloy overload relays in series with the motor branch-circuit power lines. These devices are commonly called heaters.

An **overload protective relay** is selected according to the current rating of the motor circuit to which it is connected. An identification number is used on overload protective devices. This number is used to determine the current that will cause the overload device to trip or open the branch circuit. Some typical overload protection or heater tables are given in Figures 13.23 and 13.24.

Figure 13.23 is used for melting-alloy-type devices. Table 26142 is selected according to the type and size of controller, the size of motor, and the type of power distribution (single-phase or three-phase). For example, a heater with an H37 code number is used with a motor that has a full-load current of 18.6 to 21.1 amperes.

Figure 13.24 is used with **bimetallic overload relays**. They are selected in the same manner as **melting-alloy relays**. However, this table shows the trip amperes of the heater. A heater with a K53 code number has a 13.9 trip-ampere rating. Thus, a current in excess of 13.9 amperes causes the heater element to trip. This opens the motor branch circuit by removing the power from the motor.

13.6 Selecting Motor Controllers

There are several important criteria that should be considered when selecting electric motor controllers. Among these are:

1. *Type of motor* – Induction or wound rotor; AC or DC.
2. *Motor ratings* – Voltage, current, duty cycle, and service factor.
3. *Motor operating conditions* – Ambient temperature and type of atmosphere.
4. *Utility company regulations* – Power factor, demand factor, load requirements, and the local codes.
5. *Type of mechanical load connected to the motor* – Torque requirement.

In order to become more familiar with the criteria listed above, you must be able to interpret the data on a **motor nameplate**. The information found on a typical motor i nameplate is summarized as follows:

- *Manufacturing Co.* – The company that built the motor.
- *Motor type* – A specific type of motor, e.g., split-phase AC, universal, three-phase induction.
- *Identification number* – Number assigned by the manufacturer.
- *Model number* – Number assigned by the manufacturer.
- *Frame type* – Frame size defined by NEMA. Number of phases (AC) – Single phase or three phase.
- *Horsepower* – The amount produced at rated speed.
- *Cycles (AC)* – Frequency the motor should be used with (usually 60 Hz).

- *Speed (r/min.)* – The amount at rated *hp*, voltage, and frequency.
- *Voltage rating* – Operating voltage of motor.
- *Current rating (amperes) or Amps* – Current drawn at rated load, voltage, and frequency. Thermal protection – Indicates the type of overload protection used.
- *Temperature rating (°C)* – Amount of temperature that the motor will rise over ambient temperature when operated.
- *Time rating* – Time the motor can be operated without overheating (usually continuous).

13.7 Motor-Controller Enclosures

The purpose of a **motor-controller enclosure** is obvious. A motor-controller enclosure protects an operator against accidental contact with high voltages that could cause death or shock. In some cases, however, the enclosures are also used to protect the control equipment from its operating environment. This environment can contain water, heavy dust, or combustible materials.

The categories of motor-controller enclosures were standardized by the National Electrical Manufacturers Association (NEMA). The following list, courtesy of Furnas Electric Company, summarizes various classifications.

> NEMA 1: General-purpose enclosures that protect personnel from accidental electrical contact with the enclosed apparatus. These enclosures satisfy indoor applications in normal atmospheres that are free of excessive moisture, dust, and explosive materials.
>
> NEMA 3 (3R): Weatherproof enclosures that protect the control from Neather hazards. These enclosures are suitable for applications on ship docks, canal locks, and construction work, and for applications in subways and tunnels. The door seals with a rubber gasket. They are furnished with a conduit hub and pole mounting bracket.
>
> NEMA 4: Weathertight enclosures that are suitable for application outdoors on ship docks and in dairies, breweries, etc. This type of enclosure meets standard hose test requirements. They are sealed with a rubber gasket in the door. NEMA 4X: Corrosion-resistant fiberglass

enclosures that are virtually maintenance-free. These UL-listed units are adaptable to any conduit system by using readily available metal, fiberglass, or PVC conduit hubs and are suitable for NEMA 3, 3*R*, 3*S*, 4 and 12 applications because they are dust-tight, rain-tight, water-tight, and oil-tight. NEMA 12: Industrial-use enclosure that also satisfies dust-tight applications. These enclosures exclude dust, lint, fibers, and oil or coolant seepage. The hinged cover seals with a rubber gasket.

NEMA 7: Enclosures for atmospheres containing hazardous gas. These enclosures satisfy Class 1, Group C or D applications as outlined in Section 500 of the National Electrical Code standard for electrical wiring and apparatus. These cast aluminum enclosures are for use in atmospheres containing ethyl ether vapors, ethylene, cyclopropane, gasoline, hexane, naphtha, benzine, butane, propane, alcohol, acetone, benzol, lacquer solvent vapors, or natural gas. Machined surfaces between the cover and the base provide the seal.

NEMA 9: For atmospheres containing explosive dust, these enclosures satisfy Class II, Group E, F, and G applications as outlined by the National Electrical Code for metal dust, carbon black, coal, coke, flour, starch, or grain dust. Enclosure is cast aluminum with machined surfaces between cover and base to provide a seal.

NEMA TYPES 1B1, 1B2, 1B3: Flush-type enclosures provide behind-the-panel mounting into machine bases, columns, or plaster walls to conserve space and to provide a more pleasant appearance. NEMA 1B1 mounts into an enclosed machine cavity. NEMA 1B2 includes its own enclosure behind the panel to exclude shavings and chips that might fall from above. NEMA 1B3 for plaster walls includes an adjustment to compensate for wall irregularities. Enclosures are shown in Figure 13.25.

NEMA 13: Oil-tight push-button enclosures that protect against dust, seepage, external condensation, oil, water, and coolant spray.

13.8 Other Electromechanical Power Control Equipment

There are so many types of electromechanical power control equipment used today that it is almost impossible to discuss each type. However, some of the very important types are discussed in the following paragraphs.

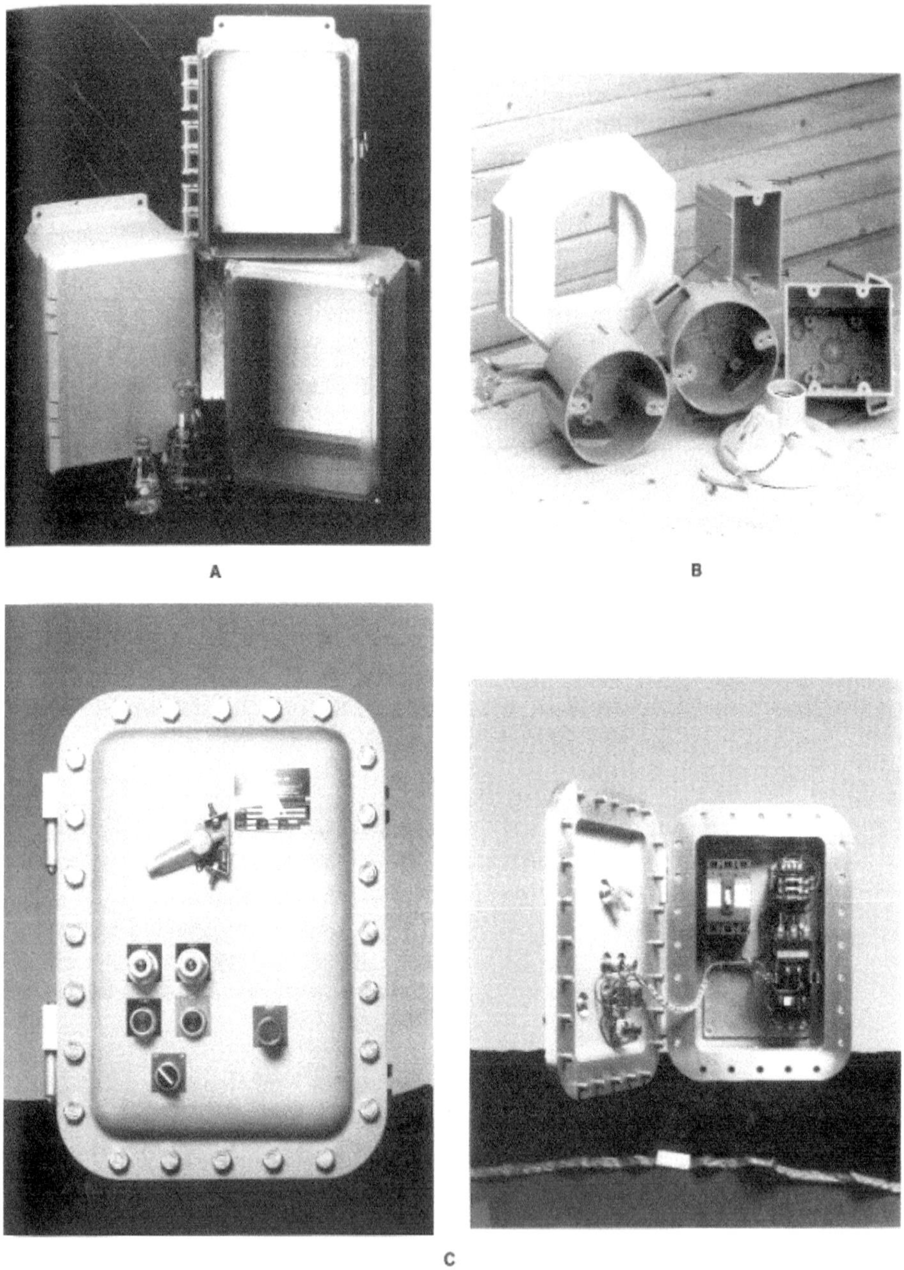

Figure 13.25 A – Plastic control enclosures. B – Enclosures used for duplex receptacles and light fixtures. C – Explosion proof enclosures for a motor starter (interior and exterior). (Carlon Co.; Allied Molded Products, Inc.; Akron Electric, Inc.)

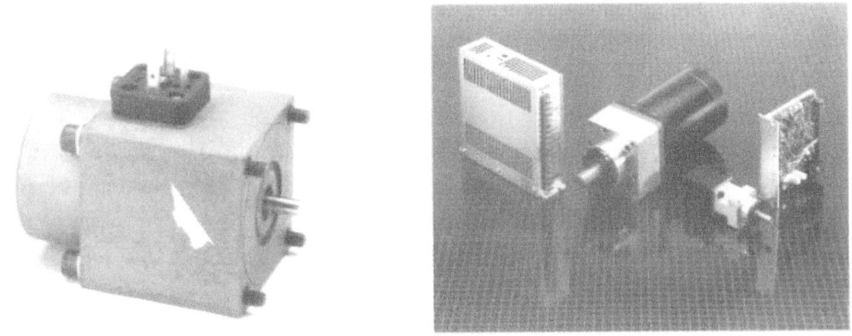

A B

Figure 13.26 A – Typical solenoid. B – Solenoids with control circuits. (Magnet-Schultz of America, Inc.; Norstat, Inc.)

13.8.1 Solenoids

Solenoids, such as those shown in Figure 13.26, are electromagnetic coils with a movable core that is constructed of a magnetic material. The core, or **plunger**, is sometimes attached to an external spring. This spring causes the plunger to remain in a fixed position until moved by the electromagnetic field that is created by current through the coil. This external spring also causes the core, or plunger, to return to its original position when the coil is de-energized.

Solenoids are used for a variety of control applications. Many gas and fuel-oil furnaces use solenoid valves to automatically turn the fuel supply on or off upon demand. Most dishwashers use one or more solenoids to control the flow of water.

13.8.2 Relays

Relays represent one of the most widely used control devices available today. The electromagnet of a relay contains a stationary core. Mounted close to one end of the core is a movable piece of magnetic material called the armature. When the coil is activated electrically, it produces a magnetic field in the metal core. The armature is then attracted to the core, which in turn produces a mechanical motion. When the coil is de-energized, the armature is returned to its original position by spring action.

The **armature** of a relay is generally designed so that electrical contact points respond to its movement. Activation of the relay coil causes the contact points to "make" or "break" according to the design of the relay.

Relays use a small amount of current to create an electromagnetic field that is strong enough to attract the armature. When the armature is attracted, it either opens or closes the contacts. The contacts, then, either turn on or turn off circuits that are using large amounts of current. The minimal current that must flow through the relay coil in order to create a magnetic field strong enough to attract the armature is known as the pickup or make current. The current through the relay coil that allows the magnetic field to become weak enough to release the armature is known as the break or dropout current.

There are two types of contacts used in conjunction with most relays – **normally open (NO) and normally closed (NC)**. The normally open contacts remain open when the relay coil is deenergized and closed only when the relay is energized. The normally closed contacts remain closed when the relay is de-energized and open only when the coil is energized.

A **relay** could be described as an electromagnetic switching mechanism. There are an almost endless number of special-purpose relays and switch combinations used for electrical power control. Figure 13.27 shows a typical assortment of relays and sockets. Figure 13.28 shows a simplified diagram of the construction of a relay that is used to control a motor.

13.8.3 Specialized relays

There are many types of relays used for electrical power control. **General-purpose relays** are used for low-power applications. They are relatively inexpensive and small in size. Many small general-purpose relays are mounted in octal base (8-pin) plug-in sockets.

Latching relays are another type of relay. They are almost identical to the relays discussed previously, but they have a latching mechanism that holds the contacts in position after the power has been removed from the coil. A latching relay usually has a special type of unlatching coil connected in series with a push-button stop switch.

Solid-state relays are used where improved reliability or a rapid rate of operation is necessary. Electromagnetic relays wear out after prolonged use and have to be replaced periodically. Solid-state relays, as do other solid-state devices, have a long life expectancy. They are not sensitive to shock, vibration, dust, moisture, or corrosion.

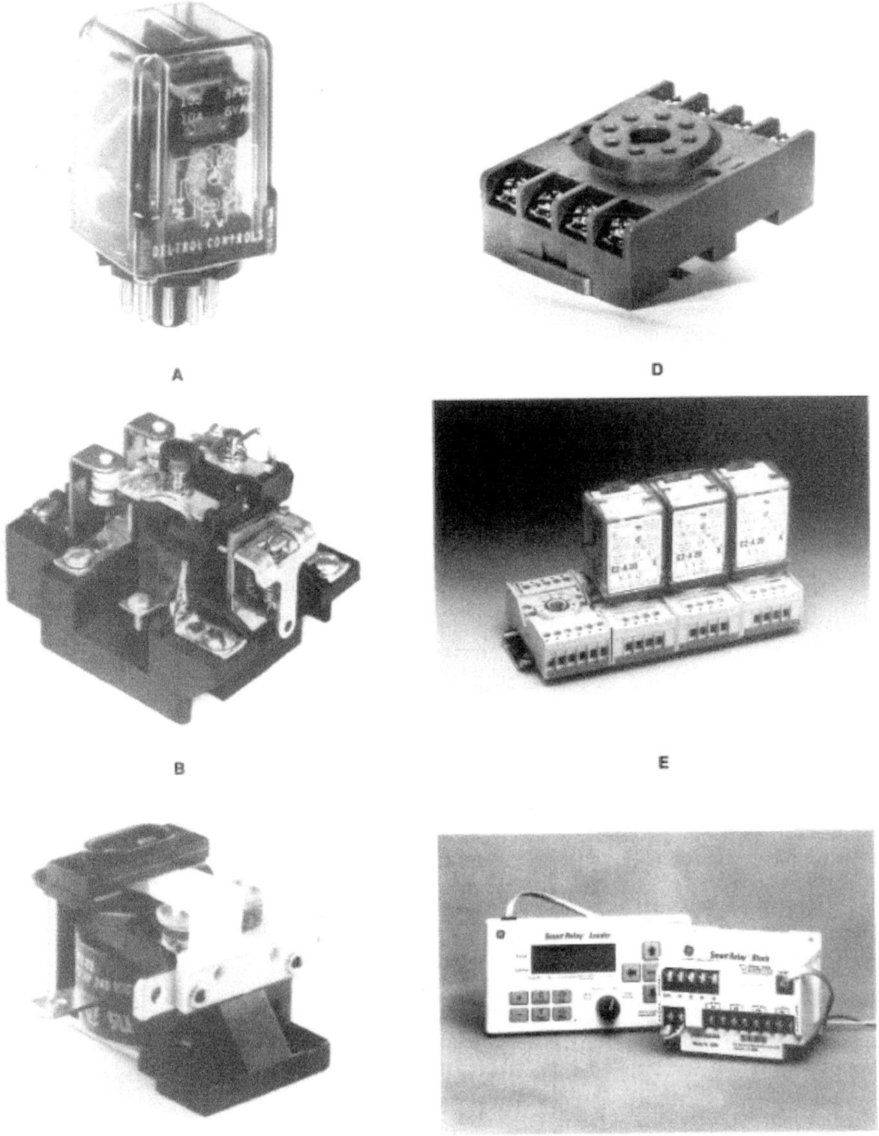

Figure 13.27 Types of relays. A – Plug-in relay. B and C – General-purpose relays. D – Plug-in relay socket for relay type shown in A. E – Plug-in relays mounted in sockets. F – "Smart relay." Eliminates the need for multi-function relays and other controls. (Deltrol Controls; Furnas Electric Co.; Releco Co.; General Electric Co.)

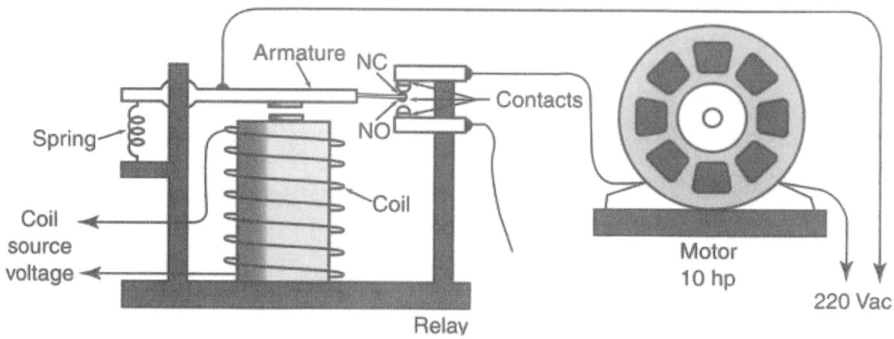

Figure 13.28 Simplified diagram of the construction of a relay that is used to control a motor.

Timing relays are used to turn a load device on or off after a specific period of time. One popular type is a **pneumatic timing relay**. The operation of a pneumatic timing relay is dependent upon the movement of air within a chamber. Air movement is controlled by an adjustable orifice that controls the rate of air movement through the chamber. The air flow rate determines the rate of movement of a diaphragm or piston assembly. This assembly is connected to the contacts of the relay. Thus, the orifice adjustment controls the air flow rate, which determines the time from the activation of the relay until a load connected to it is turned on or off.

There are several other types of **timing relays**. These relays include solid-state, thermal, oil-filled, dashpot, and motor-driven timers. Timing relays are useful for sequencing operations where a time delay is required between operations. A typical application would be as follows: (1) a start push button is pressed; (2) a timing relay is activated; (3) after a 5-second time delay, a motor is turned on.

13.9 Electronic Power Control

An example of advanced electronic control is shown in Figure 13.29. This computer controller is capable of replacing electromagnetic circuit breakers and relays used for electrical power control.

The introduction of computerized control for industrial equipment has brought about a new technology of machine control. Many industrial machines, such as automated manufacturing and robotic equipment, are now controlled by computerized circuits. Figure 13.30 shows some of the advances in machinery systems. An understanding of the basic principles of electrical control is, however, still very important for technicians.

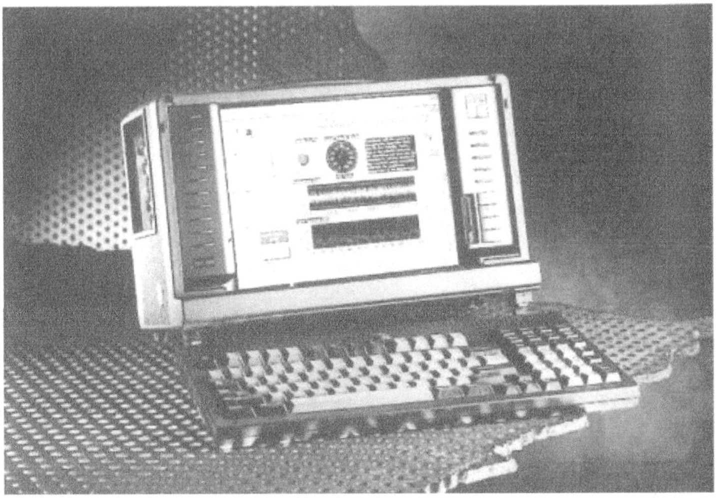

Figure 13.29 Portable computer for industrial control applications. (Dolch Computer Systems)

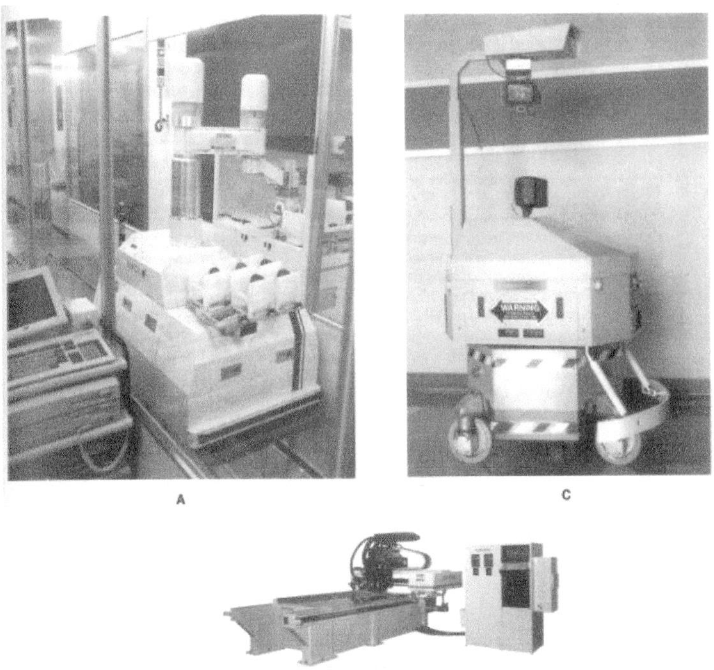

Figure 13.30 Modern machinery and equipment. A – Robotic processing system with computer controller. B – Automated machining system with controller. C – Mobile security system. (Seiko Co.; Thermwood Co.; Cybermotion Co.)

Review Questions

Answer the following questions.

1. List five types of toggle switches.

1. List four styles of push-button switches.

2. What is a drum controller?

3. What is a master switch?

4. What is a pilot device?

5. What is a manual starter?

6. What is a combination starter?

7. What is meant by pickup current?

8. What is meant by dropout current?

9. What is jogging?

10. What is the function of a motor overload protective device?

11. List five criteria to consider when selecting electric motor controllers.

12. List and describe eight types of data on a typical motor nameplate.

13. What is a solenoid?

14. What is a relay?

14

Motor Control Systems

Objectives

Upon completion of this unit, you should be able to:

- Explain the operation of the following control circuits: start/stop control, start/stop control from multiple locations, start/stop control with safe/run feature, start/stop control with run/jog feature, start/stop/jog control, forward/reverse/stop control, full voltage starting, primary resistance starting, primary reactor starting, autotransformer starting, wye-delta starting, part-winding starting, and forward-reverse starting.
- Describe dynamic braking for direct current and alternating current motors.
- Describe a universal motor speed control circuit.
- Describe frequency conversion systems.

Important Terms

across-the-line starter	jogging
autotransformer starting	part-winding starting
dynamic braking	primary-reactor starting
full voltage starting	primary-resistance starting
inching	wye-delta starting

Introduction

Chapter 12 dealt with equipment and devices that are commonly used for electric motor control. Operational motor control systems will be discussed in this chapter. Some specific applications of the equipment and devices that were studied in **Chapter 12** are investigated in this chapter.

DOI: 10.1201/9788743810056-18

14.1 Basic Control Systems

Electrical control systems are used with many types of loads. The most common electrical loads are motors, so our discussion will deal mainly with electric motor control. However, many of the basic control systems are also used to control lighting and heating loads. Generally, the controls for lighting and heating loads are less complex.

Several power control circuits are summarized in Figures 14.1 through 14.9. Figure 14.1 is a start/stop push-button control circuit with overload protection (OL). Notice that the start push button is normally open (NO) and the stop push button is normally closed (NC). Single-phase lines L1 and L2 are connected across the control circuit. When the start push button is pushed, momentary contact is made between points 2 and 3. This causes the NO contact (M) to close. A complete circuit between L1 and L2 causes the electromagnetic coil M to be energized. When the normally closed stop push button is pressed, the circuit between L1 and L2 opens. This causes contact M to open and turn the circuit off.

The circuit in Figure 14.2 is the same type of control as the circuit given in Figure 14.1. However, in the circuit of Figure 14.2, the start/stop control of a load can be accomplished from three separate locations. Notice that the start push buttons are connected in parallel and the stop push buttons are connected in series. The control of one load from as many locations as is desired can be accomplished with this type of control circuit.

The next circuit, Figure 14.3, is the same as the circuit in Figure 14.1 except that a safe/run switch is provided. The safe position ensures that the start push button will not activate the load. A safe/run switch circuit often contains a key that the machine operator uses to turn the control circuit on or off.

Figure 14.4 is also like the circuit of Figure 14.1 but with a jog/run switch added in series with the normally open contact M. In the run position, the circuit would operate just like the circuit of Figure 14.1. The jog position is used so that a complete circuit between L1 and L2 will be achieved and sustained only while the start push button is pressed. With the selector switch in the jog position, a motor can be rotated a small amount at a time for positioning purposes. **Jogging** or **inching** is defined as the momentary operation of a motor to provide small movements of its shaft.

Figure 14.5 shows a circuit that has another method of **motor-jogging** control. This circuit has a separate push button for jogging that relies upon a normally open contact (CR) to operate. Two control relays are used with this circuit.

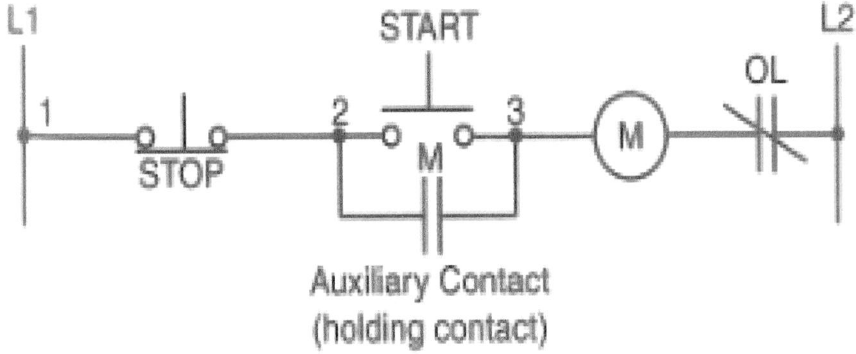

Figure 14.1 A start/stop push-button control circuit.

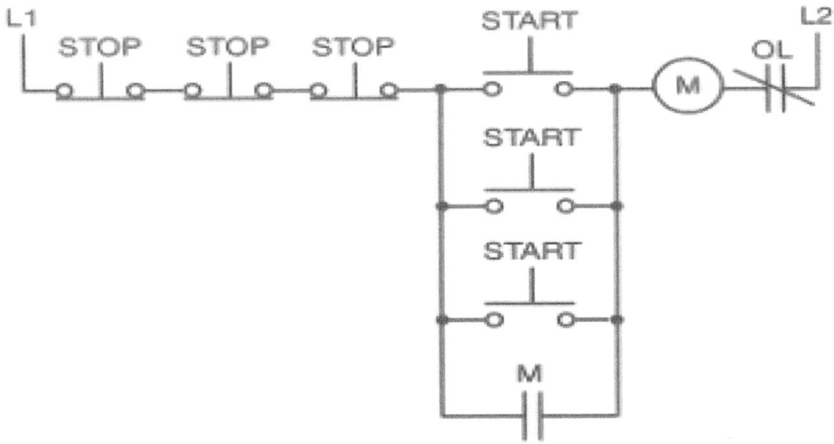

Figure 14.2 A start/stop control circuit with low-voltage protection and control from three locations.

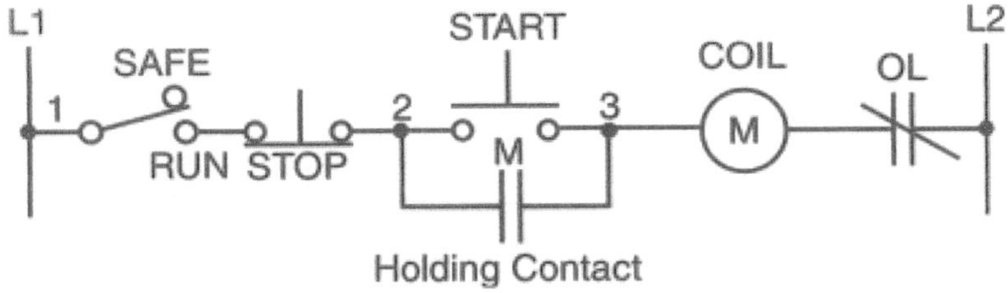

Figure 14.3 A start/stop control circuit with a safe/run selector switch.

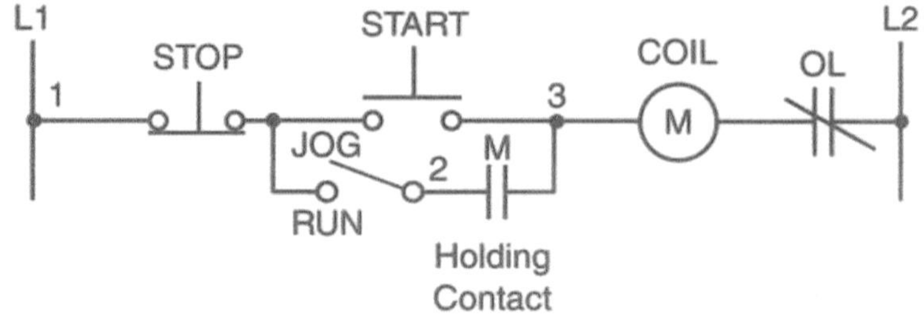

Figure 14.4 A start/stop push-button control circuit with overload protection with a jog/run selector switch.

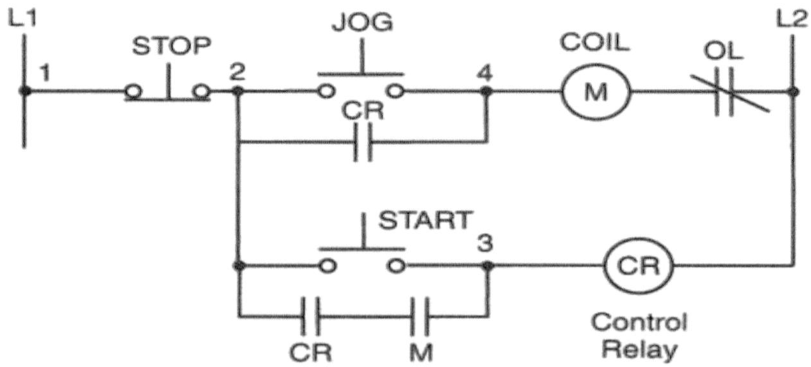

Figure 14.5 A push-button control circuit for start, stop, and jog.

The circuit in Figure 14.6 is a **forward/reverse** push-button control circuit with both forward and reverse limit switches (normally closed switches). When the *forward* push button is pressed, the load will operate until the *forward limit switch* is actuated. The load would then be turned off, since the circuit from L1 to L2 would be opened. The reverse circuit operates in a similar manner. Two control relays are needed for forward/reverse operation.

The circuit of Figure 14.7 is the same as the circuit in Figure 14.6 except for the push-button arrangement. The forward and reverse push buttons are arranged in sets. Pressing the forward push button automatically opens the reverse circuit. Pressing the reverse push button automatically opens the forward circuit. Limit switches are also used with this circuit. Their function is the same as in the circuit of Figure 14.6. In the circuit of Figure 14.**7**, when the forward push button is pressed, the top push button momentarily closes and the lower push button momentarily opens. When points 2 and 3 are

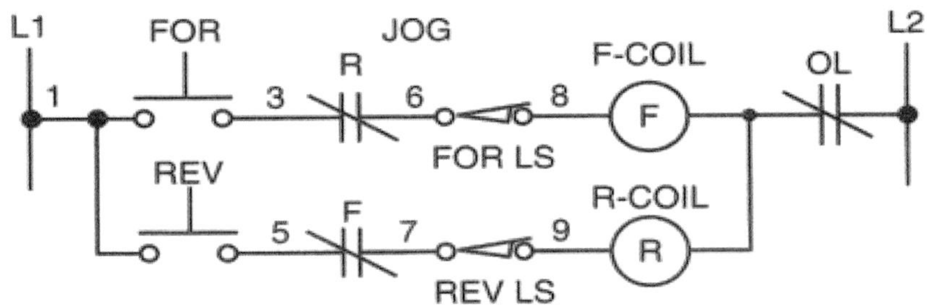

Figure 14.6 A forward/reverse push-button control circuit with forward and reverse limit switches.

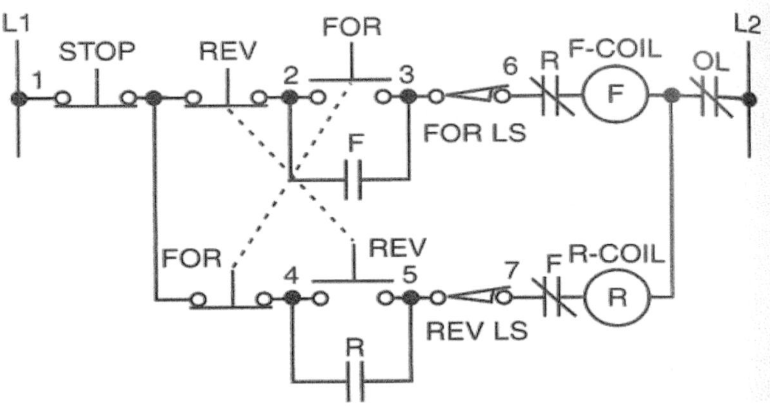

Figure 14.7 An instant forward/reverse/stop pushbutton control circuit.

connected, current flows from L1 to L2 through coil F. When coil F is energized, normally open contact F will close and the normally closed contact F will open. The forward coil will then remain energized. The reverse push buttons cause a similar action of the reversing circuit. Two control relays are also required.

The circuit of Figure 14.8 is similar in function to that of the circuit in Figure 14.7. The push-button arrangement of this circuit is simpler. When the normally open forward push button is pressed, current flows through coil F. When the forward control relay is energized, normally open contact F closes and normally closed contact F opens. This action causes a motor to operate in the forward direction. When the normally closed stop push button is pressed, the current through coil F is interrupted. When the normally open reverse push button is pressed, current flows through coil *R*. When the reverse coil

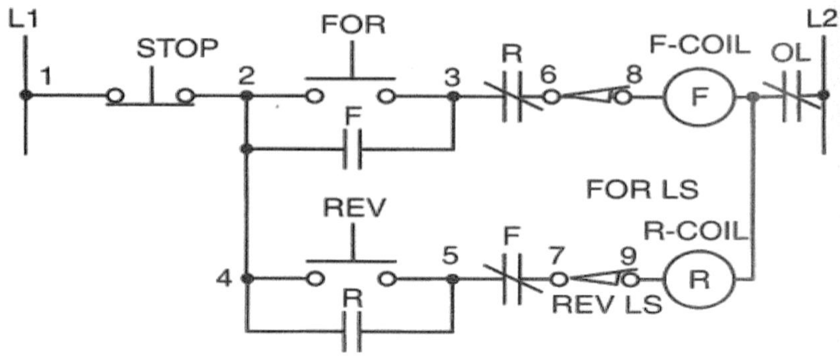

Figure 14.8 A forward/reverse/stop push-button control circuit.

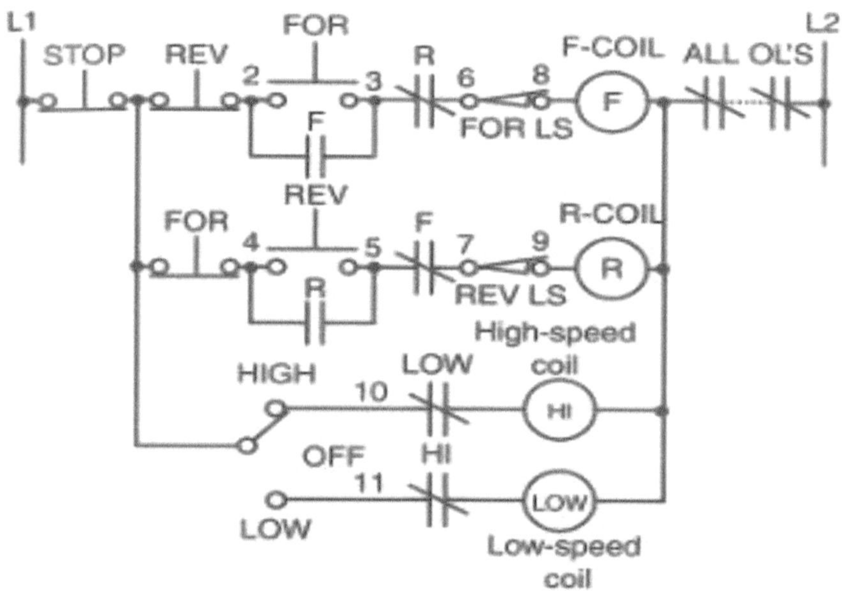

Figure 14.9 Push-button control circuit with a highand low-speed selector switch.

is energized, normally open contact F closes and normally closed contact F opens. This action causes a motor to operate in the reverse direction until the stop push button is pressed again.

Figure 14.9 shows a circuit that is another method of forward/reverse/ stop control. This control circuit has the added feature of a high- and low-speed selector switch for either the forward or reverse direction. The selector

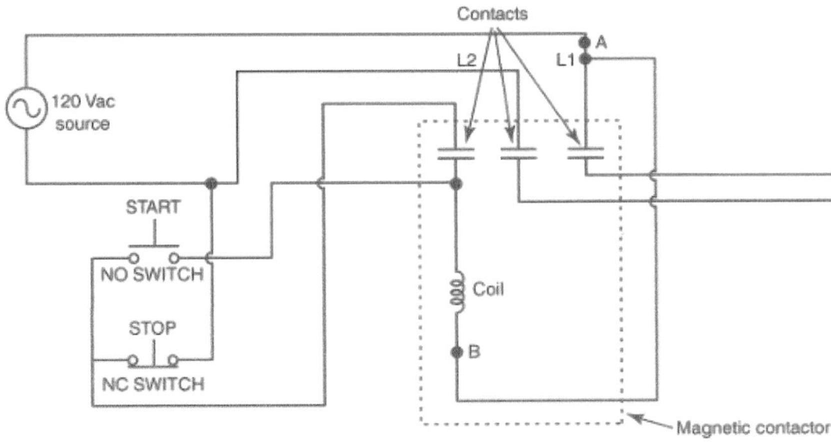

Figure 14.10 Magnetic contactor motor control circuit (schematic diagram).

switch is placed in series with the windings of the motor. When the selector is changed from the HIGH position to the LOW position, a modification in the windings of the motor can be made.

There are many other push-button combinations that can be used with control relays to accomplish motor control or control of other types of loads. The circuits discussed in this section represent some basic power control functions such as start/stop, forward/reverse/stop, jogging, and multiple speed control.

14.2 Motor-Starting Systems

The equipment used with motor-starting systems was studied in **Chapter 13**. Some specific systems used for starting electric motors are discussed in this section. Most motor-starting systems utilize one magnetic contactor. A schematic diagram of a magnetic contactor circuit used for controlling a single-phase motor is shown in Figure 14.10. Note that the magnetic contactor relies upon an electromagnetic coil that energizes when current passes through it. The activated coil performs the function of closing a set of normally open contacts. These contacts are connected in series with the power input to the motor that is being controlled.

In Figure 14.10, the START push-button switch is a normally open switch. When the start switch is pressed, current flows through the coil of the magnetic contactor. This action energizes the coil, and the solenoid of the contactor is drawn inward to close the contacts that are in series with the power lines. Once these contacts are closed, current continues to flow through the electromagnetic coil through the holding contacts (L3 in the diagram).

Current continues to flow until the STOP pushbutton switch is pressed. The stop switch is a normally closed switch. When it is pressed, the circuit to the electromagnetic coil is broken. At this time, current no longer flows through the coil. The contacts now release and cause the power lines to be interrupted. Thus, the motor is turned off.

Magnetic contactor circuits are sometimes referred to as **across-the-line starters**. The relay principle is utilized since a small current through the coil controls a larger current through a motor.

14.3 Types of Starting Systems

Motor starting, particularly for large motors, can play an important role in the efficient operation of an electrical power system. Electric motors draw a larger current from the power lines during starting than during normal operation. Motor-starting equipment should attempt to reduce starting currents to a level that can be handled by the electrical power system where they are being used. The motor-starting equipment that is used is placed between the electrical power source and the motor. There are several systems used to start electric motors. A selection of these systems will be discussed.

14.3.1 Full voltage starting

One method of starting electric motors is called **full-voltage starting**. With full-voltage starting, full power is applied to the motor initially. Since full power-supply voltage is applied to the motor initially, maximum starting torque and minimum acceleration time result. However, the power system must be able to handle the starting current drawn by the motor. This method is the least expensive and the simplest to install.

Full-voltage starting is illustrated by the diagram of Figure 14.11. In this power control circuit, a start/stop push-button station is used to control a three-phase motor. When the normally open start push button is pressed, current flows through the relay coil (M), causing the normally open contacts to close. The line contacts allow full voltage to be applied to the motor when they are closed.

When the start push button is released, the relay coil remains energized due to the holding contact. This contact provides a current path from L1 through the normally closed stop push button, through the holding contact,

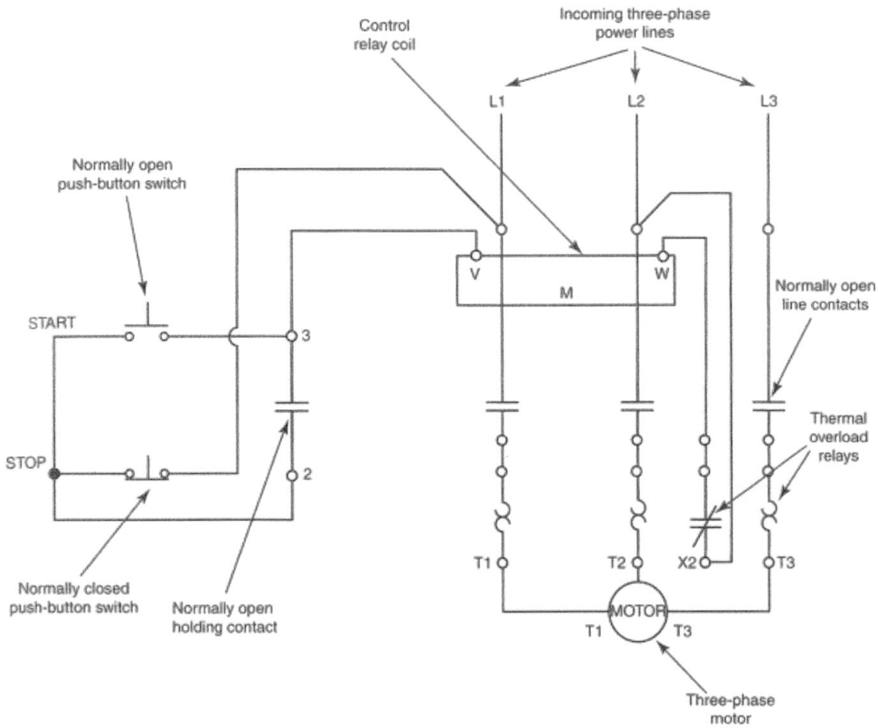

Figure 14.11 Full-voltage starting circuit for a three-phase motor.

through the coil (M), through a thermal overload relay, and back to L2. When the stop push button is pressed, this circuit is opened, causing the coil to be de-energized.

14.3.2 Primary resistance starting

Another motor starting method is called primary resistance starting. **Primary-resistance starting** uses resistors in series with the power lines to reduce the motor-starting current. Ordinarily, the resistance connected to the power lines can be reduced in steps until full voltage is applied to the motor. Thus, the starting current is reduced according to the value of the series resistance in the power lines. Since starting torque is directly proportional to the current, starting torque is reduced according to the magnitude of the current.

Figure 14.12 shows the **primary-resistance starting** method used to control a three-phase motor. When the start push button is pressed, coils S and

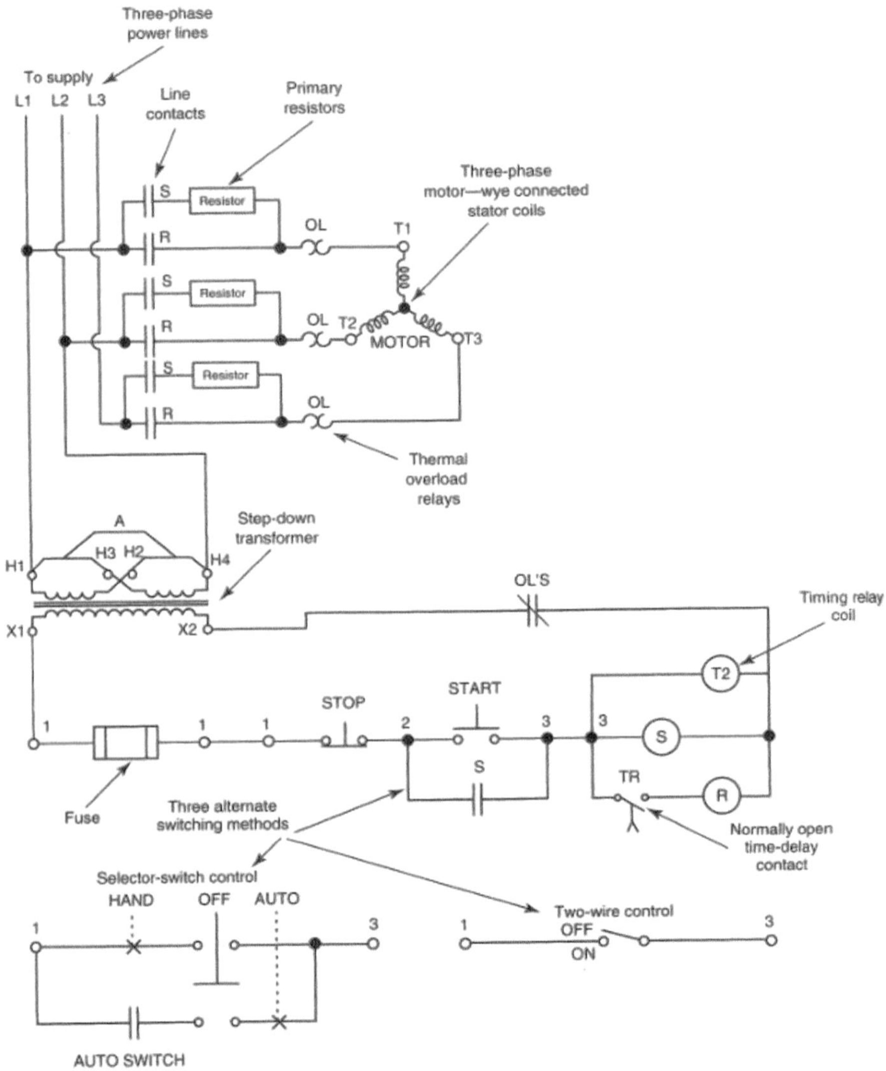

Figure 14.12 Primary resistance starter circuit.

TR are energized. Initially, the start contacts (S) will close, applying voltage through the primary resistors to the motor. These resistors (see Figure 14.13) reduce the value of the starting current.

Once the time delay period of the timing relay TR has elapsed, contact TR will close. The run contacts (R) will then close and apply full voltage to the motor. Notice that a step-down transformer is used for applying voltage

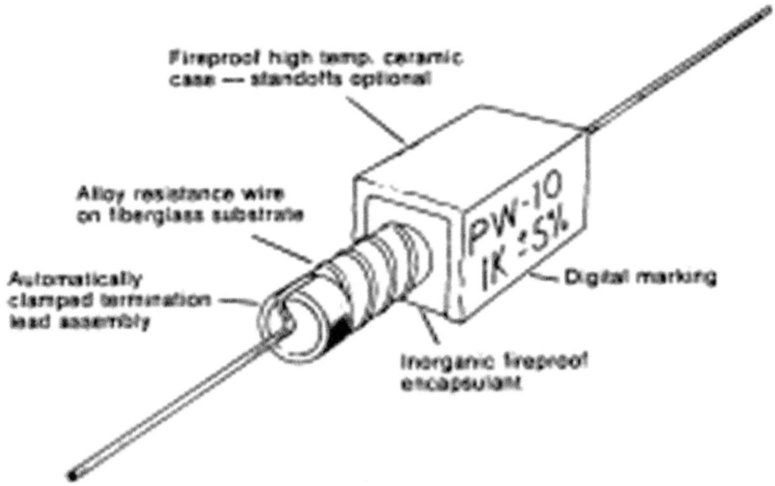

Figure 14.13 High-wattage wire-wound resistor like those used with primary-resistance starters. (TRW/IRC Corp.)

to the control portion of the circuit. This is a commonly used technique for reducing the voltage applied to the relay coils.

14.3.3 Primary-reactor starting

Another method, similar to primary-resistance starting, is called the **primary-reactor starting** method. Reactors (coils) are used in place of resistors since they consume smaller amounts of power from the AC source. Usually, this method is more appropriate for large motors that are rated at over 600 volts.

14.3.4 Autotransformer starting

Autotransformer starting is another method used to start electric motors. The **autotransformer starting** method employs one or more autotransformers to control the voltage that is applied to a motor. The autotransformers used are ordinarily tapped to provide a range of starting-current control. When the motor has accelerated to near its normal operating speed, the autotransformer windings are removed from the circuit. A major disadvantage of this method is the expense of the autotransformers.

An **autotransformer starting** circuit is shown in Figure 14.14. This circuit is an expensive type of control that uses three autotransformers and four relays. When the start push button is pressed, current flows through coils

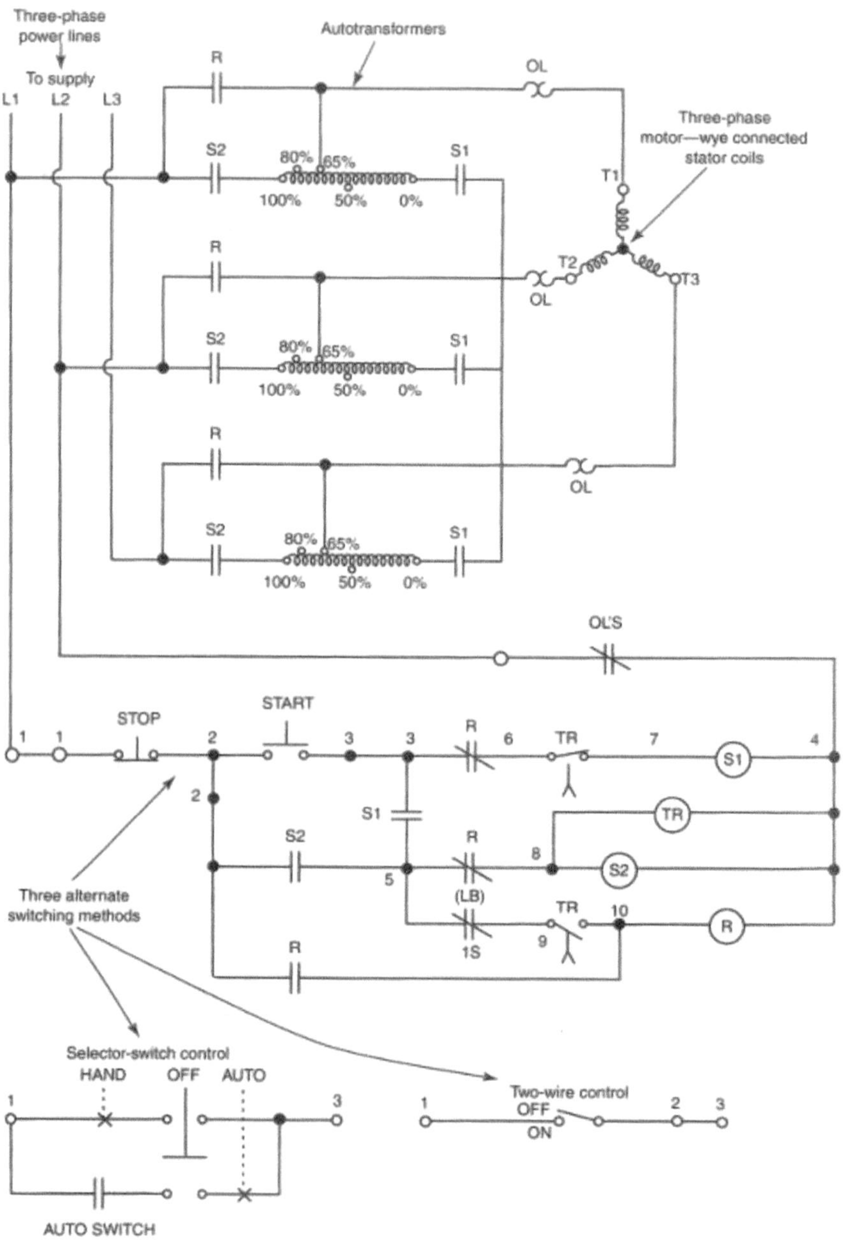

Figure 14.14 Autotransformer starter circuit used with a three-phase motor.

S1, S2, and TR. The S1 and S2 contacts will then close. Voltage will be applied through the autotransformer windings to the three-phase motor. One normally closed and one normally open contact are controlled by the timing relay TR. When the specified time period has elapsed, the normally closed TR contact opens and the normally open TR contact closes. Coil R then energizes, causing the normally open R contacts to close and apply full voltage to the motor. Normally closed R contacts are connected in series with coils S1, S2, and TR to open their circuits when coil R is energized. When the stop push button is pressed, the current to coil R is interrupted, thus opening the power line connections to the motor.

Notice that the 65% taps of the autotransformer are used in Figure 14.14. There are also taps for 50%, 80%, and 130% to provide more flexibility in reducing the motor-starting current.

14.3.5 Wye-delta starting

It is possible to start three-phase motors more economically by using the wye-delta starting method. Since, in a wye configuration, line current is equal to the phase current divided by 1.73 (or $\sqrt{3}$), it is possible to reduce the starting current by using a wye connection rather than a delta connection. This method, shown in Figure 14.15, employs a switching arrangement that places the motor windings in a wye configuration during starting and a delta arrangement for running. In this way, the starting current is reduced. Although starting torque is reduced, running torque is still high since full voltage appears across each winding when the motor is connected in a delta configuration.

In Figure 14.15, when the **start push button** is pressed, coil *S* is energized. The normally open *S* contacts will close. This action connects the motor windings in a wye (or star) configuration. It also activates the timing relay (TR) and coil M1. The normally open M1 contacts then close to apply voltage to the wye-connected motor windings. After the time delay period has elapsed, the TR contacts change state. Coil S de-energizes and coil M2 energizes. The S contacts, which hold the motor windings in a wye arrangement, open. The M2 contacts close and cause the motor windings to be connected in a delta configuration. The motor then continues to run with the motor connected in a delta arrangement.

14.3.6 Part-winding starting

Figure 14.16 shows the **part-winding starting** method, which is usually more simple and less expensive than other starting methods. However, to use this starting method, motors must be specifically designed to operate in this

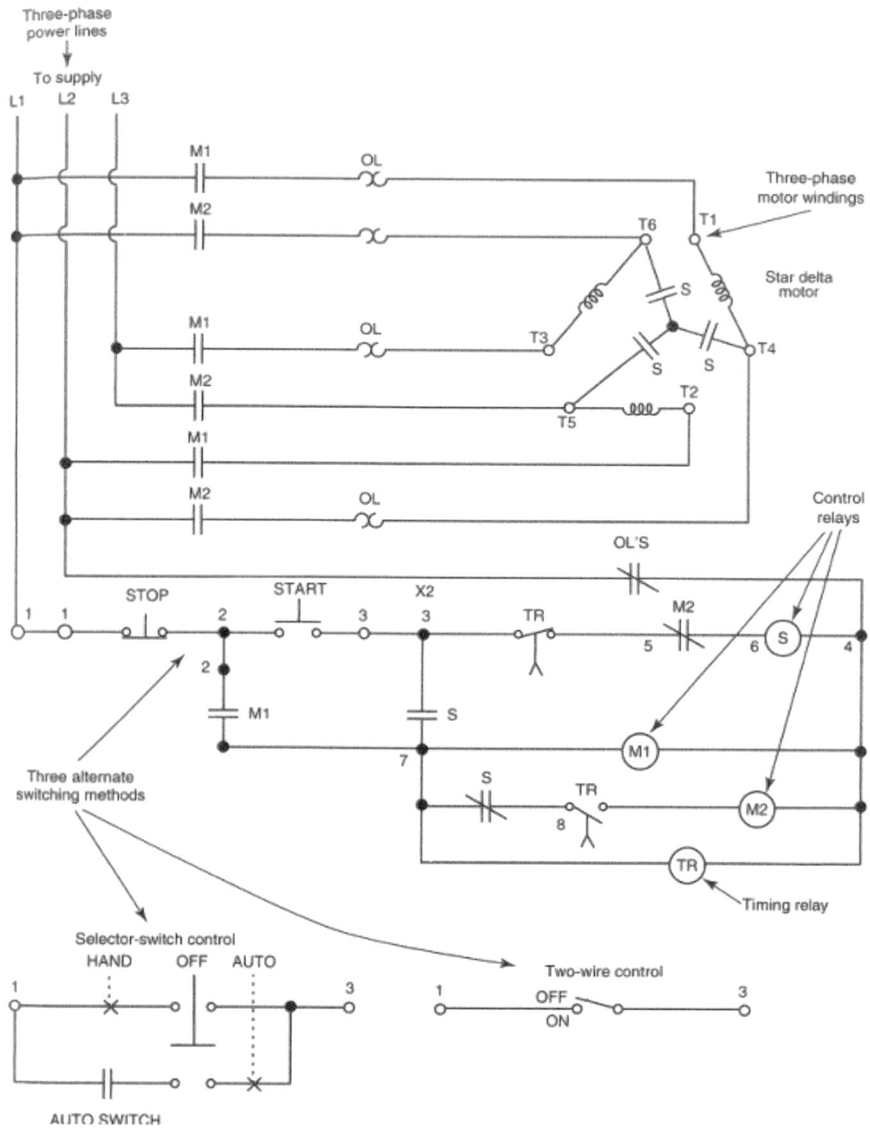

Figure 14.15 Wye-to-delta three-phase starter circuit.

manner. During starting, the power line voltage is applied across only part of the parallel-connected motor windings, thus reducing starting current. Once the motor has started, the line voltage is placed across all of the motor windings. This method is undesirable for many heavy-load applications due to the reduction of starting torque.

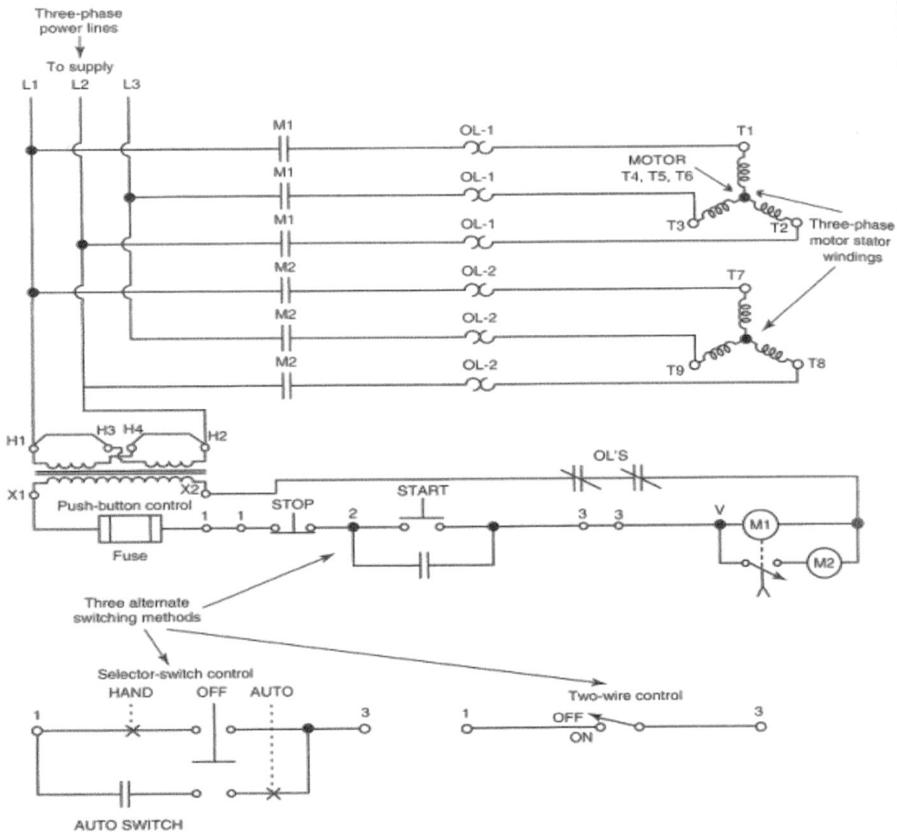

Figure 14.16 Part-winding circuit for three-phase motor starting.

In Figure 14.16, when the start push button is pressed, current flows through coil M1 of the time-delay relay. This causes the normally open contacts of M1 to close and causes a three-phase voltage to be applied to windings T1, T2, and T3. After the time-delay period has elapsed, the normally open contact located below coil M1 in Figure 14.16 will close. This action energizes coil M2 and causes its normally open contacts to close. The M2 contacts then connect the T7, T8, and T9 windings in parallel with the T1, T2, and T3 windings. When the stop push button is pressed, coils M1 and M2 are de-energized.

14.3.7 Direct-current starting systems

Since DC motors have no counterelectromotive force (CEMF) when they are not rotating (see Chapter 9), they have tremendously high starting currents.

Therefore, they must use some type of control system to reduce the initial starting current. Ordinarily, a series resistance is used. This resistance can be manually or automatically reduced until a full voltage is applied.

The four types of control systems commonly used with DC motors are:

- Current limit.
- Definite time.
- Counter-emf.
- Variable voltage.

The current-limit method allows the starting current to be reduced to a specified level and then advanced to the next resistance step. The **definite time method** causes the motor to increase speed in timed intervals with no regard to the amount of armature current or to the speed of the motor. The counter-emf method samples the amount of CEMF generated by the armature of the motor to reduce the series resistance accordingly. This method can be used effectively since cemf is proportional to both the speed and the armature current of a DC motor. The variable-voltage method employs a variable direct-current power source to apply a reduced voltage to the motor initially and then gradually increase the voltage. No series resistance is needed when the latter method is used. A solid-state reduced-voltage starter is shown in Figure 14.17.

14.4 Specialized Control Systems

Electrical power control is usually desired for some specific application. In this section, we will discuss some common types of specialized electrical power control systems that are used today.

Most types of electrical motors can be made to rotate in either direction by some simple modifications of their connections. Ordinarily, motors require two magnetic contactors, one forward and one reverse. Shown in Figure 14.18 is a solid-state starter used to accomplish forward and reverse operation. These starters are used in conjunction with a set of three push-button switches-forward, reverse, and stop. When the forward push-button switch is depressed, the forward contactor is energized. It is deactivated when the stop push-button switch is depressed. A similar procedure takes place during reverse operation.

14.4.1 Direct-current motor reversing

Direct-current motors can have their direction of rotation reversed by changing either the armature connections or the field connections to the power

Figure 14.17 Solid-state reduced voltage starters. (Eaton Corp., Cutler-Hammer Products)

source. In Figure 14.19, a DC shunt motor control circuit is shown. When the forward push button is pressed, the F coil will be energized, causing the F contacts to close. The armature circuit is then completed from L1 through the lower F contact, up through the armature, through the upper F contact, and back to L2. Pressing the stop push button will de-energize the F coil.

The direction of **rotation** of the motor is reversed when the reverse push button is pressed. This is due to the change of the current direction through

Figure 14.18 Solid-state motor reversing starter. (Payne Engineering)

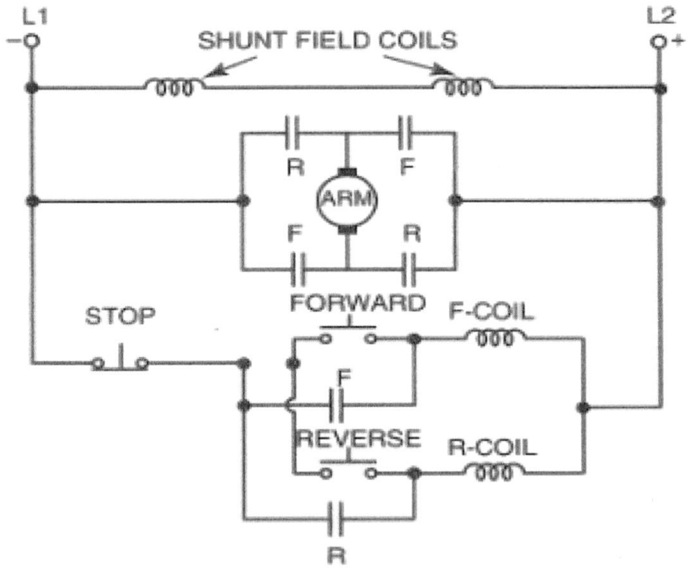

Figure 14.19 Control circuit for the forward and reverse operation of a DC shunt motor.

the armature. Pressing the reverse push button energizes the *R* coil and closes the *R* contacts. The armature current path is then from L1 through the upper R contact, down through the armature, through the lower R contact, and back to L2. Pressing the stop button will de-energize the R coil.

14.4.2 Single-phase AC induction-motor reversing

Single-phase AC induction motors that have **start and run windings** can have their direction of rotation reversed by using the circuit in Figure 14.19. If we modify the diagram by replacing the shunt field coils with the run windings and the armature with the start windings, directional reversal of a single-phase AC motor can be accomplished. Single-phase induction motors are reversed by changing the connections of either the start windings or the run windings, but not both at the same time.

14.4.3 Three-phase AC induction-motor reversing

Three-phase motors can have their direction of rotation reversed by simply changing the connections of any two power input lines. This changes the

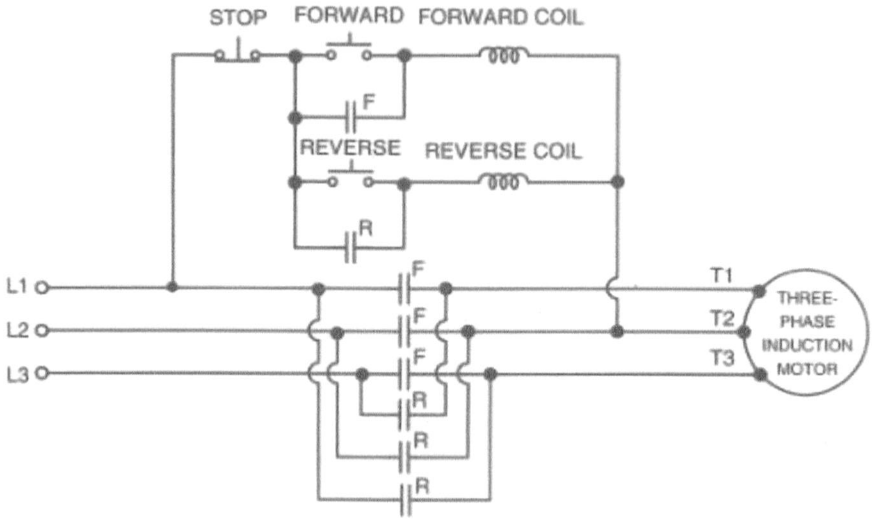

Figure 14.20 Control circuit for the forward and reverse operation of a three-phase AC induction motor.

phase sequence applied to the motor. A control circuit for three-phase induction-motor reversing is shown in Figure 14.20.

When the **forward push button** is pressed, the forward coil will energize and close the F contacts. The three-phase voltage is applied from L1 to T1, L2 to T2, and L3 to T3 to cause the motor to operate. The stop push button de-energizes the forward coil. When the reverse push button is pressed, the reverse coil is energized, and the *R* contacts will close. The voltage is then applied from L1 to T3, L2 to T2, and L3 to T1. This action reverses the L1 and L3 connections to the motor and causes the motor to rotate in the reverse direction.

14.5 Dynamic Braking

When a motor is turned off, its shaft will continue to rotate for a short period of time. This continued rotation is undesirable for many applications. **Dynamic braking** is a method used to bring a motor to a quick stop whenever power is turned off. Motors with a wound armature utilize a resistance connected across the armature as a dynamic braking method. When power is turned off, the resistance is connected across the armature. This causes the armature to act as a loaded generator. This makes the motor slow down immediately. This dynamic braking method is shown in Figure 14.21.

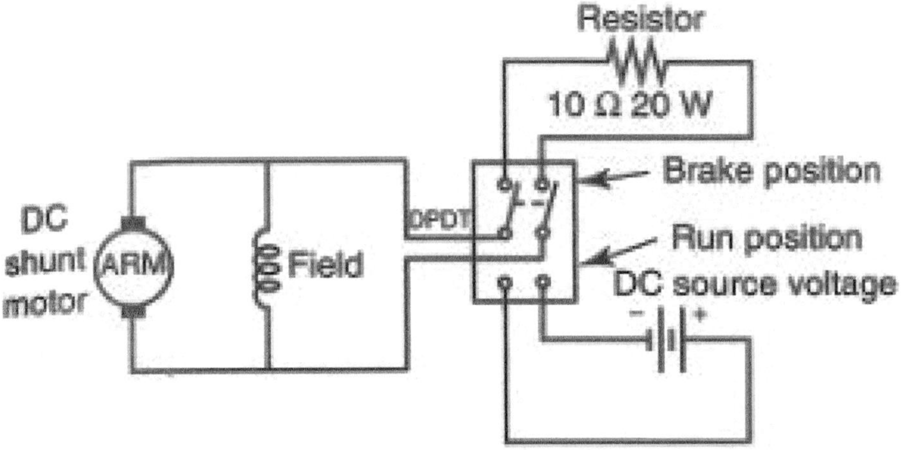

Figure 14.21 Dynamic braking circuit for a DC shunt motor.

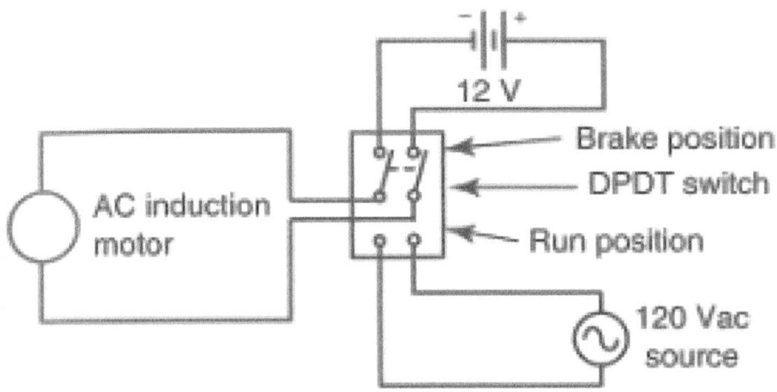

Figure 14.22 Dynamic braking circuit for a single phase AC induction motor.

Alternating-current induction motors can be slowed down rapidly by placing a DC voltage across the winding of the motor. This DC voltage sets up a constant magnetic field that causes the rotor to slow down rapidly. A circuit for the dynamic braking of a single-phase AC induction motor is shown in Figure 14.22. An electronic dynamic brake is shown in Figure 14.23.

14.6 Universal Motor Speed Control

An important advantage of a universal AC–DC motor is the ease of speed control. The **universal motor** has a brush/commutator assembly, with the armature circuit connected in series with the field windings. By varying the voltage applied to universal motors, their speed can be varied from zero to maximum.

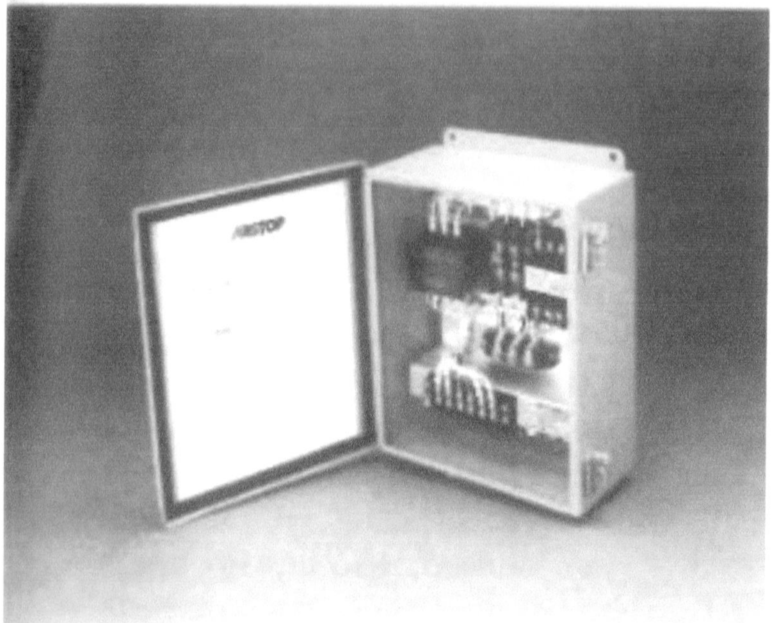

Figure 14.23 Electronic dynamic brake. (Electroid Co.)

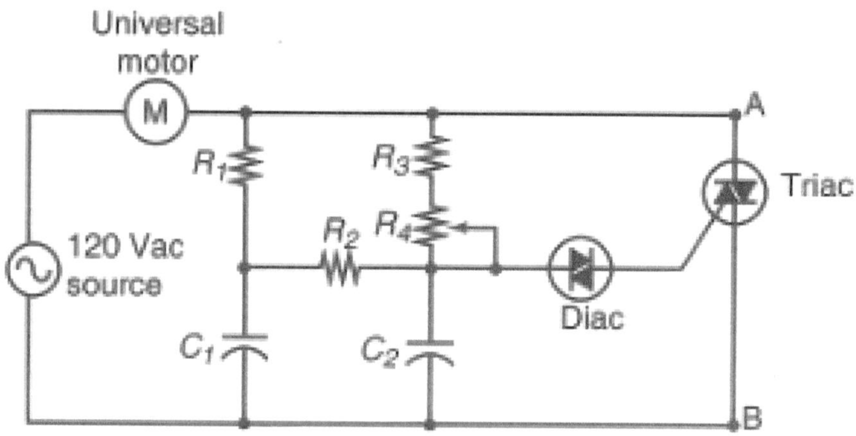

Figure 14.24 Speed control circuit for a universal motor.

The circuit used for this purpose, shown in Figure 14.24, uses a gate-controlled triac. The triac is a semiconductor device whose conduction can be varied by a trigger voltage applied to its gate. A silicon-controlled rectifier (SCR) could also be used as a speed-control device for a universal motor. Speed-control circuits like this one are used for applications such as

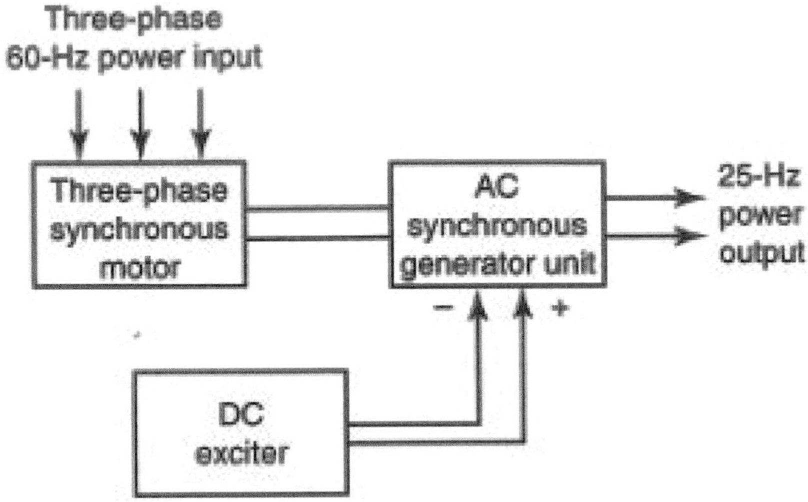

Figure 14.25 A 60 hertz to 25 hertz frequency conversion system.

drills, sewing machines, mixers, and industrial applications. Devices used for these control circuits are discussed in **Chapter 15.**

14.7 Frequency-Conversion Systems

The power system frequency used in the United States is 60 hertz, or 60 cycles per second. However, there are specific applications that require other frequencies in order to operate properly. Mechanical frequency converters are used to change an incoming frequency into some other frequency. Frequency converters are **motor-generator sets** that are connected together.

For example, a frequency of 60 Hz could be applied to a synchronous motor that rotates at a specific speed. A generator connected to its shaft could be built with the necessary number of poles to cause it to produce a frequency of 251H. Recall that **frequency** is determined by the following relationship:

$$\text{frequency (Hz)} = \frac{\text{speed of rotation (rpm)} \times \text{number of poles}}{120}$$

A frequency-conversion system is shown in Figure 14.25. Synchronous units such as the one shown are used wherever precise frequency control is required. It is also possible to design units that are driven by induction motors if some frequency variation can be tolerated.

Review Questions

Answer the following questions.

1. What is a jog control used for?

2. What is another name for magnetic contactor circuits?

3. What is meant by the following control terms?
   a. Full-voltage starting
   b. Primary resistance starting
   c. Primary-reactor starting
   d. Autotransformer-starting
   e. Wye-delta starting
   f. Part-winding starting

4. List four types of control systems used with DC motors.

5. How is the reversal of the direction of rotation of a direct-current motor accomplished?

6. How is the reversal of the direction of rotation of a single-phase AC induction motor accomplished?

7. How is reverse control of a three-phase motor accomplished?

8. What is meant by dynamic braking?

9. How is universal motor speed control accomplished?

10. What is a frequency conversion system?

UNIT FIVE

Electronic Control Basics

In **Unit V**, electronic control basics are covered. The fundamentals of electronic devices are discussed in **Chapter 15**. This chapter examines the basic concepts and applications of devices such as diodes, transistors, FETs, SCRs, triacs, diacs, op-amps, and integrated circuits.

Electronic sensors and transducers are discussed in **Chapter 16**. These devices are used in many types of control systems used with automated industrial systems today. Sensors and transducers such as light-sensitive devices, thermal devices, and proximity sensors, as well as inductive and capacitive sensors are studied.

DOI: 10.1201/9788743810056-19

15

Electronic Devices

Objectives

Upon completion of this chapter, you will be able to:

- Explain how a diode is constructed of N and P material.
- Connect a diode so that it is either forward or reverse biased.
- Distinguish between the forward and reverse characteristics of a diode.
- Identify the schematic symbol of a diode and denote the anode and cathode.
- Identify the electrodes of a diode with an ohmmeter and evaluate the condition of the device as good, shorted, or open.
- Explain the special features of the zener diode.
- Explain the special features of light-emitting diodes.
- Describe the physical construction of bipolar and unipolar transistors.
- Understand the fundamental operation of transistors.
- Identify the elements of a transistor.
- Explain how the elements of a transistor respond when it is operational.
- Properly bias NPN and PNP transistors.
- Test transistors with a multimeter to identify the leads and determine if the device is good or defective.
- Identify different electronic control devices, such as SCRs, UJTs, diacs, and triacs.
- Explain how electrical power is controlled by switching using SCRs and triacs.
- Describe operational amplifiers and integrated circuits.

DOI: 10.1201/9788743810056-20

Important Terms

anode
barrier potential
base
bias voltage
biasing
bipolar
bipolar junction transistor (BJT)
breakdown voltage
cathode
channel
collector
depletion mode
depletion zone
diac
diffusion
diode
doping
drain
emitter
enhancement mode
forward biasing
gate
hole
integrated circuit (IC)
junction
junction field-effect transistor (JFET)
light-emitting diode (LED)
majority carrier
metal oxide semiconductor field-effect transistor (MOSFET)
minority carrier
N-type crystal
NPN transistor
operational amplifier (op-amp)
P-type crystal
PNP transistor
reverse biasing
silicon-controlled rectifier (SCR)
source
transistor
triac
unijunction transistor (UJT)
unipolar
V-MOSFET (V-MOS)
zener diode
zener voltage

Introduction

Many types of electronic devices are used in automated control systems. This chapter provides an overview of many of these electronic control devices. The chapter progresses from the simplest type of device, the diode, to more complex control devices such as SCRs, triacs, and integrated circuits.

15.1 Electronic Diodes

Electronic **diodes** are semiconductor devices that allow current to flow in only one direction. Diodes are used rather extensively in nearly all phases of electronics today. Radio, TV, industrial control circuits, home entertainment equipment, computers, and electrical appliances are only a few of the applications.

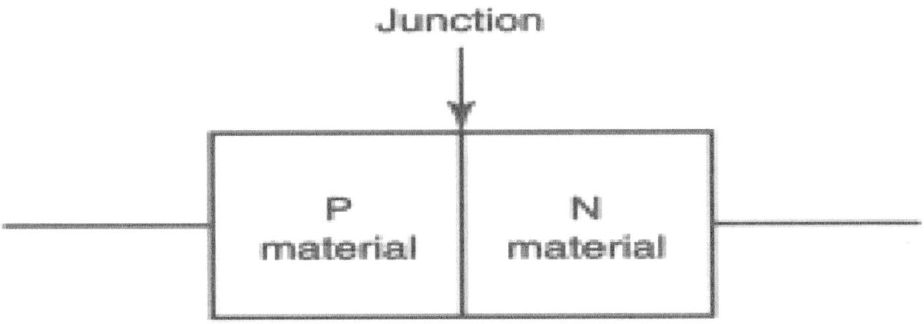

Figure 15.1 Crystal structure of a junction diode.

As a diode conducts well in only one direction, AC can be converted easily into DC with a diode. In this application, a diode is used as a rectifier. In addition to this, other functions can be performed. Some of these are the operational bases of more complex electronic devices. Diode operation is therefore extremely important. A person working in electronics must be very familiar with the characteristic operation of this device.

Diodes are probably the simplest of all electronic components. Only two leads or electrodes are used in their construction. To create a diode, a semiconductor material, such as silicon, is altered. (Recall from **Chapter 1** that a semiconductor material has four valence electrons.) Impurities are added to the material. This is called **doping**. If a **trivalent impurity** (an impurity with three valence electrons) is added, P-type crystals will result. A **P-type crystal** has an excess of holes, spaces that are looking for electrons. If a **pentavalent impurity** (an impurity with five valence electrons) is added to the semiconductor, an **N-type crystal** is created. N-type crystals have excess free electrons. Diodes are made when one P-type crystal and one N-type crystal are joined together.

The term **junction diode** is often used to describe this crystal structure made of P and N materials. A diode is generally a two-terminal device. One terminal is attached to P material and the other to N material. The common connecting point where these materials are joined is called a **junction**. A junction diode permits current carriers to flow in one direction and blocks the flow of current in the opposite direction.

Figure 15.1 shows the crystal structure of a **junction diode**. Notice the location of the P and N materials with respect to the junction. The crystal structure is continuous from one end to the other. The junction serves only as a dividing line that marks the ending of one material and the beginning of the other. This type of structure permits electrons to move readily in one direction.

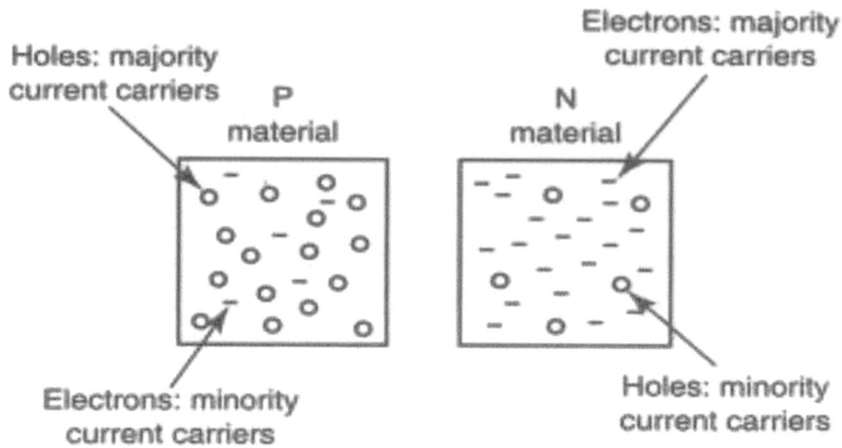

Figure 15.2 Semiconductor materials.

Figure 15.2 shows two pieces of semiconductor material before they are formed into a PN junction. As indicated, each piece of material has majority and minority current carriers called simply **majority carriers** and **minority carriers**. The number of carrier symbols shown in each material indicates the minority or majority function. **Electrons** are the majority carriers in the N material and are the minority carriers in the P material. **Holes** are the majority carriers in the P material and are the minority carriers in the N material. Both holes and electrons have freedom to move about in their respective materials.

15.1.1 Depletion zone

When a junction diode is first formed, there is a unique interaction between current carriers. Electrons from the N material move readily across the junction to fill holes in the P material. This action is commonly called **diffusion**. Diffusion is the result of a high concentration of carriers in one material and a lower concentration in the other. Only those current carriers near the junction take part in the diffusion process.

The **diffusion** of current carriers across the junction of a diode causes a change in its structure. Electrons leaving the N material cause positive ions to be generated in their place. Upon entering the P material to fill holes, these same electrons create negative ions. The area on each side of the junction then contains a large number of positive and negative ions. The number of holes and electrons in this area becomes depleted. The term depletion zone is

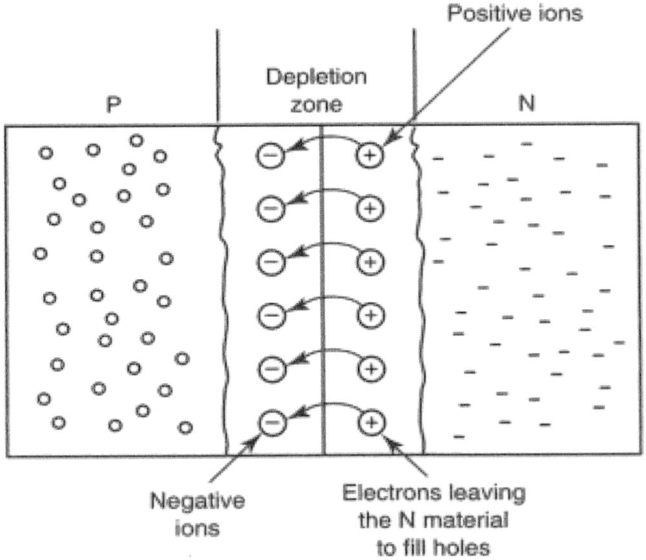

Figure 15.3 Depletion-zone formation.

used to describe this area. It represents an area that is void of majority current carriers. All PN junctions develop a depletion zone when they are formed. Figure 15.3 shows the depletion zone of a junction diode.

15.1.2 Barrier potential

Before N and P materials are joined together at a common junction, they are considered to be electrically neutral. After being joined, however, diffusion takes place immediately. Electrons crossing the junction to fill holes cause negative ions to appear in the P material. This action causes the area near the junction to take on a negative charge. Similarly, electrons leaving the N material cause it to produce positive ions. This, in turn, causes the N side of the junction to take on a net positive charge. These respective charge creations tend to drive remaining electrons and holes away from the junction. This action makes it somewhat difficult for additional charge carriers to diffuse across the junction. The end result is a charge buildup or **barrier potential** across the junction.

Figure 15.4 shows the resulting **barrier potential** as a small battery connected across the PN junction. Note the polarity of this potential with respect to P and N material. This voltage exists even when the crystal is not connected to an outside source of energy. For **germanium** semiconductor

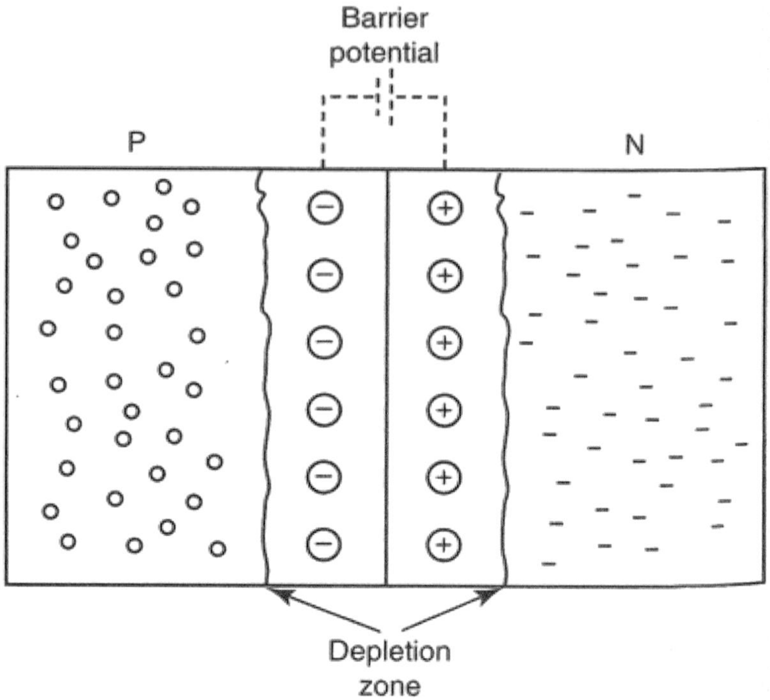

Figure 15.4 Barrier potential of a diode.

devices the barrier potential is approximately 0.3 volts. For **silicon** semiconductor devices the barrier potential is about 0.7 volts.

These voltage values cannot be measured directly. They appear only across the space charge region of the junction. However, the barrier potential of a PN junction must be overcome by an outside voltage source in order to produce current conduction.

15.1.3 Junction biasing

When an external source of energy is applied to a PN junction, it is called a **bias voltage** or simply **biasing**. This voltage either adds to or reduces the barrier potential of the junction. Reducing the barrier potential causes current carriers to return to the depletion zone. **Forward biasing** is used to describe this condition. Adding external voltage of the same polarity to the barrier potential causes an increase in the width of the depletion zone. This action hinders current carriers from entering the depletion zone. The term reverse biasing is used to describe this condition.

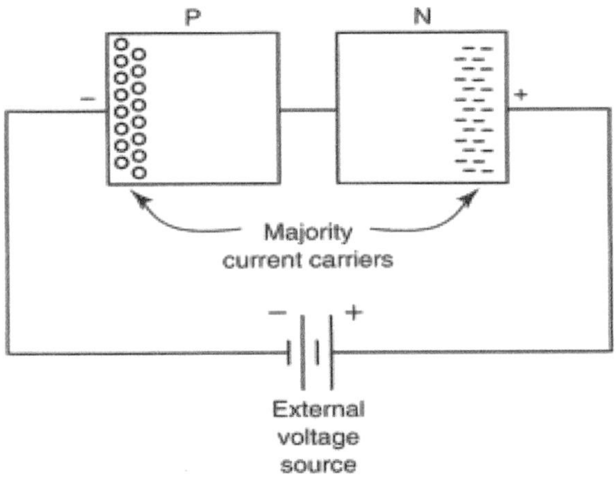

Figure 15.5 Reverse-biased diode.

15.1.4 Reverse biasing

When a battery is connected across a **PN junction** as shown in Figure 15.5, it causes reverse biasing. Note that the negative terminal of the battery is connected to the P material and the positive to the N material. Connection in this way causes the battery polarity to oppose the material polarity of the diode. Because unlike charges attract, the majority charge carriers of each material are pulled away from the junction. Reverse biasing of a diode normally causes it to be nonconductive.

Figure 15.6 shows how the **majority current carriers** are rearranged in a reverse-biased diode. As shown, electrons of the N material are pulled toward the positive battery terminal. Each electron that moves or leaves the diode causes a positive ion to appear in its place. This causes a corresponding increase in the width of the depletion zone on the N side of the junction.

The P side of the diode has a reaction very similar to the N side. In this case, a number of electrons leave the negative battery terminal and enter the P material. These electrons immediately move in and fill a number of holes. Each filled hole then becomes a **negative ion**. These ions are then repelled by the negative battery terminal and driven toward the junction. As a result of this, the width of the depletion zone increases on the P side of the junction.

The overall **depletion zone** width of a reverse biased diode is directly dependent on the value of the supply voltage. With a wide depletion zone, a diode cannot effectively support the flow of current. The charge buildup across the junction increases until the barrier voltage equals the external

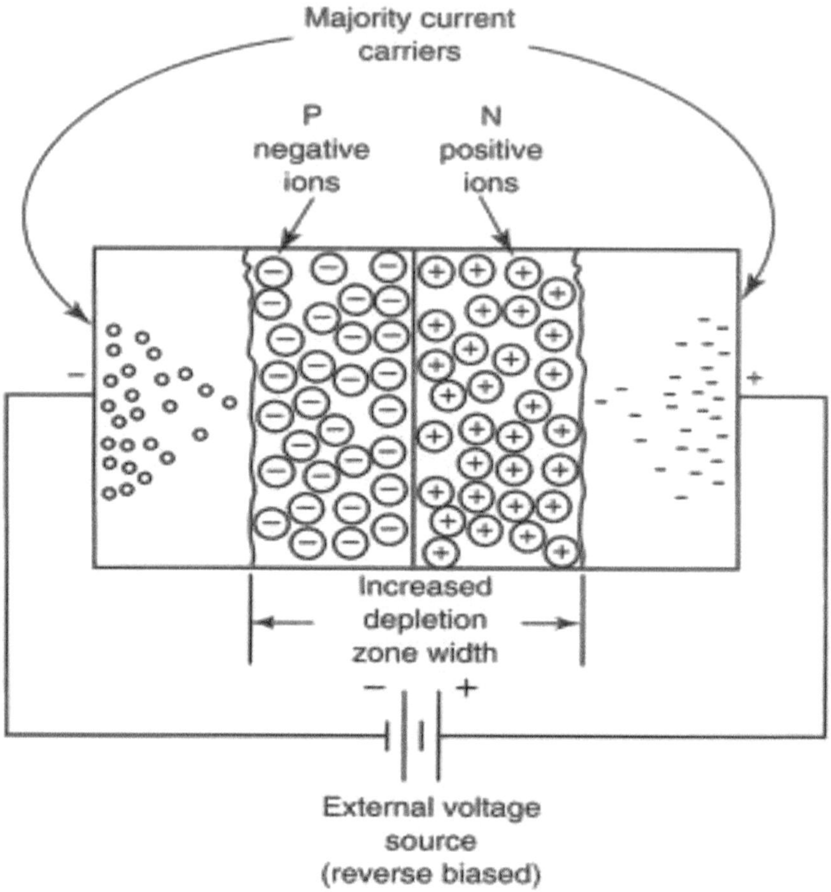

Figure 15.6 Reverse-biased PN junction.

bias voltage. When this occurs, a diode is effectively a nonconductor. A few minority current carriers are left, see Figure 15.7.

15.1.5 Forward biasing

When voltage is applied to a diode as in Figure 15.8, it causes **forward biasing**. In this example, the positive battery terminal is connected to the P material and the negative terminal to the N material. This voltage repels the majority current carriers of each material. A large number of holes and electrons therefore appear at the junction. On the N side of the junction, electrons move in to neutralize the positive ions in the depletion zone. In the P material, electrons are pulled from negative ions. This causes them to become neutral

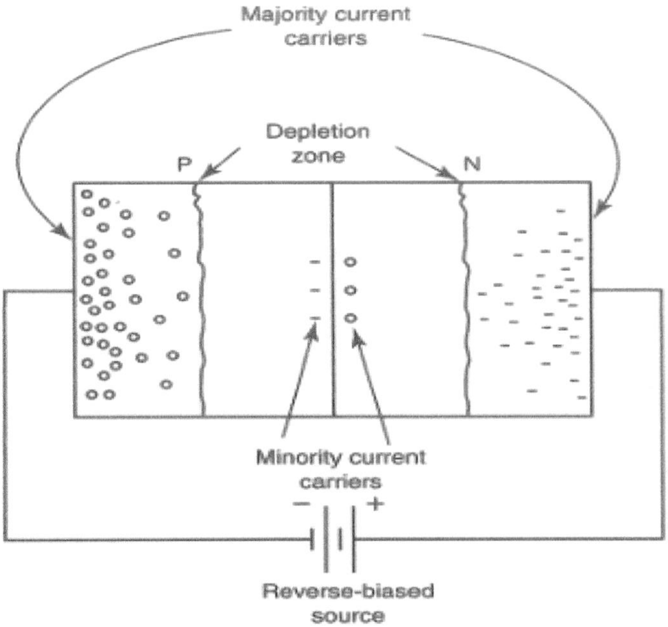

Figure 15.7 Minority current carriers.

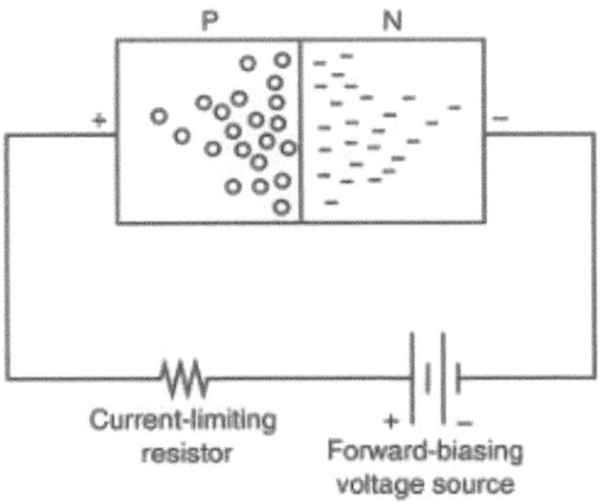

Figure 15.8 Forward-biased diode.

again. This action means that forward biasing causes the depletion zone to collapse and the **barrier potential** to be removed. The PN junction, therefore, supports a continuous flow of current when it is forward biased.

Figure 15.9 shows how the current carriers of a **forward-biased** diode respond. Since the diode is connected to an external voltage source, it has a

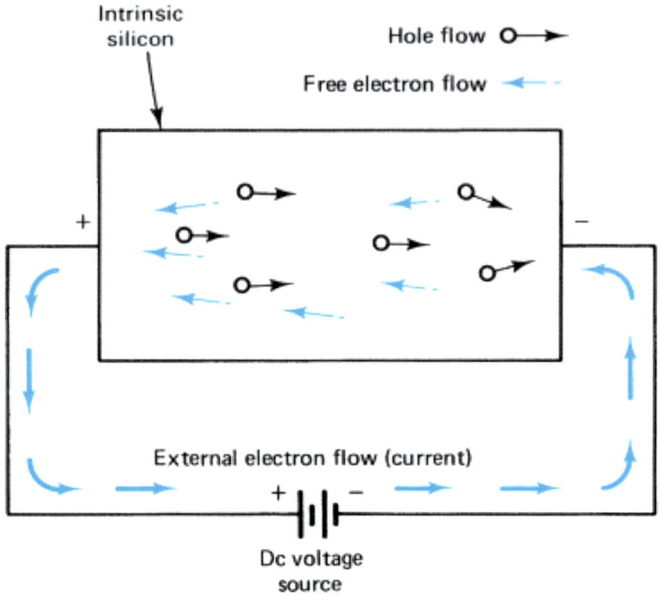

Figure 15.9 Current carrier flow in a forward-biased diode.

constant supply of electrons. Large arrows are used in the diagram to show the direction of the current outside of the diode. Inside the diode, smaller arrows show the movement of majority current carriers. Remember that electron flow and current refer to the same thing.

Starting at the negative battery terminal, assume now that electrons flow through a wire to the N material. Upon entering this material, they flow immediately to the junction. At the same time, an equal number of electrons is removed from the P material and is returned through a resistor to the positive battery terminal. This action generates new holes and causes them to move toward the junction. When these holes and electrons reach the junction, they combine and effectively disappear. At the same time, new holes and electrons appear at the outer ends of the diode. The majority of carriers are generated on a continuous basis. This action continues as long as the external voltage source is applied.

It is important to realize that electrons flow through the entire diode when it is forward-biased. In the N material, this is quite obvious. In the P material, however, holes are the moving current carriers. Remember that the hole movement in one direction must be initiated by electron movement in the opposite direction. Therefore, the combined flow of holes and electrons through a diode equals the total current.

15.1.6 Diode specifications

When selecting a diode for a specific application, some consideration must be given to **specifications**. This type of information is made available through the manufacturer. Diode specifications include such things as absolute maximum ratings, typical operating conditions, mechanical data, lead identification, mounting procedures, and characteristics. Some of the important specifications are explained in the following:

1. *Maximum reverse voltage* V_{RM}: The absolute maximum or peak reverse bias voltage that can be applied to a diode. This may also be called the peak inverse voltage (PIV) or peak reverse voltage (PRV).
2. *Reverse breakdown voltage* V_{BR}: The minimum steady-state reverse voltage at which breakdown will occur.
3. *Maximum forward current* I_{FM} : The absolute maximum repetitive forward current that can pass through a diode at 25° C(77° F). This is reduced for operation at higher temperatures.
4. *Maximum forward surge content* I_{FM} *(surge)*: The maximum current that can be tolerated for a short interval of time. This current value is much greater than I_{FM}. This represents the increase in current that occurs when a circuit is first turned on.
5. *Maximum reverse current* I_R: The absolute maximum reverse current that can be tolerated at device operating temperature.
6. *Forward voltage* V_F: Maximum forward voltage drop for a given forward current at device operating temperature.
7. *Power dissipation* P_D: The maximum power that the device can safely absorb on a continuous basis in free air at 25° C(77° F).
8. *Reverse recovery* $T_{\pi r}$ *time* The maximum time it takes the device to switch from its on to its off state.

15.1.7 Diode packaging

Diodes are manufactured today in a wide range of case styles and packages. A person working with these devices must be familiar with some of the common methods of packaging. Element identification and lead-marking techniques are essential for proper installation and testing. A diode improperly connected in a circuit can be damaged or cause damage to other parts.

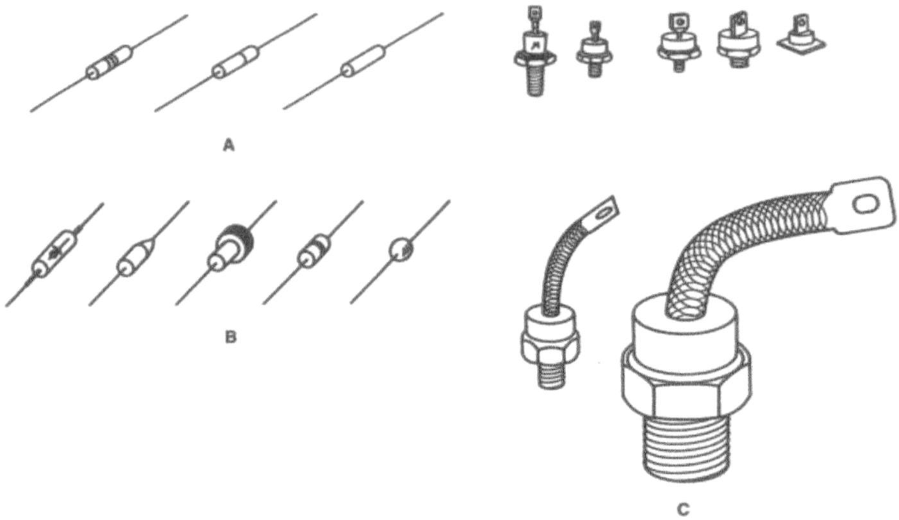

Figure 15.10 Diode packages. A – Low-current diodes. B – Medium-current diodes, C – High-current diodes.

Figure 15.10 shows three general classes of diodes arranged according to their current-handling capability. These represent some of the more popular package styles in use today. Figure 15.11 shows several types of diode devices.

15.1.8 Diode testing

Diode operation is determined by its connection in a circuit. Current conduction is directly dependent on the **polarity** of the device. Heavy conduction occurs when it is forward biased and very little current flows during reverse biasing. Connecting a diode backward in a circuit reverses its conduction. This can destroy the diode and possibly damage several other circuit parts. A person working with diodes must be absolutely certain that they are connected properly to a circuit.

The schematic symbol of a diode is commonly used to identify its polarity. Figure 15.12 shows a diode symbol, element names, and the polarity of the crystal material. The term **anode** is commonly used to describe the electrode that attracts or gathers in electrons. The P material of the crystal serves as the anode in a diode. The term **cathode** is used to denote the electrode that gives off or emits electrons. The N material of the crystal serves as the cathode. It should be noted that forward current passes through a diode from

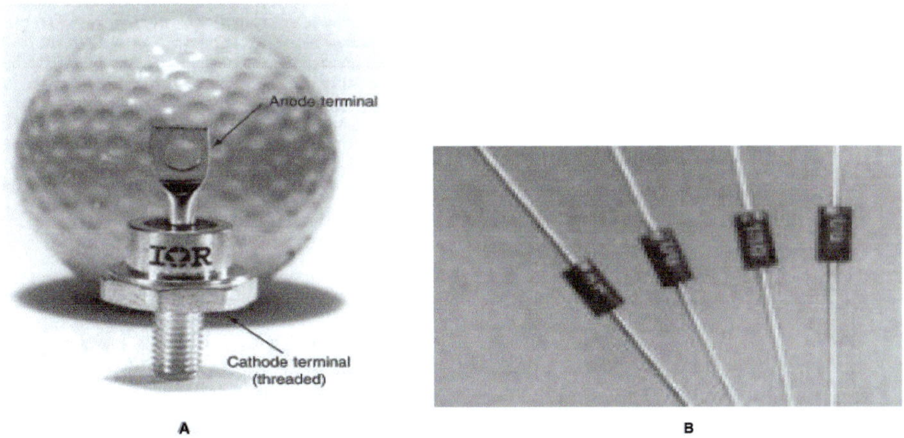

Figure 15.11 Types of diode devices. A – High-current diode. B – Low-current diodes.

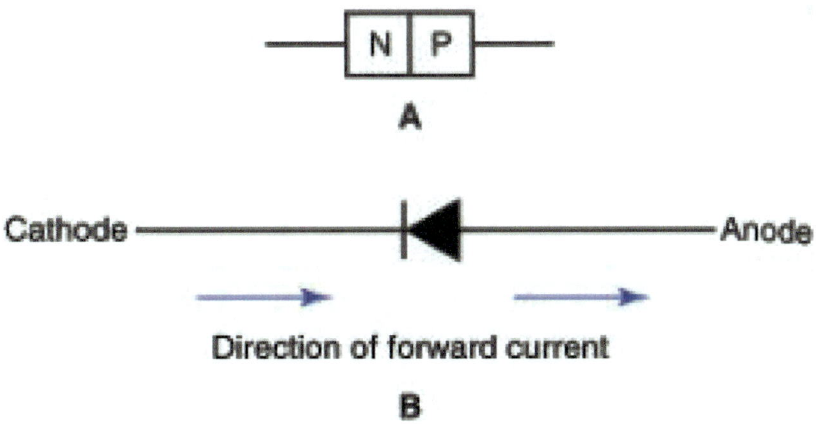

Figure 15.12 Diode. A – Crystal structure, B – Symbol and element names.

the cathode to the anode. A diode symbol shows this moving from the bar to the point of the arrow.

Diodes are available today in many different package types and styles. Because of the many differences in these devices, there is often some degree of confusion about lead identification. A person working with these devices is frequently called on to test diodes and identify leads. Multimeters are used to perform this operation. The ohmmeter section of the instrument is used to perform this test.

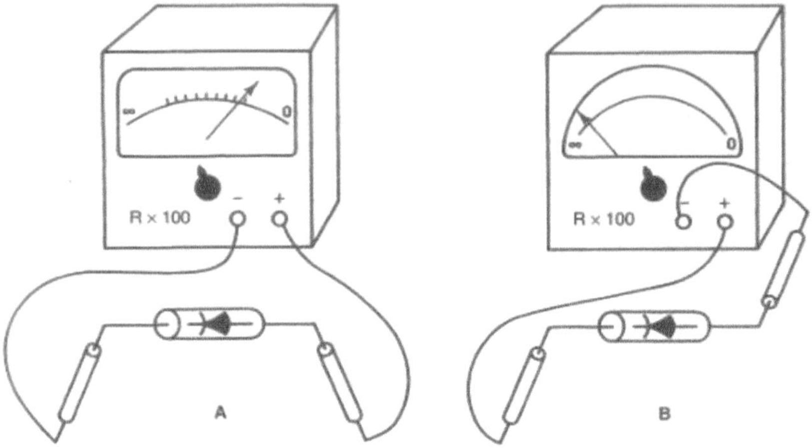

Figure 15.13 Diode testing with an analog ohmmeter. A – Forward bias test connection. B – Reverse bias test connection.

Figure 15.13 shows how an analog ohmmeter is used to test a diode. In preparation for this test, the meter should be placed in the R × 100 or 1–kΩ range. The two meter leads are then connected to the diode. Exact resistance readings are not particularly important. Only high- or low-resistance indications are used. Part A of Figure 15.13 shows how an ohmmeter is connected to ***forward bias*** a diode. This test will normally cause the meter to show a low resistance. Reverse biasing of a diode is shown in part B of Figure 15.13. An infinite resistance is indicated by the ohmmeter during this test.

A good diode will normally show low resistance when forward biased and high resistance when **reverse biased**. If a diode does not respond in this manner, it is probably defective. A diode that shows low resistance in both directions is considered to be shorted. Shorting usually occurs when the maximum current or voltage ratings of the device are exceeded. **Shorted diodes** often cause damage to other circuit components. Blown fuses, tripped circuit breakers, or overheated resistors are generally good indicators of a shorted diode. It is a good practice to test other circuit components before replacing a shorted diode.

When a diode shows **infinite resistance** in both directions, it is considered to be open. Open diodes generally cause a circuit to be nonconductive. This condition rarely causes damage to other circuit parts. The operation simply stops, and there is no current passing through the diode. Open diodes do not occur very often in electronic circuits. A quick ohmmeter check will readily detect this type of fault should it occur.

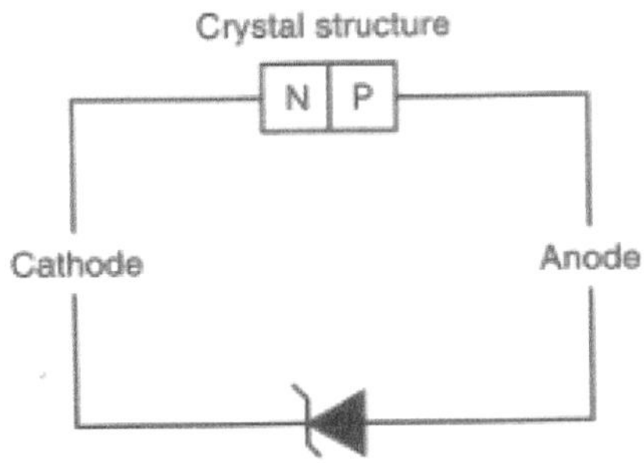

Figure 15.14 Zener diode symbol and crystal structure.

15.2 Zener Diodes

A **junction diode** is designed primarily for operation in the forward direction. **Forward biasing** will cause a large I_F with a rather small value of V_F. **Reverse biasing** will generally not cause current conduction until higher values of reverse voltage are reached. If V_R is great enough, however, breakdown will occur and cause a reverse current flow through a diode. Junction diodes are usually damaged when this occurs. However, there are special diodes designed to operate in the reverse direction without being damaged. **Zener diodes** are manufactured specifically for this purpose.

In the forward bias direction, a zener diode behaves like an ordinary silicon diode. In the reverse bias direction, there is practically no reverse current flow until the **zener voltage** or **breakdown voltage** is reached. When this occurs there is a sharp increase in reverse current. Varying amounts of reverse current can pass through the diode without damaging it. The breakdown voltage or zener voltage (V_Z) across the diode remains relatively constant. The maximum reverse current is limited, however, by the wattage rating of the diode.

Zener diodes are available commercially in a wide range of **breakdown voltages**. V_Z values range from 1.4 to 200 V with accuracies between 1% and 20%. Each V_Z value is generally specified at the minimum zener current I_Z. After this value has been reached, the V_Z remains fairly constant over a wide range of I_Z. **Power dissipation** values are therefore used to indicate the safe operating range. Typical P_D ratings are from 150 mW to 50 W. For example, when purchasing this kind of device, you would ask for a 1–W zener diode at 8.2 V_Z with an accuracy of 10%.

The symbol and crystal structure of a zener diode are shown in Figure 15.14. Notice that they are very similar to those of a regular diode. The cathode is, however, drawn with a bent line to distinguish it from a regular diode symbol. This is supposedly done to represent the letter Z. Zener diodes are normally used only in the reverse bias direction. This means that

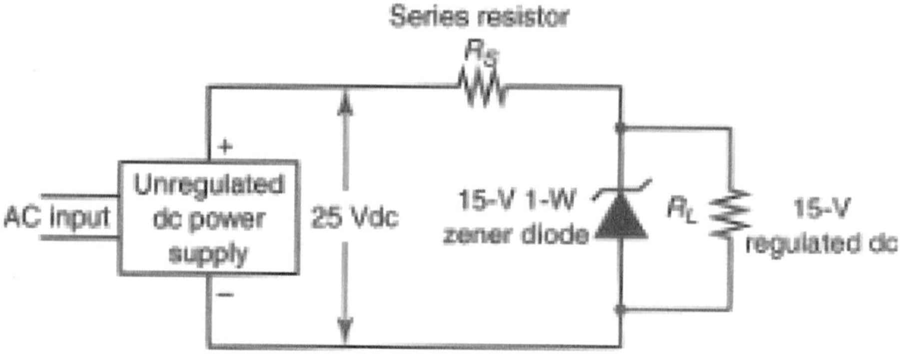

Figure 15.15 Zener diode regulator circuit.

the anode must be connected to the negative side of the voltage source and the cathode to the positive.

An important difference between zener diodes and junction diodes is in the way they are used in circuits. Because of the constant V_Z characteristic of a zener diode, it is used primarily to regulate voltages. A large change in I_Z will cause only a small change in V_Z. This means that a zener diode can be used as an alternate current path. The constant V_Z developed across the diode can then be applied to a load device. The load voltage thus remains at a constant value by altering the current through the zener diode.

Figure 15.15 shows a representative zener diode regulator circuit. Notice that the circuit contains a series resistor, zener diode, load device, and a DC power supply. The primary function of this circuit is to maintain the load voltage at a constant 15 volts. With 25 volts applied, current through R_S will cause a voltage drop of 10 volts. The output voltage or V_Z across the diode is therefore 15 volts. An increase in the supply voltage to 30 V would simply cause the zener to conduct more current. The voltage drop across would therefore increase to 15 volts. This, in turn, would allow V_Z and the load voltage to remain at 15 volts. A decrease in supply voltage to 20 volts would cause a reduction in total current. The voltage drop across R_S would now drop to only 5 volts. The load voltage across the diode would continue to be 15 volts. This means that the regulated output will remain at 15 volts for a change in supply voltage from 20 to 30 volts. Voltage regulation is an important function of many electronic circuits today.

15.3 Light-Emitting Diodes

The light-emitting diode is another important diode device. A **light-emitting diode (LED)** contains a PN junction that emits light when passing forward

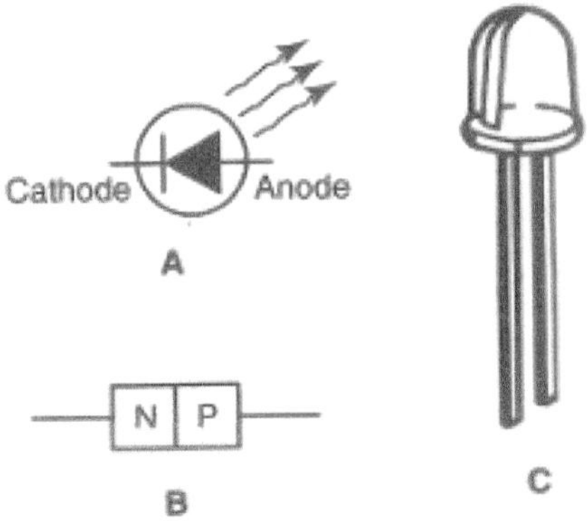

Figure 15.16 Light-emitting diode. A – Schematic symbol B – Crystal structure. C-Sketch of LED showing the transparent lens.

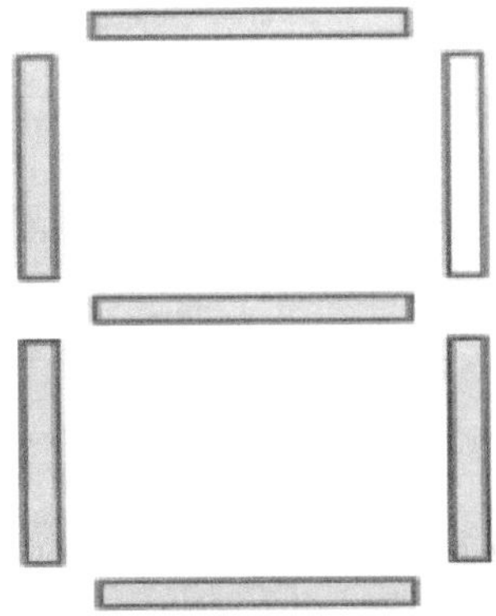

Figure 15.17 Seven-segment LED display. Segment forward biasing produces the number 6.

current. A schematic symbol and typical package of the LED are shown in parts A and B of Figure 15.16. The lens of an LED, part C of Figure 15.16, is transparent and focuses on the PN junction.

When an LED is forward-biased, electrons cross the junction and combine with holes. This action causes electrons to fall out of conduction and return to the valence band. The energy possessed by each free electron is then released. Some of this appears as heat energy and the rest of it is given off as light energy. Special materials such as gallium arsenide (GaAs) and gallium phosphide (GaP) cause this to take place when used in a PN junction. These materials are used purposely because they emit visible light energy.

Light-emitting diodes are relatively small, ruggedly constructed, and have a very long life expectancy. They can be switched on and off very quickly. Construction techniques permit this type of device to be made in various shapes and patterns. A very common application is in numeric displays indicating the numbers 0 through 9. A representative LED seven-segment display is shown in Figure 15.17. By selecting different combinations of segments, a desired number can be produced. Displays of this type are found in calculators, digital meters, clocks, and electronic billboards.

15.4 Bipolar Junction Transistors

Transistors are three-element devices made of semiconductor materials. These devices have one or more PN junctions in their construction. **Bipolar junction transistors (BJTs)** have two PN junctions. **Field-effect transistors (FETs)** have one junction in their construction. FETs are classified as

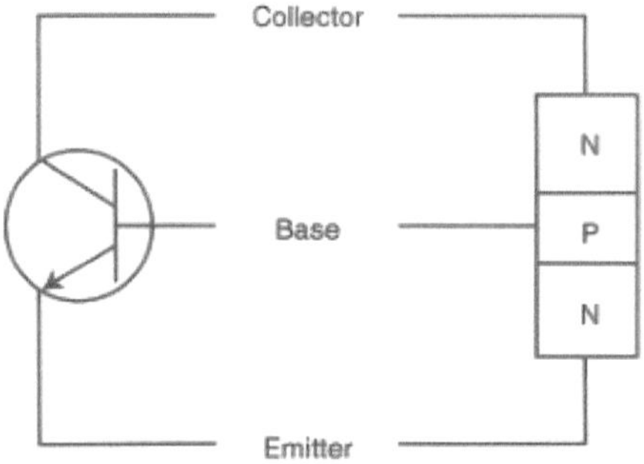

Figure 15.18 NPN transistor symbol and crystal structure.

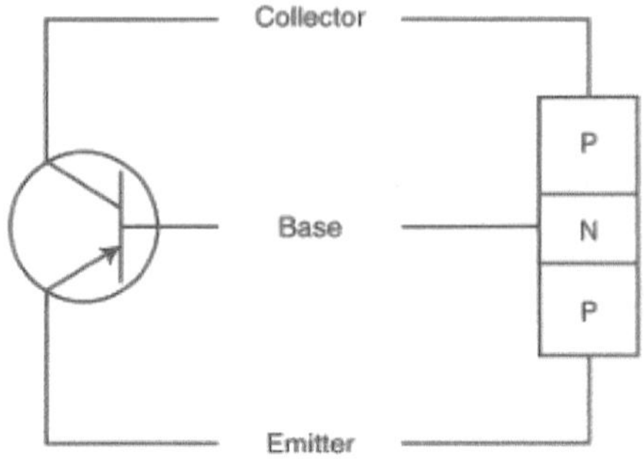

Figure 15.19 PNP transistor symbol and crystal structure.

unipolar devices. A **unipolar** device responds to one type of current carrier passing through a channel. Unijunction transistors are similar in construction to the FET. These three transistors play an important role in the field of electronics.

When the word transistor is used, it usually has some kind of letter designations to identify its type. **NPN** and **PNP** are two types of bipolar junction transistors. The letters are used to denote the polarity of the semiconductor material used in its construction. Material polarity is determined by the type of impurity added to silicon or germanium when it is manufactured. In both NPN and PNP bipolar junction transistors, the first crystal is called the **emitter**, the middle crystal is the **base**, and the third crystal is the **collector.**

An **NPN transistor** has a thin layer of P material placed between two pieces of N material. The schematic symbol, crystal structure, and lead designations of the NPN transistor are shown in Figure 15.18. The schematic symbol of this device is the same as the PNP except for the pointing direction of the arrowhead. An NPN transistor has the arrow "Not Pointing" toward the base. The polarity of the source voltage applied to each element is determined by the material used in its construction.

A **PNP transistor** has a thin layer of material placed between two pieces of P material. The schematic symbol of the PNP transistor and its crystal structure are shown in Figure 15.19. Leads attached to each piece of material are identified as the emitter, base, and collector. In the symbol, when the arrowhead of the emitter lead "Points i *N*" toward the base, it indicates that the device is of the PNP type. Lead identification is extremely important when working with transistors.

NPN and **PNP transistors** are considered to be bipolar devices. The term bipolar refers to conduction by holes and electrons. Electrons are the majority current carriers of the N material. In the P material, holes are the majority current carriers. Current conduction in the transistor is based on the movement of both current carriers. The term bipolar denotes this condition of conduction.

15.4.1 Bipolar junction transistor biasing

For a **BJT** to function properly, it must have electrical energy applied to all its electrodes. In with the semiconductor diodes, the source voltage is called bias voltage or simply bias. **Bipolar transistors** must have both junctions biased in order to function. Earlier in this chapter, we discussed diode biasing. Forward biasing of a PN junction takes place when the polarity of the source voltage is positive to the P material and negative to the N material. This condition causes a current to flow. The depletion zone of the device is reduced, and majority current carriers are driven toward the junction. One junction of a transistor must be forward-biased when it operates.

Reverse biasing of a PN junction is achieved when the polarity of the source is negative to the P material and positive to the N material. This condition does not ordinarily permit current to flow. The width of the depletion zone is increased by this action, and minority current carriers move toward the junction. In normal circuit operation, reverse bias current is extremely small. One junction of a transistor must be reverse-biased in order for the device to be operational.

15.4.2 NPN transistor biasing

Figure 15.20 shows how current carriers pass through a properly biased NPN transistor. The emitter-base junction being forward-biased causes a large amount of current I_E (emitter current) to move into the emitter-base junction. Upon arriving at the junction, a large number of the electrons do not effectively combine with holes in the base. The base is usually made very thin (0.0025 cm or 0.001 in.) and it is lightly doped. This means that the majority of current carriers of the emitter exceed the majority carriers of the base. Most of the electrons that cross the junction do not combine with holes. They are, however, immediately influenced by the positive voltage, V_{CB} (collector-base voltage), applied to the collector. A very high percentage of the original emitter current enters the collector. Typically, 95% to 99% of I_E flows into the collector junction. It then becomes collector current, I_C After passing through the collector region, I_C returns to V_{CB},V_{BE} , and ultimately to I_E. The current inside the transistor is indicated by a large arrow. Outside the transistor, the current is indicated by small arrows.

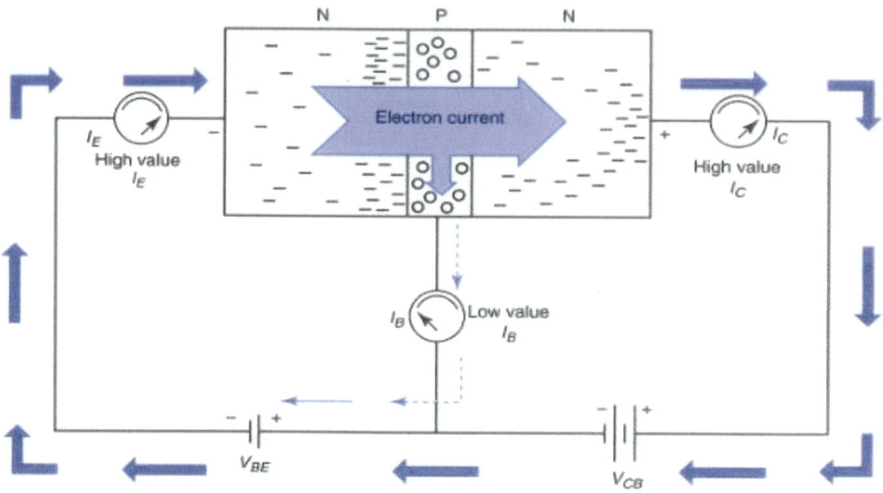

Figure 15.20 Current carriers passing through an NPN transistor.

15.4.3 PNP transistor biasing

Biasing of a PNP transistor is very similar to that of the NPN transistor. The emitter-base junction must, for example, still be forward biased and the collector-base reverse biased. The polarity of the bias voltage is, however, reversed for each transistor. The majority current carriers of a PNP transistor are holes instead of electrons as in the NPN device. Except for these two differences, the operation is primarily the same.

Figure 15.21 shows how current carriers pass through a properly biased **PNP transistor**. The emitter-base junction being forward biased causes a large number of holes to move through the junction. Upon arriving in the base region, a very small number of these holes combine with electrons. This is representative of the base current. Typically, 95% to 99% of the holes move through the base region and enter the collector junction. These holes do not find electrons to combine with in the thin base region. They are, however, immediately influenced by the negative V_{CB} voltage of the collector. This action causes hole flow through the reverse-biased base-collector junction. The end result of this is I_C passing through the collector region.

Electron flow is the opposite of hole flow. As a result, **hole flow** causes electrons to enter the collector and exit the emitter. Hole flow is indicated by the large arrow inside the transistor. Outside the transistor, current flow is achieved by electrons. Small arrows in the diagram show current outside the transistor.

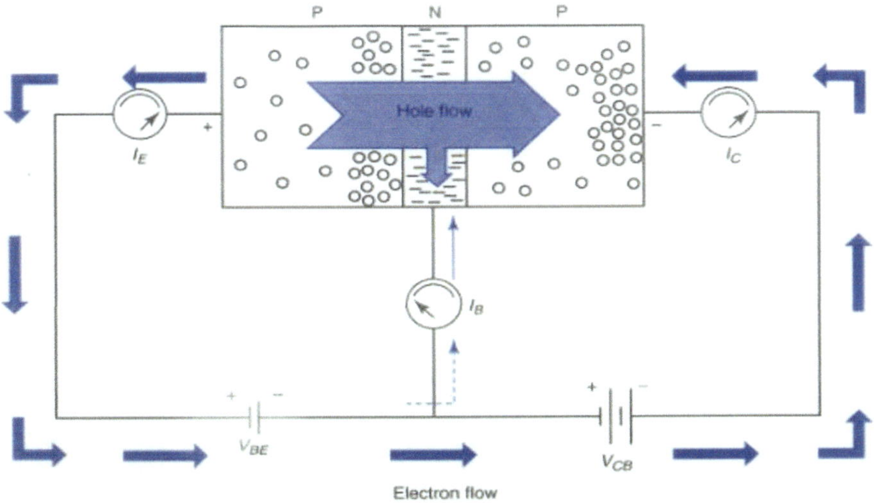

Figure 15.21 Current carriers passing through a PNP transistor.

15.4.4 Transistor packaging

Once a transistor is constructed, the entire crystal assembly must be placed in an enclosure or **package**. The unit is then sealed to protect it from dust, moisture, or outside contaminants. Electrical connections are made to each element through leads attached to the package.

Transistors in use generate heat. If this heat is not carried away from the semiconductor material, the transistor can be damaged. When metal packages are used, the housing also serves as a **heat sink**. Heat is carried away from the crystal by the metal package. A common practice in this kind of unit is to attach the collector to the metal case. This helps the collector to dissipate heat. This type of package should not be permitted to touch other circuit components when it is in operation.

Ceramic and epoxy package transistors are also available. As a rule, this type of packaging is somewhat cheaper to produce. These devices are usually not as rugged as the metal-packaged devices. Most **small-signal transistors** are housed in epoxy packages. The outside case of this device is insulated from all transistor elements. It is designed specifically for printed circuit board construction and close placement with other circuit components.

Several of the more popular transistor package styles are shown in Figure 15.22. Part A shows devices designed for small-signal applications. The letter designation **TO** stands for **transistor outline**. These same packages are also used to house other solid-state devices. Part B shows some of

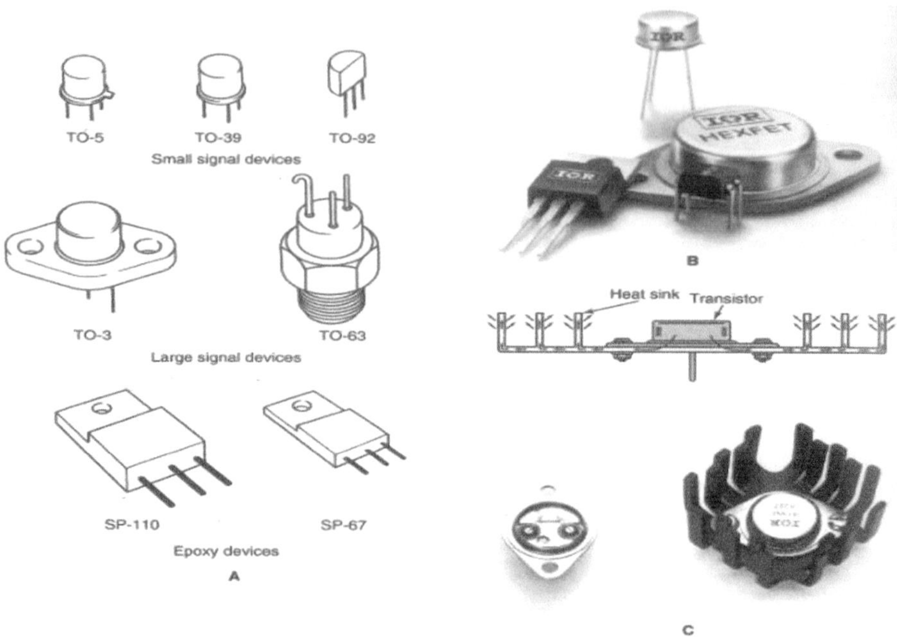

Figure 15.22 A – Sketches of transistor packages. B – Selection of transistors. C – Transistors with heat sinks. The heat sinks conduct heat from the transistors to the surrounding air.

the packages used to house large-signal or power transistors. Heavy metal cases of this type are used to dissipate heat generated by the transistor. The collector is usually attached to the metal case. Part C shows a few of the common epoxy packages. The letter designation SP stands for semiconductor package. This type of package is used to house a number of other semiconductor devices.

With the wide range of package styles available today, it is difficult to have a standard method of identifying transistor leads. Each package generally has a unique lead location arrangement. It is usually a good idea to consult the manufacturer's specifications for proper lead identification.

15.4.5 Bipolar junction transistor testing

Transistor testing is a rather common procedure that must be performed periodically when working with semiconductor devices. Lead identification, gain, open leads, shorted conditions, and leakage are some of the tests commonly performed on a transistor. Curve tracers, gain testers, and sound-producing instruments can be used effectively for these tests. However, these

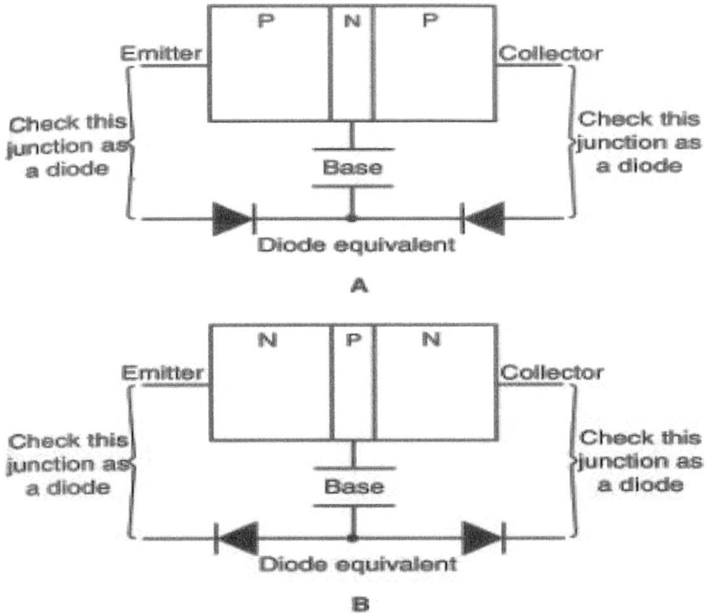

Figure 15.23 Junction polarity of P – PNP and B – NPN transistors.

instruments are not always available. An ohmmeter can be used to perform many of these tests. An ohmmeter is one of the functions of a multimeter.

A good BJT has two PN junctions in its construction. Both of these junctions can be tested with an ohmmeter. Figure 15.23 shows the junctions of PNP and NPN transistors. Each junction will respond as a diode when it is tested. Forward biasing occurs when the material polarity of the junction matches the voltage polarity of the ohmmeter. When the ohmmeter polarity is reversed, the same junction is reverse biased. Forward biasing is indicated by a low resistance reading, while reverse bias will show an extremely high resistance value. Both junctions of a transistor should respond to this test in the same way if the transistor is good.

A faulty PN junction will not show a difference in its forward and reverse resistance connections. Low resistance in both directions indicates that a junction is shorted. Excessive current generally causes this type of failure in a transistor. Infinite or exceedingly high resistance in both directions indicates an open condition. An open junction is "normally caused by a broken internal connection. This can be the result of a current overload or excessive shock. Open and shorted transistor failures, as a general rule, occur suddenly and cannot be corrected.

A rather quick test of both junctions at the same time can be made with the ohmmeter. Connect one ohmmeter lead to the emitter and the other to the collector. Ohmmeter lead polarity in this case is not important. This connection should cause an extremely high resistance reading. Then reverse the two ohmmeter leads. High resistance should also occur in this direction. A good transistor will show an infinite resistance in either direction. Any measurable resistance in either direction indicates leakage.

15.5 Junction Field-Effect Transistors

The **junction field-effect transistor (JFET)** is a three-element electronic device. Its operation is based on the conduction of current carriers through a continuous piece of semiconductor material. This piece of material is called the **channel**. Two additional pieces of semiconductor material are diffused into either side of the channel. These elements make up what is called the gate. The assembled device is housed in a package. The physical appearance of a JFET in its housing is very similar to that of a bipolar transistor.

The three leads on a JFET are the source, the drain, and the gate. The **source** is comparable to the emitter on a BJT. The **drain** is comparable to the collector. The **gate** is comparable to the base. The biasing of the JFET is accomplished with the gate. The gate produces an electric field that radiates into the channel. A stronger field restricts current from passing through the channel. If the field from the gate becomes strong enough, it will prevent all current from passing through the channel. The gate voltage that produces this field is called the **pinch-off voltage.**

15.5.1 N-channel JFETs

Figure 15.24 shows the crystal structure, element names, and schematic symbol of an **N-channel JFET**. An N-channel JFET receives a negative voltage on the source. The other side of the voltage supply is connected to the drain.

The schematic symbol of an **N-channel JFET** is somewhat representative of its construction. The bar part of the symbol refers to the **channel**. The **drain (D)** and **source (S)** are attached to the channel. **The gate (G)** has an arrow. This shows that it forms a PN junction. The arrowhead of an N-channel symbol "Points i *N*" toward the channel. This indicates that it is a PN junction. The arrowhead or gate is P material, and the channel N material. This part of the device responds as a junction diode.

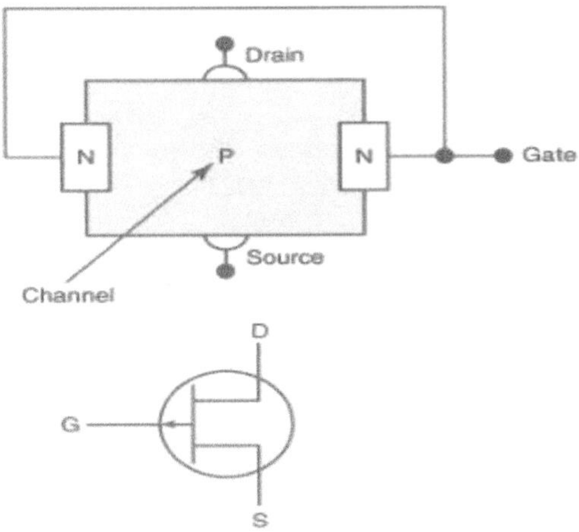

Figure 15.24 N-channel junction field-effect transistor.

15.5.2 P-channel JFETs

The crystal structure, element names, and schematic symbol of a P-channel JFET are shown in Figure 15.25. A P-channel JFET receives a positive voltage on the source. The other side of the voltage supply is connected to the drain.

The schematic symbol of a **JFET** is different only in the gate element. In a P-channel device the arrow is "Not Pointing" toward the channel. This means that the gate is N material and the channel is P material. In conventional operation, the gate is made positive with respect to the source. Varying values of reverse bias gate voltage will change the size of the PN junction depletion zone. Current flow through the channel can be altered between cutoff and full conduction. P-channel and N-channel JFETs cannot be used in a circuit without a modification in the source voltage polarity.

15.6 Metal Oxide Semiconductor Field-Effect Transistors

Metal oxide semiconductor field-effect transistors (MOSFETs) are an addition to the **unipolar** transistor family. A distinguishing feature of this device is its gate construction. The gate is, for example, completely insulated from the channel. Voltage applied to the gate will cause it to develop an electrostatic charge. No current is permitted to flow in the gate area of the device. The gate is simply an area of the device coated with metal. Silicon dioxide is used as an insulating material between the gate and the channel. Construction of a **MOSFET** is rather easy to accomplish.

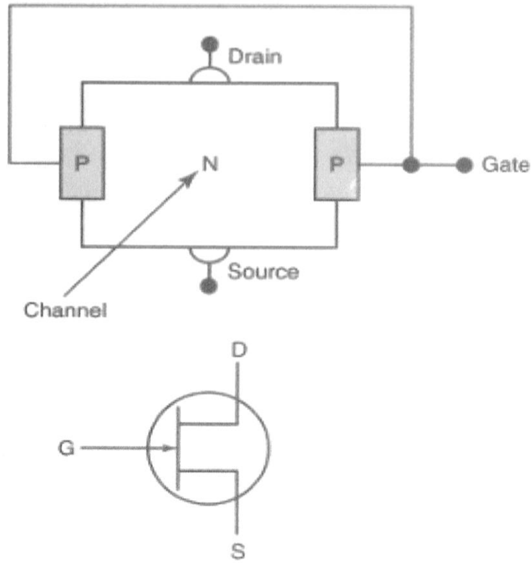

Figure 15.25 P-channel junction field-effect transistor.

The channel of a **MOSFET** has a number of differences in its construction. One type of MOSFET responds to the depletion of current carriers in its channel. These devices are called **depletion mode** MOSFETs or **D-type** MOSFETs. Depletion mode MOSFETs have an interconnecting channel built on a common substrate. Direct connection is made between the source and drain of this device. Channel construction is very similar to that of the JFET. A second type of MOSFET is called an **enhancement mode** MOSFET or E-type MOSFET. Enhancement mode MOSFETs do not have an interconnecting channel. Current carriers pulled from the substrate form an induced channel. Gate voltage controls the size of the induced channel. Channel development is a direct function of the gate voltage. E- and D-type MOSFETs both have unique operating characteristics.

Enhancement mode MOSFET construction, element names, and the schematic symbol are shown in Figure 15.26. Note in the crystal structure that no channel appears between the source and drain. The gate is also designed to cover the entire span between the source and drain. The gate is actually a thin layer of metal oxide. The conductivity of this device is controlled by the voltage polarity of the gate. A PN junction is not formed by the gate, source, and drain.

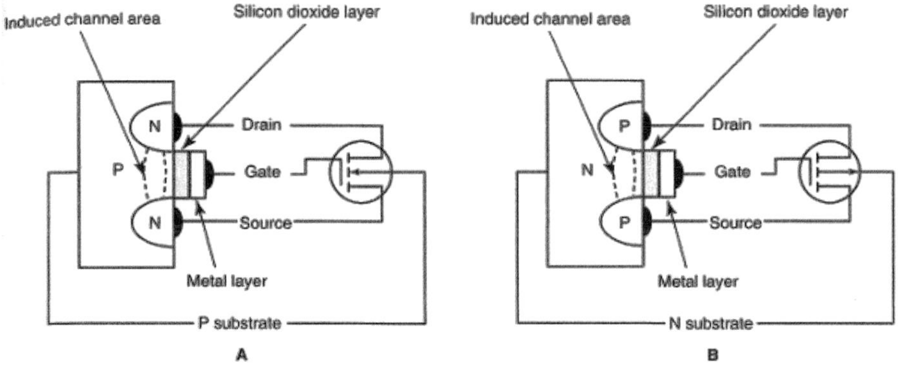

Figure 15.26 Enhancement-mode MOSFET. A – N-channel; B – P-channel.

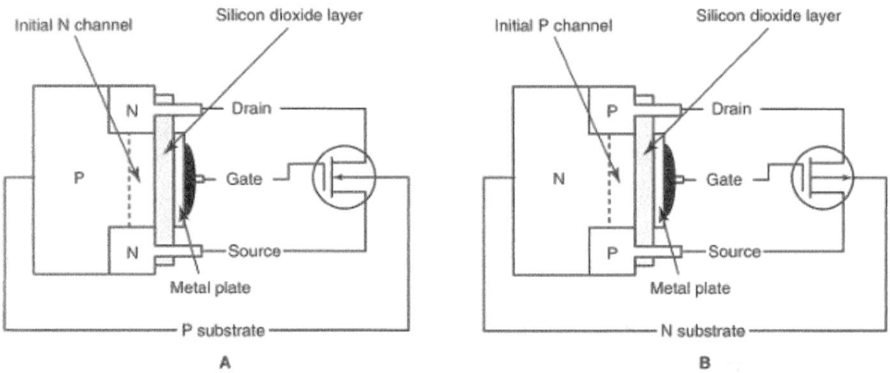

Figure 15.27 Depletion-mode MOSFET. A – N-channel. B – P-channel.

Depletion mode MOSFET construction, element names, and the schematic symbol are shown in Figure 15.27. The **N-channel** device has a thin channel of N material formed on a P substrate. Source and gate connections are made directly to the channel. A thin layer of silicon dioxide (SiO_2) insulation covers the channel. The gate is a metal-plated area formed on the silicon dioxide layer. The entire unit is then built on a substrate of *P* material. The arrow of the schematic symbol refers to the material of the substrate. When it "Points i*N*," this shows that the substrate is P material and the channel is N material.

The construction of P- and N-channel devices is essentially the same. The crystal material of the channel and substrate is the only difference. Current carriers are holes in the P-channel device, while electrons are carriers in the N-channel unit. The schematic symbol differs only in the direction of the substrate arrow. It does "Not Point" toward the substrate in a P-channel device. This means that the substrate is N material and the channel is *P* material.

The operation of a **D-type MOSFET** is very similar to that of a JFET. With voltage applied to the source and drain, current carriers pass through the channel. The gate does not need to be energized in order to produce conduction. In this regard, a D-type MOSFET is considered to be a normally-on device. Control of the drain current, I_D, is, however, determined by the polarity of gate voltage. When the polarity of V_G is the same as the channel, current carrier depletion occurs. A reversal in polarity causes I_D to increase due to the enhancement of the current carriers.

15.7 V-MOS Field-Effect Transistors

Another type of field-effect transistor is utilized in solid-state active device technology. This type of device is used mainly to replace bipolar transistors. **Power FETs,** such as these, are used in power supplies and solid-state switching applications. Collectively, these devices are called **V-MOSFETs** or **V-MOSs**. The "V" designation finds its origin in vertical-groove MOS technology. The device is constructed with a V-shaped groove etched into the substrate. Construction of this type requires less space than a horizontally assembled device. The construction of a V-MOS device permits better heat dissipation and a high density of material in the channel area.

Figure 15.28 shows a cross-sectional view of the crystal structure, the element names, and a schematic symbol of an **N-channel V-MOS transistor**. Notice that a V groove is etched into the surface of the structure. From the top, the V cut penetrates through the N+,P, and N - layers and stops near the N+ substrate. The two N+ layers are heavily doped, and the N - layer is lightly doped. A thin layer of silicon dioxide covers both the horizontal surface and the V groove. A metal film deposited on top of the groove serves as the gate. The gate is therefore insulated from the groove. The source leads on each side of the groove are connected internally. The bottom layer of N+ material serves as a combined substrate and drain. Current carriers move between the source and drain vertically.

15.8 Unijunction Transistors

The **unijunction transistor (UJT)** is often described as a voltage-controlled diode. This device is not considered to be an amplifying device.

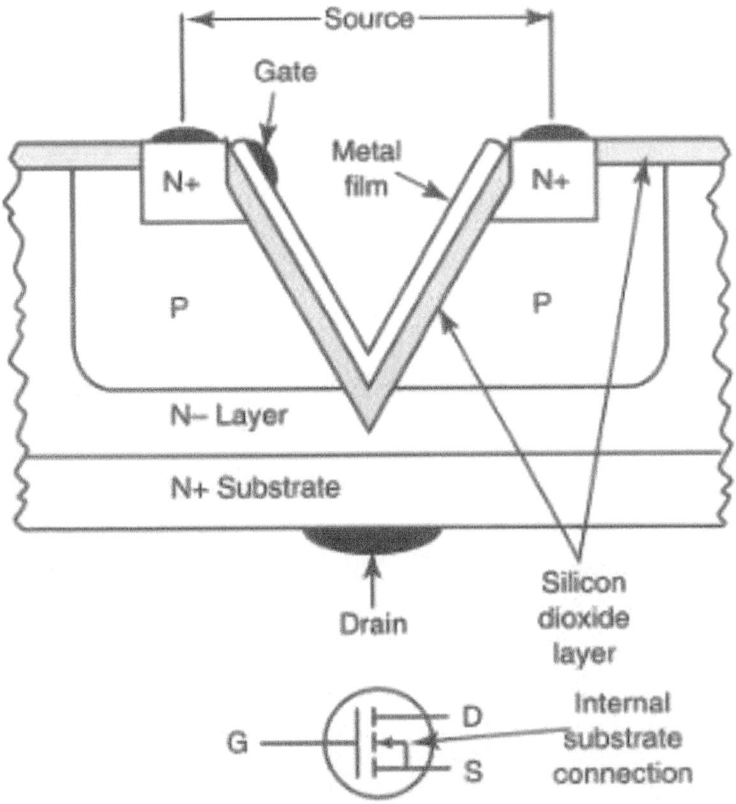

Figure 15.28 Cross-sectional view of an N-channel V-MOSFET.

It is, however, classified as a unipolar transistor. The UJT and an N-channel JFET are often confused because of the similarities in their schematic symbols and crystal structure. The operation and function of each device is entirely different.

Figure 15.29 shows the crystal structure, schematic symbol, and element names of a UJT. Note that the UJT is a three-terminal, single-junction device. The crystal is an N-type bar of silicon with contacts at each end. The end connections are called base 1(B_1) and base 2(B_2). A small, heavily doped P region is alloyed to one side of the silicon bar. This serves as the emitter (E). A PN junction is formed by the emitter and the silicon bar. The arrow of the symbol "Points iN," which means that the emitter is P material and the silicon bar is N material. The arrow of the symbol is slanted, which distinguishes it from the N-channel JFET.

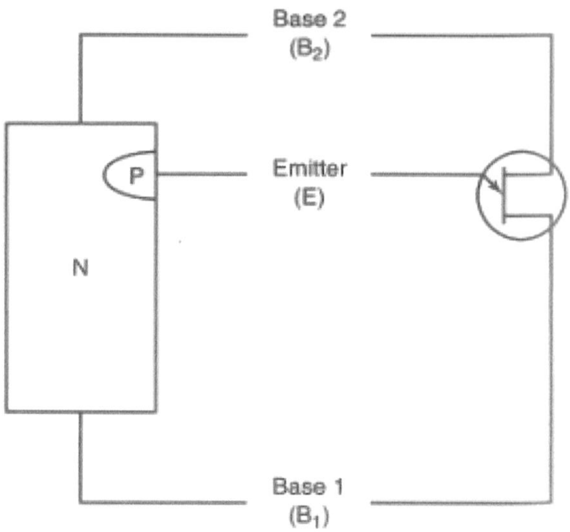

Figure 15.29 Unijunction transistor.

In circuit applications, B_1 is usually placed at the circuit ground or the source voltage negative. The emitter then serves as the input to the device. B_2 provides circuit output. A change in E to B_1 (B_1) junction voltage causes a change in R_{BB}. The output current of the device increases when E to B_1 to junction turns on. A UJT is normally used as a trigger device. It is used to control some other solid-state device.

15.9 Silicon-Controlled Rectifiers

A **silicon-controlled rectifier or SCR** is one of the most popular electronic power control devices. The SCR is used primarily as a **switching** device. Power control is achieved by switching the SCR on and off during one alternation of the AC source voltage. For 60 Hz AC the SCR would be switched on and off 60 times per second. Control of electrical power is achieved by altering or delaying the turn-on time of an alternation.

An **SCR**, as the name implies, is a solid-state rectifier device. It conducts current in only one direction. It is similar in size to a comparable silicon power diode. SCRs are usually small, rather inexpensive, waste very little power, and require practically no maintenance. The SCR is available today in a full range of types and sizes to meet nearly any power control application.

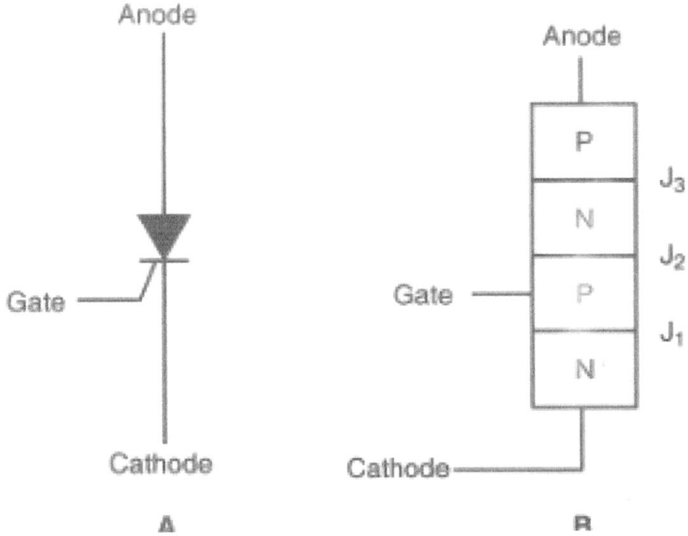

Figure 15.30 SCR crystal. A – Schematic symbol. B – Crystal structure.

Presently, they are available in current ratings from less than 1 amp to over 1400 amps. Voltage values range from 15 volts to over 2500 volts.

15.9.1 SCR construction

An **SCR** is a solid-state device made of four alternate layers of P- and N-type silicon. Three PN junctions are formed by the structure. Each SCR has three leads or terminals. The **anode and cathode** terminals are similar to those of a regular silicon diode. The third lead is called the **gate**. This lead determines when the device switches from its off to on state. An SCR will usually not go into conduction by simply forward biasing the anode and cathode. The gate must be forward biased at the same time. When these conditions occur, the SCR becomes conductive. The internal resistance of a conductive SCR is less than 1Ω. Its reverse, or off-state, resistance is generally in excess of $1M\Omega$.

A schematic symbol and the crystal structure of an **SCR** are shown in Figure 15.30. Note that the device has a PNPN structure from anode to cathode. Three distinct PN junctions are formed. When the **anode** is made positive and the **cathode** negative, junctions 1 and 3(J_1 and J_3) and are forward biased. J_2 is reverse biased. Reversing the polarity of the source alters this condition. J_1 and J_3 would be reverse biased and J_2 would be forward biased. This would not permit conduction. Conduction will occur

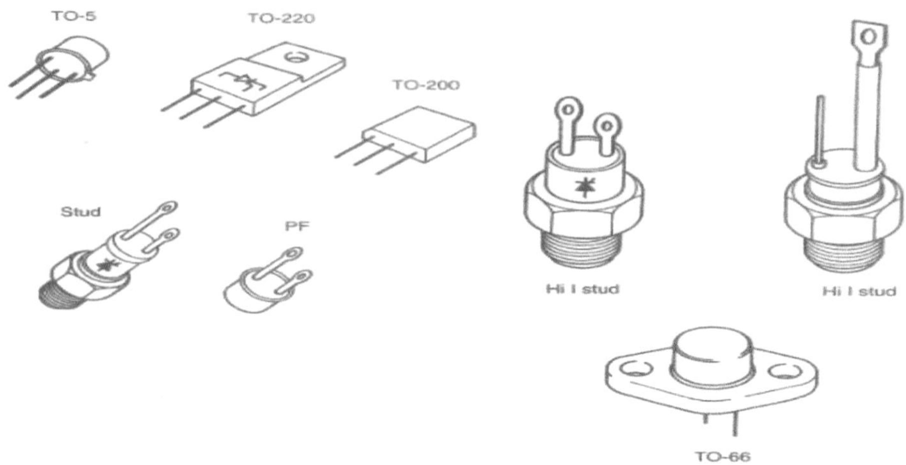

Figure 15.31 Representative SCR packages.

only when the anode, cathode, and gate are all **forward biased** at the same time.

Some representative **SCRs** are shown in Figure 15.31. Only a few of the more popular packages are shown here. As a general rule, the anode is connected to the largest electrode if there is a difference in their physical size. The gate is usually smaller than the other two electrodes. Only a small gate current is needed to achieve control. In some packages, the SCR symbol is used for lead identification.

15.9.2 SCR operation

Operation of an SCR is not easily explained by using the four-layer PNPN structure of part A of Figure 15.32. A rather simplified method has been developed that describes the crystal structure as two interconnected transistors. Part B of Figure 15.32 shows this imaginary division of the crystal. Notice that the top three crystals form a PNP transistor. The lower three crystals form an NPN device. Two parts of each transistor are interconnected. The two-transistor circuit diagram of part B of Figure 15.32 is an equivalent of the crystal structure division.

Figure 15.33 shows how the two-transistor equivalent circuit of an **SCR** will respond when voltage is applied. Note that the anode is made positive and the cathode negative. This condition forward biases the emitter of each transistor and reverse biases the collector. Without base voltage applied to the NPN transistor, the equivalent circuit is nonconductive. No current flows

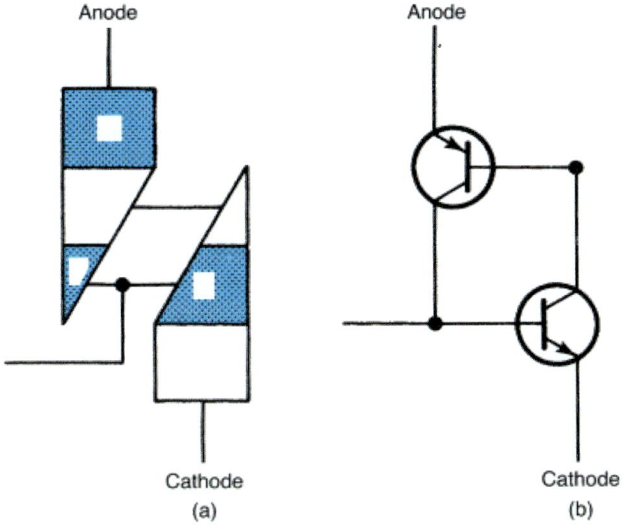

Figure 15.32 Equivalent circuit for an SCR. The two-transistor circuit of part B responds like an SCR.

through the load resistor. Reversing the polarity of the voltage source does not alter the conduction of the circuit. The emitter of each transistor is reverse biased by this condition. This means that no conduction takes place when the anode and cathode are reverse biased. With only anode-cathode voltage applied, no conduction occurs in either direction.

Assume now that the anode and cathode of an **SCR** are forward biased, as indicated in Figure 15.33. If the gate is momentarily made positive with respect to the cathode, the emitter-base junction of the NPN transistor becomes forward biased. This action immediately causes collector current to flow in the NPN transistor. As a result of this, base current will also flow into the PNP transistor. This, in turn, causes a corresponding collector current to flow. PNP collector current now causes base current to flow into the NPN transistor. The two transistors, therefore, latch or hold when in the conductive state. This action continues even when the gate is disconnected. In effect, when conduction occurs, the device will latch in its "on" state. Current continues to flow through the SCR as long as the anode-cathode voltage is of the correct polarity.

In order to turn off a conductive SCR, it is necessary to momentarily remove or reduce the anode-cathode voltage. This action turns off the two transistors. The device then remains in this state until the anode, cathode, and gate are all forward biased again. With AC applied to an SCR, it automatically

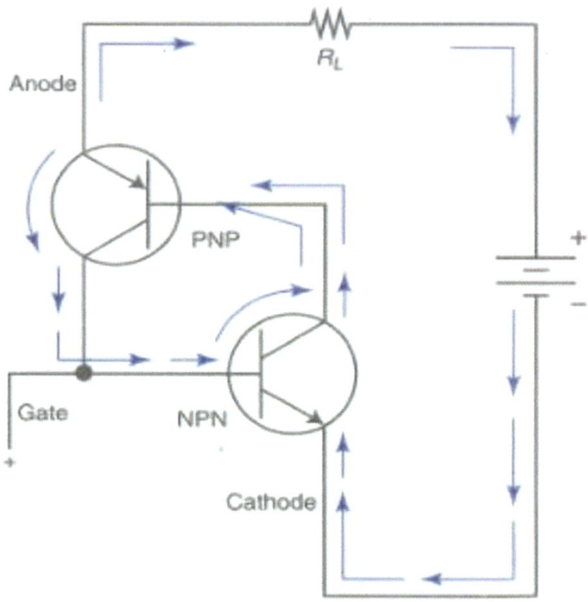

Figure 15.33 SCR response. The arrows show electron flow:

turns off during one alternation of the input. Control is achieved by altering the turn-on time during the conductive or "on" alternation.

15.10 Triacs

AC power control can be achieved with a device that switches on and off during each alternation. Control of this type is accomplished with a special solid-state device known as a triac. This device is described as a three-terminal AC switch. Gate current is used to control the conduction time of either alternation of the AC waveform. In a sense, the triac is the equivalent of two reverse-connected SCRs feeding a common load device.

A **triac** is classified as a gate-controlled AC switch. For the positive alternation, it responds as a PNPN device. An alternate crystal structure of the NPNP type is used for the negative alternation. Each crystal structure is triggered into conduction by the same gate connection. The gate has a dual-polarity triggering capability.

15.10.1 Triac construction

A **triac** is a solid-state device made of two different four-layer crystal structures connected between two terminals. We do not generally use the terms "anode" and "cathode" to describe these terminals. For one alternation, they would be the anode and cathode. For the other alternation, they would respond as the cathode and anode. It is common practice, therefore, to use the terms **main 1 and main 2**, or terminal 1 and terminal 2, to describe these leads.

The third connection is the **gate (G)**. This lead determines when the device switches from its off to its on state. The gate will normally go into conduction when it is forward biased. This is usually based on the polarity of terminal 1(T_1). If T_1 is negative, G must be positive. When T_1 is positive, the gate must be negative. This means that AC voltage must be applied to the gate to cause conduction during each alternation of the T_1–T_2 voltage.

The schematic symbol and the crystal structure of a **triac** are shown in Figure 15.34. Notice the junction of the crystal structure simplification. Looking from T_1 to T_2, the structure involves crystals N_1,P_1,N_2, and P_2. The gate is used to bias P_1. This is primarily the same as an SCR with T_1 serving as the cathode and T_2 the anode.

Looking at the crystal structure from T_2 to T_1, it is N_3,P_2,N_2, and P_1. The gate is used to bias N_4 for control in this direction. This is similar to the structure of an SCR in this direction. Notice that T_1,T_2, and G are all connected to two pieces of crystal. Conduction will take place only through the crystal polarity that is forward biased. When T_1 is negative, for example, N_1 is forward biased and P_1 is reverse biased. Terminal selection by bias polarity is the same for all three terminals.

The schematic symbol of the **triac** is representative of reverse-connected diodes. The gate is connected to the same end as T_1. This is an important consideration when connecting the triac into a circuit. The gate is normally forward biased with respect to T_1.

15.10.2 Triac operation

When AC is applied to a **triac**, conduction can occur for each alternation. For this to take place, T_1 and T_2 must be properly biased with respect to the gate. Forward conduction occurs when T_1 is negative, the gate (G) is positive, and T_2 is positive. Reverse conduction occurs when T_1 is positive, G is negative, and T_2 is negative. Conduction in either direction is similar to that of the SCR. The two-transistor operation used with the SCR applies to the triac in both directions. Figure 15.35 shows an example of the operational conduction polarities of a triac. Note that DC is used as a power source. This is used here only to denote operational polarities. In practice, DC would not be used as a power source for a triac. It is an AC power control device.

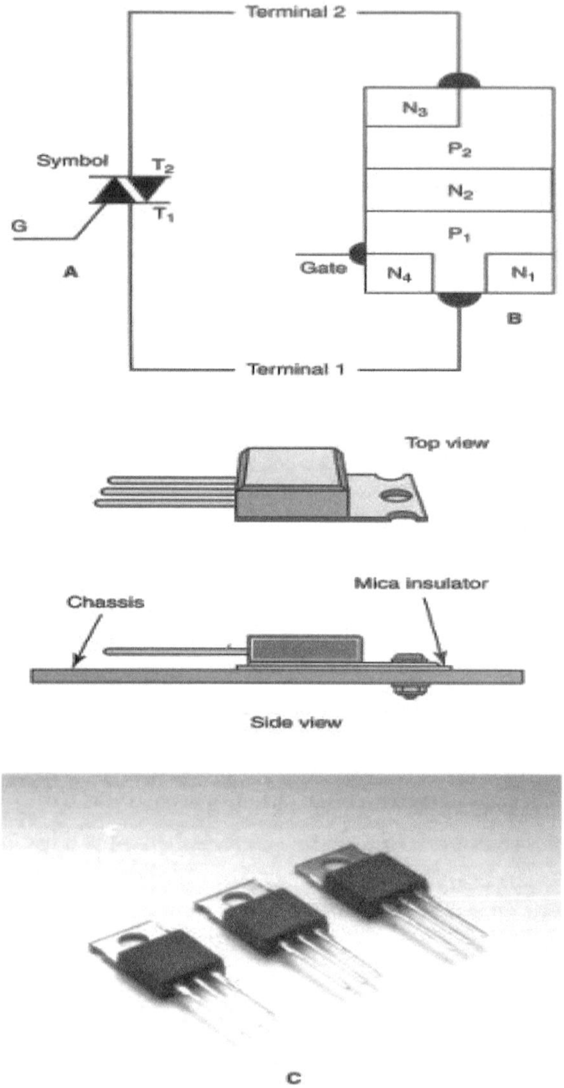

Figure 15.34 A – Triac symbol. B – Crystal structure. C – Photograph and drawings of triacs used for AC power control. (Phillips Components)

The operation of a **triac** is primarily the same as that of the **SCR**. When T_1 and T_2 are of the correct polarity, gate current must occur in order to reduce the breakover voltage. Small values of I_G cause the breakover voltage to be quite large. An increase in I_G decreases the breakover voltage. This applies equally to both forward and reverse conduction. After conduction occurs, the gate loses its effectiveness.

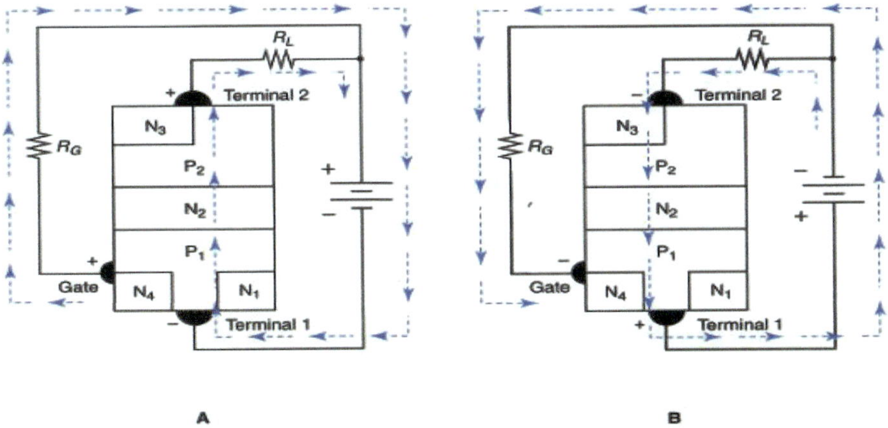

Figure 15.35 Triac condition polarities: A – Forward condition. B – Reverse conduction.

When a triac goes into conduction, its internal resistance drops to an extremely low value. Typical resistance values are less than 1Ω. The device remains in this state for the remainder of the alternation. It is essential that the current flowing through the triac be great enough to maintain conduction. This is generally called the holding current or I_H. The resistance of the load device and the value of the source voltage determine the conduction current. This current must be slightly greater than I_H to maintain conduction.

15.11 Diacs

A **diac** is a special diode that can be triggered into conduction by voltage. This device is classified as a bidirectional trigger diode. This means that it can be triggered into conduction in either direction. The word diac is derived from the terms diode for ac. This device is used primarily to control the gate current of a triac. It will go into conduction during either the positive or the negative alternation. Conduction is achieved by simply exceeding the **breakover voltage**.

Figure 15.36 shows the crystal structure and schematic symbol of a **diac**. Note that the crystal is similar to that of a transistor without a base. The N_1 and N_2 crystals are primarily the same in all respects. A diac will, therefore, go into conduction at precisely the same negative or positive voltage value. Conduction occurs only when the input voltage exceeds the breakover voltage. This particular device is a standard trigger for triac control. The voltage across the diac decreases in value after it has been triggered.

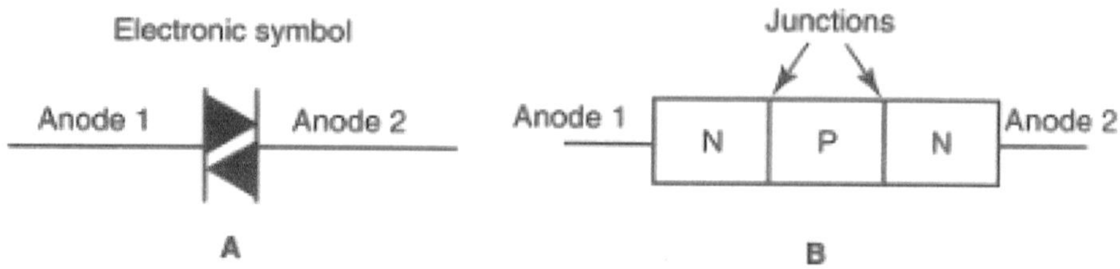

Figure 15.36 Diac construction.

15.12 Operational Amplifiers

An **operational amplifier**, or **op-amp**, is a high-performance amplifier circuit containing several transistor devices. The entire assembly is built on a small silicon substrate and packaged as an integrated circuit. ICs of this type are capable of high-gain signal amplification from DC to several million hertz.

The internal construction of an **op-amp** is quite complex and usually contains a large number of discrete components. A person working with an op-amp does not ordinarily need to be concerned with its internal construction. It is helpful, however, to have some general understanding of what the internal circuitry accomplishes. This permits the user to see how the device performs and indicates some of its limitations.

The internal circuitry of an **op-amp** can be divided into three functional units. Figure 15.37 shows a simplified diagram of the internal functions of an op-amp. Notice that each function is enclosed in a triangle. Electronic schematics use the triangle to denote the amplification function. This diagram shows that the op-amp has three basic amplification functions. These functions are generally called stages of amplification. A stage of amplification contains one or more active devices and all the associated components needed to achieve amplification.

An **op-amp** has at least five terminals or connections in its construction. Two of these are for the power supply voltage, two for differential input, and one for the output. There can be other terminals in the makeup of this device, depending on its internal construction or intended work function. On diagrams, each terminal is generally attached to a schematic symbol at some convenient location. Numbers located near each terminal of the symbol indicate pin designations.

The schematic symbol of an **op-amp** is generally displayed as a triangular-shaped wedge, denoting the amplification function. Figure 15.38 shows a typical symbol with its terminals labeled. The point, or apex, of

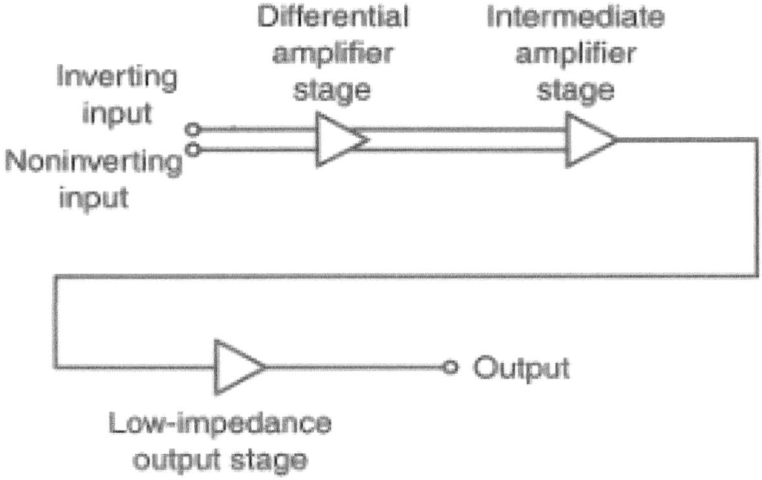

Figure 15.37 Op-amp diagram.

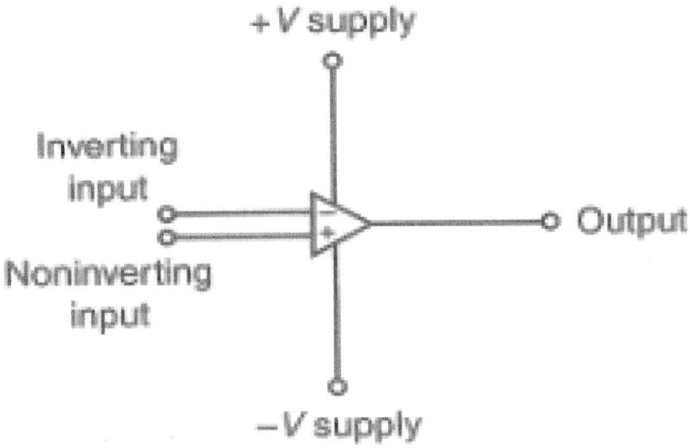

Figure 15.38 Op-amp schematic symbol.

the triangle identifies the **output**. The two leads labeled - and + identify the **differential input terminals**. The – sign indicates **inverting input** and the + sign denotes **noninverting input**. A signal applied to the – input is inverted 180° at the output. Standard op-amp symbols usually have the inverting input located in the upper-left corner. A signal applied to the + input is not inverted and remains in phase with the input. The + input is located in the lower-left corner of the symbol. In all cases, the two inputs are clearly identified as + and – inside the triangle symbol.

A B

Figure 15.39 Types of op-amps. A – Dual power opamps. (APEX Microtechnology) B – High current, high power op-amps. (Burr-Brown Corp.)

Connections or terminals on the sides of the triangle symbol are used to identify a variety of functions. The most significant of these are the two power supply terminals. Normally, the positive voltage terminal (+*V*) is positioned on the top side and the negative voltage (–*V*) on the bottom side. In practice, most op-amps are supplied by a split or divided power supply. This supply has +*V*, ground, and –*V* terminals. It is important that the correct voltage polarity be supplied to the appropriate terminals or the device can be permanently damaged. A good rule to follow for most op-amps is not to connect the ground lead of the power supply to –*V*. An exception to this rule is the current-differencing amplifier or **CDA op-amp**. These op-amps are made to be compatible with digital logic ICs and are supplied by a straight five volt source.

The actual schematic symbol of an **op-amp** generally has its terminals numbered and the element names omitted. There is a pin-out key to identify the name of each terminal. The manufacturer of the device supplies information sheets that identify the pin-out and operating data. Figure 15.39 shows some types of op-amps.

15.13 Integrated Circuits

Integrated circuits (ICs), such as the one shown in Figure 15.40, are extremely important electronic control devices. Inside each small IC package are hundreds and even thousands of individual components. Transistors, diodes, resistors and other components are built into ICs.

Figure 15.40 A-Integrated circuit-dual in-line (DIP) package. (Dallas Co.)

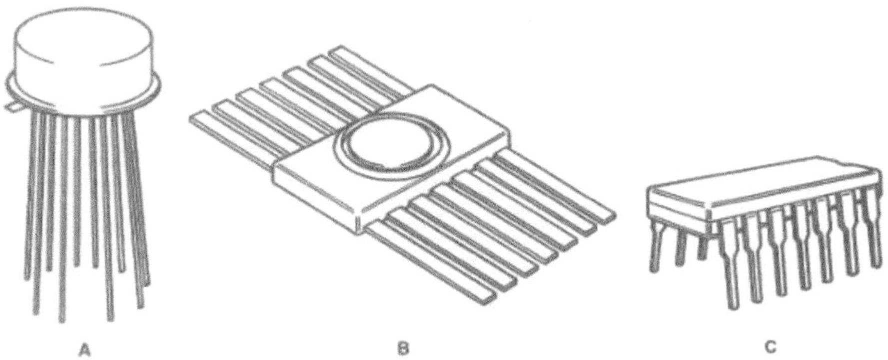

Figure 15.41 Common IC packages, A – TO-5 housing. B – Flat pack. C – Dual in-line package.

Some very common types are microprocessor and memory "chips" that are used with digital computer systems. Others include op-amps or operational amplifiers.

An important feature of these devices is their simplicity. Some self-contained small packages are shown in Figures 15.41 and 15-42.

There are a number of advantages in that integrated circuits have over standard electronic parts wired in a conventional circuit. These include:

- *Size*. ICs are becoming so small that hundreds can be on a piece of material the size of a shirt button. The industry is predicting much smaller and more complex chips in the future. Chips with a component density of 1 million are available today.
- *Cost*. The cost of an integrated circuit is based largely on its size and component density. By packing more components into less chip area, the cost per component is reduced. Additional savings can be realized

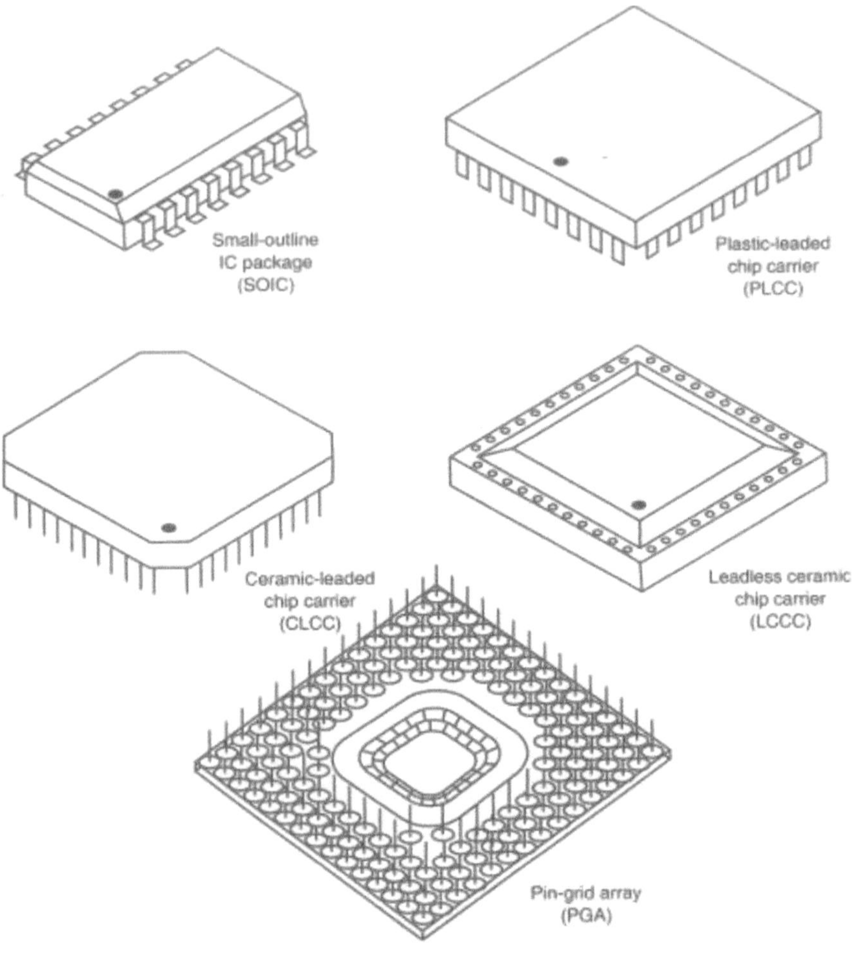

Figure 15.42 Newer IC package styles.

when using ICs because fewer parts are needed to order, inventory, and assemble into an operating system.

- *Reliability*. Perhaps the most important advantage of an IC over discrete component circuits is reliability. Increased reliability is due to a number of factors. The most significant of these is the reduction of interconnections between components. In earlier discrete-component circuits, up to 50% of circuit failures were related to interconnections between components. IC interconnections are achieved on the chip and formed at the same time during the manufacturing process. If the chip has interconnection problems they will be detected before it is placed in a completed package.

- *Power consumed.* The small size of an IC makes it well suited for low-power operation. The closeness of components in the structure reduces stray electrical pickup, allowing very small signal operation. Lowpower operation means lower internal temperature rise, which improves reliability.
- *Ease of maintenance.* Since the IC is actually a circuit containing many components, servicing can be achieved by plugging in new chips to replace faulty ones. The faulty chip is then discarded, since its repair is not practical.

Review Questions

Answer the following questions.

1. Draw the electrical symbols for the PN junction and zener diodes.

2. What type of semiconductor materials form the anode and cathode of a diode?

3. Why does the depletion area of a PN junction exist?

4. What is the difference between forward and reverse biasing?

5. How does a diode's forward resistance compare to its reverse resistance?

6. What is likely to happen when the reverse breakdown voltage of a normal PN junction diode is exceeded?

7. What manufacturer's specifications should be considered when selecting a diode for use in a circuit?

8. How can a PN junction diode's condition be checked with an ohmmeter?

9. How does the zener diode differ from the PN junction diode?

10. How does a current running through an LED produce light?

11. Draw the schematic symbol for the LED.

12. List five uses of the LED.

13. What terms are used to identify the three regions of a bipolar junction transistor?

14. Why is a transistor said to be a bipolar device?

15. How is the biasing different for NPN and PNP transistors?

16. What is a junction field-effect transistor?

17. How does the JFET differ from a PNP or an NPN transistor?

18. What is the difference between an N-channel and a P-channel JFET?

19. What is a JFET's pinch-off voltage?

20. What are the differences between the depletion mode MOSFET and the enhancement mode MOSFET?

21. How many semiconductor materials and junctions are associated with the physical structure of the UJT?

22. What is a UJT usually used for?

23. Where does the SCR get its name?

24. What role does the gate of the SCR play in triggering, or turning on, this device?

25. How does the removal of gate current affect the SCR's conduction once the device has been turned on?

26. How can the SCR be turned off once it has been triggered?

27. Why is the triac said to be like two SCRs connected in parallel?

28. What is an operational amplifier?

29. What is meant by inverting and noninverting op-amp inputs?

30. What is an integrated circuit chip?

16

Electronic Sensing Systems

Objectives

Upon completion of this chapter, you should be able to:

- Identify different types of sensing systems.
- Discuss the basic operational principle of each type of sensor and transducer.
- List the types of light-sensitive sensors.
- Describe position sensing systems.
- Discuss fiber optics and laser use for sensing systems.
- Discuss speed and mechanical movement sensors.
- Describe the Hall effect for sensing systems.

Important Terms

angstrom (Å)
Hall effect
laser
nanometer (nm)
photo conductive
photo emissive
photovoltaic
piezoelectric effect
proximity sensor
range sensor
reed switch
sensing system
stadimetry
strain gauge
stress sensor
tactile sensor
thermistor
thermocouple
thermoelectric sensor
transducer
triangulation
visual sensor

DOI: 10.1201/9788743810056-21

Introduction

The control of many industrial machines depends upon various types of electronic sensing systems. A sensing system often uses devices called transducers to convert light, heat, or mechanical energy into electrical energy. The signal output of a transducer can be used to affect the operation of a machine. Sensors give automated equipment a higher level of intelligence by improving their decision-making capability. This chapter discusses some of the sensors used with automated industrial systems in industry.

16.1 Kinds of Sensing

Sensing systems allow automated equipment to interact with their environment. It helps them deal with changes and uncertainties. A sensing system allows automated equipment to determine their own actions. If an industrial machine lacks sensing ability, support fixtures and tools must be built for it. Sensors help automated equipment deal with different positions and orientation of parts and variations in the shape and dimensions of parts. Simple sensor control is accomplished with the use of contact switches. These switches can stop the motion of a machine or modify its motion. Some machines use touch, force, or torque sensing. Advanced automated equipment has vision sensors that can recognize patterns or react to motion.

16.2 Types of Sensors

Many different sensing techniques and kinds of transducers are used with automated industrial systems. The type chosen depends on what is to be sensed, the accuracy required, and the environment. Several types of sensors are summarized in the following sections.

16.2.1 Proximity sensors

Most **proximity sensors** detect the absence or presence of an object within a certain distance. Others provide feedback about the distance between the sensor and an object, such as an end effector.

Optical proximity sensors measure the amount of light reflected from an object. They can respond to either visible or infrared light. Incandescent lights can be used as the light source. However, **light-emitting diodes (LEDs)** are generally preferred, since they are more reliable than lamps and are not sensitive to shock and vibration. As discussed in **Chapter 15**, LEDs

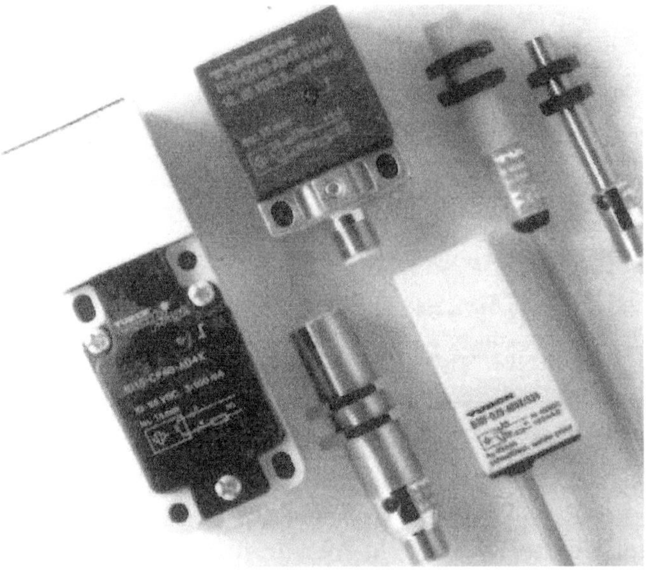

Figure 16.1 Proximity sensors. These devices are limit switch style, solid-state, non-contact sensors. (Turck, nc.)

(or solid-state lamps) are small, lightweight optoelectronic devices. The term *optoelectronic* refers to the combination of optics and electronics. LEDs are easily used with digital and other miniaturized systems.

Eddy current proximity sensors produce a magnetic field in the small space of a detector unit, which can be mounted in a probe. The magnetic field induces eddy currents into any conductive material that is near the probe. A pick-up coil senses a change in magnetic field intensity when an object enters the field. Some types of proximity sensors are shown in Figure 16.1.

16.2.2 Reed switches

Reed switches can either make or break contact in response to a magnetic field. Two flat metal strips, or reeds, are housed in a hollow glass tube filled with an inert gas. When the reeds are exposed to a magnetic field, they are forced together, completing an electrical circuit. When the magnetic field is turned off, the reeds spring open, breaking the circuit.

Because the contacts are housed inside a hermetically sealed glass tube, contact sparks are isolated from the outside. This type of construction is useful in explosive environments. Contacts are also isolated from outside dust and corrosion. This means improved life expectancy for the switch.

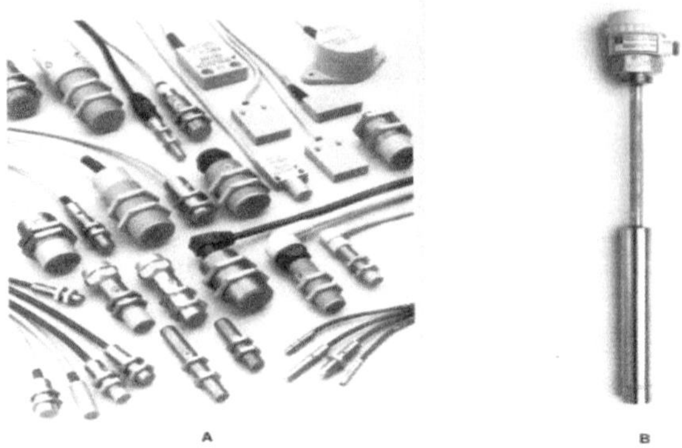

Figure 16.2 Touch-sensitive proximity sensors. A – Capacitive probes (sensors) used for liquid level control. B – Pressure level transmitter used to sense the hydrostatic pressure of liquids in tanks. (Pepper! & Fuchs, Inc.)

16.2.3 Touch-sensitive proximity sensors

Touch-sensitive proximity sensors (detectors) operate on capacitance developed by a large conductive object (such as the human body or a liquid). The capacitance developed changes the frequency of an electronic circuit (see Figure 16.2). A conductive plate or rod can be used to sense contact.

Acoustic proximity sensors react to sound. Standing sound waves are generated within a cylindrical, open-ended cavity inside the detector. The presence of a nearby object interferes with these waves, altering the detector's output. A microphone can be used to detect a change in sound pressure and measure the distance of the object from the sensor.

16.2.4 Range sensors

Range sensors determine the precise distance from the sensor to an object. Such devices are useful for locating objects near a workstation. One range sensing system, called a **laser interferometric gauge**, is sensitive to humidity, temperature, and vibration. Another range-sensing system is a television camera that operates on the sonar (sound) principle.

16.2.5 Tactile sensors

Tactile sensors indicate the presence of an object by touch, while **stress sensors** produce a signal that indicates the magnitude of the contact made. A simple touch sensor is a microswitch. Limit switches also respond to contact with an object. Devices called strain gauges are often used as stress sensors.

16.2.6 Visual sensors

Visual sensors can be used to recognize objects or to measure their characteristics. Camera-equipped computer systems are an example. These identify a part by means of a television camera. They can distinguish that part from any other part. An object can be identified by its shape, outline, or area, regardless of its orientation to the camera. Computer vision systems are used to sense spatial relationships. Computer vision provides depth information by means of stadimetry and triangulation. **Stadimetry** determines the distance to an object based on the apparent size of a camera image. **Triangulation** involves measuring angles and the baseline of a triangle to determine the object's position.

Position detection can also be done by solid-state TV cameras. A camera is placed on a machine and feedback is used to guide the movement of the machine to a specific location. This process, which is used for moving material from one spot to another, is referred to as visual servoing. Servo movement can be used with either stationary or moving objects.

16.2.7 Light sensors

Light-sensing systems respond to changes in light energy using various optoelectronic devices. These devices, along with **lasers** and **X-ray** devices, are becoming increasingly important in machine control circuits.

Light is a visible form of radiation. It occupies a narrow band of frequencies along the vast electromagnetic spectrum. As shown in Figure 16.3, the **electromagnetic spectrum** includes frequencies for radio, television, radar, infrared radiation, visible light, ultraviolet light, X-rays, and gamma rays. The different types of radiation, such as light, heat, radio waves, and X-rays, differ only in their frequency and wavelength.

The human eye responds only to light in the visible frequencies. See Figure 16.4. Each color of light has a different frequency and wavelength. In order of increasing frequency and decreasing wavelength, colors range as follows: red, orange, yellow, green, blue, and violet.

The **wavelengths** of light are measured in nanometers. A **nanometer (nm)** is equal to 1000 micrometers (millionths of a meter, more commonly called microns). Visible light wavelengths are in the 400-nm (violet) to 700-nm (red) range. Angstroms are also used for light measurement. An **angstrom (A)** is one-tenth of a nanometer. Thus, visible light ranges from 4000 to 7000 A. The human eye is most sensitive to the frequency near 5500 A. It is the least sensitive around 4000 A on the lower wavelengths and 7000 A on the higher wavelengths. The human eye perceives various degrees of brightness because of its response to the wavelengths of light. The normal

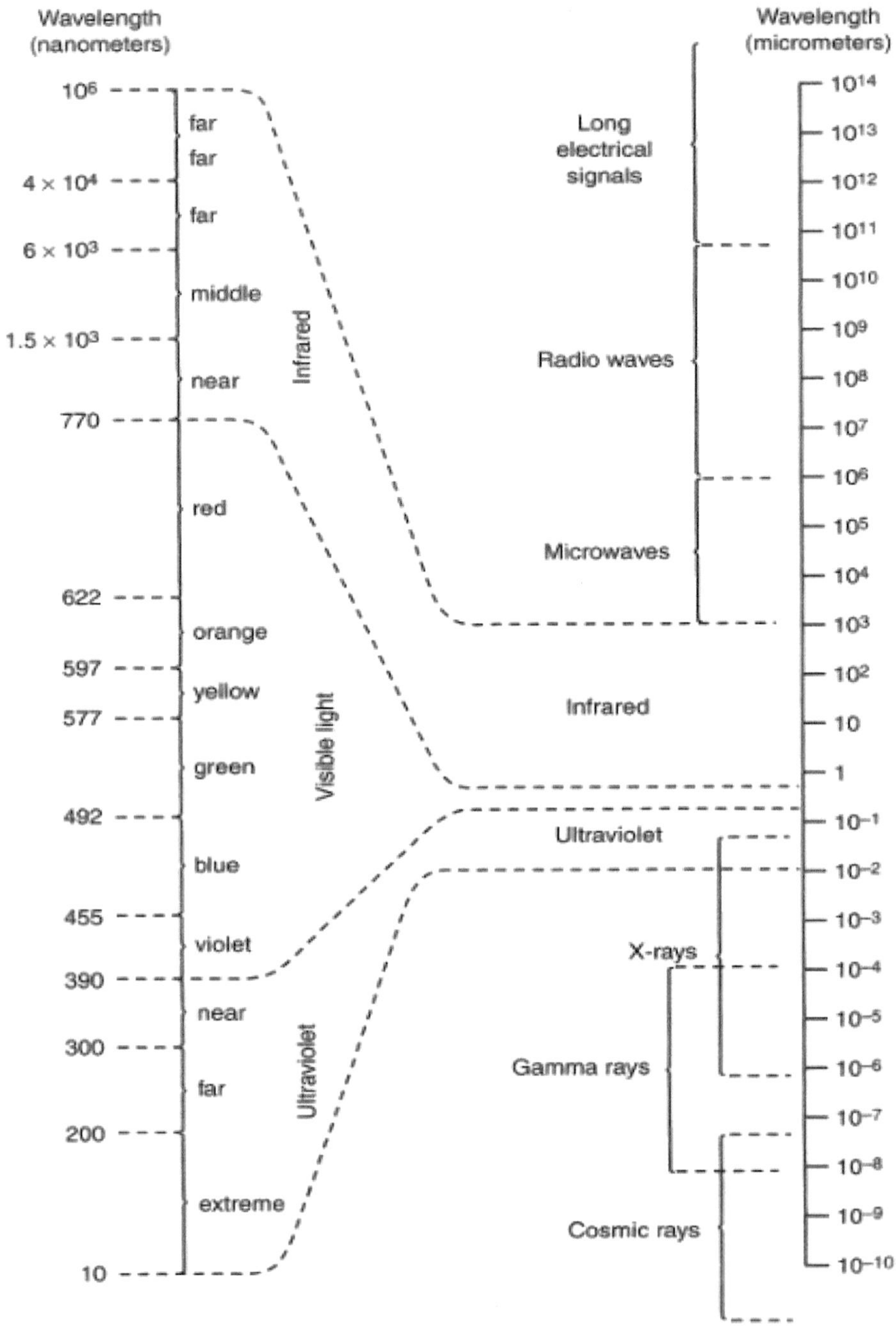

Figure 16.3 The electromagnetic spectrum includes a narrow band of visible light.

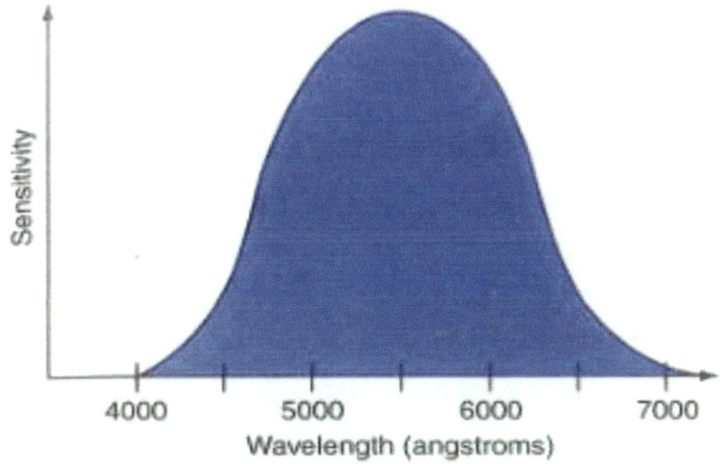

Figure 16.4 The human eye responds to visible light at these wavelengths. It is most sensitive to light in the middle wavelengths, such as shades of green and yellow.

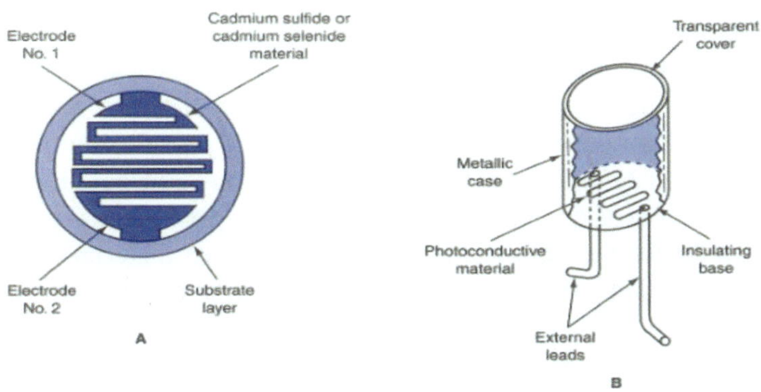

Figure 16.5 This cadmium sulphide cell will vary in conductivity with the amount of light energy striking it. A – Top view. B – Cutaway view.

human eye cannot see a wavelength of less than 4000 A or more than 7000 A (400–700 nm).

Optoelectronic devices fall into three categories: photoemissive, photoconductive, and photovoltaic. **Photoemissive** devices emit electrons in the presence of light. Phototubes are a type of photoemissive device. **Photoconductive** devices vary in conductivity according to fluctuations in light. Their electrical resistance decreases when light is more intense and increases when light intensity decreases, Figure 16.5. **Photovoltaic devices,**

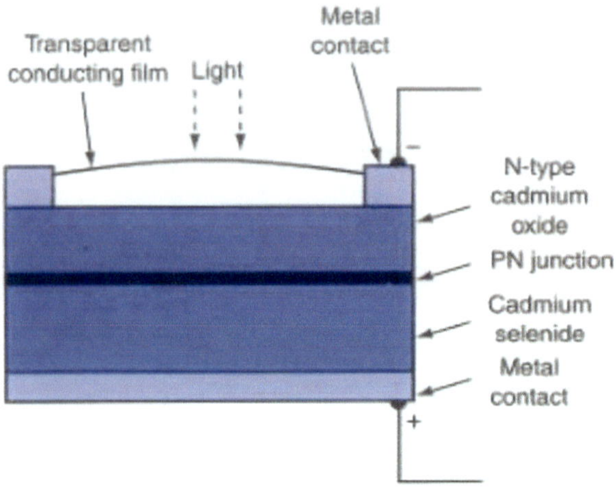

Figure 16.6 Electrons in this selenium photovoltaic cell move from layer to layer, generating electrical energy.

or solar cells, convert light energy into electrical energy. When light energy falls on a photovoltaic device, it creates an electrical voltage.

The electrical output of a **solar cell** is proportional to the amount of light energy falling on its surface. The cell in **Figure 16.6** has layers of selenium and cadmium deposited on a metal base. A layer of cadmium selenide and another layer of cadmium oxide are produced. A transparent conductive film and a section of conductive alloy are placed over the cadmium oxide. External leads are then connected to the conductive material. When light strikes the cadmium oxide layer, electrons are emitted and move toward the leads. A deficiency of electrons is now created in the cadmium oxide. This deficiency is compensated for by electrons from the selenium material. Electrons then move from the metal base into the selenium. As a result, a voltage develops between the two external terminals.

Selenium cells are relatively inefficient in producing electricity; **silicon cells** are now more frequently used. A **silicon photovoltaic cell** is shown in Figure 16.7. Photovoltaic cells have a variety of applications. Although their electrical output is low, they are used with amplifying devices to drive a load.

16.2.8 Optoelectronic position sensors

Various types of optoelectronic devices have been designed to sense the position of a light beam. They are used with digital control systems to produce electrical outputs, Figure 16.8. The output is based on patterns created by the photoconductive material onto which the light beam is focused.

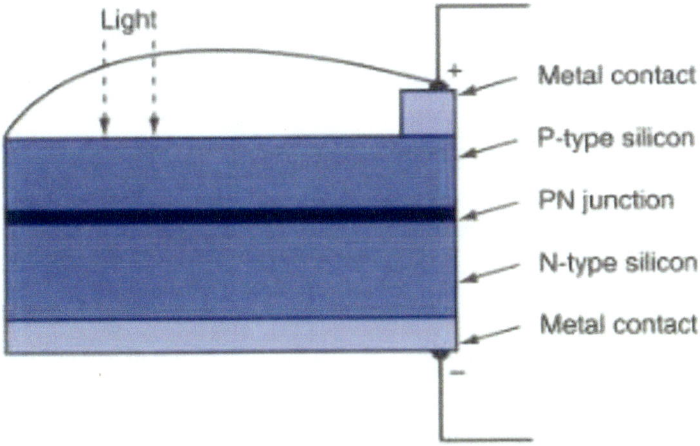

Figure 16.7 Silicon photovoltaic cells like this one produce electrical energy more efficiently than selenium cells.

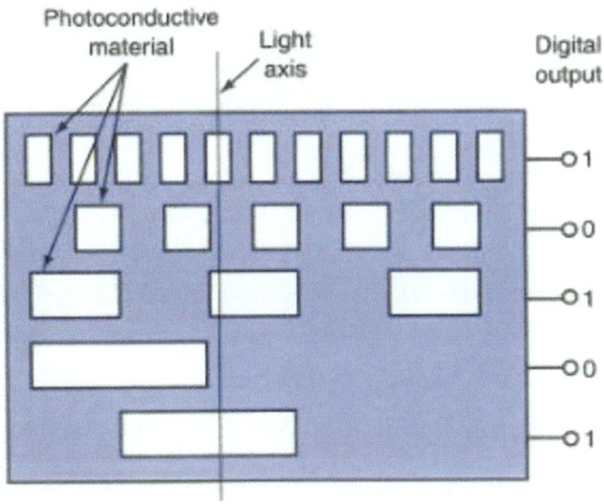

Figure 16.8 This optoelectronic digital readout indicates the position of a light beam.

16.2.9 Fiber-optic sensors

Optical fibers made of glass or plastic can transmit light from one point to another, **Figure 16.9.** The light can travel around corners, within a limited space, or over long distances. The light travels through the fiber-optic material regardless of how the material is bent or shaped. The cladding (cover) material is reflective, so the light bounces from side to side within the optical fiber light pipe as it moves. Fiber optics is used

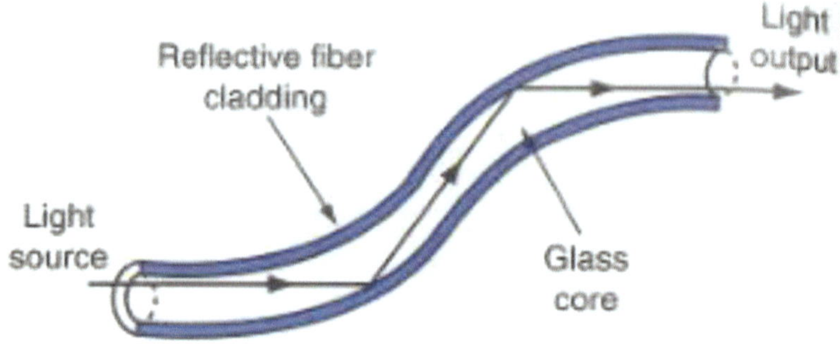

Figure 16.9 Light can be "bent" around corners through the use of an optical fiber, sometimes called a "light pipe."

Figure 16.10 Fiber-optic sensor. This is a noncontact sensor with sensitivity adjustment for fine discrimination of objects. (Ramco Electric Co.)

for numerous industrial applications. A fiber-optic sensor is shown in **Figure 16.10.**

16.2.10 Infrared sensors

Another type of device that responds to radiant energy is an **infrared detector**. These devices pick up radiation in the infrared region of the electromagnetic spectrum. All objects emit infrared thermal radiation. Infrared camera systems can detect such radiation even in darkness. Industrial uses include heat-sensitive control systems, optical pyrometers, and infrared spectroscopy for gas analysis.

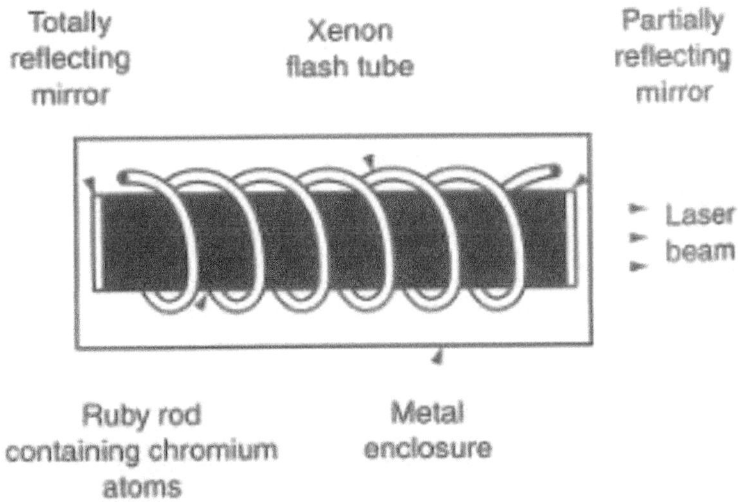

Figure 16.11 The ruby laser shown in this simplified drawing produces a beam of photons that do not spread out| and scatter as ordinary light does.

16.2.11 Ultraviolet sensors

Ultraviolet sensors respond to electromagnetic radiation in the ultraviolet range. However, design problems have been encountered with such sensors. They are thus used less frequently than detectors of visible light or infrared energy.

16.2.12 Laser sensors

The development of the laser has had a significant continuing impact on industrial control systems. The major advantage of the **laser** is that its tightly focused beam can travel long distances with relatively little spreading. The term **laser** means light amplification by stimulated emission of radiation.

A **ruby laser** is shown in Figure 16.11. When the xenon tube flashes, the chromium atoms in the ruby rod absorb photons of light. The chromium atoms then emit their own photons. Many of these travel along the axis of the ruby rod, and are reflected by the mirrors on each end. This causes additional photons to be released, which amplifies the light. Eventually, the photons fall into step with one another so that they produce only one wavelength (color) of light. The result is an intense beam of light energy.

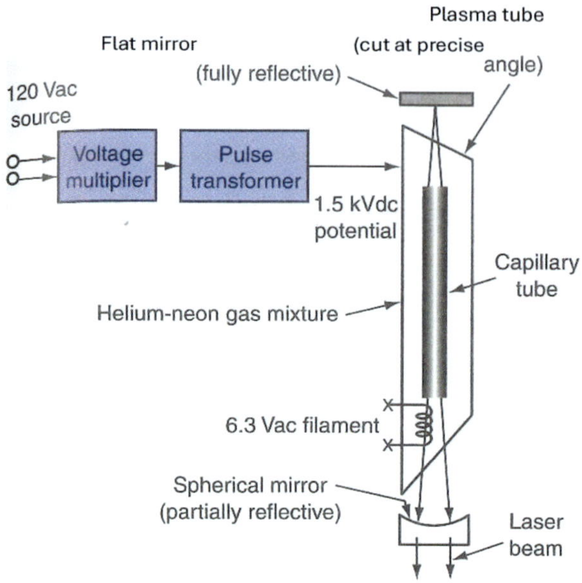

Figure 16.12 A laser using helium-neon gas is one of the types used in industry for sensing and control applications

16.2.13 Gas lasers

Gas lasers are often used with sensing and control systems. A popular type is the helium-neon laser, Figure 16.12. Many other types are available. All use basically the same operating principle, as described below.

A high DC potential is applied to the plasma tube by means of a voltage multiplier circuit and a pulse transformer. The filament within the plasma tube is heated by a 6.3-volt AC source. As electrons from the filament are accelerated, they strike helium-neon gas atoms in the tube and cause them to ionize. The ionized gas emits light. This is similar to the action of a fluorescent tube used for household lighting. The light reflects from a flat, fully reflective mirror at the top. The plasma tube is cut at a precise angle to control reflection. The light is reflected back and forth several times to the partially reflective spherical mirror at the bottom. Here, it is concentrated into a laser beam that is emitted through the spherical mirror.

16.2.14 Semiconductor lasers

It is also possible to generate laser beams by means of semiconductors. These lasers have a resonant cavity similar to other lasers, except that it is formed on a chip of semiconductor material.

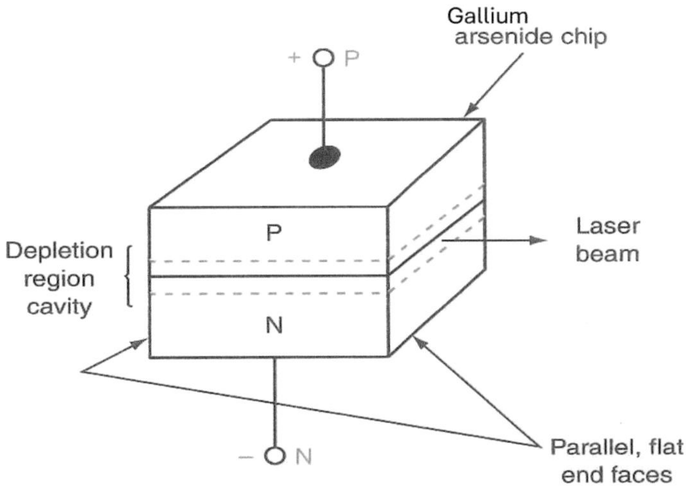

Figure 16.13 This semiconductor injection laser requires no mirrors and is tiny and very efficient.

Semiconductor injection lasers, Figure 16.13, are very efficient and extremely tiny in size, compared to other lasers. The end faces of a gallium arsenide chip are made parallel and flat. Since gallium arsenide is a reflective material, no mirrors are needed. This is a distinct advantage in terms of complexity and cost.

As current flows through the chip, light is emitted from the **gallium arsenide**. The atoms collide near the PN junction of the material and cause the release of additional photons. Due to the reflective properties of the gallium arsenide, a wave of photons is developed between its flat surfaces. The back-and-forth movement of this wave creates the resonant action required to produce a laser beam. A **laser beam sensor** is shown in Figure 16.14.

16.2.15 X-ray sensors

X rays make up one band of frequencies in the electromagnetic spectrum. The element **radium** emits three kinds of these invisible rays: alpha, beta, and gamma rays. Some of these rays can pass through the human body, a property that makes them valuable in medical treatment and analysis.

It is possible to use a vacuum tube, Figure 16.15, to produce rays similar to those emitted by radium. This X-ray tube has a cathode that is heated by means of a filament voltage. The anode is constructed of a heavy metal. A high positive potential is applied to accelerate the electrons emitted from the cathode at a very rapid rate. The electrons strike the anode with such velocity

Figure 16.14 Laser beam sensor. These sensors provide safe, accurate, and fast response sensing. (Ramco Electric Co.)

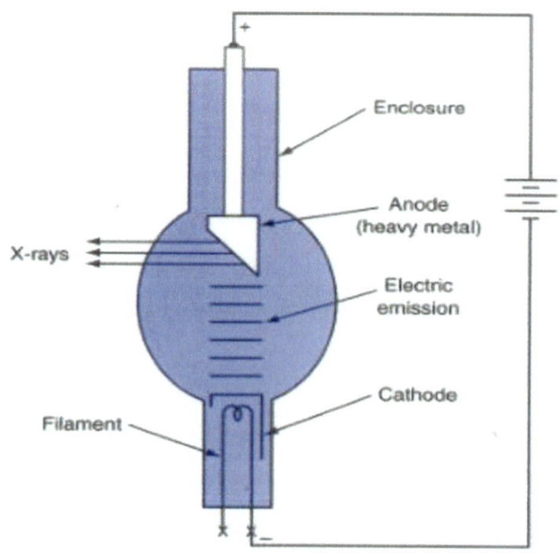

Figure 16.15 X-rays generated by tubes like this one are used to probe the structure of metals.

that *X* rays are created. If the potential is increased, the frequency of the *X* rays also increases, while the wavelength decreases. *X*-ray tubes can operate with potentials in excess of one million volts dc. The resulting *X* rays are similar to the high-frequency gamma rays emitted by radium.

In industry, *X* rays are used to control processes that involve metals. The short wavelength of rays allows them to pass through metals and reveal structural characteristics.

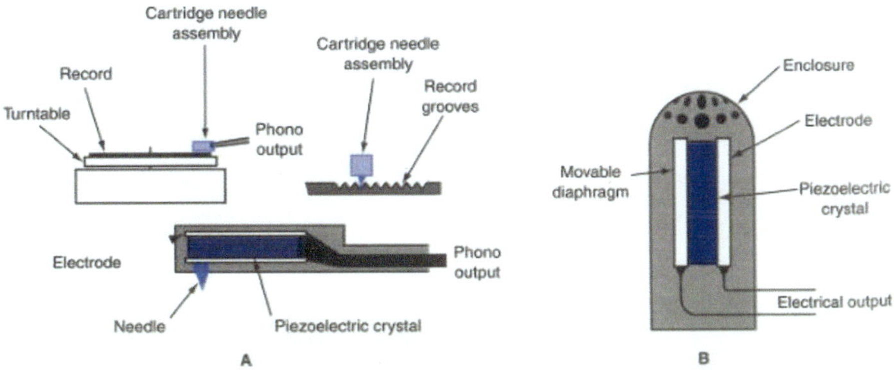

Figure 16.16 Applications of the piezoelectric effect. A – A phonograph cartridge/needle (stylus) assembly. B – A crystal microphone.

16.2.16 Sound sensors

Sound sensing systems rely on the piezoelectric effect to convert sound to electrical energy. The piezoelectric effect occurs in certain crystals, such as quartz and Rochelle salt. When these crystals are subjected to mechanical stress, an electrical potential is developed in them.

An application of the **piezoelectric effect** is the cartridge/stylus needle assembly of a phonograph, part A of Figure 16.16. The cartridge contains a crystal that vibrates as the stylus attached to the cartridge rides through the record grooves. This creates an electrical potential across the crystal. The small electrical variations are then amplified by a control system. Thus, mechanical energy (vibration) is converted to electrical energy. The same principle is applied by crystal microphones, in which sound waves (vibrations in air) strike a piezoelectric crystal. See part B of Figure 16.16. A voltage is developed across the crystal and is amplified by the control system. Some piezoelectric switches are shown in Figure 16.17.

16.2.17 Temperature sensors

Sensors that produce a change in electrical output due to a change in temperature are referred to as **thermoelectric sensors**. They are used to control processes in which temperature must be held at a given level or not allowed to exceed a specific point.

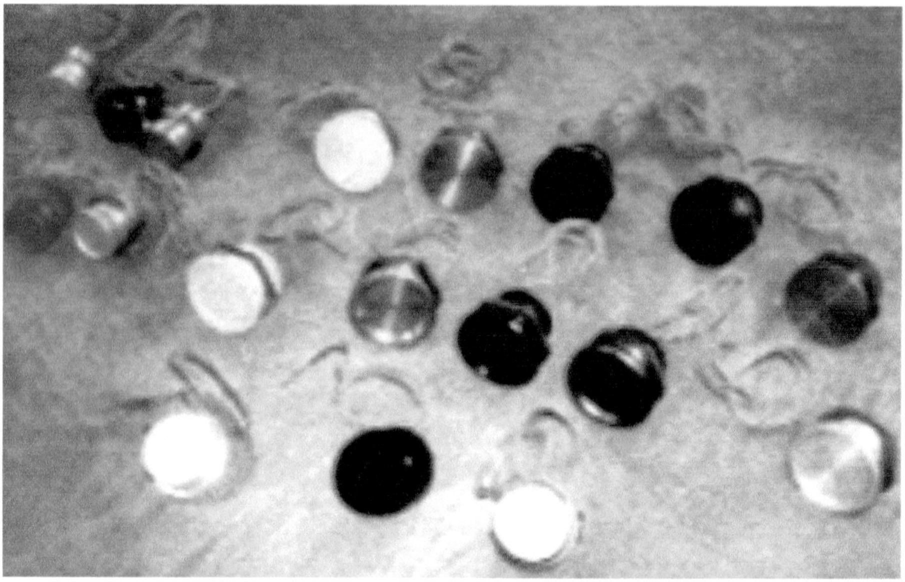

Figure 16.17 Piezoelectric switches. These switches are used to withstand temperature extremes, moisture, dust, chemicals, and oils. They are very durable sensors with no moving parts. (Everswitch, USA)

One device that is commonly used for heat sensing is the **thermistor**. The resistance of a thermistor, a temperature-sensitive resistor, decreases as temperature increases (and vice versa). This phenomenon is referred to as a negative temperature coefficient of resistance. Various metal-oxide semiconductor materials are used to construct thermistors, Figure 16.18. Thermistors are manufactured in a wide range of resistance characteristics and temperature coefficients.

Thermocouples are devices that convert heat energy into electrical energy. A thermocouple, Figure 16.19, consists of two dissimilar metal strips fused together at one end. The metals are usually combinations of iron-constantan, copper-constantan, and platinum-rhodium. Different metal combinations respond to different ranges of temperature. When the fused end is heated, a voltage develops at the unconnected ends. This voltage exists due to the differing coefficients of expansion of the two metals. The voltage produced by a thermocouple is usually in the millivolt range.

16.2.18 Displacement sensors

Displacement (movement) can be detected by various types of sensors. Figure 16.20 illustrates how resistive, capacitive, and inductive sensors can

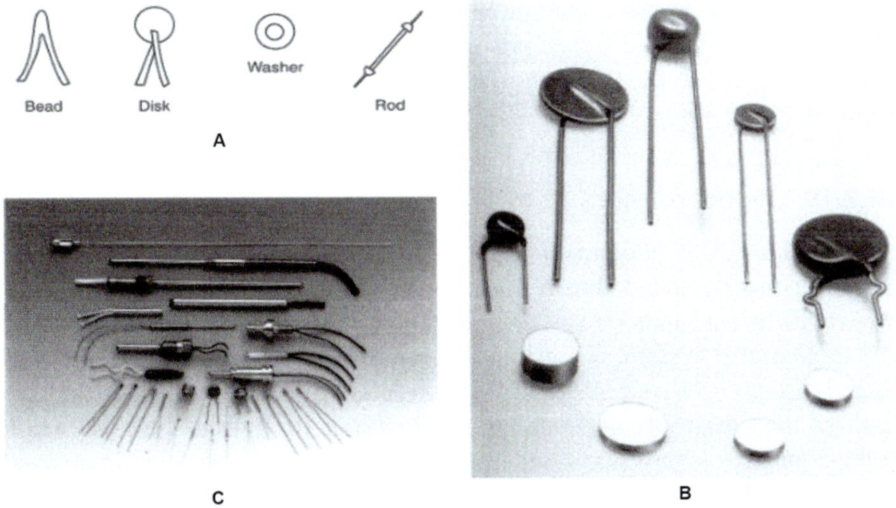

Figure 16.18 In a thermistor, a decrease in temperature causes increased resistance. A – Sketches of some of the common forms of the device. B – Thermistors used to protect sensitive components from over-current conditions. C-Thermistor applications include probes such as those shown here, which are used to measure high temperatures. (Advanced Thermal Products, Inc.; Thermometrics Co.)

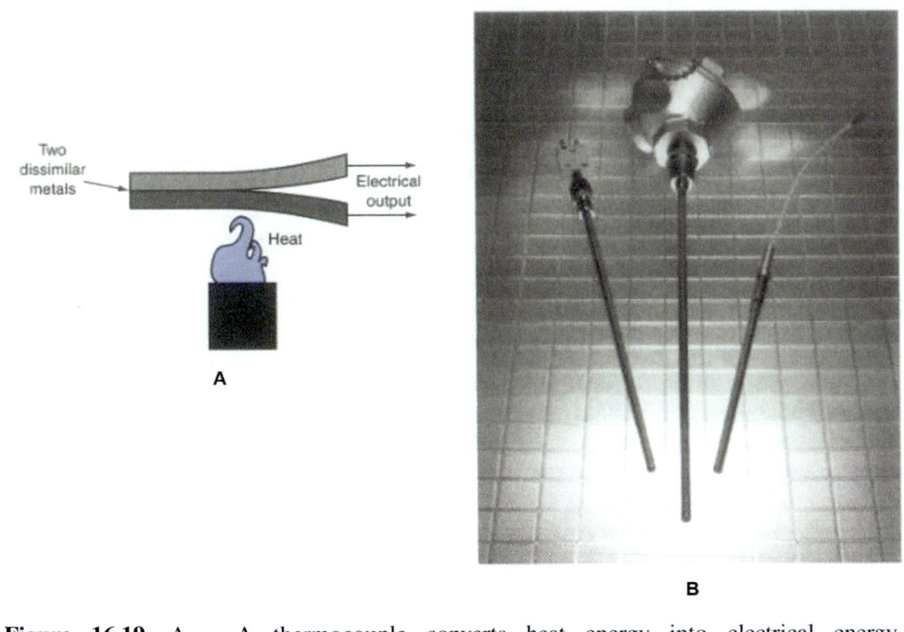

Figure 16.19 A – A thermocouple converts heat energy into electrical energy. B – Thermocouple probe sensors for heat measurement and control. (Tudor Technology Ltd., An Athena Control Co.)

be used to create an electrical output that shows displacement. Displacement is sensed in reference to a fixed position or to the force required to move from one position to another. It is possible to sense a displacement of less than 1mm (0.04 in.).

16.2.19 Speed sensors

Because industrial equipment has shafts, gears, pulleys, and other rotating components, speed sensing usually involves measuring rotary motion. Several different methods can be used for **speed sensing.** One of these, the **DC tachometer** system, is illustrated in Figure 16.21. The tachometer is connected directly to a rotating piece of equipment. As its movement causes the shaft of the small DC generator to rotate faster or slower, the generator's voltage output increases or decreases. Voltage output is translated into speed changes or used to control equipment operation.

Electronic tachometers are now popular because of their precision and ease of use. Reflective material is placed on the surface of the rotating portion of the equipment. The tachometer has a light source that is reflected back to a photocell as the material passes in front of it. The photocell converts the pulses of reflected light energy into electrical signals that are used to measure the speed of rotation.

16.2.20 Mechanical movement sensors

Mechanical movement (a change in dimension resulting from strain or stress) can be sensed using a **strain gauge**. In a typical application, strain gauges emit electrical signals (feedback) based on the amount of pressure the mechanical fingers of a robot are exerting to lift an object.

As shown in Figure 16.22, the gauge is made of fine-gauge resistance wire about 0.001in. (0.025 mm) in diameter mounted on a strip of insulation. The wire is elastic so that it can easily change dimensions. When subjected to stress, it is stretched. As the wire stretches, its cross-sectional area is reduced and its resistance changes. The wire's resistance can be expressed mathematically as:

$$R = \frac{\rho \times L}{A}$$

where R is the resistance of the wire conductor, ρ is the resistivity constant of the conductor, L is the length of the conductor, and A is the cross-sectional area of the conductor.

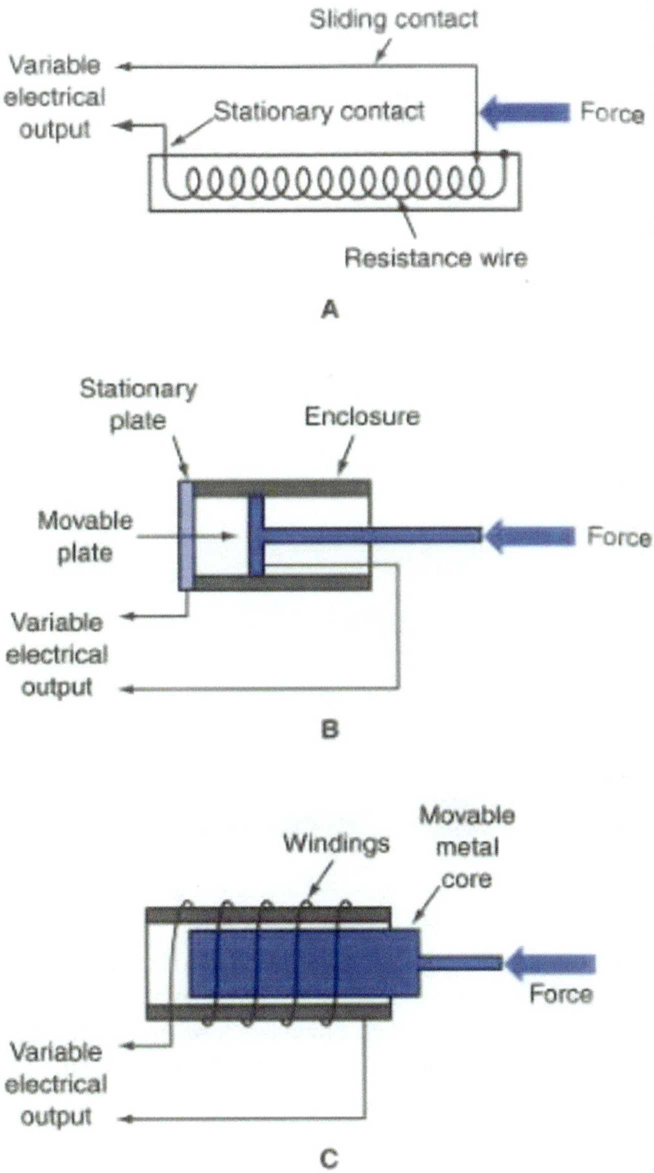

Figure 16.20 Displacement sensing systems. A – Resistive sensor. B – Capacitive sensor. C – Inductive sensor.

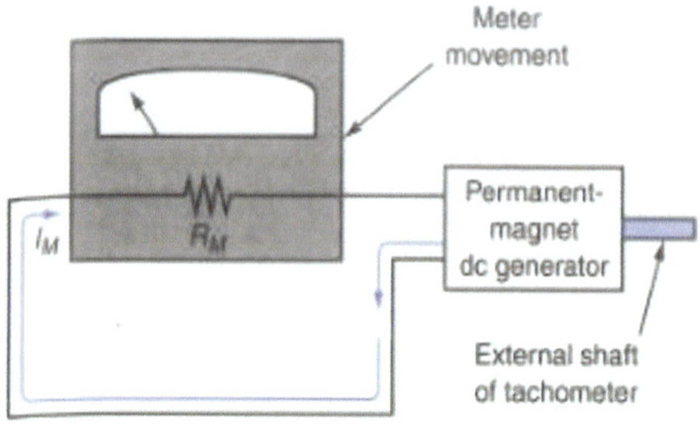

Figure 16.21 A DC tachometer senses speed changes and displays them as meter fluctuations.

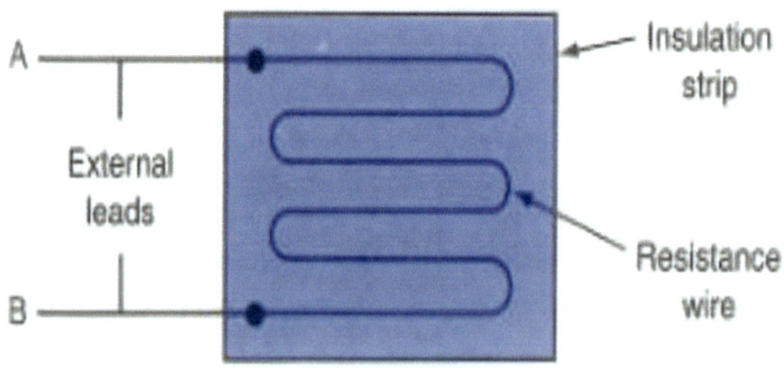

Figure 16.22 A strain gauge responds to mechanical energy by changing its resistance.

Semiconductors are now often used in place of traditional metal **strain gauges**. Their rate of change of resistance is approximately 50 times higher, and they are more sensitive to small changes. A semiconductor strain gauge is as stable as the metallic type, but has a much higher output.

16.3 Types of Transducers

The conversion of physical quantities to electrical quantities is a basic sensing function. **Transducers** are used to make this conversion. For example, a thermocouple is a transducer that converts heat energy into electrical energy. A microphone is a transducer that converts sound energy into electrical energy.

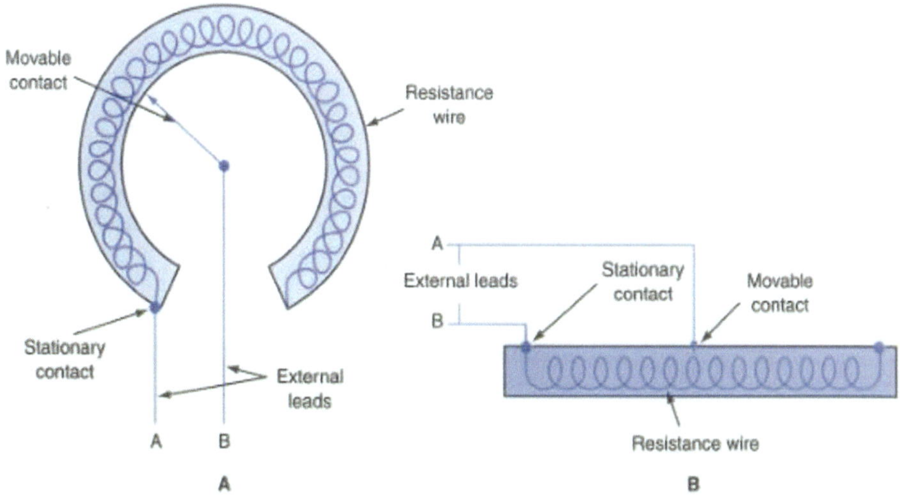

Figure 16.23 Resistive transducers. A – Rotary potentiometric transducer. B – Flat potentiometric transducer.

Numerous other examples of transducers can be found in the home as well as in industry. Transducers can be resistive, capacitive, or inductive.

Resistive transducers convert resistance variations into electrical variations. One type works on the potentiometer principle, as shown in Figure 16.23. This transducer changes resistance when the position of its movable contact is changed. By increasing the length of the wire between terminals A and B, the resistance between those two points is increased. This transducer is often used to sense physical displacement. Displacement causes movement of the sliding contact, and thus a resistance change in the control circuit.

Capacitive transducers measure a change in capacitance, or the ratio of charge to the potential difference between conductors. Capacitance exists when two conductive materials (plates) are separated by an insulating material. Capacitance can be increased by increasing the area of the plates or by decreasing the thickness of the insulation.

One use for capacitive transducers is for sensing **fluid pressure**, Figure 16.24. As shown, the transducer is placed into a fluid line. Plate 1 is a conductive diaphragm that senses any variation in fluid pressure. Plate 2 is stationary, while plate 1 varies position due to changes in pressure. When the pressure of the fluid in the line increases, plate 1 moves closer to plate 2. When the distance between the capacitor plates decreases, the capacitance

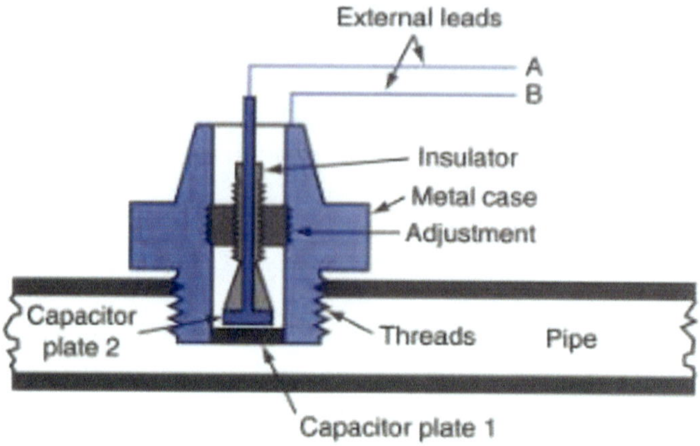

Figure 16.24 This capacitive transducer senses fluid pressure. Pressure changes cause movement of plate 1, changing the capacitance of the circuit and affecting the control signal that is emitted by the transducer.

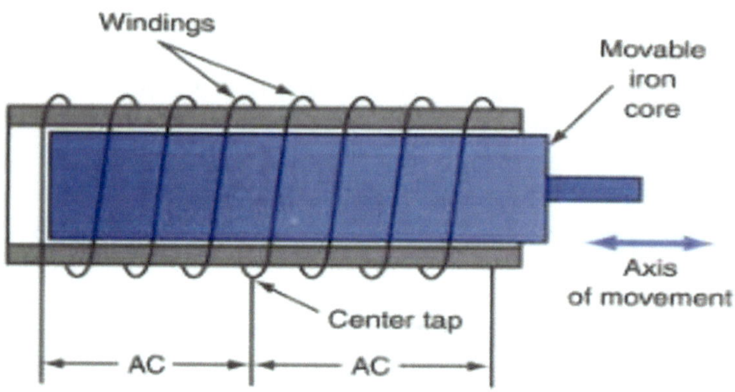

Figure 16.25 A movable-core inductive transducer permits measurement of movement.

between terminals A and B increases. When pressure decreases, the distance between plates increases, and capacitance decreases.

An **inductive transducer**, such as the one shown in Figure 16.25, usually has a stationary coil and a movable core. The movable core is connected to an object whose movement is to be measured. As the core changes position within the coil, the inductance of the coil flow through the coil drops increases ($X_L = 2\pi f L$ and $I = V$ One common type of inductance sensor is the linear variable differential transformer (LVDT), is illustrated in Figure 16.26. **A movable** metal core is placed in a tubular housing that has three windings.

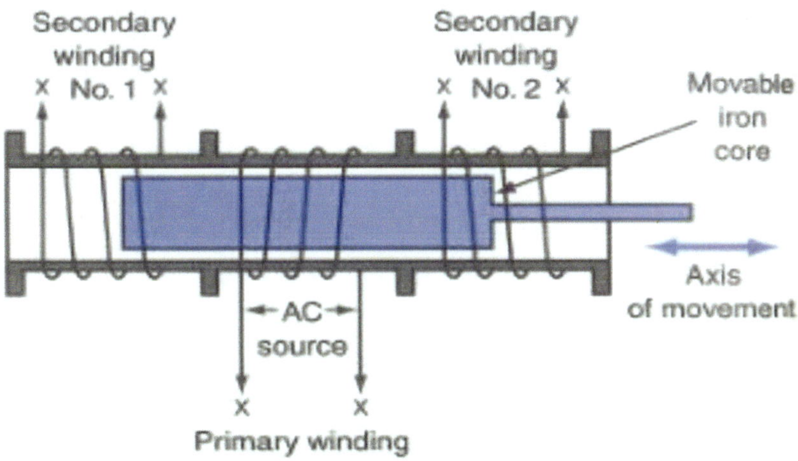

Figure 16.26 A linear variable differential transformer (LVDT) measures movement based on changes in the voltage induced in its secondary windings.

The center winding (primary) is connected to an AC source. Voltage is induced in the two outer windings by the primary winding. Initially, the voltages induced in the two outer windings are equal. Any movement of the core will cause one induced voltage to increase and the other induced voltage to decrease. The difference between the induced voltages depends upon the amount of core movement.

16.4 Hall Effect Applications

Magnetism can be used for purposes other than in generators to produce voltage. When a magnetic field is placed at right angles to a current-carrying conductor, a small voltage is produced across the conductor. This is known as the **Hall effect**. This effect has many applications in switching and measurement circuits. Voltages produced by this process are small (in the microvolts) when conductors are used. When a semiconductor material such as indium arsenide is used, the voltage produced can be as much as 100 mV .

The **Hall effect** is illustrated in Figure 16.27. A block of indium arsenide has a small current flowing through it. If a magnetic field is placed perpendicular to the direction of the current, a voltage is produced across the width of the semiconductor. The amount of voltage produced is directly proportional to the flux density of the magnetic field.

An application of the **Hall effect** is as a sensor in instruments designed to measure the strength of magnetic fields. One such device is known as a

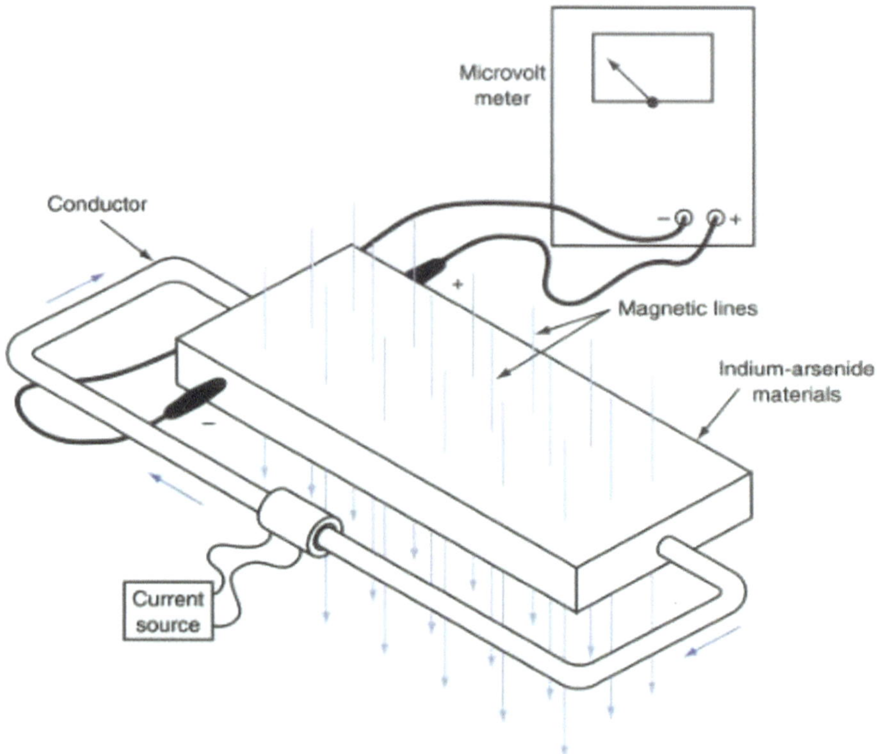

Figure 16.27 Illustration showing the Hall effect.

Gauss meter. Within its measuring probe is a Hall effect device. When placed within a magnetic field, it produces a voltage directly proportional to the flux density of the field. This voltage is sensed and displayed on the meter, which is calibrated in gauss. There are many other uses for Hall effect devices, including position sensors for machines, switches for computer keyboards, and in electronic ignition systems of automobiles.

Review Questions

Answer the following questions.

1. What do electronic sensing systems give automated equipment?

2. List the two types of proximity sensors and describe their detection methods.

3. How does a reed switch sensor operate?

4. List a use for each of the following sensors.

   a. Range sensor.

   b. Tactile sensor.

   c. Visual sensor.

5. List the colors of the visible portion of the electromagnetic spectrum from shortest to longest wavelength.

6. List the three major categories of optoelectronic sensors.

7. List five optoelectronic sensing devices.

8. How can X-rays be used for industrial sensing applications?

9. What is the piezoelectric effect?

10. List and briefly describe the three types of thermoelectric sensors.

11. List a use for each of the following sensors.
    a. Displacement sensor.
    b. Speed sensor.
    c. Mechanical movement sensor.

12. How does a strain gauge operate?

13. Explain the basic operating theory of resistive, capacitive, and inductive transducers.

14. What is an LVDT?

15. Describe an application of a Hall effect sensor.

UNIT SIX

Industrial Electronic Control Systems

In **Unit VI**, industrial electronic control systems will be studied. The fundamentals of industrial timing systems are discussed in **Chapter 17.** This chapter examines the basic concepts and applications of timing systems. These applications include delay, integral, repeat, and reset timers.

Optoelectronic control systems are discussed in **Chapter 18**. These systems are used to control automated industrial systems. Light-emitting diodes, photodiodes, phototransistors, and other light-activated devices are studied.

DOI: 10.1201/9788743810056-22

17

Industrial Timing Systems

Objectives

Upon completion of this chapter, you should be able to:

- Identify different types of timing systems.
- Discuss the basic operational principle of each type of timing system.
- List the types of timing systems.
- Describe methods of accomplishing timing.

Important Terms

actuating speed
analog timer
astable multivibrator
bimetal-strip timer
comparator
delay timer
digital timer
electromechanical timer
electronic timer
hydraulic timer
interval timer
isolation
mercury displacement timer
monostable multivibrator
motor-driven cycle timer
motor-driven delay timer
motor-driven interval timer
operational life expectancy
pneumatic timer
pulse stretcher
repeat-cycle timer
repeatability reset timer
RC time constant (τ)
solid-state timer
thermal expansion timer
timer

Introduction

Industrial timing systems are an important part of many automatic process control operations. These systems are designed to provide the necessary

DOI: 10.1201/9788743810056-23

time needed to perform a number of different operations. This ranges from a simple delay to some rather complex event timing that is used to establish sequential steps. With increased emphasis on automatic process control, timing becomes an even more essential part of industrial control.

17.1 Timers

The word timer is more commonly used in industrial applications than the term timing system. **Timers**, in a strict sense, are often complete operating systems in a self-contained unit. They can be energized by an outside source and used to control the operation of a device outside the enclosure. Timers are typically small in physical size and are often permanently attached to the machine they are controlling. Timers can be classified as either separate systems or as the control element of a much larger electrical system.

17.2 Timer Functions

Industrial applications of the timer are so numerous today that it is difficult to place them in classifications. One category places all timers in classifications related to the type of functional control achieved. This includes timers that delay a starting cycle, control intervals, repeat operational cycles, and those that reset themselves after a cycle has been completed. Some timers have the capability of achieving two or more of these functions. Others can be designed to achieve only one specific function. The function of a timer is an important concept.

17.2.1 Delay Timers

In machine tool operations and a variety of applications, **delay timers** are used to alter the starting time of a load device or controlled circuit. When the timer is initially energized, it sees an adjustable time setting value and delays the turn-on operation according to this setting. It eventually closes the power switch, which energizes the system load. Many delay timers are designed to de-energize their own circuitry after the control cycle has been completed.

Figure 17.1 shows typical **delay timers** of the solid-state type. These have adjustable time ranges with potentiometer control. Initially, the unit is adjusted to a specific delay time setting. When electrical energy is applied, the load device or output circuits remain off until the delay time has elapsed. The load then turns on and latches in the on position. The timer remains actuated without consuming a great deal of electrical energy from the source.

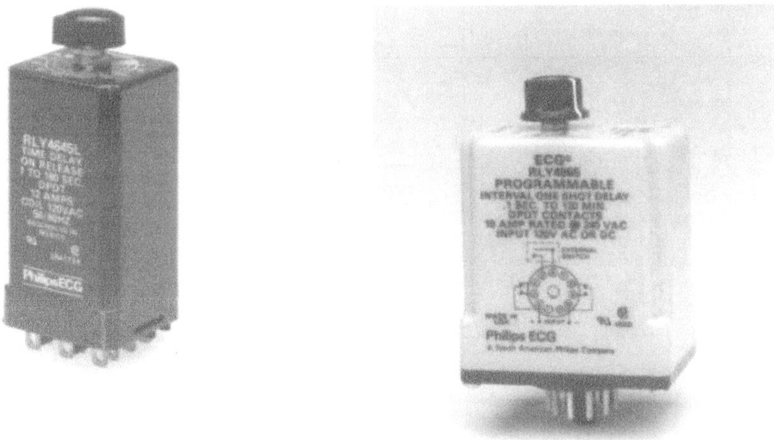

Figure 17.1 Delay timers. (Philips ECG)

17.2.2 Interval timers

Interval timers are used in applications where the load is turned on at the start of an operational cycle and kept on for a specific time period. At the end of this interval, the load is turned off automatically. Industrial applications of this timer are found in heating units, photographic light source control, mixing valves, and automatic machine tool operations. Interval timers can be adjusted to a range of different settings. The type of adjustment needed to change the interval varies with different timer mechanisms. In some applications, a number of different intervals and control possibilities can be achieved by ganging two or more timers together.

Figure 17.2 shows **interval timers**. To operate, the digit dials are adjusted to a desired interval. When energized, the unit counts the interval. When the interval is reached, the load turns off. The same interval can be repeated by actuating the unit again.

17.2.3 Repeat-cycle timers

Repeat-cycle timers are designed to perform their operating cycle continually until the operation is stopped. Timers of this type can be stopped by an automatic overriding program, by a safety interlock, or through manual switch action. Some type of resetting operation must take place after the timer has been stopped. Both manual and automatic resetting devices are available.

A large number of **repeat-cycle timers** used in industrial timer applications are of the motor-driven type. A synchronous AC motor is

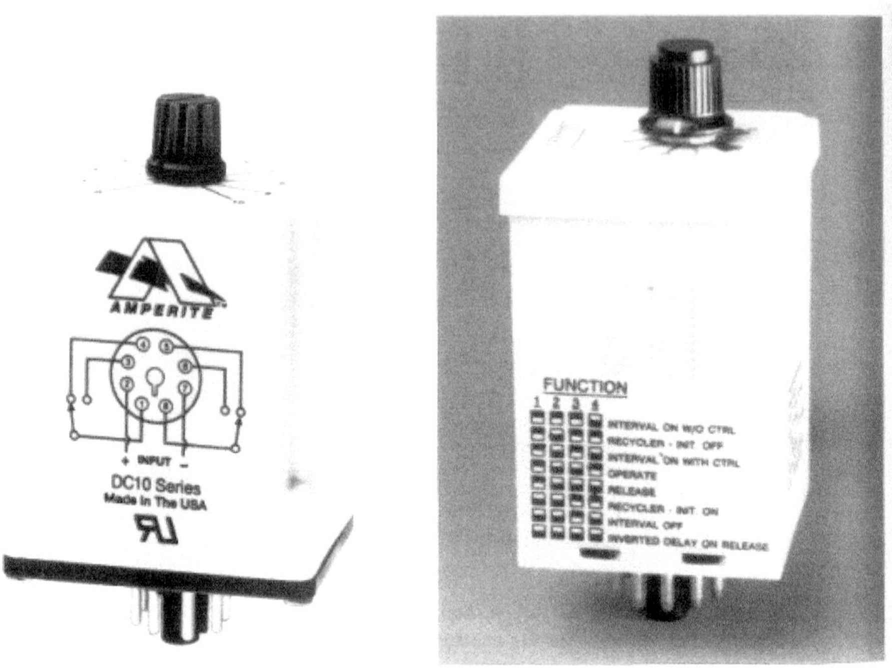

A B

Figure 17.2 Interval timers. (Amperite Co., Inc.; Philips ECG)

commonly used to drive a camshaft or a rotating dial. Contacts are opened or closed according to cam action or by switch-tripping tabs attached to the dial. A specific cam or tab can be used to turn on a switch and an alternate one can be used to turn it off. Several sets of tabs or cams can be attached to the drive mechanism of these timers. Cycle changes can range from 10 seconds to 24 hours by changing the cam notches.

The rotational speed of a **repeat-cycle timer** is determined by the gear ratio of the drive mechanism and the operating speed of the motor. When 60 Hz of AC is used to energize the drive motor, power line accuracy can be achieved for long periods of time. Timers of this type are typically found in electrical energy-saving applications in industry. Some wiring diagrams are shown in Figure 17.3.

17.2.4 Reset timers

Reset timers represent a variation of the cycling timer operation just discussed. This type of timer employs a motor-driven gear assembly that is

Bold Lines Are Internal Wiring

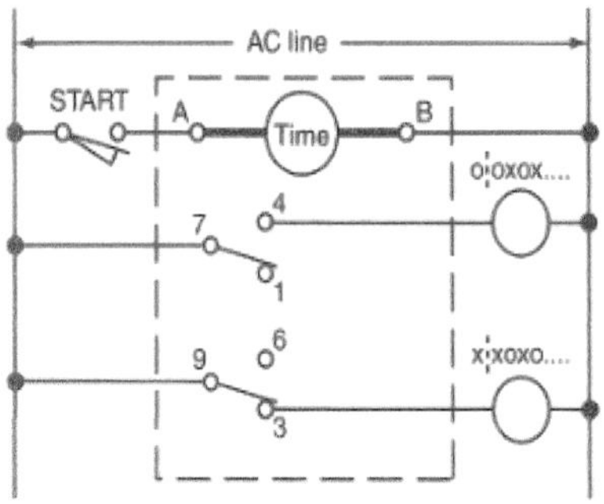

Repeat Cycle Timing, OFF Time First. Outputs connected to the normally open relay contacts will be OFF when the timer is reset, and during the first timing period after timer start. Close control switch to start timing cycle. Open control switch to reset.

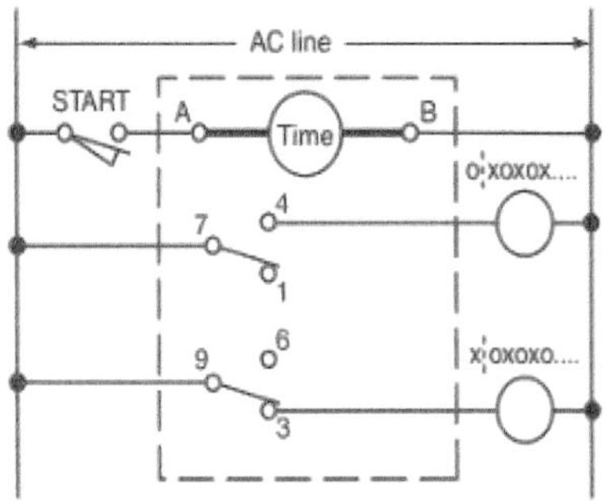

Repeat Cycle Timing, ON Time First. Timer contacts are shown in reset and during the OFF-Time. Contacts transfer during ON-Time. Contacts will transfer as soon as control switch is closed. Timing cycle continues as long as control switch is closed. Timer resets when control switch is opened.

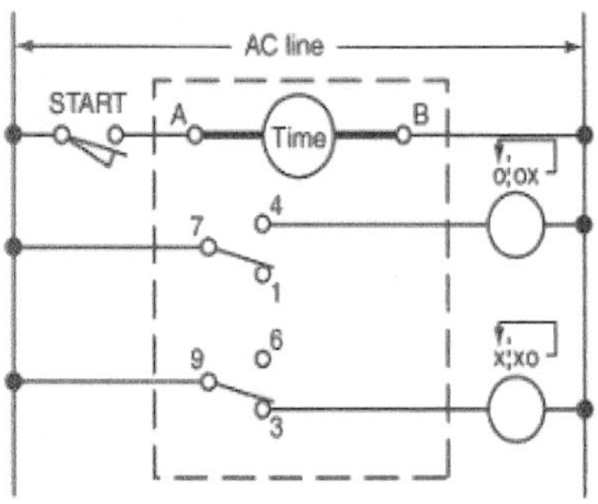

One Cycle Timing. OFF Time First. Closing the control switch starts one OFF and one ON timing period. The timer resets after the ON period has been completed. To initiate another timing cycle, the start switch must be opened and reclosed.

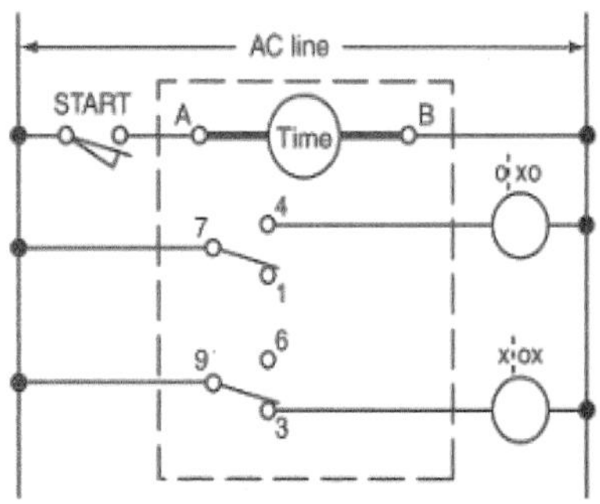

One Cycle Timing. ON Time First. Closing the control switch starts one ON timing period. The timer times the OFF period after the ON period has been completed. To initiate another timing cycle, the start switch must be opened and reclosed.

Figure 17.3 Wiring diagrams for repeat-cycle timers. (Eagle Signal Controls)

actuated by a clutch mechanism. As a result, the cycle starts and goes through an interval. At a certain time, the clutch mechanism is tripped. This action disengages the drive motor from the gear mechanism and permits it to return to the starting position by spring action. Upon its return to the starting point, this timer can turn off, go into an idling condition, or repeat the cycle again automatically. The cycle can be repeated by manually depressing a push button without making a new dial setting. The clutch mechanism is a unique part of this timer.

Figure 17.4 shows **reset timer** wiring diagrams. Timers of this type are designed to energize the load at a preset interval. Interval setting is achieved by turning the control to the desired time interval. The cycle then starts when a push button is depressed. When the time interval is reached, the load is turned off and the pushbutton pointer returns to the starting position. The cycle is repeated again by simply depressing the push button.

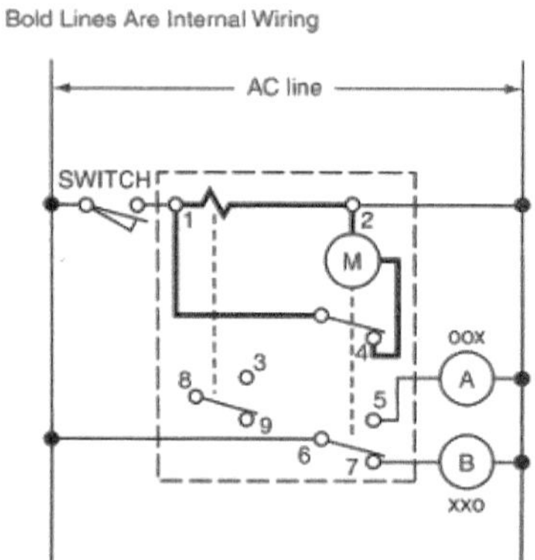

Figure 1

Standard Start Substained Control Switch—Close to start. Open to reset. Simple delayed closing and opening of load circuits.

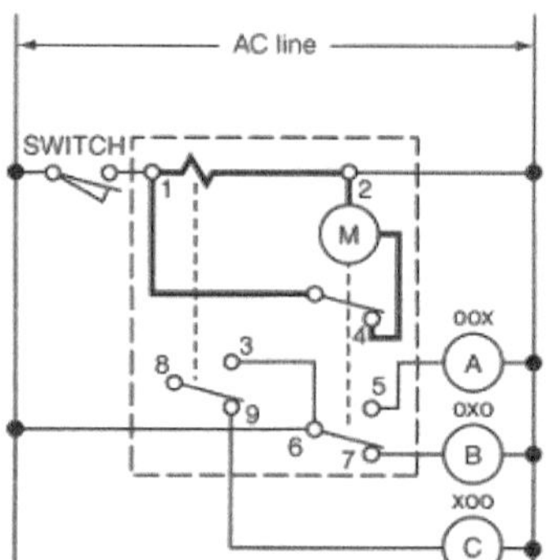

Figure 2

Standard Start—Sustained control switch. Close to start. Open to reset. Various control circuit operations obtained by connecting contacts in series.

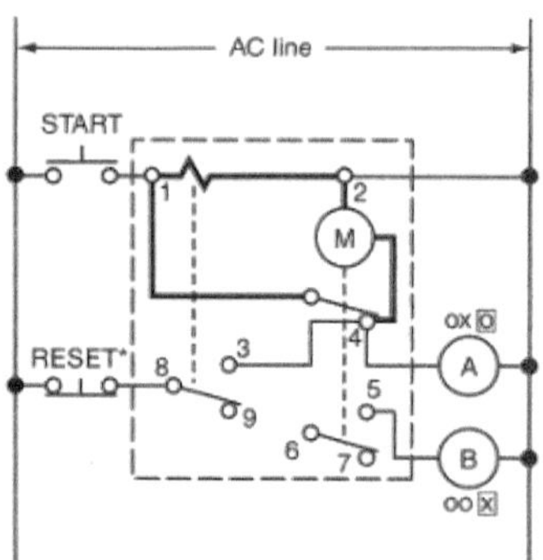

Figure 3

Momentary Start, Automatic Reset—Switch 6-5-7 transfers to timed out condition indicated by X for ¾% of maximum dial setting.

*Optional RESET switch—open to reset timer during timing. If reset switch is not used, connect terminal 8 directly to line.

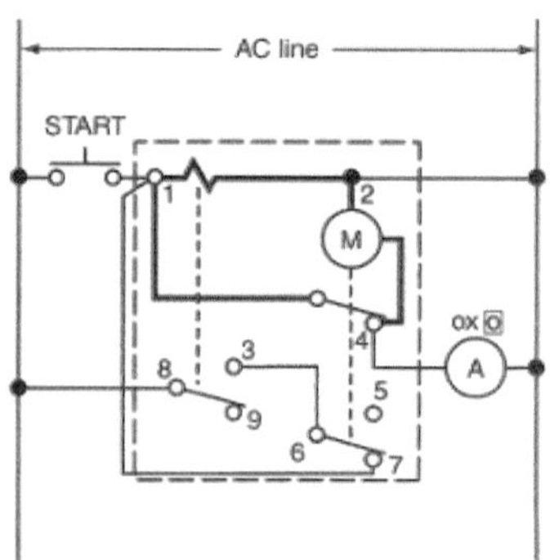

Figure 4

Momentary Start Automatic Reset—With motor shut-off protection. Use this circuit when the start switch may not always be opened before end of timing. This ensures that power is removed from motor at end of timing.

Figure 17.4 Wiring diagrams for reset timers. (Eagle Signal Controls)

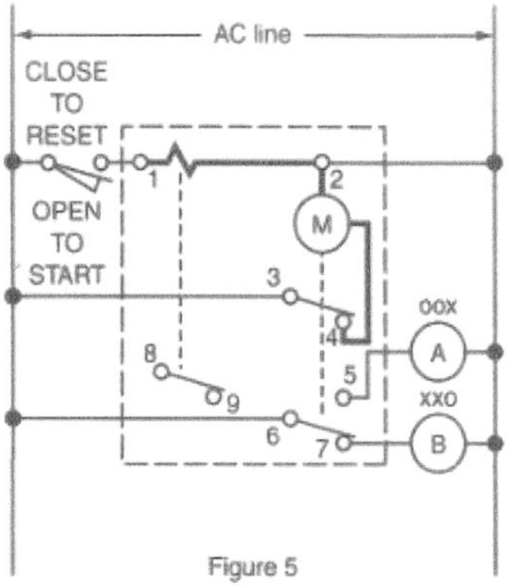

Reverse Start. Close switch to reset open to start timing. Timer remains in timed out condition until switch is closed to reset.

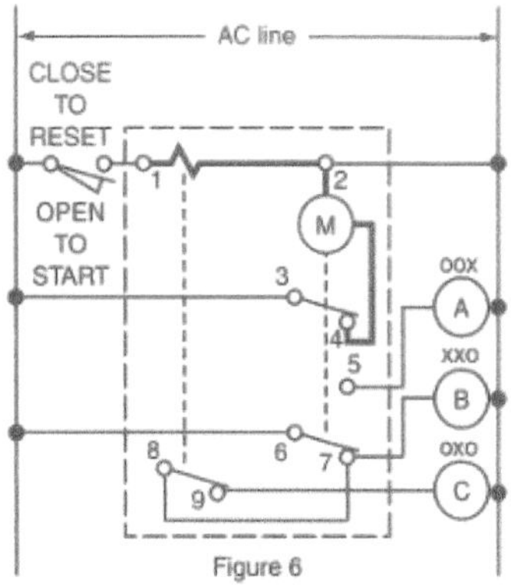

Reverse Start. Close switch to reset. Various control circuit operations obtained be connecting controls in series.

Figure 17.4 Continued.

17.3 Timer Classifications

A broad range of techniques is used to achieve industrial timing. This includes thermal changes, electromechanical action, electrochemical changes, electronic signals, mechanical action, and digital operations. For convenience, these techniques are first placed into groups that respond to either analog or digital changes. **Digital timers** are designed to count pulses generated by an electronic clock circuit. **Analog timers**, by comparison, respond to physical quantities or changes.

Expansion of the analog classification finds the general terms thermal, electromechanical, electrochemical, electronic, and mechanical in common use. Division of these terms leads to some specific kinds of timers. The block diagram of Figure 17.5 shows the relationship of these timer classifications.

17.4 Thermal Timers

Thermal energy has been used in industrial applications for as long as any of the other basic principles of timing. The principle that metal expands when heated and contracts when cooled is used in the thermal timing element. Each type of metal has unique linear expansion characteristics. When dissimilar metal strips are connected together, they tend to bend or flex when heat is applied. As a result of this action, the bending property of a bimetal strip is used to force electrical contacts together. A majority of

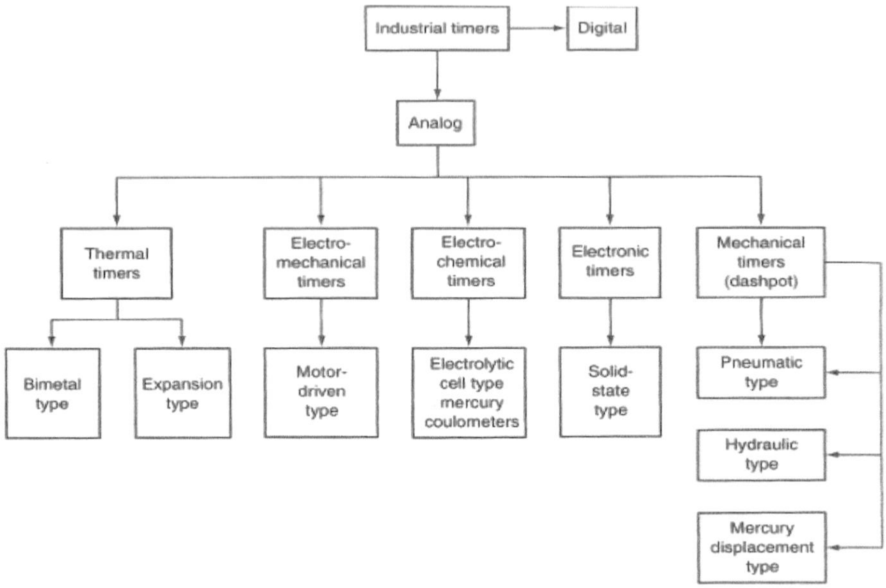

Figure 17.5 Block diagram of timer classifications.

thermal timers respond to an applied heat source and contract naturally by heat reduction.

17.4.1 Bimetal-strip timers

Part A of Figure 17.6 shows a delay type of thermal timer that has been used in industrial applications for a number of years called a **bimetal-strip timer**. The electrical part of the timer, diagrammed in part B of Figure 17.6, employs two self-supporting metal strips with electrical contacts. When electrical energy is applied to the heating element, it causes the bimetal strip to bend or flex. The side of the strip near the heating element expands faster than the contact side. As a result, the bending action causes the contacts to be forced together after a short delay period.

Bimetal-strip timers are designed in a variety of different styles and types. The glass-enclosed type of part A of Figure 17.6 is designed for a fixed delay time. The delay ranges from 2 to 180 seconds (s). Timers of this type must be replaced periodically, which accounts for the plugin-type structure. Thermal delay timers of this type may be used to permit proper warm-up periods for other control devices before full source voltage is applied.

A **dual-bimetal-strip timer** is shown in Figure 17.7. Timers of this type are designed to correct the contact point position that occurs with ambient

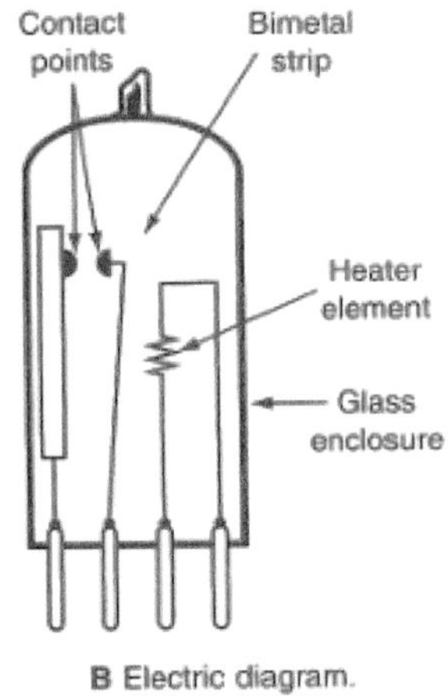

Figure 17.6 Thermal delay timer. (Amperite Co., Inc.)

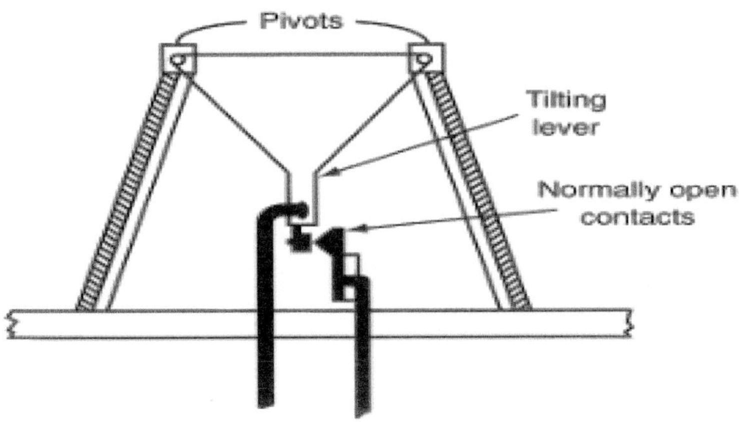

Figure 17.7 Dual bimetal-strip time delay unit.

temperature changes. In this construction, any change in temperature alters both supports equally. As a result, the contacts maintain a constant gap with normal temperature changes. This means that the delay action stays fairly consistent over a wide range of temperature changes. The time accuracy of this device is more consistent than that of the timer shown in part A of Figure 17.6.

The operation of a dual-bimetal-strip timing element is similar to that of the single-element type. Heat is applied to only one strip when the element is energized. After a period, the heated strip bends outward due to its expansion. The alternate strip does not change shape significantly. As a result of this action, the normally open contacts are forced to close, thus completing the delay action.

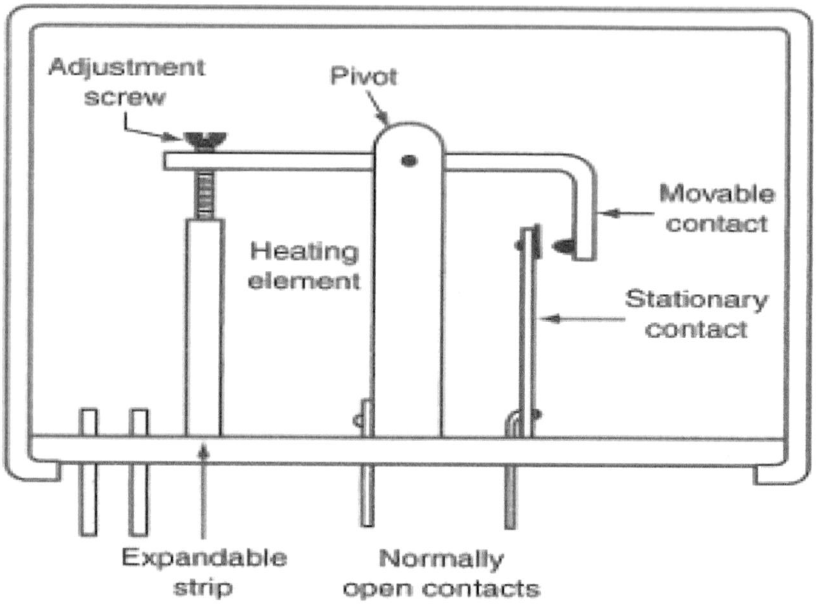

Figure 17.8 Expansion-type timer.

17.4.2 Thermal expansion timers

The **thermal expansion timer** shown in Figure 17.8 is a very common industrial device. Its operating principle is also based on the expansion characteristic of metal. In this device, metal is made to expand instead of bend or warp. The end result of this device is, however, the same as that of the bimetal-strip timer.

An **expansion timer** is energized by an external source. Either an AC or DC source of energy can be applied to the heating element. After a short period, the wide upright piece of metal tends to expand. The amount of expansion that takes place before the contacts are closed is based on the position setting of the tension compression adjustment screw. A variable degree of time delay is therefore permitted with this device. Delays ranging from one to three minutes are possible with the timer of Figure 17.8. In general, delay timers have a rather limited range of delay capabilities.

17.4.3 Thermal timer application

Thermal timers usually suffer from a rather serious problem when they are used independently as control devices. The heating element must be

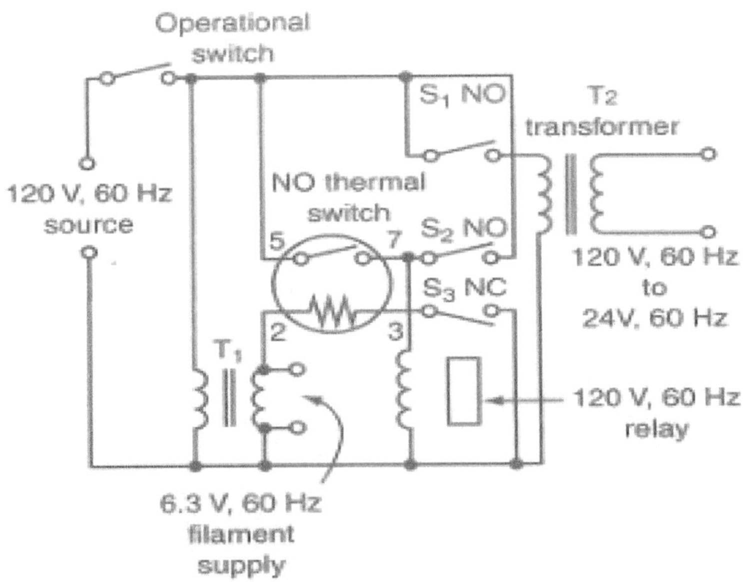

Figure 17.9 Thermal delay timer circuit. (Amperite Co., Inc.)

constantly energized in order to continue contact conduction after the delay action takes place. For long-duration operating cycles after delay, a great deal of electricity is needed to keep this device operational. A great deal of heat is also created. To solve this problem, thermal timers are frequently used in latching relay circuits for applications other than short operational periods.

An example of this can be seen in the **thermal delay-relay timer** circuit shown in Figure 17.9. In this circuit, when the operational switch is turned on, 6.3 Vac is applied to the heating element of the thermal delay unit. After a short delay period, the contacts come together and activate the relay coil. The normally closed (NC) relay contacts break, and the two normally open (NO) contacts make a connection. Note that the NC contacts are placed in series with the heating element. As a result of this action, the heater element source is disconnected. The S_2 NO contacts are now closed, bypassing the contacts of the delay unit. This creates a latching condition that holds the relay coil actuated until the operational switch is turned off. Control energy to the load device is completed through the S_1 NO contacts by the same relay action.

Through the action of a delay-relay circuit, the heater source is removed after the load has been actuated. The operational efficiency of this circuit is improved, and electrical energy waste is reduced to a very low level. As a general rule, the operational life of a thermal delay unit is increased when it is used in a circuit of this type.

17.5 Electromechanical Timers

Electromechanical timers are used for several industrial timer applications. Primarily, the electromechanical timer classification refers to those devices driven by a motor mechanism. Since 60 Hz of AC is readily available at most industrial sites, it typically serves as the energy source for these timers. The drive mechanism is usually actuated by an inexpensive synchronous motor. These low-cost timing devices are commonly used to achieve delay, interval, repeat cycle, and reset functions with power-line frequency accuracy. As a general rule, the life expectancy of this kind of timer is quite good. The operational life of electromechanical timers ranges from 100,000 to 10 million cycle completions.

Practically all **motor-driven timers** conform to the basic system structure outlined in **Chapter 1**. This refers to the energy source, transmission path, control, load device, and indicator. Specifically, the primary energy source is AC electricity to drive a small synchronous motor. Electromagnetic energy developed by the motor is then used mechanically to turn a gear train or transmission. Control of this energy can then be achieved by a gear mechanism, clutch assembly, or spring. The end result of the system sees mechanical energy being applied to a drive mechanism that actuates switches or an electrical relay coil. Calibrated dials with time pointers are also attached to the actuator drive mechanism to achieve the indicator function. Each timer can be classified as a unique system or as a control element for a larger system.

17.5.1 Motor-driven delay timers

Motor-driven delay timers are used in industrial applications where a specific amount of time is needed before a load device is energized. Timers of this type are used to provide warm-up time for some electrical components or to ensure proper sequencing steps in production-line assembly applications.

Part A of Figure 17.10 shows a schematic diagram of a **motor-driven delay timer**. When the operational switch is turned on, power is applied directly to the drive motor through the normally closed (NC) relay contacts. This energizes the motor and causes rotation until it strikes an actuator mechanism that trips the relay switch. Actuation of the relay armature automatically closes the normally open (NO) contacts and opens the NC contacts. Through this same action, the load device is energized electrically. With the NO contacts now closed, the relay and load continue to operate. Through this

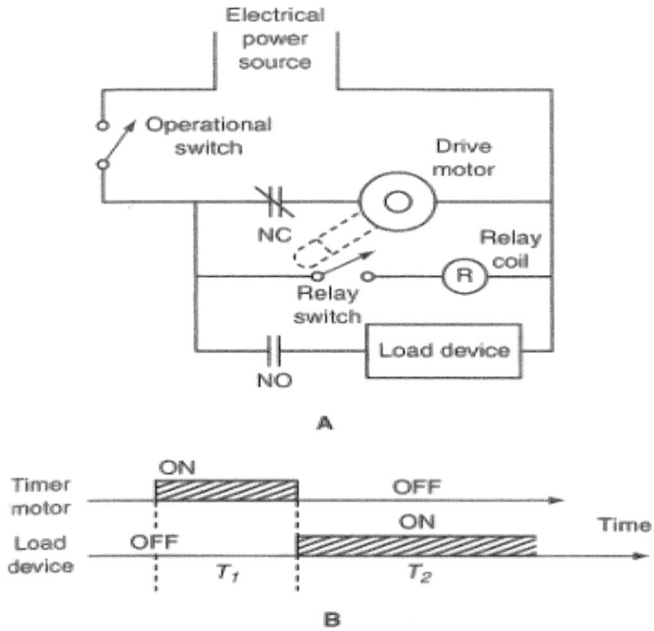

Figure 17.10 Motor-driven delay timer. A – Circuit. B – Timing.

same relay action, the drive motor energy is removed by opening the NC contacts. As a result of this action, the timer remains in this operating state until the electrical power source is turned off by the operational switch.

The bar graph in part B of Figure 17.10 shows the operational sequence of the timer motor and load device. Note that the motor is in operation only during T_1 and off during T_2. The load responds in reverse order to this action.

17.5.2 Motor-driven interval timers

Motor-driven interval timers are commonly used in industrial applications where the load is turned on immediately and left on for a certain interval of time. After this interval, the load is automatically turned off. It then remains in the off state until the timer is reset and started again. Figures 17.11 to 17.14 show some application diagrams of the interval timer.

17.5.3 Push-button interval with stop feature

In Figure 17.11, the start button of the interval timer needs to be depressed to start the operational cycle. When this occurs, the load is energized

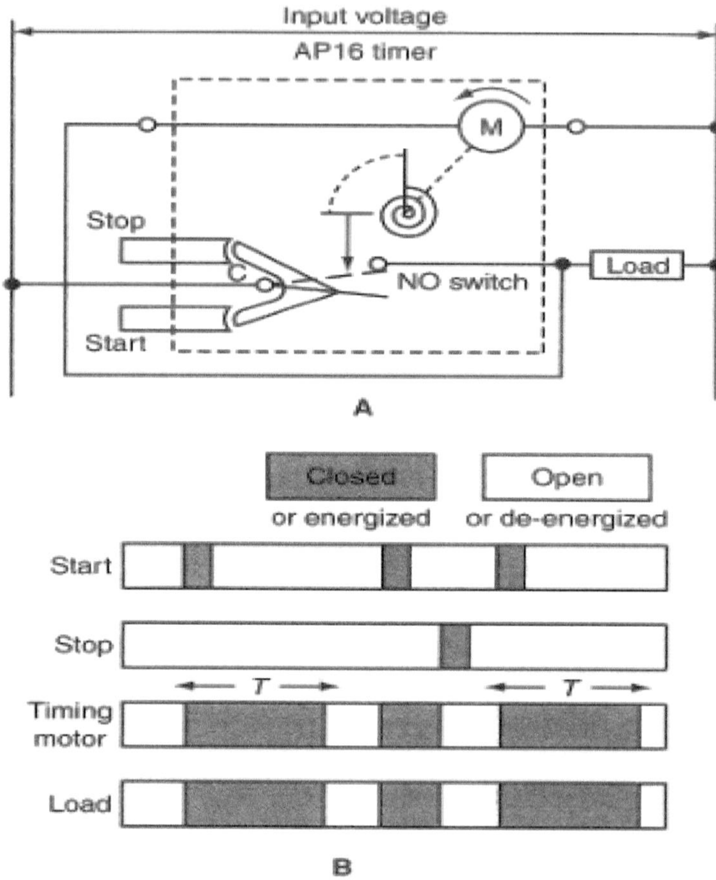

Figure 17.11 Interval timer with push-button stop. A – Circuit. B – Timing.

immediately for a set time period. Depressing the stop button de-energizes the operation before it completes the cycle if the need arises.

When the NO start button is depressed, it completes the energy path from the source to the load and timing motor. With the motor in operation, its drive mechanism turns the switch-tripping arm. After an elapsed time, the switch-tripping arm actuates the timing switch, which turns the motor and load off. This action de-energizes the entire unit. To start the cycle again in the same sequence, the timer must be reset to its starting position.

17.5.4 Push-button interval ON or interval OFF

Figure 17.12 shows an **interval timer** that causes an on or off load control condition to be initiated automatically by push-button action. Depressing the start

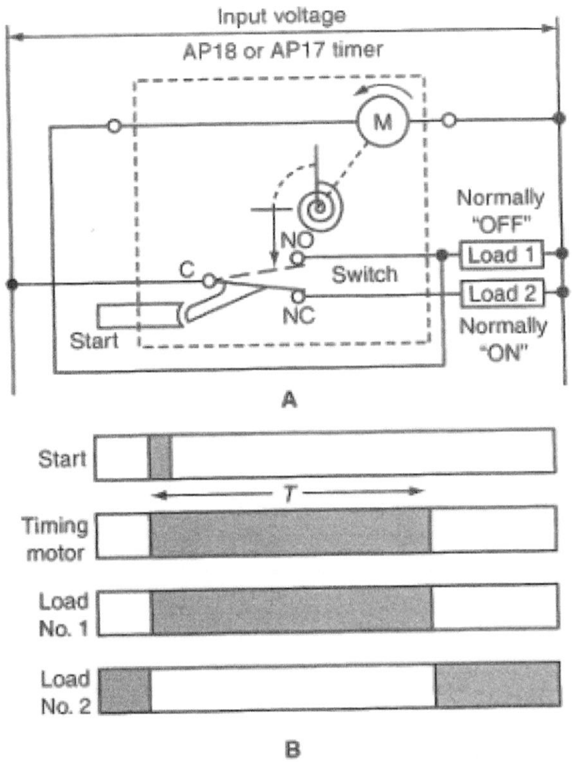

Figure 17.12 Interval timer with push-button on and off interval. A – Circuit. B – Timing.

button actuates the motor and energizes load 1. This same action causes load 2 to change from its normally on state to an off condition. After an elapsed time interval, the switch-tripping arm changes the SPDT switch back to its original starting position before actuation. Load 2 is, therefore, on again, and load 1 is off. The timing motor must be reset to ready itself for the next operational cycle.

17.5.5 Limiting control with memory

The primary function of the timer of Figure 17.13 is to limit the time that a load is energized. Exceeding a prescribed time setting causes this device to interrupt both the load and timer motor. Through this action, the timer cannot be recycled until the reset button or memory switch is pressed.

When this timer is in the reset state, the load and timer motor are both turned off as indicated. Depressing the control push button starts the operation cycle by turning on both the load and the motor. Holding the push button closed for a time longer than the set time interval causes the switch-tripping

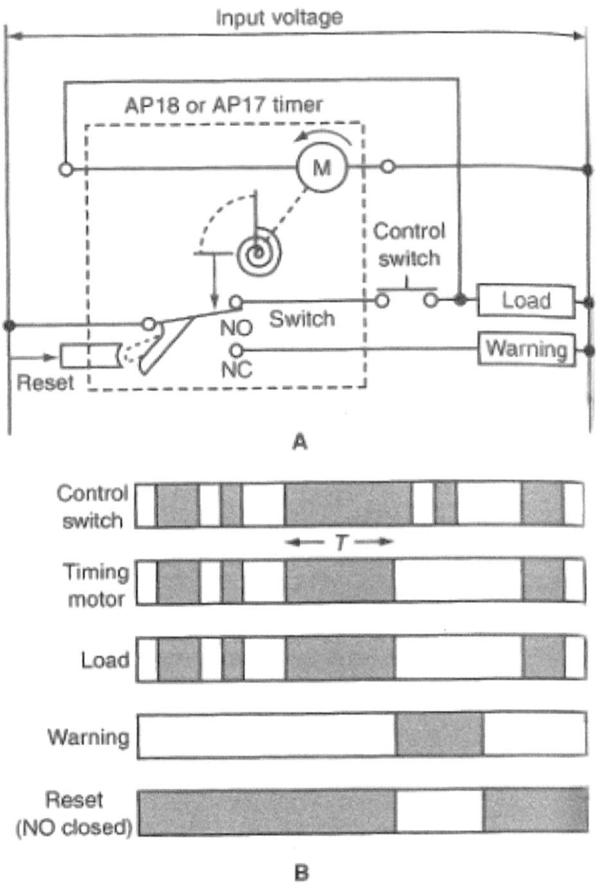

Figure 17.13 Limiting control interval timer with (reset) memory. A – Circuit. B – Timing.

arm to change the SPDT reset switch. This de-energizes the load and timer and actuates the warning light. When this occurs, the timing motor will not reset upon release of the control switch until it is reset manually. Holding the control push button in for a time less than the prescribed time interval will not actuate the tripping switch. As a result, the motor resets to its original starting point when the push button is released. Through this action, the timer cannot be turned on again until its timing cycle is reset.

17.5.6 Hand-set interval timer

With a hand-set interval timer, Figure 17.14, the load and motor are energized immediately when the timing dial is manually turned to a set-time position.

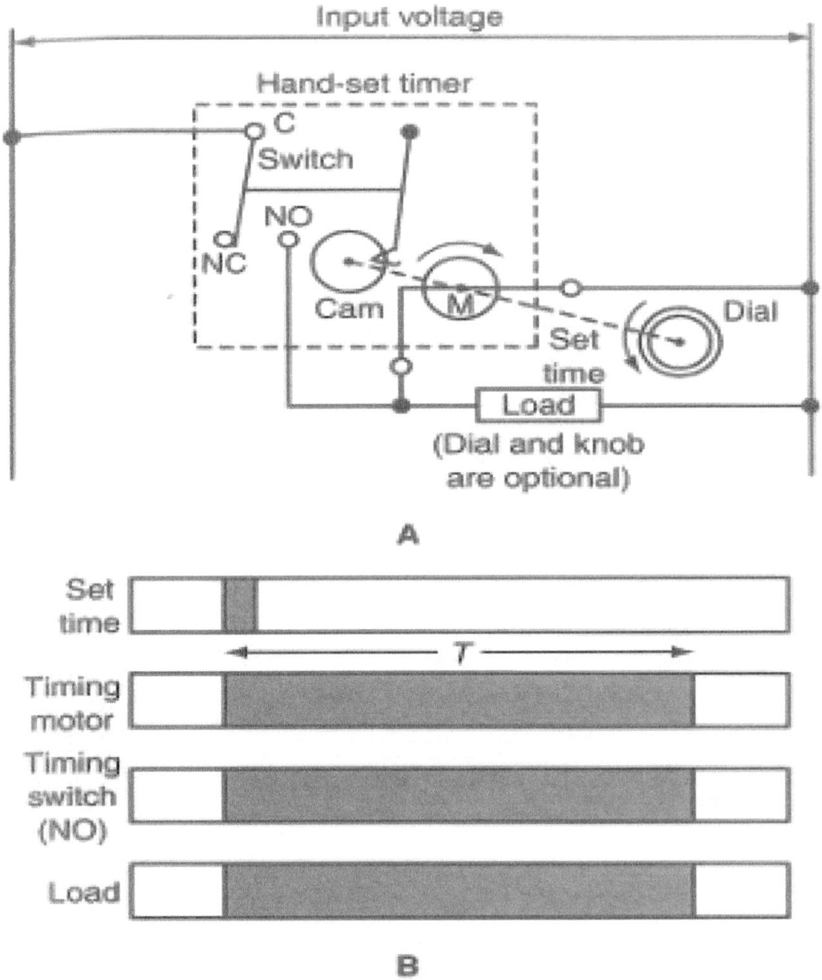

Figure 17.14 Hand-set interval timer. A – Circuit. B – Timing.

The motor operates for this set time period until its timing switch drops into a notch on the cam drive wheel. When this occurs, the SPDT switch turns the motor and load off. To initiate another operating cycle, the calibrated dial is adjusted to a new time interval manually.

17.5.7 Motor-driven cycle timers

Motor-driven cycle timers play a very important role in sequence control operations in industry today. Primarily, this type of timer is used to control

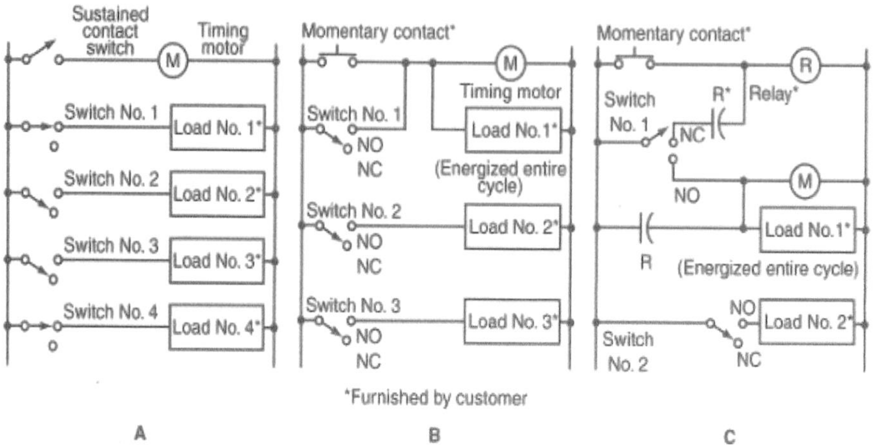

Figure 17.15 Typical circuit applications of motor-driven on-off cycle timers. A – Four-load circuit. B – Three-load circuit. C – Two-load circuit.

one or more different on and off operations on a continual basis. Figure 17.15 shows three different circuit possibilities of multiple-load control applications. Each switch of this timer is controlled by a separate cam that can be adjusted to any position according to the needs of the system. Special switch-actuating arms that ride on the cam of a motor-driven timer are used to trigger the switch when a physical change in cam diameter occurs. The screws on the cams are tightened in place when the cam is adjusted to a desired position.

17.6 Electronic Timers

A wide range of electronic devices has been used through the years as industrial timing devices. Initially, vacuum tubes served as trigger devices for relays. In some low-current load control applications, direct control was achieved by the device without a relay. High-current load applications often included vacuum-tube triggering of gaseous-tube thyratrons.

Solid-state devices were then added to the list of electronic timers. Now, many different timing applications are achieved by these devices. Bipolar transistors, FETs, and UJTs are a part of this classification. These devices can be used to drive a relay or, by triggering an SCR or a triac, they can achieve direct control of the load device.

Solid-state timers offer a number of advantages over the earlier vacuum- and gaseous-tube units. A solid-state timer, for example, consumes very little energy. Vacuum tubes and gaseous devices, in contrast, usually required heater voltage to be made operational. Solid-state timing can be achieved without moving parts. This improves device reliability and operational life expectancy. Solid-state timing can have electrical isolation between the input and output. This reduces safety hazards and removes any potential breakdown problems. In general, solid-state timing is physically smaller, less fragile, and has a substantially longer operational life span than **other types of timers**.

The most recent trend in electronic timer design is to include integrated circuits in the construction of timers. Both linear amplifiers and digital counting timers are included in this category. A number of advantages, such as long time delay and accuracy, are the result of the addition of ICs to the timer family.

17.6.1 Electronic timer classifications

The term **electronic timer** is a rather broad classification used today to describe a number of devices employing electronic components. To make this classification a little more workable, we have divided these timers into two major groups according to the type of timing element employed. These are resistor-capacitor or RC elements and digital timers. The RC element produces a time delay by the charge or discharge action of a capacitor through a resistor. Digital timers, by comparison, employ oscillators or pulse generators that switch on and off at a precise RC rate. These pulses are ultimately counted and used to trigger the load device into operation. The end result is basically the same for both units. Timing accuracy, repeat accuracy, and setability are the major differences between these two major groups.

Further classification of the **electronic timer** is directed toward the triggering element. This component detects changes produced by the timing element and processes them for application to the load control element. The triggering element can be a bipolar transistor, FET, UJT, or operational amplifier. In general, this component amplifies or changes the signal from the timer element in such a way that the load control element can be actuated by it.

The final division of **electronic timer** classification refers to the load control switching element. In this regard, timers achieve either direct switching control of the load device or actuate it by relay action. In both cases, the end result is the same. There are, however, some specific differences in the two methods of control. These differences include such things as repeatability time, actuating speed, isolation between input and load, and operational life expectancy.

Repeatability refers to the rate or number of times that the load can be switched on and off in a given period. Direct device switching has an excellent repeatability factor, while relay switching is inherently very slow. Actuating speed refers to the time it takes for the load to be energized after the control signal has been applied. Repeatability and actuating speed are primarily influenced by the same factors.

Isolation is an expression of the breakdown voltage that separates the input and load. Relay isolation can be very high because of the physical distance separating the actuating coil and the load. Direct switching often has a lower isolation breakdown factor. This has been improved rather significantly, however, with the inclusion of optoelectronic isolation devices. The operational life expectancy of the switching element is certainly a distinguishing characteristic. Relay switching is generally limited by moving contacts, pivot points, and spring tension. Direct actuation does not involve any moving parts. This is obviously more desirable than the moving parts of a relay. The functional application of the timer usually dictates these features when it is selected to do a specific job.

Figure 17.16 shows a block diagram of the three basic elements of an **electronic timer**. Some of the possible components used by each element are represented by schematic symbols.

17.6.2 RC timing

For a number of years, the timing elements of industrial timers have been of the **resistor-capacitor** or **RC** type. In this component network, when a DC voltage is applied, the capacitor gradually builds up a charge potential or voltage. In general, this is described as the *RC* time constant (τ). The formula $\tau = R \times C$ refers to the time in seconds that it takes a capacitor to build up to 63% of the full charge potential of the applied source.

In electronic **RC timing** elements, the voltage buildup of a capacitor is used to trigger an active device into conduction. Transistors, UJTs, and ICs are commonly used to sense this voltage level change. Ultimately, the sensing device triggers the control element, which actuates the load device.

An **RC timing** element has two major factors that can be used in timing applications. These factors are the charge and discharge of the capacitor voltage. If the charging condition is used as the timing factor, it is usually limited to one time constant. The **universal RC time curve** is fairly linear during this **charging time**. Figure 17.17 shows a universal RC time constant curve. Notice the nearly linear characteristic from 0 to 1 along the horizontal axis. From this point on, the curve is nonlinear due to its exponential

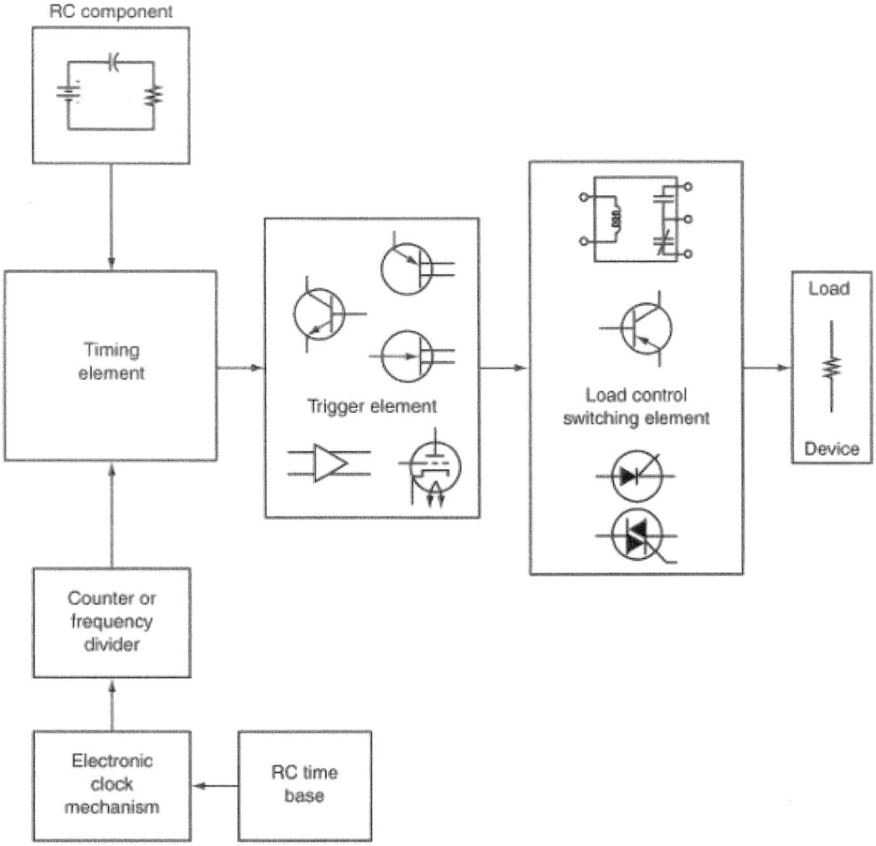

Figure 17.16 Block diagram of the basic elements of an electronic timer.

characteristic. RC timing applications are generally restricted to the first time constant because the values can be accurately predicted.

The **discharge action** of the RC network must also be taken into account in a timer element. In one RC time constant, the capacitor discharges 63% of its full charge potential. This means that the charge voltage drops down to 37% of its charge in one time constant. This part of a universal *RC* time curve is also linear. Time element accuracy is primarily based on operation on the linear part of the RC curve.

A large number of **electronic timers** are designed to respond to the discharge condition of an RC network. The capacitor must therefore charge very quickly and discharge gradually. Large fixed capacitors and variable resistors are typically employed to make timing elements adjustable. Ranges

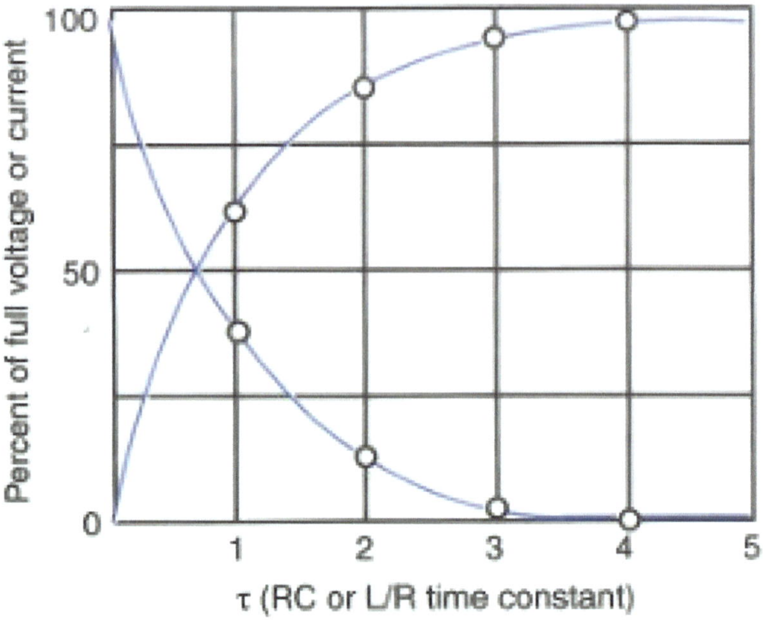

Figure 17.17 Universal RC time constant curve.

from a few milliseconds to approximately two hours are possible today. A large number of all industrial timer applications operate in a range of approximately 10 minutes or less.

If the **discharge action** of a capacitor is used to determine timing action, then the charging cycle is also very important. In this case, the charging action determines the reset interval. This is the smallest time interval that is needed to repeat a cycle after its completion. Ranges from 1 to 30 milliseconds are typical in industrial timers. As a rule, long delay times tend to lengthen the reset time period.

A **solid-state timer** uses capacitor discharge time to achieve delay of energization at the start of its cycle. An adjustable time delay control can be used to alter the value of a resistor in the RC network. The accuracy of this timer is restricted to operation on the linear part of the universal RC time curve. Beyond this point, nonlinearity occurs very quickly and the timing action loses its accuracy.

Another method to accomplish **solid-state timing** is to extend the linear portion of the **RC time constant** well into the normal nonlinear area. This is achieved in a number of ways. Switching devices can be used to turn the source voltage on and off for short periods. As a result, the charge or

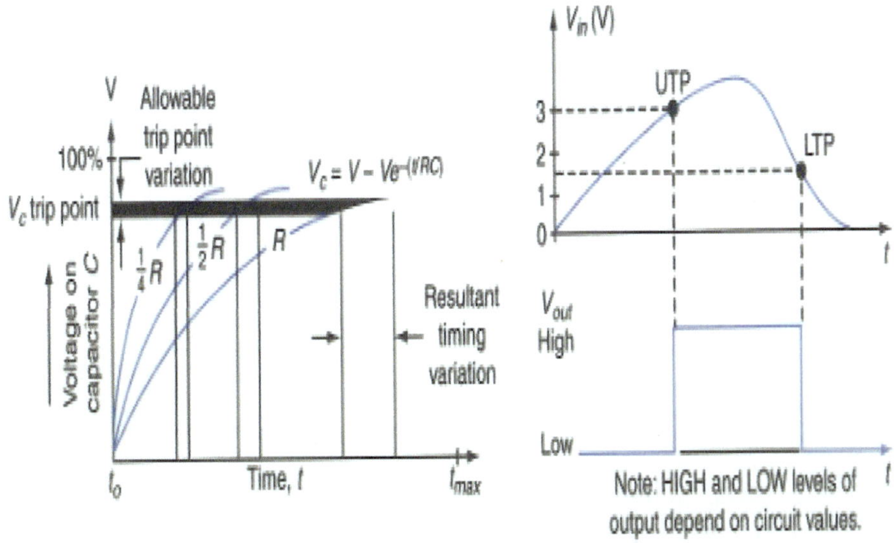

Figure 17.18 Extended RC curve using solid-state timers.

discharge action is delayed beyond its normal time period. In addition, it can be extended by distorting in a direction that offsets the normal distortion. Through these methods and others, it is possible to extend the RC curve to achieve 30 minutes or more of delay very accurately. Standard component values can be readily employed to achieve this circuitry. This reduces production cost for longer-time-delay units. Figure 17.18 shows an extended **RC curve** used in some solid-state timers.

17.6.3 555 timing systems

Special **integrated circuit chips** are available that have the capability of producing a variety of timing functions. This development opened the way for a number of new and unusual approaches to electronic timing. Through this device, timing operations can now be achieved with a minimum of components. The 555 IC series was one of the first developments in this area. It was first introduced by Signetics Corporation. Several variations of the basic timer are now available through other manufacturers. Figure 17.19 shows a functional block diagram of the timer chip.

The operation of an **SE/NE 555 timer** is directly dependent on its internal functions. The three resistors, for example, serve as an internal voltage divider for the source voltage. One-third of the source appears

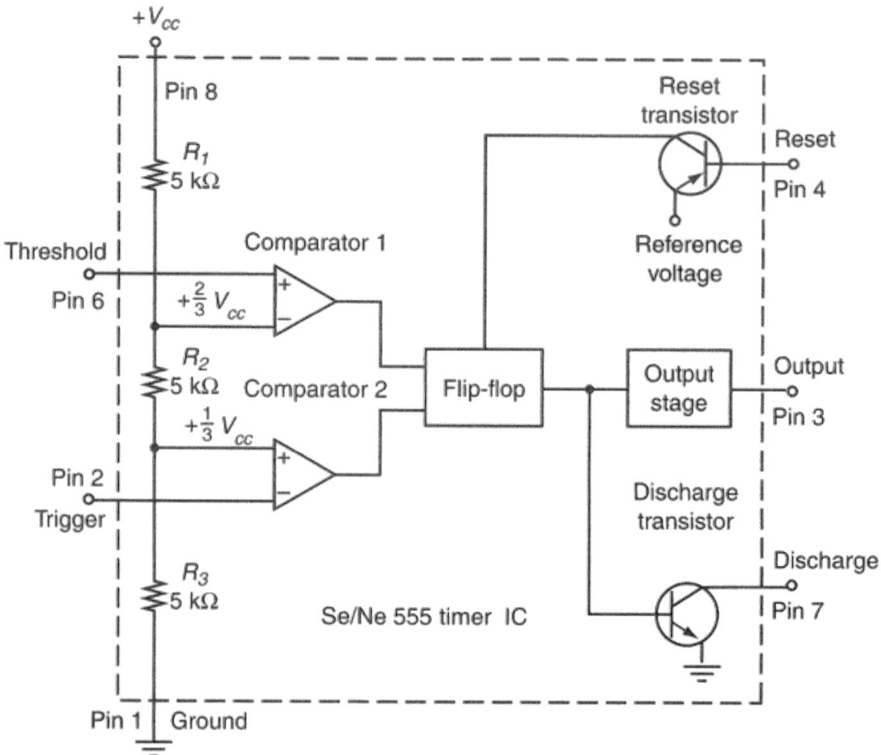

Figure 17.19 Functional block diagram of an SE/NE 555 IC.

across each resistor. Connections made at the 1/3 and $1/3V_{CC}$ points serve as reference voltages for the two comparators. A **comparator** is an op-amp that changes states when one of its inputs exceeds the reference voltage. Comparator 2 is referenced at $+\frac{1}{3}V_{CC}$. If a trigger voltage applied to the negative input of this comparator drops below $+1/3V_{CC}$ it causes a state change. Comparator 1 is referenced at $+3/V_{CC}$ If the voltage at the threshold terminal exceeds this reference voltage, the comparator goes through a state change. Note that the output of each comparator is connected to an input of the flip-flop.

The **flip-flop** of the SE/NE 555 timer is a **bistable multivibrator**. It changes states according to the voltage value of its input. If the voltage value of the threshold terminal rises in excess of $+1/3V_{CC}$, it causes the flip-flop to change its state. A decrease in trigger voltage below $+1/3V_{CC}$ also causes comparator 2 to change its state. This means that the output of the flip-flop is controlled by the voltage values of the two comparators. A state change occurs when the trigger input drops below $+1/3V_{CC}$ or when the threshold rises above $+1/3V_{CC}$.

Note that the **flip-flop output** of the SE/NE 555 is used to drive the discharge transistor and the output stage. A high or positive flip-flop output is used to turn on both the discharge transistor and the output stage. This condition causes the discharge transistor to be conductive and causes it to respond as a low-resistance short to ground. The output stage responds in the same manner. A reversal of this action takes place when the flip-flop output changes to a low or zero state. The discharge transistor then responds as an open circuit or becomes infinite. The output state swings to its high or positive V_{CC} state. In effect, the operational state of the output and discharge transistor is based on the voltage applied to the trigger or threshold input terminals.

17.6.4 555 timer operation

The **555 timer** series has two fundamental modes or methods of operation. Practically all timing applications are in some way a variation of these operational modes. One very important application deals with the generation of digital time-base waveforms. Used in this manner, the 555 responds as an astable multivibrator. Only two resistors and one capacitor are needed to make an astable multivibrator. The other mode of operation is as a monostable multivibrator. In its simplest form, a monostable multivibrator can be made with one resistor and one capacitor. Monostable operation permits the 555 IC to be used in delay and interval applications.

17.6.5 Astable multivibrator mode of operation

Figure 17.20 shows an **SE/NE 555 IC timer** connected as an astable multivibrator. In this time base generator, pins 8 and 4 are connected to the DC energy source of somewhere between 5 to 15 volts DC. When pins 2 and 6 are connected, the 555 triggers itself and operates as a free-running multivibrator. The external capacitor (*C*) charges through both resistors R_A and R_B and discharges through R_B and the internal parts of the IC. The duty cycle, which is total cycle time divided by the on time, is precisely set by the ratio of these two resistors.

In actual circuit operation, the capacitor charges from one-third to two-thirds of the V_{CC} voltage. It then discharges through R_B and the IC to a value $1/3V_{CC}$. As shown by the waveform graph of part B of Figure 17.20, as C starts to charge, the output goes to a high value. It remains at this level for

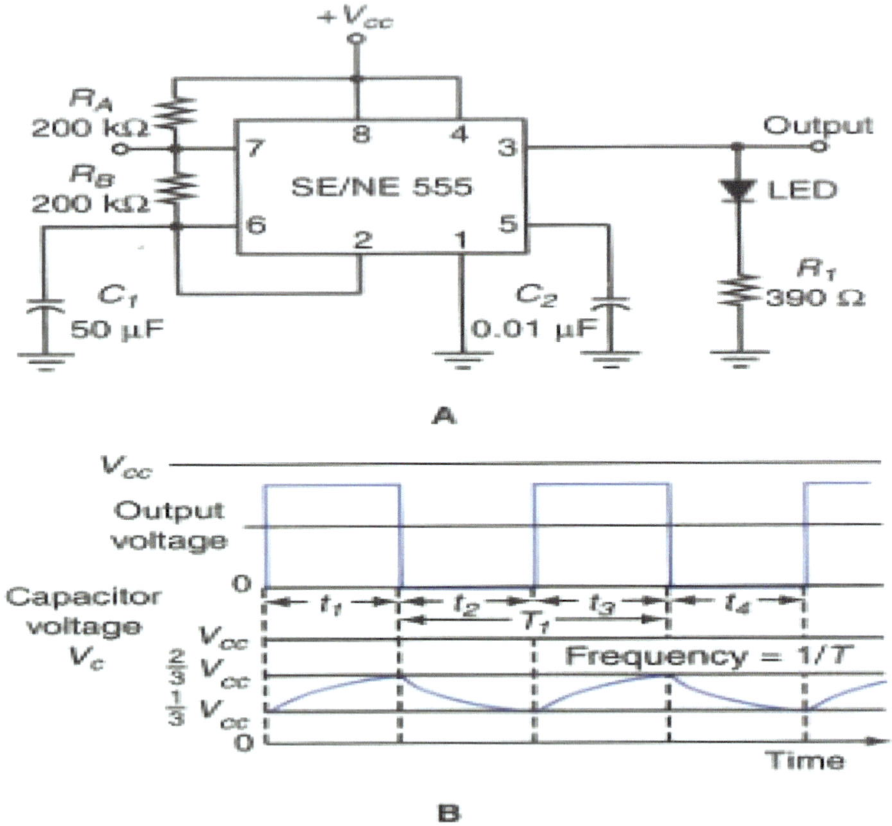

Figure 17.20 Astable multivibrator using an SE/NE 555. A – Circuit. B – Waveform.

the time period t_1. In seconds, this represents 0.693 of $(R_A + R_B) \times C$. When C begins to discharge through R_B, the voltage drops to one-third of the value of V_{CC} This is represented by the time that the output drops to a low value, or t_2. In seconds, t_2 is $0.693R_BC$.

The **output frequency** of an **astable multivibrator** is represented by the total time required to charge and discharge C. The combined period of $t_1 + t_2$, $t_2 + t_3$, or $t_3 + t_4$ is represented by a capital letter such as T. The frequency of a total operational cycle is, therefore, a function of time. In practice, the formula $f = 1/T$ shows this relationship. For a multivibrator, the following formula is used:

$$\text{Frequency in hertz} = \frac{1.44}{(R_A + R_B) \times C}$$

The resistance ratio of R_A and R_B of an IC multivibrator is quite critical. If R_B is more than half the value of R_A the circuit will not operate. Essentially, this would not permit pin 2 to drop in value from $2/3V_{CC}$ to $1/3V_{CC}$. This would not allow the IC to retrigger itself, which prepares it for the next time period. IC manufacturers usually supply design data information of this type that can be used to select proper resistor-capacitor ratios for a desired **operating frequency**.

17.6.6 Monostable multivibrator mode of operation

Not all timing applications require a continuous repetitive wave, such as that developed by the free-running or **astable multivibrator**. Many timing situations call for operations that last for a specified length of time. The monostable multivibrator permits the 555 timer series to be used for these applications. Part A of Figure 17.21 shows the 555 IC connected in a monostable circuit configuration. Note that only one resistor (R_A) and one capacitor (C) are needed in a monostable circuit.

A **monostable multivibrator** is primarily designed to generate a single output pulse in response to an input trigger pulse. The duration of the output pulse depends entirely on the time constant of C and R_A. When a negative-going pulse is applied to the trigger input (pin 2), it causes the output (pin 3) to go high or to $+V_{CC}$ The trigger pulse causes comparator 2 to drop below its referenced value of $+1/3V_{CC}$. This action causes the flip-flop to go to its low state. A negative voltage to the discharge transistor causes it to become infinite. This removes the short to ground for capacitor C. The voltage across C begins to rise in value according to the time constant of R_A and C. When the voltage of C exceeds $+2/3V_{CC}$, it causes comparator 1 to change states. This, in turn, causes the discharge transistor to again become conductive. C then discharges very quickly to ground through pin 7. The output stage follows this action and drops to its low or ground state. In effect, the output stage follows the change in the trigger input level. Part B of Figure 17.21 compares the trigger input and output waveforms.

The duration of the trigger pulse of a **multivibrator** can be either longer or shorter than the generated output pulse. The resulting width of the output pulse is based on the values of R_A and C. The time (T) of the output is based on the formula:

$$\text{Time (high)} = 1.1 \times R_A \times C$$

This means that the **time duration** of the output can be altered according to the application. A short-duration input pulse can therefore be used to produce

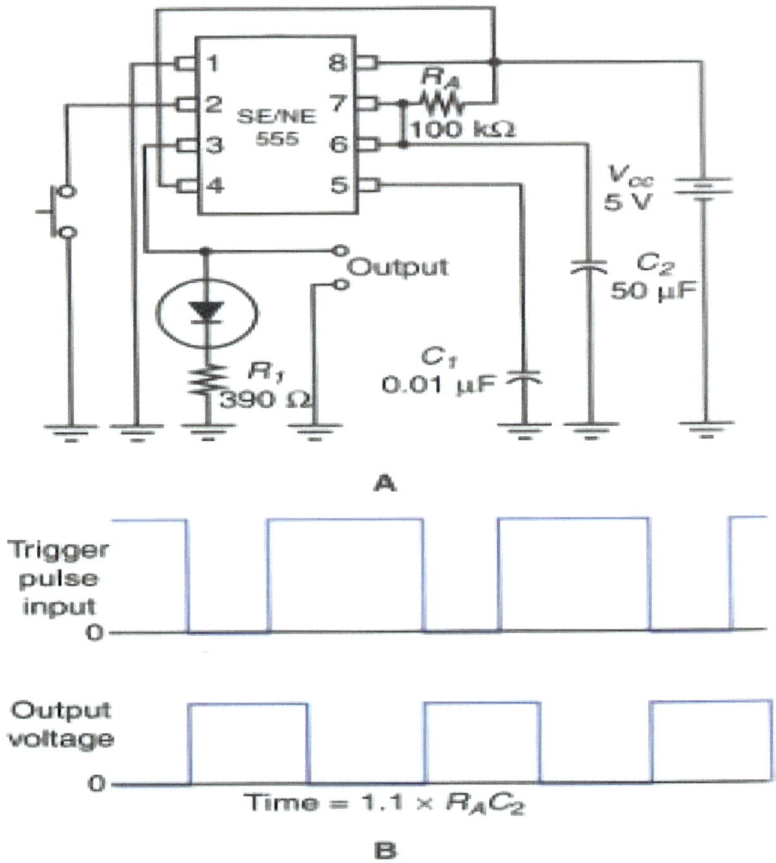

Figure 17.21 Monostable multivibrator operational mode. A – Monostable multivibrator operation. B – Trigger-output waveforms.

a rather long duration output pulse. Because of this, monostable multivibrators of this type are often called pulse stretchers. The monostable mode of operation is also used to produce delay and interval timing operations. Note that the starting of a monostable multivibrator is based on the application of a trigger pulse.

17.7 Electrochemical Timers

An unusual alternative for industrial timing applications is achieved by **electrochemical action**. This method of timing uses specially designed electrochemical devices called electrolytic timing cells and mercury

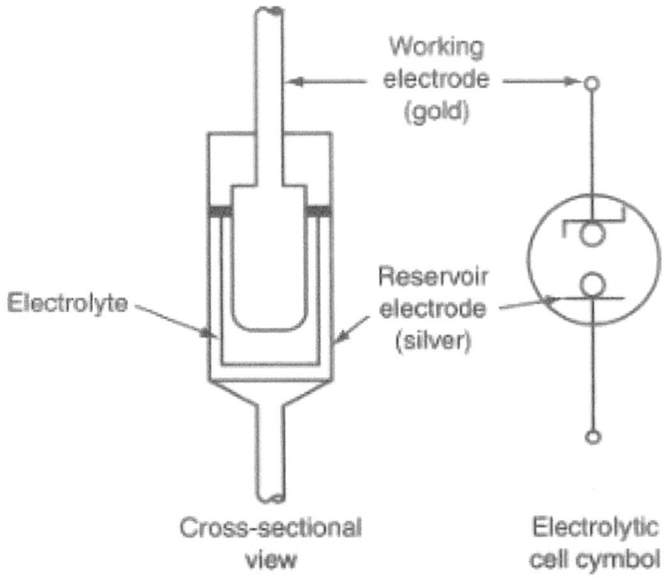

Figure 17.22 Electrolytic cell for timing.

coulometers. In these devices, a form of electroplating takes place that is based on the amount of current passing through the cell. During the plating process, ions and electrons serve as the current carriers. The resistance of the cell and the voltage developed across it are quite small during this operational time. As the cell nears completion of the plating process, its internal resistance and voltage rise rather abruptly. This change in voltage can then be applied to a trigger element, and in some cases, it can be applied directly to the load switching element.

Figure 17.22 shows an **electrolytic timing cell**, a cross-sectional view of its interior, and the electrical symbol. Typically, the working electrode of this cell is made of gold, and the reservoir electrode holding the electrolyte solution is silver. The electrolyte solution transfers current carriers between the two electrodes.

17.8 Mechanical Timers

Many **industrial timers** use mechanical energy as a means to achieve different timing operations. Devices in this category are often called dashpot timers because of the way they operate. The action of these timers is based on the amount of time that it takes a fluid or air to pass through a small opening or orifice. The physical size of the orifice can either be fixed or adjustable according to the desired design. Through this type of construction, many types of industrial applications are possible.

Three types of **dashpot timers** are used. These types are pneumatic, hydraulic, and mercury displacement timers. These classifications are based primarily on the material being processed by the timer.

17.8.1 Pneumatic timers

The operational cycle of a **pneumatic timer** is based on the amount of time it takes for air to be forced through a small hole in an enclosed container. Timers of this type are used to achieve delay or interval timing operations. These timers often have an adjustable time setting capability.

17.8.2 Hydraulic timers

Hydraulic timers are similar in appearance and operation to the pneumatic timer. Hydraulic fluid is used in a dashpot unit to achieve the desired delay action. These units retain fluid in an enclosed area where it is used each time the device is actuated. Silicone fluid is typically used because of its wide range of viscosity. Timers of this type are very reliable and have long life expectancies.

17.8.3 Mercury displacement timers

Mercury displacement timers are an important type of mechanical timer. They depend on the displacement of a pool of mercury that makes a connection between two electrodes. The electrical contacts are hermetically sealed, making them ideal for high-current switching in hazardous environments.

Review Questions

Answer the following questions.

1. How are timing systems used with industrial machine control?

2. Describe the following types of timing operations.

   a. Delay.

   b. Interval.

   c. Repeat cycle.

   d. Reset.

3. How does an analog timer differ from a digital timer?

4. Briefly describe the operation of the following timers.

   a. Bimetal-strip.

   b. Thermal expansion.

5. List three types of electromechanical timers.

6. What is a solid-state timer?

7. Define the following factors.

   a. Repeatability.

   b. Actuating speed.

   c. Isolation.

   d. Operational life expectancy.

8. Briefly describe the operation of an RC timer.

9. Describe the astable and monostable modes of operation of a 555 timing system.

10. List three types of mechanical timers.

18

Optoelectronic Control

Objectives

Upon completion of this chapter, you should be able to:

- Explain the difference between interruptible and non-interruptible optoelectronic systems.
- Discuss the different radiation sources used with optoelectronic devices.
- List the differences between through-beam detection, reflex detection, and proximity detection.
- Explain the important factors in optoelectronics system design.
- List the three different types of detector systems and discuss their differences.
- List and describe nine uses for optoelectronic control systems.
- Identify different types of optoelectronic systems.
- Discuss the basic operational principle of at least six types of optoelectronic systems.
- List at least six types of optoelectronic systems.
- Describe methods of accomplishing industrial control using optoelectronic systems.

Important Terms

candela (cd)
convergence
divergence
dynode
heterochromatic
infrared
detector
interruptible system
lens
light-activated silicon-controlled rectifier (LASCR)
light-dependent resistor (LDR)

DOI: 10.1201/9788743810056-24

lumen (lm)
lux (lx)
mirror
modulated
monochromatic
non-interruptible system
optical filter
panchromatic
photo-field-effect transistor (photo-FET or PFET)
photoconductive device
photodiode
photoelectric emission
photo emissive device
photometer
photometric
photo resistive device
phototransistor
photovoltaic device
prism
proximity detection
radiometric
rectilinear propagation
reflection factor
reflex detection
retroreflective
solar cell
through-beam detection
window

Introduction

Optoelectronic systems are designed to respond to light signals or to transfer information from one point to another by light energy. These systems are unique in that they employ a number of rather specialized light transducers. There must be a source of light energy and a device that will detect light. Changes in light are then used to actuate the final control element of the system. On and off control and partial variations in the load can be achieved through an optoelectronic system.

18.1 Light Characteristics

When dealing with light, there are several characteristic terms that should be understood. The unit of luminous intensity of a standard light source is called a **candela (cd)**. The intensity of light is therefore expressed in candelas. The amount of light falling on a surface of which all points are a unit distance from a uniform point source of one candela is called a **lumen (lm)**. The illumination of a surface is the number of lumens falling on it per unit area. The unit of illumination is the **lux (lx)** or lumens per square meter.

We see only light that is reflected. Reflected light is measured in candelas per square meter of surface. A term called the **reflection factor** is the percent of light reflected from a surface expressed as a decimal. Therefore, the light reflected from a surface is equal to the illumination of the surface times the reflection factor.

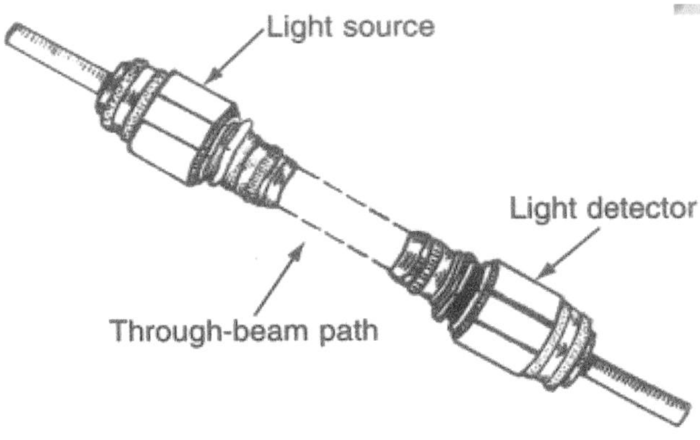

Figure 18.1 Photoelectric system with independent source and detector.

18.2 System Classifications

The method used to couple light energy from the source to the detector is often used to distinguish one optoelectronic system from another. Two general classifications are made in industry. These can be described as interruptible systems and non-interruptible systems.

18.2.1 Interruptible systems

Interruptible systems are the most common of all optoelectronic applications. Systems of this type have a source of electromagnetic radiation and a detector or receiver. A system with an independent source and detector is shown in Figure 18.1. Electromagnetic energy emitted from the source travels through the air to the receiver. Systems of this type are used in industry to achieve a number of different control operations. Response time is considered to be very fast. Actuation is accomplished by breaking or interrupting the radiation path between the source and detector.

Interruptible systems are designed to detect the presence or absence of light in the transmission path. In this type of system, the light source and detector are usually positioned so that they will respond to changes in the transmission path. Counting, automatic door control, motion or product movement, and object position adjustments are made by this type of system. A large number of these systems are in use today.

18.2.2 Non-interruptible systems

Non-interruptible systems are designed to respond to a constant flow of light through a slightly unusual medium. This type of system can be used to detect changes in light intensity, respond to variations in object colors, and transmit data. The end result of the system could be film exposure, the slicing of an object by a laser beam, or circuit isolation. Operation is not necessarily based on the breaking of a light beam. Data transmission, for example, is achieved by turning the light source on and off according to some predetermined coded signal. Electrical isolation between different circuit components can be achieved by interconnecting them through a light beam. Communication signals can be transmitted by light between different locations through fiber-optic cables. Non-interruptible systems represent some of the most innovative developments in the optoelectronics field.

There is very little difference between the general operation of an **interruptible** and a **non-interruptible** system. A source of radiant energy and a receiver or detector are usually needed in both system types. The end result of the system or the resulting work produced by its load is used to distinguish one system from the other. In general, the most significant advances in optoelectronics have occurred in non-interruptible system applications.

18.3 System Functions

An **optoelectronic system** has two important operational conditions that are used to distinguish it from other systems. First, the source must generate some form of radiant energy. Normally, this energy is in or near the visible portion of the frequency spectrum. Second, the system must have a detector or receiver of radiant energy. A variety of devices can be used to detect this energy. The design of the system and its application have a great deal to do with the selection of its energy source and detector. System operation is primarily dependent on the function of these two components.

18.3.1 Radiation sources

A variety of **radiation sources** are now used in optoelectronic systems. Light-emitting diodes, tungsten lamps, neon lamps, fluorescent lamps, and xenon tubes are some of the sources of electromagnetic radiation. As a rule, these sources radiate energy in the visible part of the electromagnetic spectrum, see Chapter 17. Some light-emitting diodes (LEDs), however, are purposely designed to radiate only infrared energy.

Emitters that generate infrared wavelengths or energy outside of the visible part of the spectrum are said to be **radiometric**. If the radiation source produces some form of visible light, however, it is described as **photometric**. Radiometric terms are usually preceded by the word radiant.

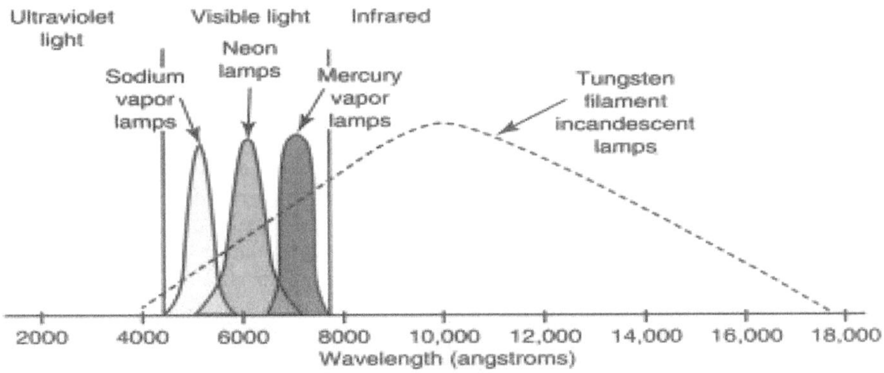

Figure 18.2 Spectral response of radiation sources.

Common expressions are radiant flux, radiant intensity, and radiant incidence. Photometric terms are preceded by the word luminous. Luminous flux, luminous intensity, and luminous incidence are typical photometric expressions.

Photometric sources are frequently classified by the amount or range of visible light emitted. In this regard, sources are described as being panchromatic, heterochromatic, or monochromatic. The term chromatic in these expressions refers to a specific color or various colors of the spectrum. An incandescent lamp is an example of a panchromatic source. Its radiation extends over a very large portion of the optical spectrum. A wide range of visible colors can be produced by an incandescent source.

Heterochromatic sources are different because they produce a very limited number of different colors. Mercury arc lamps, for example, produce colors that are predominantly red or orange. Specifically, this particular source generates light in the range of 6500 angstroms (Å).

A **monochromatic** source radiates energy of only one specific wavelength or a very narrow part of the spectrum. A sodium-vapor lamp, for example, radiates energy at 5000 Å. Neon lamps emit energy at 5500 Å. Figure 18.2 shows the spectral response of several different light sources. Notice the chromatic differences in these sources.

18.4 Incandescent Lamp Sources

Incandescent lamps can be used as the radiation source of an interruptible system. This type of source is readily available, is very reliable, and has a consistent level of operation. Tungsten filament lamps are widely used in industrial applications. The radiation output of this source peaks at approximately

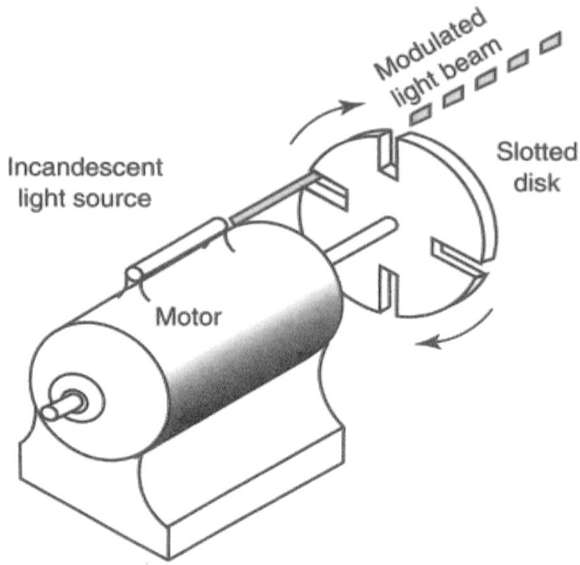

Figure 18.3 Rotating disk modulation source.

10,000 Å (see Figure 18.2). It also radiates electromagnetic energy in the visible light region.

The **incandescent lamps** of an interruptible system are generally operated at the lowest voltage value that is consistent with good performance. Low-voltage operation extends the life span of the lamp. Lamp life in an optoelectronic system ranges from 5000 hours (h) to something in excess of 50,000 h, depending on the lamp and its operating voltage.

Many **optoelectronic radiation sources** employ modulated light signals. A modulated light signal is one that is interrupted or pulsed on and off several times per second. The modulation rate ranges from 100 pulses per second (pps) to several thousand.

Modulation can be accomplished mechanically or electrically. Mechanical pulsing is normally achieved by a slotted rotating disk. The disk is rotated by a motor. Light passes through each slot and is blocked by the remainder of the disk. In a sense, the light beam is turned on and off according to the rotational speed of the disk. Modulation of this type can be very high in accordance with motor speed and the number of slots in the disk. Figure 18.3 shows an example of a rotating disk modulation source.

Electrical switching of an incandescent lamp is usually achieved by an **SCR** or a **triac**. The lamp is simply made to be conductive or turned off at a constant rate. As a rule, electrical modulation of an incandescent lamp

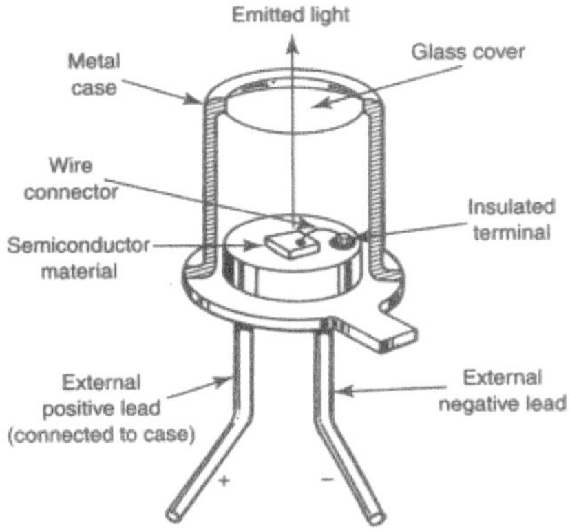

Figure 18.4 Light-emitting diode.

can only be achieved at a rather slow rate. There is a time factor involved in this operation. When electrical power is applied initially, it takes a short time for the lamp to produce filament heat and ultimately light. When the power is removed, there is a short time delay before the filament cools off and stops emission. **Modulation** by switching is generally limited to a few hundred pulses per minute. Modulation of the light source is only used where the effects of strong ambient light are objectionable to normal system operation.

18.5 Light-Emitting Diode Sources

A semiconductor optoelectronic device that is used extensively as a light source is the light-emitting diode (LED), sometimes called a **solid-state lamp (SSL)**. A typical LED is shown in Figure 18.4.

In general, the **LED** is small and lightweight. This makes it very desirable for use with digital circuitry and other miniaturized applications. The PN junctions of LEDs are commonly made of gallium arsenide phosphide (GaAsP) and gallium phosphide (GaP). Light-emitting diodes have fast response times, and they present a low impedance that is compatible with solid-state circuitry.

An **LED** is operated in a forward-biased condition. The semiconductors used in its fabrication have the property of producing electromagnetic radiation when a forward-biased potential is applied. Thus, electrical energy causes the radiation of visible light energy.

The **spectral response** of LEDs varies according to the type of semiconductor materials used in their design. LEDs are made to produce a number of different colors. Typical LED spectral response curves are shown in

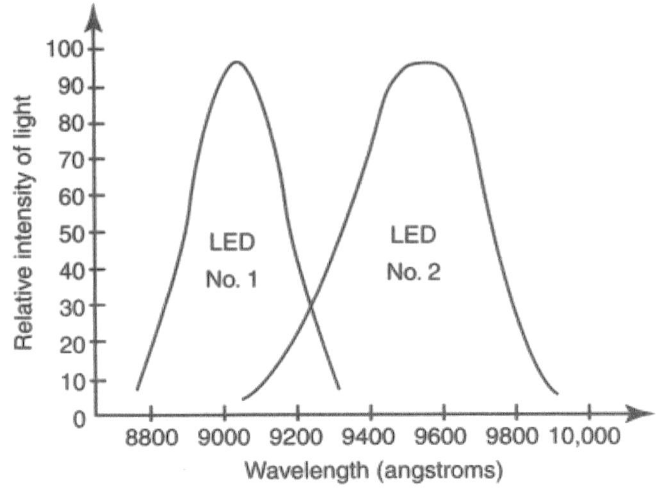

Figure 18.5 LED spectral response curves.

Figure 18.5. Compare the curves of two LEDs and their peak response to the color response curve shown in Figure 18.2. This comparison shows that the wavelength of radiated energy from an LED is well beyond the visible light range. Therefore, phosphors are added to produce the desired color effects. LEDs are very sensitive to temperature changes. Any change in temperature will cause the spectral response of an LED to shift its wavelength.

A unique feature of the **LED** as a light source is its rapid switching time. Compared with an incandescent lamp, the LED can be switched on and off very rapidly. Switching rates of several thousand hertz (Hz) can be achieved. The LED does not have a significant delay time for its switching action. Radiation occurs when the diode is forward biased. No heating of an element is required to produce light. Modulation of an LED source is commonly achieved by a high-speed switching device. Electronic switching is widely used for this type of modulation.

18.6 Radiation Detection

Interruptible optoelectronic systems are generally classified according to the method of light detection. The mode or method of detection selected for a particular installation is often determined by the amount of contamination in the air. Any buildup of foreign matter on a lens or in the air has a great deal to do with the response of the system. Three modes of detection are used in interruptible systems today. These are described as through beam, reflex, and

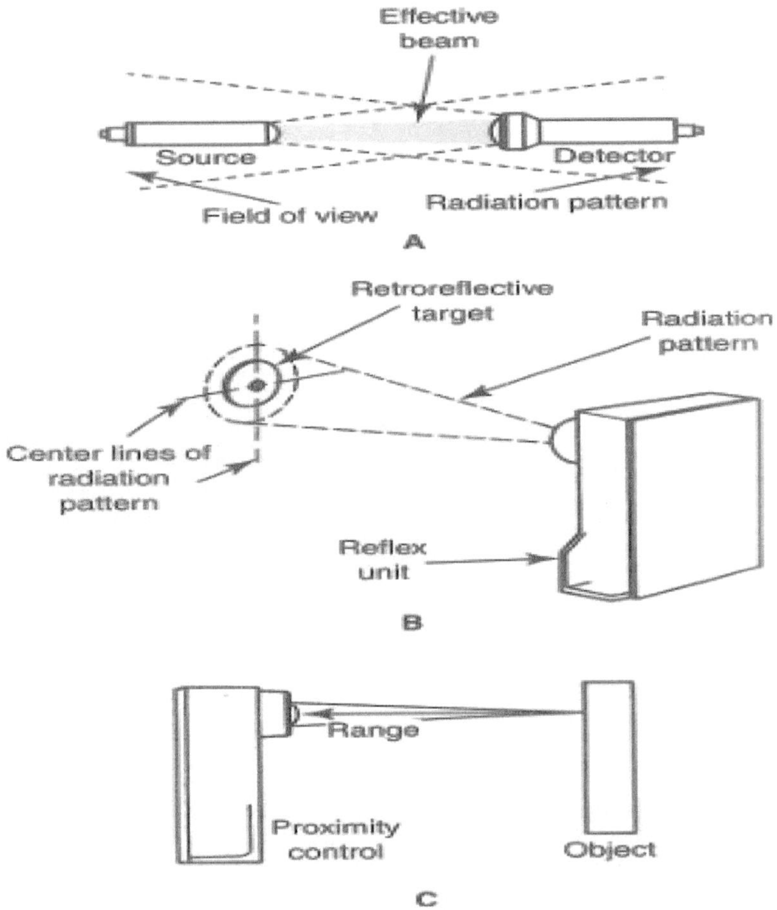

Figure 18.6 Modes of light detection. A – Through beam. B – Reflex. C – Proximity.

proximity modes of detection. Each **detector** is unique and has a set of characteristics that distinguishes it from the others. These characteristics have a great deal to do with the installation of a particular system. Figure 18.6 shows the three detection modes.

18.6.1 Through-beam detection

Through-beam detection is the most common detection mode. Through-beam detection has the light source and the detector both mounted in independent units. The source sends out a light signal from its location. The detector then responds to this signal at some remote location. Object detection occurs

Figure 18.7 Retroreflective-type reflex detection. (Scientific Technologies, Inc.)

when the light beam is broken. The light source and detector both require electrical energy for operation.

18.6.2 Reflex detection

Reflex detection has the light source and detector unit both mounted in the same assembly. The source can be either an incandescent light or a modulated LED. The operation is based on the reflection of an emitted light beam back to the detector. A light beam sent out by the source strikes a retroreflective target and reflects back to the detector. A **retroreflective** target reflects or turns back radiant waves of energy so that the reflected waves are parallel to the original wave. Retroreflective surfaces are generally less sensitive to misalignment than plane-surfaced mirrors. Detection then occurs when an object crosses between the light beam and the target. This type of system only requires electrical power to one unit.

Figure 18.7 shows a **reflex detector**. A block diagram of the reflex detector system is shown in Figure 18.8. Note that only one lens is needed for this assembly. A special beam-splitting mirror permits light emission and detection on the same axis. Alignment is generally not critical because the retroreflective target returns light to the detector even when it is tilted 15° at

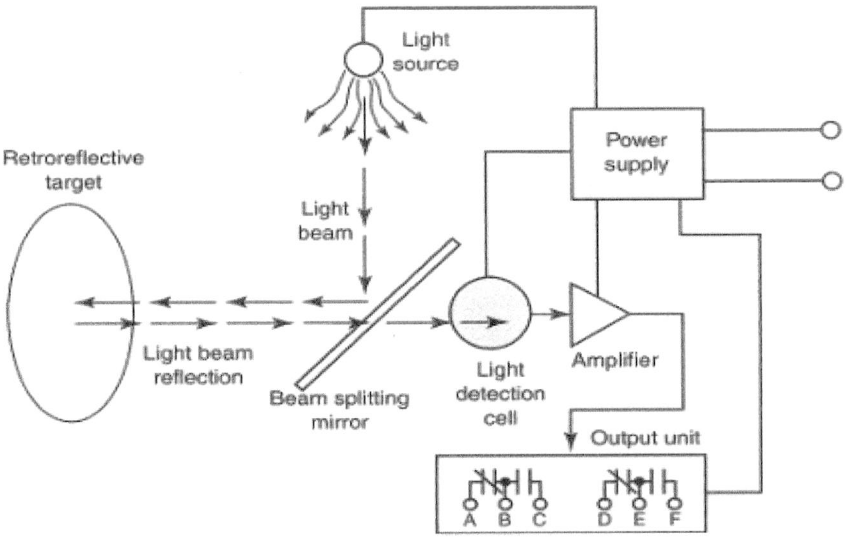

Figure 18.8 Block diagram of a reflex detector system.

the rated operating range. The detector in this unit is a silicon photovoltaic cell. Amplification of the detected signal is achieved by high-speed silicon transistors. The output of this system has a number of different options. It can be on/off solid-state switching, relay control, or one-shot pulser switching. It can also have a time-delay feature.

18.6.3 Proximity detection

Proximity detection is achieved with a combined light source and detector in a single unit. The specific differences between proximity and reflex detection are the light source and the operating range. The light source and detector in proximity detection usually respond to visible light. The object being detected simply reflects the light signal back to the detector. Electrical energy is only supplied to the combined source/detector unit.

Figure 18.9 shows a **proximity detector** unit. Systems of this type usually employ a visible light LED source. The source is normally energized by a high-frequency square wave. Frequencies of 2.5kHz are widely used. This wave is generated by an oscillator that is built into the supply unit.

The operational range of a **proximity detector** system is limited when compared with other methods of detection. The operation is based on the material surface being detected. Smooth, glossy surfaces produce more reflection than rough-textured surfaces. The maximum range of detection is around

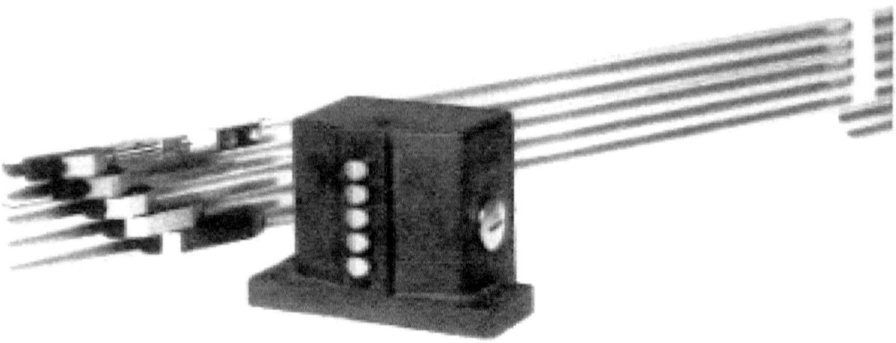

Figure 18.9 Proximity sensor. (Norstat, Inc.)

24 inches or 600 mm. System response is also influenced by the condition of the air. Most installations of this type are found in clean environmental areas.

18.7 System Design

In the design of an **optoelectronic system** for a specific application, a number of considerations must be taken into account. These include such things as component matching characteristics, the transmission medium selection, and a variety of optical factors. The operation of the system is directly dependent on these considerations. As a rule, the considerations should not be altered after a system has been designed for a specific application.

18.7.1 Component matching

Component matching of an **optoelectronic system** is an important system design consideration. In this regard, matching refers to the spectral response of the light source and the detector. These two parts must have compatible characteristics in order for the system to be effective. A graphic indication of the component matching characteristic is shown in Figure 18.10. Part A shows a complete mismatch between the source and the detector. A characteristic of this type shows that the detector will not respond to any of the light source energy. This would be similar to tuning a radio receiver to a frequency at one end of the band for reception of a station at the other end of the band.

When the **spectral response** of the source and detector overlap, operational efficiency is improved. Part B of Figure 18.10 shows a characteristic

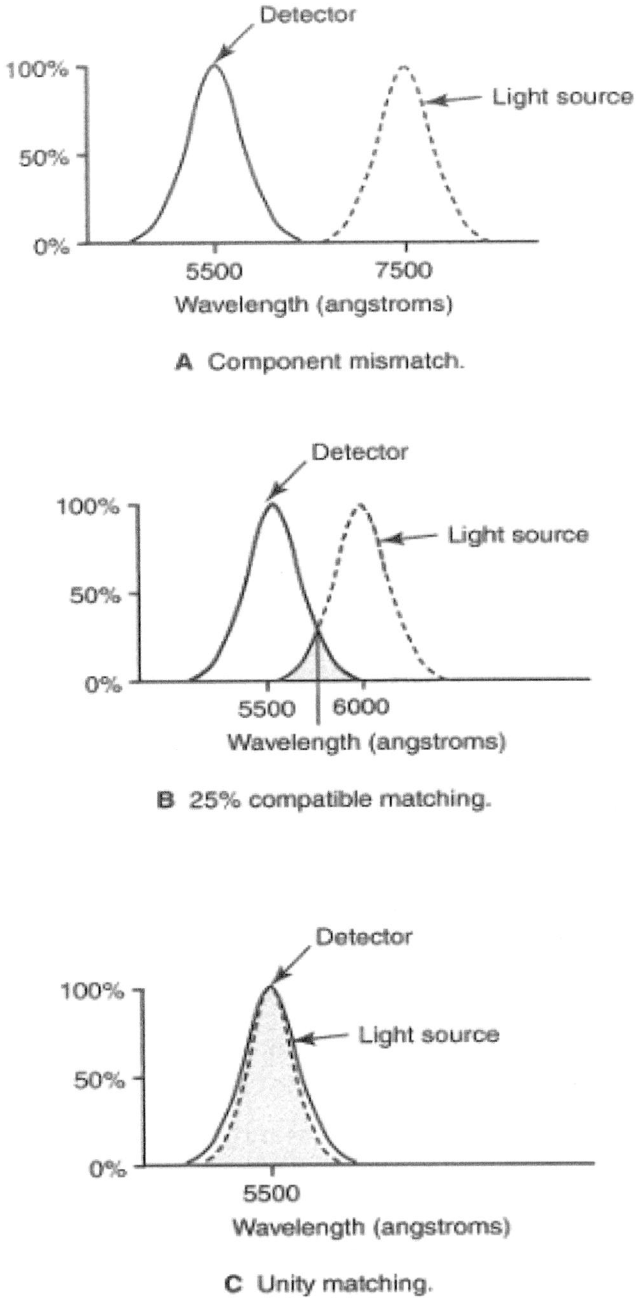

Figure 18.10 Component-matching characteristics.

Source of Light Energy	Eye	Light sensitive silicon PN device
Sun	0.16	0.5
2,200K tungsten lamp	0.007	0.19
2,600K tungsten lamp	0.02	0.24
3,000K tungsten lamp	0.04	0.3
Neon lamp	0.35	0.7
Light-emitting diode gallium arsenide 9000A	0.0	1.0
Light-emitting diode gallium phosphide 7000A	0.08	0.7

Figure 18.11 Light-matching relationship table.

where the devices are approximately 25% compatible. Matching is indicated by the shaded region where the two curves overlap. Matching, in this case, is very effective around 5750 angstroms (Å).

Unity component matching is shown in part C of Figure 18.10. This condition occurs when all the source energy falls on the detector. Ideally, this means that the two components are perfectly matched. Their matching factor is based on the product of the two response curves. The matching factor is the shaded region where the two curves overlap.

The **matching factor** of an optoelectronic system is an extremely important operational characteristic. The table of Figure 18.11 shows representative matching factor relationships for different light sources. Note that the human eye and a light-sensitive PN silicon crystal are compared. Examination of the table shows that the light-sensitive crystal is a much better detector than the human eye in all examples. The human eye is most sensitive to the sun and the neon lamp. In both cases, the light-sensitive detector is a much more efficient detector.

Sensitivity, in this case, is not an indication of brightness. It primarily refers to a higher percentage of a specific light wavelength that is detectable by the device. Optoelectronic systems do not necessarily need to respond to light energy that the human eye responds to. In practice, many systems employ ultraviolet (UV) or infrared (IR) light instead of visible light.

18.7.2 Transmission medium

The **medium** through which light energy is transferred from the source to the detector is an extremely important operational consideration. Regardless of the type of system, there is always a time when there must be some transmission of light energy. The medium must first be capable of passing light at a desired frequency with minimal loss or attenuation. In an interruptible system, blocking the light source must also cause a corresponding change in the output. The medium must therefore be capable of responding to both the presence and absence of light in its operation.

Air is a very common transmission **medium** for optoelectronic systems. Light travels through air very rapidly. Typical indications of the speed of light are 300,000 kilometers per second or 186,000 miles per second. Light travels in this medium through waves. These waves are of the transverse type. This means that their motion is in a plane that is at right angles to the direction of travel. We believe that light waves travel in a straight line. This condition is called **rectilinear propagation**. Most optoelectronic systems respond to light wave propagation through air for only a limited distance.

Optical fibers are also frequently used in many systems as a transmission medium. These fibers are made from glass or plastic. The fiber material serves as the actual transmission medium. Glass fibers are used frequently and are available in diameters from two micrometers to more than six millimeters.

Each fiber has a central core that has its outside coated or clad with a reflective surface. The cladding material has a reflective index slightly lower than that of the core. To a **light wave**, the interface between the core and the cladding material acts like a mirror. Light striking the interface is reflected back into the core. It moves and remains in the core until it strikes the core cladding interface again. The process is then repeated. As many as 500 or more reflections can occur in 30 centimeters of the fiber core.

The principle of **fiber optics**, illustrated in Figure 18.12, utilizes optical fibers made of glass or plastic to transmit light from one point to another. Light can be transmitted in very unconventional ways, such as around corners, in limited space, or over long distances by using the fiber optics principle. The light will transmit through the fiber-optic material regardless of how it is bent or shaped. Advances in fiber-optic design have made it possible to use low-loss fiber lengths for numerous applications. Optical communication is the most common use of fiber optics.

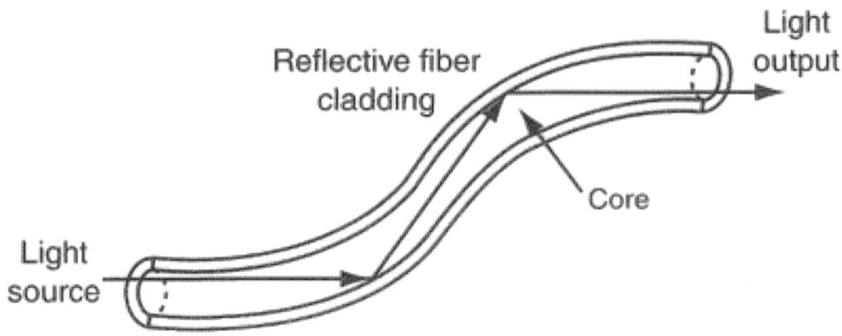

Figure 18.12 Cutaway view of a fiber-optic cable.

18.7.3 Optical coupling considerations

Optical coupling considerations take into account those things that influence the transmission of light other than the conduction medium. Such things as windows, optical filters, lenses, mirrors, and prisms are included in this design consideration. Each item can be used in a system singly or in combination with others.

18.7.4 Windows

Windows are a very common optical coupling consideration. A window is the covering or closure through which light passes from a packaged light source to the outside or from the outside to the inside of a packaged detector. Figure 18.13 shows window types for a number of discrete optoelectronic devices.

Windows are normally made of glass, plastic, quartz, or sapphire. The design of the window has a great deal to do with the wavelength of the light energy radiated by the source or received by the detector. Such things as refraction and diffusion must be taken into account. Light waves passing from one medium into another have a natural tendency to bend. Refraction of this type could cause light to be emitted at an unusual angle or be improperly directed toward the active device. Window selection must take this into account.

Diffusion is primarily dependent on the intended function of the device. In some applications, diffusion is needed to disperse the light source over a large area. In other cases, it is important for light to be developed in a finely concentrated beam. The surface of the window has a great deal to do with its diffusion factor. Smooth, well-polished windows reflect light with a minimum of distortion. Rough-surfaced windows cause light to be scattered or dispersed. Selection of the window surface is a very important system consideration.

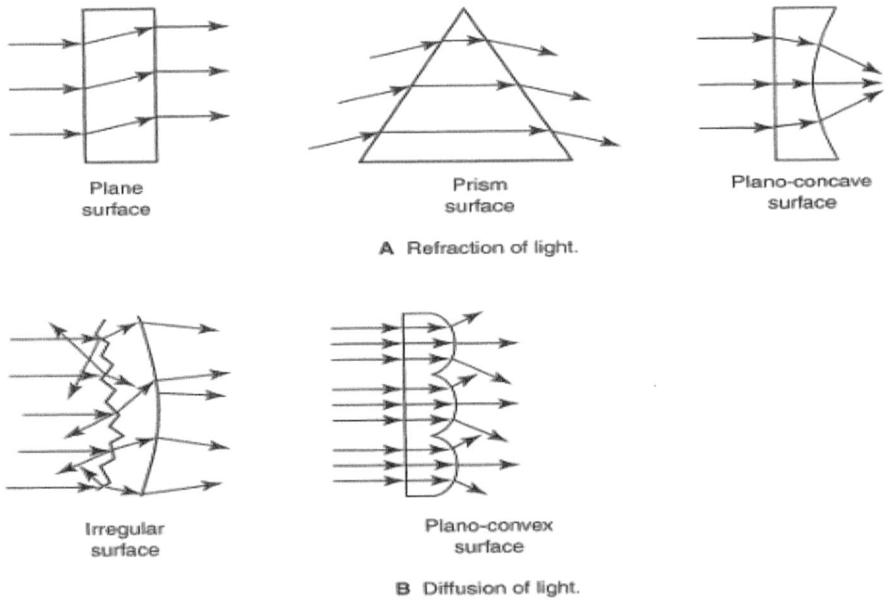

Figure 18.13 Optoelectrical window surfaces.

18.7.5 Optical filters

Optical filters are purposely placed in a light path to restrict the passage of specific frequencies. In many applications, only selected bands of frequencies are permitted to pass through the system. Optical filters are designed to pass only desired wavelengths and to block out unwanted frequencies. A red optical filter, for example, passes only the frequencies of light that the system detects as the color red. All other frequencies are attenuated.

A wide range of **optical filters** is used today in optoelectronic systems. Typically, these filters are made by depositing very thin layers of metal on a base of glass or quartz. The operational frequency of a system usually dictates the need for an optical filter. In most industrial applications, filters are used to reduce or eliminate light that can produce an erroneous indication in the output.

18.7.6 Lenses

Lenses are primarily used in an optical system as a window. In this regard, they must pass light waves without causing a great deal of attenuation. They are purposely used to focus or send light energy to a specific point or in a particular direction. Directing light waves into a single or common point is

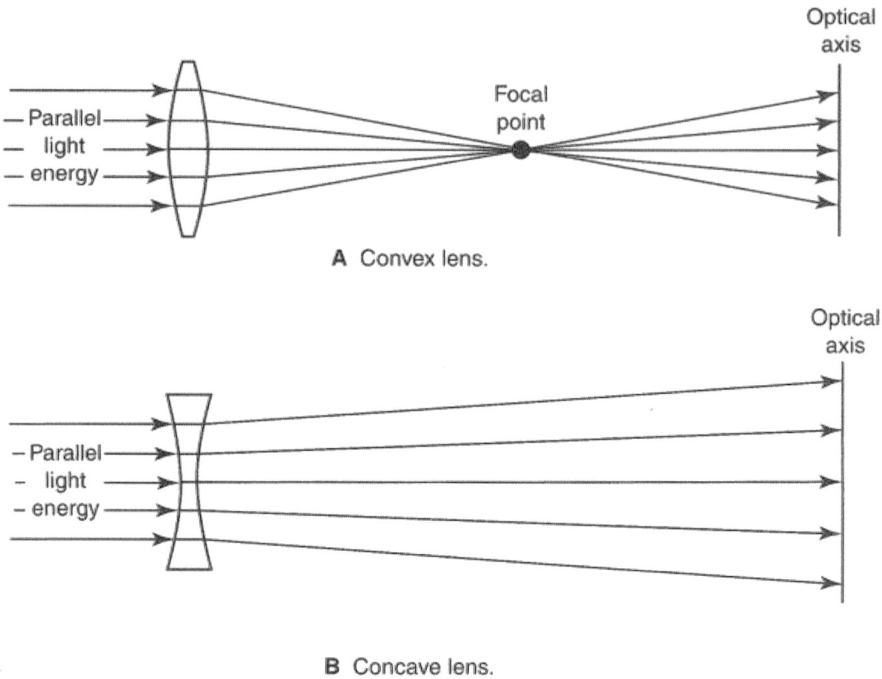

Figure 18.14 Convex and concave lens action.

called convergence. A lens can also be used to cause a light wave to spread out. This function is called divergence. The physical structure of the lens determines how it will respond when light energy is applied.

The ability of a lens to modify light energy is primarily dependent on refraction. Figure 18.14 shows how light energy is altered by a lens. Parallel light waves entering a convex lens are bent toward a common point or axis. A convex lens is considered to be positive and causes light to converge at a common point. The concave lens has a negative function and causes a divergence of light energy. Parallel light waves entering this lens are bent away from the axis. The function of an optical system primarily dictates the type of lens being used.

18.7.7 Mirrors

Mirrors are designed to reflect light and in most applications do not pass light energy. They respond by directing light beams in one direction which can cause the beam to either converge or diverge. The structure of the mirrored surface determines its response.

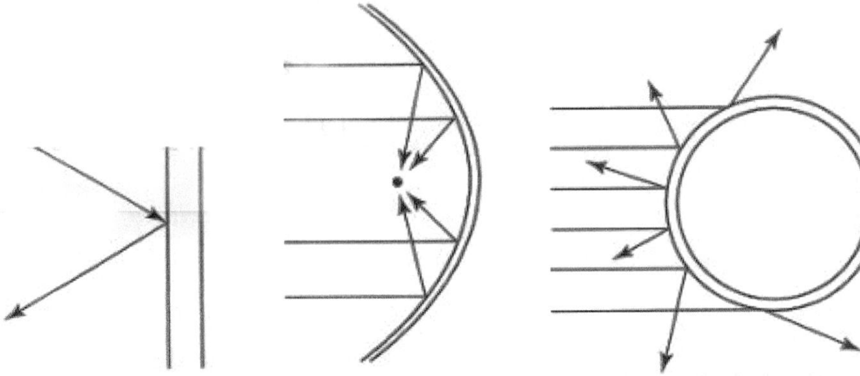

Figure 18.15 Three mirror surfaces.

Optoelectronic **mirrors** generally consist of a thin metal surface deposited on a glass substrate. The reflecting properties of the mirror are based on the reflectance of the metal being used. Aluminum is widely used today because it has excellent reflectance throughout the visible and infrared regions of the optical spectrum.

The surface of a **mirror** is primarily responsible for its response to light energy. Figure 18.15 shows three common mirror surface configurations. Plane mirrors have a smooth, flat surface. This type of surface is used to reflect all or part of an optical wave in a new direction. Curved mirrors respond as a lens. A curved parabolic surface reflects light to a common point. This type of mirror permits light to be collected at a distant location and be directed into a narrow beam. A spherical surface causes light to diverge. This type of surface causes light to scatter in many directions.

18.7.8 Prisms

Prisms respond in an optoelectronic system as a lens. Operation, in general, is based on the refraction of light. The velocity of a light wave passing through glass depends on its wavelength. For this reason, prisms are used to divide a light beam into its component wavelengths. Figure 18.16 shows how incident light is separated into distinct components.

A system employing a **prism** could be used to detect a particular color from a common light source. An alarm could be turned on at a remote location if an unusual amount of one color appeared in the output of a specific

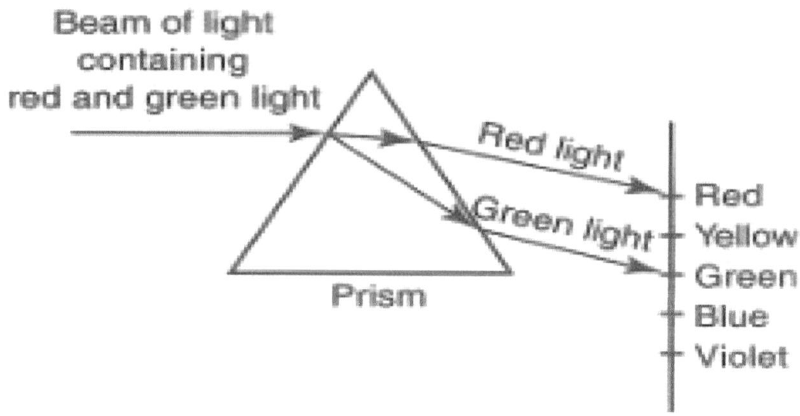

Figure 18.16 Incident light separation by a prism.

detector. Systems of this type are frequently used in **color mixing** operations in the textile field.

18.8 Optoelectronic Devices

Optoelectronic or photoelectric devices are by far the most unique part of a composite light-actuated system. These devices, in general, are classified as either sources or detectors. A source is responsible for changing electrical energy into some form of radiant energy. Incandescent lamps, LEDs, gaseous or vapor lamps, and lasers are typical sources of radiant energy. These sources can be responsible for the emission of visible light or some form of radiation in another part of the spectrum. The source, in a strict sense, is a transducer. Electrical energy is changed into radiant energy.

The **detector** of an optoelectronic system is also classified as a transducer. Detectors are designed to change radiant energy into electrical energy. In effect, detectors are the reverse or have an opposite function to that of the source. A number of detection devices are available today. Detectors are divided into three general categories:

- Photoemissive.
- Photoconductive.
- Photovoltaic.

Photoemissive devices emit electrons in the presence of light. Vacuum and gaseous phototubes are photoemissive devices. **Photoconductive devices** are

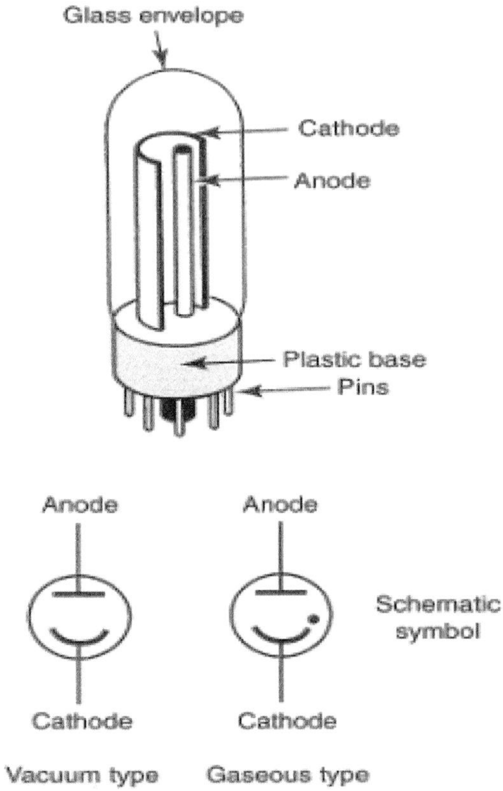

Figure 18.17 Photoemissive cell.

designed so that their resistance decreases when light becomes more intense and increases when light intensity decreases. Photodiodes, phototransistors, and photoresistive devices are common examples of photoconductive devices. **Photovoltaic devices** convert light energy into electrical energy. When a photovoltaic device is illuminated, an electrical potential is created by the device. Most optoelectronic detectors fit into one of these categories. However, there is such a diversity of new semiconductor optoelectronic devices used in industry today that it is more desirable to discuss each device individually. The sections that follow deal with various types of optoelectronic devices and related systems that have applications in industry.

18.8.1 Photoemissive devices

The phototube shown in Figure 18.17 is similar in appearance to a vacuum tube or gaseous tube. This particular type of tube has a cathode that emits electrons when it is struck by light. This effect is called **photoelectric**

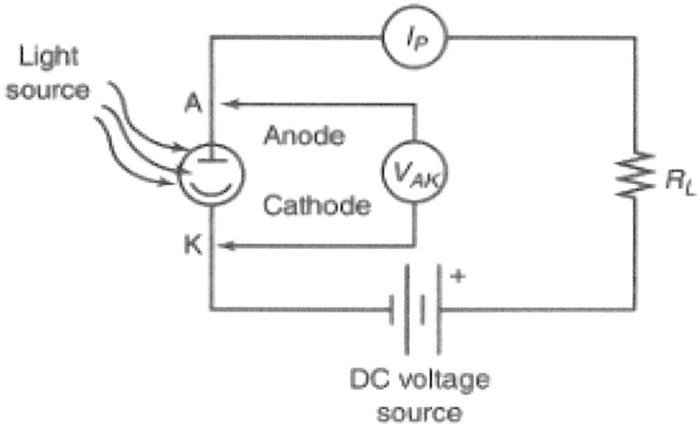

Figure 18.18 Phototube circuit action.

emission. The cathode emits electrons, whose energy depends on the wavelength of the light striking it.

Light energy exists in the form of **photons**, which are discrete bundles of energy. When photoemission takes place in a phototube, photons of light are absorbed by the surface of the cathode. The absorption of this energy causes electrons on the surface of the cathode to gain enough energy to leave the cathode. The energy possessed by the electrons that leave the cathode is based on the frequency of the applied light. Electron emission is based on the intensity and wavelength of light.

A **phototube** can be connected in a circuit as shown in Figure 18.18. At one time, nearly all industrial control by light was achieved by photoemissive tubes. The unusually high voltage needed to energize a phototube is a disadvantage. This type of device has been replaced by solid-state photodiodes. Circuit control can be achieved more efficiently with lower voltage values.

Phototubes have a very small output when the light intensity is low. The multiplier phototube, shown in Figure 18.19, overcomes this disadvantage. Note the internal construction of this phototube. When photons of light energy strike the cathode, electrons are emitted from its surface. These electrons then strike successive positive plates called **dynodes**. Through the principal secondary emission, a larger number of electrons is produced by each successive dynode. Most multiplier phototubes use from 9 to 14 dynodes to produce a high output at low light levels.

Light that strikes the cathode causes electrons to be emitted and focused toward the second dynode. Each dynode possesses a successively higher positive charge to attract these electrons. The anode has the highest positive

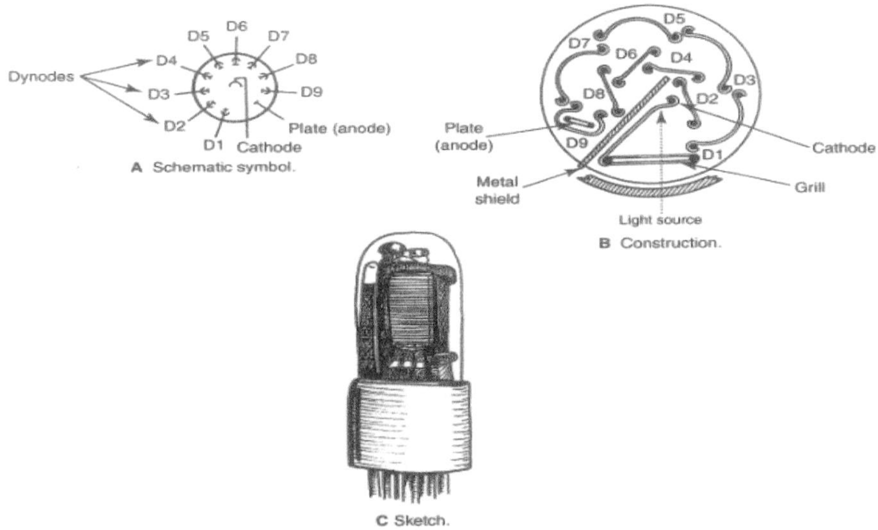

Figure 18.19 Multiplier phototube. (RCA)

potential. Thus, a small amount of light can cause enough electrons to be emitted to produce a significant output. Photomultiplier tubes have a fairly high level of signal multiplication, which is important in sensing low levels of light.

18.8.2 Photoconductive devices

The **photodiode**, shown in Figure 18.20, is a specialized semiconductor diode that is operated in the reverse-biased condition. With the photodiode, there is a linear increase in reverse current as light intensity increases. When light of the proper wavelength is focused toward the PN junction of the diode, current carriers move across the junction. This causes a current through the external load that is proportional to the amount of light falling on the junction.

The photodiode is usually followed by an amplifying device. If the amplifier is made by the manufacturer on the same silicon chip, the device is called a **phototransistor**. A phototransistor equivalent circuit and symbol are shown in Figure 18.21. The collector-base junction is reverse-biased. When light is focused on the collector-base junction, base current flows. This base current is amplified by the transistor. It is possible to bias the base to affect the sensitivity of the device.

A modification of the basic phototransistor is the **Darlington-connected phototransistor** shown in Figure 18.22. This device is similar to the basic

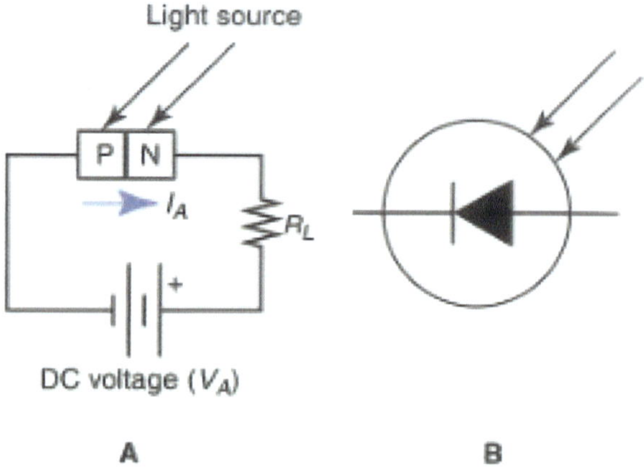

Figure 18.20 Photodiode circuit and symbol. A – Circuit. B – Schematic symbol.

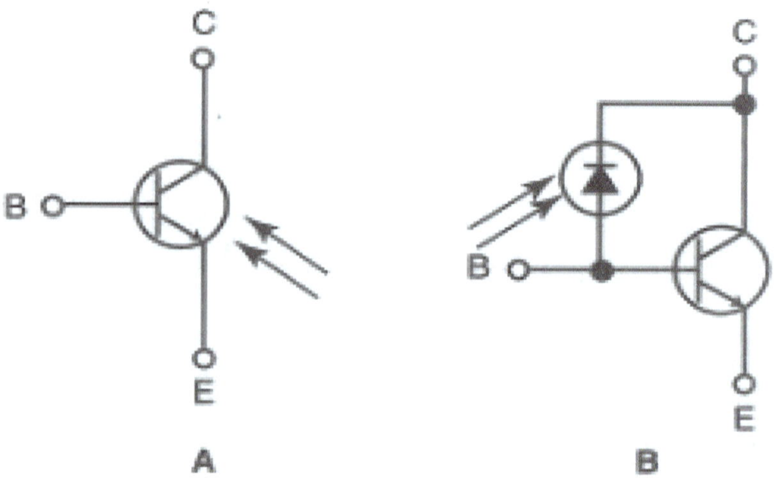

Figure 18.21 Phototransistor

phototransistor except it has much higher current gain. The additional gain is largely due to a directly coupled transistor configuration built on the same silicon chip. Low light levels can be easily detected by a high-gain Darlington phototransistor. Light energy is applied to both transistor sections at the same time. This configuration has the advantage of higher input impedance, lower output impedance, and less current drawn from the source.

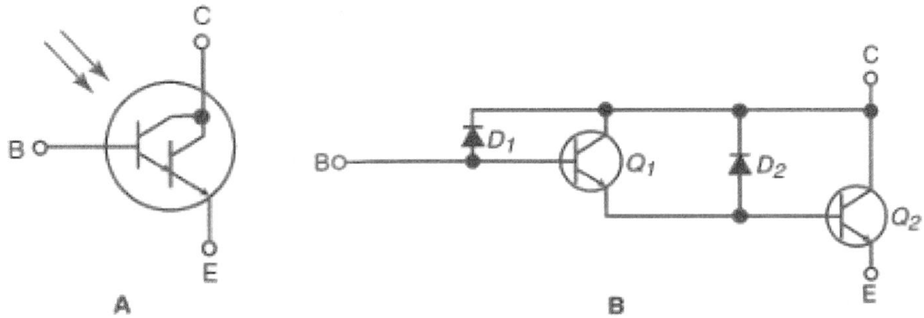

Figure 18.22 Darlington-connected phototransistor.

There are numerous control applications for photodiodes, phototransistors, and Darlington-connected phototransistors. Most of these applications use the light-sensitive device as a control element to activate some type of load. Load activation is normally accomplished through the amplifying action of transistors or other electronic devices. The load can be activated by either the presence or absence of light. The specific function of the load determines when the control circuit must be actuated.

The circuit of Figure 18.23 is a basic **photodiode relay** control circuit. When light shines on the reverse-biased photodiode, it causes an increase in the current through R_B. The voltage across R_B is such that the top of it is positive with respect to ground. This voltage (V_B) forward biases the base of Q_1. An increase in V_B produces a corresponding increase in I_B. An increase in I_B, in turn, causes an increase in collector current. Since the relay coil is connected in the collector circuit, it will actuate. This causes the incandescent light load to be energized. Removal of the light source causes the relay to drop out or change back to its original state. This type of circuit is connected in the forward direction. An increase in light, therefore, causes an increase in load current.

Another **photodiode** application is shown in Figure 18.24. The two photodiodes are used to control the input signal of an operational amplifier. The output of the op-amp is used to position items that are passing along an assembly line. This output is used to drive a servomotor system that controls assembly line movement. The principle of operation for this circuit is that when the photodiodes conduct equally, the op-amp output will cause the assembly line to stop for a predetermined period. This circuit application provides exact positioning control.

A **phototransistor** is often more desirable to use in control circuitry than a photodiode because of its built-in amplifying capability. The

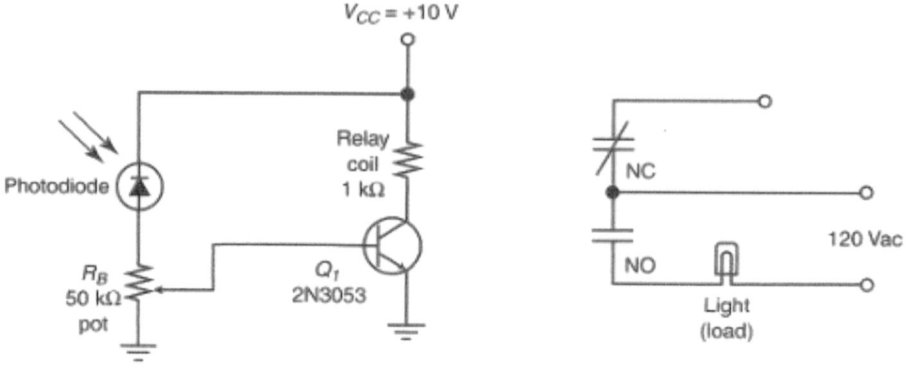

Figure 18.23 Photodiode relay control.

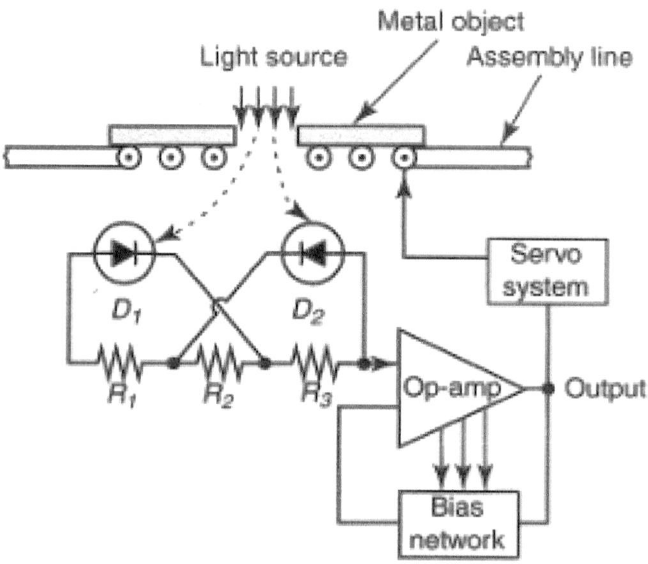

Figure 18.24 Photodiode application.

phototransistor can be used in circuits that are controlled by either the presence or absence of light. Figure 18.25 shows a circuit in which light causes the SCR to turn on. When the phototransistor (Q_1) has no light applied, base current through Q_2 is minimum. However, when light strikes the phototransistor, it becomes conductive and causes Q_2 to go into conduction. When Q_2 conducts, gate current flows into the SCR, causing it to be triggered. Conduction of the SCR causes the light to be energized. Once the load

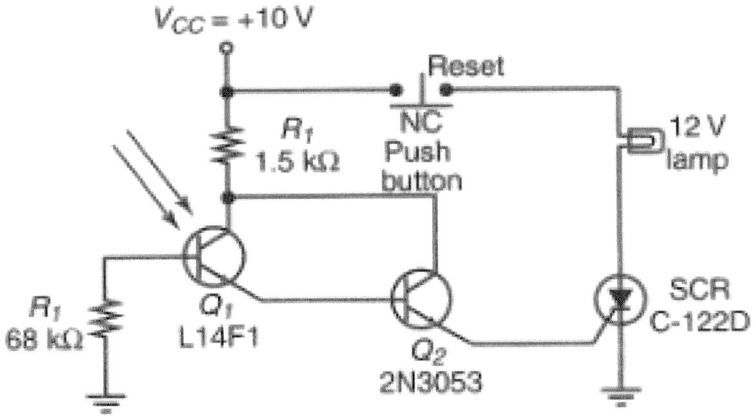

Figure 18.25 Phototransistor control circuit.

has been energized, it continues to remain in the on state even when the light source has been removed. To reset the SCR, the reset button must be momentarily depressed. This prepares the circuit to accept the next change in light energy.

A **phototransistor** can be used to energize a load with the absence of light. This type of circuit is frequently used to energize an emergency light source in the event of a power failure. It could also be described as an automatic nightlight. Figure 18.26 shows a phototransistor driving a power transistor for direct control of a 6-volt lamp. Specifically, the circuit is reverse connected. This means that, with light applied, Q_1 is on and Q_2 is cut off. The load or lamp L_1 is off. The circuit remains in this condition as long as light of a suitable intensity is applied to Q_1.

Assume now that the light intensity applied to Q_1 is suitable enough to cause conduction. This condition causes current carriers (holes and electrons) to return to the depletion region of the reverse-biased emitter-base junction of Q_1. As a result of this condition, there is an increase in base current (I_B). An increase in I_B causes a corresponding increase in I_C. The collector-emitter voltage of Q_1 then drops to a rather small value. With Q_1 being directly coupled to Q_2, the value of V_{CE} will not provide a suitable I_B for Q_2 to be conductive. As a result, Q_2 is cut off, and no current is supplied to lamp L_1. The lamp is off and will remain in this state.

If the **light intensity** applied to Q_1 drops below an ambient value based on the value of R_1, the transistor is driven to cutoff. This condition causes the V_{CE} of Q_1 to rise in value. With an increase in V_{CE}, the base current of Q_2 rises. This causes a corresponding increase in I_C, which is then great enough to

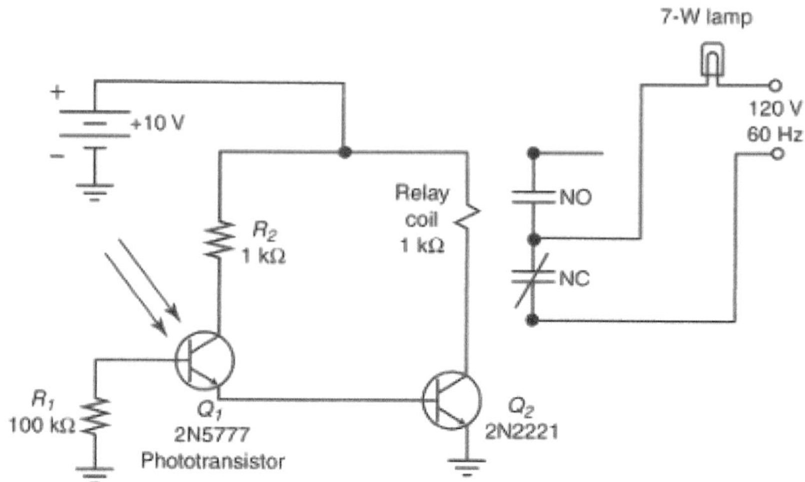

Figure 18.26 Phototransistor control circuit.

energize L_1. The lamp continues to be conductive as long as the light intensity of Q_1 remains at a low level. At a low level of light, the lamp is on. A high level of light causes the lamp to be off. The conduction of lamp current, in this case, is based on the current-handling capability of Q_2. All the load current must pass through Q_2. It must, therefore, have a power dissipation rating that will handle the lamp used in the circuit.

Phototransistors are not very widely used as discrete component control devices. Other optoelectronic devices can achieve control equally as well. Phototransistors are commonly used in opto-coupling devices today. This application permits the interfacing of a computer output to a high-current or high-voltage source with excellent isolation. Light energy of radiation is coupled from an LED to a phototransistor without a direct electrical connection between the two devices. This permits electrical signal transfer between equipment without the hazards of a direct electrical connection. Many different **opto-coupling devices** are available today. These devices are discussed in a later section of this chapter.

18.8.3 Photo-field-effect transistors

A **photo-field-effect transistor** or **photo-FET (PFET)** with its crystal structure is shown in Figure 18.27. This optoelectronic device is similar in many ways to a conventional junction field-effect transistor. Its construction has

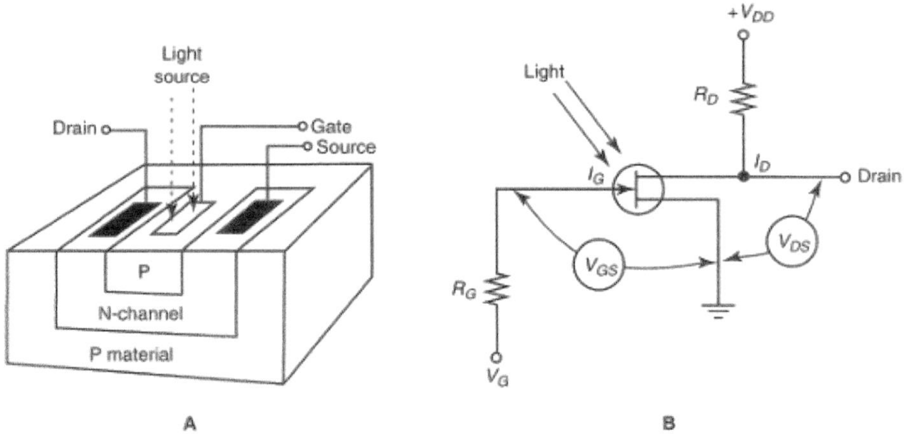

Figure 18.27 Photo-FET, A – Structure. B – Photo-FET circuit.

a lens that focuses light on a sensitive photodiode gate. This diode is normally operated in the reverse-biased direction. Examining the circuit of part B Figure 18.27, when light energy strikes the photodiode gate-channel junction, it changes the value of the reverse-biased voltage appearing across resistor R_G A very small amount of reverse-biased gate current lowers the value of the gate voltage. A reduction in the reverse-biased gate voltage V_G causes a corresponding increase in drain current through the channel.

A **photo-FET** is quite different in operation from that of a conventional JFET. The gate impedance is extremely high when the photodiode junction is dark. An increase in light intensity tends to lower the gate impedance. Some photocurrent must flow in the gate circuit of a PFET. This could be compared with the leakage gate current of a JFET. The gate leakage current of a JFET and the gate photocurrent of a PFET are both temperature dependent. PFETs, as a rule, do not produce a very significant amount of drain current. The drain current can, however, be controlled by light instead of voltage.

A **PFET** used as a light control is shown in Figure 18.28. In this circuit, very little power is consumed in darkness owing to the high impedance of the channel. The output voltage will be at its highest value, and the drain current will be extremely low or zero. The gate voltage will have an extremely high reverse-biased voltage. When the gate is exposed to light, there is a corresponding increase in drain current and a reduction in drain voltage. This is reflected in the output voltage that is developed across the source-drain. The power output of this circuit is very small and is primarily used to drive another active device.

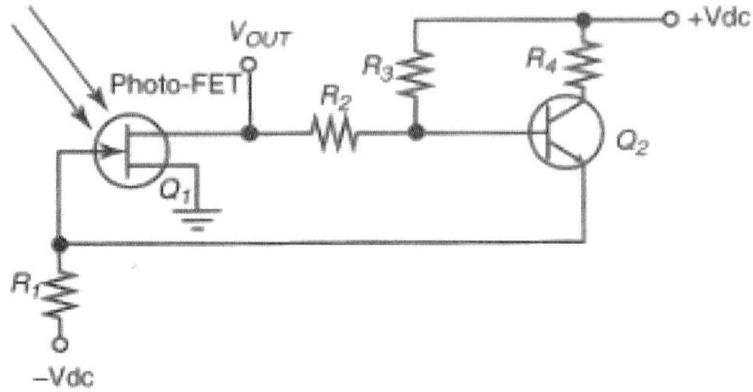

Figure 18.28 Photo-FET applications.

18.8.4 Light-activated SCRs

An optoelectronic device that has become a very popular control element is the **light activated silicon-controlled rectifier**, or **LASCR**. The LASCR is shown in Figure 18.29. We know that SCRs are four-layer PNPN devices that act as semiconductor switches for handling large amounts of current. The SCR is similar to a latching relay that changes from a high-resistance state to a low-resistance state when a trigger pulse is applied to its gate terminal. The LASCR is a semiconductor optoelectronic switch that has a lens that focuses light on its gate. It is triggered into conduction by light rather than by an electrical trigger pulse. Its power-handling capability far exceeds that of the other semiconductor optoelectronic devices discussed. LASCRs can be used as switching devices in systems for counting, sorting, or similar functions where a circuit must be switched on and off. The LASCR uses a small amount of light energy to control a much larger amount of electrical energy.

An application of the ***LASCR*** is shown in Figure 18.30. Here the LASCR serves as a control device that alters the conduction of a conventional SCR. When the conventional SCR goes into conduction, current flows into the load device. In this circuit, the LASCR serves as part of a voltage divider network that controls the AC source voltage to the gate of the SCR. The circuit shown here is described as being reverse connected. With no light applied to the LASCR, current flows through the SCR and the load device. When light is applied, the load becomes nonconductive and switches to its off state. With the SCR energized by an AC source, it turns off without a reduction in gate current after one alternation.

When no light is applied to the **LASCR** of Figure 18.30, it functions as a nonconductive device. Essentially, this means that an open condition exists between its anode–cathode terminals. As a result of this, full gate current is

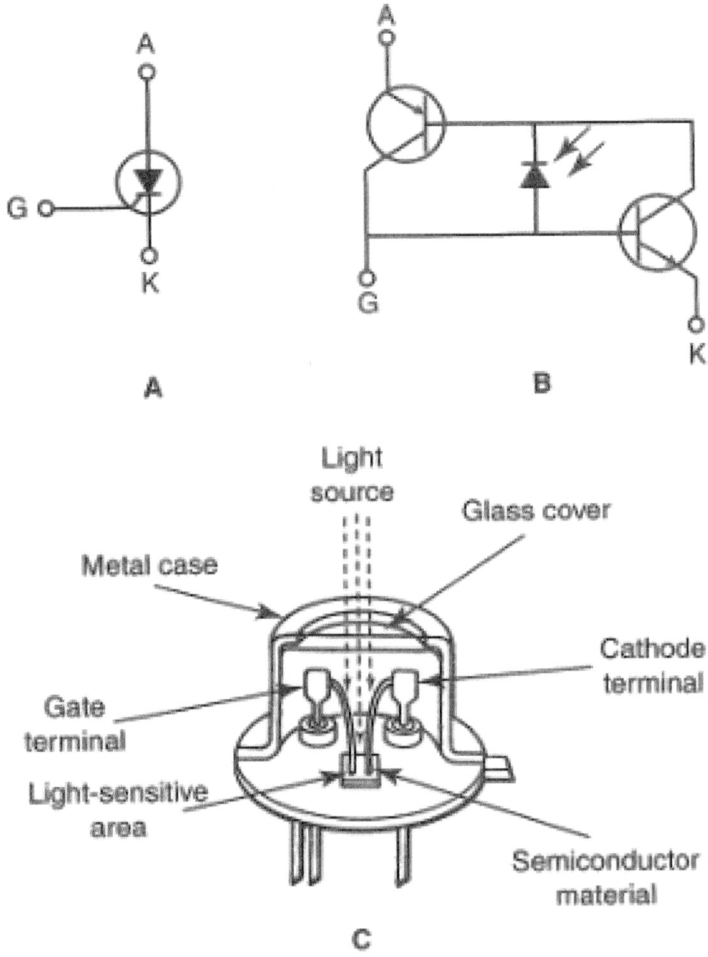

Figure 18.29 Light-activated SCR (LASCR). A – Symbol. B – Equivalent circuit. C – Construction.

applied to the SCR through R_1 and D_1. The SCR becomes fully conductive, and the load is energized. This condition continues as long as AC power is supplied to the circuit.

When light is applied to the **LASCR**, it becomes conductive. Its anode–cathode drops to an extremely low resistance value. In a sense, the LASCR responds as a *low resistance short* for the gate current of the SCR. Hence the I_G of the SCR drops in value. With AC energizing the SCR, a reduction in gate current causes the SCR to be nonconductive. The load, which is

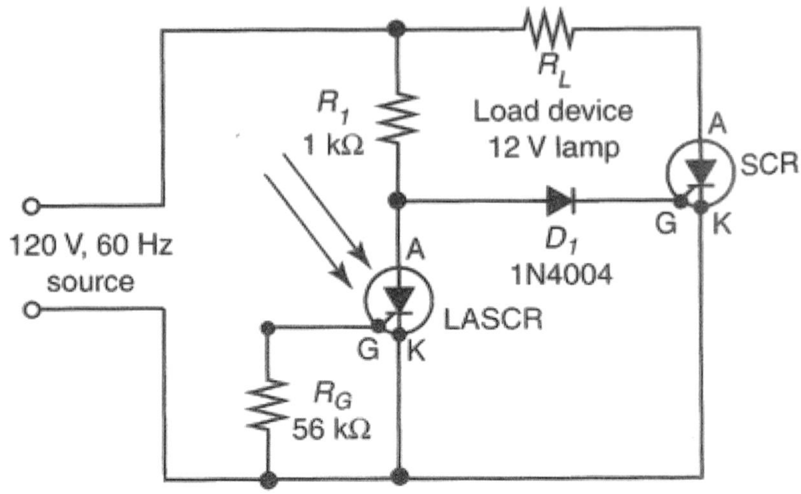

Figure 18.30 Reverse-connected LASCR control circuit.

controlled by the SCR, then becomes inoperative. This means that no light will cause full conduction of the load, and light applied to the LASCR will cause the load to be nonconductive. Reverse conduction of this type can be used to detect the absence of an object in a production line.

A **LASCR** circuit with forward conduction is shown in Figure 18.31. In this circuit, the locations of the LASCR and resistor R_1 have been transposed. Conduction of the SCR will not occur unless light is applied to the LASCR. With light applied, the LASCR becomes conductive and I_G flows into D_1 and the gate of the SCR. The load device then becomes conductive or turns on. Without light, the LASCR is nonconductive. No gate current flows into the SCR, and the load is in an off state. Forward conduction means that light causes conduction of the load, and no light causes the load to be nonconductive. Forward conduction of this type could be used to detect "log jamming" of a conveyor belt assembly line. A large accumulation of objects at one point in the production line would shut down the conveyor motor drive assembly.

18.8.5 Photoresistive devices

Photoresistive devices are designed to change their electrical conductivity when variations of light energy occur. Such devices are also called **light-dependent resistors (LDRs)**, since their resistance varies in inverse proportion to their conductivity. The cadmium sulfide (CaS) cell, shown in Figure 18.32 is a common type of photoresistive device that is called a

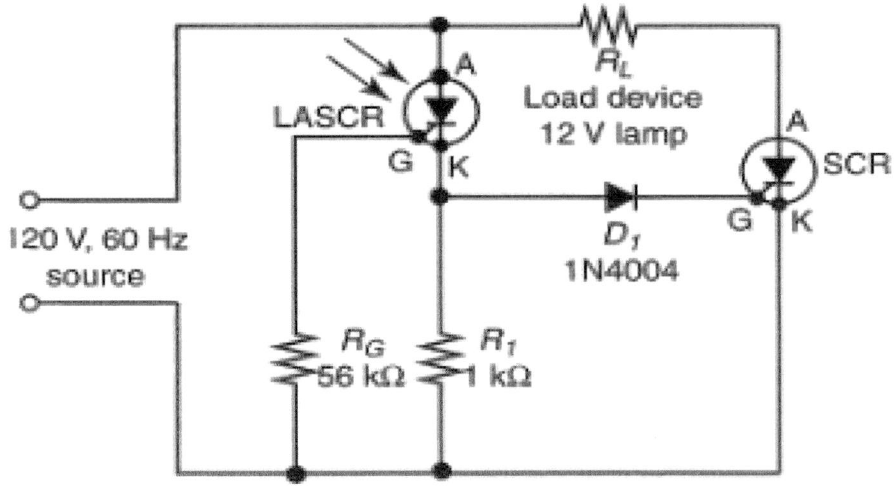

Figure 18.31 Forward-connected LASCR control circuit.

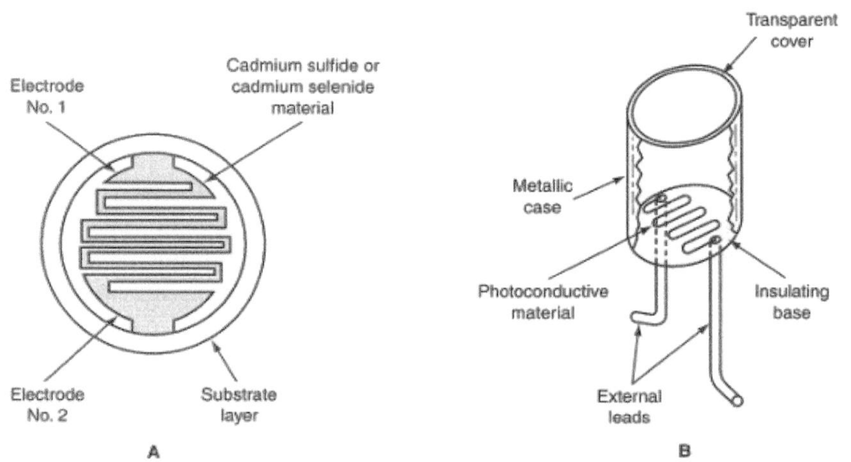

Figure 18.32 Cadmium sulfide photoconductive cell. A – Top view. B – Cutaway view.

photoconductive cell. When exposed to varying intensities of visible light, the cadmium sulfide cell changes resistance. An increase in light energy falling onto its surface increases the cell's conductivity. The cell is highly sensitive to variations in light intensity. These devices are typically used in alarm and relay control systems.

The operational principle of a **photoconductive cell** is that when light strikes its surface, valence electrons of the semiconductor material are released

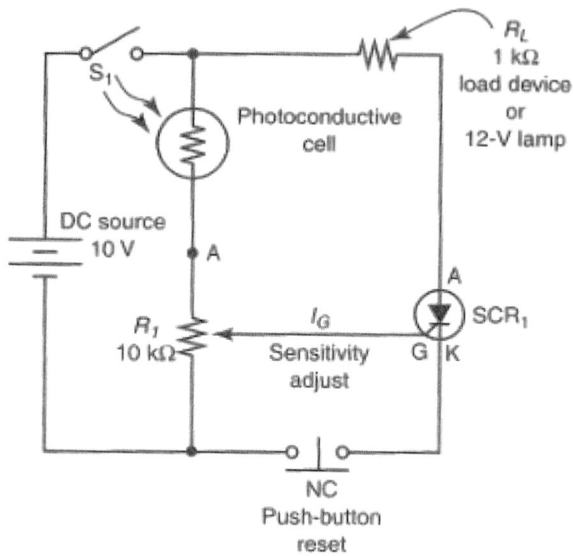

Figure 18.33 Photoconductive cell applications.

from their atomic bonds. When electrons are released, the resistance of the material decreases. We can also say that the material becomes more conductive. If the light intensity increases, more electrons are released, and the material becomes more conductive. The resistance of photoconductive devices can range from several megohms in darkness to between 50 and 100Ω in fairly intense light. The spectral response curve of a cadmium sulfide photoconductive cell very closely resembles the spectral response of the human eye.

An application of the **photoconductive cell** is shown in Figure 18.33. In this circuit, SCR_1 conducts when light is focused onto the photoconductive cell. When light strikes the photocell, its resistance decreases. The potential at point *A* becomes more positive and causes gate current (I_G) to flow. A sufficient amount of gate current triggers the SCR into conduction. When SCR_1 conducts, the load device is activated. The SCR then conducts until its anode circuit is opened. Variable resistor R_1 is used as a sensitivity adjustment to control the level of light required to cause the SCR to conduct.

18.8.6 Photovoltaic cells

The **photovoltaic cell**, commonly called a **solar cell**, is used to convert light energy into electrical energy. Since this process represents a direct energy conversion, a great deal of research has been conducted in an attempt to convert

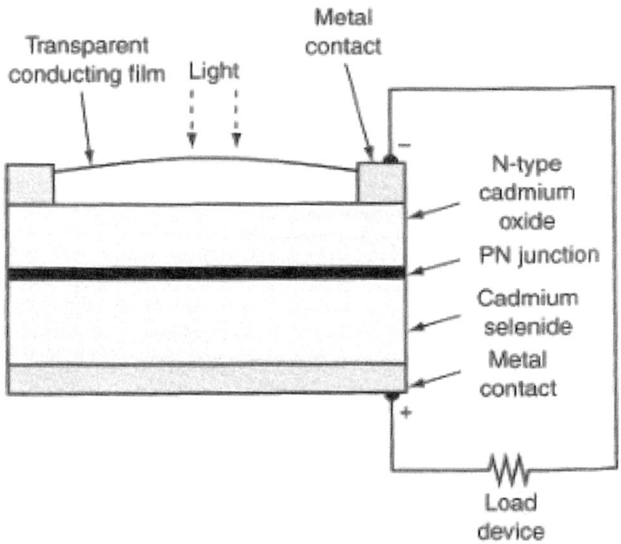

Figure 18.34 Selenium photovoltaic cells.

large amounts of light energy into electrical energy. A common application of the solar cell is for photographic light-exposure meters. The electrical output of the solar cell is proportional to the amount of light falling onto its surface. The output of the solar cell is used to energize a light-intensity meter.

The construction of a **photovoltaic cell** is shown in Figure 18.34. This selenium cell has a layer of selenium deposited on a metal base, followed by a layer of cadmium. In the fabrication, one layer of cadmium selenide and another layer of cadmium oxide are produced. A transparent conductive film is placed over the cadmium oxide, and a section of conductive alloy is then placed on the film. External leads are connected to the conductive material around the cadmium oxide layer and the metal base. When light strikes the cadmium oxide layer, electrons are emitted and move toward the external load device. A deficiency of electrons is then created in this region, which is filled by electrons from the cadmium selenide. Electrons from the metal contact then flow into the cadmium selenide. The metal contact plate then becomes positive. Thus, light energy causes a difference in potential between the two external terminals.

Selenium cells have very low efficiency for converting light energy into electrical voltage. Typical efficiency ratings are less than 1%. A great deal of light and an extremely large active area are needed to produce electrical energy of any significant value. Selenium photovoltaic cells, however, have

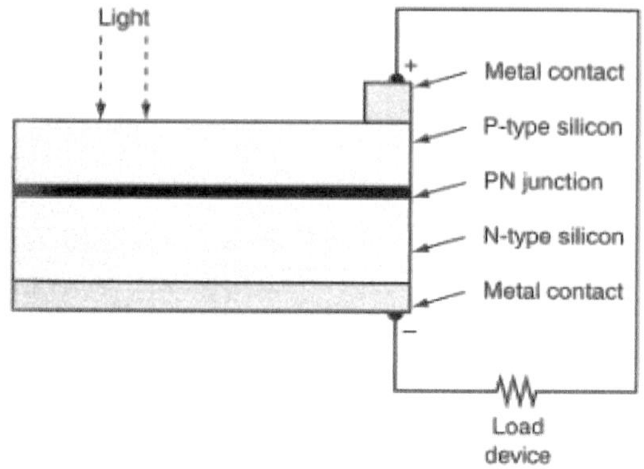

Figure 18.35 Silicon photovoltaic cell.

a spectral response that is very similar to that of the human eye. In many applications, the spectral response is much more significant than a high level of output efficiency.

Silicon photovoltaic cells are used much more readily today than the selenium units. Silicon cells have efficiencies as high as 15%. The more common silicon photovoltaic cell is shown in Figure 18.35. When no light is focused onto the silicon cell, it operates similarly to a conventional PN junction diode. When light strikes the cell, a voltage is developed across the external leads. The more intense the light, the greater the potential difference across the cell.

Photovoltaic cells are used for a variety of applications. Although their electrical output is low, they can be used with amplifying devices to develop an output that will drive a relay or some other load device. One such application is shown in Figure 18.36. In this circuit, the output of the photovoltaic cell is amplified by transistor Q_1 so that, when light strikes the photovoltaic cell, the base current (I_B) of the transistor increases. The increase in I_B is amplified by Q_1, causing a collector current (I_C) that is of sufficient value to cause the relay to activate. The load connected across the relay terminals could be turned either off or on by the presence of light on the photovoltaic cell.

Another application of photovoltaic cells is light meters, such as shown in Figure 18.37.

18.8.7 Infrared detectors

Another type of device that responds to radiant energy is referred to as an infrared detector. These devices respond to radiation in the infrared region of the electromagnetic spectrum (Figure 18.2).

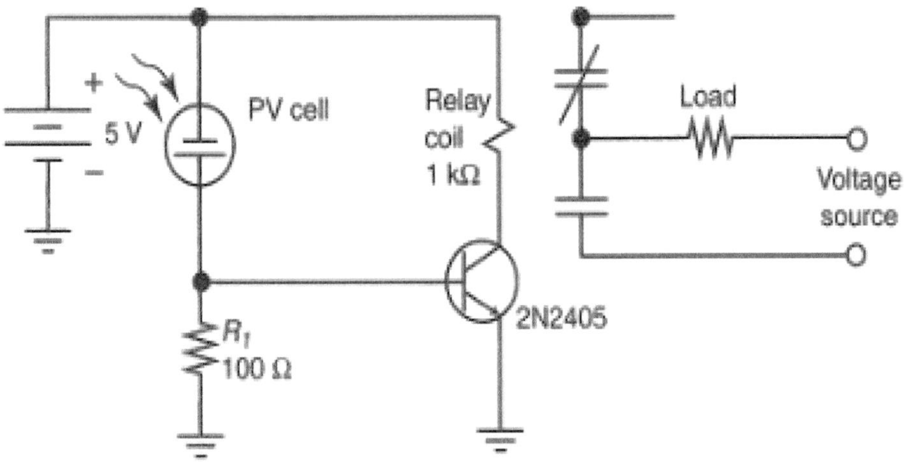

Figure 18.36 Photovoltaic cell relay control circuit.

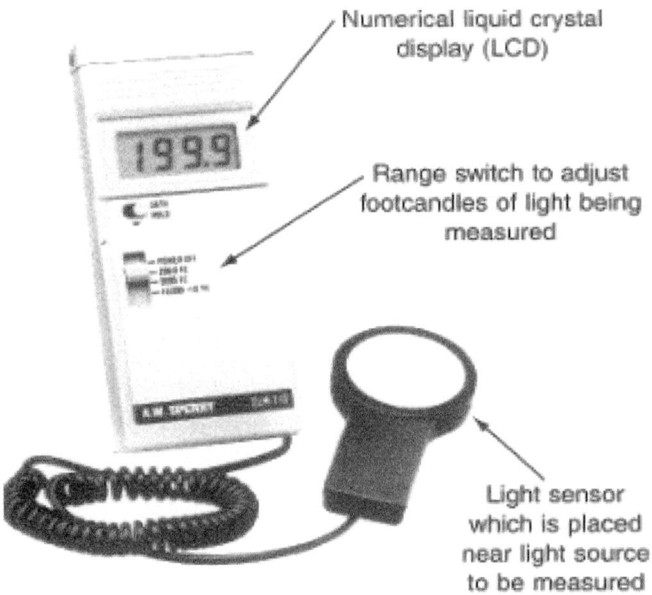

Figure 18.37 Digital light meter. (A.W. Sperry Instruments)

Infrared detectors have been used for military applications for many years. These applications include guided missile systems, night detection scopes, haze detection, and infrared photography. Infrared detectors are common in many industrial applications now. These applications include heat-sensitive control systems, optopyrometers, and infrared spectroscopy for gas analysis.

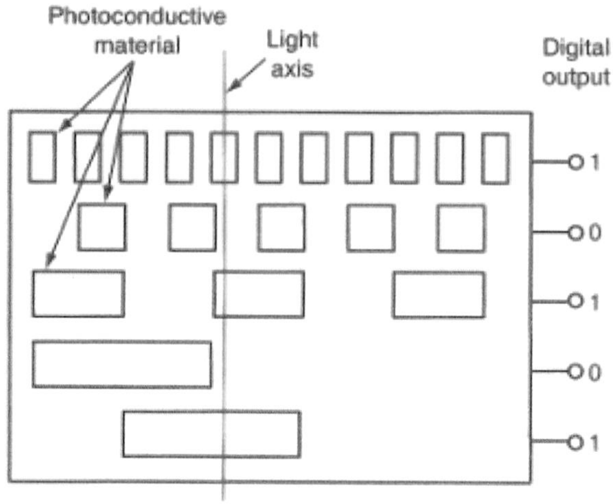

Figure 18.38 Digital readout position indicator.

An important principle of infrared detection is that all objects emit infrared radiation. Infrared camera systems have been devised that can produce images in darkness by detecting infrared thermal radiation. Since the human body emits thermal infrared radiation, infrared detectors can be used to detect individuals in darkness without a light source. This application is used in security applications.

18.8.8 Ultraviolet detectors

Detectors have been designed to respond to electromagnetic radiation in the ultraviolet range. Many design problems have been encountered with ultraviolet detectors. Consequently, they are not used as often as detectors for the visible light and infrared regions of the spectrum. Two types of optoelectronic devices that can be designed for use in the ultraviolet region are multiplier phototubes and cadmium selenide photoconductive cells.

18.8.9 Optoelectronic position detectors

Various types of **optoelectronic devices** have been designed to sense the position of a light beam. These devices ordinarily produce an electrical output based on the position of the light beam. These devices can be used with digital systems to produce electrical outputs as shown in Figure 18.38. The electrical signal output is based on the patterns of the photoconductive material onto which the light line is focused.

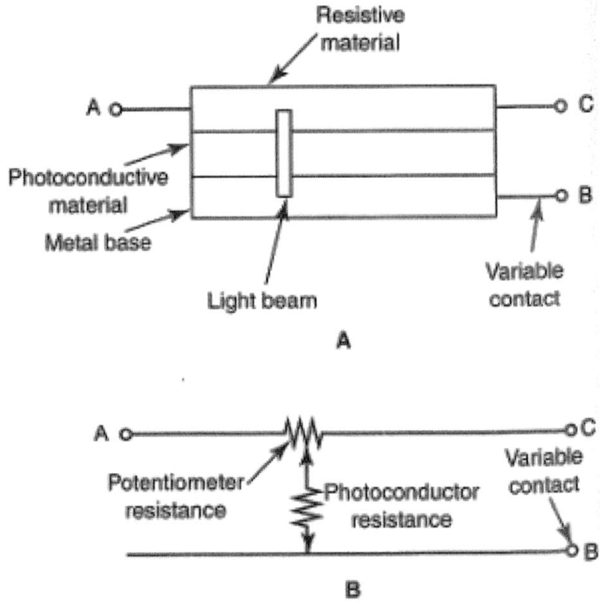

Figure 18.39 Photopotentiometer. A – Construction. B – Electrical equivalent.

18.8.10 Photo potentiometric devices

We previously discussed photoresistive devices. The photoresistive principle can be extended to design photo potentiometric devices. A **photo potentiometer** uses a light beam rather than a wiper contact such as a conventional potentiometer does. Thus, many of the problems associated with potentiometers, such as wear and linearity, can be avoided by using a photo potentiometer.

The operational theory of a **photo potentiometer** is shown in Figure 18.39. A precision wire-wound or composition resistive material is used to establish the potentiometer's maximum resistance between points A and C. Point B is the variable contact. A light beam is focused between the resistive material and a metal base. Between the metal base and the resistive material, a photoconductive material is used. The resistance of the photoconductive material varies according to the light exposed to its surface area. Conduction from either terminal A or terminal C to terminal B of the photo potentiometer depends on the position of the light beam. The light beam can be focused anywhere along the entire length of the resistive material. Thus, a variable resistance between points *A* and *B* and between points *C* and *B* can be accomplished by adjusting the position of the light beam.

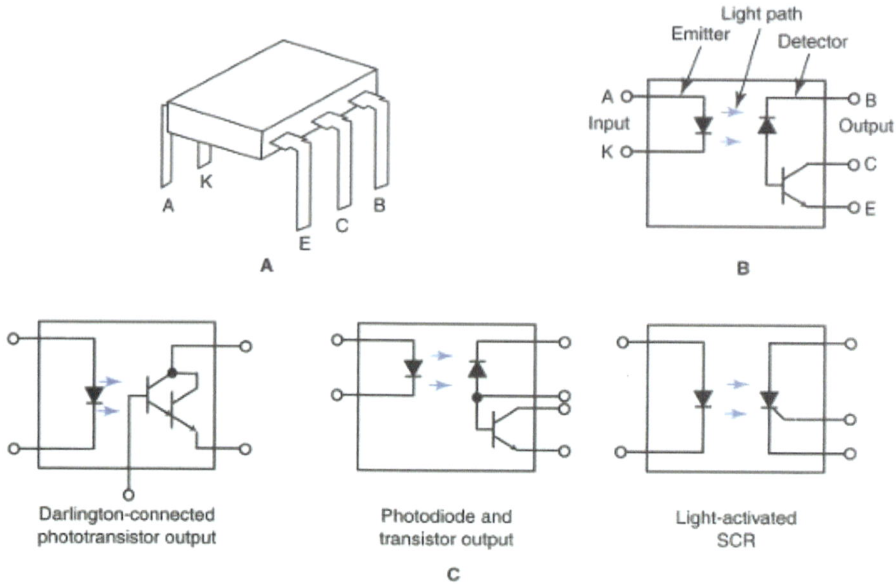

Figure 18.40 Optoelectronic couplers. A – Package. B – Equivalent circuit. C – Other configurations.

18.8.11 Optoelectronic couplers

Optoelectronic couplers, which are sometimes referred to as optoisolators, provide a one-way transfer path for electrical signals. Optoelectronic couplers are commonly manufactured in dual inline packages as shown in part A of Figure 18.40. Note the equivalent circuit of the optoelectronic coupler (part B Figure 18.40). A light-emitting diode (LED) or infrared-emitting diode (IRED) is combined with a transistor amplifier in the same package. The input of this device is the LED. The output is the transistor circuit. The collector current of the transistor varies in direct proportion to the current through the LED. Therefore, when an input signal is applied to the LED, it emits light in proportion to the applied signal voltage. The light causes the transistor to conduct, thus increasing the collector current. This is achieved with no electrical connection between the devices that compose the coupler. Isolation of the two circuits is therefore achieved.

Optoelectronic couplers can be used for many coupling requirements. They are not as bulky as transformers, and they last for longer periods. Their input/output characteristics are relatively linear in terms of LED current versus transistor collector current. They can be used for rapid-response applications, as well as for isolation between a power source and the load and isolation of noise from electronic circuits.

The amount of isolation between the two portions of the **optoelectronic coupler** is determined by the material in the light path and the distance between the LED and the transistor detector. As the distance between these portions increases, greater isolation is achieved. However, as the distance increases, the current transfer ratio between the two devices decreases, causing signal transfer to become poorer. Therefore, a compromise must be made between isolation and current transfer for optimum results. Other electronic couplers are available for voltage isolation requirements as high as 50,000 volts.

Optoelectronic couplers can be manufactured in a variety of ways. By adding additional electronic elements in combination with the detector, better sensitivity and response time can be achieved. The equivalent symbols for optoelectronic couplers using (left) a Darlington-connected phototransistor output detector, (middle) a photodiode and transistor output detector, and (right) a light-activated silicon-controlled rectifier detector output are shown in part *C* of Figure 18.40. Other configurations of emitters and detectors can be used for specialized coupling or isolation applications. Some typical applications of optoelectronic couplers are in telephone communications, relay control, DC power control, and high-voltage AC switching.

Optoelectronic couplers have numerous industrial control applications. Part A of Figure 18.41 shows an optocoupler time-delay circuit. The time delay is brought about by the combination of R_1 and C_1. The LED is shunted by C_1, and therefore turn-on of the LED is delayed by the time constant of R_1C_1. After the time interval required to activate the LED, an output is produced at the collector of the transistor detector. In this circuit, the time-delay elements are isolated from the controlled circuit.

Part B of Figure 18.41 shows an **optocoupler switch** that uses an SCR detector. A trigger pulse applied to the LED causes the SCR to conduct. The conductive SCR then causes current to flow through an external load. Again, the detector is electrically isolated from the circuit that controls it.

Part C of Figure 18.41 illustrates a pair of **LASCR optoelectronic couplers** used as an AC relay. The optocouplers are connected in a back-to-back configuration so that a trigger pulse of proper polarity applied to either LED element causes the corresponding SCR detector to conduct. SCR conduction causes the load, which is connected to an AC source, to be activated.

18.9 Optoelectronic Control Examples

Several applications of **optoelectronic systems** in industry have been discussed in the previous sections. An understanding of how these devices

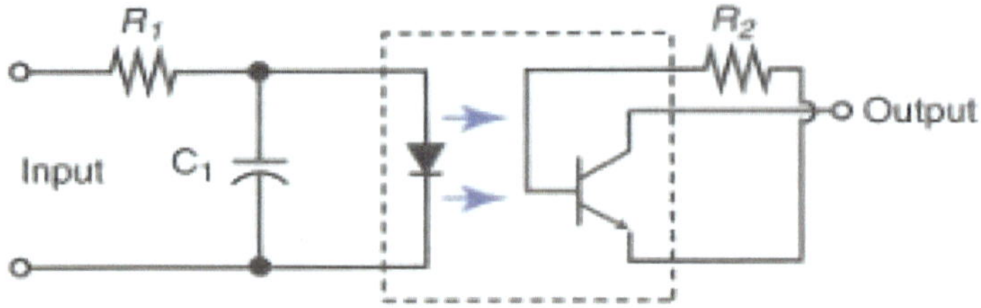

A Optocoupler time delay circuit.

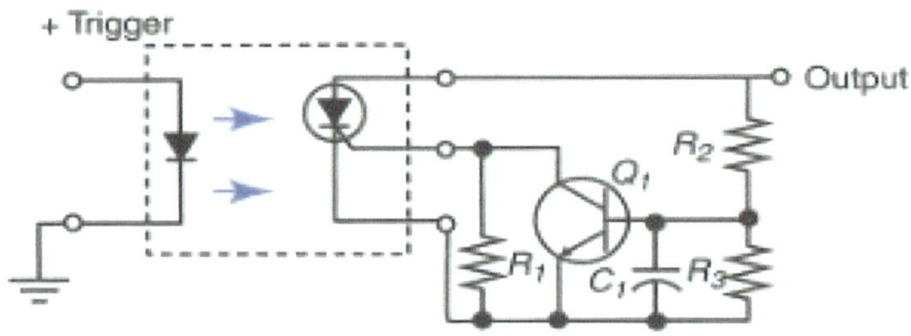

B Optocoupler switch.

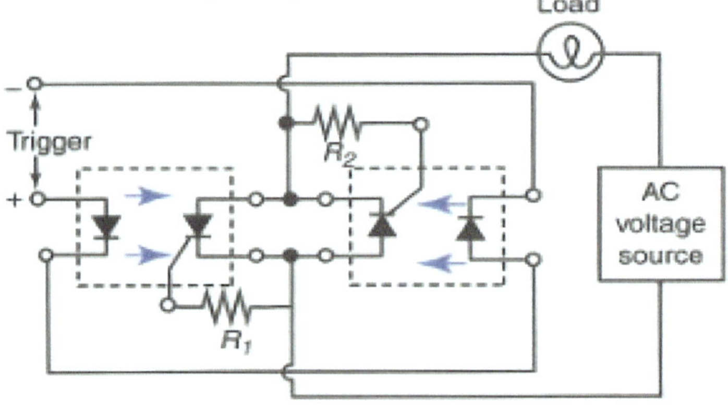

C Optocoupler ac relay.

Figure 18.41 Optoelectronic coupler applications.

operate is basic to the design of circuits for a specific application. The previous sections have attempted to show some of the basic operational characteristics of typical optoelectronic devices.

Optoelectronic control systems in industry are becoming more prevalent each day. As new optoelectronic devices are developed, new ways of utilizing them will result. In the following sections, we will discuss some of the general types of controls that utilize optoelectronic devices and some circuits that can be used for optoelectronic control.

There are so many applications of **optoelectronic control** today that it is difficult to categorize all of them. We will therefore attempt to discuss only those applications that are widely used. Some of the more popular applications are the following:

- Level control.
- High-speed counting.
- Sorting and routing.
- Detecting position.
- Automatic inspection.
- Preventing conveyor jamming.
- Safety hazard detection.
- Automatic stacking.
- Measurement.

18.9.1 Level control

Level control is accomplished by placing the light source at the same height above the conveyor line as the desired level of fill of a container. In the illustration of Figure 18.42, containers on a conveyor line are positioned under a material dispenser. When the container is in position, the actuator causes the dispenser to allow material to pass into the container. When the material reaches the level of the light source, the light beam is interrupted. With no light striking its surface, the detector causes a relay to activate. The activated relay, in turn, causes the actuator on the dispenser to close. When another container is in position, the actuator opens once more. This control system ensures a uniform level of material in each container.

18.9.2 High-speed counting

High-speed counting is also accomplished by the interruption of a light source. In this application, as shown in Figure 18.43, a precise beam of light is focused onto a detector. When the light beam is broken, the electronic counter is activated. High-speed counters can be used to count items at a rate of over 100 per second.

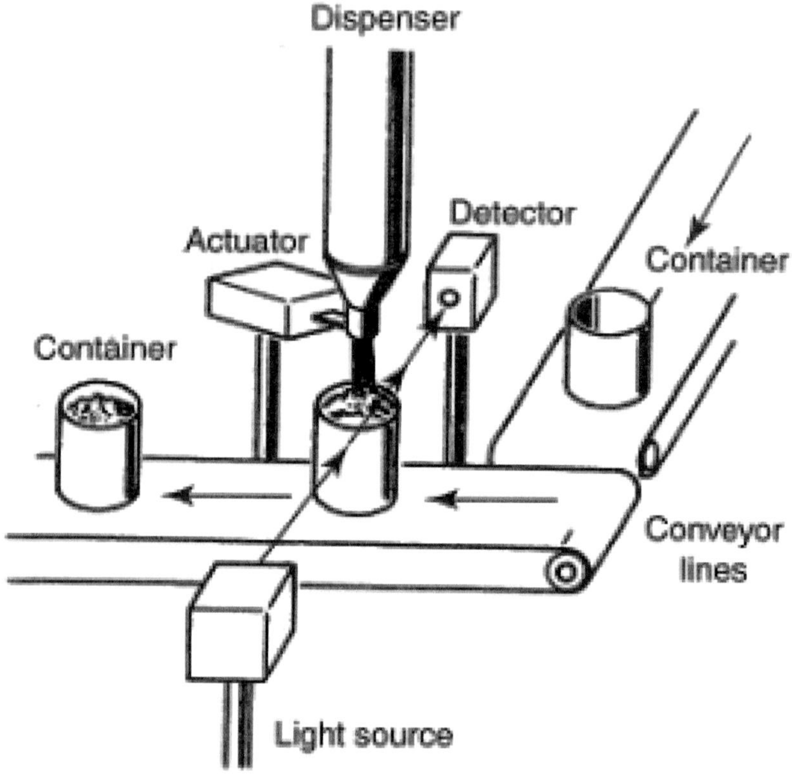

Figure 18.42 Level control.

18.9.3 Sorting and routing

Sorting and routing are conveyor processes that involve the sorting of unlike items and their routing to different locations. Items are typically sorted by size or by register marks. Figure 18.44 shows how cartons filled with different items can be sorted and routed by using register marks. A register mark is reflective and therefore causes a light-sensitive device to change condition when light focuses onto it. In the example shown, some cartons have one register mark and others have two marks. Cartons with one register mark activate the control, which pushes them onto conveyor line 1. Those cartons with two marks are pushed onto line 2. This system could be used for any number of conveyor lines.

18.9.4 Detecting position

The detection of **position** is important for many industrial processes. Figure 18.45 shows one method of detecting position. In this case, the

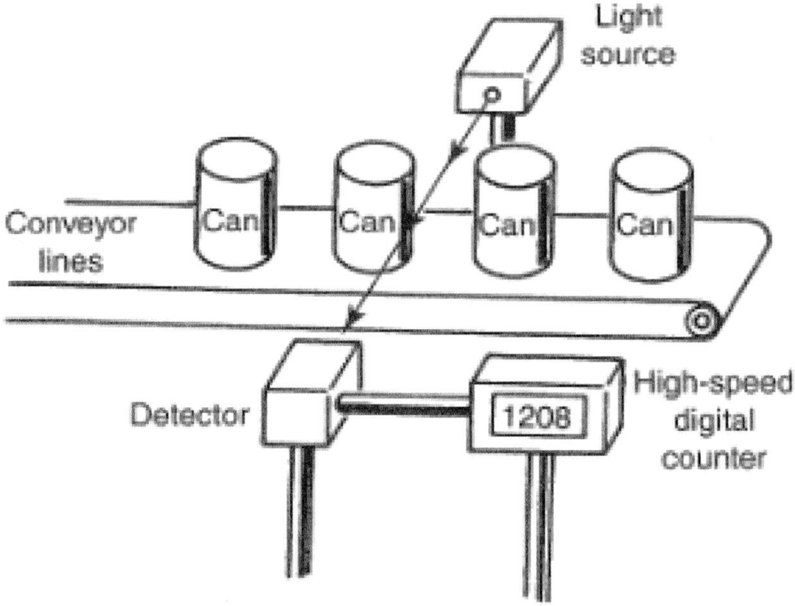

Figure 18.43 High-speed counting.

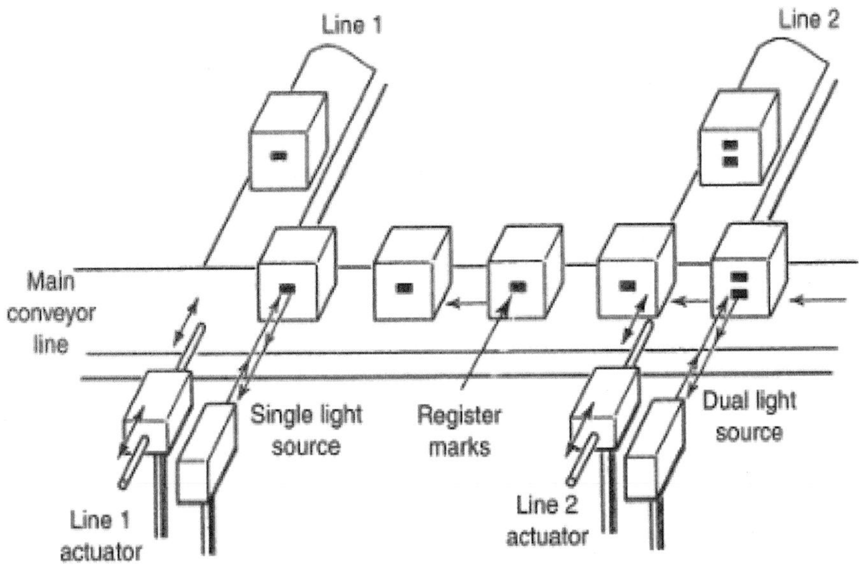

Figure 18.44 Sorting and routing.

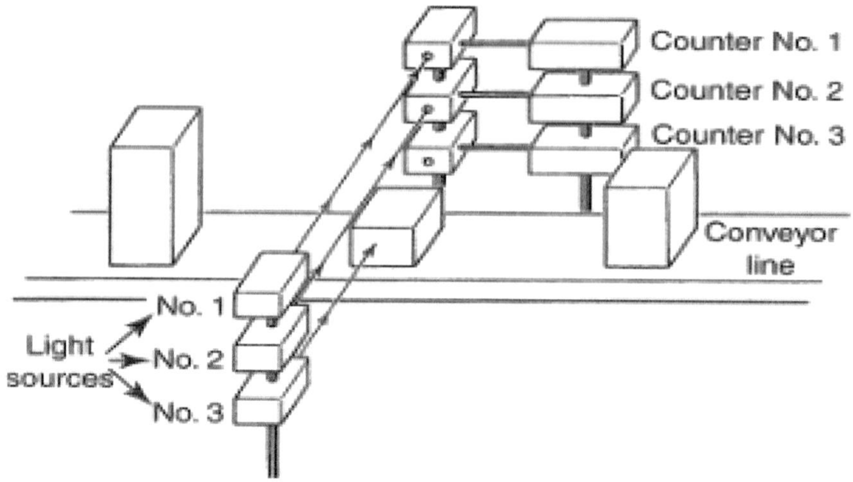

Figure 18.45 Detecting position.

height of the items is detected. Items of each height are counted by separate optoelectronic counting circuits. The counting circuits are designed so that counter 1 counts the smaller items. Counter 2 counts only when both source 1 and source 3 are interrupted. Counter 3 counts when sources 1, 2, and 3 are interrupted. Therefore, the position, in this case height of the object, determines which counter is activated.

18.9.5 Automatic inspection

Many industrial operations must be continuously or periodically inspected to ensure the accuracy of construction or that no parts are defective or missing. Several types of **inspection** can be accomplished by **optoelectronic systems**. Figure 18.46 shows an example of how workpieces can be inspected for possible missing parts. As a workpiece passes down the conveyor line, source 1 is interrupted when the workpiece passes. Source 1 serves as a reference or locator. At the instant source 1 is interrupted, source 2 should also be interrupted. Source 2 is used to check the location of the portion of the workpiece located in the center.

If the center portion is missing, source 2 will not be interrupted. A control circuit is designed to cause an actuator to remove the workpiece from the conveyor line if the center portion of the workpiece is missing. This method could be used to make several inspections of more complicated workpieces.

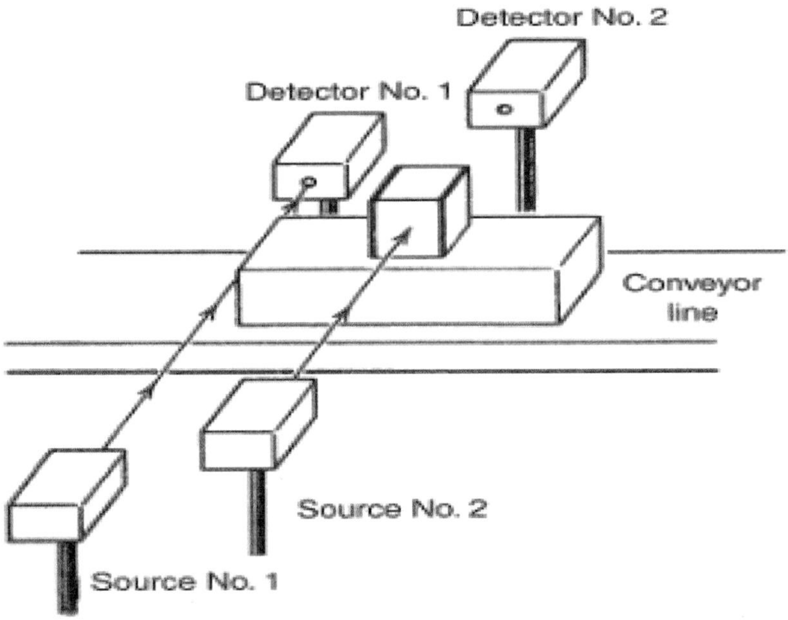

Figure 18.46 Automatic inspection.

18.9.6 Preventing conveyor jamming

Another common use of optoelectronic systems is to prevent **conveyor jamming**. Figure 18.47 shows an example of how this operation could be accomplished. Each part that passes through the light source interrupts the light beam and causes a timer to start. If the light beam remains interrupted by parts that are jammed together, the length of the timing period causes the conveyor motor to turn off or slow down so that jamming does not continue. Another timing period controls the amount of time that the conveyor motor remains off or slower in speed.

18.9.7 Safety hazard detection

Some machines used in industry are dangerous to use if not operated properly. Machine operators must be constantly alert to avoid accidents. It is possible to use optoelectronic systems to rapidly switch a machine off when the operator makes a mistake that might injure him or her in some way. An example of safety hazard detection is shown in Figure 18.48. When the machine operator places his or her hands on the table while the machine is running, a light beam is interrupted. Interruption of the light beam causes the machine

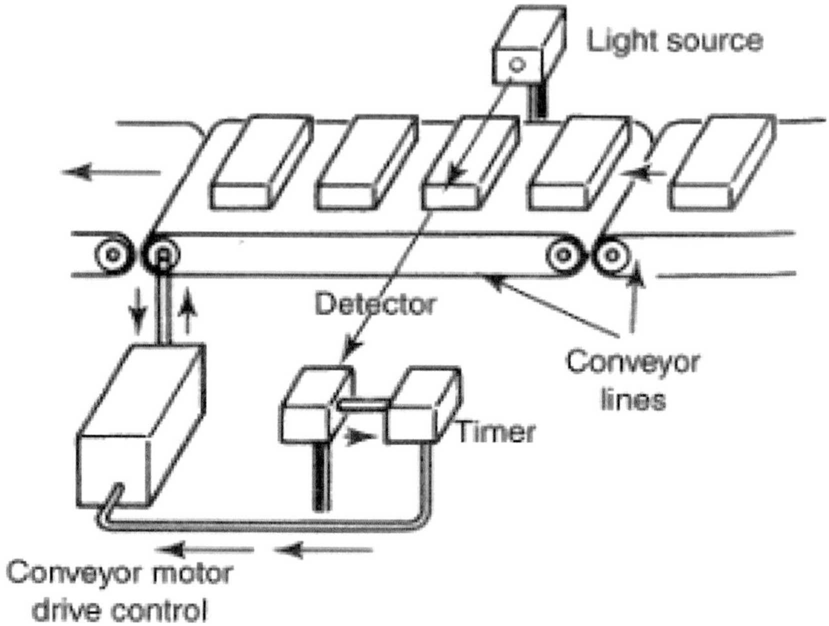

Figure 18.47 Preventing conveyor jamming.

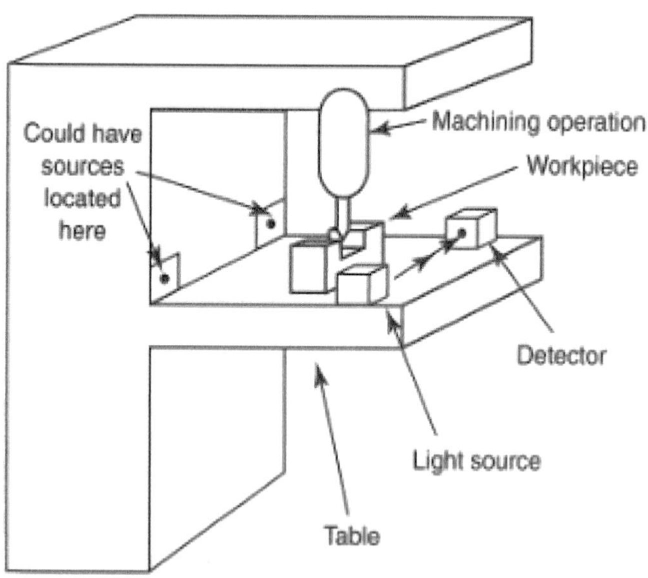

Figure 18.48 Safety hazard detection.

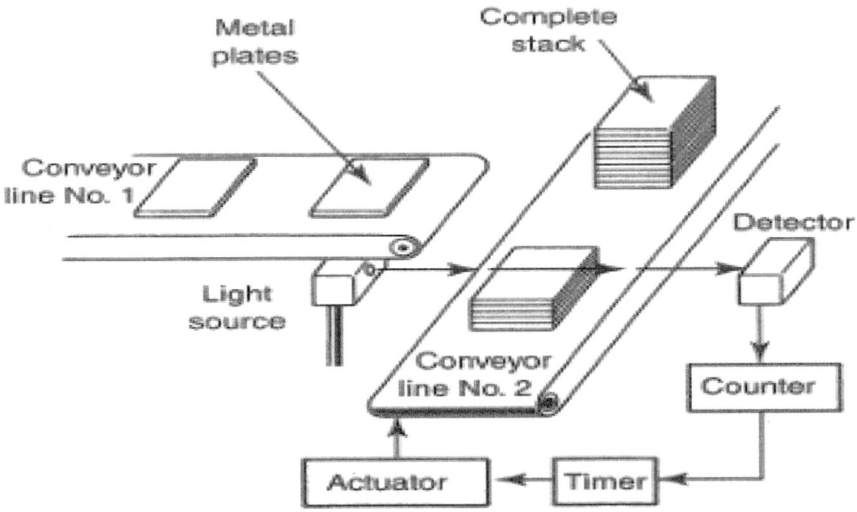

Figure 18.49 Automatic stacking.

to turn off. Thus, if the operator gets too near the machining operation, the machine automatically stops. This method helps eliminate many **industrial accidents**.

18.9.8 Automatic stacking

Figure 18.49 shows a conveyor line situation in which metal plates are being stacked onto a conveyor. It is possible to use optoelectronic systems to stack the desired amount of objects. In this case, a light source is interrupted as each metal plate passes through it from conveyor line 1. When the light source is broken, the counter connected to it is activated. When the desired number of metal plates has been stacked, conveyor line 2 is moved so that another stack can be started. **Automatic stacking** could also be accomplished by positioning the light beam at the desired height of the stack. When the objects are stacked to this height, conveyor line 1 stops and conveyor line 2 moves so that another stack can be started.

18.9.9 Measurement

Optoelectronic systems are readily adaptable to several types of **measurement**. An instrument called a **photometer** is used to measure illumination. This meter can be used to measure the luminous intensity of light

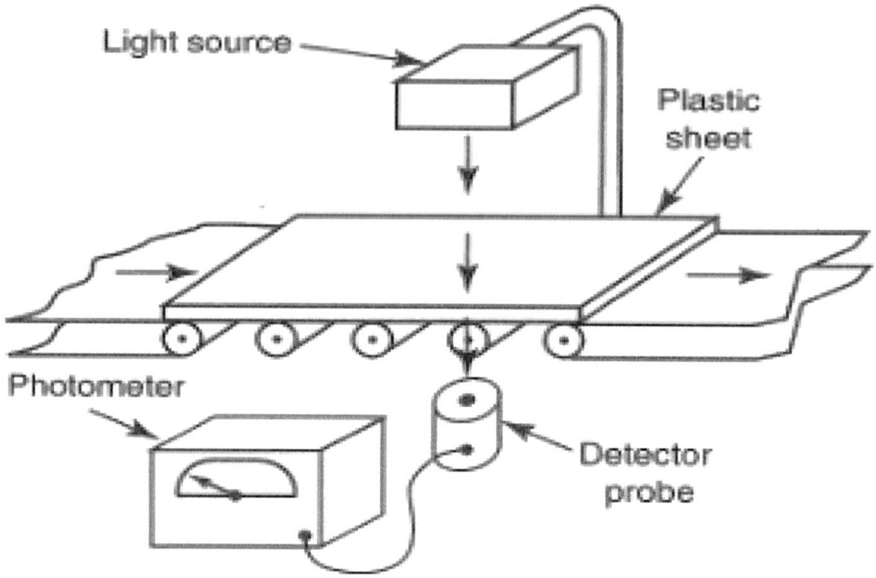

Figure 18.50 Optoelectronic measurement.

sources, to set light levels, and for other optical measurements. An example is shown in Figure 18.50 of an industrial measurement application using a photometer. The photometer uses an optoelectronic detector in its probe to serve as a transducer for the instrument. In the example shown, the density of sheet plastic is being checked by the deflection produced on the photometer. The probe detector is a photovoltaic cell that changes its electrical output with variations in light. The density of the sheet plastic should be consistent and, therefore, cause uniform deflection on the photometer. If a sheet of plastic does not have uniform density, the meter will indicate this. Thus, a photometer can be used to sense changes in the density of a translucent material.

Review Questions

Answer the following questions.

1. What are optoelectronic systems designed to do?

2. Compare interruptible and non-interruptible optoelectric systems.

3. What is an incandescent light source used for in an optoelectronic system?

4. What advantages do LEDs have over incandescent light sources in optoelectronic systems?

5. Define the following terms.

   a. Radiometric.

   b. Photometric.

   c. Heterochromatic.

   d. Candela.

   e. Reflection factor.

   f. Lumen.

   g. Lux.

6. What is meant by the terms through beam, reflex, and proximity detection? What is the purpose of each?

7. What are the three general categories of optoelectronic detector devices?

8. List some applications for each of the following devices.

   a. Phototube.

   b. Multiplier phototube.

   c. Photodiode.

   d. Phototransistor.

e. Darlington-connected phototransistor.

f. LASCR.

g. Photovoltaic cell.

h. Infrared detector.

i. Optoelectronic coupler.

9. List five industrial applications of optoelectronic systems for process control. Briefly discuss each application.

10. What is a photometer?

UNIT SEVEN

Digital Control Basics and Applications

In **Unit VII**, control systems are studied with regard to digital and programmable logic controllers (PLCs), and applications such as robotics, troubleshooting, and preventive maintenance (PM). These topics are discussed in the last five chapters.

Chapter 19 examines the basic concepts of digital control. Computers and microprocessors are discussed in **Chapter 20**. **Chapter 21** provides coverage of programmable logic controllers (PLCs). Robotics is studied in **Chapter 22**. **Chapter 23** is a logical last chapter. It deals with the troubleshooting and preventive maintenance of the equipment with which you are now familiar.

DOI: 10.1201/9788743810056-25

19

Basics of Digital Control

Objectives

Upon completion of this chapter, you will be able to:

- Describe digital electronic information processing.
- Convert numbers between decimal, binary, octal, binary coded decimal, and hexadecimal systems.
- Discuss the basic operational principles of binary AND, OR, and NOT gates.
- Explain the operation of flip-flop circuits.
- Explain how a binary and decimal counter works.
- Discuss numerical control (NC) systems.

Important Terms

analog
AND gate
base
binary
binary-coded decimal (BCD),
binary counters
binary point
bistable
bit
byte
closed-loop system
comparator
decade counter
decimal point
digital
feedback
flip-flop
hexadecimal
inverter
logic gate
NAND gate
negative logic
NOR gate
NOT gate
numerical control (NC) system
octal
open-loop system
OR gate
place value
positive logic
standard, truth table

DOI: 10.1201/9788743810056-26

Introduction

The electronic circuitry used with automated machines supplies commands that control motors and other industrial systems. These circuits also store information, count, encode, and decode. The circuits that perform these functions are called **logic circuits.** The technology that controls automated systems is called **digital electronics**. In this chapter, the basics of digital control will be discussed.

19.1 Electronic Information Processing

The term **computer** is used to cover a number of functions, but primarily refers to a system that performs automatic computations. Computers range from pocket calculators to complex central units that serve an entire organization. Information is processed in two states-digital or analog.

Analog information varies continuously. An example is taking the temperature with a common thermometer. The mercury indicator in an analog thermometer is, at times, between one degree marking and the next. **Digital** information, by contrast, occurs in separate full units. With a digital clock, for example, the time is always displayed as one unit (second) or the next. It is never between units.

Digital computers use two numbers that represent either the presence or absence of voltage. A voltage pulse is usually represented by one (1). Absence of voltage is usually indicated by a zero (0). A single pulse is described as a bit of information. The term bit is a contraction of the phrase binary digit. A group of 8 bits produces a **byte**, sometimes called a **binary word**.

19.2 Digital Number Systems

The **number system** we use in everyday life is the base 10 (decimal) system. In this system, ten digits 0,1,2,3,4,5,6,7,8, and 9 are used for counting. The number of digits used in the system is called its base. The decimal system, therefore, has a base of 10.

Number systems give digits a **place value**. This refers to the position of a digit with respect to a reference. In the decimal system, this reference is called the **decimal point**. The largest digit that can be used in a specific place is determined by the base of the system. In the decimal system, the first position to the left of the decimal point is called the units place. Any digit from 0 to 9 can be used in this place. When number values greater than 9 are used, they must be expressed in two or more places. The position to the left of the

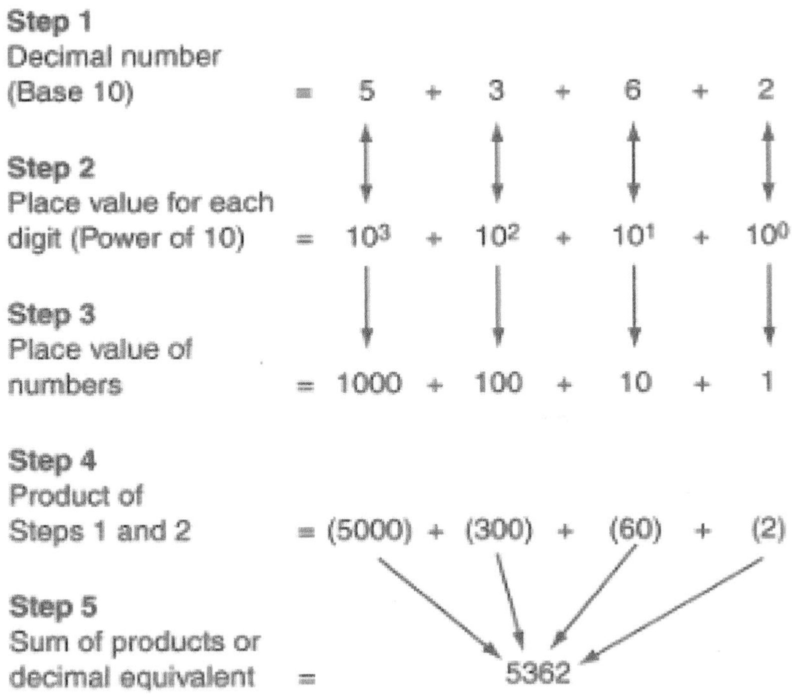

Figure 19.1 Expressing a base 10 (decimal) number.

units place is the 10s place. The number 99 is the largest value that can be expressed by two places in the decimal system. Each place added to the left extends the number value by a power of 10.

A **number value** in any system can be read by adding numbers in each place. The decimal value 2583, for example, can be expressed as: $(2 \times 1000) + (5 \times 100) + (8 \times 10) + (3 \times 1)$. The values have increased by a power of ten for each place to the left of the decimal point. In the decimal system, this is 10^3, 10^2, 10^1, 10^0. Mathematically, each place value is the number times a power of the system base (Figure 19.1). The decimal number system is difficult to use with digital control systems. Instead, the binary system is typically used.

19.2.1 Binary number system

The **binary** number system uses 2 as its base. The largest value that can be expressed in a specific place is the number 1. As a result, only the numbers 0 and 1 are used. The first place to the left of the reference (in this case, called the **binary point**) represents units, or 1s. Places to the left of the binary point

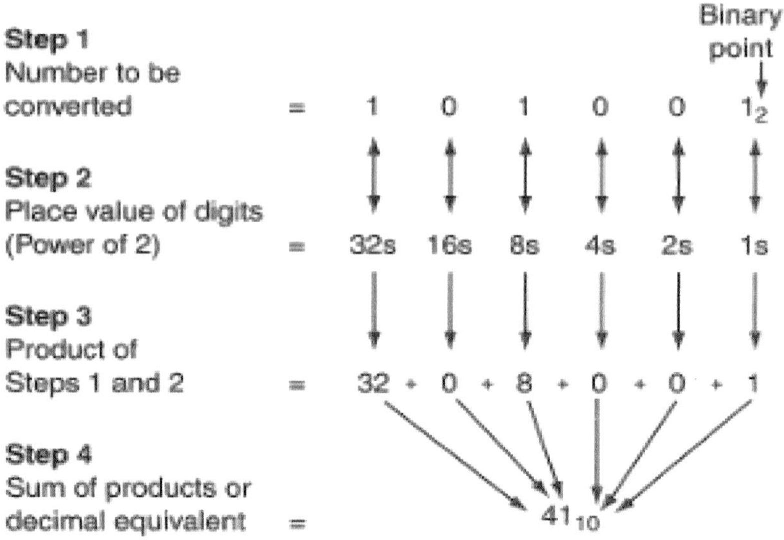

Figure 19.2 Making a binary-to-base 10 conversion.

are expressed in powers of 2. These include: $2^0 = 1$, $2^1 = 2$, $2^2 = 4$, $2^3 = 8$, $2^4 = 16$, $2^5 = 32$, $2^6 = 64$, and so on.

When different number systems are used, a **subscript** number is used to identify the base. The number 100_2 indicates that the binary (base 2) system is being used. In the binary system, 100 is read as one-zero-zero, instead of one hundred. Start at the first digit to the left of the binary point. This number would have place values of $(0 \times 2^0) + (0 \times 2^1) + (1 \times 2^2)$ or $0 + 0 + 4$. Thus, the binary number 100 is equal to 4 in the decimal number system.

The conversion of a **binary number** to an equivalent **decimal number** is shown in Figure 19.2. When converting, write down the binary number first. Starting at the binary point, indicate the decimal equivalent for each place location where a 1 is indicated. For each 0 in the binary number, leave a blank space or indicate a 0. Add the place values and then record the decimal equivalent.

The conversion of a **decimal number** to its **binary equivalent** is shown in Figure 19.3. As shown, the conversion is done by repeatedly dividing by 2. When the quotient is even with no remainder, the value is 0. When the quotient has a remainder, the value is 1. Division continues until the quotient is zero.

Electronically, the value of 0 is expressed as low voltage or no voltage. The number 1 is used to indicate a voltage larger than 0. Binary systems that use these values are said to have positive logic. (**Negative logic**, by comparison, assigns a voltage to the 0 and no voltage to the number 1.)

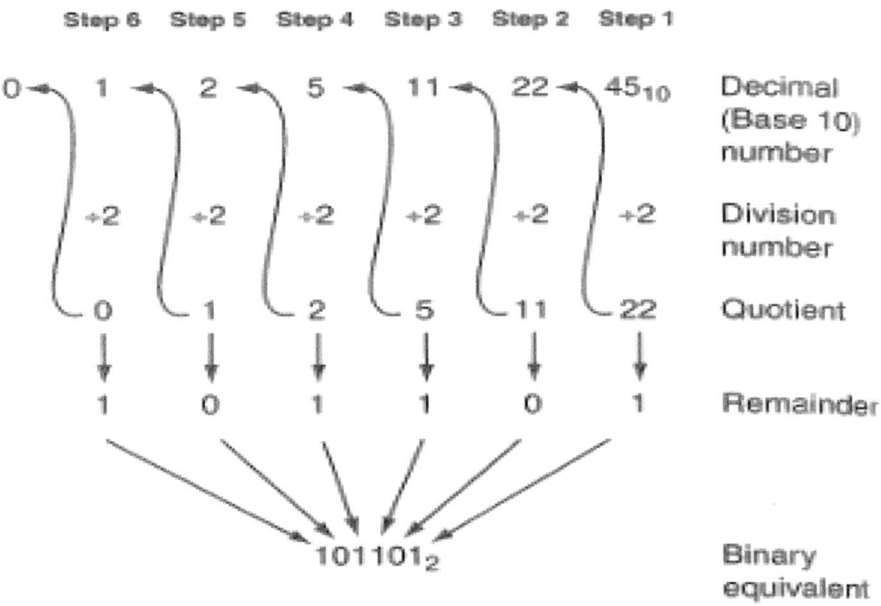

Figure 19.3 Decimal-to-binary conversion is achieved by dividing by 2.

Positive logic will be used for all discussions and examples in this chapter.

The two **operational states** of a binary system, 1 and 0, are natural circuit conditions. When a circuit is turned off or has no voltage applied, it is in the 0 state. A circuit that has voltage applied is in the 1 state. This makes it possible to change states in less than a microsecond. Electronic devices can thus manipulate millions of 0s and 1s per second, and process information very quickly.

19.2.2 Binary-coded-decimal number system

When large numbers are written using the binary method, they are difficult to use. For example, $1101001_2 = 105_{10}$. The **binary-coded decimal (BCD)** number system of counting was devised to overcome this problem.

With the **BCD system**, the base 10 number is first divided into digits according to place value, Figure 19.4. The number 105_{10} is made up of digits 1-0-5. Each digit is then converted to a four-digit binary number. Thus, 105_{10} results in 000100000101_{BCD}. The largest digit expressed by any group of *BCD* is 9. Decimal numbers up to 999_{10} can be converted using only three four-digit binary numbers. The space between each group of digits is important to include.

Given decimal number		105_{10}	
Step 1 Group the digits	(1)	(0)	(5)
Step 2 Convert each digit to binary group	(0001)	(0000)	(0101)
Step 3 Combine group values	0001 0000 0101 BCD		

Figure 19.4 When making a BCD conversion, four binary digits represent each decimal digit.

19.2.3 Octal number system

The **octal number system**, or **base 8 system**, can be used to process large numbers. The octal system is based on the same principles as the decimal and binary systems. The digits 0,1,2,3,4,5,6, and 7 are used in the place positions. The place values of digits, moving to the left of the octal point, are powers of eight: 8^0 = units, or 1 s, $8^1 = 8$ s, $8^2 = 64$ s, $8^3 = 512$ s, $8^4 = 4096$ s, and so on.

The process of converting an **octal number** to a **decimal number** is similar to the process used for binary-to-decimal conversion (using powers of 8 instead of 2). Suppose that the number 372_8 is to be changed to an equivalent decimal number. The procedure is outlined in Figure 19.5.

Converting an **octal number** to an equivalent **binary number** is very simple. It is similar to *BCD* conversion. The octal number is first divided into digits according to place value. Each digit is then converted into an equivalent binary number using only three digits, Figure 19.6.

Converting a **decimal number** to an **octal number** means dividing by the number 8. After the quotient has been determined, the remainder is brought down as the place value. When the quotient is even with no remainder, the value is 0. The procedure for converting into 4098_{10} octal is outlined in Figure 19.7.

If a **binary number** (110100100_2, for example) is to be changed to an equivalent **octal number**, the digits must first be divided into groups of three binary numbers starting at the binary point, Figure 19.8. Each binary group

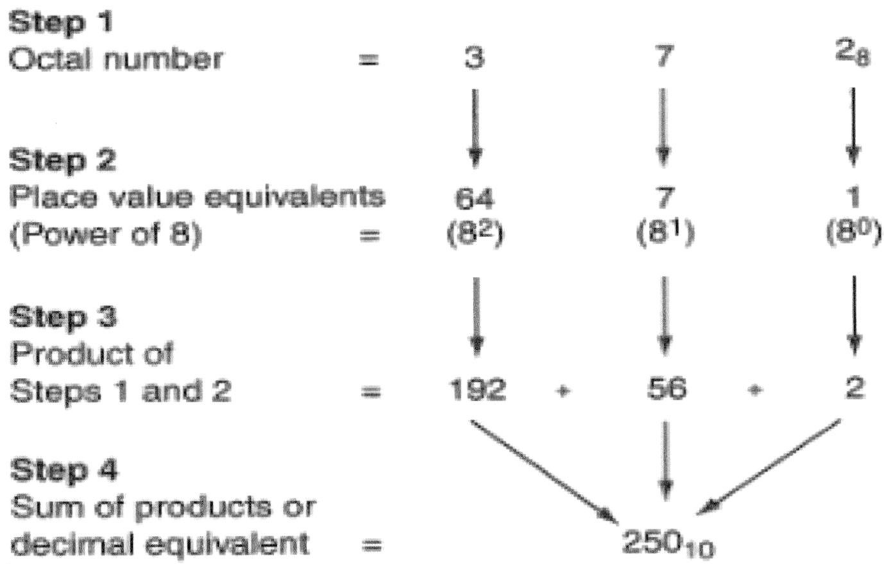

Figure 19.5 Octal-to-decimal conversion requires the same basic process as binary-to-decimal conversion.

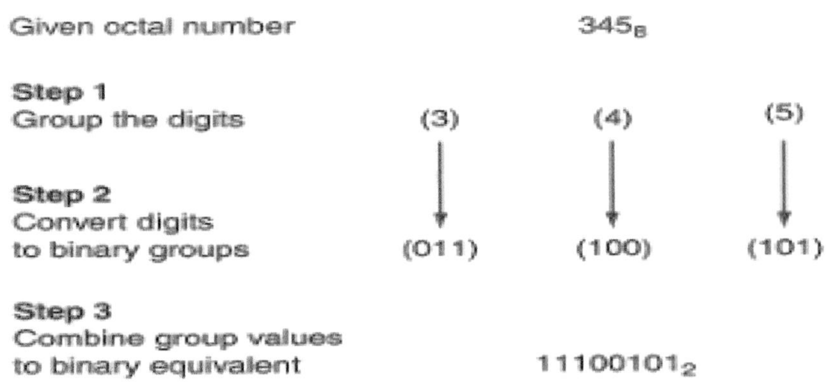

Figure 19.6 In octal-to-binary conversion, a maximum of three binary numbers is used per digit.

is then converted into an equivalent octal number. These numbers are then combined, while remaining in their same respective places, to represent the equivalent octal number. Converting a binary number to an octal number is often done for digital circuits. Binary numbers are first processed at a very high speed. An output circuit then accepts this signal and converts it to an octal number displayed on a readout device.

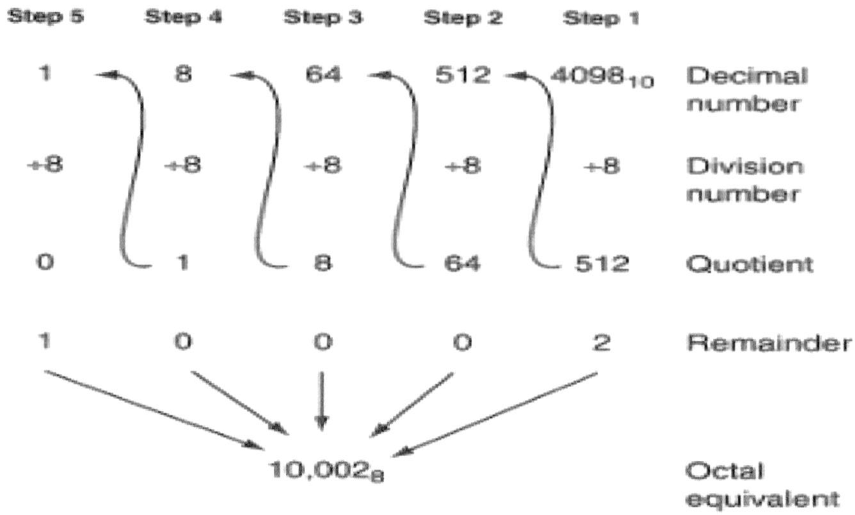

Figure 19.7 A decimal-to-octal conversion involves dividing by 8.

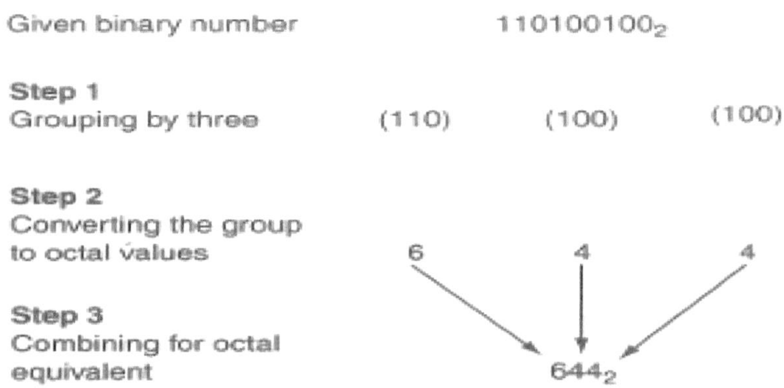

Figure 19.8 Binary-to-octal conversion is often done for digital circuits.

19.2.4 Hexadecimal number system

The **hexadecimal number system** is also used to process large numbers. The base of this system is 16. The largest number used in a place is 15. Digits used are the numbers 0–9 and the letters A–F. The letters A–F stand for the digits 10–15, respectively. The place value of digits to the left of the hexadecimal point is powers of 16:$16^0 = 1$, $16^1 = 16$, $16^2 = 256$, $16^3 = 4096$, $16^4 = 65{,}536$, and so on.

Changing a **hexadecimal number** to a **decimal number** uses a process similar to those used for other conversions. The hexadecimal number is

Step 1: Hexadecimal number	=	1		2		C		D_{16}
Step 2: Place values (Power of 16)	=	4096s		256s		16s		1s
Step 3: Conversion of letters to numbers	=	1		2		12		13
Step 4: Product of Steps 2 and 3	=	4096	+	512	+	192	+	13
Step 5: Sum of products or decimal equivalent	=	4813_{10}						

Figure 19.9 Hexadecimal-to-decimal conversion involves an extra step, converting the letters to numeric values.

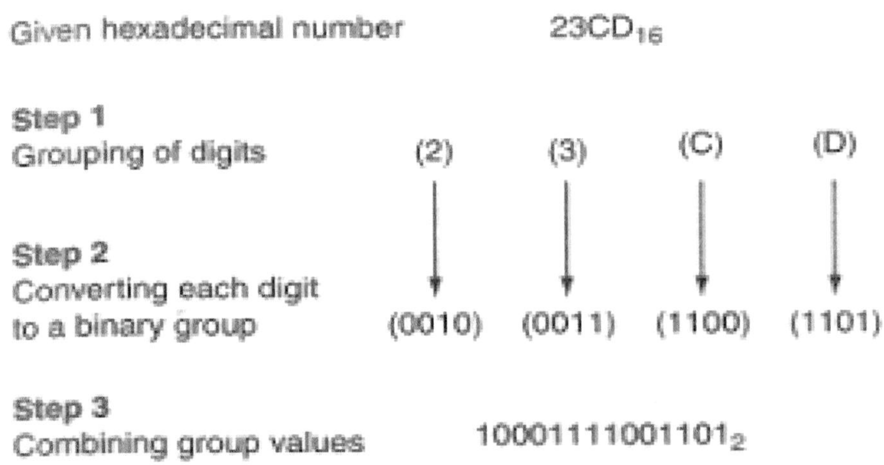

Figure 19.10 A hexadecimal-to-binary conversion requires grouping the binary numbers in Step 2.

written in proper digital order, Figure 19.9. The place values are then positioned under each respective digit in Step 2. In Step 3, the letters are converted to numeric values. Each of the values in Step 3 is then multiplied by its corresponding place value from Step 2. In Step 4, the products are added together. The resulting sum (Step 5) is the decimal equivalent of the hexadecimal number.

The process of changing a **hexadecimal number** to a **binary equivalent** is a simple grouping operation, Figure 19.10. Initially, the hexadecimal number is separated into digits. Each digit is then converted to its equivalent in binary form (D in hexadecimal notation equals 13 in decimal form, or 1101 in binary). Step 3 shows the groups combined to form the equivalent binary number. Note that the leading zeros from the binary number 0010 are

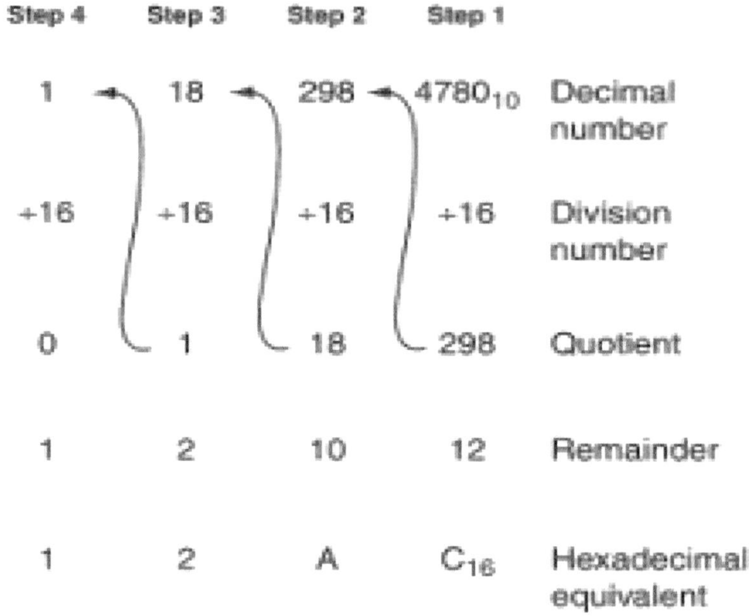

Figure 19.11 A decimal-to-hexadecimal conversion uses a divisor of 16.

dropped. Also note that, unlike converting to , there are no spaces between each block of four binary digits.

Decimal number-to-**hexadecimal number** conversion is done by dividing by 16 . Remainders can be as large as 15, Figure 19.11.

Converting a **binary number** to a **hexadecimal equivalent** is the reverse of the hexadecimal-to-binary process, Figure 19.12. Initially, the binary number is divided into groups of four digits, starting at the binary point. Each group is then converted to the equivalent hexadecimal value, and the groups are combined to form the answer.

19.3 Binary Logic Circuits

Binary signals are far superior to octal, decimal, or hexadecimal signals for use in logic circuits. They can be processed easily because they can be represented by two stable states of operation: on or off, 1 or 0, up or down, voltage or no voltage, right or left, and so on. There is no in-between state.

The **symbols** used to represent an operational state are very important. In positive binary logic, voltage, on, or true results from the 1 operational state. **No voltage**, **off**, or **false** results from the 0 condition. A circuit can be set to either state and will remain at that setting until a change occurs.

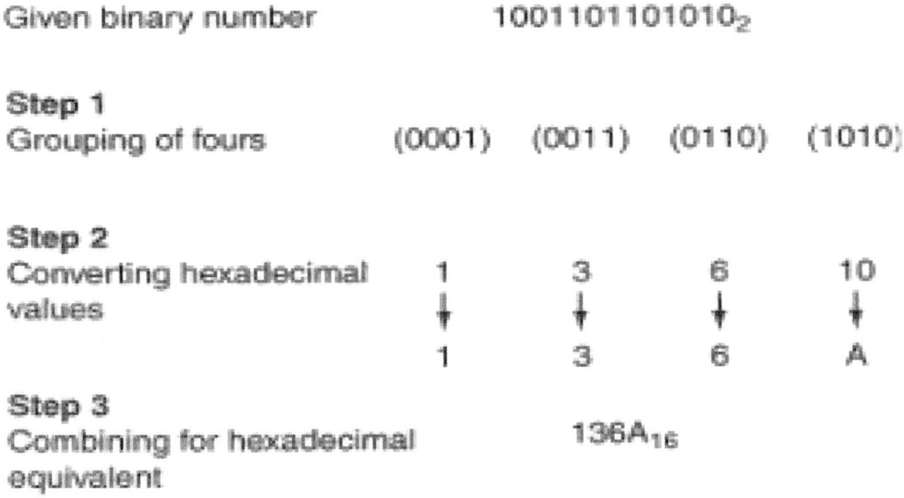

Figure 19.12 Binary-to-hexadecimal conversion is done by first creating groups of four binary numbers, then converting them to hexadecimal values.

Any electronic device that can be set in one of these two operational states by an outside signal is said to be **bistable**. These devices include relays, lamps, switches, transistors, and diodes. A bistable device can store one binary digit or bit of information. By using many of these devices, it is possible to build a binary logic circuit that will make logical decisions based upon input signals. The basic binary logic circuits are the AND circuit, the OR circuit, and the NOT circuit. The decision made by each type of circuit is unique. These circuits are also called **logic gates**. The term gate refers to the circuit's ability to pass or block certain signals. An IF ... THEN sentence is often used to describe the basic operation of a logic gate. For example, IF the inputs applied to a gate are all 1, THEN the output will be 1.

The **logic gates** discussed in the following examples illustrate basic operation using a simple switch and lamp input/outputs. In actual applications, logic gates are typically integrated circuits (ICs).

19.3.1 AND gates

An **AND gate**, Figure 19.13, has two or more inputs and one output. In part A of Figure 19.13, when a switch is turned on, it represents a 1. Off represents a 0. The lamp displays a 1 when it is on and 0 when it is off. If both inputs are in the 1 state (closed) simultaneously, then a 1 will be output (the lamp will light). A **truth table** shows combinations of inputs and the resulting outputs of a logic gate. The symbol for an AND gate is shown in part B of Figure 19.13.

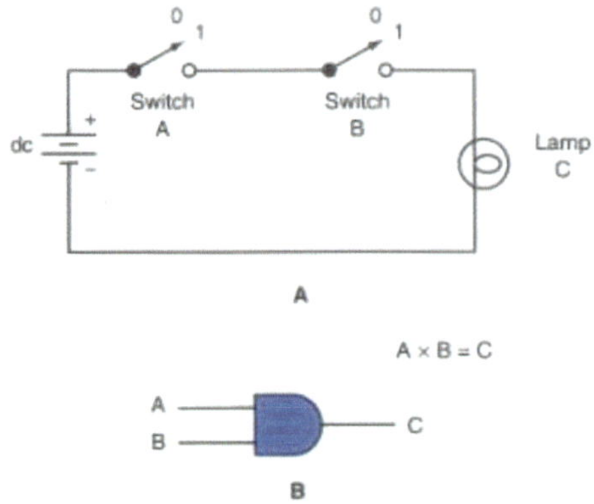

Input switch	Input switch	Output lamp
A	B	C
0	0	0
0	1	0
1	0	0
1	1	1

C

Figure 19.13 The AND gate circuit. A – Simple circuit example. B – The AND gate symbol. C – The AND gate truth table.

The operation of the gate is simplified by describing the input/output relationships in the truth table. The table in part C of Figure 19.13 shows the predictable alternatives. The AND gate produces a 1 output only when switches A and B are both 1. Mathematically, this action is described as A × B = C.

Each input to an **AND gate** can create two operational states: 1 and 0 . A two input AND gate has 2^2, or 4, possible combinations that influence the output. A three-input gate has 2^3, or 8, combinations, while a four input gate has 2^4, or 16, combinations. These combinations are normally placed in the truth table in binary order. For a two input gate this is 00,01,10, and 11. This represents the binary count of 0,1,2, and 3 in order.

19.3.2 OR gate

An **OR gate**, Figure 19.14, has two or more inputs and one output. Like the AND gate, each input to the OR gate produces two possible states: 1 or 0. *A*1 is output when both switches are 1 or when either switch A or B is 1. Mathematically, this action is described as $A + B = C$. This gate is used to make decisions as to whether or not a 1 appears at either input. The truth table, part *C* of Figure 19.14, shows that if any input is a 1, the output will be a 1.

19.3.3 NOT gate

A **NOT gate**, Figure 19.15, has one input and one output. The output of a NOT gate is opposite to the input. When the switch in part A of Figure 19.15

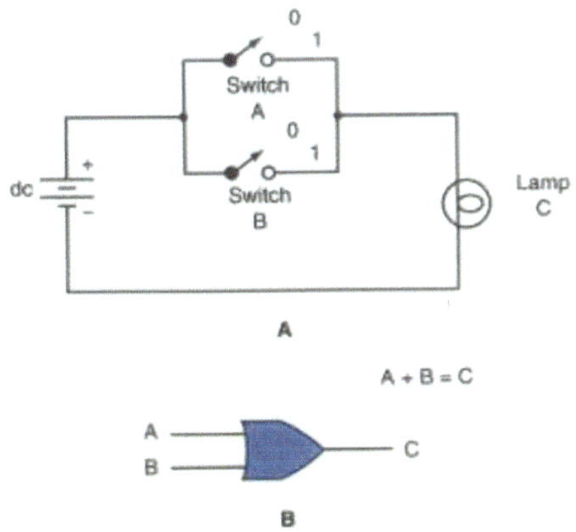

Input switch A	Input switch B	Output lamp C
0	0	0
0	1	1
1	0	1
1	1	1

C

Figure 19.14 The OR gate circuit. A – Simple circuit example. B – The OR gate symbol. C – The OR gate truth table.

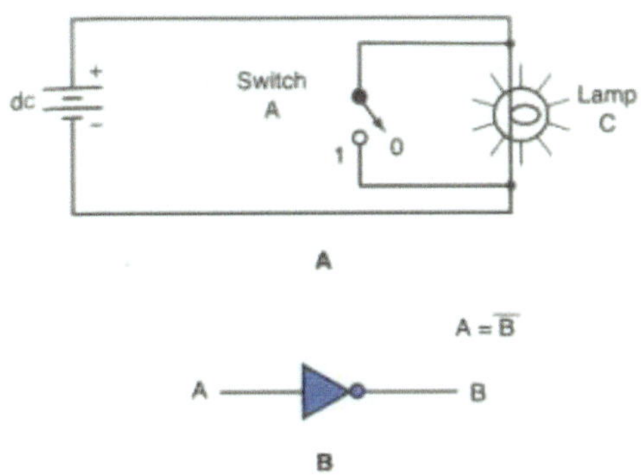

Input switch A	Output lamp B
0	1
1	0

C

Figure 19.15 The NOT gate circuit. A – Simple circuit example. B – The NOT gate symbol. C – The NOT gate truth table.

is on (in the 1 state), it shorts out the lamp. Placing the switch in the off condition (0) causes the lamp to be on (1). The NOT gate is also called an inverter.

19.3.4 Combination logic gates

When a **NOT gate** is combined with an **AND gate**, it is called a **NAND gate**. This is an inverted AND gate, Figure 19.16. When switches A and B are both on (in the 1 state), lamp C is off (0). When either or both switches are off, lamp is on (in the 1 state). Mathematically, the operation of a NAND gate is $A \times B = \overline{C}$. The bar over the C denotes the inversion, or negative function, of

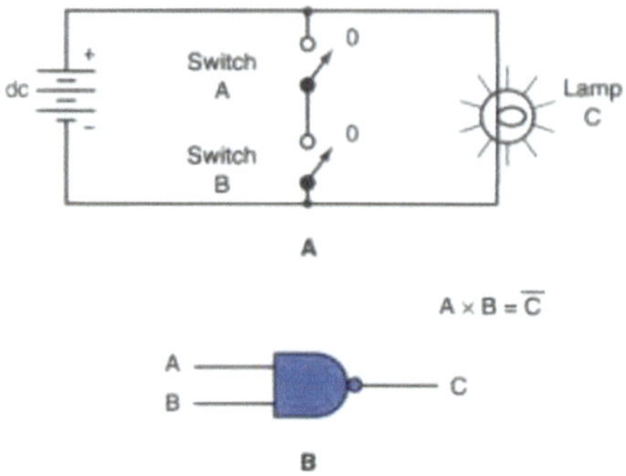

Input switch A	Input switch B	Output lamp C
0	0	1
0	1	1
1	0	1
1	1	0

C

Figure 19.16 The NAND gate circuit. A – Simple circuit example. B – The NAND gate symbol. C – The NAND gate truth table.

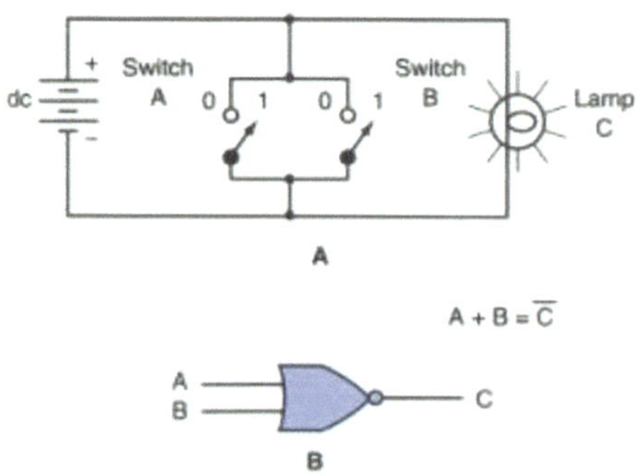

Input switch A	Input switch B	Output lamp C
0	0	1
0	1	0
1	0	0
1	1	0

C

Figure 19.17 The NOR gate circuit. A – Simple circuit example. B – The NOR gate symbol. C – The NOR gate truth table.

the gate. The schematic symbol for a NAND gate is the symbol for an AND gate with a small circle at the output end.

The combination of a **NOT gate** and an **OR gate** results in a **NOR gate**, Figure 19.17. A NOR gate is the opposite of an OR gate. Mathematically, the operation of a NOR gate is $A + B = \overline{C}$. A1 will appear at the output only when *A* is 0 and *B* is 0. The schematic symbol for a NOR gate is the symbol for an OR gate with a small circle at the output end.

19.4 Flip-Flops

Flip-flops are memory devices used in digital circuits. They are often the basic logic element for counting, temporary memory, and sequential switching operations. They can be used to hold an output state even when the input is completely removed. They can also change their output based on an appropriate input signal.

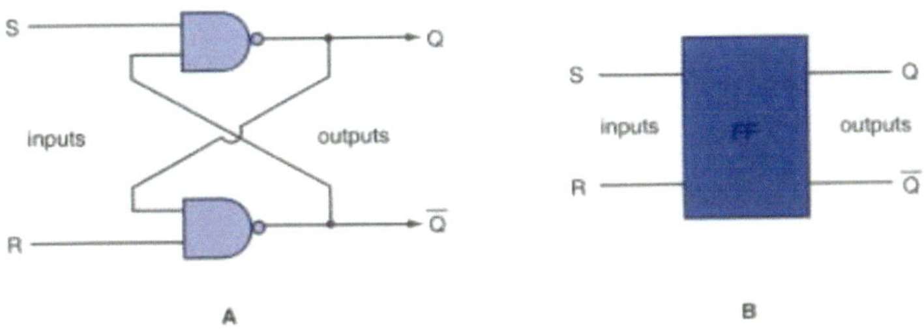

Applied inputs		Previous outputs		Resulting outputs		
S	R	Q	$\overline{Q}$	Q	$\overline{Q}$	
0	0	1	0	1	0	
0	1	1	0	0	1	
1	0	1	0	1	0	
1	1	1	0	-	-	Invalid
0	0	0	1	0	1	
0	1	0	1	0	1	
1	0	0	1	1	0	
1	1	0	1	-	-	Invalid

C

Figure 19.18 The R-S flip-flop. A – Logic diagram with NAND gates. B – Symbol. C – Truth table.

The **reset-set (R-S) flip-flop** is shown in Figure 19.18. Its logic diagram, symbol, and truth table are more complicated than those for a simple logic gate. The truth table includes the different states before an input occurs and how they change afterwards. Two of the operating conditions produce an unpredictable output.

Flip-flops often must be set and cleared at specific times with respect to other circuits. They operate in step with a clock pulse. The appropriate R-S inputs and clock pulse must all be present to cause a state change. A device called an R-S triggered flip-flop, or simply an **R-S-T flip-flop**, is shown in Figure 19.19. The truth table for an R-S-T flip-flop is basically the same as that for an R-S flip-flop. A state change occurs only when the clock pulse arrives at the T input. A two-input AND gate is added to the set and reset inputs for this purpose.

The **J-K flip-flop**, Figure 19.20, is unique. It has no unpredictable output states. It can be set by applying a 1 to the *J* input and cleared by feeding a

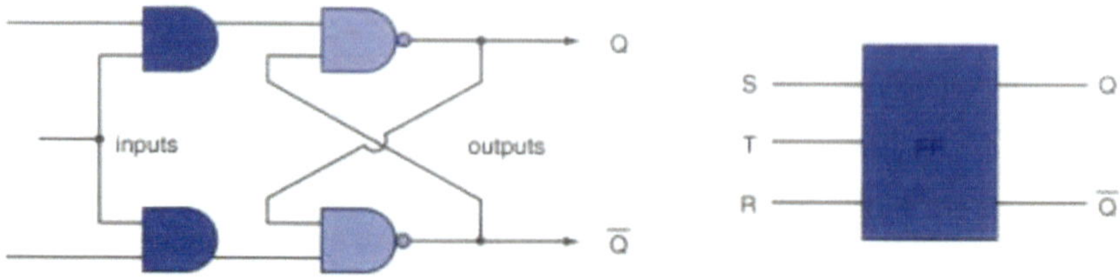

A B

Applied inputs			Previous outputs		Resulting outputs		
S	T	R	Q	$\bar{Q}$	Q	$\bar{Q}$	
0	0	0	1	0	1	0	
0	1	1	1	0	0	1	
1	0	0	1	0	1	0	
1	1	1	1	0	–	–	Invalid
0	0	0	0	1	0	1	
0	1	1	0	1	0	1	
1	0	0	0	1	0	1	
1	1	1	0	1	–	–	Invalid

C

Figure 19.19 The R-S-T flip-flop. A – Logic diagram. B – Symbol. C – Truth table.

1 to the K input. A 1 signal applied to both and inputs simultaneously causes the output to change states, or toggle. A 0 applied simultaneously to both J and K inputs does not initiate a state change. The inputs of a J-K flip-flop are controlled directly by clock pulses. Several modifications of the basic J-K flip-flop are available. Some have preset and preclear inputs that are used to establish sequential operations at precise times.

19.5 Digital Counters

One of the most versatile and important logic devices is the counter. **Counters** are used to count a wide variety of objects in various applications. However, they really count only one thing: electronic pulses. These pulses can be produced mechanically, acoustically, with a clock mechanism, or by a number of other processes. The two common forms of this device are the binary counters and the decade (BCD) counter.

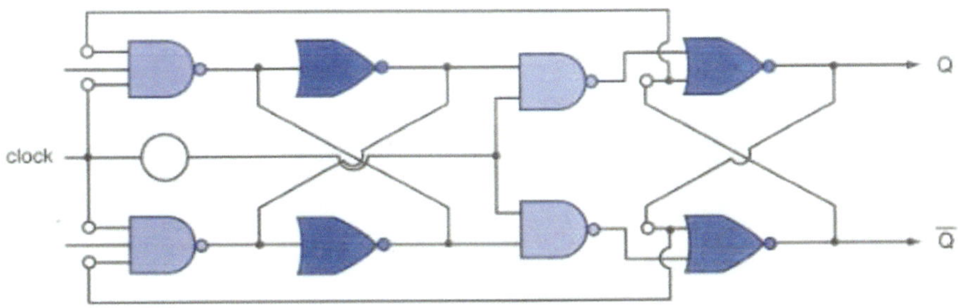

A

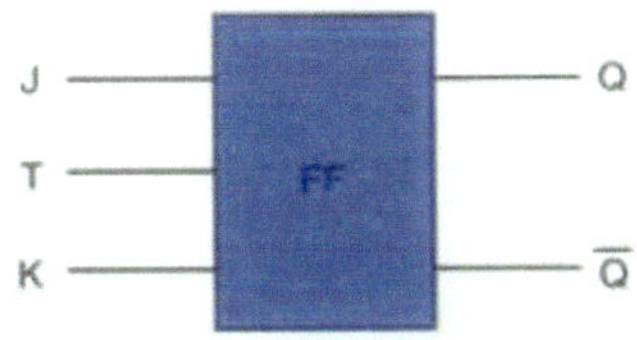

B

Applied inputs			Previous outputs		Resulting outputs		
J	T	K	Q	$\overline{Q}$	Q	$\overline{Q}$	
0	0	0	0	1	0	1	
0	0	1	0	1	0	1	
0	1	1	0	1	0	1	
1	0	0	0	1	0	1	
1	1	0	0	1	1	0	
1	1	1	0	1	1	0	← Toggle state
0	0	0	1	0	1	0	
0	0	1	1	0	1	0	
0	1	1	1	0	0	1	
1	0	0	1	0	1	0	
1	1	0	1	0	1	0	
1	1	1	1	0	0	1	← Toggle state

C

Figure 19.20 The J-K flip-flop. A – Logic diagram. B – Symbol. C – Truth table.

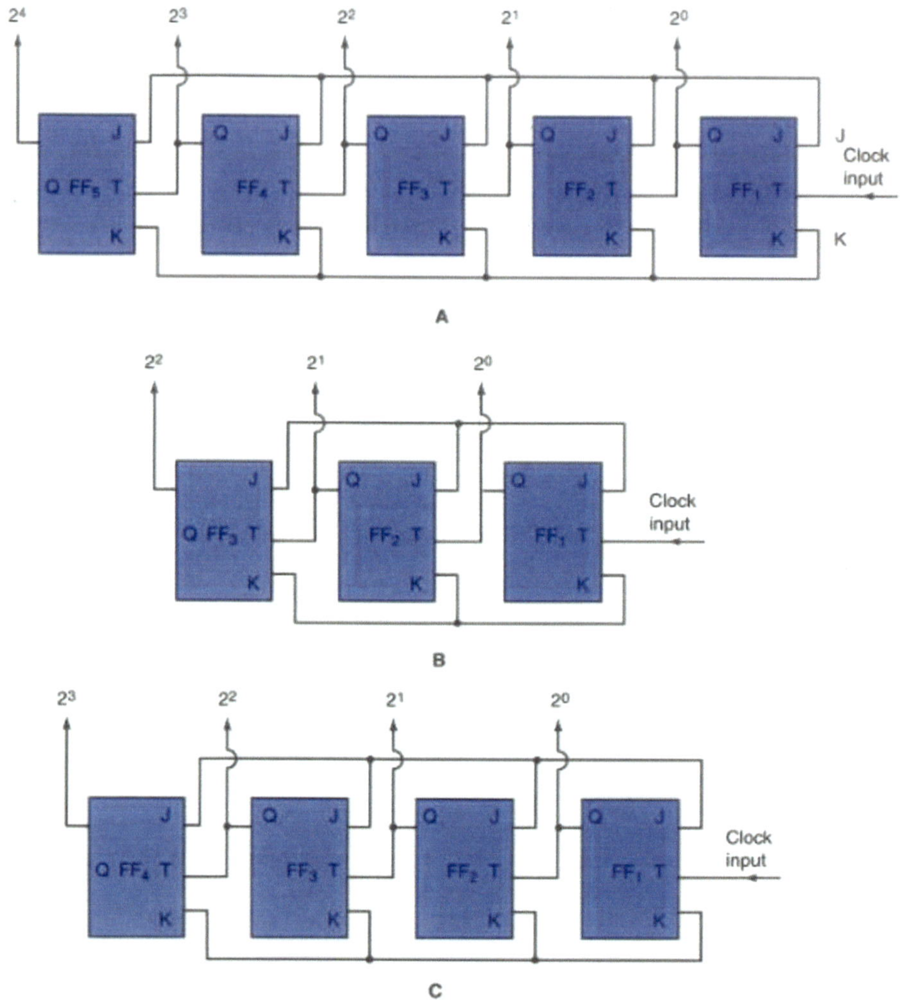

Figure 19.21 Some digital counters that use a J-K flip-flop. A – Binary ripple counter (32 count). B – Octal counter (8 count). C – Hexadecimal counter (16 count).

19.5.1 Binary counters

In **binary counters**, flip-flops are connected so that the Q output of the first circuit drives the trigger, or clock input, of the next circuit. See Figure 19.21. Each flip-flop, therefore, has a divide-by-two function.

The **counter** in part A of Figure 19.21 is called a **binary ripple counter**. The J and K inputs for each flip-flop are held at a logic 1 level. Each clock pulse applied to the input of FF 1 then causes a change in state. Since the flip-flops trigger only on the negative-going part of the clock pulse, the output of FF 1 will alternate between 1 and 0 with each pulse. Thus, a 1 output will appear at *Q* of FF 1 for every two input pulses. The output of 1 for every two input pulses produces the divide-by-two function at each flip-flop. Five flip-flops connected in this manner produce a 2^5, or 32, count. The largest count in this case is 11111_2 (31_{10}). The next applied pulse clears the counter so that 0 appears at all the Q outputs.

By grouping three flip-flops together (part B of Figure 19.21), a **binary-coded-octal (BCO) counter** can be created. The number 1111_2 represents the seven count, or seven units of an octal counter. Two groups of three flip-flops connected in this manner produce a maximum count of 111111_2, which represents 77_8, or 63_{10}.

It is possible to develop the units part of a binary-coded-hexadecimal (BCH) counter by placing four flip-flops together in a group. See part C of Figure 19.21. Thus, 1111_2 is used to represent F_{16} or 15_{10}. Two groups of four flip-flops could produce a maximum count of 11111111_2, which would represent FF_{16} or 255_{10}. Each succeeding group of four flip-flops raises the count to the next power of 16.

19.5.2 Decade counters

Since most of the mathematics that we use today is based upon the decimal (base 10) system, it is important to be able to count by this method. The output of a binary counter must be changed into a decimal form before it can be used in this way. The first step is to change binary signals into a binary-coded-decimal (BCD) form. This is done by using a **decade counter**.

A **four-bit binary counter** is shown in Figure 19.22. In this counter, 16 counts are achieved by the four flip-flops. The counts are listed below the binary counter, reading right to left. To convert this counter into a **decade counter**, it must skip some of its counts, as shown in Figure 19.23. Four J-K flip-flops are built into a chip. The Q output of each flip-flop appears at terminals A, B, C, and D. These serve as the 2^0, 2^1, 2^2, and 2_3 or 1, 2, 4, and 8 outputs.

The first seven counts occur naturally. At the seventh count, 1s appearing at the *Q* outputs of FF_B and FF_C are applied to the AND gate. This action produces a logic 1 and applies it to the J input of FF_D. Arrival of the next clock pulse triggers FF_A, FF_B, and FF_C into the off state and turns on FF_D. This represents the eighth count.

When FF_D is in the on state, Q is 1 and $\bar{Q}$ is 0. This causes *a* to be fed to the *J* input of $FF_{B'}$ which now prevents it from triggering until cleared.

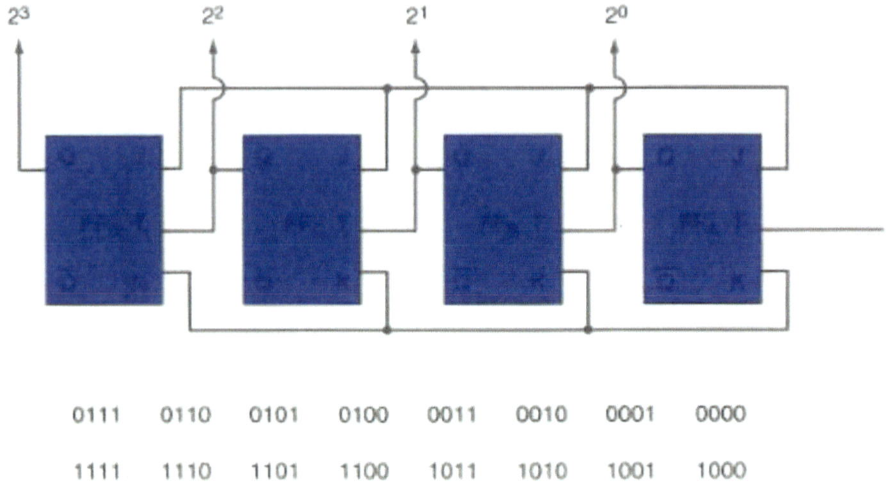

Figure 19.22 A four-bit binary counter uses four flip-flops.

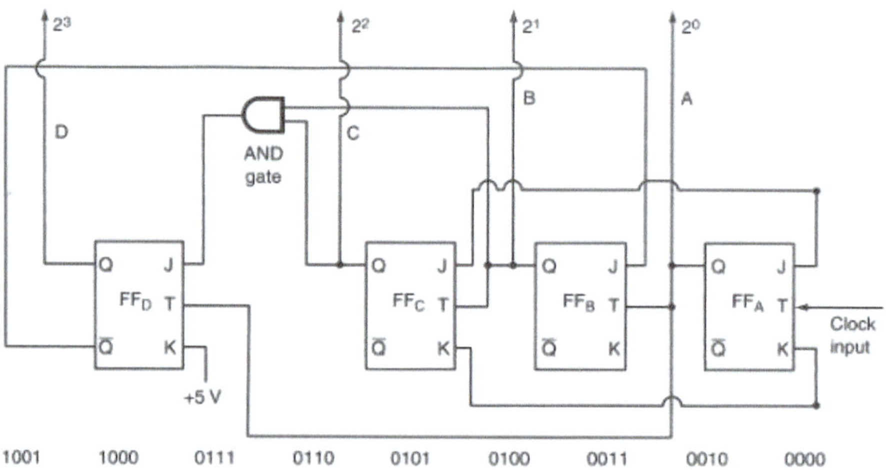

Figure 19.23 A four-bit binary counter is converted into a decade (BCD) counter.

Arrival of the next clock pulse causes FF_A to be set to a 1. This registers a 1001_2, which is the ninth count. Arrival of the next count clears FF_A and FF_D instantly. Since FF_B and FF_C were previously cleared by the seventh count, all 0s appear at the outputs. The counter has, therefore, cycled through the ninth count and returned to zero, ready for the next input.

19.6 Digital Control Systems

Automatic fabrication methods, packaging, and machining operations have been improved through advances in digital systems. Digital (numeric) instructions over the years have been supplied by perforated paper tape, punched cards, magnetic tape, or variations in pressure, temperature, or electric current. These instructions are then changed into a series of on/off electrical signals. The signals are processed by the logic gates of a computer and are directed to specific subsystems that then perform the necessary operations.

Electrical energy is used to energize the load device, which performs the work. The load of a system can be electrical actuators or fluid-power cylinders designed to move parts of the machine. When appropriate signals are received, these parts move an object to a specific location. Machine positions and movements, clamping operations, and material flow can all be controlled by digital systems.

19.6.1 Control systems

Control of an automated manufacturing system is accomplished by human input or automatically by a physical change. During production, **control systems** are continually at work, making adjustments that alter machine operation. The complexity of the operation determines the number of control functions needed. In many cases, varying numbers of components are used in different parts of the system.

The **control** unit of an industrial system determines its flexibility and efficiency. Some machines have only mechanical stops on each axis. Others have microprocessor (computer) control with memory to store position and sequence data. Some important factors in the selection of a control unit are speed of operation, repeatability, accuracy of positioning, and the speed and ease of reprogramming.

Non-servo, or **open-loop**, control systems are the most basic. They typically use sequencers and mechanical stops to control the machine.

19.6.2 Open-loop control

Open-loop systems are used almost exclusively for manual control operations. There are two variations of the open-loop system: **full control** and **partial control**. **Full control** simply turns a system off or on. For example, in an electrical circuit, current flow stops when the circuit path is opened. Switches, circuit breakers, fuses, and relays are used for full control. **Partial control**

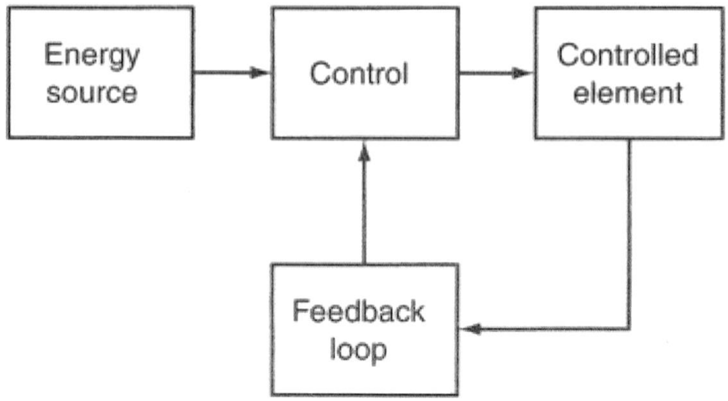

Figure 19.24 Block diagram of a closed-loop system.

alters system operations, rather than causing it to start or stop. Resistors, inductors, transformers, capacitors, semiconductor devices, and integrated circuits are commonly used to achieve partial control.

19.6.3 Closed-loop control

To achieve automatic control, interaction between the control unit and the controlled element must occur. In a **closed-loop system**, this interaction is called **feedback**. Feedback can be activated by electrical, thermal, light, chemical, or mechanical energy. Both full and partial control can be achieved through a closed-loop system. See Figure 19.24.

Many of the **automated systems** used in industry today are of the closed-loop type. Figure 19.25 shows the block diagram of a closed-loop system with automatic correction control. In this system, energy goes to the control unit and the controlled element. Feedback from the controlled element is directed to a **comparator,** which compares the feedback signal to a reference signal or **standard**. A correction signal is developed by the comparator and sent to the control unit. This signal alters the system so that it conforms to the data from the reference source. Systems of this type maintain a specified operating level regardless of external variations or disturbances.

Automated control has gone through many changes in recent years with the addition of control devices that are not obvious to the casual observer. These include devices that change the amplitude, frequency, waveform, time, or phase of signals passing through the system.

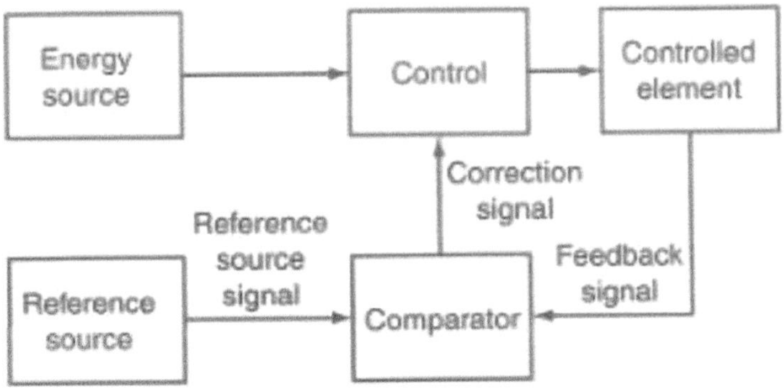

Figure 19.25 Block diagram of a closed-loop system with automatic correction.

19.7 Numerical Control Systems

A very important digital system application found in industry today is called numerical control. With **numerical control (NC)** systems, a series of numerical-coded instructions is used to control the operation of a machine automatically. Duplicate parts can be produced by this process, or specific assembly operations can be performed automatically. Manufacturing consistency is a major advantage of machinery controlled by this process.

The term **numerical control** actually refers to two distinct functions that are interrelated. The term numerical refers specifically to the instructional information applied to the machine. This generally represents digital information that is read by the input part of the system. Two-state conditions (1s and 0s) are used to instruct the machine. The digital console is primarily responsible for this operation.

The control function of an **NC system** refers to the physical changes that take place in the machine after it has been commanded to do something. Timing, sequencing, positioning, guiding, directing, altering speed, and clamping are typical machine control operations. Electromechanical, hydraulic, and pneumatic operations are used to achieve this function independently or through a combination of all three operations working together.

Figure 19.26 shows a typical **NC milling machine** unit. The operator's control console is located on the right side of the machine. This unit employs a drive unit and the digital logic components needed to control different machine operations. Electrical disconnect switches, control transformers, spindle speed power control, and auxiliary operating functions are controlled by the console of this system.

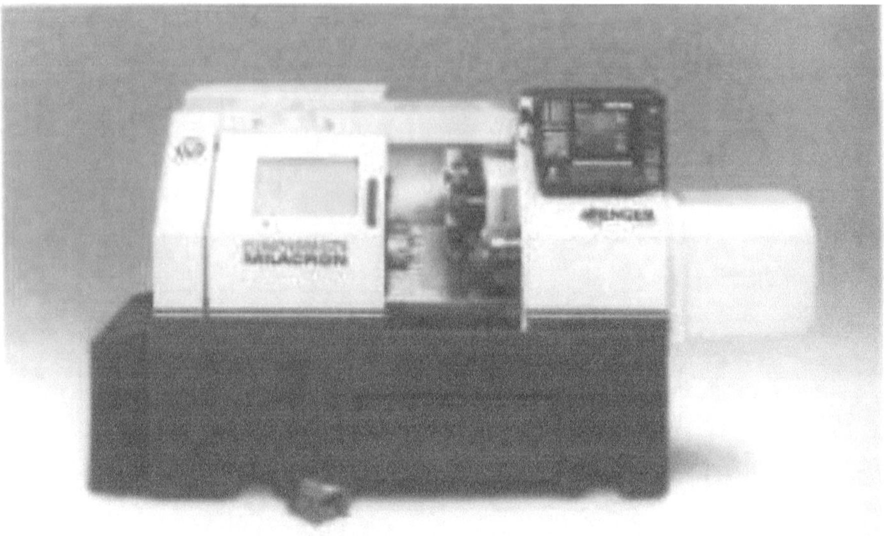

Figure 19.26 CNC machine. (Cincinnati Milacron)

Physical changes that take place in the operation of an NC machine are achieved electrically, hydraulically, or by pneumatic actuators. DC stepping motors and AC servomotors are commonly used to drive the different axes of the worktable electrically. Table position can then be monitored by synchro motors to relay accurate position information back to the control console for precise reference correction. Vertical position of the worktable, tool selectors, coolant circulation, spindle feed, and turret rotation are typical hydraulic and pneumatic control operations. Control of this type can be energized electrically or mechanically and actuated hydraulically or pneumatically.

The operation of an **NC machine** can be simplified somewhat by placing it in the framework of a basic system structure. Figure 19.27 shows two distinct types of NC machine control systems. The open-loop system (part A of Figure 19.27), in its simplest form, only necessitates a logic processor and amplifier to drive stepping motors to achieve specific control operations. The theory behind the operation of this system is that the stepping motor is usually reliable enough to act on a command without fail. Systems of this type are less expensive and generally easier to maintain and operate than closed-loop systems. The key to this type of system is a reliable stepping motor. A number of basic milling machines and machine lathes are controlled by this type of system.

Traditionally, **NC machines** have nearly all been of the closed-loop type (part B of Figure 19.27). Commands applied to the input are processed

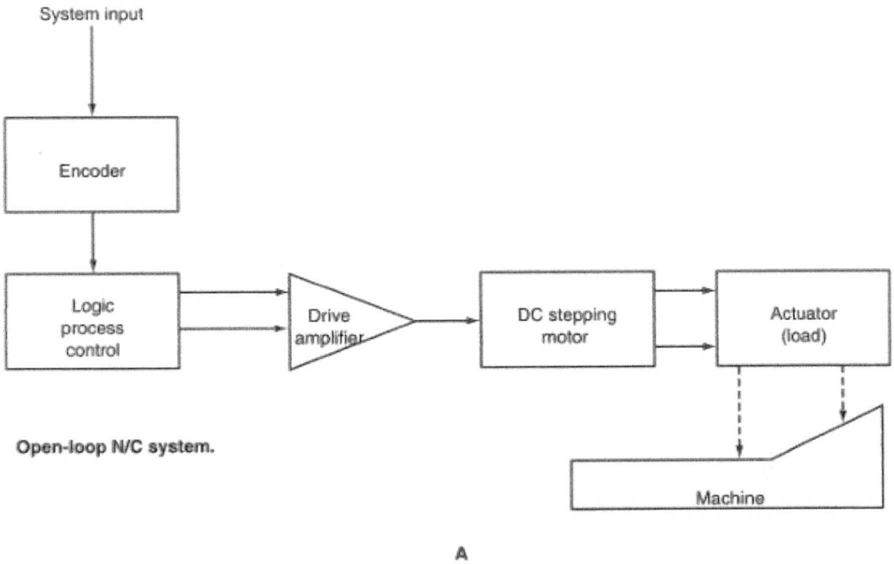

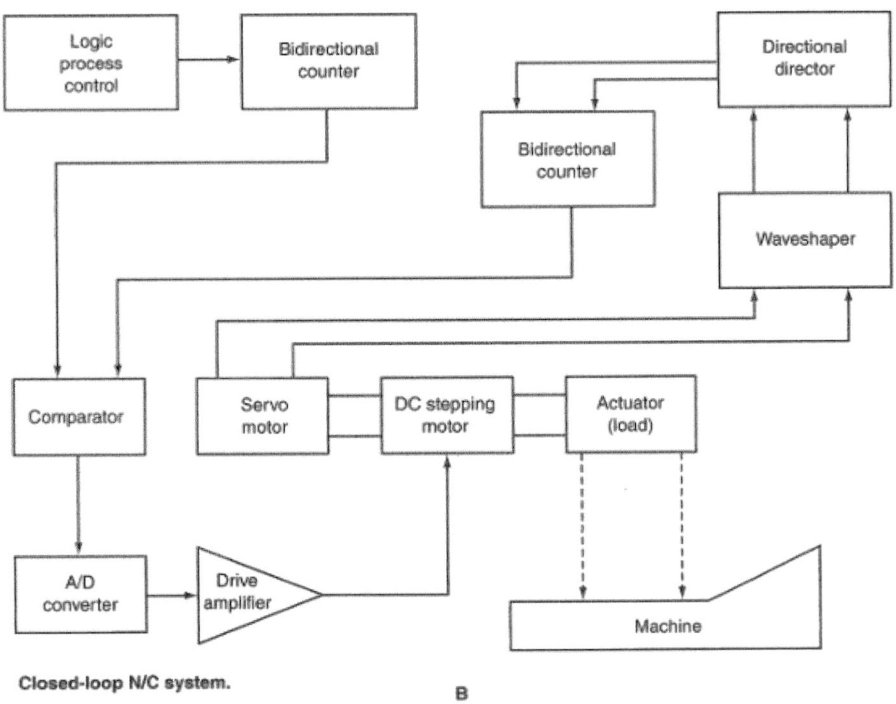

Figure 19.27 NC system diagrams.

and applied to the actuators. Feedback signals developed by the actuator are then returned to the processor for comparison. Input and feedback signals are then compared, and a correction signal is used to modify the control process should the need arise. Systems of this type are quite sophisticated when compared with the open-loop type of system. The intended work function of the system is primarily used to decide the need for open- or closed-loop control.

19.8 Numerical Control Input

For an **NC machine** to operate, it must receive digital information from its input. Coded digital instructions are usually prepared in advance and can be placed on perforated paper tape, punched cards, or magnetic tape. Perforated paper tape was used as a standard input medium for NC systems. Instructions placed on this tape are punched in eight horizontal channels that run the length of the tape. Holes punched in different channel locations are done in accordance with an established code. Figure 19.28 shows representative codes of the EIA 244A and ASCII. The EIA code was developed by the Electronics Industry Association. ASCII refers to a code developed by the American Standards Institute for Information Interchange. Similar operations are now performed using magnetic systems.

19.8.1 Tape readers

The primary input of an NC machine was previously a tape reading device. This device was needed to interpret the information prepared in the program of instructions and direct it to the process control unit. Essentially, a tape reader gathers digital information from the tape. In a strict sense, a tape reader detects two-state (1 or 0) or binary-coded information and changes it into a digital signal. This process is the basis for systems that have now been developed.

Figure 19.29 shows an NC tape reader. In the example shown, holes punched in the tape pass light from the source to a specific photocell above each channel. When light reaches a specific cell, it generates an output signal. Readers of this type are well-suited for high-speed operation and show very little part wear for long hours of operation.

19.8.2 NC machine control unit

The **machine control unit or MCU** of an NC system is the primary link between the programmed information placed on tape and actual machine

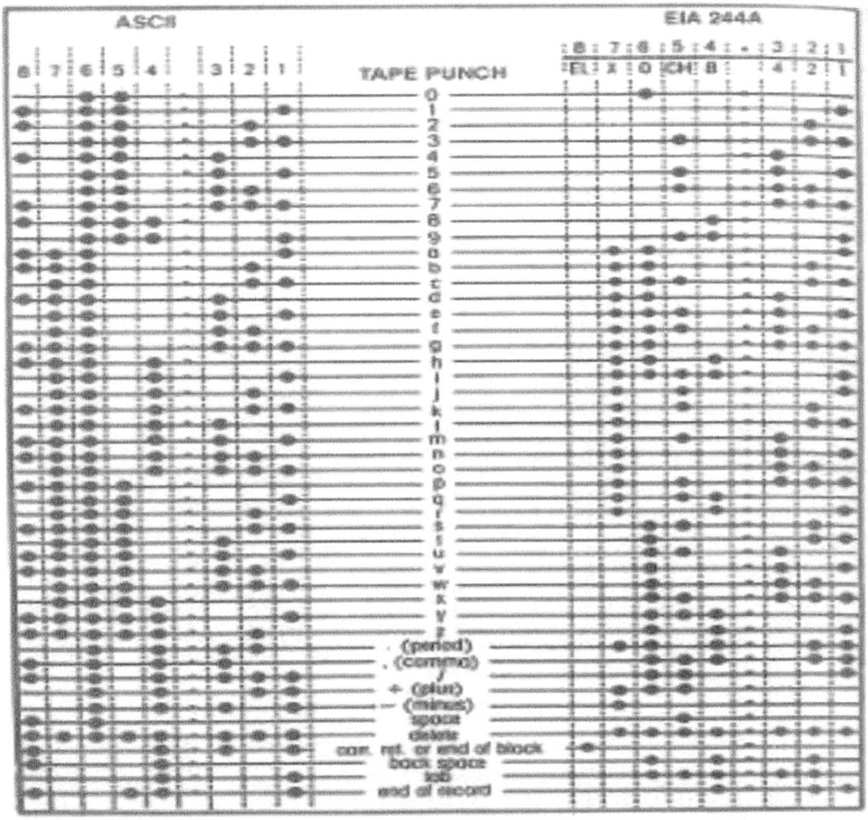

Figure 19.28 EIA 244A and ASCII tape codes.

operations. This part of the system acts on the digital input signal from the tape reader and converts it into actual machine motions through actuators.

One thing that makes the MCU of an NC machine somewhat difficult to discuss in general terms is the large number of different system options available today. Each option obviously changes the operation and circuitry of a system a great deal. A rather simple unit is designed for two-axis, point-to-point positioning and straight-line machine operations. This is a rather compact unit with logic circuits, a control panel, motor drive transistors, and a self-contained power supply. Control can be achieved by tape drive or by manually overriding switches. Systems of this type are typically of the open-loop type.

Another MCU has two-, three-, or four-axis control of the contouring type. Cutting, welding, and milling operations can be applied to contouring

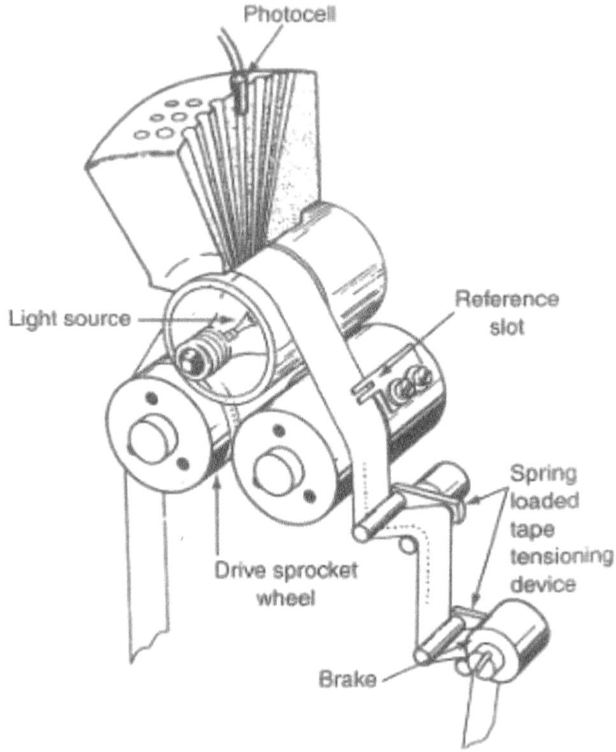

Figure 19.29 Drawing of NC tape reader equipment.

surfaces automatically by this unit. It can be operated by a paper tape or by its own built-in **keyboard**, or it can be interfaced with a computer. This unit is a **closed-loop system** that samples the position of each axis and corrects itself according to the tolerance level controlled by the program. The MCU of an NC machine is designed around the specific needs of the machine and its intended applications.

19.9 Numerical Control Load Actuators

The end result or work function of an NC machine is commonly applied to the system load through an actuator. Rotary actuators such as **DC stepping motors** or AC **servomotors** are commonly used to achieve this operation.

The rotary motion of a **DC stepping motor** is commonly changed into linear motion by driving a lead screw. When the motor moves one step, it can cause the screw to turn 1/200 of a revolution. Two hundred continuous

Figure 19.30 Positioning tables. (Parker-Hannefin, Compumotor Division)

steps can be used to rotate the lead screw one complete revolution. Typical NC stepping motor accuracy is ± 0.09° of the desired shaft position. This level of accuracy is usually not cumulative, which makes it quite accurate for low-tolerance machining operations.

Figure 19.30 shows an NC machine **positioning table**. The horizontal or *x*-axis of a table is controlled by a servomotor attached to the end of the table. Actuation of this motor causes the accurate rotation of a lead screw. Each revolution of the motor causes the lead screw to advance 0.100 in or 1 mm, according to the design of the unit. Dials attached to the lead screw are commonly divided into 100 equally spaced slots out into its periphery. Rotation of the lead screw produces coded information that is relayed back to the control console as an indication of position.

Front-to-back or **y-axis** movement of the positioning table is also controlled by a servomotor. This motor is attached to the front part of the table. A lead screw driven by this motor produces control that is similar to that of the *x*-axis actuator.

Machine spindle feeds are often controlled automatically by hydraulic or pneumatic actuators. This type of load control can drive the spindle fast initially and at a slower rate when it nears the work and during its operation.

Special toolholder turrets are often attached to an NC machine to permit multiple tool machining capabilities. Specific load actuators must be designed to be compatible with the machine and the MCU.

Review Questions

Answer the following questions.

1. Define the terms bit and byte.

2. What is meant by the term base, and how does it apply to numbering systems?

3. Why is the two-state operating condition of an electronic device important in digital systems?

4. Define the following.

    a. Binary number system.

    b. BCD number system.

    c. Octal number system.

    d. Hexadecimal number system.

    e. Positive logic.

    f. Negative logic.

5. Compare the input-output characteristics of two-input AND, OR, NAND, and NOR gates.

6. List three uses for flip-flops.

7. What is a binary counter? Decade counter?

8. What is the difference between an open-loop and a closed-loop systems?

9. What is a numerical control (NC) system?

10. What two devices are commonly used as load actuators?

All sensory input data must be converted into a digital form if it is to be used with computer analysis. This thermal recording and analysis system records and examines temperature data over time. (FLIR Systems Inc.)

20

Computer Systems and Microprocessors

Objectives

Upon completion of this chapter, you will be able to:

- Define the terms bit, word, and byte.
- List the basic parts of a computer.
- Explain the purpose of the arithmetic logic unit.
- Compare and contrast read-only and random-access memories.
- Describe the role of read-only memory in a computer system.
- List the functions that are basic to a microprocessor.
- Differentiate between 16-, 32-, and 64-bit microprocessors.

Important Terms

accumulator
address
address register
arithmetic logic unit (ALU)
bus
bus network
cache
computer
data register
firmware
inherent-mode instruction,
instruction decoder
machine language
microprocessor unit (MPU)
operand, program
program counter
random-access memory (RAM)
read-only memory (ROM)
sequence controller
software

Introduction

Computer technology has brought significant changes to industrial machine control. Centralized computers are used to perform a wide range of control

DOI: 10.1201/9788743810056-27

operations in automatic process applications. Elaborate calculations can be performed quickly, data can be stored and retrieved at a moment's notice, and deductions can be made from data that will influence a process. All of this has caused the industry to become more dependent on computers and has placed a higher premium on computer operational time.

Around 1963, **minicomputers** appeared on the scene to provide computer technology for industrial applications not requiring the capacity of a large-scale computer. These units were significantly less costly than the big computers. Instant success demonstrated the usefulness of this computer with a wide range of applications.

Microcomputers were introduced by a number of different manufacturers in the early 1970s. These computers were an extension of the technology employed by the minicomputer. **Large-scale integration (LSI)** technology was used to place thousands of discrete solid-state components on a single **integrated circuit (IC) chip**. The entire **central processing unit (CPU)** of a microcomputer is now achieved by a single IC chip called a microprocessor unit (MPU). A **microprocessor** coupled with memory, input/output interface chips, and a power source forms a microcomputer system. See Figure 20.1. This combination of components can be built on a single printed circuit board and easily housed in the cabinet of the instrument being controlled. Microcomputer technology has revolutionized industrial control by providing inexpensive, small-capacity computers that can be built into machines.

20.1 Computer Glossary

In an investigation of computers and microprocessors, one frequently encounters a number of new terms. They are quite important and in many cases serve as the basis of a control principle to the operation of a computer. A few of these terms have been singled out for study before proceeding with the chapter. Reviewing these terms will make the reading of the text more meaningful.

Accumulator: A temporary register that is designed to store data that is to be processed.

Address register: A microprocessor function that temporarily stores the address of a memory location that is to be accessed for data.

Architecture: The physical component layout of a microprocessor or computer assembly.

Arithmetic/logic unit (ALU): A circuit capable of performing a variety of arithmetic or logic functions.

Figure 20.1 Microcomputer controller. (Ziatech Co.)

Assembly language: A machine-oriented language in which mnemonics are used to represent each machine language instruction.

Base: The number of symbols in a number system. A decimal system has a base of 10, a binary system has a base of 2, an octal system has a base of 8, and a hexadecimal system has a base of 16.

Basic: A high-level programming language. The acronym for Beginner's All-purpose Symbolic Instruction Code.

Baud: A unit of data transmission speed equal to the number of bits that can be manipulated per second.

Binary: A system of numerical representation that uses only two symbols, 1 and 0.

Bit: A contraction of the term binary digit.

Bus: Parallel lines used to transfer signals between devices or components.

Byte: A special term used for an 8-bit word. A byte usually consists of 8 bits.

Clock: A flip-flop input that controls timing or a pulse generator that controls computer operations.

Data register: A group of flip-flops that store data.

Digital: A value or quantity related to numbers or discrete values.

Enable: To make a computer operation possible, such as permitting data to be transferred from one location to another by applying a signal to the enable input.

Encoding: The process of converting an input signal into binary data or of changing one digital code to another.

Erasable programmable read-only memory (EPROM): A read-only memory in which information has been entered by applying electrical pulses to selected cells. This information can be erased by applying ultraviolet light to the device through a quartz window.

Execute: One of the operational cycles of a central processing unit.

Fetch: An operational cycle of the central processor.

Firmware: Permanently installed instructions in computer hardware that control operation.

Flowchart: A symbol diagram used to aid in program development.

Gate: A circuit that performs special logic operations such as AND, OR, NOT, NAND, and NOR.

Instruction: Information that tells a microprocessor or computer what to do.

Interface: A process or piece of equipment in a computer that brings things together.

Large Scale Integration (LSI): The combining of several circuits or components on a single IC chip.

Logic: A decision-making capability of computer circuitry.

Machine language: Instructions that are written in binary form that a computer can execute directly.

Memory: A process that can store logic 1 and logic 0 bits in such a manner that they can be accessed or retrieved.

Microprocessor: An integrated circuit or set of ICs that can be programmed to process data.

Mnemonic: An abbreviation that resembles a word or reminds one of a word.

Network: A group of computers that are connected to each other by communication lines to share information and resources.

Nonvolatile memory: Memory that is capable of retaining data after the power supply voltage has been removed.

Operand: A quantity being operated on or performed in a processing system.

Operational code (opcode): The part of a computer instruction word that designates the task to be performed.

Program: A list of tasks in an order that are used to control the operation of a processor or computer system.

Programmable read-only memory (PROM): A memory device that can be programmed by the user.

Radix: The base of a numbering system.

Random-access memory (RAM): Memory that can be both read and changed during computer operation. Unlike other semiconductor memories, RAM is volatile. If power is lost, all data stored in RAM is lost.

Read-only memory (ROM): A form of stored data that can be sensed or read. It is not altered by the sensing process.

Read/write (R/W): A memory operation that can be read from or written into with equal ease.

Registers: A digital circuit that is capable of storing or moving binary data.

Software: Instructions, such as a program that controls the operation of a computer.

Substrate: A piece of N or P semiconductor material that serves as a foundation for other parts to be formed on in the construction of an integrated circuit.

Volatile memory: A type of memory that necessitates electrical power in order to store information. If the power is removed, all information will be lost.

Write: A memory operation where a new word or bit of data is placed in a particular memory location.

20.2 Computer Basics

The term **computer** is a general term that is used to describe a number of operating systems. It refers to a device that will perform automatic arithmetic computations when provided with information. It is much more than a fast calculator. A computer can choose, copy, move, compare, and perform other nonarithmetic operations on alphabetic, numeric, and other symbols that are used to represent different things. The computer manipulates these symbols through a routing map called a program. The **program** is a detailed set of instructions that directs the computer to function in a way that will produce a desired result. Computers are fast and accurate systems that are designed to automatically accept and store input data, process it, and produce a resulting output.

Computers use numbers represented by the presence or absence of voltage at a particular level. A high voltage level or pulse is usually taken to represent a 1 state, while no voltage indicates the 0 state. A voltage value or signal pulse is described as a "bit" of information. The term bit is a contraction of the letters bi from the word binary and the letter *t* from digit. A group of four bits, pulses, or voltage level changes, produces a word. A byte consists of eight bits.

A **computer system**, regardless of its complexity, has a number of fundamental parts that are the basis of its operation. These parts can be arranged in a variety of different ways and still achieve the same function. The internal organization and design of each block of a computer differ considerably between manufacturers. Figure 20.2 shows a block diagram of a computer.

A digital computer consists of input and output devices, arithmetic logic and control circuitry, and some form of memory. Before the computer is placed into operation, a program written in a language that the computer understands must be provided. The program can be supplied to the input unit through a magnetic disk or tape, a compact disc, a keyboard, punched tape, or punched cards.

Input data is translated into some type of numeric code called **machine language**, which the control unit then manipulates according to programmed instructions. After all the internal operations are complete, the coding process is reversed. The machine language is then translated back into another language. This information is used to actuate the output device.

Computer memory stores the operating instructions that direct the central processing unit (CPU). Coded data in the form of 1 s and 0 s are written into memory, based on directions provided by a program. The CPU then reads these instructions and uses them to carry out operations. When the program is logical, processing proceeds in the correct manner with useful results.

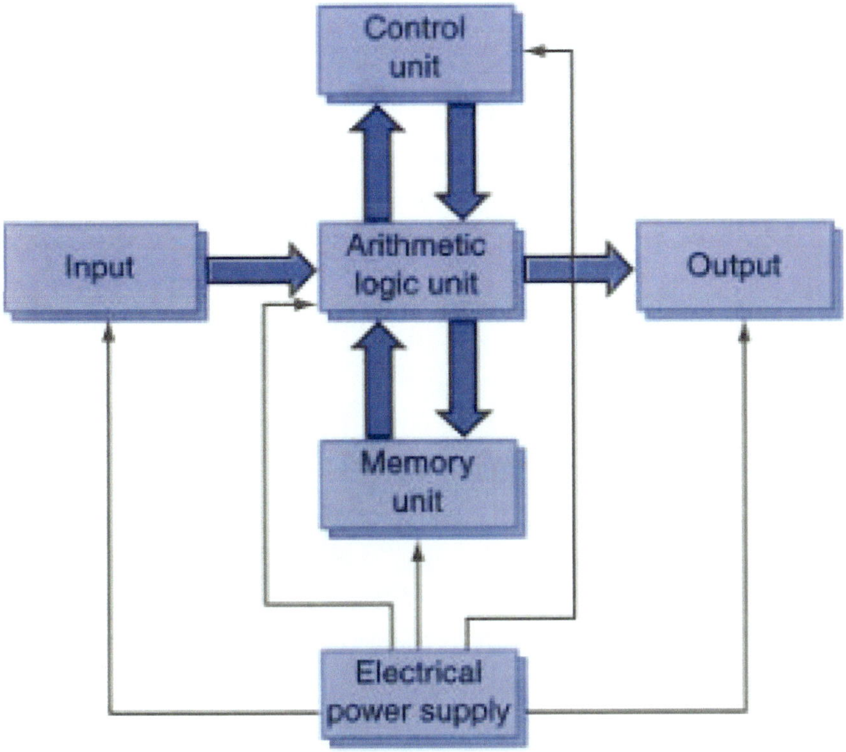

Figure 20.2 The design of a computer system can vary from one type or size to another, but all computers have the same basic parts.

Computers have one or more output ports that permit the CPU to communicate with other devices. Monitors (sometimes called video display terminals or VDTs), printers, magnetic disk or tape drives, and modems (interface devices allowing communication with other computers over telephone lines) are typical output devices. Operating speed is an important characteristic of the output device.

The parts of a **computer** are basically the same for all sizes and types. The differences among units are in physical size, amount of memory, and processing speed and capacity. The largest computers, called mainframes, have literally millions of components included in their circuits. They are used for complex and large tasks, such as scientific computations or operating industrial plants or government departments. Very large mainframes, dubbed supercomputers, have been built.

Minicomputer systems are primarily designed for applications that do not require the capacity of a mainframe computer. They have a smaller

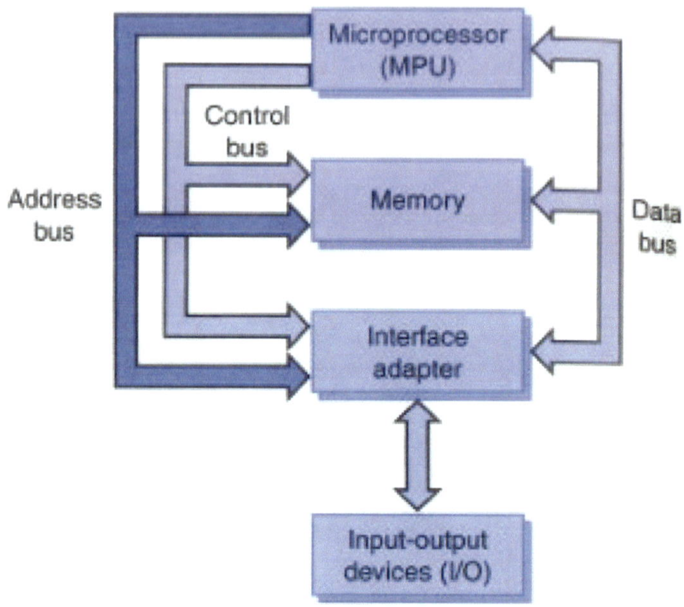

Figure 20.3 Simplified block diagram of a personal computer.

memory capacity and process data less rapidly, but are also less expensive. Minicomputers are used for many business, manufacturing, and educational applications.

Personal computers, usually referred to as **PCs**, are small desktop or portable units built around a single integrated circuit. Called microcomputers when they were first produced for home and small business use in the early 1980s, these units have achieved considerable popularity. They are widely used today in applications ranging from handling office correspondence and bookkeeping tasks to architectural drafting to controlling a machining cell in a factory. This type of computer is often a dedicated system. It generally serves a specific need, solves a specific type of problem, or handles one application. While most PCs are used as standalones, many others are wired together in networks to enable them to share information.

A **personal computer** includes a microprocessor, memory, an interface adapter to handle input and output, and several distribution paths called buses, Figure 20.3. The microprocessor contains both the arithmetic logic unit and control unit of the basic computer system shown in Figure 20.4.

Information processed by a computer is of two types: **instructions** and **data**. A simple addition problem, such as 9 + 2 = 11, can be used as an

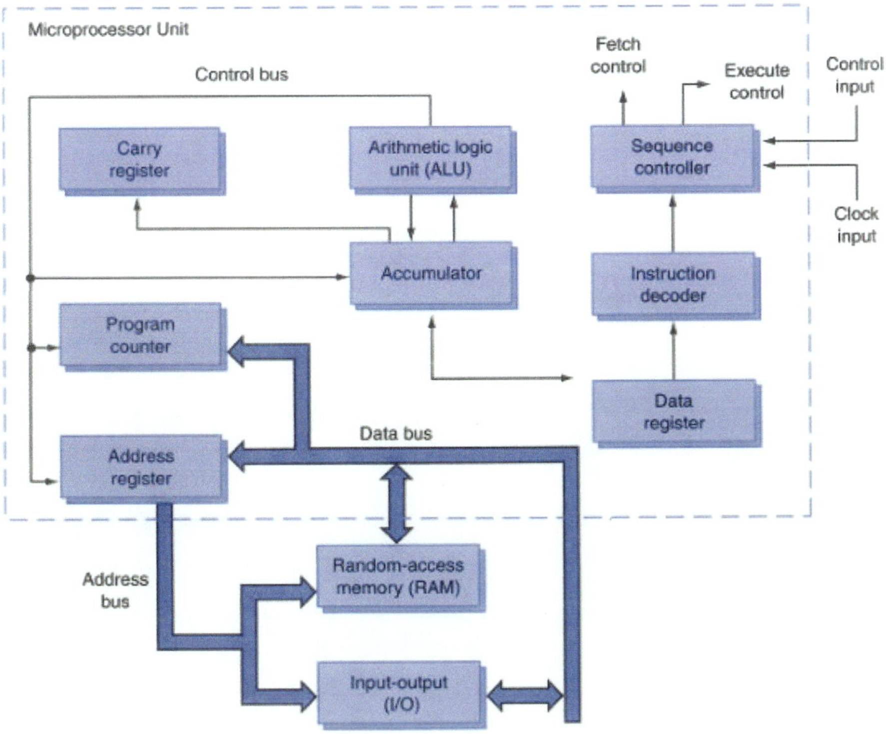

Figure 20.4 This block diagram of a microprocessor has been simplified for clarity.

analogy to illustrate the distinction. In the problem, the numbers 9,2, and 11 are data, while the plus sign is an instruction. The data is distributed by the data bus to all parts of the system. Instructions (the program) are distributed by the control bus through a separate path. The address bus of the unit forms an alternate path for the distribution of address data. It is normally used to place information into memory at the correct address. The address is that spot in memory where certain information is located. With the appropriate command, data can be removed from memory and distributed as output.

20.3 Microprocessors

A **microprocessor unit (MPU)** combines the arithmetic logic unit and control section of the computer. Today, a microprocessor fits into a single integrated circuit, or chip. A typical chip, approximately 0.5 cm (0.20 in) square, contains thousands of transistors, resistors, and diodes. Many companies

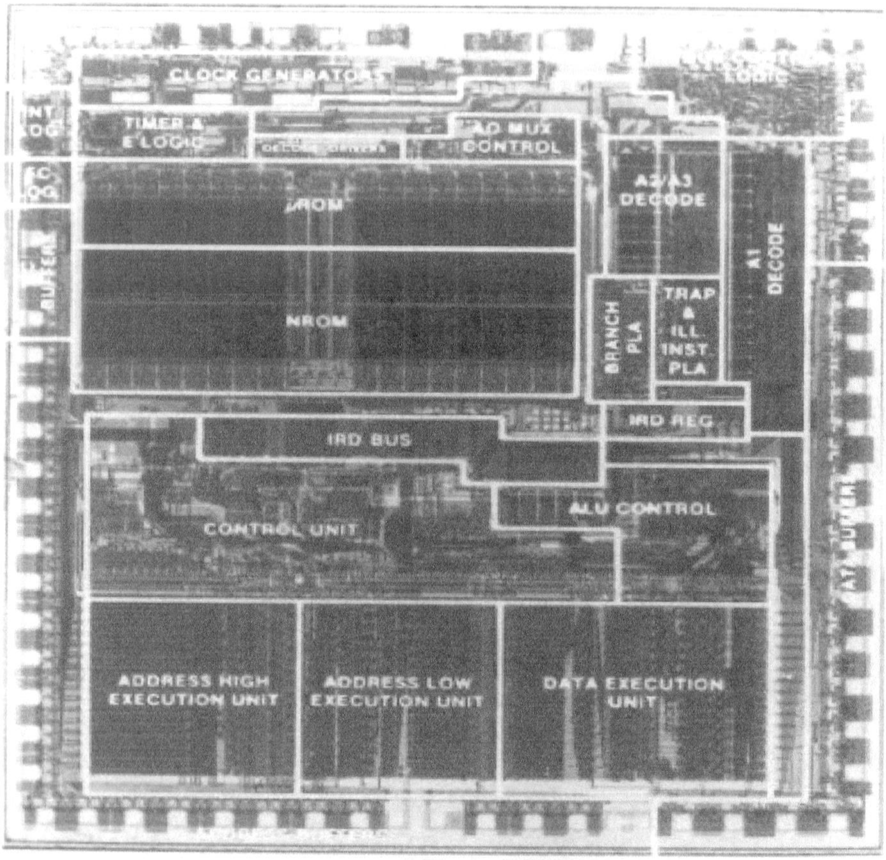

Figure 20.5 An expanded view of a microprocessor chip. (Motorola, Inc.)

manufacture microprocessors in different designs and for different prices. Applications for them are increasing at a remarkable rate.

A microprocessor receives data in the form of 1s and 0s. It can then store this data for future processing, perform arithmetic and logic operations, and deliver the results to an output device. In a sense, a microprocessor is a computer on a chip. The block diagram of a typical microprocessor, Figure 20.4, shows that it contains a number of components. Included are the arithmetic logic unit (ALU), an accumulator, a data register, address registers, a program counter, an instruction decoder, and a sequence controller. One type of microprocessor chip is shown in Figure 20.5.

All microprocessors contain an **arithmetic logic unit (ALU).** The ALU is a calculator that performs mathematical and logic operations. It works automatically by means of signals sent from the instruction decoder.

The ALU combines the input from the data register and the accumulator. Addition, subtraction, and logic comparisons using binary numbers are its primary operations.

The data supplied to the inputs of an **ALU** is normally in the form of 8-bit binary numbers (bytes). This data is combined by the ALU using binary arithmetic. Since a mathematical operation is performed on the two data inputs, they are often called **operands**.

Assume that two numbers are to be added 6_{10} and 8_{10}. Initially, the binary number $00000110_2(6_{10})$ is placed in the accumulator. The second operand, $00001000_2(8_{10})$, is then placed into the data register. When a proper control line to the ALU is activated, binary addition is performed. This produces an output of 00001110_2, or 14_{10} which is the sum of the two operands. This value is then stored in the accumulator, replacing the operand that appeared there originally.

Accumulators temporarily store operands that are to be processed by the ALU. The output of the ALU is also stored in the accumulator after processing. A typical instruction for a microprocessor is LOAD ACCUMULATOR. This means that the contents of a particular memory location are to be placed into the accumulator. A similar instruction is STORE ACCUMULATOR. This causes the contents of the accumulator to be placed in a selected memory location.

The **data register** temporarily stores information applied to the data bus. For example, it stores operands for ALU input. It can also hold an instruction while that instruction is being decoded or hold data prior to storage in memory.

Address registers temporarily store the address of a memory location that is to be accessed. In some units, this register is programmable. This means that new instructions can alter its contents. Such a program can also build an address in the register prior to giving an instruction referring to memory.

The **program** counter is a memory device that indicates the location in memory of the next instruction to be executed. This unit counts the instructions in sequential order. When the MPU receives instructions addressed by the counter, the count advances to the next location. The numbering sequence can be modified so it does not follow strict numerical order. The counter can jump from one point to another in a routine.

Each operation that the **MPU** performs is identified by binary numbers known as instruction codes. Eight-bit words are commonly used. There are 2^8 (256_{10}) separate operations that can be represented by this code. After an instruction code is pulled from memory and placed in the data register, it must be decoded. The **instruction decoder** examines the coded word and decides which operation is to be performed by the ALU. The output of the decoder is then sent to the sequence controller.

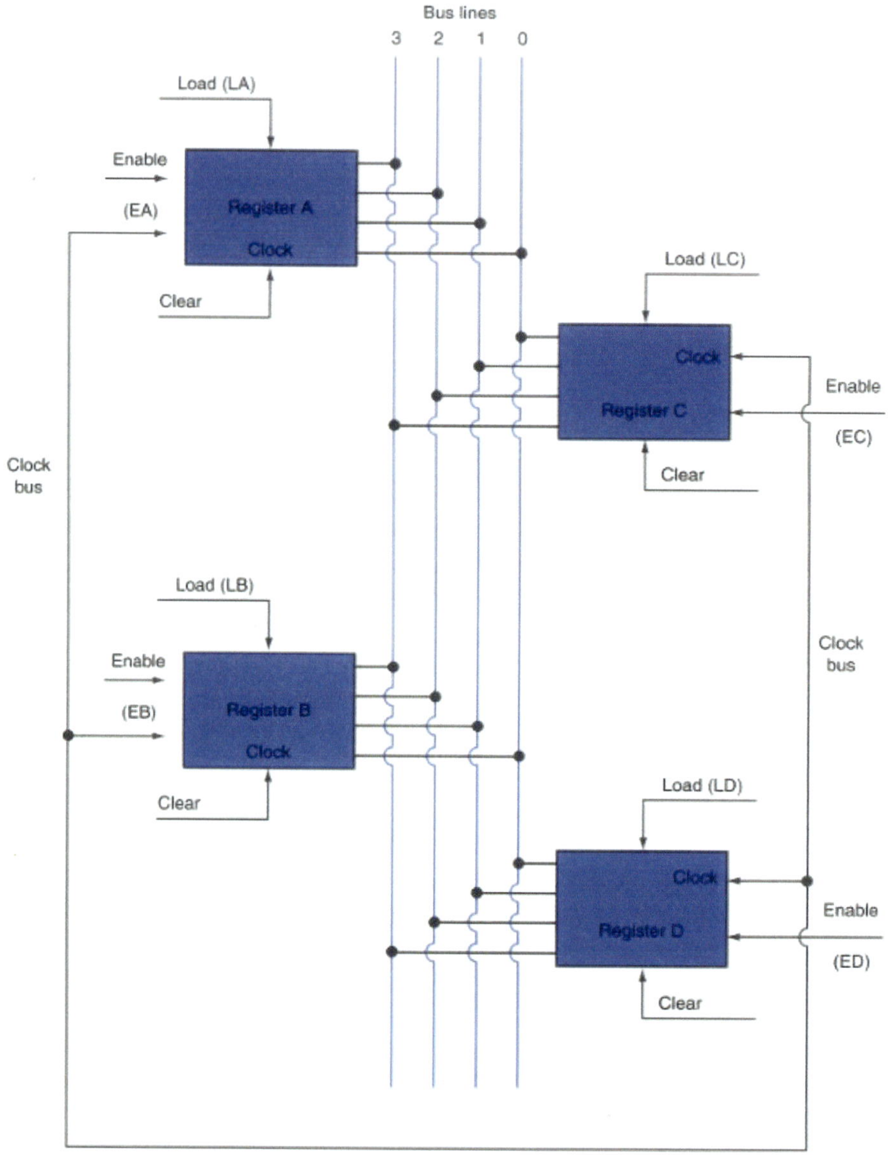

Figure 20.6 In a computer, registers are connected to buses that serve as paths for data transfer.

Using **clock inputs**, the sequence controller maintains the proper sequence of events required to perform a task. After instructions are received and decoded, the sequence controller issues a signal that starts the proper action. In most units, the controller can respond to external signals.

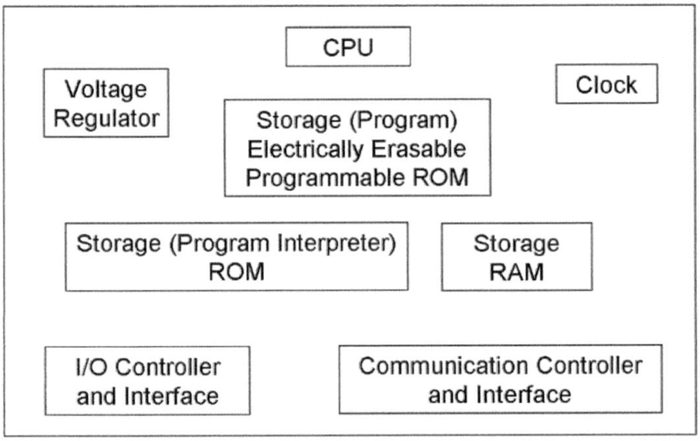

Figure 20.7 Main components of a microcontroller.

20.4 Microcontrollers

Just as a microprocessor combines all the functions of a CPU on a single chip, a microcontroller goes a step further and combines all the functions of an entire computer on a single chip. It thus has a **CPU, internal clock, storage (both ROM and RAM)**, **Input/Output** control capabilities, and could even include power regulation components needed for circuit operation. This is shown in Figure 20.7.

Microcontrollers are a remarkable advance in the field of digital electronics. These devices make it possible to control a wide variety of systems, ranging from car ignition, temperature control, washing machines, ATMs, security systems, and digital signal processing

The **processor core** itself consists of control circuitry on a chip that fetches each instruction sequentially from the program memory. The processor then initiates the specific steps that the instruction requires on an **Arithmetic and Logic Unit (ALU)**. This uses both the value and the status (such as positive, negative, zero, etc.) of the result stored in memory registers for making decisions. Microcontrollers include registers, small high-speed storage in the core itself, connected to different **I/O pins** for determining the purpose each serves, along with suitable software mechanisms to either write data to or read data from specific hardware pins.

Readily accessible, online microcontroller platforms are available for use by engineers, designers, hobbyists, artists, and others interested in creating interactive objects or environments, audio/video applications, microwave ovens, and numerous others. Figure 20.8 shows a representative microcontroller.

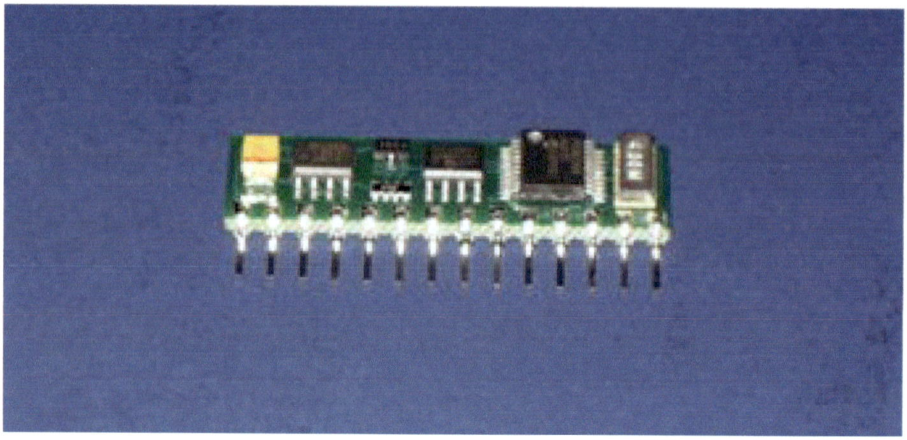

Figure 20.8 Microcontroller

The control code or program for the microcontroller may be written in assembly language, in which case it must be assembled and then downloaded to the storage on the microcontroller. Other alternatives include using an interpreted language, such as **BASIC (Beginners All-purpose Symbolic Instruction Code),** wherein the program is written in a higher-level language, for processing on a per-line basis by the microcontroller. It is then downloaded to the microcontroller. Each line of the program is interpreted, i.e., converted into machine language for the microcontroller to process and execute. This process of interpreting each line of the program slows down the overall execution time, and as such, interpreted programs run slower than compiled ones. It should be noted that prior to download, the program should be checked for proper syntax, keywords, or constructs. Most microcontrollers will not permit the code to be downloaded if it contains such errors. However, as yet, there are no checks for logic errors in the program. Such errors are commonly identified by running the program and applying sample inputs (called test vectors) to activate all portions of the program code.

20.4.1 Introducing Arduino Microcontrollers

Tinkercad by Autodesk, Inc. a free, three-dimensional (3D) modeling program that runs in any modern web browser, permits users to publish designs related to solid geometry, architecture, and more recently, electronic circuits using microcontrollers. These are made available for others to adapt and use worldwide, under a Creative Commons license. This modeling program is a

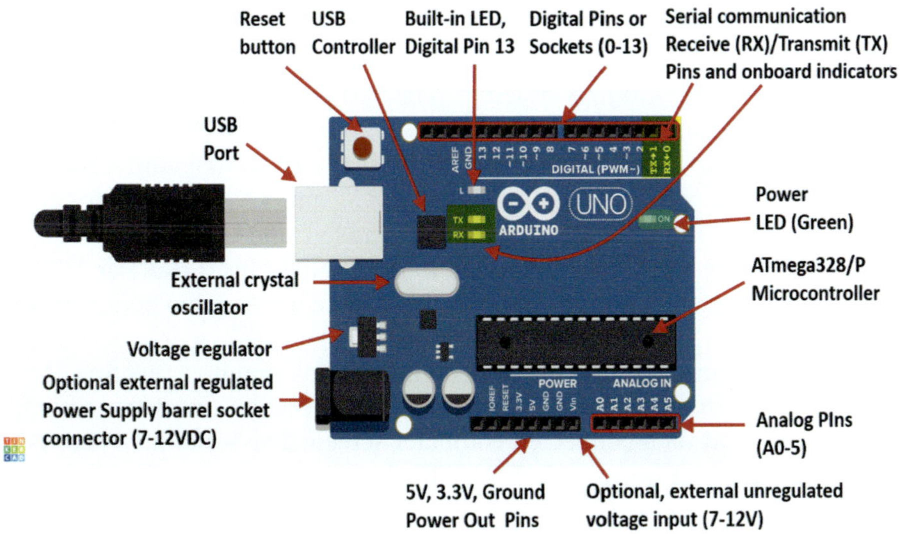

Figure 20.9 Arduino Uno board and a USB cable for interfacing it with a computer.

wonderful avenue to learn about microcontrollers, their wiring and programming, even if one does not have physical access to these devices.

Arduino Microcontrollers, such as shown in Figure 20.9, focus on creating digital projects, called "sketches." These projects can be developed for those in the creative arts who may not be familiar with electronics. The methodology includes an **Integrated Development Environment (IDE)** based on the processing program development platform. This allows the user to learn programming within a more intuitive, graphical setting. The IDE includes an easy-to-use library of functions for completing typical tasks. Arduino systems are based on open-source hardware and open-source software, allowing users to customize and use the hardware and software as they choose to, even for commercial purposes.

20.4.2 The Arduino Uno Board

The **Arduino Uno**, version R3 utilizes one of the AVR family of microcontrollers, with flash memory for storing programs. The **Arduino Uno** consists of the ATmega128 microcontroller by Atmel, now Microchip Technology, Inc., on a Printed Circuit Board (PCB). It includes both digital and analog sockets or pins, power, and ground pins, arranged along the longer edges, as shown in Figure 20.9. Serial communications monitoring capabilities for reporting messages from the microcontroller to the computer are also shown.

The Arduino Uno connects to a Windows, Linux, or Mac computer using a **Universal Serial Bus (USB)** that is also used for uploading programs. The functionality of the Arduino system can be easily expanded by connecting it to a **shield**.

On the top right-hand corner of the Arduino Uno board shown in Figure 20.9, the two sockets (pins) labeled RX_0 and TX_1, are used for communicating with the computer system connected via the USB cable. No input or output devices should be connected to the Digital Pins 0 (RX) or 1 (TX), as this may interfere with communication between the Arduino board and the computer.

The Arduino **IDE** includes a built-in serial monitor function. Live streaming input/output status information and messages from the Arduino can be displayed on a connected computer screen. The serial hardware port of the Arduino, known as the **USART**, short for Universal Synchronous/Asynchronous Receiver/Transmitter, enables device communications using serial protocols. Using specific software commands, we can read in the status of digital inputs as Boolean-value signals, logic Low or logic High. We can then display these on the computer screen. Digital inputs to the Arduino are often provided by switches or sensors, which can be in either open or closed states. Based on how the input is wired to the Arduino board, using any of the pins 2–13, shown in the top row of Figure 20.9, the status of a single digital input bit can be either Logic Low or Logic High.

The digital pins (2–13) in the top row of Figure 20.14 can be configured by the user to function as either inputs or outputs. For ease of wiring Arduino circuits, it may be convenient to use the lower-number digital pins (2–7), on the Arduino board as **inputs**, and the higher-number digital pins (8–13) as digital **outputs**. So, for a foundational digital system, with one input and one output, we may designate pin 2 as the input, and pin 13 as the output, using specific commands in the software. In fact, there is a built-in **LED** connected to pin 13 of the Arduino Uno board, shown in Figure 20.9.

20.4.3 Arduino Uno Features and Specifications

The **Arduino Uno** is often pre-loaded with a program that turns the built-in LED on and off at regular intervals once the device is first powered on. Specific Digital pins on the **Arduino Uno** board can provide **pulse-width modulation (PWM)** outputs. The average value of the output voltage developed by a pin can be adjusted by varying the duration of the output pulse at that pin. If it is on all the time (100%), the output voltage will be 5V, and if it is on for half the time (50%), the output voltage will average to 2.5V. The specific Digital Pins that permit the use of **PWM**, when used as output are designated by a tilde (~) symbol next. On the Uno, the six Digital pins can be configured to provide PWM outputs: 3, 5, 6, 9, 10, and 11.

While the **Arduino Uno** cannot directly provide continuously varying analog output voltage signals, specific **PWM Digital Pins** on the Arduino board have the capability to mimic analog outputs. These pins, when set to function as **PWM** outputs, use 8 bits to encode the different possible voltage levels, yielding 28 = 256 possible output voltage levels, varying from 0 to 255. Thus, a value of 0 corresponds to 0V, 127 to approximately 2.5V, and 255 to 5VDC.

20.4.4 ATmega328/P Microcontroller

The brainpower for the Arduino is provided by the **ATmega328/P** integrated circuit (IC) from Microchip Technology. It includes a Reduced Instruction Set Computer (**RISC**) 8-bit processor for carrying out instructions, flash memory for holding instructions and data, along with communication capabilities. In essence, it is an economical computer on a chip. While the microcontroller cannot directly connect to peripheral devices, when integrated into the Arduino development board, using suitable software, it can read the status of digital and analog input pins, run a pre-loaded sketch, set the status of digital output pins, and communicate using serial protocols with connected devices. The pinout of the Arduino Uno development board, based around the ATmega 328/P microcontroller, is shown in Figure 20.10 (a) and (b).

20.5 Working with Arduino Software

20.5.1 The Arduino Integrated Development Environment (IDE)

The microcontroller requires the support of additional technologies, both software and hardware, for leveraging its capabilities. The **Arduino Integrated Development Environment (IDE)** includes programming tools for Windows, Linux, and Mac computers, allowing us to create programs that run on the microcontroller. Older Arduino Uno boards included a USB controller chip for communications between the computer and the Uno. This **communications controller chip** has been replaced with an ATmega16U2. Once a program for the Arduino has been written using the IDE, it can be uploaded to the flash memory of the microcontroller. Figure 20.11 shows the

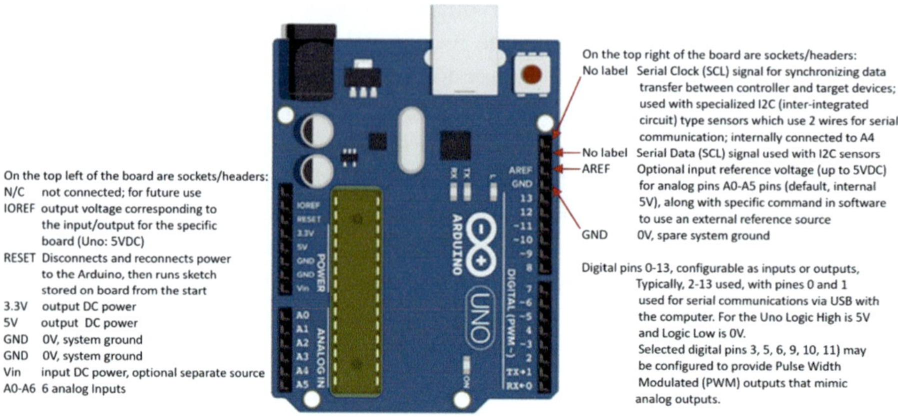

(a). Arduino with the ATmega328/P microcontroller highlighted.

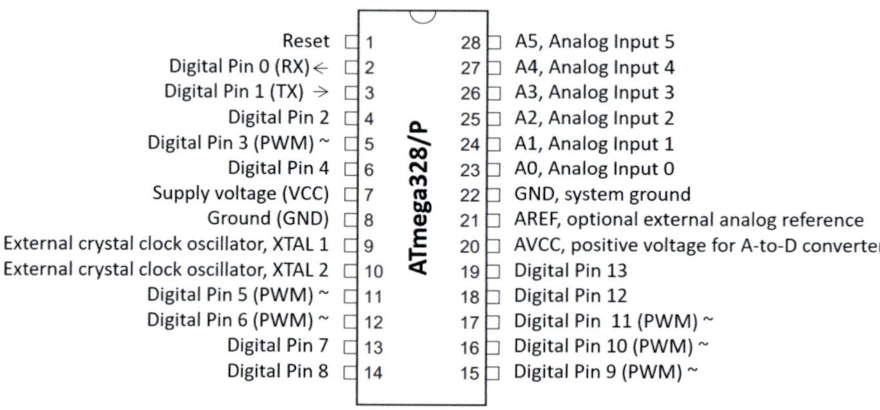

(b). Pin-out of ATmega328P microcontroller.

Figure 20.10 Sockets on the Arduino development board and pinout of the ATmega328/P Microcontroller.

Arduino IDE that is typically used to create a sketch, verify, and then upload it to the Arduino Uno.

20.5.2 Circuit Design Suite: Tinkercad Emulator

Emulators such as the Circuit Design suite of *Autodesk Tinkercad* allow the sketch to be directly run and view the status of inputs and outputs reports using the Serial Monitor function. To control outputs such as lights, buzzers,

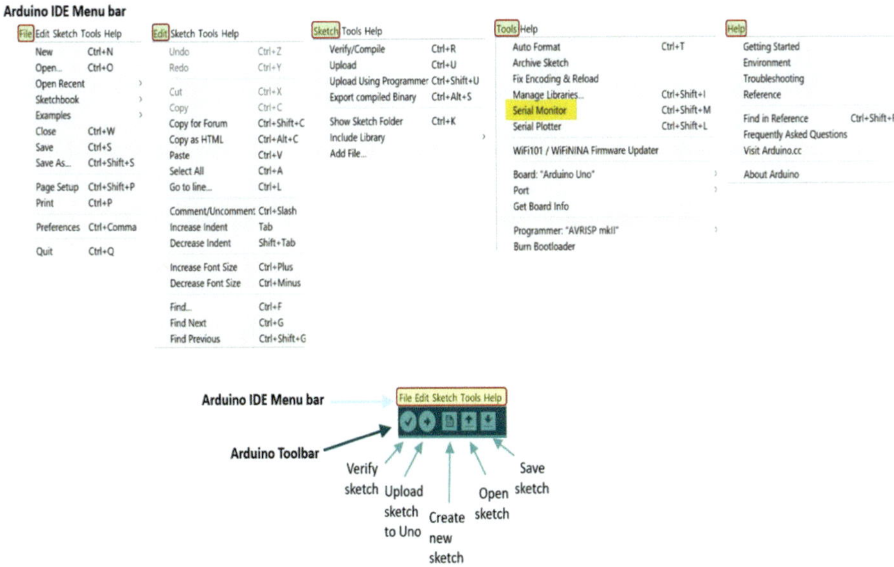

Figure 20.11 Menu and toolbars for creating, editing, verifying, and uploading sketches using the Arduino IDE.

and motors, based on inputs provided from switches and sensors, procedures called Arduino sketches are developed. Procedures allow sketches to be readily customized based on personal interest and the availability of specific input or output devices.

The procedure for developing a new Arduino-based design circuit is shown in Figure 20.12 (a), (b), (c), and (d). A free **Autodesk Tinkercad** account can be obtained at *https://www.tinkercad.com/login*.

20.5.3 Understanding Microcontroller Code (Arduino Sketches)

Viewing the underlying microcontroller code for blinking the onboard LED can be done by using the "Code" button shown in Figure 20.13(a) to see the underlying code "Blocks". The drop-down arrow for changes the code to "Text" as shown in Figure 20.13(b). The code comprises two functions: setup and loop. The setup function is run when the Arduino is first powered on, then continues with the loop function.

20.5.4 Arduino Sketches

Arduino sketches are programs that use a relatively small number of commands that can be combined in various ways to perform specific tasks. For

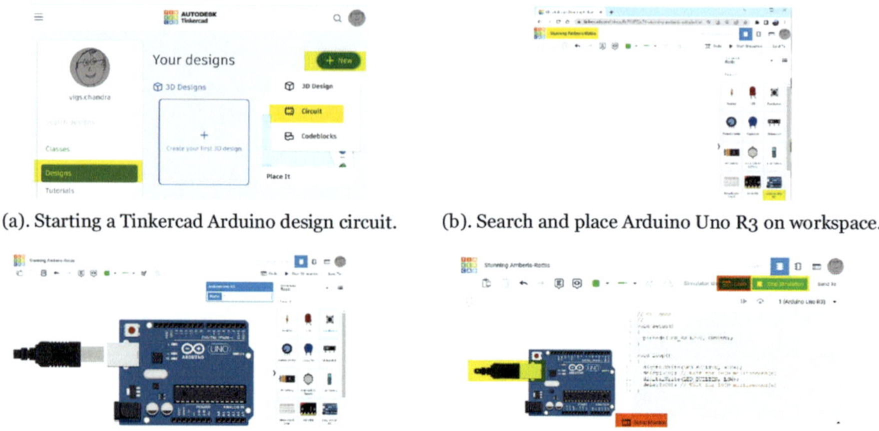

(a). Starting a Tinkercad Arduino design circuit.

(b). Search and place Arduino Uno R3 on workspace.

(c). Arduino placed on the workspace.

(d). Press the "Start Simulation" button to connect the USB power cable and run the selected program.

Figure 20.12 Starting an Arduino-based circuit using Autodesk Tinkercad. Note – the Serial Monitor is available under the Code section of the screen on the right-hand side.

example, these may allow us to evaluate conditions and expressions, perform math operations, or make decisions regarding continuing processing. Sketches are performed step-by-step at the current location in memory, or jump to a different section of the code and continue processing, starting from that location.

Arduino sketches are converted into code that can be understood and executed by the Arduino board, with the Arduino **IDE** acting as a front-end, adding pre-processing and compilation. The Arduino **IDE** uses a compiler to convert the higher-level source code into machine code (binary instructions) that the CPU on the Arduino board can execute.

20.5.5 Flowchart Structures

Flowcharts use logical control structures for depicting different types of operations, ranging from step-by-step sequences, making decisions, and repeating a task either a given number of times, while a specific condition is met, or continually.

Sketches in an Arduino use three main types of control structures:

- **sequence** (shown using a series of rectangular shapes within which we indicate the action being performed, or a parallelogram shape if it relates to interacting with inputs or outputs)
- **choice** (if-then-else, shown using a diamond shape, with outgoing paths based on the specific conditions that are met)

(a). Press the Code button to see the underlying coding "Blocks."

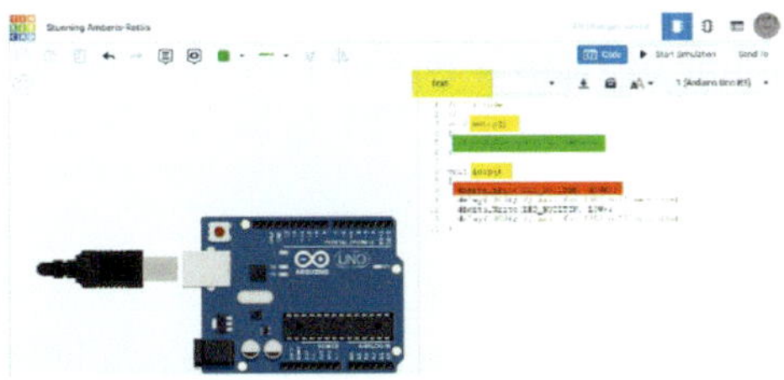

(b). Use the drop-down next to "Blocks" to change to "Text" view.

Text ▾ | 1 (Arduino Uno R3) ▾

```
// C++ code
//
void setup()
{
  pinMode(LED_BUILTIN, OUTPUT);
}

void loop()
{
  digitalWrite(LED_BUILTIN, HIGH);
  delay(1000); // Wait for 1000 millisecond(s)
  digitalWrite(LED_BUILTIN, LOW);
  delay(1000); // Wait for 1000 millisecond(s)
}
```

(c). A closer look at code.

Figure 20.13 Underlying Microcontroller code in Block and Text forms.

- **repetition** (a more advanced structure that includes a decision regarding performing a task or tasks several times in a loop that may either repeat indefinitely, or repeat a specific number of times, in turn keeping track of how many times the task has been performed, either counting up/down to a specific value, then terminating the loop; with provision to break out of a loop under certain conditions)

20.6 Buses

The circuits of most microprocessors are connected together by a common bus network. A **bus network** is a series of registers connected together.

See Figure 20.6. An advantage of a bus network is the ease with which data can be transferred. Each register has inputs labeled clock, enable, load, and clear. When the load and enable input lines are at 0, each register is isolated from the common bus line. To transfer a word from one register to another, the appropriate inputs must convert to the 1 state. For example, to transfer the data of register *A* to register *D*, enable *A(EA)* and load *D(LD)* inputs are both placed in the 1 state. This causes the data of register A to appear on the bus line. When a clock pulse arrives at the common inputs, the transfer is completed. The amount of **data** a bus can carry is based on the number of conductor paths. Buses for 8,16, and 32 bits are commonly used

20.7 Memory

The uses to which a **microcomputer** can be put depend largely on the amount of memory it has available. Memory refers to the ability of the MPU to store data so that a single bit or a group of bits can be easily accessed or retrieved. Additional memory can be added by means of auxiliary chips. The two most common types of memory, **ROM** and **RAM**, are usually handled by semiconductor circuits on a single IC chip.

Most **microcomputers** make use of some permanently stored or rarely altered data. Examples include math tables and programs. Storage of this type is provided by **read-only memory (ROM)**. This information is often placed in the chip when it is manufactured. ROM data is permanent. It cannot be changed, and (unlike RAM, as described in the next section) it is not lost when the power source is turned off.

A variation of ROM is the **read-mostly memory**. This is used where read operations are needed more frequently than write operations. **Programmable read-only memory (PROM)** is another type. It can be erased by exposure to an ultraviolet light source. After light exposure, each memory cell goes to a 0 state. New data can then be written into the chip. Another alternative is the **electrically altered read-only memory (EAROM)** or the **erasable programmable read-only memory (EPROM)**. These types of chips permit erasures of individual cells. Memory cell structure is similar to that of the PROM. The EAROM and EPROM are very useful for certain applications.

Read/write memory is memory that can be written to, as well as read. In other words, it can be altered. **Random-access memory (RAM)** is a type of read/write memory commonly found in microcomputers because it is very fast. Read/write chips can store large quantities of data. The structure of this chip includes a number of separate circuits. Each can store binary data in an

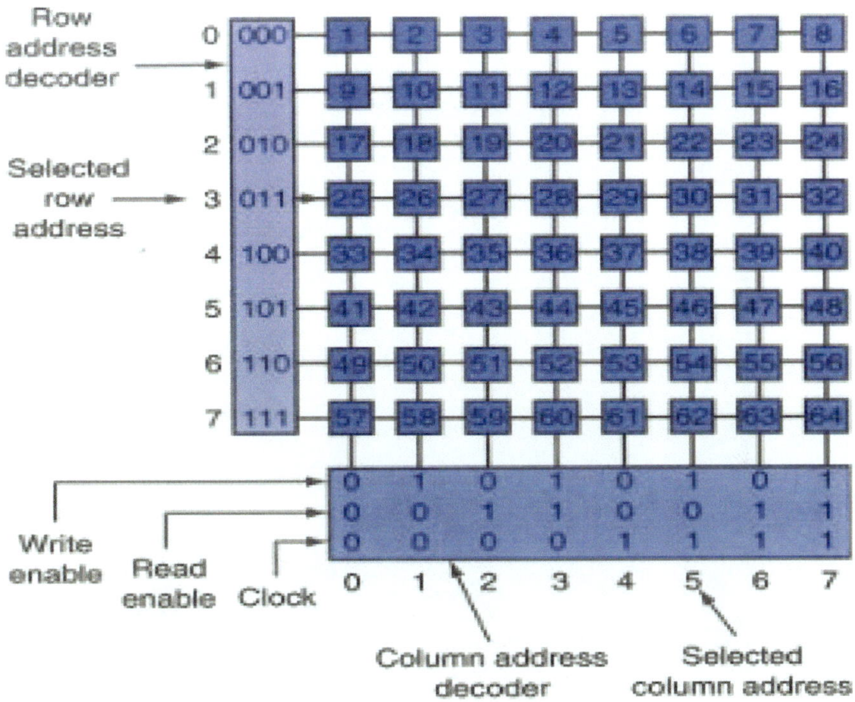

Figure 20.14 Data is stored in rows in a memory unit, such as this basic 8 × 8 array.

organized pattern. Access to each location is provided by coded information from the address bus. Data can be placed into memory or retrieved at the same rate.

Memory ICs are usually organized in a rectangular pattern of rows and columns. Figure 20.14 shows a memory array of eight rows that can store 8-bit words, or 64 single bits. Some read/write memory units available today have much larger capacities. To select a particular memory address, a 3-bit binary number indicates the specific row, and three additional bits indicate the column. In this example, the row address is 3_{10}, or 011_2, and the column address is 5_{10}, or 101_2. The selected memory cell address is 30. Eight-bit word storage is achieved by energizing one row and all eight columns simultaneously. The row and column decoders perform this operation.

To **write** a word into memory, data sent along the address bus selects a specific address. Reading data from memory does not destroy the data. Data reading can take place as long as the unit is turned on. A loss of power, or turning the unit off, destroys the data, however. Random-access memory is

Figure 20.15 An expanded view of a RAM (random access memory) memory chip for a microcomputer system. (Motorola, Inc.)

volatile, and disappears when power is interrupted. A type of RAM is shown in Figure 20.15.

20.8 Computer Functions

Many functions are basic to most computer systems. Among these are timing, fetch, execute, read memory, write memory, input-output transfer, and interrupt.

The operation of a computer involves a sequence of **instructions**. The sequence is controlled by a timing signal. The MPU fetches an instruction, executes the required operations, fetches the next instruction, executes it, and so on, in a cycling pattern. This means that all actions occur at or during a precise time interval. Such an orderly sequence requires a free-running electronic oscillator or clock. In some systems, the clock can be built into the MPU. In others, timing units feed the system through a separate control bus. Operations such as fetch and execute are achieved within a period called the NPU cycle.

Fetch takes the same amount of time in each instance for each instruction. **Execute**, however, can consist of many events and sequences, depending upon the instruction. Execute timing varies a great deal.

The interval taken for a timing pulse to pass through a complete cycle from beginning to end is called a **period**. One or more clock periods can be needed to complete an instruction.

After programmed information is placed into memory, its action is directed by a series of **fetch and execute operations**. The sequence is repeated until the entire program has cycled to its conclusion.

The **start signal** actuates the control section of the MPU, which automatically starts the sequence of operations. The first instruction is to fetch the next instruction from memory. The MPU can then issue a read instruction. The contents of the program counter are then sent to memory, which returns the next instruction. The first word of this instruction is then placed into the instruction register. If more than one word is included in an instruction, a longer cycling time is needed. After the complete instruction is in the MPU, the program counter records one count. The instruction is then decoded. This prepares the unit for the next fetch instruction.

The **execution phase** is based on which instruction is to be performed. It can be to read memory, write to memory, read the input signal, or transfer to output, among others. The timing of an operation depends upon the programmed information. A microprocessor will take a number of clock periods to perform an operation. As an example, we will use a chip that can be operated at a clock rate of 33MHz (33 million cycles per second). A single cycle has a period of 1/33,000,000 or 0.00000003 second. Periods this small are best expressed in microseconds (millionths of a second) or nanoseconds (billionths of a second). At a clock rate of 33 MHz, the length of a single cycle would be 0.03 microseconds (μs) or 30 nanoseconds (ns). If the fastest instruction change that this chip could execute requires four clock cycles, the change would take 0.12μs or 120 ns. If the slowest instruction would require 18 periods, it would require 0.54μs or 540 ns. The operating time of an MPU is a good measure of its effectiveness and power.

Read memory calls for data to be read from a specific location. The MPU issues a read operational code (opcode) and sends it to the proper memory address. The read/write memory unit then sends the data into the data bus. This number is fed to the MPU, where it is placed in the accumulator after the timing pulse has been initiated.

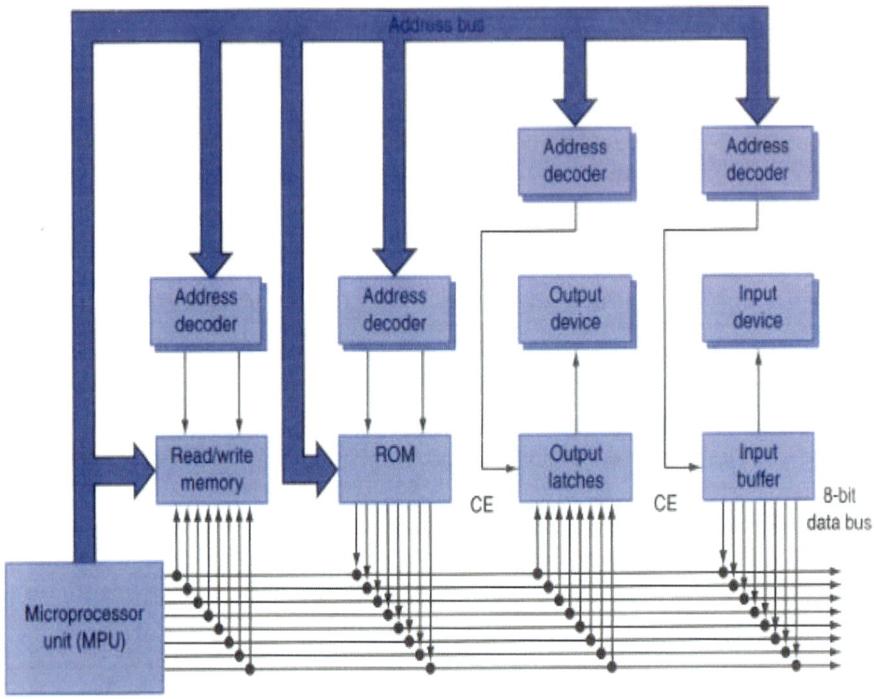

Figure 20.16 This simplified view of an MPU address bus and data bus illustrates input-output transfer. The 8-bit data bus has been superseded by 16-bit, 32-bit, and 64-bit buses in newer personal computers.

Write memory calls for data to be stored at a specific location. The MPU issues a write operation code and sends it to a selected read/write memory unit. Data is sent via the data bus and placed into the selected location.

Input-output (I/O) transfer operations are similar to read/write operations. The major difference is the operational code number used to call up the operation. This code actuates an I/O port, which either receives data from the input or sends it to the output device. In the system shown in Figure 20.16, data moves in either direction to the read/write memory and flows from the ROM into the data bus. The output flows from the data bus to the output device.

Interrupt operations are often used to improve efficiency. Interrupt signals come from peripheral equipment such as keyboards, displays, modems, or printers. They inform the MPU that a peripheral device needs some type of attention. Assume that a system is designed to process a large volume of data, and the output is a printer. The MPU can output data much faster than the printer can print a character. This means that the MPU has to remain idle while waiting for the printer to complete its task. If an interrupt is used, the MPU can output a byte and then return to processing data while the printer works. When the printer is ready for the next data byte, it requests an interrupt. The MPU stops and automatically goes to a subroutine that outputs the next data byte. The MPU then continues processing information until it receives another interrupt.

In **microcomputer systems**, some programs are in read-only memories called firmware, and others are in the form of software. **Firmware** is placed on ROM chips. Program changes can be made by changing the ROM. Software has the greatest flexibility. **Software** is transferred to the system by keyboard, magnetic disk or tape, CDROM, or (in some older systems) by punched cards or punched tape. Instructions are stored in read/write memory.

Instructions normally appear as a set of characters or symbols that define a specific operation. These symbols are similar to those on a typewriter keyboard. Included are decimal digits 0 – 9, the letters A – Z, and in some cases, punctuation marks and special characters. Instructions can also appear in the form of binary numbers, hexadecimal numbers, or mnemonic codes.

Each type of **MPU** is designed to understand a certain set of instructions. These instructions are in the form of binary data, normally held in a read-only memory unit. The unit is address-selected and connected to the MPU through the common data bus. Instructional sets are an example of firmware. They are fixed at a specific location and cannot be changed.

Instructions usually consist of 1,2, or 3 bytes of data. This type of data must follow the commands in successive memory locations. These commands are usually called addressing modes.

One-byte instructions are also called **inherent-mode instructions**. They are designed to send data to the accumulator registers of the ALU. No address code is needed because it is an implied machine instruction. The instruction CLA, for example, is a 1-byte opcode that clears the contents of accumulator A. No address or specific definition of data is needed. Inherent mode instructions differ a great deal among manufacturers.

Representative **opcode instructions** are shown in Figure 20.17. Note that the code is in hexadecimal form, and a mnemonic is given for each one. Each computer system has a number of instructions of this 1-byte type that contain only an opcode.

Immediate addressing is done with a 2-byte instruction that contains an opcode and an operand. The opcode appears in the first byte, followed by the 8-byte operand. A common practice is to place intermediate addressing instructions in the first 256 memory locations. Since this is the fastest mode of operation, these instructions can be retrieved very quickly.

Mnemonic	Opcode	Meaning
ABA	1B	Add the contents of accumulators A and B. The result is stored in accumulator A. The contents of B are not altered.
CLA	4F	Clear accumulator A to all zeros.
CLB	5F	Clear accumulator B.
CBA	11	Compare accumulators: Subtract the contents of ACCB from ACCA. The ALU is involved but the contents of the accumulators are not altered. The comparison is reflected in the condition register.
COMA	43	Find the ones complement of the data in accumulator A, and replace its contents with its ones complement. (The ones complement is simple inversion of all bits.)
COMB	53	Replace the contents of ACCB with its ones complement.
DAA	19	Adjust the two hexadecimal digits in accumulator A to valid BCD digits. Set the carry bit in the condition register when appropriate. The correction is accomplished by adding 06,60, or 66 to the contents of ACCA.
DECA	4A	Decrement accumulator A. Subtract 1 from the contents of accumulator A. Store result ACCA.
DECB	5A	Decrement accumulator B. Store in accumulator B.
LSRA	44	Logic shift right, accumulator A or B. 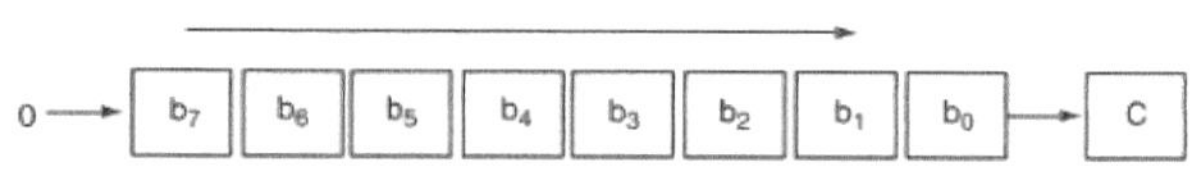
SBA	10	Subtract the contents of accumulator B from the contents of accumulator A. Store results in accumulator A.
TAB	16	Transfer the contents of ACCA to accumulator B. The contents of register A are unchanged.
TBA	17	Transfer the contents of ACCB to accumulator A. The contents of ACCA are unchanged.
NEGA	40	Replace the contents of ACCA with its twos complements. This operation generates a negative number.
NEGB	50	Replace the contents of ACCB with its twos complements. This operation generates a negative number.
INCA	4C	Increment accumulator A. Add 1 to the contents of ACCA and store in ACCA.
INCB	5C	Increment accumulator B. Store results in AACB.
ROLA	49	Rotate left, accumulator A.

Figure 20.17 An example of opcode instructions. The code is in hexadecimal form.

Relative addressing instructions transfer program control to a location other than the next consecutive memory address. Transfer is often limited to a number of locations in front or in back of the present location. The 2-byte instruction contains an opcode in the first byte and a memory location in the second byte. The second byte points to the location of the next instruction that is to be executed.

Indexed addressing is similar to relative addressing. The second byte of the 2-byte instruction is added to the contents of the index register to form a new, or effective, address. This address is obtained during execution rather than being held at a predetermined location. It is held in a temporary memory address register to ensure that it will not be altered or destroyed during processing.

In **direct addressing**, the address is located in the byte of memory following the opcode. This permits addressing the first 256 bytes of memory, from to $00FF_{16}$.

Extended addressing increases the ability of direct addressing to accommodate more data. It is used for memory locations above $00FF_{16}$ and requires 3 bytes of data for the instruction. The first byte is a standard opcode. The second byte is an address location for the most significant 8 bits of data. The third byte holds the address of the least significant 8 bits of data being processed.

A computer cannot solve even the simplest problem without the help of a well-defined **program**. The system follows this program to accomplish a task. This is why programming is an essential part of all computer applications. The programmer must be fully aware of the instructions used by the system. The instruction set of a computer is the basis for all programs.

A **programmer** should be able to decide what specific instructions are needed to perform a given task. A limited number of operations can usually be developed without the aid of a diagrammed plan of procedure. Complex problems, however, require a specific plan in order to avoid confusion or the loss of an important step. Flowcharts are commonly used to aid in this type of planning. See Figure 20.18.

20.9 16-Bit Microprocessors

To improve the data handling and addressing capabilities of their products, microprocessor manufacturers introduced a number of new devices in the early 1980s. These chips were designed to manipulate 16 bits of data at a time. This brought about a whole new generation of personal computers. Most of the **16-bit PCs** are built around a few of the popular microprocessors. Intel's 8088, for example, was found in the IBM personal computer and a dozen or more clones. The data path between the 8088 and the primary storage is 8 bits wide. This permitted all operations to and from memory to be done 8 bits at a time. Once the data was retrieved, it was processed internally 16 bits at a time. The 8088 is described as an 8/16-bit microprocessor. It functions as a fast 8-bit device with an extended set of instructions. Intel's 8086 has internal and external data paths that are all 16 bits wide. The 8086 is classified as a 16/16-bit microprocessor. The address bus of this chip has 20 lines, which permit it to identify 1 megabyte of memory.

Motorola's 68000 is another **16-bit microprocessor**. The 68000 uses a 16-bit external bus to move data between storage and the chip. Data

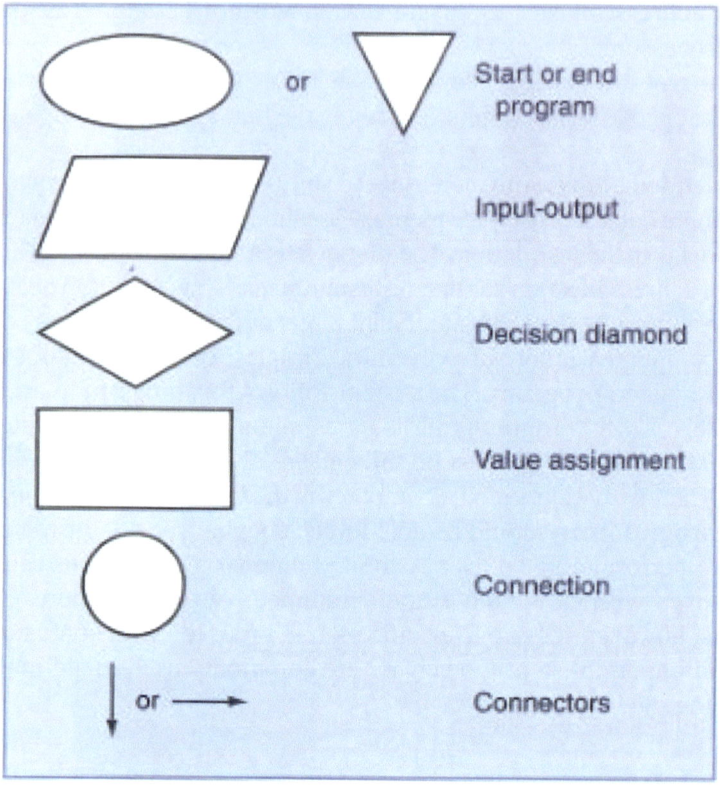

Figure 20.18 These symbols are used by programmers when constructing flowcharts.

processing is done internally 32 bits at a time. This chip is therefore classified as a 16/32-bit microprocessor. The address bus has 24 lines and permits a primary storage capacity of up to 16 megabytes.

The 68000 family of microprocessors has two modes of operation called the user and the supervisor. The user mode of operation is reserved for executing programs written by the user. The supervisor mode of operation is designed to handle all other programs presented to the chip.

20.10 32-Bit Microprocessors

The **32-bit microprocessor** is an enhancement or extension of the 16-bit devices. Motorola's 68030 is an enhanced version of the 68000 chip. The MC68030 has 300,000 transistors in its circuitry, and it is classified as a 32/32-bit device. A base operating frequency of up to 20 MHz makes this chip compatible with the most demanding applications in process and numerical control, robotics, workstations, communications, and environmental control systems.

This chip has parallel address and data buses in its construction. Each bus can handle 16 bits of data. By being parallel, these bus lines can handle 16 + 16 or 32 bits of data. The chip also has a memory management unit and two memory caches. The term cache refers to a localized block of read/write memory that is internally accessible by the chip in which it resides. The 68030 has a data cache and an instruction cache. A memory management unit, or MMU, is used to keep track of where memory is and is not, as well as which portions of physical memory have been used the least or farthest in the past. The 68030 has a paged memory management unit in its construction. Paged addressing has the length of its memory addresses fixed to a specific length.

Intel's 386SX is a full **32-bit microprocessor**. It has a very high-performance ALU that ensures short average instruction execution times. This unit is capable of execution at sustained rates of 2.5 to 3.0 million instructions per second. The memory management unit of this chip includes an address translation cache, multitasking hardware, and a four-level hardware-enforced protection mechanism. The circuitry also offers on-chip testability and debugging features.

The 386SX chip consists of a central processing unit, memory management unit, and a bus interface. The CPU has execution and instruction units in its construction. The execution unit consists of the register file, multiply/divide, barrel shifter, control, and instruction decoder to manipulate its operations. The instruction unit decodes the opcode signals and stores them in the decoded instruction queue for immediate use by the execution unit. The MMU has segmentation and paging units. Segmentation is used to isolate and protect operations. The paging unit is transparent to segmentation and manages the address space. This chip also has real mode and protected modes of operation. This permits the circuit to respond as a fast 8086 by using the paging unit. Finally, the bus interface serves as a direct address pipe for each byte of applied data. The 386 has a 32-bit CPU with a 16-bit external data bus and a 24-bit external address bus that brings high-performance software to midrange systems.

20.11 64-Bit Microprocessors and Beyond

Progress in **microprocessor** development has not stopped with the 32-bit chips. Manufacturers began producing 64-bit devices. The Intel i860 is a representative 64-bit microprocessor. This unit has balanced integers, floating decimal points, and graphics performance for applications such as engineering workstations, scientific computing, three-dimensional graphics, and networking capabilities. Its parallel architecture permits high speed throughout for graphic techniques, pipeline processing units, wide data paths, on-chip memory caches, over a million transistors in its construction, and one-micron fast operation. Microprocessors of this type make the small personal computer equivalent to the supercomputer of a few years back. This innovation permits the computer to serve as the building block of industrial automation. Continued progress in the production of microprocessors will impact automation and control systems. The development of the Pentium processor series has been a more recent example.

Review Questions

Answer the following questions.

1. What is another name for a personal computer?

2. What is the basic organizational pattern of a computer system?

3. What does the ALU of a microprocessor do?

4. What is an accumulator, and what function does it perform?

5. Describe the purpose of a data register.

6. What is the difference between read/write and ROM chips?

7. When is data written into ROM?

8. How are data optically erased from a ROM?

9. What is an EAROM?

10. What happens to data in RAM when power is lost?

11. The sequence of instructions a computer runs through is controlled by what computer function?

12. Explain the fetch and execute operations of a microcomputer.

13. What are the sequential steps of a read memory operation?

14. What is the purpose of the interrupt operation of the microcomputer system?

15. Explain the differences between: immediate addressing, relative addressing, indexed addressing, direct addressing, and extended addressing.

16. What is the difference between a microprocessor and a microcontroller?

17. What are the three main types of control structures used while programming an Arduino microcontroller?

21

Programmable Logic Controllers

Objectives

Upon completion of this chapter, you should be able to:

- Describe a PLC system.
- Discuss the development of PLC systems
- Discuss the basic operational principle of PLCs.
- Discuss the basic methods of relay logic for PLCs.
- Describe methods of accomplishing programming for PLCs.
- Discuss sequencing and arithmetic operations using a PLC.

Important Terms

assembler
assembly language
central processing unit (CPU)
mini-PLCs mnemonics
programmable logic controller (PLC)
relay contact
relay ladder
relay logic
sequencer
sequencing
table

Introduction

For many years, machines were controlled by electromechanical devices, such as relays, solenoid valves, motors, linear actuators, and timers. Although they sometimes performed some basic logic functions, most controllers simply turned the machine on or off. Production line sequences were controlled by motors with timers. Nearly all controllers were permanently hard-wired into the system, making changes difficult and expensive. In industries where frequent changes were needed, this type of control was costly. However, it

DOI: 10.1201/9788743810056-28

was in many cases not simply the best way, but the only way that control could be achieved with any success.

Things changed when digital electronics came on the scene. It made possible the devices known as programmable logic controllers (PLCs). PLCs have numerous applications in controlling machines as part of computer-integrated and flexible manufacturing systems. This chapter discusses the development of programmable logic controllers, their components and operation, and how they are programmed.

Programmable logic controllers (PLCs) are evolving in complexity as the available technologies advance. New PLC technologies are continually emerging to augment the current capabilities and applications. The following will detail recent trends in PLC technology.

PLCs provide control for complex industrial systems, coordinating the movements, interactions, and behavior. In robotics, PLCs are responsible for controlling the movements of robotic arms, grippers, and other mechanical components. Executing precise motions, coordinated on multiple axes, and ensuring accurate positioning, PLCs also perform sensor Integration functions in an industrial setting. Vision systems, face sensors, and proximity sensors gather information about a machine's surroundings. PLCs can be used to sequence and coordinate tasks and operations. This can be anything from picking and placing objects to assembly processes. Using sensor data and programmed logic in conjunction, PLCs make decisions about actions. PLCs allow machines the ability to choose the best path to take, select the appropriate tool, or respond to unexpected situations.

The Internet of Things (IoT) is a common term that denotes a network of physical objects that are embedded with sensors, software, and other technologies, enabling them to interact with other devices over an internet connection. These interactions can include data transfer, connectivity, and control. This allows systems to function together under unified control.

PLCs and IoT function together to optimize machine operation and integration. Machines can communicate, share data, and have centralized control in real time through the connectivity allowed using IoT and PLCs together. IoT provides connectivity, and PLCs provide computational control for the system to connect to larger networks and communicate with a wider range of devices. IoT-enabled sensors provide constant data to PLCs, allowing for real-time control of manufacturing processes.

In the future, PLCs will include more IoT integration, increased processing power, and improved integration between information technology (IT) and operational technology. Allowing PLCs to handle more complex tasks and larger amounts of data.

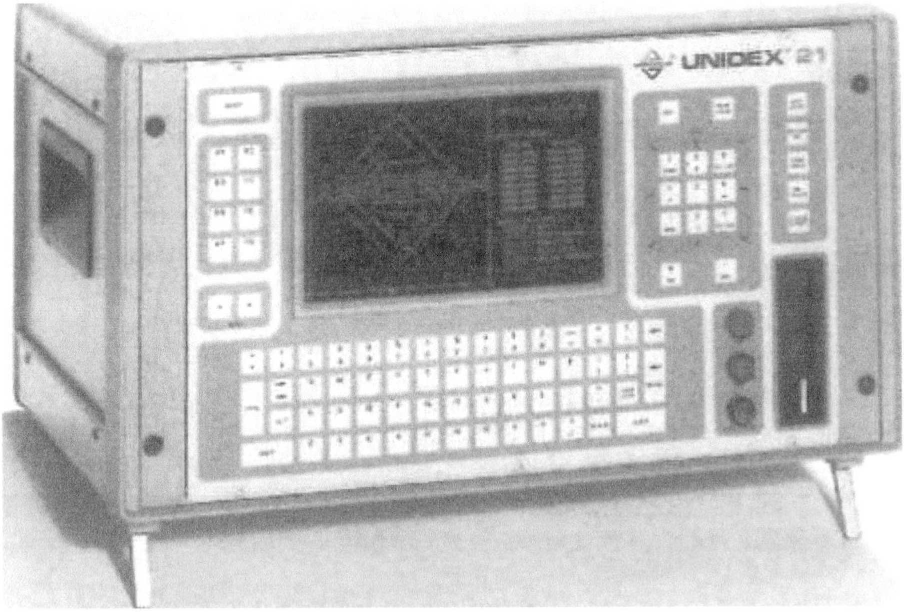

Figure 21.1 Programmable logic controllers (PLCs) are used to control machine tools. There are many different sizes and types available. (Aerotech, Inc.)

21.1 Development of PLCs

In the late 1960s, digital electronic devices began to replace the electromechanical equipment used to control machines. Further refinement of these devices led to the development of the **programmable logic controller (PLC)**, Figure 21.1. Early PLCs could perform only AND and OR functions. Today, they can make a variety of calculations. In fact, their ability to do simple arithmetic is the main thing that distinguishes them from electromagnetic controllers. Today, they perform logical operations compatible with the traditional relay logic formerly used. They are very versatile and can be easily changed to meet the demands of complex situations. They have reduced downtime when making changeovers, occupy very little space, and have improved efficiency.

A **PLC** can respond to numerical data, such as size, weight, temperature, or pressure. For example, it can accept a temperature signal, multiply it by a constant, and print out the results in degrees Fahrenheit. The resulting action can be retained in memory, recorded, displayed on a monitor, or used to set off an alarm. PLCs can also handle trigonometric functions, square roots, and logarithms. These calculations are important in process

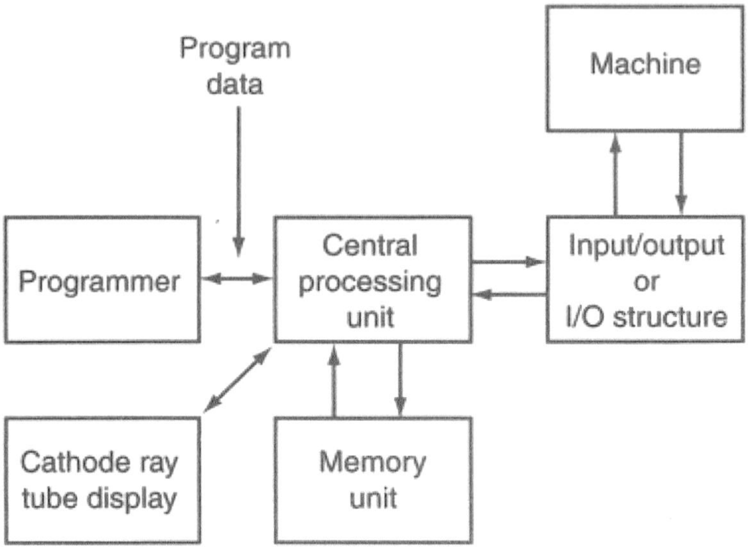

Figure 21.2 This block diagram represents the parts of a PLC.

control, process modelling, real-time error correction, and production line control.

In a **programmable logic controller**, instructions can be entered or changed without using complex languages and routines. The operator needs only a sound understanding of relay logic and its design. Functions can be altered by simply depressing a few buttons or keys to change the logic sequence. In many cases, an entire system change can be made. The new procedure can be completely different from the original. Once the system has been modified, the program is retained in memory.

21.2 Controller Systems

A block diagram for a simplified **PLC system** is shown in Figure 21.2. Its parts include the input/output structure, the processor, the memory unit, the monitor, and the programmer. In general, the devices making up the PLC are more complex than a single block can indicate. Most programmable logic controllers, in fact, are considered dedicated computers. Their degree of sophistication or power depends on the application in which they are used. Many PLCs interface with a mainframe computer. Others are independent and respond only to those things needed to control a specific machine.

21.3 Components

PLCs typically have both a standard power supply and a backup in case of power failure. The **central processing unit (CPU)** is its nerve center. A microprocessor is the basic logic element. It has address outputs and address buses, stores and handles data, and monitors the status of input and output signals. The output is based on information being processed through the system.

The **arithmetic processing unit (APU)** is part of the microprocessor. It computes basic arithmetic, logarithmic, and trigonometric functions. The APU generally functions independently once the initialization (starting sequence) is complete. The PLC typically has **memory modules** that are expandable. Possible choices are an ultraviolet erasable programmable read-only memory (LTV-PROM) or semiconductor random access memory (RAM). Both types can be intermixed in the PLC assembly. A display monitor and keyboard (buttons) are also components. Control buttons are also generally found on the front panel and are used to select such functions as timer control, parts count, or temperature.

Interface modules can be installed to connect the PLC with a network, or with such peripheral devices as printers, monitors, modems, or data storage devices. As a rule, each type of connection requires its own interface module.

The plug-in **input/output (1/O) modules** are housed in the assembly. For output, the module can be a solid-state relay, electromechanical relay, timer, or power transistor. It makes a direct connection between the PLC and the load device. This type of construction eliminates hard-wired circuits and reduces interfacing problems between the system and the load. Input modules are designed to interface signals. Their circuitry accommodates such things as thermocouples, light sensors, and practically any type of analog input signal. Communication with a programmable logic controller can be done using an external monitor (CRT or cathode ray tube). Designed for operation in environments where electromagnetic noise, high temperature, humidity, and mechanical shock occur, such a monitor is typically housed in a rugged case. The CRT can be easily moved to the worksite where the controller is located. When the environment is not harsh, a standard CRT can be used.

21.4 Mini Systems

Continuing improvements in integrated circuits and power transistor technology are responsible for the development of *mini-PLCs*. These systems can control simple machine operations and many types of manufacturing

Figure 21.3 Mini-PLC system. (Eagle Signal Controls)

processes. A number of companies are now producing them. One such system is shown in Figure 21.3.

Mini-PLCs can economically replace as many as four electromechanical relays in a control application. They provide timer and counter functions, as well as relay logic, and are small enough to fit into a standard 19-inch rack assembly. Most have fewer than 32 I/O ports or modules. Some units, however, can be expanded to drive up to 4001/O devices. As a rule, the I/O modules respond to digital signals. However, some units can handle analog information. This makes it possible for the system to respond to changes in temperature, pressure, flow, level, light, weight, and other analog data.

In general, mini-PLCs can achieve the same control as larger PLCs. However, they are usually smaller, less expensive, simpler to use, and more efficient. In the future, these systems will play a greater role in the control of industrial processes.

21.5 Programming

A PLC is programmed (given instructions to store and execute) through the use of a keyboard, push buttons, magnetic disks or tape, or a network connection to a computer. How the PLC performs is based on its programming. In general, PLCs are programmed using assembly language or relay logic.

21.5.1 Assembly language

Assembly language is a machine-oriented instructional set for a specific microprocessor. In assembly language, the operation code is in the form of symbols or words called *mnemonics*. A mnemonic is easier to recognize and remember than binary code. An advantage of assembly language is that it makes keeping track of loops and variables easier than other programming methods. It also makes it easier for someone else to look at the program and see how it works. However, assembly language programs are long and complicated. Many errors can occur. Also, the mnemonics must ultimately be translated into binary data before they can be used by the machine. This can be done by writing down each one, line by line, and then entering the translation into memory. An easier (and more usual) method is to use a program called an *assembler* that does the translation automatically. The data is then usually input directly into memory.

For some microprocessors, **assembly language** uses bit and byte I/O modes and provides a number of special control functions. For example, a common microprocessor has *244* instructions with 1-3 bytes of code per instruction (1 byte of operand and 0-2 bytes of data). Sample instructions include:

- LAL – Load the accumulator with the content of the L register.
- AD – Add 1 to the contents of the accumulator and store the result in the accumulator.
- LLE – Load the L register with the contents of the E register.

Assembly language is generally considered a low-level language. It is efficient with respect to the amount of memory and execution time required. However, it is somewhat difficult to use and debug unless the programmer works with the language frequently. Other languages used by PLCs are typically variations of assembly language.

21.5.2 Relay logic

Some PLCs are programmed using **relay logic**. A **ladder diagram** is entered into the system by means of the keyboard. Figure 21.4 shows the layout of a PLC keyboard for relay logic. It has network control, relay control, and a numerical entry section. When using relay logic, the basic programming element is the *relay contact.* This contact can be normally open or normally closed.

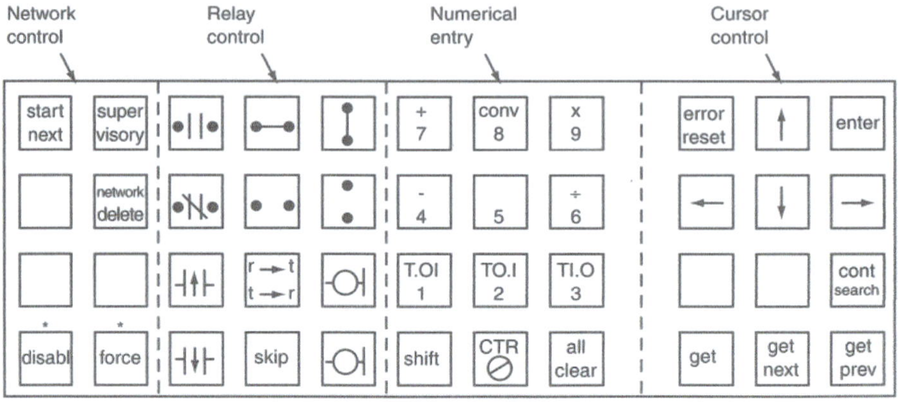

Figure 21.4 This diagram shows the layout of a typical relay logic PLC keyboard.

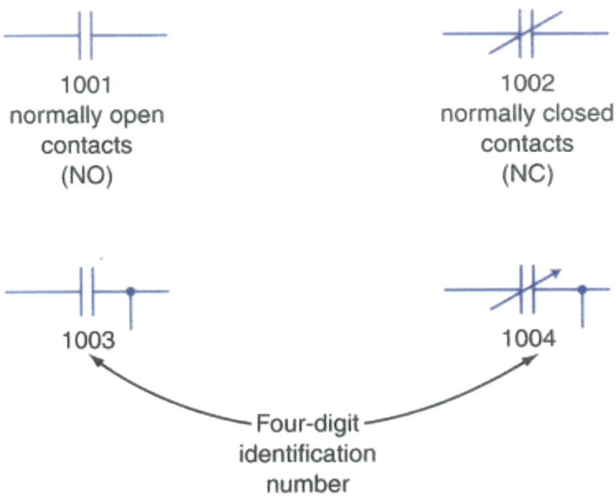

Figure 21.5 Relay contacts may be of the normally open or normally closed type.

In Figure 21.5, the normally open contacts are on the left and the normally closed contacts are on the right. The line to the right of the two lower contacts is for connection to optional branch circuits. Below each contact is a four-digit reference number. This number identifies specific contacts used in the system. Contacts are connected, in either series or parallel, to form the horizontal rungs of a *relay ladder.*

Figure 21.6 shows a typical **relay ladder diagram**. This format allows for up to ten elements in each horizontal rung. The rungs are connected into networks. A network can be a single rung or a combination of several, as

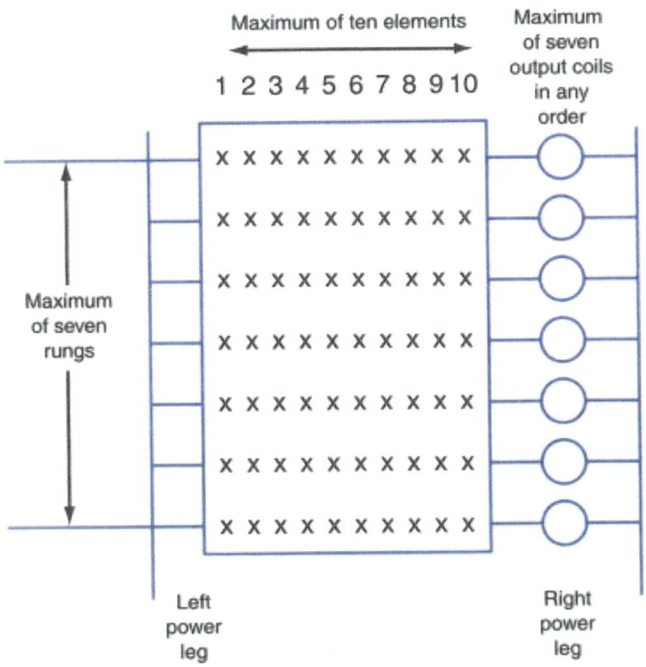

Figure 21.6 This relay ladder format allows for 10 elements in each rung.

long as the elements in each rung are connected. Up to seven rungs can be contained in a network. The left rail of the ladder is the common connecting element. On the right rail, up to seven coils are connected in random order. These coils are given an identification number. The number can be used only once during the operational sequence. The number of devices and registers available for use depends on the capacity of the system. Power flow is only from left to right. The operating sequence occurs vertically from top to bottom. In loop operations, it can be recycled from the bottom back to the top.

Once a **relay program** has been entered into the PLC, it is monitored on the CRT and can be modified using the keyboard. On the monitor, the path of electrical current through the network is brighter than the other components. Changes can be made by placing a cursor on the device to be altered and pressing the correct key. Contact status can be changed or bypassed and different outputs can be turned on or off by this process. All components and registers are identified by numbers. For example, each 1/O module has a reference number or address that identifies its location in the system. Each manufacturer has its own numbering system. An example is shown in Figure 21.7. Notice that the outputs are identified by a number beginning with zero, while

I/O Reference Configuration

Module number (top to bottom)	Circuit number	Channel one housing number							
		One		Two		Three		Four	
		output	input	output	input	output	input	output	input
	1	0001	1001	0033	1033	0065	1065	0097	1097
1	2	0002	1002	0034	1034	0066	1066	0098	1098
	3	0003	1003	0035	1035	0067	1067	0099	1099
	4	0004	1004	0036	1036	0068	1068	0100	1100
	1	0005	1005	0037	1037	0069	1069	0101	1101
2	2	0006	1006	0038	1038	0070	1070	0102	1102
	3	0007	1007	0039	1039	0071	1071	0103	1103
	4	0008	1008	0040	1040	0072	1072	0104	1104
	1	0009	1009	0041	1041	0073	1073	0105	1105
3	2	0010	1010	0042	1042	0074	1074	0106	1106
	3	0011	1011	0043	1043	0075	1075	0107	1107
	4	0012	1012	0044	1044	0076	1076	0108	1108
	1	0013	1013	0045	1045	0077	1077	0109	1109
4	2	0014	1014	0046	1046	0078	1078	0110	1110
	3	0015	1015	0047	1047	0079	1079	0111	1111
	4	0016	1016	0048	1048	0080	1080	0112	1112
	1	0017	1017	0049	1049	0081	1081	0113	1113
5	2	0018	1018	0050	1050	0082	1082	0114	1114
	3	0019	1019	0051	1051	0083	1083	0115	1115
	4	0020	1020	0052	1052	0084	1084	0116	1116
	1	0021	1021	0053	1053	0085	1085	0117	1117
6	2	0022	1022	0054	1054	0086	1086	0118	1118
	3	0023	1023	0055	1055	0087	1087	0119	1119
	4	0024	1024	0056	1056	0088	1088	0120	1120
	1	0025	1025	0057	1057	0089	1089	0121	1121
7	2	0026	1026	0058	1058	0090	1090	0122	1122
	3	0027	1027	0059	1059	0091	1091	0123	1123
	4	0028	1028	0060	1060	0092	1092	0124	1124
	1	0029	1029	0061	1061	0093	1093	0125	1125
8	2	0030	1030	0062	1062	0094	1094	0126	1126
	3	0031	1031	0063	1063	0095	1095	0127	1127
	4	0032	1032	0064	1064	0096	1096	0128	1128

Figure 21.7 Reference numbers are used to identify input/output modules. In this example, all output numbers begin with 0, and all input numbers begin with 1.

input numbers begin with one. The identity of each I/O device is extremely important during programming. For this reason, the numbers cannot be altered or used more than once.

Once a **relay diagram** has been entered, the PLC examines the network and solves the logic in numerical order. Network 1 is solved first, then networks 2,3,4, and so on. Each network is solved by a scanning process. Scanning consists of a series of high-speed pulses. Each pulse must pass through the network, following the proper sequence. Scanning moves from the left rail to the right rail and from the top to the bottom. It occurs the instant power is applied to the processor and continues for as long as it remains energized.

Scanning usually occurs so rapidly that the PLC appears to solve the logic steps instantly. The result is that each network is immediately available again to all following networks, regardless of any change in its status or in numerical value. This ensures that each network is solved in its correct numerical order and not by the numerical value assigned to a specific coil or contact.

21.6 Programming Methods

During **programming**, the appropriate component number and function are entered. The procedure is then placed in memory for the example that follows; programming is done using a relay ladder that can accommodate up to ten elements in each rung. As many as seven of these rungs can be connected to form a network (see Figure 21.6**).** When programming a relay ladder diagram into a PLC, the separate devices and registers are numbered (Figure 21.7).

The **numbering system** used depends on the memory size of the system. In a low-capacity system, number assignments might be 0001 to 0064 for output coils and 0258 to 0320 for internal coils. A larger system might use number assignments of 0001 to 0256 for output coils and 0256 to 0512 for internal coils. Any coil can be used only once. However, references to contacts controlled by a specific coil can be used as many times as needed. Output coils that are not used to drive a specific load can be used in the programming procedure.

When **programming** the response of a particular input, the input can be identified as a **normally-closed (NC) or normally-open** (NO) relay contact. The coil for the selected contact takes on the same number. While the coil is identified as a circle on the diagram, the contacts are identified by the standard contact symbol. See Figure 21.8**.** The contacts can be programmed for the NO or NC condition, depending on their intended functions.

Any external input that is normally closed, such as a safety switch, overload switch, or stop push button, must be treated differently. These are programmed on the CRT as normally open. An external NC push button, for example, would be entered on the CRT as an open contact. By switching the external contact function and its input signal, a double inversion occurs, causing the desired condition.

Assume that the *logic* shown in Figure 21.8 is used for a simple start-stop motor controller. The start and stop buttons are located externally. Pushing the start button closes contacts CR1 and energizes the relay coil. Contacts CR2 close at the same time, completing the circuit to the motor and turning it on. The motor will continue to run as long as power is supplied. Pushing the stop button turns off the motor and removes the latch from the start button.

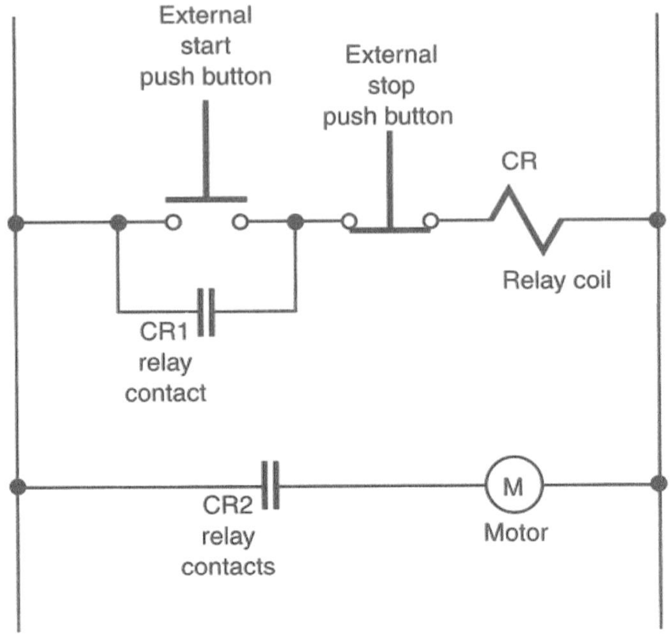

Figure 21.8 These relay ladder symbols are used for a start-and-stop controller.

The PLC equivalent of this motor controller is shown in Figure 21.9. Numbers refer to the specific components of the PLC. Input devices are numbered 1001 and 1002. This includes the input module and its resulting switching operation. The output device is numbered 0049. The start and stop buttons are connected externally, so they do not have a number.

An actual PLC circuit for starting and stopping a motor is shown in Figure 21.10. The 1/O modules interface with the microprocessor. The input and output modules are independent parts of the system, but both are controlled by the microprocessor. This circuit would be displayed on the CRT and could be modified with a few simple keystrokes. It is somewhat more complex than the ladder diagram equivalent. However, it is more versatile and can be quickly changed.

21.7 Sequencing Operations

Sequencing is the control of a load device in each of its stepping positions. A sequencer is basically a counting register. Sequencers are widely used in automatic assembly operations and in the control of production lines. Each

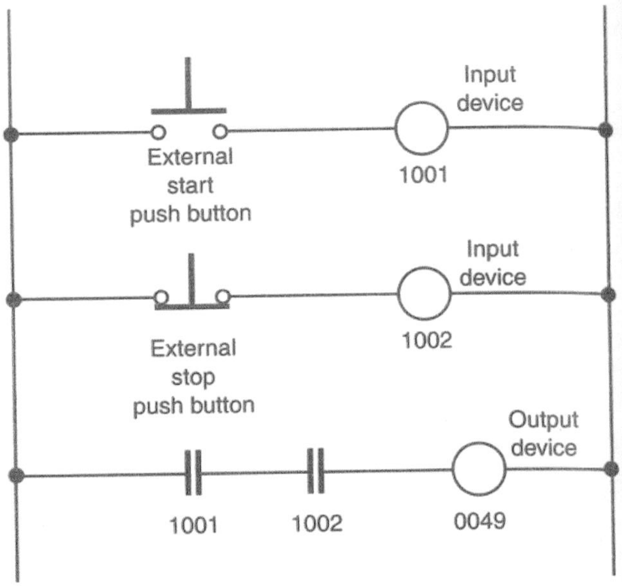

Figure 21.9 The **PLC** equivalent of the motor starter circuit shown in Figure 21.8.

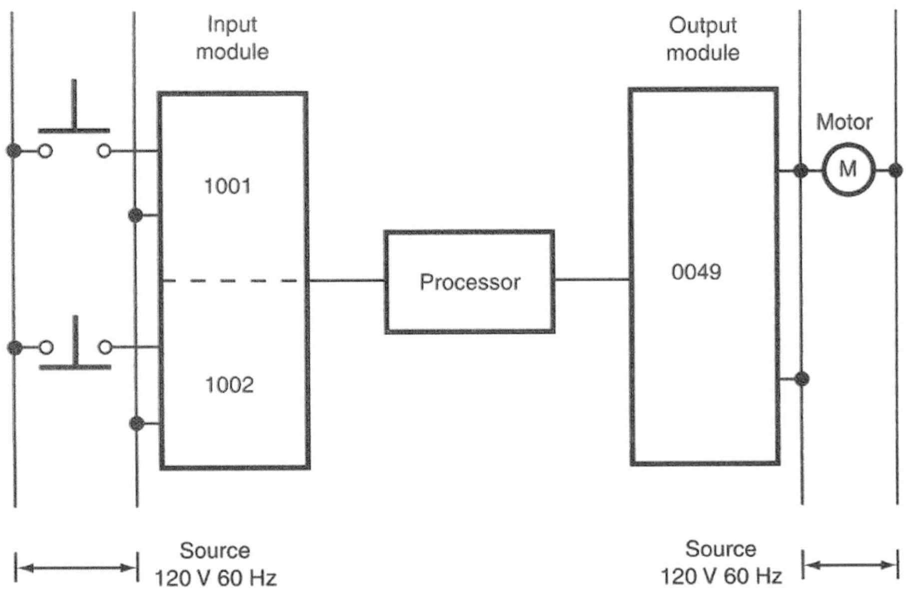

Figure 21.10 This diagram shows the actual PLC circuit of a motor controller.

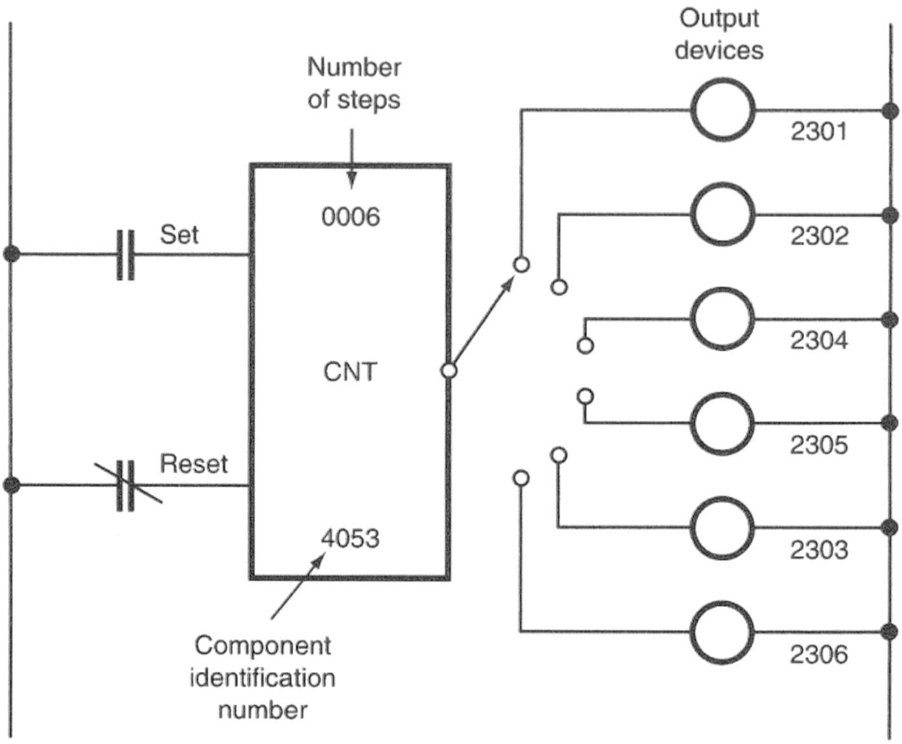

Figure 21.11 An example of a sequencing circuit.

PLC manufacturer has a unique way of identifying the operation, its location in memory, the actuating procedure, and the numbering of components. Generally, a sequencer is represented as a block in ladder diagrams and as a rectangle on a CRT.

In Figure 21.11, the number of steps being controlled is indicated by the top four-digit number. The number 0006 indicates six steps. In general, sequencers can achieve up to 32 steps (shown as 0032). This number can be altered using the keyboard. Counting information is stored in a specific register. **Registers** are identified by four-digit numbers. In Figure 21.11, the number is 4053.

Control of a sequencer is like that of other registers. It has two inputs, **set** and **reset**. The output is attached to six independent devices. **Each** time the set input is energized, the number value of the register increases by 1. This is the equivalent of moving a stepper relay one step.

When three steps are achieved, the third output device is actuated. In this case, a four-digit numbering system is used to identify the output device.

Output device 2304 would be actuated by the third input. When the **reset input** is energized, the stepping count goes to zero and the sequence returns to the starting position. This is not affected by the location of its present count or step position. The sequencer will not energize any of the outputs during the transition period.

Each **sequencer** responds in an independent manner. It is controlled by separate register values. Sequencer references can be used as often as necessary in a PLC circuit. They can appear anywhere that a relay contact is appropriate. By using calculated logic, it is possible to get the sequencer to skip steps, advance forward, or move backward according to some prescribed procedure. Intermediate references are not energized during this unusual procedure. Sequencers are updated as soon as a change is made in the content of a sequence register. One independent network can be used to drive a sequencer, and a second network can be used to update the most recent value changes. Sequencers in general cannot be used on transitional contacts in control applications.

21.8 Arithmetic Functions

In addition to the normal relay functions, a **programmable controller** is capable of several data-manipulation functions. These functions are capable of performing some arithmetic operations. An **arithmetic operation** has three elements, usually registers, placed vertically one above the other in a block or rectangular box. The top and middle elements can be either a fixed or three-digit variable value not exceeding a maximum of 0999. The bottom element is a holding register. The registers are all assigned reference numbers for identification. A particular number assignment would vary between different manufacturers. In this explanation, the numbers used have been developed by the manufacturer of a PLC system. They would be different when applied to a system made by a different manufacturer. We are only using the number assignments here to serve as general identification of the particular registers being discussed.

Figure 21.12 shows the **programming format** of an arithmetic element. The upper element represents data or the numerical content of a register. This normally has a register identification number where its content is stored. The enable input on the left side of the block is used to achieve control of the function. The middle element is usually the same as the upper element. It represents data entry or the contents of data stored in a register. It is usually identified by a different numbering sequence to distinguish it from the upper register. The lower element is a holding register that derives its data from the

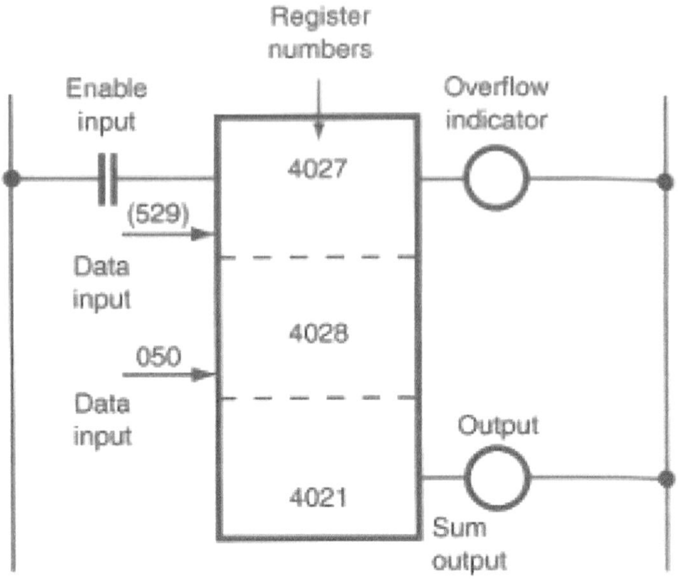

Figure 21.12 Arithmetic format.

upper two registers. These data can be the sum, difference, product, or quotient of the data in the upper two registers.

The **arithmetic function** of the unit determines the data value that appears in the lower register. Each arithmetic unit has up to three output possibilities depending on the function of the unit. These appear on the right side of the block. The outputs are updated after each scan of the enable input. The organization of the data and how it is manipulated is a programmable function that can be altered by a keystroke operation of the programming unit.

21.8.1 Addition

The **addition function** is achieved by adding the contents of the upper register to the contents of the middle register and placing the sum in the holding register. The upper and middle registers can have fixed or variable values from 000 to 999. In the addition function, generally, only one output is developed. This is used to represent the content of the holding register. In most units, the result of the addition function cannot be greater than the three-digit number 999. Values greater than this will only indicate the three least significant digits. In some cases, an overflow indication will appear when the sum is greater than 999. The overflow can be stored in a separate register, with the three least significant digits stored in another register.

Figure 21.12 shows an example of the **addition function** and some representative identification numbers used in a programmable controller. The four-digit numbers are register identification numbers. The numbers in brackets are the data values to be added and the sum of the two values. In this case, the sum is less than 999, so the overflow indicator is not actuated. The addition function, in this example, is updated each time that the system is scanned when the enable input is in its active or on state. The addition function of a programmable controller is generally limited to rather small number values. This is purposely done so that the function can be achieved with a small number of electronic components.

21.8.2 Subtraction

The **subtraction function** of a programmable controller is achieved by the same type of circuitry as that of the addition function. The difference between the two functions is the data value placed in the holding register. In the adder, the sum of the top and middle registers is placed in the holding register. In a subtractor, the difference between the upper and middle registers is placed in the holding register. Furthermore, a subtractor generally has three output conditions. The upper output passes power to the load when the upper register is greater than the middle register. The middle output passes power to the load when the upper and middle registers are equal. The lower output is actuated when the upper register is less than the middle register. The three outputs can be selected or used according to the demands of the control function being utilized.

Figure 21.13 shows an example of the **subtraction function** and some of the component numbers that can be identified with the programming procedure. When the enable input is energized, it causes data stored in the respective registers to be applied to the subtractor unit. Notice that each register is identified by a four-digit number. The bracketed numbers represent data that appear at each register. Respective outputs are represented by four-digit numbers of the 0030 series. With the indicated data applied, output coil 0030 would be energized because the upper register is greater than the middle register. The data value placed in the holding register is the actual difference between the upper and middle registers.

The **subtraction function** occurs and is updated each time that the enable input is scanned. This takes place very rapidly, and the entire process

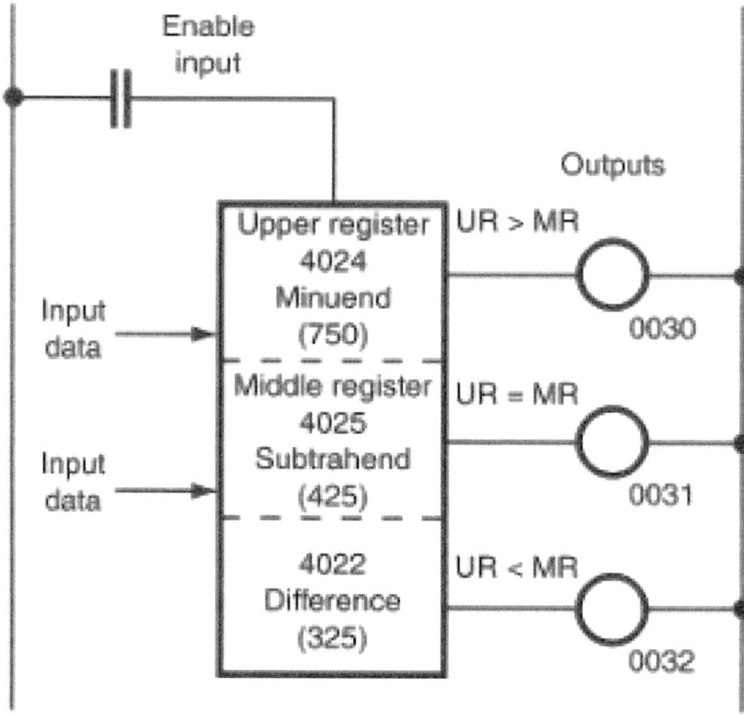

Figure 21.13 Subtraction format.

appears to be functioning on a continuous basis. The largest data number that can appear in any of the three registers is 999.

21.8.3 Multiplication

The **multiplication function** of a programmable controller is very similar to that of the other arithmetic functions. Three registers are involved in the function. This function has data applied to the upper and middle registers. The product of the two register values appears in the holding register. As a rule, the holding register is divided into two sections that accommodate the numbers 999 and 999. The three least significant digits are considered to be *low-order values.* The three most significant digits are considered to be *high-order values.*

The combined high- and low-order values permit a six-digit value to be produced as the product of the two registers. The product is an actual data value that is placed in a combination holding register. A **multiplie**r generally has only one output. This indicates when the multiplication process is being performed by the assembly.

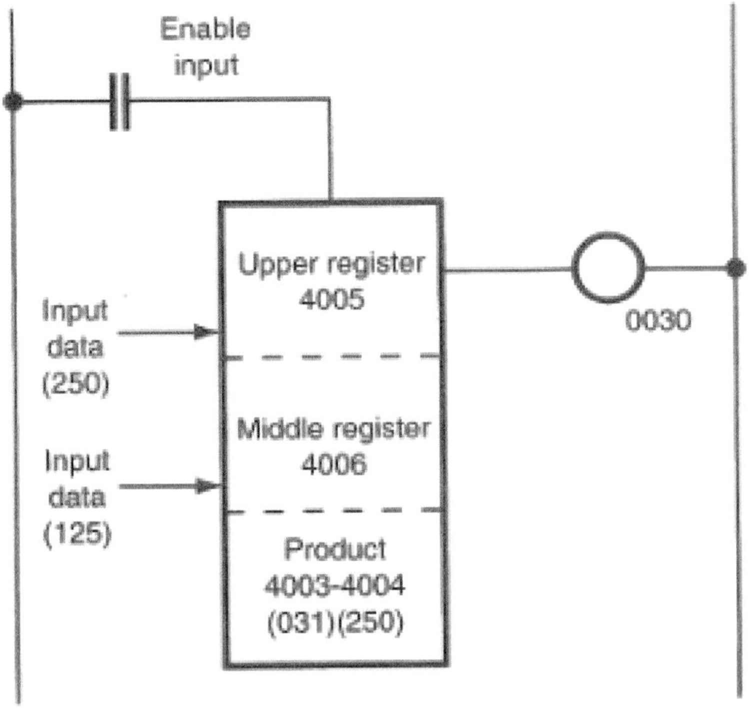

Figure 21.14 Multiplier format.

Figure 21.14 shows the **multiplication programming** format of a programmable controller. Four-digit numbers are used to identify the specific registers involved in this function. The enable input is actuated by contacts 1018. Each time that the enable input is energized, it causes the multiplication function to be performed and updated with each scanning cycle. In general, this function appears to take place instantly and occurs on a continuous basis. The bracketed numbers are representative of the data applied to each register. Note that the upper register is 250 and the middle register is 125. The product is 031 250 and is divided between two registers. The multiplication output can be used if the system requirements demand some indication that the function is being performed. The product of the two values appears as an actual data value.

21.8.4 Division

The **division function** of a programmable controller is similar to that of the multiplication function. The holding register indicates the quotient of the two numbers appearing in the upper and middle registers. The data value of

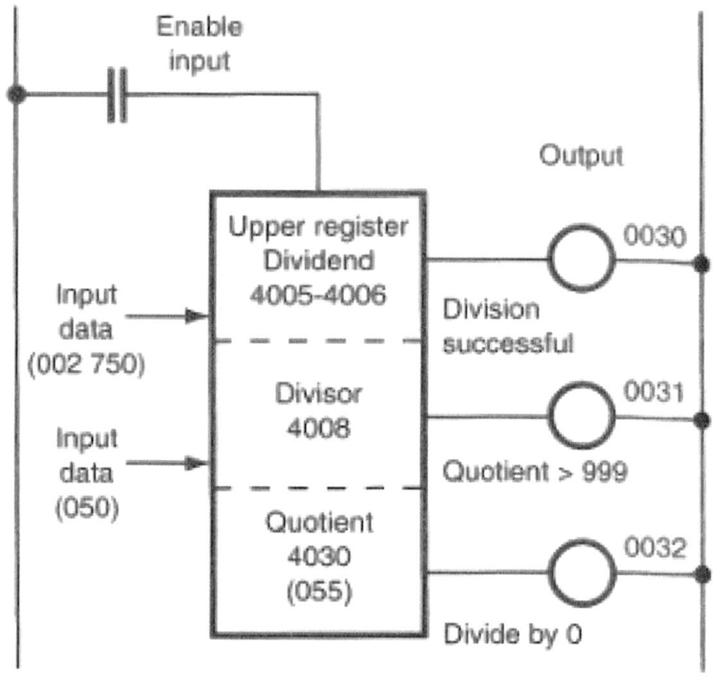

Figure 21.15 Divider format.

the upper register is the dividend, and the middle register is the divisor. The dividend is divided by the divisor, and the quotient appears in the holding register. The dividend generally takes two consecutive registers to accommodate its value. Data values of up to 999,999 can be used as the dividend. The divisor and quotient utilize a single register with a maximum value of 999. A divider generally has three outputs. These indicate when the division operation is successful, if the quotient is greater the 999, and a divide by zero indication.

The programming format of a **divider** is shown in Figure 21.15. The four-digit numbers are used to identify the register. Registers 4005 and 4006 form the upper register or dividend. Register 4008 serves as the divisor, and 4030 is the quotient. The bracketed numbers are representative of the data being manipulated by the divider. In this case, 002 750/50 = 55. The dividend is held in two consecutive registers. Output 0030 would be energized by this operation, indicating that the division function has been successfully performed. The quotient appears in register 4030 as an actual data value.

The entire operation can be performed only when the **enable input** is energized. When enabled, the operation is performed on a continuous basis and appears to take place instantly. Only a limited value of division can be achieved by a programmable controller.

21.8.5 Move function

The transfer of data between different registers in a programmable controller is described as the **move function**. Essentially, data in an input or holding register can be transferred or moved to another holding register within the controller through this function. The data are not altered during the transfer operation. Moving takes place very quickly because it is initiated by the scanning signal.

Two types of moving are achieved by a programmable controller**:** register to table and table to register**.** The format of the move function is primarily the same for both operations. A **table** is a group of consecutive registers all holding data. Moving the data of a table is achieved by a pointer register. The pointer identifies each register **of** the table and causes its data to be transferred during the scanning operation. The table data are not altered during the transfer operation. The data at the new location is destroyed and replaced with new data from the table.

Figure 21.16 shows symbolic representations of the **move function**. When a scan pulse arrives at the set input, it enables the function. The contents of the register at location 4003 are transferred to the output. The lower register, identified as 4004, is used as the pointer. Its content determines the transfer location of the data moved from register 4003. Through this procedure, the content of a register can be transferred to a table, or the content of a table can be transferred to a register. The letter designation R-T denotes register to table, and T-R refers to the table-to-register moving function.

21.8.6 Conversion

The **conversion function** of a programmable controller is used to change BCD data into binary form. This permits a variety of different devices to be connected to discrete input/output modules. Each input module, having four circuits, is capable of interfacing one BCD digit. Up to three consecutive modules are combined to provide coverage of numbers from 000 to 999 for a single register. Each data value will be converted and applied to or from the I/O module during a scanning cycle.

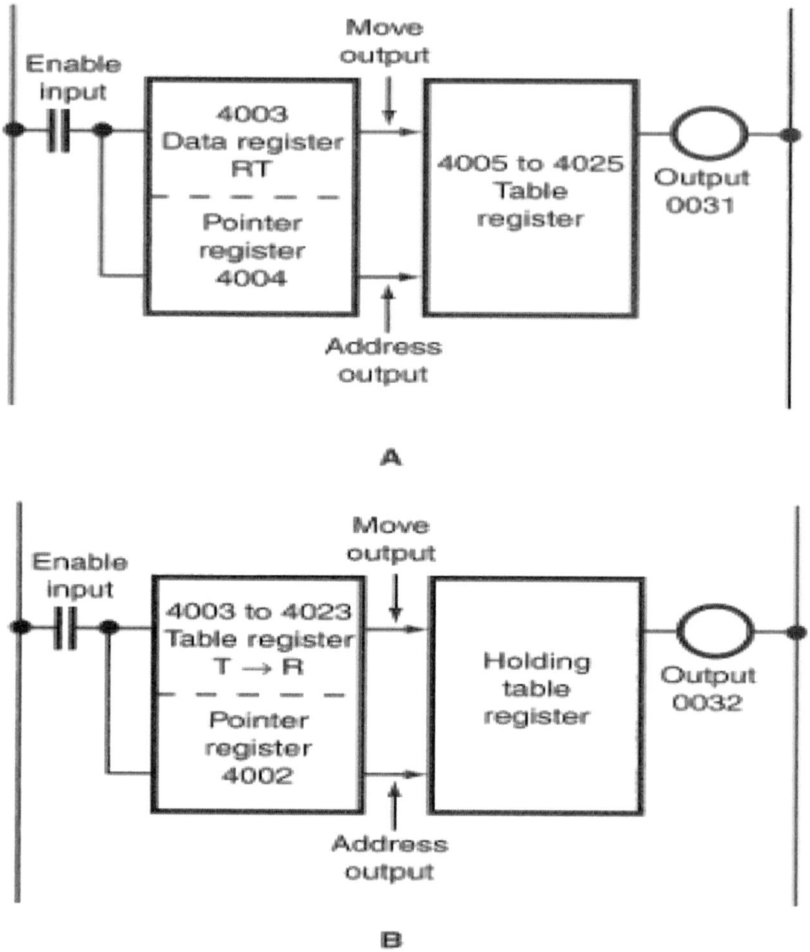

Figure 21.16 Data-moving function. A – Register to table moving. B – Table to register moving.

Figure 21.17 shows the general format of a **converter**. Assume in this application that the output of the digital meter is in BCD form. The BCD value of 357 is shown being applied to the input register. Each of the three BCD values is represented by a combination of four output lines. These lines are connected to an input module that has four lines for each input digit. The input module is divided into three sections according to digital values. When an enable signal is applied to the set input, it causes the BCD value of the input register to be applied to the digital ranked values of the corresponding

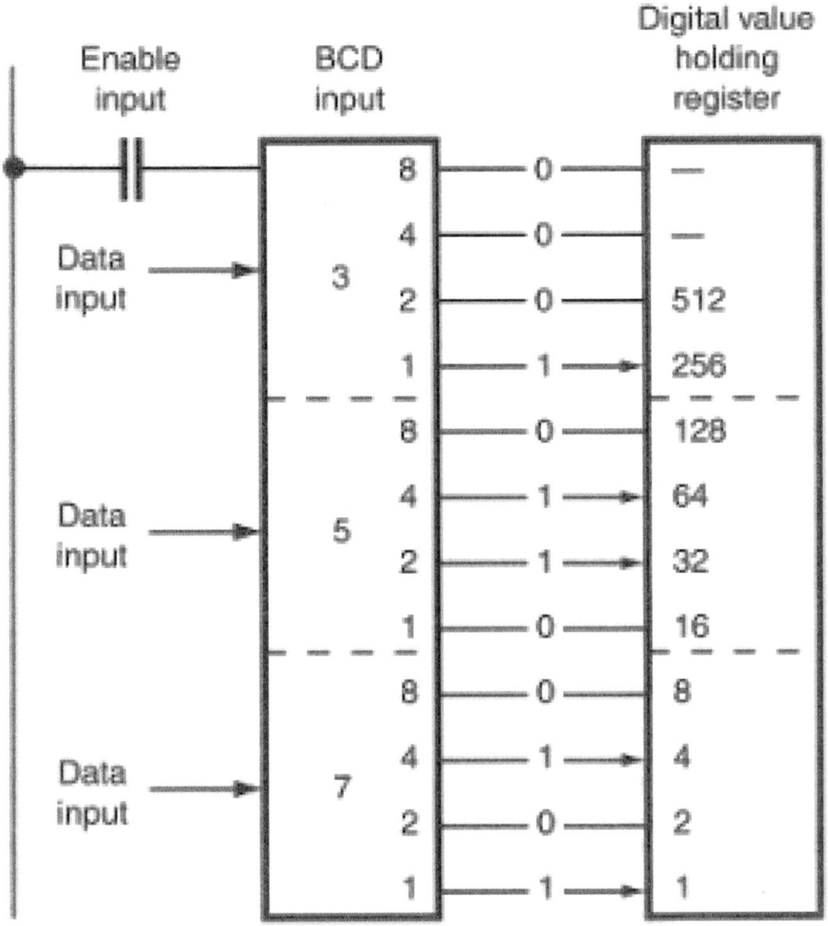

Figure 21.17 BCD-to-digital converter format.

input holding register. The binary equivalent is 0001-0110-0101. This digital value is then placed in the holding register for storage.

Conversion is also used to change digital data into BCD data to actuate an output device such as a meter or LED display. The conversion process is the reverse of the **input conversion** function of Figure 21.17. Figure 21.18 shows the format of an **output converter**. In this situation, the input is a digital signal that is divided into three sections of four lines each that represent the digital value of 0010-0001-1000. This input signal generally comes from a holding register. When an enable input signal is applied, it causes the digital

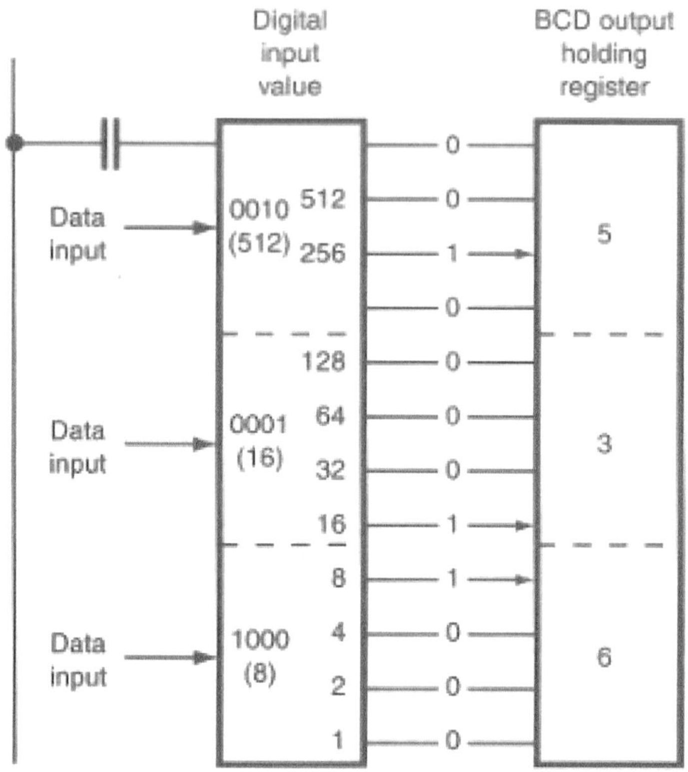

Figure 21.18 Digital-to-BCD converter format.

signal to be transferred to three consecutive connected holding registers. The resulting output is the BCD equivalent or 536, which is the original digital input value. Conversion of this type is used to interface the PLC with an instrument that has a digital display or response.

Review Questions

Answer the following questions.

1. What are the fundamental parts of a programmable logic controller?

2. Briefly explain the function of each fundamental part of a programmable logic controller.

3. What is a mini-PLC, and how does it differ from a full-scale PLC?

4. What is the meaning of the expression assembly language?

5. What is relay logic for a PLC?

6. What is sequencing?

7. How often can sequencer references be used in a PLC circuit?

8. How is adding achieved in a programmable logic controller?

9. How is division achieved in a programmable logic controller?

10. What is the purpose of the move function of a PLC?

22

Industrial Robotics

Objectives

Upon completion of this chapter, you should be able to:

- List and describe the parts of a robotic system.
- Define what is meant by degrees of freedom of a robot.
- Discuss methods of programming industrial robots.
- Describe various robot configurations that are used for automated manufacturing.
- Identify and describe various methods of interfacing robots with computer systems or other types of control systems.
- Describe machine vision systems used with automated manufacturing applications.

Important Terms

actuator
angular actuators
anthropomorphic
Cartesian configurations controller
cylindrical configurations
degrees of freedom
direct-drive electric motors
electric drive actuators
end effector
hierarchical control
hydraulic drive system
image acquisition
image analysis
image interpretation
image preprocessing
input ports
interface
joint
linear actuators
machine vision systems
manipulator
non-servo robot
output ports
parallel ports
parallel transmission
pitch

DOI: 10.1201/9788743810056-29

pixels
pneumatic drive systems
power supply
radial traverse
revolute configurations
roll
rotational traverse
SCARA
serial ports
serial transmission
servo robot
spherical configurations
vertical traverse
work envelope
yaw

Introduction

The development of the computer has created what some experts have called the Second Industrial Revolution. Many consider robots to be the prime movers of this revolution. Their use in our society is increasing and will continue to increase. The need to know more about robots has become more important for technicians. To fully understand automated industrial control, you should know the technical aspects of robotics.

22.1 Parts of a Robot

Robots come in many shapes and sizes. The industrial robots illustrated in Figures 22.1 and 22.2 resemble an inverted human arm mounted on a base. Industrial robots consist of a number of subsystems that work together. These subsystems, labeled in Figure 22.1, are the controller, the manipulator, the power supply, the end effector, and a means for programming. Figure 22.3 is a diagram of the relationships among these five major systems.

The **controller** is the part of a robot that coordinates all movements of the mechanical system. It also receives input from the immediate environment through various kinds of sensors. The **manipulator** consists of segments that can be jointed and that move about, allowing the robot to do work. In Figure 22.2, for example, the manipulator is the robot's arm. The **power supply** may convert AC line voltage to the DC voltages required by the robot's internal circuits, or it may be a pump or compressor providing hydraulic or pneumatic power. The end effector is the robot's hand. The area that the robot's end effector can reach is called its work envelope. The means for programming is used to record movements into the robot's memory.

22.1.1 Controller

The heart of the robot's controller is generally a **microprocessor** linked to input/output and monitoring devices. The commands issued by the controller

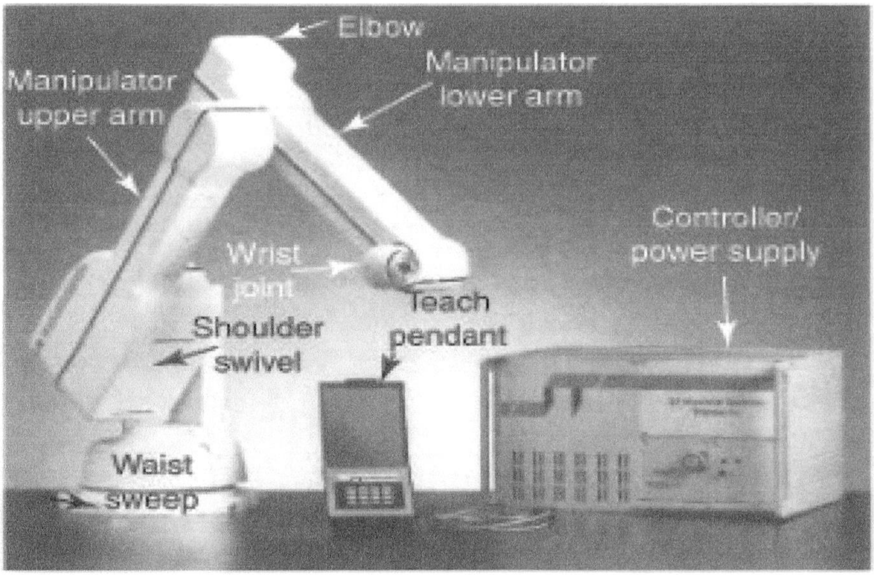

Figure 22.1 This robot clearly illustrates the systems of a typical industrial robot. This electric robot can be used in a variety of industrial applications. (S.T. Monforte Robotics, Inc.)

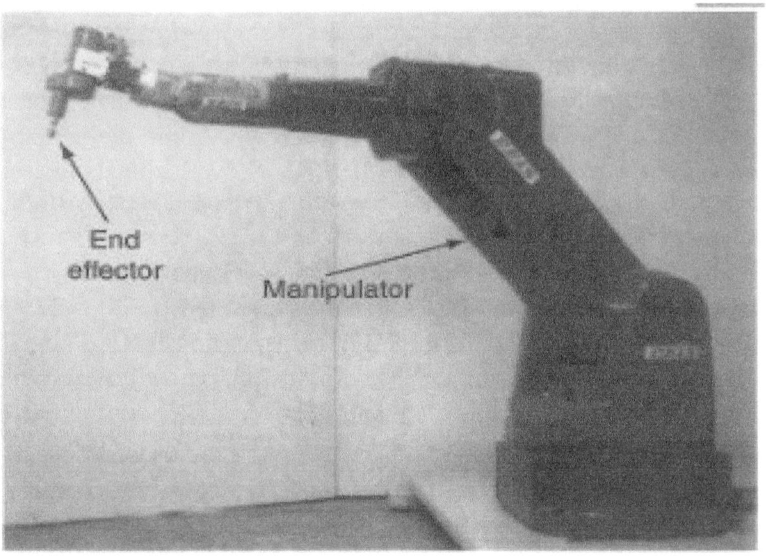

Figure 22.2 This robot has been designed expressly for use in precise path-oriented tasks such as deburring, milling, sanding, gluing, bonding, cutting, and assembly. (Reis Machines, Inc.)

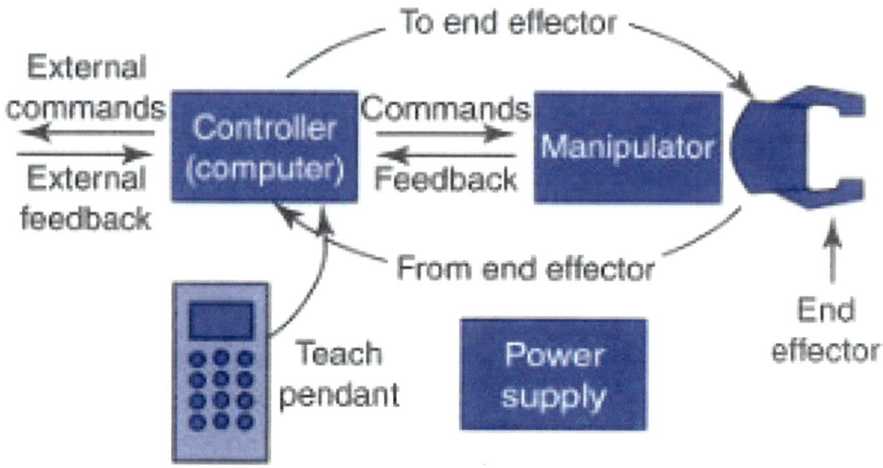

Figure 22.3 The relationships among the five major systems that make up an industrial robot are shown in this diagram.

activate the motion control mechanism, consisting of various controllers, amplifiers, and actuators. An actuator is a motor or valve that converts power into robot movement. This movement is initiated by a series of instructions, called a program, stored in the controller's memory. Figure 22.4 shows a Kawasaki **controller/power supply** and **teach pendant**.

The **controller** has three levels of hierarchical control. A hierarchical arrangement is one in which a given level is lower (more elementary) than the one above it, and is dependent on the level above it, for its instructions. Figure 22.5 illustrates the relationships among the basic levels of hierarchical control. The three levels are:

- *Level I:* **Actuator Control**. This is the most elemental level, at which the separate movements of the robot along various planes, such as the X,Y, and Z axes are controlled. These movements will be explained in detail in the section of this chapter that covers robot configurations.
- *Level II* **Path Control.** The path control (intermediate) level coordinates the separate movements along the planes determined in Level I into the desired trajectory (path).
- *Level III:* **Main Control.** The primary function of this highest control level is to interpret the written instructions from the human programmer as to the tasks required. The instructions are then with various

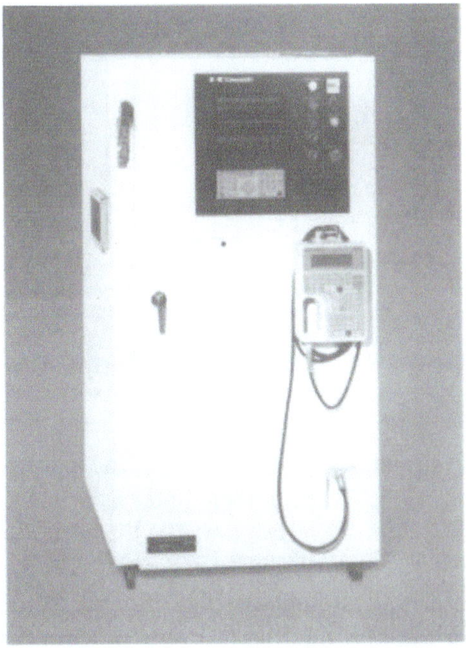

Figure 22.4 A controller/power supply with a teach pendant. (Kawasaki)

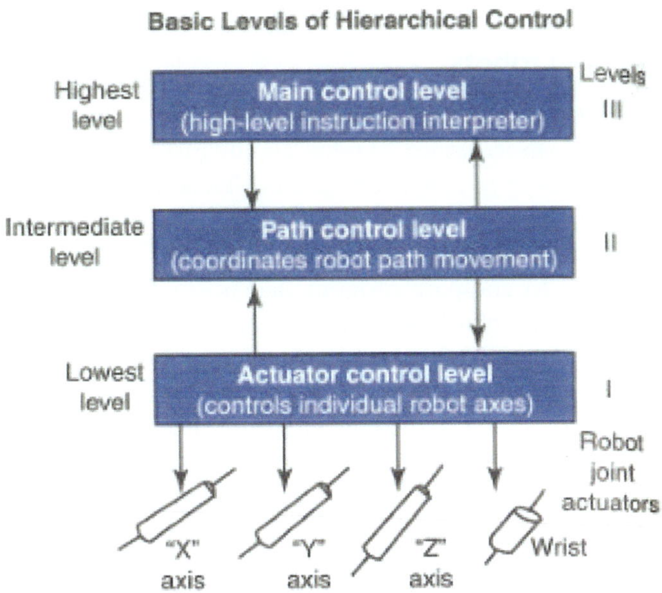

Figure 22.5 The three basic levels of hierarchical control.

Comparison of Robots by Definition			
Item No.	Japanese Definition	U.S. Definition	
		Nonsophisticated	Sophisticated
1	Manual manipulator		
2	Fixed sequence		
3	Variable sequence	Nonservo	
4	Playback		Servo—intelligent
5	Numerically controlled		Servo—intelligent
6	Intelligent		Servo—highly intelligent

Figure 22.6 Differing definitions of robot types by Japanese and U.S. industry are shown in this table.

environmental signals and translated by the controller into the more instructions that Level II can understand.

Robots can be classified according to the type of control system used. The **non-servo robot** is a non-intelligent robot. **Servo robots** are classified as either intelligent or highly intelligent. The primary difference between an intelligent and a highly intelligent robot is its level of awareness of its environment.

22.1.2 Non-servo robots

Non-servo robots are the simplest robots, according to the definition used in the United States. (See Figure 22.6 for a comparison of U.S. and Japanese classifications.) Non-servo robots are often referred to as limited sequence, pick-and-place, or fixed-stop robots.

The **non-servo robot** is an open-loop system. That is, no feedback mechanism is used to compare programmed positions to actual positions. A good example of an open-loop system is the operating cycle of a modern washing machine. Figure 22.7 is a block diagram of the steps performed by a typical washing machine. At the beginning of the operation, the dirty clothes and the detergent are placed in the machine's tub. The timer/control is set for the proper cleaning cycle, and the machine is activated by the start button. The machine fills with water and begins to go through the various washing, rinsing, and spinning cycles. The machine finally stops after the set sequence is completed. The washing machine is considered an open-loop system for two reasons. First, because the clothes are never examined by sensors during the washing cycle to see if they are clean. Second, the length of the cycle is not automatically adjusted to compensate for the amount of dirt remaining in

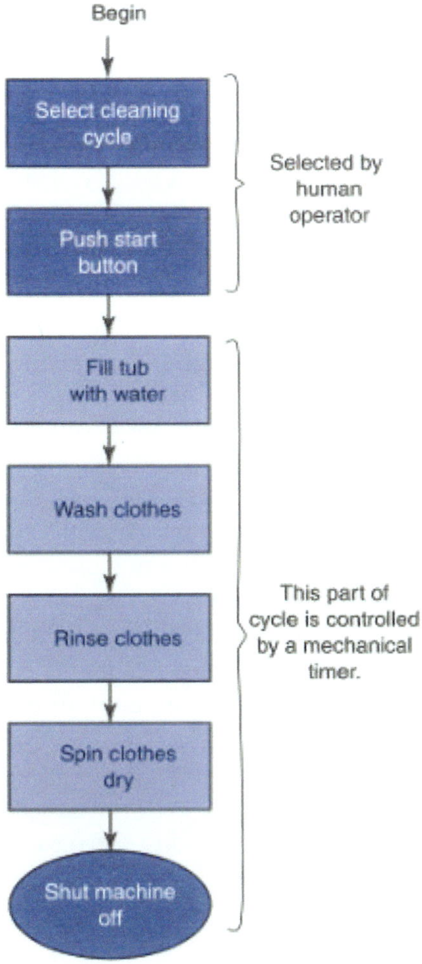

Figure 22.7 In this block diagram depicting the sequence of steps performed by a washing machine, no feedback is used. In such an open-loop control system, the condition of the clothes during the washing operation is not monitored and used to alter the process.

the clothes. The cycle and its time span are determined by the fixed sequence of the timer/control.

Study the diagram in Figure 22.8, which represents a **three-axis** pneumatic (air-controlled) robot. The three axes allow movement along certain planes. For the sake of simplicity, only one axis is shown. At the beginning of the cycle, the **controller** sends a signal to the control valve of the manipulator. As the valve opens, air passes into the air cylinder, causing the rod in the cylinder to move. As long as the valve remains open, this rod continues to

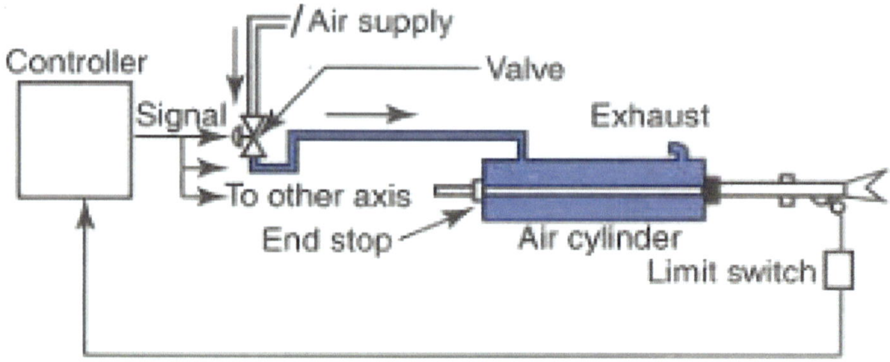

Figure 22.8 In a non-servo system, movement is regulated by such devices as a limit switch, which signals the controller when it is activated.

move until it is restrained by the end stop. After the rod reaches the limit of its travel, a limit switch tells the controller to close the control valve, and the controller sends the valve a signal to close. The controller then moves to the next step in the program and initiates the necessary signals. If the signals go to the end effector, for example, they might cause the gripper to close in order to grasp an object. The process is repeated until all the steps in the program have been completed.

- **Non-servo robots** are:
  - Relatively inexpensive compared to servo robots.
  - Simple to understand and operate.
  - Precise and reliable.
  - Simple to maintain.
  - Capable of fairly high speeds of operation.
  - Small in size.
  - Limited to relatively simple programs.

22.1.3 Servo robots

The **servo robot** is a closed-loop system because it allows for **feedback**. The feedback signal sent to the servo amplifier affects the output of the system. In a sense, a servomechanism is a type of control system that detects and corrects errors. Figure 22.9 shows a block diagram of a servo-controlled robotic system.

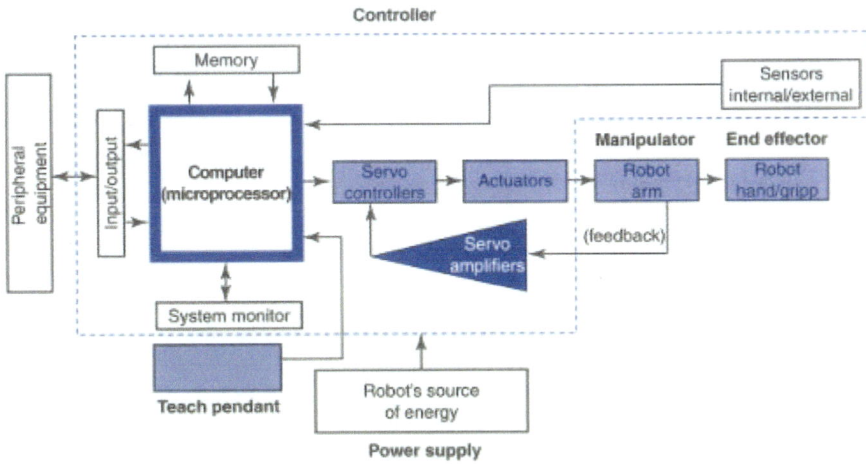

Figure 22.9 A servo-controlled industrial robotic system such as the one depicted in this block diagram might be classified as "intelligent" or "highly intelligent," depending on the level of sensory data it can interpret.

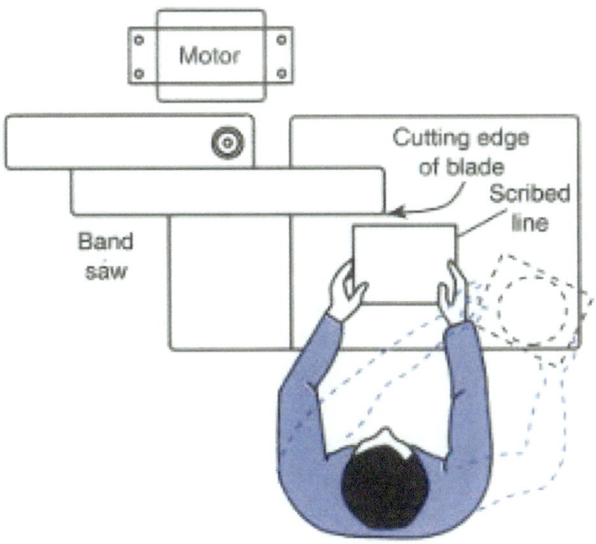

Figure 22.10 Human beings employ the principle of the servomechanism for many tasks, such as cutting a circle on a bandsaw.

The principle of **servo control** can be compared to many tasks performed by human beings. One example is cutting a circle from a piece of stock on a power bandsaw, shown in Figure 22.10. The machine operator's eye studies the position of the stock to be cut in relation to the cutting edge

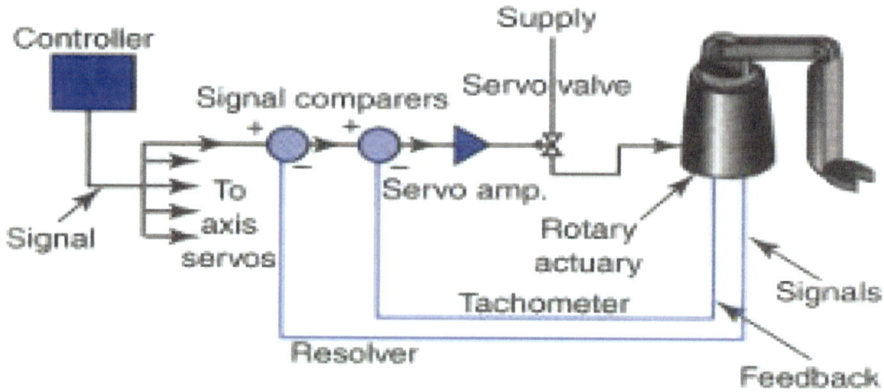

Figure 22.11 Feedback signals from the tachometer and resolver sensing systems allow the system to make corrections whenever the actual speed or position of the robot does not agree with the values contained in the robot's program.

of the blade. The eye transmits a signal to the brain. The brain compares the actual position to the desired position. The brain then sends a signal to the arms to move the stock beneath the cutting edge of the blade. The eye is used as a feedback sensing device, while the brain compares desired locations with actual locations. The brain sends signals to the arms to make necessary adjustments. This process is repeated as the operator follows the scribed line during the sawing operation.

The diagram in Figure 22.11 helps to explain the operation of a six-axis **hydraulic robot**. Only one axis is shown in detail. When the cycle begins, the controller searches the robot's programming for the desired locations along each axis. By means of feedback signals, the controller determines the actual locations on the various axes of the manipulator. The desired locations and actual locations are compared. Error signals are fed back to the servo amplifier. The greater the error, the higher the intensity of the signal. These error signals are amplified (increased) by the servo amplifier and applied to the servo control valve on the appropriate axis. The valve opens in proportion to the level of the command signal received. The opened valve admits fluid to the proper actuator to move the various segments of the manipulator.

New signals are generated as the **manipulator** moves. When there are no more error signals, the servo control valves close, shutting off the flow of fluid. The manipulator comes to rest at the desired position. The controller then addresses the next instruction in the program. It can be to move to another location, or to operate some peripheral equipment. The process is repeated until all steps of the program are completed. The output of a tachometer, a

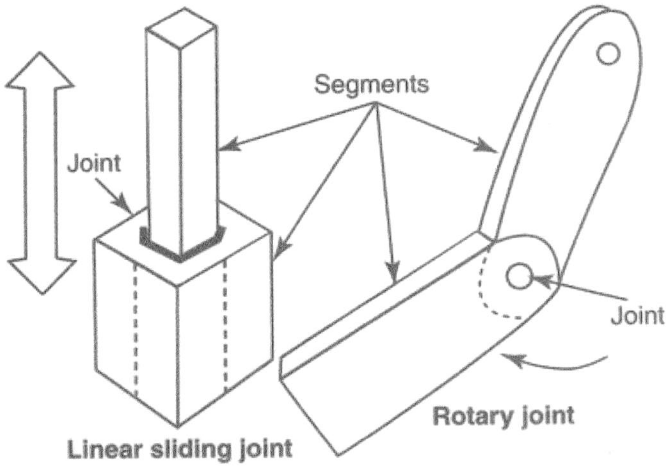

Figure 22.12 Both linear and rotary joints are commonly found in robots.

speed-measuring device, is used to control acceleration and deceleration of the manipulator's movements.

22.1.4 Servo robots

- They are relatively expensive to buy and cost more than non-servo robots to operate and maintain.
- They use a sophisticated, closed-loop controller.
- They have a wide range of capabilities.
- They can perform multiple point-to-point transfers as well as transfer along a controlled, continuous path.
- They can respond to very sophisticated programming.
- Use a manipulator arm that can be programmed to avoid obstructions within the work envelope.

22.1.5 Manipulator

The **manipulator** must move materials, parts, tools, or special devices through various motions to provide useful work. The manipulator is the arm of the robot and is made up of a series of segments and joints, much like those found in the human arm. Joints connect two segments together and allow

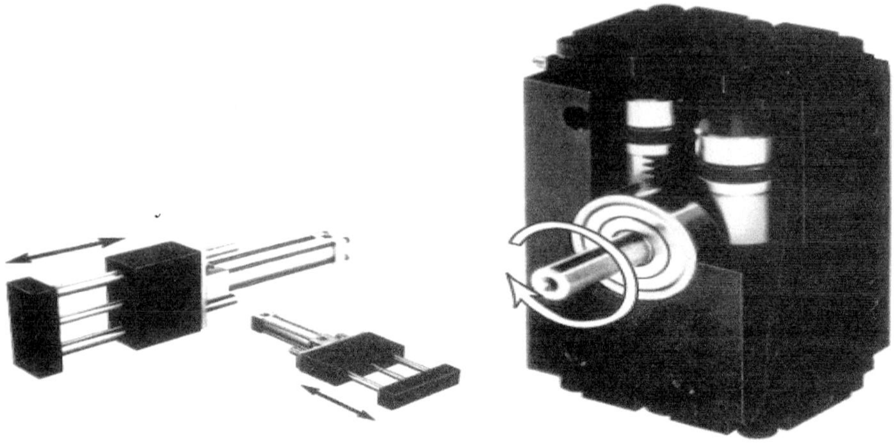

A B

Figure 22.13 A – Linear actuators provide straight-line movement. B – Rotational movement around an axis is provided by the angular (rotary) actuator. Actuators can be powered by electric motors, pneumatic (air) cylinders, or hydraulic (oil) cylinders. (Phd, Inc.)

them to move relative to one another. The joints provide either linear (straight line) or rotary (circular) movement. Figure 22.12 shows simple linear and rotary joints commonly found in manipulator arms.

The muscles of the human body supply the driving force that moves the various body joints. Similarly, a robot uses actuators to move its arm along programmed paths and then to hold its joints rigid once the correct position is reached.

There are two basic types of motion provided by actuators. **Linear actuators** provide motion along a straight line. They either extend or retract their attached loads. **Angular actuators** provide rotation. They move their loads in an arc or circle. Rotary motion can be converted into linear motion using a lead screw or other mechanical means of conversion. Part A of Figure 22.13 shows a linear actuator. Part B shows an angular (rotary) actuator. These types of actuators are also used outside the robot to move workpieces and provide other kinds of motion within the work envelope.

22.1.6 Types of actuator drives

One common method of classifying robots is by the type of **drive** required by the actuators. Electrical actuators use electric power. Pneumatic actuators use pneumatic (air) power. Hydraulic actuators, shown in Figure 22.14, use hydraulic (fluid) power.

Figure 22.14 A large hydraulic actuator provides up-and-down motion to the manipulator arm of this industrial robot.

22.1.7 Electric drive

Three types of motors are commonly used for electric drive actuators: AC servo motors, DC servo motors, and stepper motors. Both AC and DC servo motors have built-in methods for controlling the exact position. Many newer robots use DC servo motors rather than hydraulic or pneumatic ones. DC servo motors are commonly found on small- and medium-sized robots. AC servo motors are found in heavy-duty robots because of their high torque capabilities (Figure 22.15). A stepper motor is a type of incrementally controlled DC motor. Stepper motors are rarely used in commercial industrial robots, but are commonly found in educational robots (Figure 22.16).

Conventional **electric-drive motors** have several advantages. They are quiet, simple, and can be used in clean-air environments. Electric robots generally require less floor space, and their energy source is readily available. However, the conventionally geared drive causes problems of backlash, friction, compliance, and wear. These problems cause inaccuracy, poor dynamic

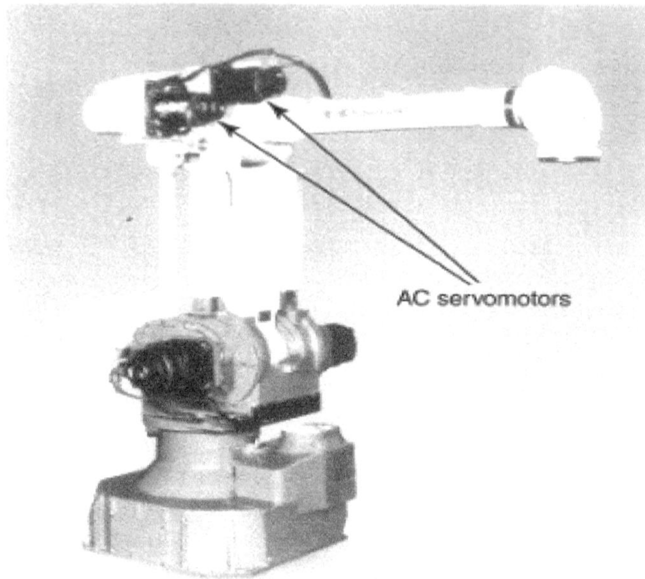

Figure 22.15 The arrows indicate the AC servomotors on this heavy-duty industrial robot.

response, need for regular maintenance, poor torque control capability, and limited maximum speed on longer moves. Loads that are heavy enough to stall (stop) the motor can cause damage. Conventional electric motors also have poor output power compared to their weight. This means that a larger, heavier motor must be mounted on the robot arm when a large amount of torque is needed.

The rotary motion of most **electric drive motors** must be geared down (reduced) to provide the speed or torque required by the manipulator. However, manufacturers are beginning to offer robots that use direct-drive electric motors, which eliminate some of these problems. Direct-drive electric motors are high-torque motors that drive the arm directly, without the need for reducer gears. The prototype of a direct-drive arm was developed by scientists at Carnegie-Mellon University in 1981.

The basic construction of a **direct-drive motor** is shown in Figure 22.17. Coupling the motor with the arm segment to be manipulated eliminates backlash, reduces friction, and increases the mechanical stiffness of the drive mechanism. Figure 22.18 shows a direct-drive robot arm. Compare its design to the non-direct-drive robot in Figure 22.15. Simplicity of design results when direct-drive motors are used. Maintenance requirements are also reduced. Direct-drive robots provide higher speeds, greater flexibility, and greater accuracy than non-direct-drive types.

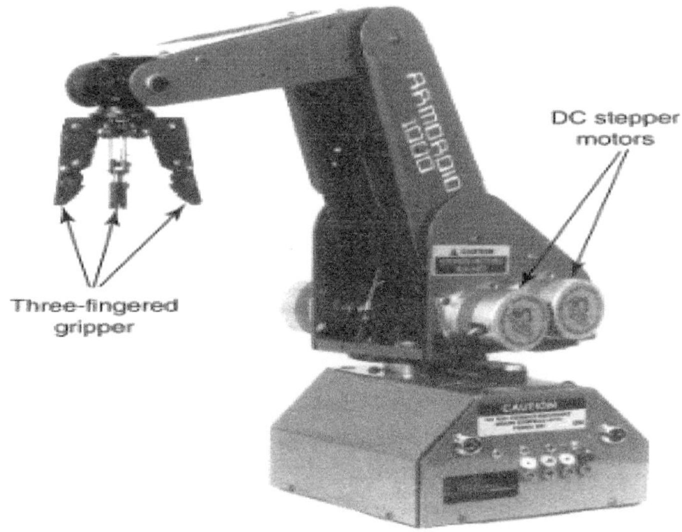

Figure 22.16 DC stepper motors are used on this tabletop educational robot. Note the three-fingered gripper. (D & M Computing, Inc.)

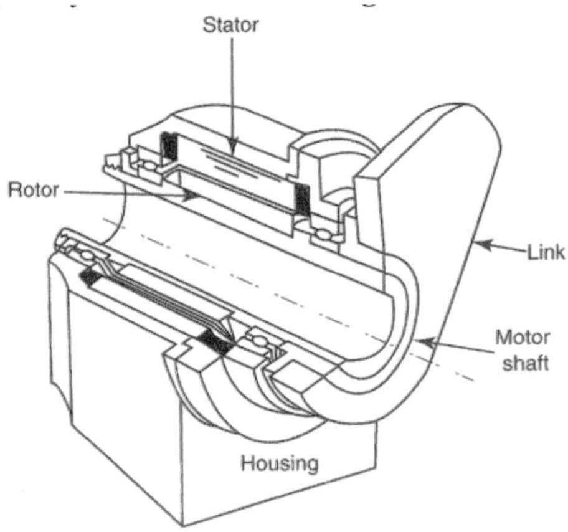

Figure 22.17 This drawing shows the basic construction of the type of DC motor commonly used in direct drive robots.

Applications currently being performed by **electric direct-drive** robots are mechanical assembly, electronic assembly, and material handling. These robots will increasingly meet the demands of advanced, high-speed, precision applications. Such applications include laser cutting of sheet metal, which requires speeds of over one meter per second and tolerances of less than 0.1millimeter, and fiber optic assembly, which requires accuracy within 0.1 to 0.5 microns (millionths of a meter).

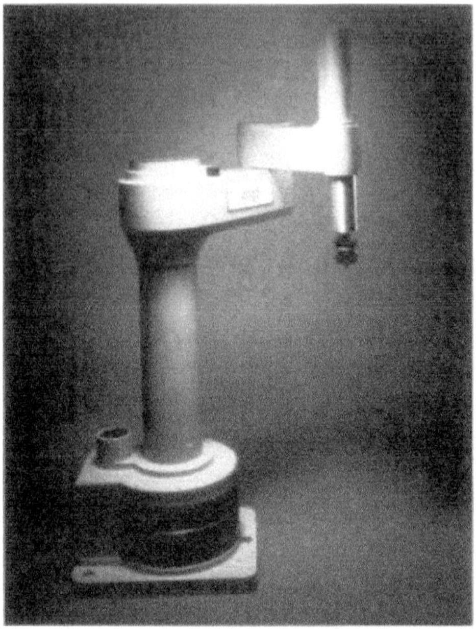

Figure 22.18 Note the simplified mechanical design of this direct-drive robot as compared to the non-direct drive robot shown in Figure 22.15.

22.1.8 Hydraulic drive

Many earlier robots were driven by hydraulic drive systems. A **hydraulic drive** system consists of an electric pump connected to a reservoir tank, control valves, and a hydraulic actuator. Hydraulic systems provide both linear and rotary motion using a much simpler arrangement than non-direct-drive electrical systems, Figure 22.19. The storage tank supplies large amounts of instant power not available from electric systems.

Hydraulic-drive units have several advantages. They provide precise motion control over a wide range of speeds. They can handle heavy loads on the end of the manipulator arm, can be used around high explosives, and are not easily damaged when quickly stopped while carrying a heavy load. However, they are expensive to purchase and maintain and are not energy efficient. They are also noisier than electric units. Because of hydraulic fluid leaks, they are not recommended for clean environments.

22.1.9 Pneumatic drive

Pneumatic drive systems make use of air-driven actuators. Since air is also a fluid, many of the same principles that apply to hydraulic systems apply

Figure 22.19 A large hydraulically operated robot being used for demonstration at a manufacturing trade show.

to pneumatic systems. Pneumatic and hydraulic motors and cylinders are very similar. Since most industrial plants have a compressed air system running throughout assembly areas, air is an economical and readily available energy source. This makes the installation of pneumatic robots easier and less costly than that of hydraulic robots. For lightweight pick-and-place applications that require both speed and accuracy, a pneumatic robot, like the one in Figure 22.20, is potentially a good choice.

Pneumatic-drive units work at high speeds and are most useful for small- to medium-sized loads. They are economical to operate and maintain and can be used in explosive atmospheres. However, since air is compressible, precise placement and positioning are more difficult to control. It is also difficult to keep the air as clean and dry as needed by the control system. Pneumatic robots are noisy and vibrate as the air cylinders and motors stop. Non-direct electric, pneumatic, and hydraulic systems are compared in Figure 22.21.

22.1.10 Power supply

The robot **power supply** provides the energy to drive the controller and actuators. The three basic types of power supplies are electrical, hydraulic, and pneumatic.

The most common **energy source** available where industrial robots are used is electricity. The second most common is compressed air, and the least common is hydraulic power. These primary sources of energy must

Figure 22.20 This twin-arm assembly robot is controlled by a pneumatic servo system. Note the dual grippers, which are placing components on electronic circuit boards. (Rexroth-Pneumatic)

	Nondirect	Pneumatic	Hydraulic
Cost	poor	good	poor
Cleanliness	good	good	poor
Controllability	good	poor	good
Noise	good	poor	poor
Weight	poor	good	good
Strength	fair	poor	good
Speed	poor	good	good
Stiffness	good	poor	good

Figure 22.21 Characteristics of robots with three types of drives are compared on this chart.

be converted into the form and amount required by the type of robot being used. The electronic part of the control unit and any electric motor actuators require electrical power. A robot containing hydraulic actuators requires the conversion of electrical power into hydraulic energy using an electric motor-driven hydraulic pump. A robot with pneumatic actuators requires compressed air, which is usually supplied by a compressor driven by an electric motor.

22.1.11 End effector

End effector is the technical term for the end-of-arm tooling on the robot. It is often referred to as the hand or gripper. An end effector is better defined as a device attached to the wrist of the manipulator for the purpose of grasping, lifting, transporting, maneuvering, or performing operations on a workpiece.

22.1.12 Means for programming

A robot can be programmed using any of several different methods. The **teach pendant**, also called a teach box or hand-held programmer, is one commonly used device. It teaches a robot the movements required to perform a useful task. The human operator uses a teach pendant to move the robot through the series of points that describe its desired path. The points are recorded by the controller for later use. The teach pendants shown in Figures 22.1 and 22.4 are typical of those found with most industrial robots.

22.2 Degrees of Freedom

Although robots have a certain amount of dexterity, theirs is nothing compared to human dexterity. The movements of the human hand are controlled by 35 muscles. Fifteen of these muscles are in the forearm. The arrangement of muscles in the hand provides great strength to the fingers and thumb for grasping objects. Each finger can act alone or together with the thumb. This enables the hand to do many intricate and delicate tasks. In addition, the human hand has 27 bones. Figure 22.22 shows the bones found in the hand and wrist. This bone, joint, and muscle arrangement gives the hand its dexterity.

Degrees of freedom is a term used to describe a robot's dexterity, or freedom of motion. For each degree of freedom, a joint is required. A robot requires six degrees of freedom to be completely versatile. Its movements are clumsier than those of a human hand, which has 22 degrees of freedom.

The number of degrees of freedom defines the robot's configuration. For example, many simple applications require movement along three axes X, Y, and Z. See Figure 22.23. These tasks would require three joints, or three degrees of freedom. The three degrees of freedom in the robot arm are the

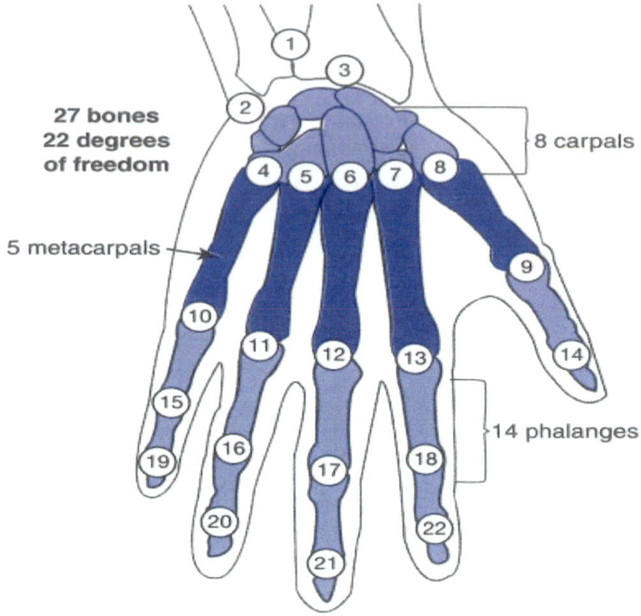

Figure 22.22 The arrangement of bones and joints found in the human hand provides dexterity. Each joint represents a degree of freedom; there are 22 joints, and thus, 22 degrees of freedom in the human hand.

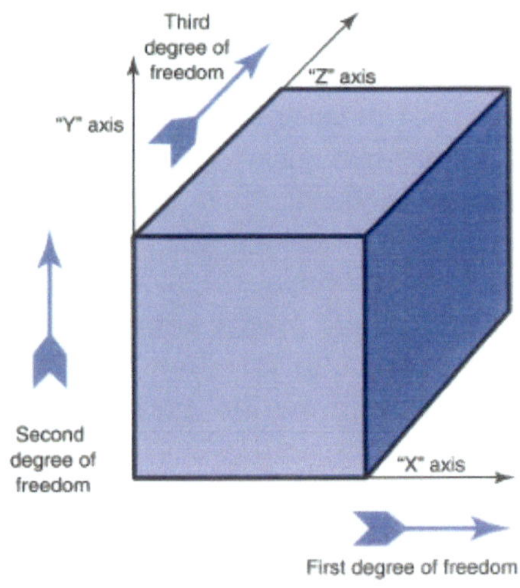

Figure 22.23 The three basic degrees of freedom associated with movement along the , and axes of the Cartesian coordinate system.

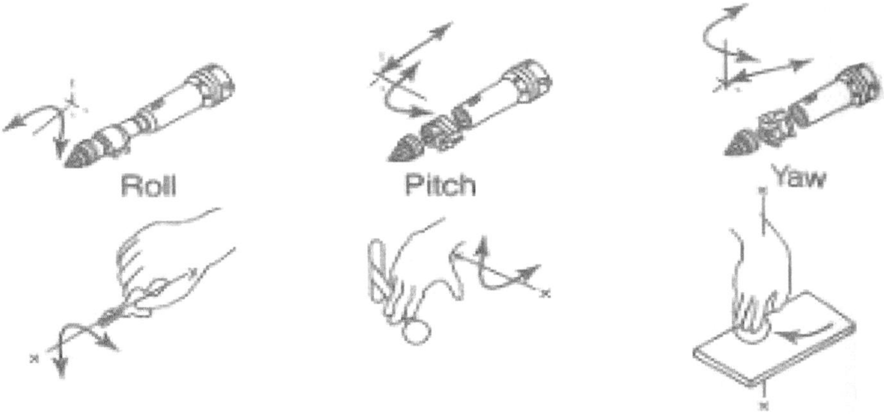

Figure 22.24 Three additional degrees of freedom: roll, pitch, and yaw – are associated with the robot's wrist. (Mack Corporation)

rotational traverse, the radial traverse, and the vertical traverse. The rotational traverse is a movement about a vertical axis. This is the side-to-side swivel of the robot's arm about its base. The radial traverse is the extension and retraction of the arm, creating in-and-out motion relative to the base. The vertical traverse provides up-and-down motion.

For applications that require more freedom, additional degrees can be obtained from the wrist, which gives the end effector its flexibility. See Figure 22.24. The three degrees in the wrist bear aeronautical names-**pitch, yaw**, and **roll**. The **pitch**, or bend, is the up-and-down movement of the wrist. The **yaw** is the side-to-side movement, and the **roll**, or swivel, involves rotation.

A robot requires a total of six **degrees of freedom** to locate and orient its hand at any point in its work envelope, Figure 22.25. Although six degrees of freedom are required for maximum flexibility, most applications require only three to five. The more degrees of freedom required, the more complex the robot's motions and controller design must be. Some industrial robots have seven or eight degrees of freedom. These additional degrees are achieved by mounting the robot on a track or moving base, as shown in Figure 22.26. The track-mounted robot shown in Figure 22.27 has a total of seven degrees of freedom. This addition also increases the robot's reach.

Although the robot's freedom of motion is limited in comparison with that of a human, the range of movement in each of its joints is considerably

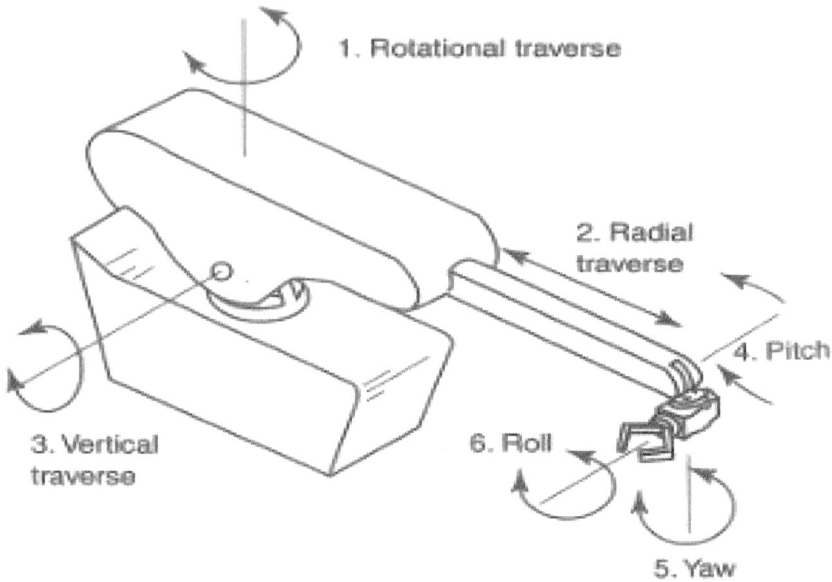

Figure 22.25 Six degrees of freedom provide maximum flexibility for an industrial robot.

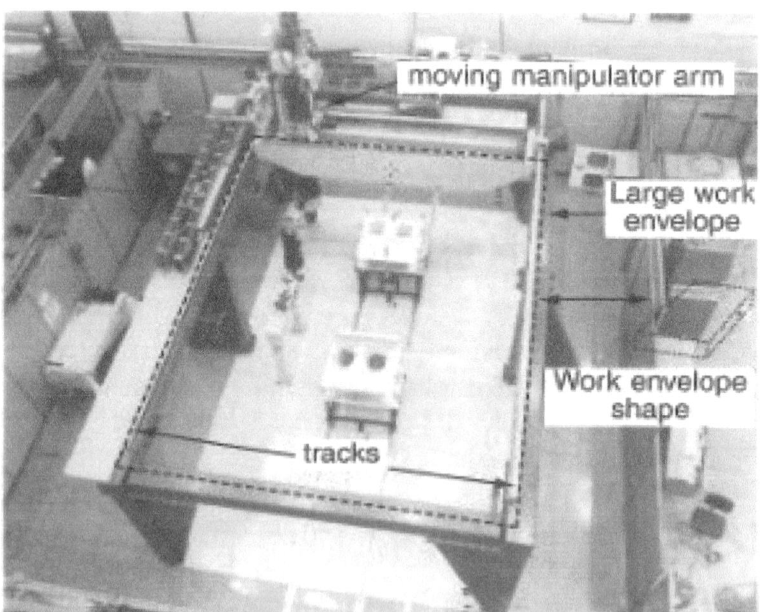

Figure 22.26 This gantry robot, viewed from above, illustrates the large work envelope obtained by mounting the manipulator arm on tracks. (Cybotech Automation Systems)

Figure 22.27 Mounting this robot on tracks gives the system seven degrees of freedom, six from the robot/gripper system and one additional degree from the track mount.

greater. For example, the human hand has a bending range of only about 165 degrees. Figure 22.28 shows the six major degrees of freedom by comparing the terms of a robot to a person using a spray gun.

22.3 Robot Configurations

Robots come in many sizes and shapes. They also vary as to the type of coordinate system used by the manipulator. The type of coordinate system, the arrangement of the joints, and the length of the manipulator's segments all help determine the shape of the **work envelope**. To identify this maximum area, a point on the robot's wrist is used, rather than the tip of the gripper or the end of the tool bit. As a result, the work envelope is slightly larger when the tip of the tool is considered.

Work envelopes vary from one manufacturer to another, depending on the exact design of the manipulator arm. Combining different configurations in a single robot can result in another set of possible work envelopes. Before choosing a particular robot configuration, the application must be studied carefully to determine the precise work envelope requirements.

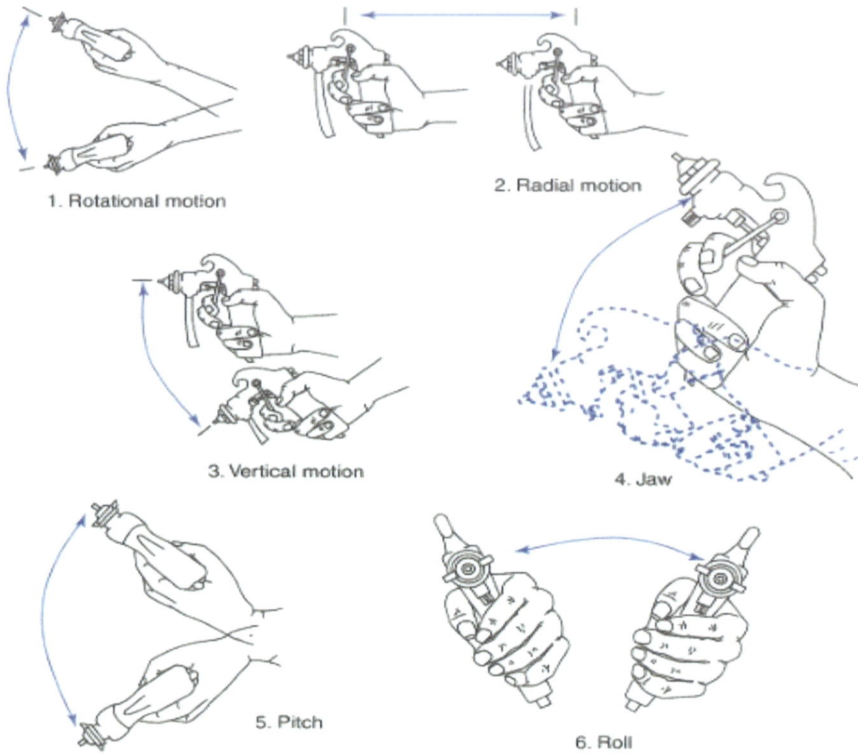

Figure 22.28 The six degrees of freedom, illustrated by a person using a spray gun. Degrees 1,2, and 3 are arm movements; 4,5, and 6 are wrist movements.

Some **work envelopes** have a geometric shape; others are irregular. One method of classifying a robot is by the geometric configuration of its work envelope. Some robots can employ more than one configuration. The four major ones are revolute, Cartesian, cylindrical, and spherical. Each configuration is used for specific applications.

22.3.1 Revolute configuration

The jointed-arm or **revolute configuration** is the most common. These robots are often referred to as anthropomorphic (humanlike) in form because their movements closely resemble those of the human body. Rigid segments resemble the human forearm and upper arm. Various joints mimic the action of the wrist, elbow, and shoulder. A joint called the sweep represents the waist.

A **revolute coordinate** robot performs in an irregularly shaped work envelope. There are two basic revolute configurations: vertically articulated and horizontally articulated.

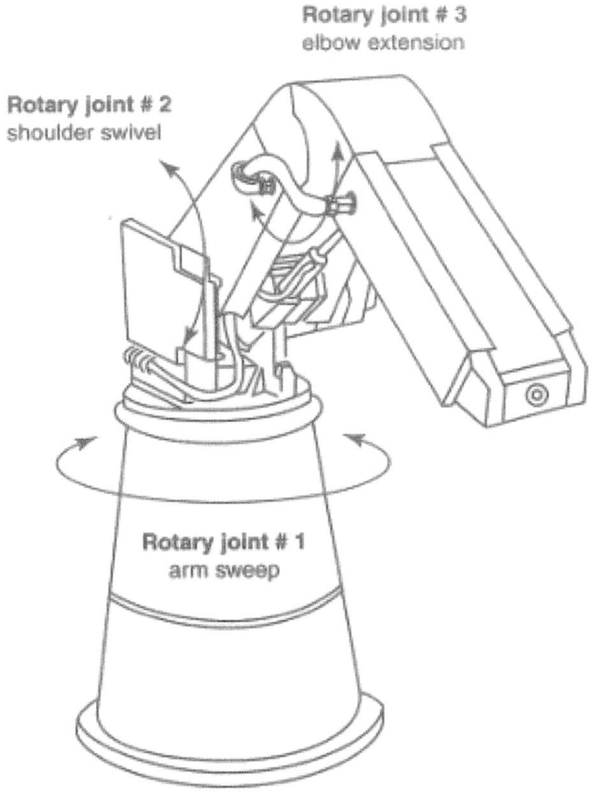

Figure 22.29 These three revolute (rotary) joints are associated with the basic manipulator movements of a vertically articulated robot.

The vertically **articulated configuration** shown in Figure 22.29 has three revolute (rotary) joints. Part A of Figure 22.30 shows a vertically articulated robot. Part B shows the vertical plane and top view of its work envelope. The jointed-arm, vertically articulated robot is useful for painting applications because of the long reach this configuration allows.

The **horizontally articulated** configuration generally has one vertical (linear) and two revolute joints. Also called the SCARA (Selective Compliance Assembly Robot Arm) configuration, its primary objective is a configuration that is fairly yielding in the horizontal motions and rather rigid in the vertical motions. The basic SCARA configuration is illustrated in part A of Figure 22.31. It is an adaptation of the cylindrical configuration. Part B shows the work envelope. The SCARA robot shown in Figure 22.32 is designed for clean-room applications, such as wafer and disk handling in the electronics industry.

A

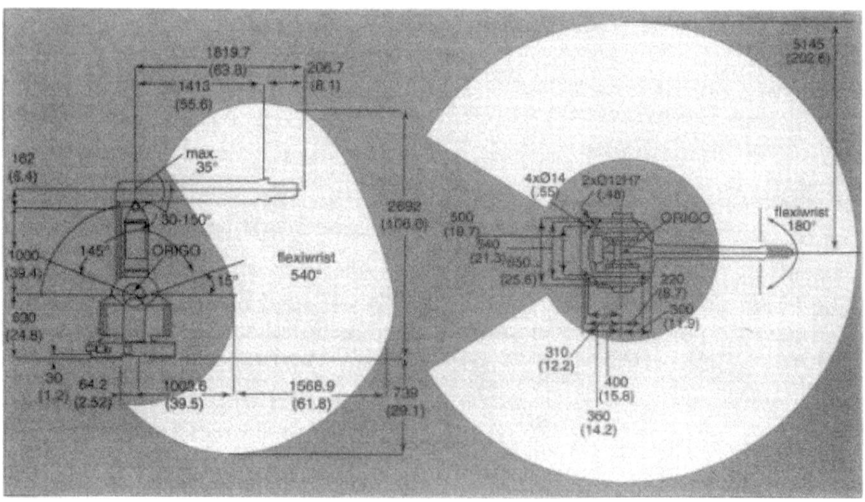

B

Figure 22.30 A – This electrically actuated painting robot is vertically articulated. B – Shaded areas show the work envelope for the vertically articulated painting robot. The numbers shown are the work envelope dimensions in millimeters and inches. (ABB Graco Robotics, Inc.)

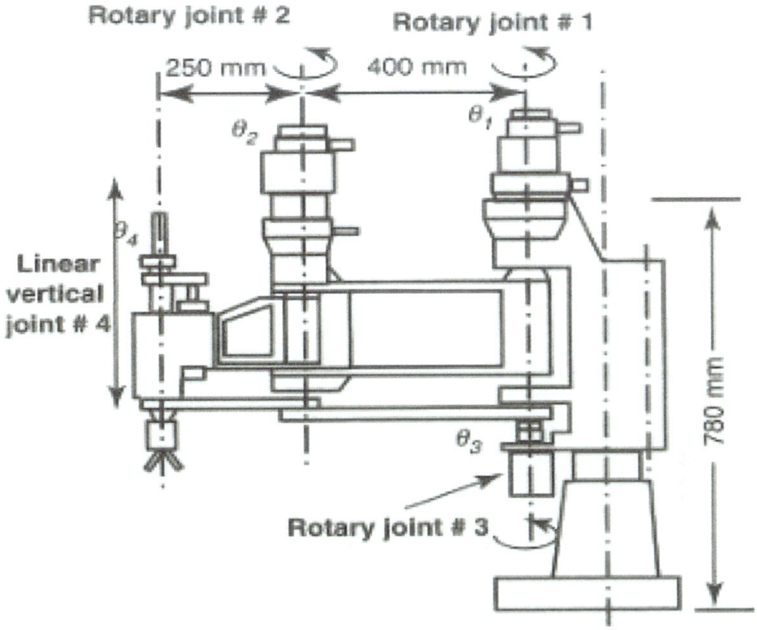

A

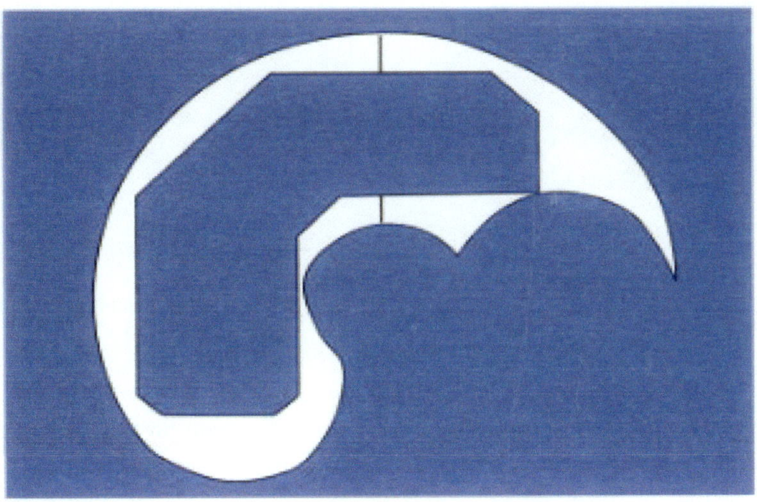

B

Figure 22.31 A – This diagram shows the basic SCARA robot configuration. Note the two rotary joints and the single vertical joint used in this horizontal articulated configuration. B – This is a top view of the work envelope of a typical SCARA horizontally articulated robot configuration. This work envelope is sometimes referred to as the “folded book configuration.”

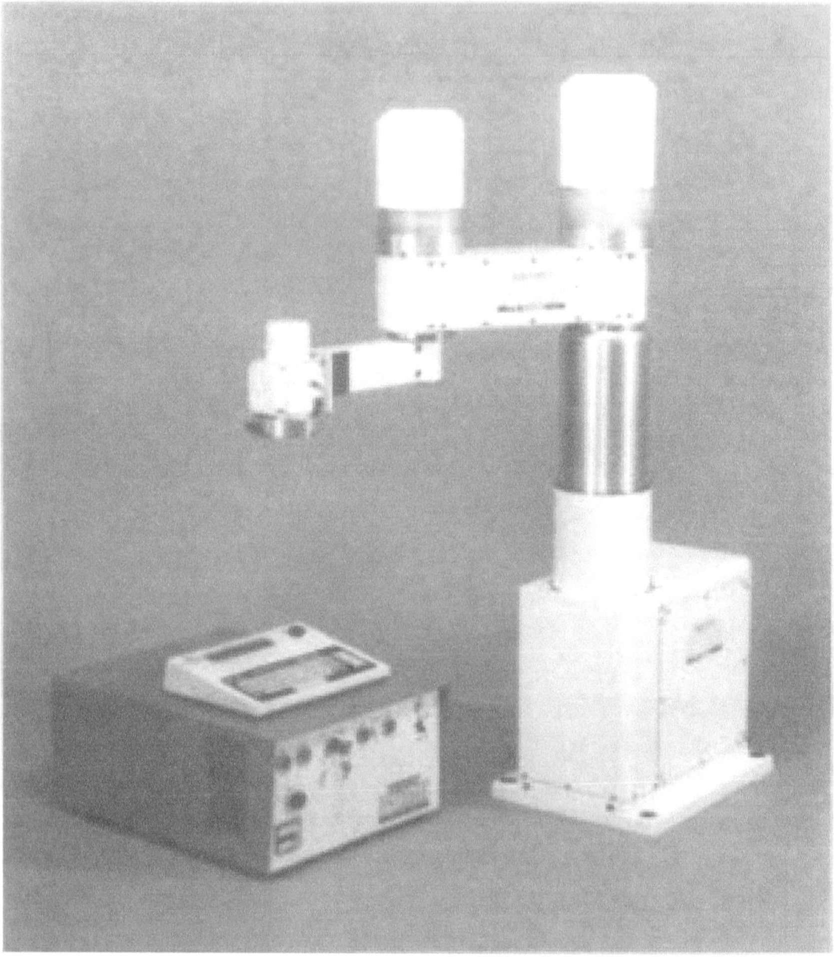

Figure 22.32 This Seiko SCARA robot is specifically designed for clean-room applications. (Seiko Instruments)

SCARA robots are ideally suited for operations in which the vertical motion requirements are small compared to the horizontal motion requirements. Such an application would be assembly work in which parts are picked up from a parts holder and moved along a nearly horizontal path to the unit being assembled.

The **revolute configuration** has several advantages. It is by far the most versatile configuration and provides a larger work envelope than the Cartesian, cylindrical, or spherical configurations. It also offers a more flexible reach than the other configurations, making it ideally suited to welding and spray-painting operations.

However, there are also disadvantages to the **revolute configuration**. It requires a very sophisticated controller. Programming is more complex for this configuration than for the other three configurations. In addition, different locations in the work envelope can affect accuracy, load-carrying capacity, dynamics, and the robot's ability to repeat a movement accurately. This configuration also becomes less stable as the arm approaches its maximum reach.

Typical applications for **revolute configuration** robots include:

- Automatic assembly.
- Parts and material handling.
- Multiple-point light machining operations.
- In-process inspection.
- Palletizing.
- Machine loading/unloading.
- Machine vision.
- Material cutting.
- Material removal.
- Thermal coating.
- Paint and adhesive application.
- Welding.
- Die casting.

22.3.2 Cartesian configuration

The arm movement of a robot using the **Cartesian configuration** can be described by three intersecting perpendicular straight lines, referred to as the **X, Y** and **Z axes**. See Figure 22.33. Because movement can start and stop simultaneously along all three axes, the motion of the tool tip is smoother. This allows the robot to move directly to its designated point, instead of following trajectories parallel to each axis. See Figures 22.34 and 22.35.

Figure 22.36 illustrates the rectangular **work envelope** of the typical **Cartesian configuration**. (Refer to Figure 22.26 for a photograph of

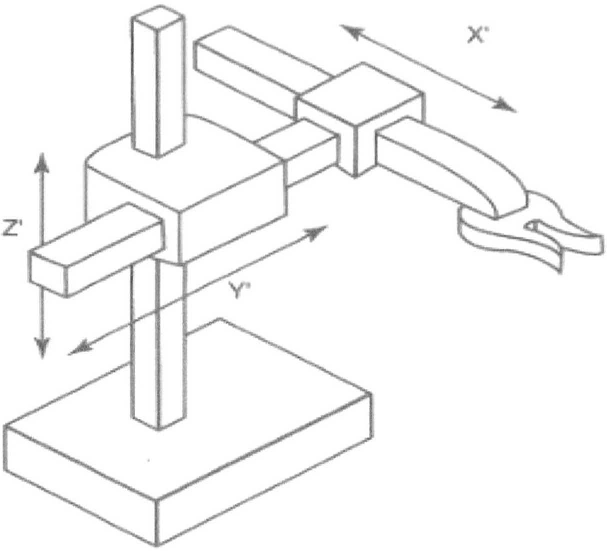

Figure 22.33 A robot with a Cartesian configuration moves along *X*, *Y*, and *Z* axes.

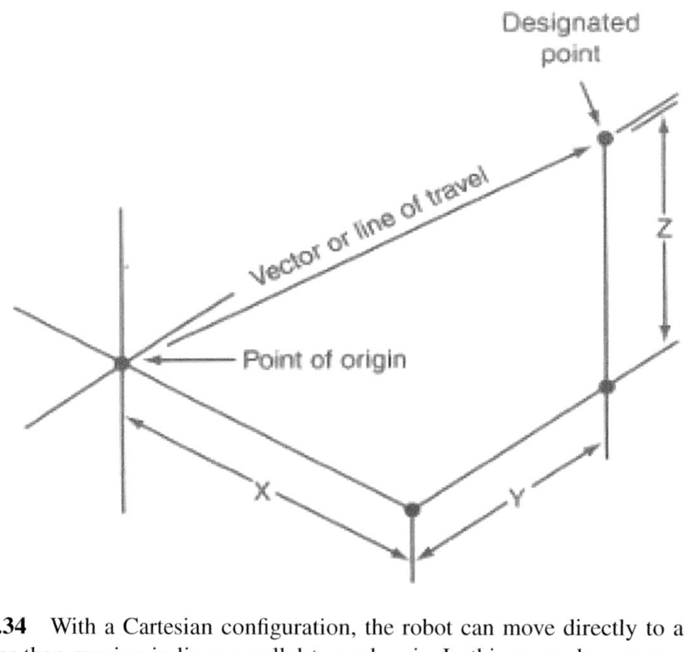

Figure 22.34 With a Cartesian configuration, the robot can move directly to a designated point, rather than moving in lines parallel to each axis. In this example, movement is along the "vector" connecting the point of origin and the designated point, rather than moving first along the *X* axis, then *Y*, then *Z*.

Figure 22.35 This robot has a Cartesian configuration and is used for high-precision jobs. (Seiko Instruments)

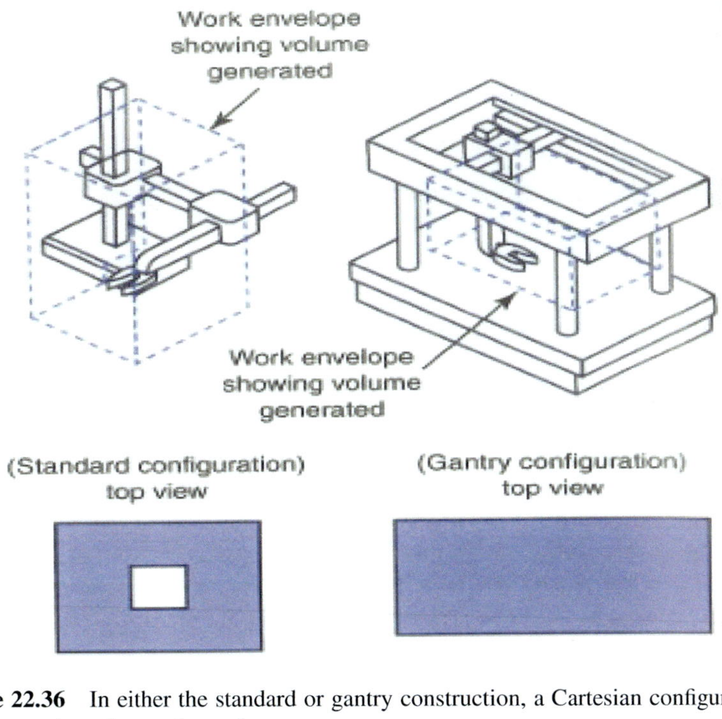

Figure 22.36 In either the standard or gantry construction, a Cartesian configuration robot creates a rectangular work envelope.

a Cartesian gantry robot.) One advantage of robots with a Cartesian configuration is that their totally linear movement allows for simpler controls. They also have a high degree of mechanical rigidity, accuracy, and repeatability. They can carry heavy loads, and this weightlifting capacity does not vary at different locations within the work envelope. As to disadvantages, **Cartesian robots** are generally limited in their movement to a small, rectangular workspace.

Typical applications for **Cartesian configuration** robots include:

- Assembly.
- Machining operations.
- Adhesive application.
- Surface finishing.
- Inspection.
- Waterjet cutting.
- Welding.
- Nuclear material handling.
- Robotic X-ray and neutron radiography.
- Automated CNC lathe loading and operation.
- Remotely operated decontamination.
- Advanced munitions handling.

22.3.3 Cylindrical configuration

The **cylindrical configuration** consists of two orthogonal (at a 90-degree angle) slides mounted on a rotary axis, Figure 22.37. Reach is accomplished as the arm of the robot moves in and out. For vertical movement, the carriage moves up and down on a stationary post, or the post can move up and down in the base of the robot. Movement along the three axes, as shown in Figure 22.38, traces points on a cylinder.

A **cylindrical configuration** generally results in a larger work envelope than a Cartesian configuration. These robots are ideally suited for pick-and-place operations. However, cylindrical configurations have some disadvantages. Their overall mechanical rigidity is lower because robots with a rotary axis must overcome the inertia of the object when rotating. Their repeatability and accuracy are also lower in the direction of rotary movement.

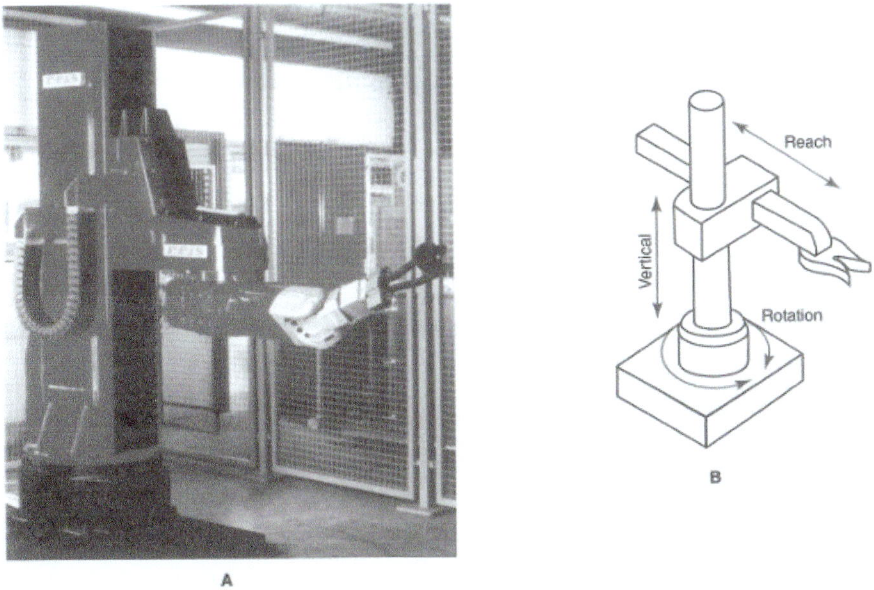

Figure 22.37 A – This cylindrical robot has two slides for movement up-and-down or in-and-out, and is mounted on a rotary axis. (Reis Machines, Inc.) B – The basic configuration for a cylindrical robot.

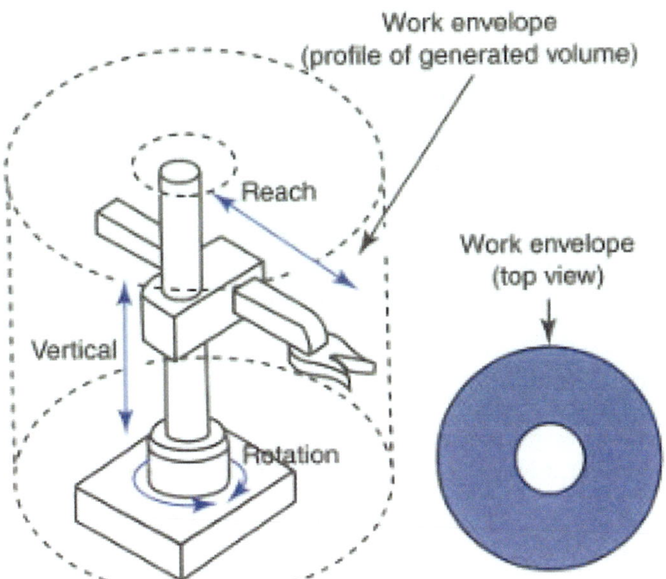

Figure 22.38 Motion along the three axes traces points on a cylinder to form the work envelope.

The cylindrical configuration requires a more sophisticated control system than the Cartesian configuration.

Typical applications for **cylindrical configuration** robots include:

- Machine loading/unloading.
- Investment casting.
- Up setter loading.
- Conveyor pallet transfers.
- Foundry and forging applications.
- General material handling and special payload handling and manipulation.
- Meat packing.
- Coating applications.
- Assembly.
- Injection molding.
- Die casting.

22.3.4 Spherical configurations

Many **industrial robots** use combinations or special modifications of the four basic configurations. One example is shown in part A of Figure 22.42. This robot uses an articulated configuration, but its base does not rotate horizontally. It is designed to literally bend over backwards to grasp objects behind it. This feature makes it possible to install these robots very close to other equipment to minimize space requirements, while maintaining a large, effective work envelope. See part B of Figure 22.42. This robot has been used in such applications as spot welding and material handling.

22.4 Robot Interfacing

Robots do not work in isolation. They interact continuously with other production equipment, Figure 22.43. This is called interfacing.

22.4.1 Interfacing

An **interface** is the common point at which two or more systems communicate with each other. External end effector sensors, work-cell limit switches, external relays, operator alarms, safety equipment, bar code readers, conveyors,

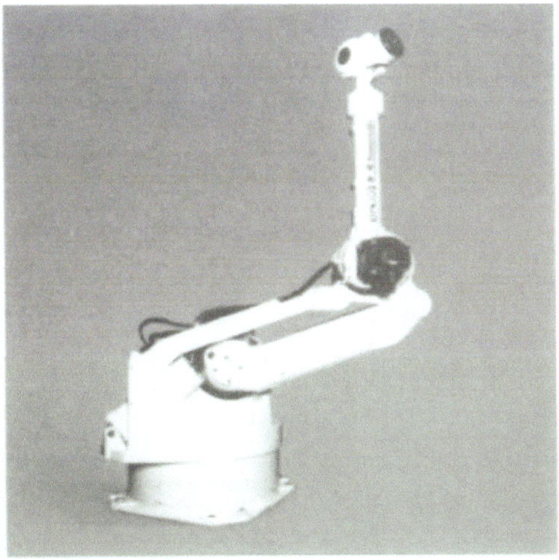

A

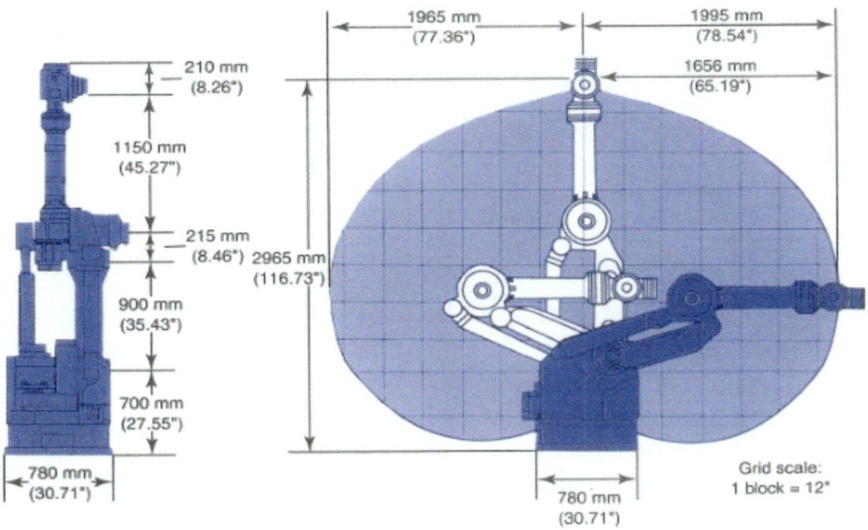

B

Figure 22.42 A – This heavy-duty robot literally bends over backward. B – The work envelope for the Kawasaki robot is large. (Kawasaki)

transfer lines, programmable logic controllers (PLCs), and machine vision systems all interface with robots.

For a robot to communicate, it must be provided with **input ports and output ports**. The digital signals used for communication travel through these ports. There are four basic types: dedicated digital input and output ports, serial ports, and parallel ports.

Figure 22.43 In a robotic work cell, robots interface with other equipment.

22.4.2 Dedicated digital input ports

Through a **digital input port**, the robot receives a signal conveying the state (on or off) of an external switch. Limit switches, such as the one shown in Figure 22.44, communicate information about such things as the presence or absence of a part at a given location. This information is used by the robot's computer program to initiate some action. For example, Figure 22.45 shows a limit switch connected to a typical digital input port. When no part is present, the limit switch is open. The base of the input transistor Q_1 is held at ground potential by resistor R_1. In this condition, transistor Q_1 is in a nonconducting state. When a part is present, the limit switch is closed, supplying the base of input transistor Q_1 with 24 V direct current. In this condition, the transistor is in a fully conducting state. The controller is programmed to detect these two conditions (presence of a part and a change of transistor state). This information is then passed on to the robot's computer program, and the appropriate action is carried out.

Another use of **digital input ports** is signals from safety devices. Safety devices , such as carpet switches, light curtains, motion detectors, or door switches on safety fences, can be connected to a digital input port. The controller shuts down the robot when an operator or other person, or object enters

Figure 22.44 Limit switches like this one communicate information via input ports. (Euchner-USA, Inc.)

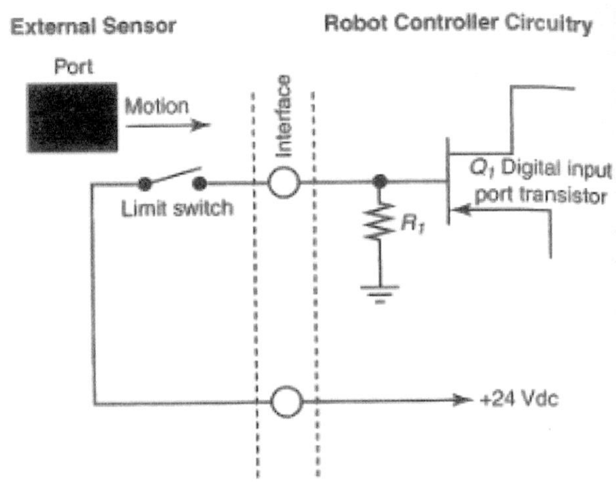

Figure 22.45 When a part moves past the limit switch, the contacts close and 24 volts DC is applied to the base of the digital input port transistor (Q_1). The controller circuitry detects this change in condition and initiates the appropriate action.

the work envelope. Although Figure 22.45 shows only one digital input port, most robots come equipped with 20 or more.

22.4.3 Dedicated digital output ports

One way that a robot interacts with external equipment, such as a conveyor belt drive motor, is through **digital output ports**. Output ports permit

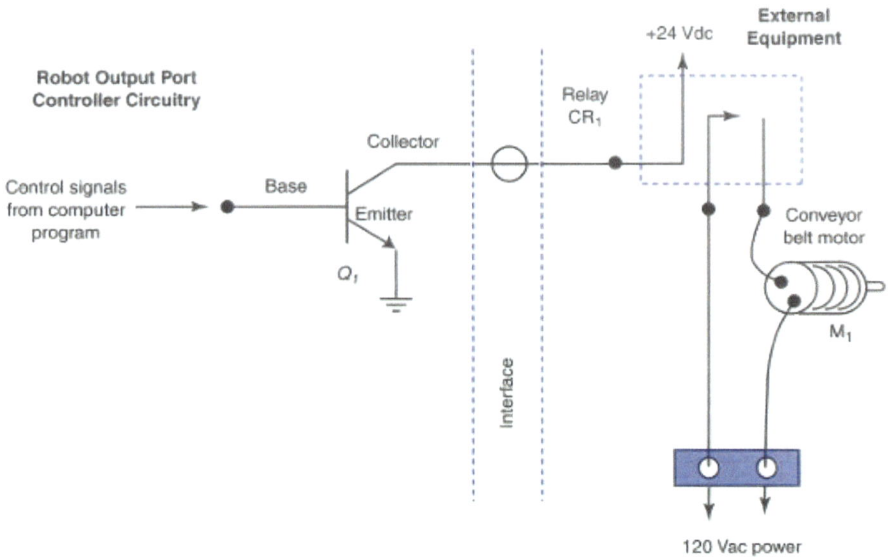

Figure 22.46 When enough voltage is applied to digital output port transistor (Q_1), the lower end of relay CR_1 is grounded. Grounding the lower end of the relay coil causes the relay contacts to close. This supplies AC voltage to the conveyor belt motor (M_1).

digital signals to be sent from the robot controller to equipment controls. See Figure 22.46.

Whenever the **control signal** from the robot's computer program is at a low voltage level, the base of transistor Q_1 is nonconducting. The output current between the emitter and collector of Q_1 is zero. The circuit is open, or in a state of high resistance. In this condition, the motor control relay CR_1 will not be energized. The motor will be off. When the computer sends a high-level signal (12 V or 24 V) to the base of Q_1, however, the current will be at maximum. The resistance between the emitter and collector of Q_1 will be very low, and the motor control relay will be grounded. Under these conditions, relay CR_1 is energized, and the motor is switched on. Most robot controllers come equipped with 20 or more **digital output ports** to match the number of input ports.

22.4.4 Serial ports

Serial transmission delivers digital data over long distances, but at a slow rate. The rate is slow because the data must be transmitted one bit at a time. Two computers can communicate with each other over telephone circuits via

Current Loops	Teletype to computer
RS232C	Computer to terminal
RS422 & RS423	Peripheral to computer (distances greater than 20 m)

Figure 22.47 Typical computer serial interface standards.

serial ports connected to modems. Existing transmission lines can be used, which is an advantage.

The more common serial interface used in robotics is the standard RS232C, Figure 22.47. It was developed to standardize the interface between data terminal equipment and data communications equipment. The RS232C interface is used when binary data is to be transmitted over short distances, such as between two computers or between a computer and a piece of peripheral equipment. The circuit performs three functions: data transfer, timing, and control.

Part A of Figure 22.48 defines the connections of the RS232C. Applications that do not involve long-distance communications require only a few connections. These are the signal ground (AB), transmitted data (BA), and received data (BB). **Long-distance communications** (those involving modems) require the use of many of the other connections listed. Part B of Figure 22.48 shows how AB, BA, and BB connections communicate between the microprocessor in a robot's controller and an external video terminal.

Another common use of the RS232C **serial port** is to connect a robot to PC. The PC can be used offline to write and debug the initial program. Then the PC can be connected to the robot via the RS232C port, and the robot be taught the exact moves needed to run the application. Once the program is complete, it can be downloaded into the robot's memory.

The most common connector used with an RS232C cable is a 25-pin DB25. When preparing cables to be used with the RS232C port, do not exceed a total cable capacitance of 2500 picofarads (pF). The capacitance limits the distance between the two pieces of equipment. Another limitation is a 20-kilobit maximum data transfer rate. Although the RS232C standard does not specify the data format, the most common is 8-bit ASCII.

Standards have been developed to allow serial transmission over distances up to 1000 meters. These newer standards are the RS422 for unbalanced transmission lines and the RS423 for balanced transmission lines.

			Ground	Data		Control		Timing
Interchange Circuit	Pin Number	Description		From DCE	From DCE	From DCE	To DCE	To DCE
AA	1	Protective ground	X					
AB	7	Signal ground/common return	X					
BA	2	Transmitted data			X			
BB	3	Received data		X				
CA	4	Request to send					X	
CB	5	Clear to send				X		
CC	6	Data set ready				X		
CD	20	Data terminal ready					X	
CE	22	Ring indicator				X		
CF	8	Received line signal detector				X		
CG	21	Signal quality detector				X		
CH	23	Data signal rate selector (DTE)					X	
CI	23	Data signal rate selector (DCE)				X		
DA	24	Transmitter signal element timing (DTE)						
DB	15	Transmitter signal element timing (DCE)						X
DD	17	Receiver signal element timing (DCE)						X
SBA	14	Secondary transmitted data			X			
SBB	16	Secondary received data		X				
SCA	19	Secondary request to send					X	
SCB	13	Secondary clear to send				X		
SCF	12	Secondary received line signal detector				X		

A

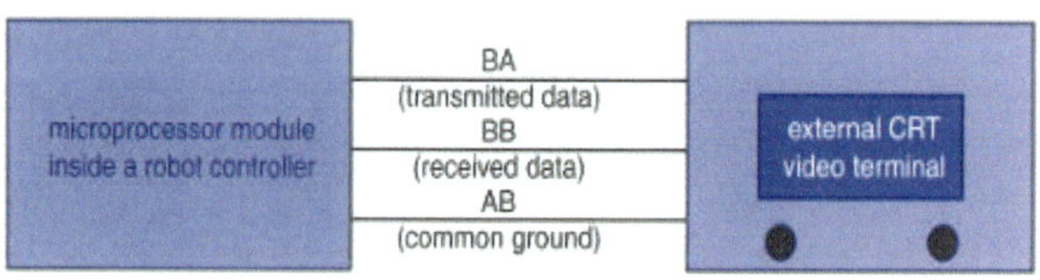

B

Figure 22.48 RS232C interfacing. A – This chart lists pin assignments and interchange circuit functions for the RS232C serial interface standard. B – The external CRT video terminal is interfaced to a robot controller's microprocessor module via RS232C connections.

22.4.5 Parallel ports

Another way that digital equipment can communicate with peripheral equipment is by means of **parallel transmission**. Multiple bits of data are sent at the same time, following side-by-side paths like the lanes of a highway. Parallel transmission is done through **parallel ports**. Although parallel transmission is faster, it is generally considered too costly to use over long distances.

One **parallel port** standard is the IEEE 488. It was developed in 1975 by the Institute of Electrical and Electronic Engineers to reduce the amount of time needed to set up test equipment. (At that time, test equipment had to have its own independent interface. Interfacing more than two devices to a computer could get complicated and expensive.) The IEEE 488 standard allows computer-controlled test equipment to be set up in a matter of hours. These advantages make it useful for any industry that uses automated testing.

The IEEE 488 standard is used when a computer is connected to measuring equipment by means of a **general-purpose interface bus (GPIB)**. The GPIB is a cable that interfaces system controllers with programmable instruments. Message length over the 8-bit serial bus is variable, but the rate should not exceed 1 megabyte per second. A maximum of 15 devices can be supported by the bus. Total cable length is 20 meters, or 2 meters times the number of connected devices (whichever is less). Cable between any two devices must not exceed 4 meters.

Using the GPIB to connect test instruments to a **system controller** or is shown in Figure 22.49. The bus carries data in both directions, 8 bits at a time. The data and the direction are controlled by the control bus. The bus can connect various kinds of devices to the controller: equipment that can only talk, equipment that can only listen, and equipment that can either talk or listen.

The **controller** sets the direction of data transfer. Many controllers can be attached to the GPIB, but only one controller at a time can have a command. If a second controller needs to communicate with the devices on the bus, the first must surrender its control. The controller is generally a computer with a GPIB interface card. Both microprocessor-based computers and minicomputers work well as controllers.

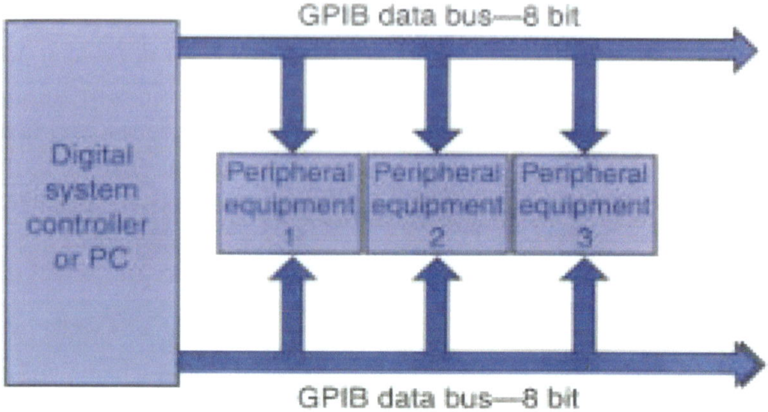

Figure 22.49 The GPIB parallel interface bus is used here to connect external test equipment to a digital system controller or to a PC.

22.5 Machine Vision

Machine vision systems give the robot "eyes" and allow it to perform in a more intelligent manner. They can be used for such applications as robot guidance, determining part orientation, measuring and inspecting parts, and identifying images. See Figure 22.50. Machine vision is becoming an essential requirement in modern automated work cells. When robotic work cells are not equipped with machine vision, all parts must be pre-oriented before the robot is able to grasp them and perform its task.

Machine vision can help eliminate the need for many machine operators. Machines equipped with vision can perform many complex tasks once done only by humans. Used for assembly and inspection, they have had a major impact on the automotive and electronics industries. For example, today's printed circuit boards often contain thousands of individual parts. If even a tiny fraction of these parts is incorrectly placed on the board, the cost of finding and repairing the problem is enormous. When human workers perform circuit board assembly, fatigue can cause as much as a 20 percent error rate. Using robots with machine vision is much more successful.

In the automotive industry, such simple tasks as determining the presence or absence of parts in subsystems coming off the assembly line can be done with machine vision. General Motors has stated that it would like to install tens of thousands of vision systems.

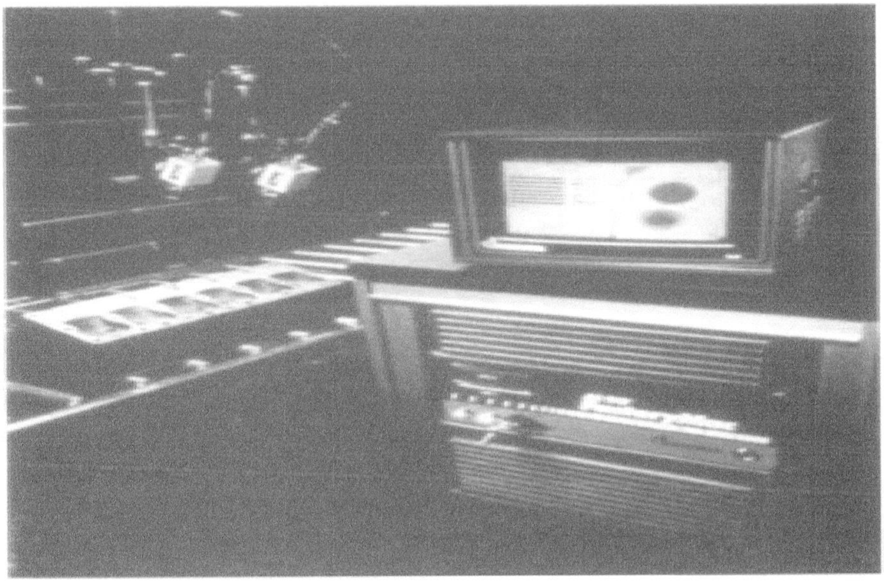

Figure 22.50 The video camera "eyes" of this system are suspended over the conveyor. (Automatix)

22.5.1 Fundamentals of machine vision

Machine vision systems use video cameras and computers to translate light energy into an image. These images are then used to determine part orientation or some other task. Even the most effective machine vision systems cannot come close to achieving the human eye's level of perception and sophistication.

Machine vision technology is relatively new. Some of the earliest systems were introduced in the late 1970s. These early systems were rather expensive, due to the high cost of computer memory needed to handle the images. They were also inefficient, because all but the mainframe computers contained slow processors. Today, however, faster computers and less expensive memory chips are making vision systems practical. The data obtained can be used to make important decisions regarding action taken by either human operators or a robot. Vision systems can also be used in conjunction with material inventory and material flow within a factory.

Four functions take place during image processing. These functions are acquisition, preprocessing, analysis, and interpretation, Figures 22.51

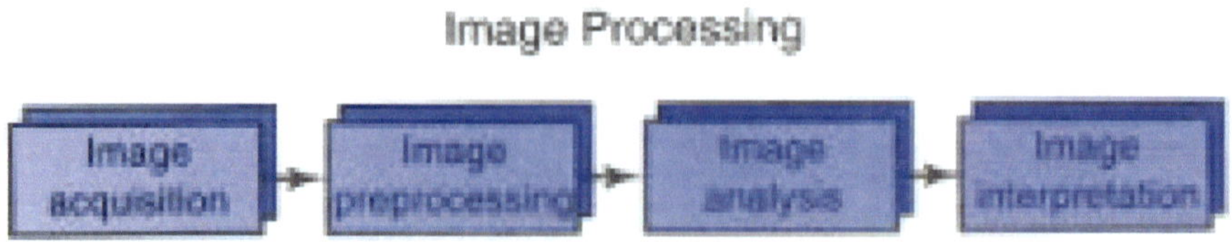

Figure 22.51 A typical machine vision system carries out four image processing functions.

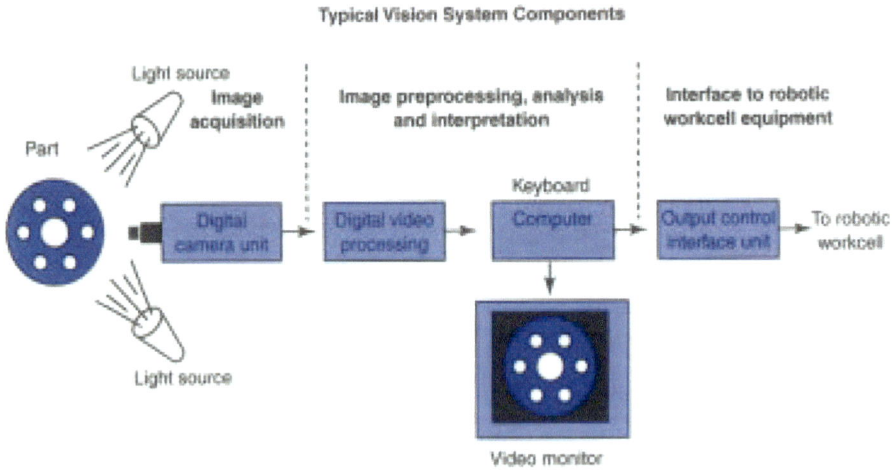

Figure 22.52 These digital electronic subsystems are found in a typical machine vision system.

and 22.52 shows the relationship among the subsystems that carry out these functions.

22.5.2 Image acquisition

Image acquisition consists of illuminating the workpiece and digitally scanning its image. Fluorescent lamps, incandescent bulbs, strobe lights, or arc lamps usually provide the illumination. The type and amount of light needed depend on the application. Front lighting is used to enhance surface features, such as bar codes or labels. Side lighting is used when three-dimensional (3D) images are desired. Back lighting provides a silhouette of the object.

Scanning is done by a video camera. The solid-state digital cameras commonly used with vision systems have either a charge-coupled device (CCD) or a charge-injected device (CID). At the heart of the camera is a silicon chip used with an array of photosensitive elements. The light reflected into the camera lens from the workpiece falls onto this photosensitive array and is converted into an analog electrical signal. The image is broken down into dots of light called picture elements, or pixels. Pixels are the means used to form the image on a TV screen. A typical digital camera can produce an image consisting of 256 by 256 pixels.

Older **machine vision systems** often used vidicon tube cameras. However, the newer solid-state cameras have several advantages over vidicon cameras. They are more rugged, smaller in size and weight, and more sensitive to low levels of light. They require fewer adjustments, and the image is less distorted. They also operate at a higher speed, which creates less image bloom (glare from light-colored objects).

22.5.3 Image preprocessing

During **image preprocessing**, an analog-to-digital (AD) converter is used to change the analog signal into an equivalent digital signal. The digital signal represents light intensity values over the entire image. These values are stored in memory, permitting the digital image to be analyzed and interpreted.

22.5.4 Image analysis and interpretation

Image analysis involves gathering information from the image that will enable the robot to do a task. Analysis is performed by computer software. Using algorithms (special codes), the software identifies and measures features of the digital image. After all the features are analyzed, the information is interpreted. The goal of a machine vision system is image interpretation. This enables the robot to make decisions about the work that needs to be done.

Image analysis computer programs employ various techniques. One example of a state-of-the-art program is Automatix's Image Analyst. "See Figure 22.53. Image Analyst allows the computer to take measurements and other data from video images. First, the user defines regions of interest (ROIs). An ROI can be as small as a single pixel, or it can include the entire image. The positions and sizes of the ROIs are determined using a graphical display. Figure 22.54 is a block diagram showing the major components and features of the Image Analyst program.

Another example of analysis is shown in Figure 22.55. This **vision system** provides 100 percent inspection of plastic film as it is manufactured. The system uses a state-of-the-art computer and software. Any defects found in the plastic film are automatically reported by type, location, and size on a high-resolution color display.

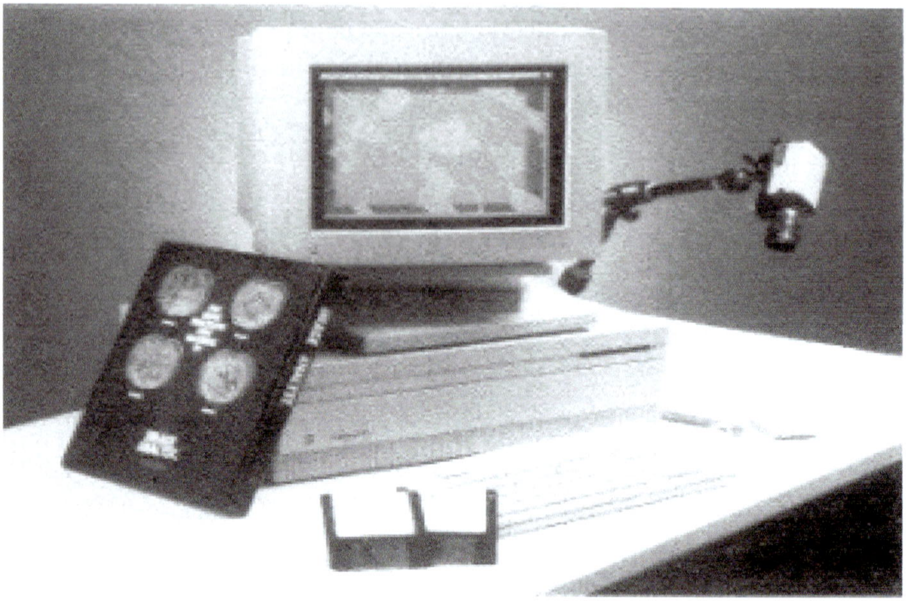

Figure 22.53 Regions of interest can be selected using this image analysis software. (Automatix)

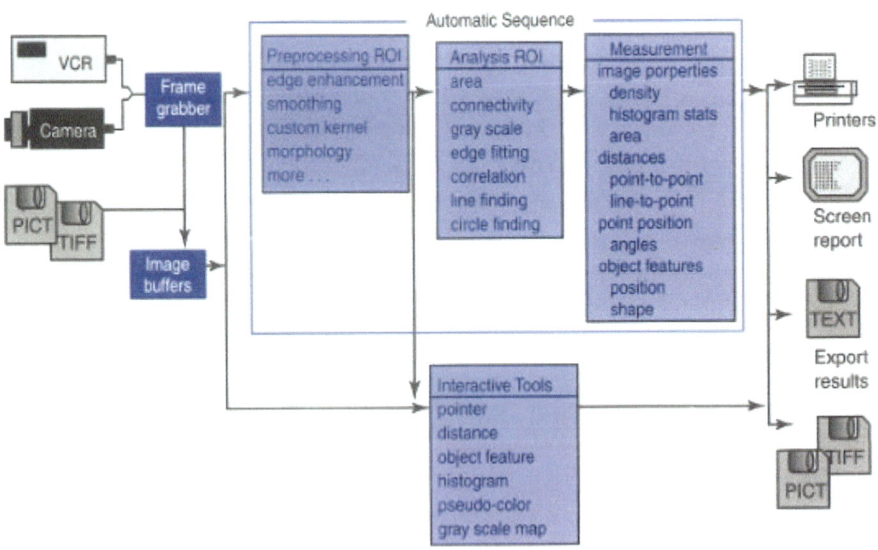

Figure 22.54 This block diagram shows the functions of the Image Analyst software. (Automatix).

Figure 22.55 This vision system provides 100 percent inspection of similar but different parts as they are being manufactured. (Adept Technology)

Review Questions

1. List the five major systems that make up an industrial robot.

2. What is the difference between servo and non-servo robots?

3. What three types of power supplies are used with industrial robots?

4. What is meant by the degrees of freedom of a robotic system?

5. Which has more degrees of freedom, a typical robotic arm or a human hand?

6. What are the four robotic system configurations?

7. What is an interface?

8. Compare serial ports and parallel ports.

9. What is meant by machine vision?

10. What are the four functions of machine vision?

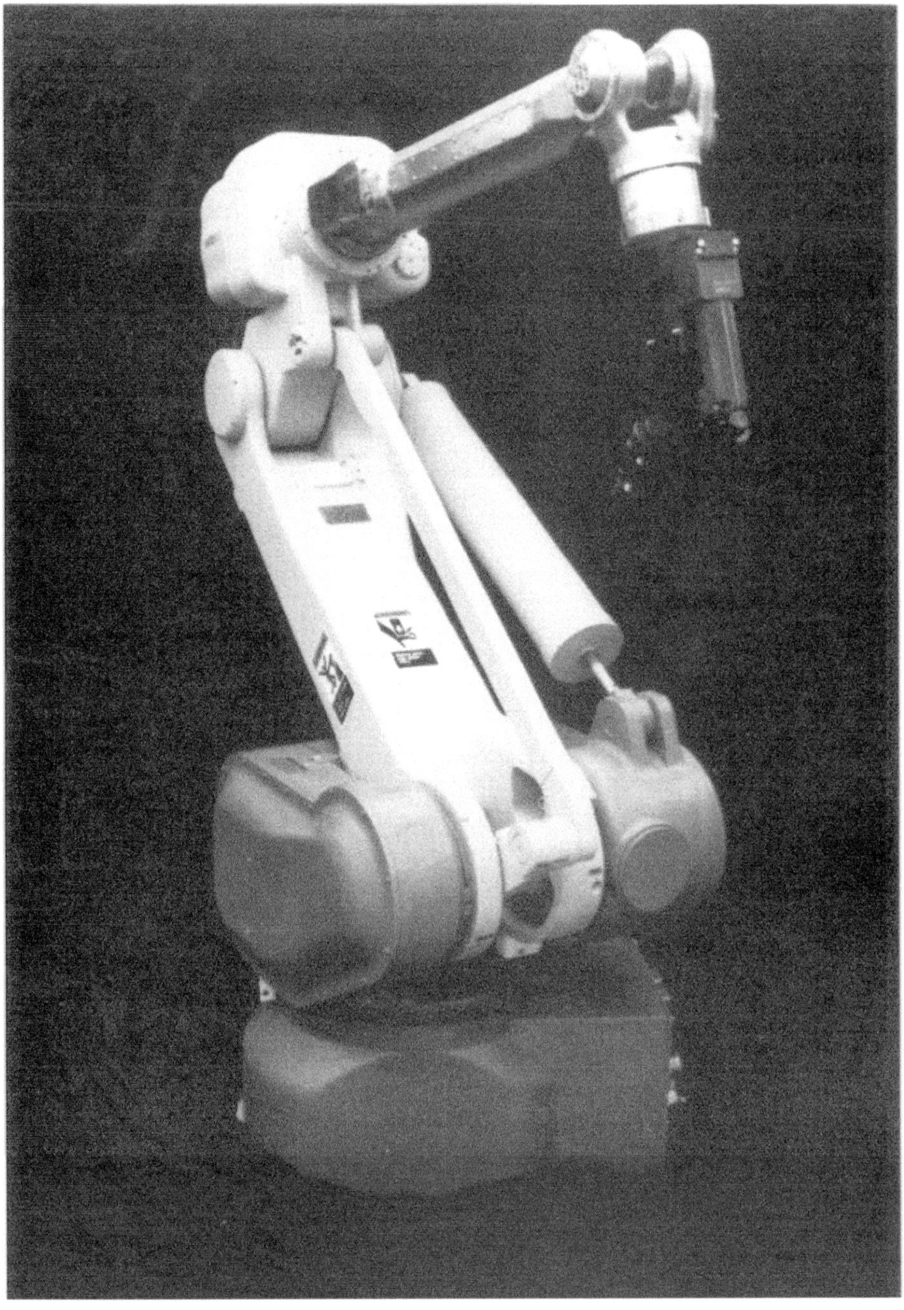

This robot was designed for painting and spray finishing applications. (Motoman, Inc.)

23

Preventive Maintenance and Troubleshooting

Objectives

Upon completion of this chapter, you should be able to:

- Describe general servicing techniques.
- List the seven steps for servicing equipment.
- Discuss troubleshooting basic principles.
- Use a troubleshooting chart.
- Explain the concept of preventive maintenance (PM).
- Troubleshoot basic series and parallel circuits.
- List at least three common maintenance problems.
- List several types of equipment that can be used for troubleshooting.
- List at least six measurement quantities and the type of equipment used to measure them.

Important Terms

assignment sheet
capacitor analyzer
curling
digital storage oscilloscope (DSO)
digitizing
dynamometer
feeler gauge
inductance analyzer
power analyzer
preventive maintenance (PM)
prony brake assembly
rework, scrap
troubleshooting
watthour meter
wattmeter

DOI: 10.1201/9788743810056-30

Introduction

Automated industrial systems are complex because they involve many different areas of technology. These include hydraulic, pneumatic, and electrical power systems, as well as electronic control systems, and digital computer systems. This makes their maintenance complex as well.

In this chapter, automated industrial system maintenance is discussed. Emphasis is placed on general servicing techniques, a well-organized approach to troubleshooting, preventive maintenance, and developing and implementing a preventive maintenance program.

23.1 General Servicing Techniques

Machines seldom require servicing of individual parts. Many repairs to systems take place at the subassembly level. Parts are not repaired. Instead, the module (unit) is simply replaced. You should become familiar with certain general servicing techniques. They are outlined in the following seven steps.

Step 1. *Initial observation and inspection*
See if the problem is electrical or mechanical in nature. Never accept anyone's word as to what the machine is or is not doing. Check it yourself! Inspect the power system. Determine if the proper current single-phase AC, three-phase AC, or DC is being used to power the equipment. Magnetic contactors, transformers, switches, circuit breakers, fuses, and other electronic controls should all be checked. Observe the physical condition of the equipment. Look for such things as fluid or air leaks, unusual noises, odors, or poor electrical connections or loose wires.

Step 2. *Gathering information*
If possible, acquire a schematic, a service or maintenance manual, and an operating manual for the machine. Such tools are essential to most equipment servicing. If these are unavailable, consult with someone who has had experience in servicing the machine. If available, an identical automated industrial system can be used to determine specific values (such as voltage, current, and waveforms). These values can be compared to those that exist in the machine being serviced.

Step 3. *Analyzing the equipment*
After gathering the information, you should have a good idea of how the system functions. Mentally divide or block out the different equipment according to purpose. For example, input power systems, rectifiers,

transformers, filters, feedback loops, amplifiers, interface/control circuits, and other devices can be sorted out. Then, identify the input and output of each block.

Step 4. *Isolating the problem*
You are now ready to trace the malfunction to a specific subsystem within the equipment. Use a multimeter or an oscilloscope to read measurements or waveforms. The secret here is knowing how to use test equipment and how to interpret the measured values. Compare your readings to those indicated on the schematic or in the service manual. When you compare the test readings to the recommended values, it is usually easy to isolate the problem.

Sometimes, verifying a malfunction is difficult. Generally, if a subsystem's input is correct and its output is not, you will assume the subsystem itself to be the problem. This is not true in all cases. In some cases, the output of one subsystem can become distorted because of another subsystem to which it is connected. When this is suspected, the subsystem under inspection should be disconnected and the voltages and waveforms checked again.

Step 5. *Narrowing the problem to a component, circuit, or module*
To isolate the malfunction of a specific component or components, use the procedure just described, making a series of measurements. Once the problem has been isolated to a specific component, circuit, or module, you must then check it to identify the problem. This checking can be completed in various ways with a number of different test instruments.

Step 6. *Replacing a component or device*
Any component used to replace another must be equal to, or better than, the original. Important considerations are voltage, maximum current, and power rating. Always replace a component with one that has an equal or higher maximum value. For example, a one-watt resistor can be replaced with a two-watt resistor, but never with a half-watt resistor.
Desoldering is also an important step in replacing faulty components. Use care in selecting the proper iron and soldering tip. Additionally, care must be taken not to damage components located nearby. A solder sucker should be used when working with a printed circuit board to properly remove a defective component without damaging the board. Care must also be taken when soldering a new component into the circuit. The replacement part must not be damaged by overheating. You must also guard against cold solder joints. Resin core (never acid core) should be used for all electronic circuit work.

Step 7. *Checking repairs and keeping records*
After the system has been repaired, operate it for several minutes to be sure it is working properly. When you are satisfied that the operation is correct, document the problem and the solution. Make a record of the original problem, what caused it, and what you did to correct it. This information should be kept with the service manual or in a special file so that it is on hand for future reference.

23.2 Troubleshooting Basics

Troubleshooting is the process of finding out why something does not work properly. If you follow the steps outlined earlier, you will be able to locate most problems. Some problems are easy to solve and require little time. Others are more difficult: they occur intermittently or are so complex that they require many hours of concentration and work.

Methods that most technicians find helpful in **troubleshooting** include:

- Using common sense.
- Following a logical sequence.
- Knowing how automated industrial systems work.
- Knowing how to use test equipment.
- Knowing how to read and use schematic diagrams effectively.

To begin any kind of **troubleshooting**, organize your thoughts and identify possible courses of action. Without a well-organized approach, troubleshooting can become a time-consuming guessing game. You must use a planned approach in order to save time. As you become more experienced and more familiar with the equipment, less time is required. Remember, no system is perfect.

An important part of troubleshooting is the **initial inspection**. When this inspection is done properly, many problems can be diagnosed without a lot of unnecessary steps. This means looking for the obvious. Use your senses of sight, touch, smell, and hearing. If you suspect a specific part, turn off the equipment, then examine it carefully, smell it, touch it, listen to it, and look closely at it.

A good **visual inspection** should be done before you get into actual circuit or system testing. Look for such things as:

- *Burned parts*. These are often obvious. They can be charred, blistered, or discoloered. They can even have holes.

Symptom List	
Applications 1 through 5 on at the same time	10–10
Axis does not move	10–11
Axis moves erratically or runs away	10–13
Control panel failures	10–15
Data error (DE) LED on or blinking	10–17
DI/DO failure (Standard)	10–19
DI/DO failure (Extended)	10–21
Does not home	10–23
Manipulator power does not work (No other indications)	10–25
MTCB indicator failure	10–26
MTCB switches do not work	10–27
Overtime error (OT) LED on	10–28
Overrun (OR) LED on	10–29
Overrun detection failure	10–32
Overrun reset failure	10–33
Power failure (PF) LED on	10–34
Remote operator panel interface failure	10–35
Repeatability varies	10–36
Servo error (SE) LED on	10–39
Servo error (SE) LED blinks and operator panel beeps	10–40
Transmission error (TE) LED on or communication error	10–42

Figure 23.1 This system list has page references to indicate where detailed information can be found.

- *Broken parts*. Breaks can appear in the form of cracks, wires pulled out of parts, or parts that have been completely destroyed.
- *Broken wires and poor connections.*
- *Smoke or heat damage*. Parts can smoke when the equipment is turned on. This identifies damaged part-but not the cause of the damage.
- *Oil, air, or water leaks.*
- *Loose, damaged, or worn parts.* These can be located visually or by touch.
- *Noisy parts, such as motor bearings*. Uncommon noises often indicate defective parts.

Keep in mind that most problems are **component failures**. If you know what each component is supposed to do, you will be aware of the troubles it can cause. Use the proper tests. To avoid duplicating effort, make a list of circuit or system operations you have already tested.

Good troubleshooting involves knowing how the system operates and how to use test equipment. Most of all, be patient and systematic.

Defective Battery
Note: Do not dispose of the battery in a fire. With the system off, check the voltage across the battery. The voltage should be between +3.2 and +4.2 Vdc. A discharged battery may load down the power supply while charging. To test for this, unplug the battery and reload the applications. If the battery is suspected of only being discharged, leave the controller powered up for eight hours to sufficiently charge the battery. If the voltage is still low after charging, replace the battery card.
Low Power Supply Voltages
Check for 24 Vdc, –12 Vdc, 12 Vdc, 5 Vdc at the power supply. If any of the voltages differ by more than 5%, adjust or replace the power supply. See power supply adjustment (4009). **Note:** The power supply may be left powered on with no load attached.
Defective Motor Control Board
If the battery is good and the voltages are good, suspect the Motor Control Board.

Figure 23.2 Detailed troubleshooting information helps to identify the actual problem.

Troubleshooting equipment and specific troubleshooting examples are discussed later in this chapter.

23.2.1 Troubleshooting charts

Most manufacturers of automated industrial equipment provide troubleshooting manuals or online information with charts and diagrams to aid in repair. Use this information to help locate the faulty module or part. A

AXIS DOES NOT MOVE

Defective Servo Pack Fuse

Swap the servo pack fuse for the failing axis. Do not rely on the indicator.
Theta 1 is labeled SV1 in the controller. (15 amp.)
Theta 2 is labeled SV2 in the controller. (10 amp.)
Z-Axis is labeled SV3 in the controller. (7.5 amp.)
Roll is labeled SV4 in the controller. (7.5 amp.)

Defective Z-Axis Brake

The brake should remain locked until manipulator power is brought up. Then, the Z-Axis servo pack takes over by controlling the power to the servo Z-Axis.
If the brake remains locked after manipulator power is up, then suspect the magnetic brake or circuit to the brake (brake is approximately 85 ohms when good). Also, the brake can be tested by disconnecting connector CN16H with manipulator power on, this makes a clicking noise.

Out of Adjustment Zero

Perform the "Servo Pack Zero Adjustment."

Defective Motor Control Board

1. If the Z-axis only does not move, set controller power to 0 (off), then ground CN16H-1 (WD16) to release the brake. Set the power switch to 1 (on). Move the axis to check for binds. If the axis cannot be moved, isolate the bind. Remove the jumper.
   For any other axis, set the power switch to 0 (off). Move the axis to check for binds. If the axis cannot be moved, isolate the bind.
2. Set the controller power switch to 1 (on). Press the Manip Power key on the control panel.
   DANGER:
   THE NEXT STEP REQUIRES YOU TO PUSH ON THE AXIS WITH MANIPULATOR POWER ON. DO NOT ALLOW YOUR BODY TO ENTER THE WORKSPACE.
3. Attempt to push or turn the axis. If the axis moves easily, then suspect the Servo Pack or wiring from the Servo Pack to the motor (WD07-WD10).

Figure 23.3 Manufacturers usually provide complete directions for diagnosing and solving problems. Note the safety warning included as part of a diagnostic procedure.

typical symptom list is shown in Figure 23.1. This list itemizes problems associated with a specific automated industrial system. It then directs the maintenance person to detailed information, such as the material shown in Figures 23.2 and 23.3. Such detailed instructions help to specifically identify the problem.

Some manufacturers use a flow chart, Figure 23.4, to guide you to the problem. If you understand how the circuit, device, or system functions and know how to use test equipment, troubleshooting and testing are relatively easy to accomplish. This is true for the simplest circuit or the most complex system. Practice is the best way to learn the procedures.

23.3 Preventive Maintenance

Unfortunately, some companies practice management by crisis: they avoid thinking about a problem until one occurs. In terms of maintenance, this means not fixing equipment until it breaks down. This leaves the production line always vulnerable to equipment failure.

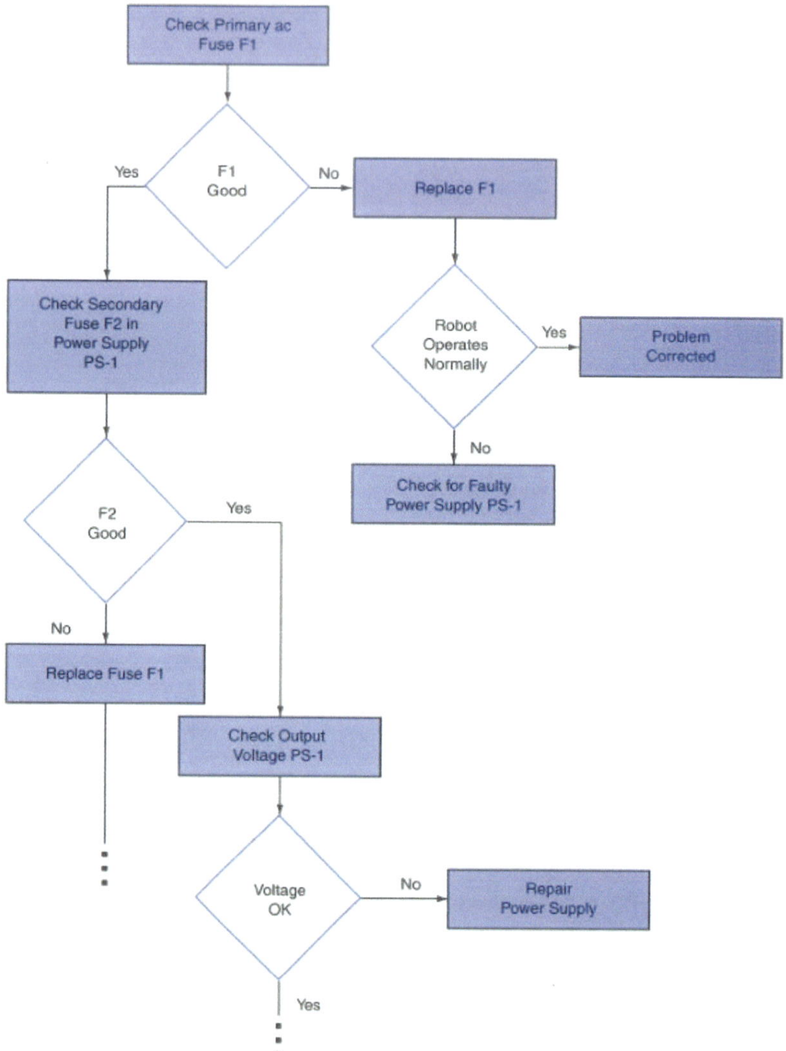

Figure 23.4 A flow chart like this one guides you, step-by-step, through diagnosing and correcting problems.

Equipment failures are costly, since they can create production bottlenecks. When work in process backs up at a workstation shut down by equipment failure, it means the operation must later run additional hours to catch up. This causes the operations that follow to need additional production time as well. Idle time results when operators or machines are not being utilized. If the order is not produced in time, the customer may go to another source.

Equipment failure also causes increased rework and scrap because the equipment had excessive wear prior to breakdown. Rework is the process of fixing parts that do not meet product specifications. Scrap is a part that cannot be fixed. Inventory costs are also increased when materials must be held beyond the normal time it takes for them to cycle through the manufacturing process.

Preventive maintenance is the best solution for avoiding the costly results of equipment failure. Preventive maintenance (PM) is the process of regularly checking equipment, cleaning and maintaining it, and replacing worn parts before breakdowns occur. It can play an important part in an organization's management strategy. Preventive maintenance is essential to robots and all other types of industrial equipment. It can increase flexibility, help maintain production flow, and allow for a continuous analysis of equipment. It decreases the need for a large on-call maintenance crew. Used in combination with mathematical analysis, it aids in predicting when equipment needs to be adjusted or replaced to avoid breakdowns.

Preventive maintenance can be performed at convenient times, such as when a robot is not being used. This reduces overtime for skilled operators and balances the workload among shifts. When performed conscientiously, preventive maintenance ensures consistent quality, reduces product costs, and reduces downtime of critical production equipment.

23.3.1 Developing a maintenance program

To develop a preventive maintenance program, a few basic steps should be followed:

Step 1. *Establish a schedule*
Each shift must share in maintenance responsibilities. An example of a maintenance schedule is shown in Figure 23.5. Such a schedule can be altered to accommodate the appropriate number of shifts. However, it must be followed carefully for maximum effectiveness. If one shift fails to follow the schedule, it can cause problems for the entire program. Usually, the schedule is easy to maintain and should be posted on the machines.

Step 2. *Use an assignment sheet*
An assignment sheet describes tasks that must be performed on a regular basis, Figure 23.6. Each assignment sheet identifies the unit to be maintained, the maintenance intervals, and the type of action required. The steps can be

Daily-Weekly Preventive Maintenance Schedule

Machine Type: IBM-7547 — **Serial Number: 257698ZMW**
Department: 179 — **Building Number: 5**

Shift	Week	Check as completed Mon	Tues	Wed	Thurs	Fri	Weekly	Operator	Special Comments
Shift 1	1	X	X	X	X	X	X	John Wasson	Air Leak
Shift 2	2	X	X	X	X	X	X	Robert Towers	Bad Relay
Shift 3	3	X	X	X	X	X	X	Stephen Fardo	No Problems
Shift 1	4	X	X	X	X	X	X	John Wasson	No Problems
Shift 2	5	X	X	X	X	X	X	Robert Towers	Air Leak
Shift 3	6	X	X	X	X	X	X	Stephen Fardo	Bolt Loose
Shift 1	7	X	X	X	X	X	X	John Wasson	Motor Noise
Shift 2	8	X	X	X	X	X	X	Robert Towers	No Problems
Shift 3	9	X	X	X	X	X	X	Stephen Fardo	Bad I/O Port
Shift 1	10	X	X	X	X	X	X	John Wasson	Loose Clamp
Shift 2	11	X	X	X	X	X	X	Robert Towers	Air Leak
Shift 3	12	X	X	X	X	X	X	Stephen Fardo	No Problems
Shift 1	13	X	X	X	X	X	X	John Wasson	No Problems
Shift 2	14	X	X	X	X	X	X	Robert Towers	Bolt Loose
Shift 3	15	X	X	X	X	X	X	Stephen Fardo	No Problems
Shift 1	16								
Shift 2	17								
Shift 3	18								
Shift 1	19								
Shift 2	20								
Shift 3	21								
Shift 1	22								
Shift 2	23								
Shift 3	24								
Shift 1	25								
Shift 2	26								
Shift 3	27								
Shift 1	28								
Shift 2	29								
Shift 3	30								
Shift 1	31								
Shift 2	32								
Shift 3	33								
Shift 1	34								
Shift 2	35								
Shift 3	36								
Shift 1	37								
Shift 2	38								
Shift 3	39								
Shift 1	40								

Figure 23.5 A preventive maintenance schedule helps find and eliminate small problems before they become large ones.

carried out in any order, but each should be done completely. Assignment sheets should be posted on the machines.

Step 3. *Keep work areas clean*
Once a preventive maintenance program is begun, machinery and work areas must be kept very clean. This will allow any leaks or visible damage or wear to be noticed more easily by the operator.

IBM-7547 Preventive Maintenance Assignment Sheet			
Unit	**Action**	**Interval**	**Lubrication**
Manipulator	Clean	Daily	—
	Check Oil Level	Daily	—
	Check for Air Leaks	Daily	—
Controller	Check Air Filters	Monthly	—
Roll Axis Belts	Check Tension	Monthly	—
Roll Axis Gears	Lubricate	Monthly	Molycoat G
Theta 1 Axis	Change Oil	Semi-Annually	Number 10 Oil
Theta 2 Axis	Change Oil	Semi-Annually	Number 10 Oil
Bearings	Pack Bearings	Every Replacement	Number 23 Grease

Figure 23.6 Assignment sheets should be posted on the machines to identify maintenance tasks and the interval at which they should be performed.

Step 4. *Make the operator part of the program*
Preventive maintenance places a great deal of responsibility on the operator. That person must be dedicated to the program for it to work properly. This does not mean that the operator is personally responsible for maintenance. It does mean, however, he or she must keep maintenance personnel informed of any problems.

While a good **preventive maintenance program** will definitely reduce the rate of unexpected equipment failure, such programs are not without their drawbacks. These include:

- Increased need for dedicated workers.
- Increased responsibility and paperwork for operators.
- Access time needed to work on equipment.
- Initial cost.

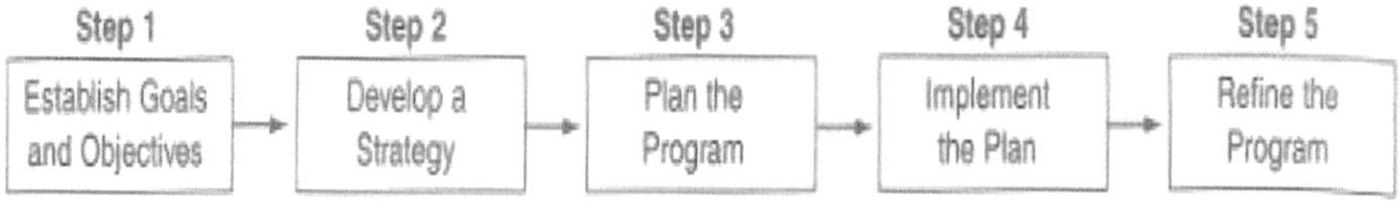

Figure 23.7 These steps should be taken when implementing a preventive maintenance program.

23.3.2 Implementing a new program

Implementing a new preventive maintenance program involves a number of steps, Figure 23.7. Because operators are so important to preventive maintenance, they should be involved early in the planning process.

23.4 Equipment Troubleshooting

Electronic systems are highly reliable, but still subject to failure. A technician must be capable of quickly and accurately analyzing a problem and making necessary repairs. It is necessary that a technician develop a logical, step-by-step approach to locate faults.

The first step in **troubleshooting** equipment is to determine that a fault really exists. A technician must know the purpose of the system and look for symptoms of trouble. Once the symptom is determined, possible causes must be considered and tested until the actual fault is located. The old saying that "if it is not broken, do not fix it" applies in troubleshooting. If no symptoms exist, testing should not be done.

Troubleshooting equipment requires certain steps in the process. These steps consist of (1) determining from the operation of the equipment that a fault does exist; (2) checking the source voltage for its proper value; (3) determining the cause of zero, high, or reduced current; (4) making repairs to the system; and (5) checking the system for proper operation.

You should take an analytical approach to **troubleshooting**. Through Ohm's law, you know the relationships between voltage, current, and resistance in a circuit. The power equations make it possible to determine the amount of power developed in a circuit or in a component. Mathematical calculations make it possible to determine the theoretical circuit operation. Measurements in the circuit make it possible to determine if these values are true operational values.

In problems involving **power** conversion, two of the senses (sight and smell) are helpful. Overheated wiring or printed circuit boards often have a darkened appearance. This discoloration can be an indication of trouble.

This appearance should be carefully examined as a possible effect of a circuit problem. Resistors can overheat to the point where the surface looks charred. This makes it impossible to determine the color code value. A burned resistor can even break completely. The burned resistor may not be the actual trouble in a circuit, but an effect of the trouble.

Normal **heating** in an electronic system does not produce charring or discoloration. A technician must learn the difference between normal heating and overheating. Some circuits operate at high temperatures.

When **troubleshooting** electronic equipment, the sense of smell sometimes gives the first indication of trouble. Overheated insulation or resistors have a distinct odor which technicians will recognize.

When this **odor** is perceived, turn the equipment off as quickly as possible! If this is not done, serious damage to the equipment can occur. Components burning or smoking are usually the result of a trouble, rather than the problem.

Since power is equal to the product of the current squared and the resistance ($I^2 \times R$), overheating occurs in a circuit when an overcurrent flows. An overcurrent is produced when either the source voltage is too great or the resistance is too small. The source voltage can be measured to determine its value. If it is correct, then something has happened to reduce the resistance. A possibility is a short circuit path developed around a resistor. Once the trouble is determined, it must be corrected. Then, the circuit must be checked.

23.4.1 Basic series circuits troubleshooting

Circuits operate through the control of electric current. Faults occur that cause the current to be too low, too high, or zero. Once the fault is determined, possible causes must be analyzed. Ohm's law, Kirchhoff's laws, and the behavior of series circuits are used in analysis.

The first step in **troubleshooting** circuits is checking the voltage source. Current flows as a result of the source. No circuit will operate without the proper source voltage. Always check first that the power switch is on. For circuits with batteries used as a voltage source, make sure that the batteries are tested under full-load conditions.

In a **series circuit**, zero current indicates that the circuit is open. An open circuit can be found using a voltmeter. A voltage (IR) drop is developed when current flows through a resistor. So, with no current, all IR drops are zero. The voltage across an open resistor will equal the source voltage. An open circuit can be caused by a broken wire or printed circuit board strip. It can also be caused by a component that has overheated. This trouble is corrected by replacing the open component or repairing the break in the printed circuit board.

If the **current** is too high in a series circuit, the resistance is too low. Low resistance can be caused by a shorted component or an alternate current path. A low-resistance alternate path can be caused by corrosion, solder bits, or wire clippings. A shorted resistor will have no voltage drop, while the other series components have voltage drops larger than normal. This problem can be corrected by replacing the shorted resistor or removing the alternate path. Solid-state components, such as transistors, which are discussed later, are more likely to short than resistors. Resistors are more likely to burn open due to high currents.

If the current in a series circuit is too low, the resistance is too high. A possible cause is a resistor that has increased in value. This increase is usually caused by overheating. If a resistor increases in value, it has a larger voltage drop than normal. This produces lower voltage drops across other resistors. This trouble is remedied by replacing the component that has changed value. The replacement should be exactly like the original. Never replace a resistor with one that has greater tolerance. The replacement should also have an equal or higher wattage rating than the original.

23.4.2 Basic parallel circuits troubleshooting

Basic techniques of **troubleshooting** for series circuits can also be applied to parallel circuits. A parallel circuit failure also results in a current that is too high, too low, or zero. The first step in troubleshooting is to check the voltage source. Check for blown fuses, a tripped circuit breaker, or a disconnected power cord. No circuit will operate properly if the voltage source is not proper.

Zero current in a parallel circuit is the result of an **open circuit**. The only other cause of zero current in a parallel circuit is if all loads are open.

High current is caused by a resistance that is too low. The resistance of a parallel circuit decreases as paths for current are added. If any component in a parallel circuit shorts, the short is across all components. The current would then become extremely high. The circuit protection device (circuit breaker or fuse) would then open to protect the source from damage. **Shorted** resistors are uncommon. A cause of high current could be a resistor that has decreased in value. In this case, one parallel branch conducts a much greater current than normal. This increases the total current. It is necessary to measure the branch currents to determine which resistor has changed value. The faulty resistor can then be replaced in order to correct the trouble.

Low current is likely to be caused by a resistor that has opened. The full source voltage is measured across an open resistor. This measurement is useless in a parallel circuit since the source voltage is across all resistors. It is necessary to make current measurements to locate the fault.

Remember, when measuring the **resistance** of a branch, it is necessary to disconnect one end of the resistor. Otherwise, the parallel combination of other branches gives a false reading on an ohmmeter.

23.5 Some Common Maintenance Problems

Some simple techniques can be used to check common sources of electrical failures. Three common sources of failure are: capacitors, transformers, and contact points. Special tools are available to help check these components.

23.5.1 Capacitor testing

Capacitor problems are of three types: **opens, shorts**, or **leaks**. An open capacitor is often caused by one of the connecting wires burning open where it connects to the plate. Shorts occur when current arcs through the dielectric, leaving a current path that allows excess current to flow. A leaky capacitor is one that is not shorted but does allow current through the dielectric. Thus, it never fully charges. Leaky capacitors are more likely to occur as equipment ages.

An **analog scale** ohmmeter can be used to test capacitors for opens or shorts. A normal capacitor, when connected across the ohmmeter, initially shows a low resistance. Then, as the capacitor charges, the needle moves toward infinity. If the capacitor is shorted or leaky, the resistance reading stays low and does not move toward infinity. The normal resistance should be between 500kΩ and 1MΩ for capacitors above 1μf. Smaller capacitors charge too quickly for the ohmmeter to respond and should read infinite on all scales. When checking electrolytic capacitors, the polarity of the meter must match the polarity of the capacitor leads.

A better check of a capacitor is made with a **capacitor analyzer**. This instrument checks the value of the capacitor and measures the leakage current under the normal working voltage of the capacitor.

23.5.2 Transformer testing

The most likely problems to occur in a transformer are shorts or opens. The instruments used for testing transformers are an ohmmeter and an inductance analyzer. The rated DC resistance of each winding of a transformer is normally known. An ohmmeter can be used to measure these resistance values. They can then be compared to the rated values. With an inductance analyzer, the inductance of each winding can be measured. Once again, these values can then be compared to rated values.

Short circuits sometimes occur in transformer windings. A direct short across the primary of a transformer removes all the inductance. With only the very low resistance of the short to oppose current, the primary current greatly increases. The protection device (fuse or circuit breaker) will operate to protect the voltage source. A shorted primary can be detected using an ohmmeter. The ohmmeter will read nearly zero across a direct short. The primary resistance of a power transformer, however, can be less than one ohm. Thus, the ohmmeter must be on a scale low enough to indicate a small value of resistance. A transformer with a shorted primary winding requires replacement.

A **short** across the secondary of a transformer, either in the winding or load, will be "reflected" into the primary. The primary current increases greatly. The protection device should operate. If the protection device does not operate, the heat produced by the high current could destroy the primary. If disconnecting the load removes the short, then the transformer is probably good. If the short exists when the load is removed, the secondary winding is at fault, and the transformer must be replaced.

A **partial short** means some, but not all, of the windings have shorted. A loss of windings on the primary or secondary results in a change in the turns ratio. Thus, the secondary voltage would be different from its designed value. These conditions could be detected using an ohmmeter, but it would be necessary to know the resistance value of each winding.

If an **open** occurs in the primary winding, no energizing current flows, and thus no magnetic field is created. No voltage is induced in the secondary under this condition. This condition could be detected by an ohmmeter indicating infinite resistance. Also, an ammeter check would show zero current flow.

23.5.3 Contact checking and maintenance

Among potential causes of **maintenance problems**, one of the most common is worn electrical contacts. It is easy to forget about contacts because they tend to perform reliably for long periods. Cadmium silver oxide contacts are used in modern equipment. They last far longer than the copper contacts previously used. And definite-purpose (DP) rated contacts, used in many electrical systems, can survive for up to 250,000 electrical operations. Like any wearing part, contacts require attention. You should make their inspection a part of your regular maintenance program.

During **maintenance inspections**, you want to recognize contact problems so you can replace them in time to prevent a system failure. You also want to avoid throwing away contacts that still have service life left. Cadmium silver oxide contacts show clear signs of wear. You can learn to recognize common wear patterns. You can then visually inspect to determine when to replace them.

Motor control contacts have both electrical and mechanical wear. **Mechanical wear** is usually insignificant. However, electrical wear, caused by arcing, will gradually erode contacts. Each time arcing occurs, it vaporizes and blows away a small part of the contact. As contacts continually make and interrupt the current, the contact surfaces change in color, contour, and smoothness. These changes lead to discoloration, surface irregularity, and pitting.

- **Color**. A new contact normally has a uniform silver color. As electrical wear occurs, the color may change to blue, brown, or black. The black results mainly from silver oxidation, which is beneficial to the contact operation.

- **Contour**. Irregularity of the surface is caused by uneven electrical wear. This does not necessarily mean that a contact is worn out. Contacts have a considerable amount of wear allowance. Irregularity of contour usually involves a slanted wear pattern. For example, one corner of a contact might wear more quickly than the others.

- **Smoothness**. Pitting is caused when the contact material melts and vaporizes. The size of a pit mark, or crater, is determined by the arcing amperage. On a high-amperage contact, the pit marks will be larger than on a low-amperage contact. When it appears that chunks are being torn away from the contact's surface, you probably should replace the contact.

A more problematic condition is curling. **Curling** occurs when the corners of the contact material separate from the backing plate. Curling usually results from manufacturing processes that produce high heat. Curling can also result when the contact has been exposed to fault conditions from short circuits or ground faults in the system. You should replace contacts when curling occurs. If curling continues to occur each time you replace a contact, check the application carefully to determine if the contactor is rated properly for the application. Also, check that other associated equipment is operating properly.

Regardless of surface appearance, do not file cadmium silver oxide contacts. Filing does not improve performance. It will only remove useful service life.

Contact inspection is a simple maintenance operation. For safety's sake, the first step is to disconnect the power at the branch circuit disconnecting device. After performing this step, simply remove the contact block cover and slide out the movable contact bar. Inspect the movable contacts. If the movable contacts are in reasonably good condition, the stationary contacts probably are also. If the movable contacts appear questionable, however, check the stationary contacts and measure the wear. Wear allowance is the total thickness of contact material that can wear away before the contact no longer carries the rated current. Generally, a contact is serviceable as long as the remaining wear allowance exceeds a specific minimum value. When measuring wear allowance, use a wide **feeler gauge**.

In checking **contacts**, avoid replacing contacts prematurely. Contact kits, which contain all the contacts and springs you need to make these replacements, are readily available. Look on the contactor nameplate to get the proper kit part number. These maintenance procedures can extend the life of motor control systems and prevent unnecessary downtime.

23.6 Electrical Machinery Measurement and Testing Systems

The operation and maintenance of electrical machines rely on several types of measurement and testing systems. Some of the types of equipment commonly used for electrical machine measurements and tests are as follows.

23.6.1 Basic test equipment

There are a number of pieces of **test equipment** that you will find useful for testing electrical equipment. The tools used depend on the task to be accomplished. The multimeter, megohmmeter, and oscilloscope are three of the most useful.

23.6.2 Testing with multimeters

One of the most valuable and often used types of equipment is the **multimeter**. Some types are shown in Figure 23.8. These instruments are used to measure AC voltage, DC voltage, DC current, and resistance. Basic tests on machines can be accomplished. It is a very versatile meter that can be used easily for maintenance tests.

Clamp-type meters, such as shown in Figure 23.9, are popular for measuring current in power lines. These indicators can be used for periodic

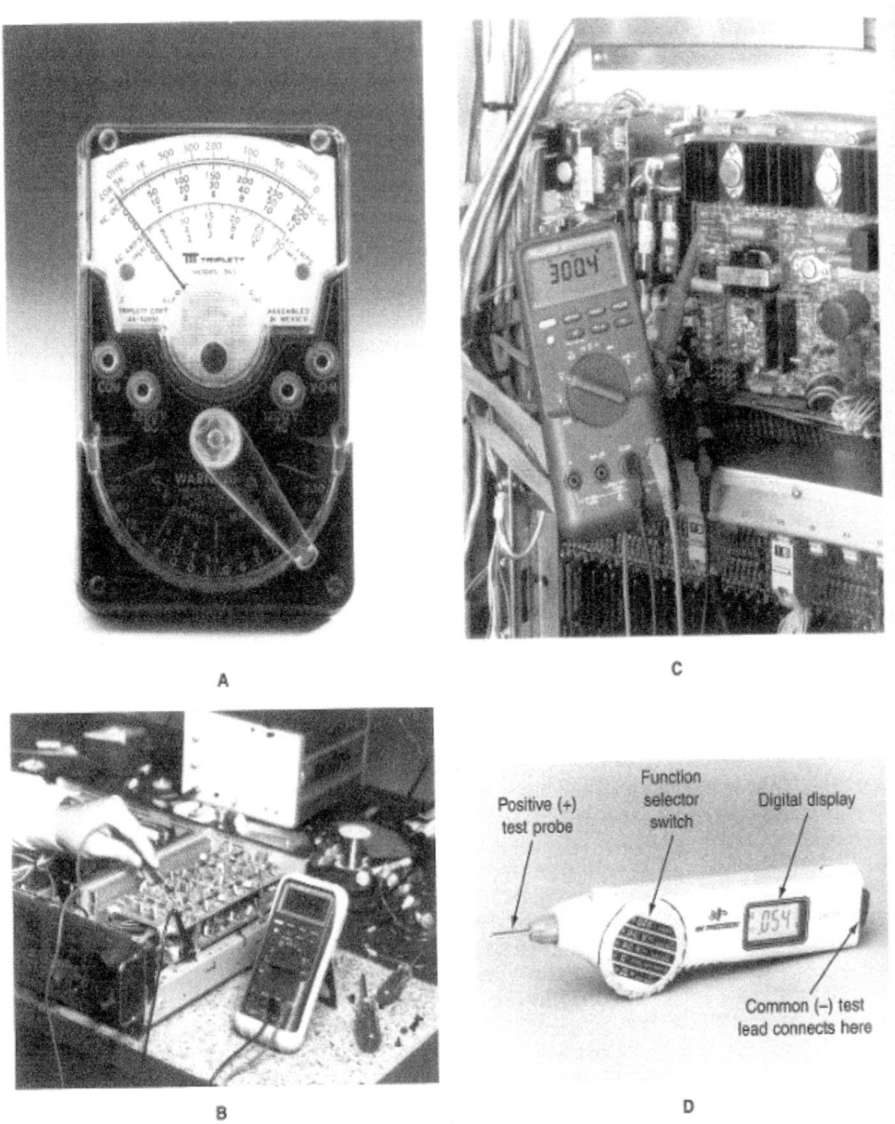

Figure 23.8 A – Portable analog meter. Notice the scale and pointer, which indicates the meter reading for ohms, volts, or amps. B – Measuring with a digital multimeter. Notice the common test lead connected to the equipment chassis. This provides voltage readings with reference to the equipment ground. C – Digital multimeter used for electrical testing. This is an auto-ranging meter. Notice that there are no number settings on the range switch. D – Portable hand-held digital multimeter. (Triplett Corp., Fluke Mfg. Co., Inc., Maxtec International Corp.)

A

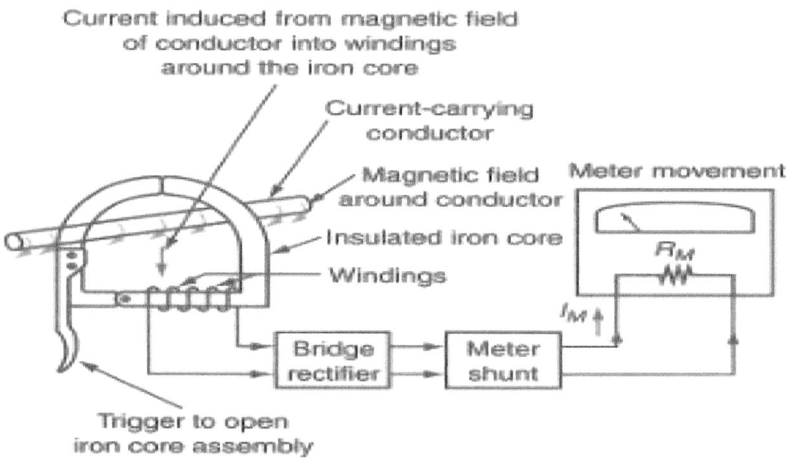

B

Figure 23.9 A – Clamp-type meter. B – Clamp-on multimeters to measure voltage, current, and resistance. (Fluke Mfg. Co., Inc.)

checks of current by clamping them around power lines. They are easy to use and convenient for maintenance and to test instruments.

The simplified circuit of a **clamp-type current meter** is shown in part B of Figure 23.9. Current flowing through a conductor creates a magnetic field around the conductor. The varying magnetic field induces a current into the iron core of the clamp portion of the meter. The meter scale is calibrated so that when a specific value of current flows in a power line, it will be indicated on the scale. Of course, the current in the power line is proportional to the current induced into the iron core of the clamp-type meter. Ordinarily, the clamp-type meter also has voltage and resistance functions that utilize external test leads.

23.6.3 Testing with digital storage oscilloscopes

An oscilloscope provides a time graph of a voltage amplitude versus time. This means that you see the change in voltage as it happens. In most applications, a real-time display is adequate. There are times when this type of graph is not desirable. If a voltage change is short in duration and only occurs after long periods of time, it would be difficult to study on a real-time display. Another example is when a voltage change is only a small part of a much longer, complex waveform. Here, getting the one small portion would be difficult on a real-time oscilloscope. In these cases, it is best to store a waveform sampled by an oscilloscope for future analysis. Such storage can be accomplished using a digital storage oscilloscope (DSO).

With a DSO, the signal viewed can be stored in memory and later retrieved for viewing and analysis. This method of storage digitizes the waveform. Digitizing means to sample an analog signal and assign each sample a number. In other words, the analog (continuously changing) voltage is converted to a series of digital (binary 1 or 0) numbers. Each digital number is then stored for either immediate or future display. To display the waveform, the binary numbers of the digitized waveform are used to restore the points on the graph. Figure 23.10 shows a typical digital storage oscilloscope. Since DSOs have basically the same circuitry and input probes as other scopes, most function as a conventional oscilloscope.

23.6.4 Testing with megohmmeters

Megohmmeters, such as shown in Figure 23.11, are used to measure high resistances that extend beyond the range of a typical ohmmeter. These indicators are used primarily for checking the quality of insulation on electrical machines. The quality of insulation of equipment varies with age, moisture

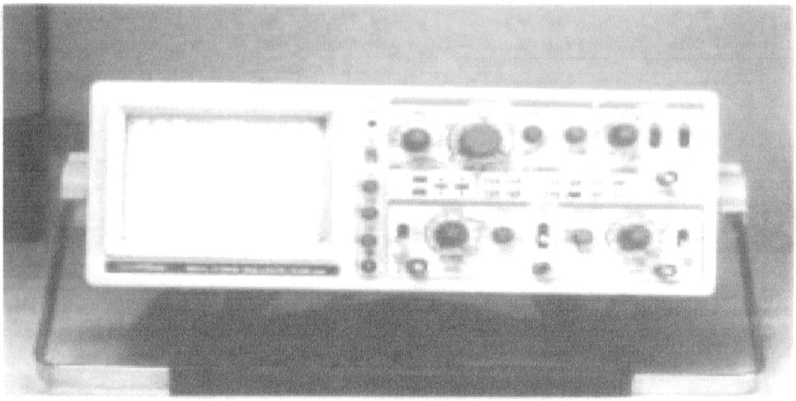

Figure 23.10 Digital storage oscilloscope (DSO). (Gold Star Precision Co.)

content, and applied voltage. The megohmmeter is similar to a typical ohmmeter except that it uses a high-voltage source of energy.

Periodic **insulation tests** should be done on rotating electrical machines and transformers. As insulation breaks down with age, the equipment starts to malfunction. A good method is to develop a periodic schedule for checking and recording insulation resistance so that it can be predicted when a piece of equipment needs to be replaced or repaired. A resistance versus time graph can be made, and the trend on the graph noted. A downward trend (decrease in insulation resistance) over a time period indicates that a problem in insulation exists. Additional types of insulation testers are shown in Figure 23.12.

23.6.5 Measurement of power quantities

Several types of machinery **measurement** applications involve equipment used to measure electrical power and related quantities. The day-to-day operation of machines relies on the use of electrical power. Therefore, to use electrical machines more effectively, their power conversion should be monitored.

23.6.6 Measuring electrical power and energy

Wattmeters, which monitor both voltage and current in a circuit, are ordinarily used to measure the power converted by electrical machines. For instance, the amount of electrical energy converted into mechanical energy by a motor can be measured by properly connecting a wattmeter to the motor circuit.

A

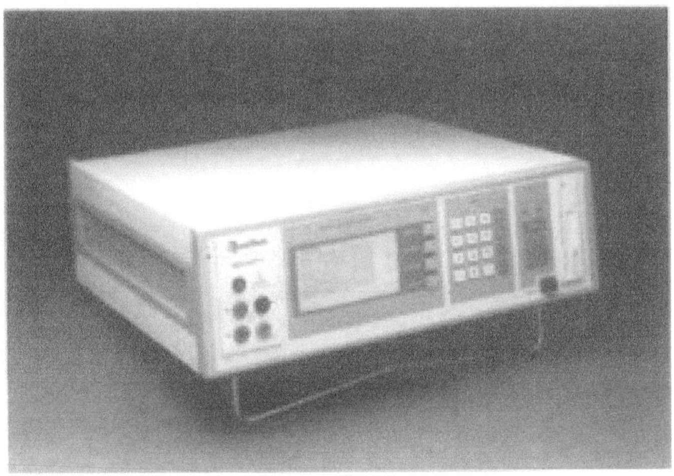

B

Figure 23.11 A – Analog megohmmeter. B – Digital megohmmeter. (AVO Biddle Instruments, Quad Tech)

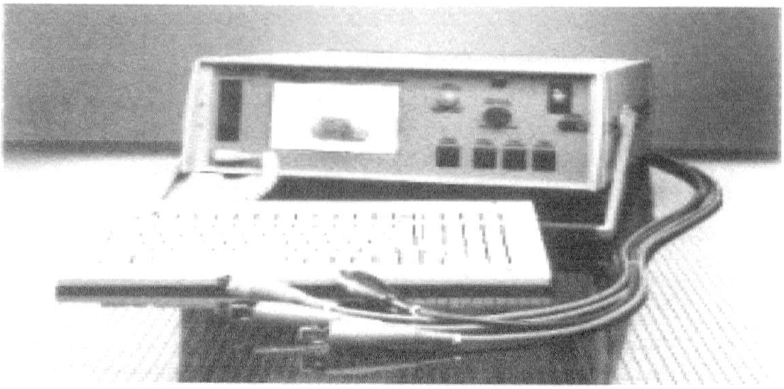

A

B

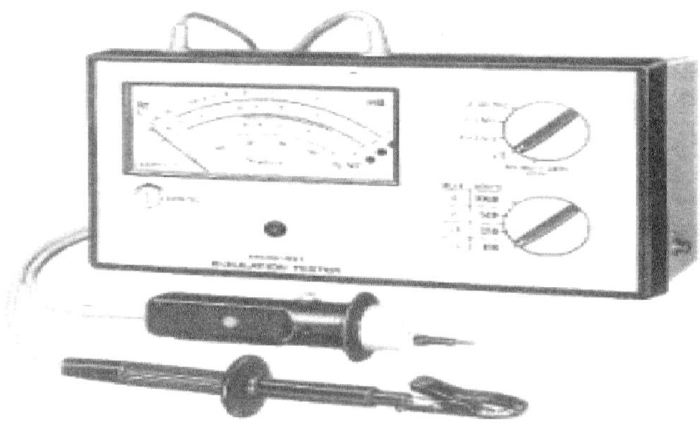

C

Figure 23.12 A – Advanced winding analyzer. This device checks resistance and other characteristics of equipment with windings. B – This Megger(i) unit analyzes insulation resistance. C – Analog insulation tester. (Baker Instrument Co, AVO Biddle Instruments, Simpson Co.)

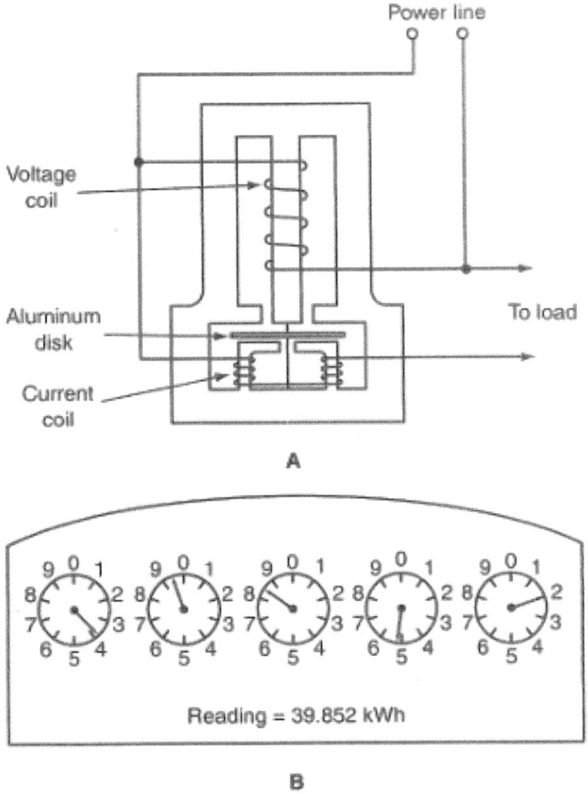

Figure 23.13 Watthour meter. A – Internal construction. B – Dial-type register.

The amount of electrical energy used over a period of time can be measured by using a watt-hour meter. A **watt-hour meter**, illustrated in Figure 23.13, relies on the operation of a small motor mounted inside its enclosure. The speed of the motor is proportional to the power applied to it. The internal construction is shown in part A of Figure 23.13. The rotor is an aluminum disk that is connected to a register, which usually indicates the number of kilowatt hours of electrical energy used. Part B of Figure 23.13 shows the dial type of register that is frequently used on watt-hour meters. Other types have a direct numerical readout of kilowatt hours used.

A **watt-hour meter** is connected between the voltage feeder lines and the branch circuits of a power delivery system. In this way, all electrical energy used must pass through the watt-hour meter.

The electrical operation of a **watt-hour meter** relies on a potential coil connected across the power lines to monitor voltage, while a current coil is placed in series with the line to measure current. The voltage and current of the system affect the movement of an aluminum-disk rotor. The operation of the watt-hour meter is similar to that of an AC induction motor. The stator is an electromagnet that has two sets of windings: the voltage windings and the current windings. The field developed in the voltage windings causes eddy currents to be induced into the aluminum disk. The torque produced is proportional to the voltage and the in-phase current of the system. Therefore, the watt-hour meter will monitor the true power converted.

23.6.7 Measuring three-phase electrical power

For certain applications, it is usually necessary to monitor the **three-phase power** converted by electrical machines. It is possible to use a combination of single-phase wattmeters to measure total three-phase power, as shown in Figure 23.14. The methods shown are ordinarily not very practical since the sum of the meter readings would have to be found in order to calculate the total power of a three-phase system.

Three-phase power meters, see Figure 23.15, are designed to monitor the true power of a three-phase system by using one meter. The reading developed on a power meter is dependent on the voltage and current of the three phases acting on the system. Three-phase watt-hour meters are also available that use three aluminum disks on the same shaft to monitor the three-phase power used by a system. The total three-phase power is a combination of the power values developed by each phase. Power analyzers simplify the measurement of three-phase power. See Figure 23.16.

23.6.8 Measuring power factor

Power factor is the ratio of the true power of a system to the apparent power (volts × amperes). To determine the power factor, a wattmeter, a voltmeter, and an ammeter can be used. The relationship PF = W/VA of each measurement can then be calculated. However, it is more convenient to use a power-factor meter in situations where the power factor must be monitored.

The principle of a power-factor meter is illustrated in Figure 23.17. The power-factor meter is similar to a wattmeter, except that it has two armature coils that rotate due to their electromagnetic field strengths. The armature coils are mounted on the same shaft so that their alignment is about 90° apart. One coil is connected across the AC line in series with a resistance, while

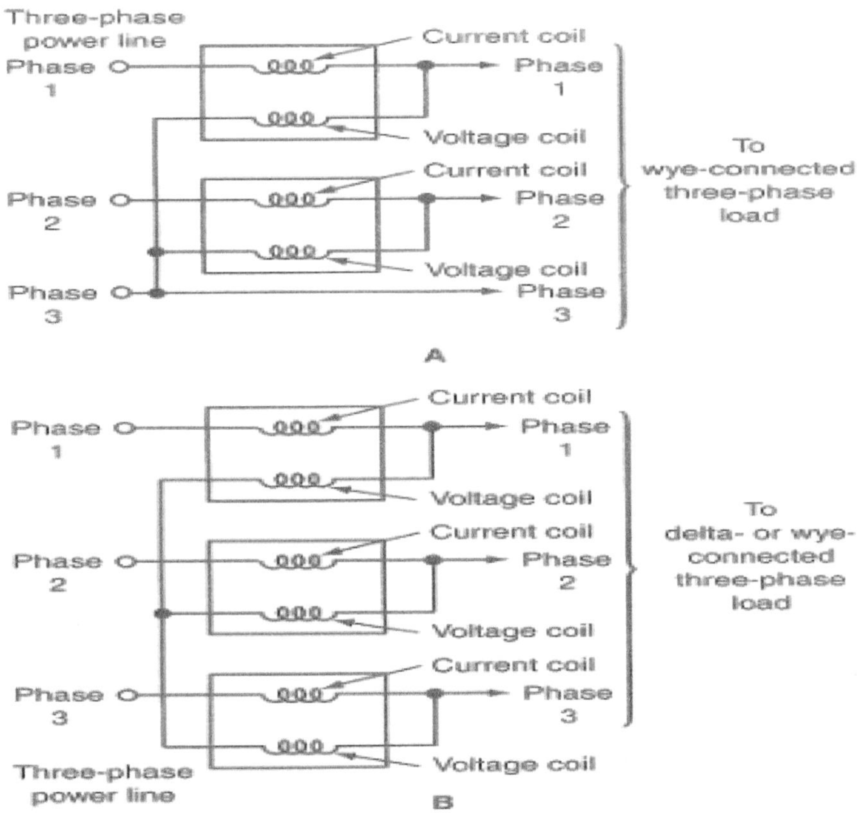

Figure 23.14 Measuring three-phase power. A – Two-watt-meter method. B – Three-watt-meter method.

the other coil is connected across the line in series with an inductance. The resistive path through the coil reacts to produce a flux proportional to the in-phase component of the power. The inductive path reacts in proportion to the out-of-phase component of the power.

If a **unity (1.0) power-factor** load is connected to the meter, the current in the resistive path through coil A should develop full torque. Since there is no out-of-phase component, no torque would be developed through the inductive path. The meter movement would now indicate full-scale, or unity power factor. As power factor decreases below 1.0, the torque developed by the inductive path through coil B would become greater. This torque would be in opposition to the torque developed by the resistive path. Therefore, a power factor of less than 1.0 would be indicated. The scale must be calibrated to measure power-factor ranges from zero to unity.

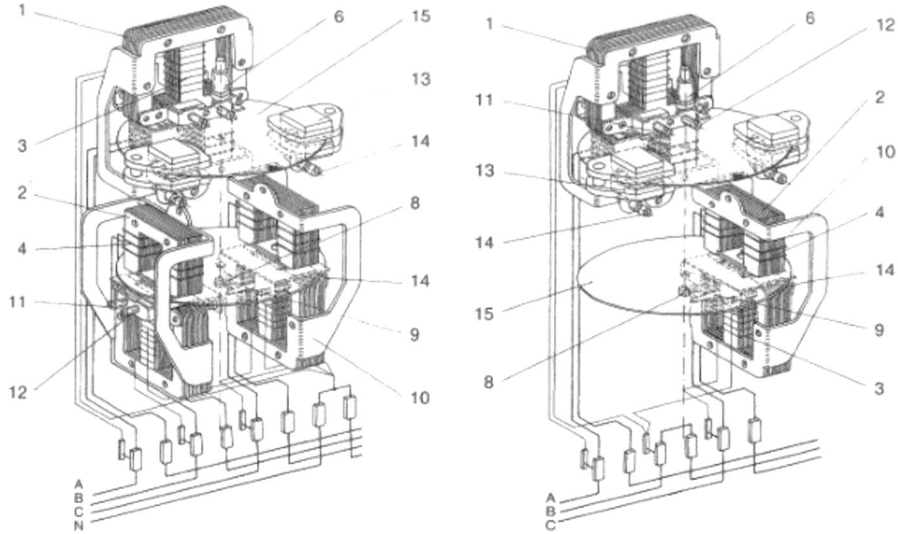

Schematic arrangement of three measuring systems

1—Voltage core; 2—Current core; 3—Voltage coil; 4—Current coil; 5—Rotor; 6—Worm-gear; 7—Brake magnet; 8—Brake flag; 9—Brake tongue; 10—Base plate; 11—Phase-angle regulator; 12—Torque regulator; 13—Brake magnet regulator; 14—Low-load regulator

Schematic arrangement of two measuring systems

1—Voltage core; 2—Current core; 3—Voltage coil; 4—Current coil; 5—Rotor; 6—Worm-gear; 7—Brake magnet; 8—Brake flag; 9—Brake tongue; 10—Base plate; 11—Phase-angle regulator; 12—Torque regulator; 13—Brake magnet regulator; 14—Low-load regulator

Figure 23.15 Three-phase power meters. (Electric Metering Corp.)

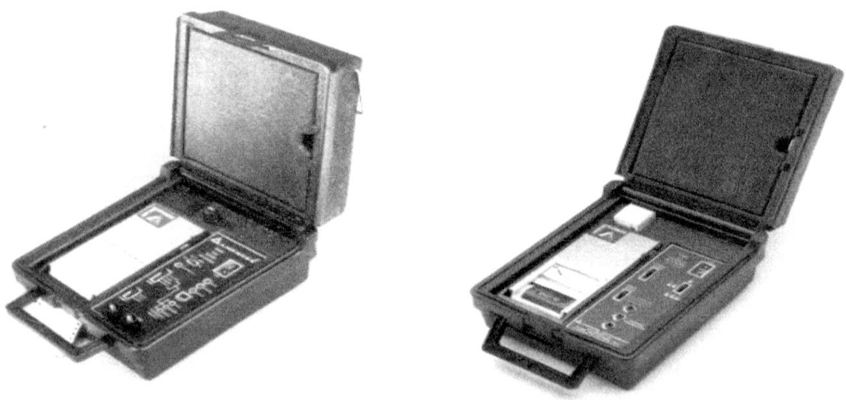

A B

Figure 23.16 Chart recorders for monitoring AC power quantities. A – Recorder for monitoring AC power line disturbances. B – Recorder to monitor true rms AC voltage. (Amprobe Instruments)

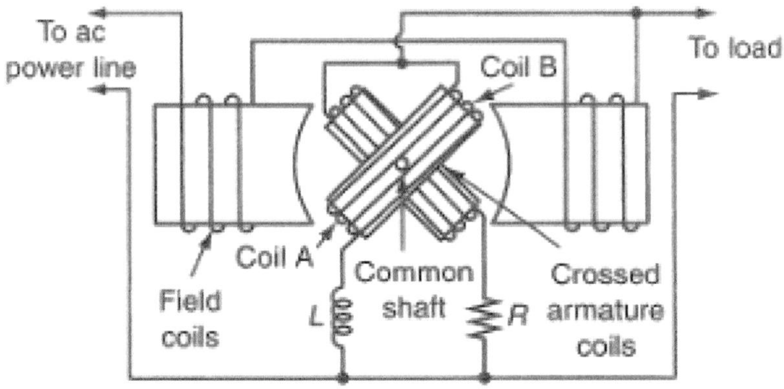

Figure 23.17 Schematic diagram of a power-factor meter.

Various other types of **instruments** are commercially available for measuring three-phase power and associated quantities. Power-monitoring instruments, which can be used for analysis of true three-phase power, reactive power, current, and voltage, are sometimes referred to as power analyzers. Such equipment usually has several instrument circuits housed in a single enclosure. In this way, the related three-phase quantities can be observed easily to better understand how the power system or machinery is functioning.

23.6.9 Measurement of frequency

Another important measurement is **frequency**. The frequency of the power source must remain stable, or the operation of many types of machines will be affected. Frequency refers to the number of cycles of voltage or current that occur in a given period of time. The international unit of measurement for frequency is the hertz (Hz). A frequency of one hertz is one cycle per second. The standard power frequency in the United States is 60 Hz.

Frequency can be measured with several different types of indicators. An **electronic counter** is one type of frequency indicator. Vibrating-reed frequency indicators are also common for measuring power frequencies. An oscilloscope, such as the one shown in Figure 23.18, can also be used to measure frequency. Graphic recording instruments can be used to provide a visual display of frequency for different time periods.

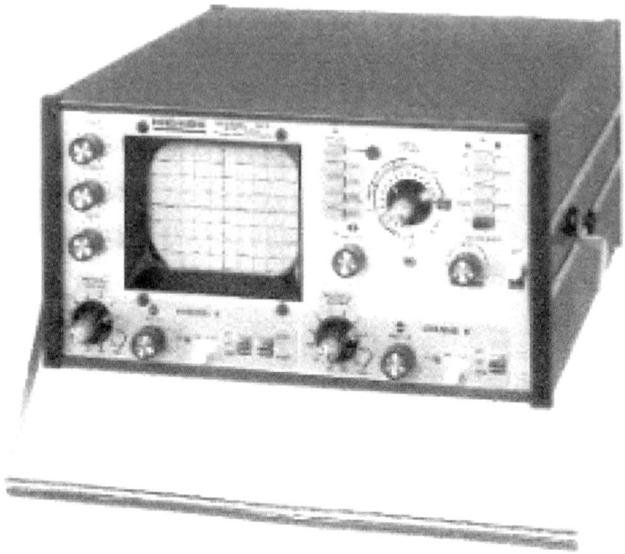

Figure 23.18 Oscilloscope used to measure AC voltage and frequency. (Hickok Electrical Instrument Co.)

23.6.10 Measurement of speed

A common type of electrical machine measurement is the measurement of **speed**. Usually, speed is measured as a rotary movement with a **tachometer**. Several different principles can be used for measuring speed. One method is referred to as a DC tachometer system and is illustrated in part A of Figure 23.19. This tachometer is connected directly to a rotating machine or piece of equipment. The principle involved in this system is that as the shaft of the small permanent-magnet DC generator rotates faster, the voltage output increases in proportion to the speed of rotation. Voltage output increases can then be translated into speed changes and indicated on a calibrated tachometer scale.

Electronic tachometers, such as shown in part B of Figure 23.19, are now used extensively due to their increased precision and ease of usage. In this type of tachometer, movement is measured by providing a reflective material on the surface of the equipment or machine subject to measurement. The tachometer has a light source that is interrupted by the passage of the reflective material. A photocell converts the changes of light energy into electrical impulses. The electrical impulses control the movement of an indicator calibrated in revolutions per minute, or the pulse rate of a

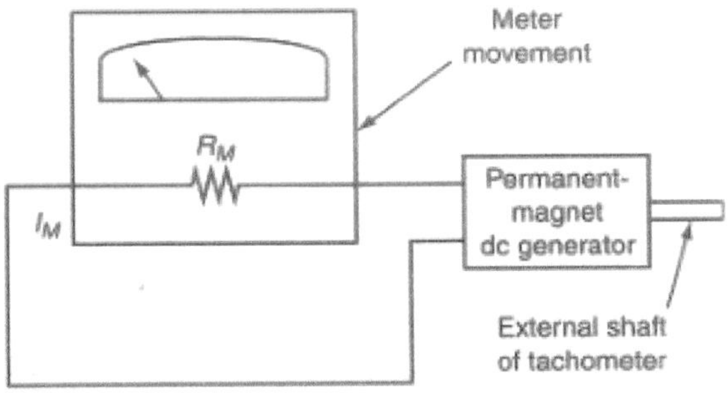

Figure 23.19 A – Diagram of a DC tachometer system. B – Tachometer-microprocessor controlled digital hand-held noncontact instrument for measuring speeds from 6 to 30,000rpm. (Shimpo Instruments)

digital counter that provides a direct numerical readout of revolutions per minute.

23.6.11 Measurement of vibration

Vibration is a factor that must be controlled in electrical machines. In order to measure vibration in a piece of equipment, a vibration meter with a special transducer can be used. The reduction of vibration will improve the operating characteristics of machinery. The transducer used to measure vibration has an output that is proportional to the amplitude of the vibrations that are sensed while a machine is in operation.

23.6.12 Motor testing

Electric motors convert electrical energy into mechanical energy. The **efficiency** of this energy conversion is critical in terms of motor operation.

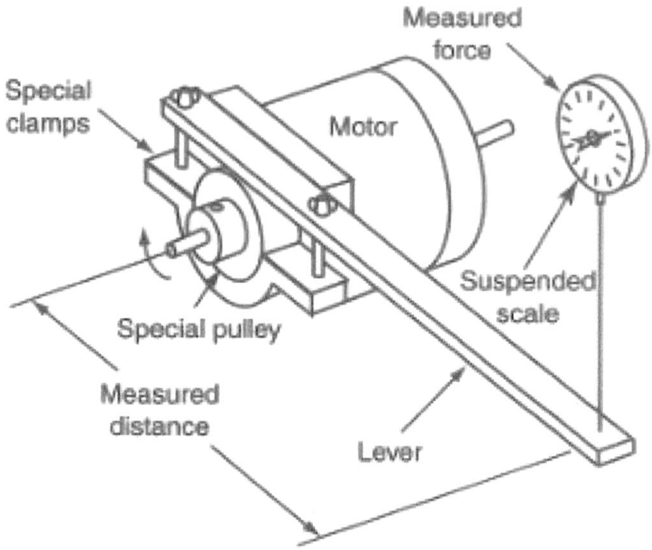

Figure 23.20 Representation of a prony brake.

Recall that efficiency is calculated as the ratio of power output to power input of a machine. For electric motors:

$$\text{efficiency} = \frac{\text{power output (hp)}}{\text{power input}\left(\text{W}\right)} \text{ or } \frac{hp \times 746}{W}$$

The **horsepower** of a motor is determined by the **torque** and **speed** of the machine. Torque can be determined by measurement with a **prony brake assembly**. A representation of a prony brake is shown in Figure 23.20. Tightening the bolts onto the special pulley causes the brake unit to turn with the motor and apply force to the suspended scale. The motor's torque is calculated as the product of the scale reading (pounds, ounces, or grams) and the measured distance shown (feet, inches, or centimeters). The prony brake method is a friction-type brake that would be impractical for large motors.

The principle of the **prony brake** is used for commercial units called **dynamometers**, which are used extensively for motor testing. Dynamometer testing has become very important with increased awareness of energy consumption and rising power costs. A dynamometer is used to load a motor and measure its power output directly. A load cell is commonly used to provide an accurate measure of force exerted by a motor on the dynamometer.

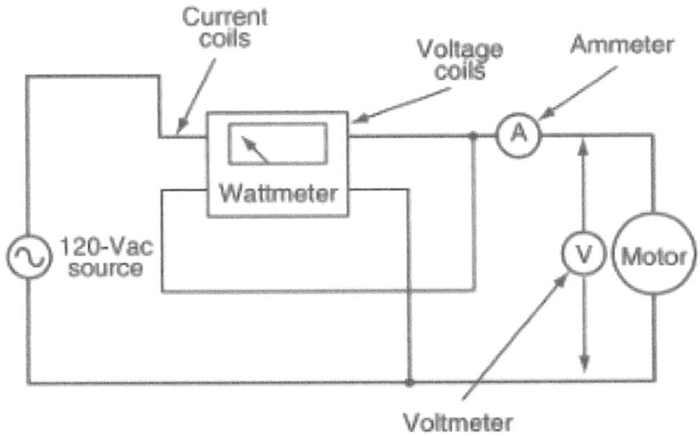

Figure 23.21 Wattmeter, ammeter, and voltmeter are used to measure motor input values.

Figure 23.22 Phase sequence indicator. (Syntel, Inc.)

Measurement of input power, current, and voltage of a motor can be accomplished by using a wattmeter, ammeter, and voltmeter as shown in Figure 23.21. The **wattmeter** measures the true power converted by the motor. The **power factor** of a motor can be calculated as follows:

$$\text{PF} = \frac{\text{true power}\left(\text{W}\right)}{\text{apparent power}\left(\text{VA}\right)} = \frac{\text{watts}}{\text{volts} \times \text{amperes}}$$

23.6.13 Three-phase motor phase sequence

For **three-phase motors**, an important type of test is **phase sequence** indication. A simple type of equipment used for such tests is shown in Figure 23.22. A reversed phase sequence will cause a three-phase motor to rotate in the wrong direction. The correct direction of rotation often does not lend itself to trial-and-error testing. Applications such as conveyor lines, pumping

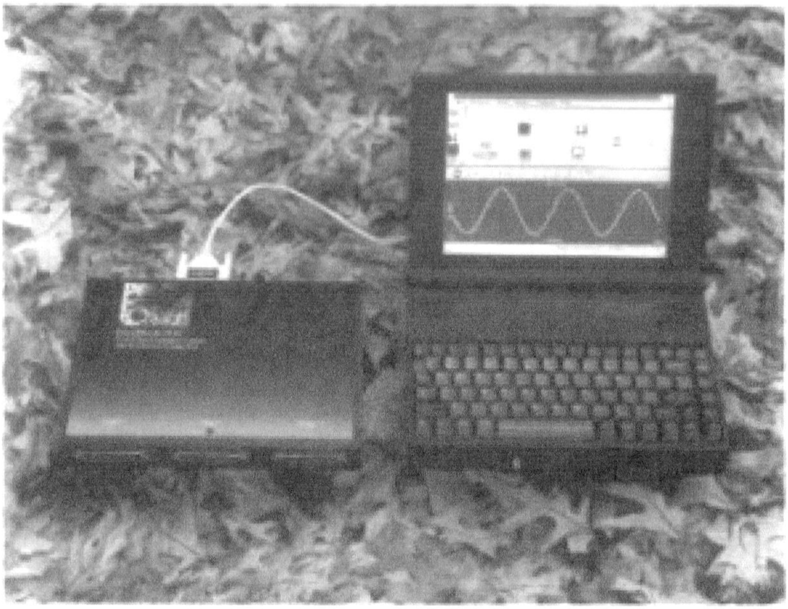

Figure 23.23 Computer interface. Now commonly used to provide data acquisition to monitor the status of equipment. (IOtech, Inc.)

systems, and interconnected drive systems require a definite direction of rotation. Thus, phase sequence testing can be used to avoid problems associated with the reversed direction of rotation.

23.7 Preventive Maintenance Equipment

Preventive maintenance equipment is extremely important in industry. Preventive maintenance consists of inspecting, cleaning, and performing tests on equipment and components before failure. Poor preventive maintenance results in longer periods of downtime for equipment and costs industries in terms of reduced production and increased costs. In today's competitive world economy, companies cannot afford to have poor preventive maintenance programs.

In theory, **preventive maintenance** is simply keeping machines and equipment operational so that there is minimum downtime. Trained technicians who are familiar with test and preventive maintenance equipment help to keep companies running smoothly. Several types of test and preventive maintenance equipment are illustrated in Figure 23.24.

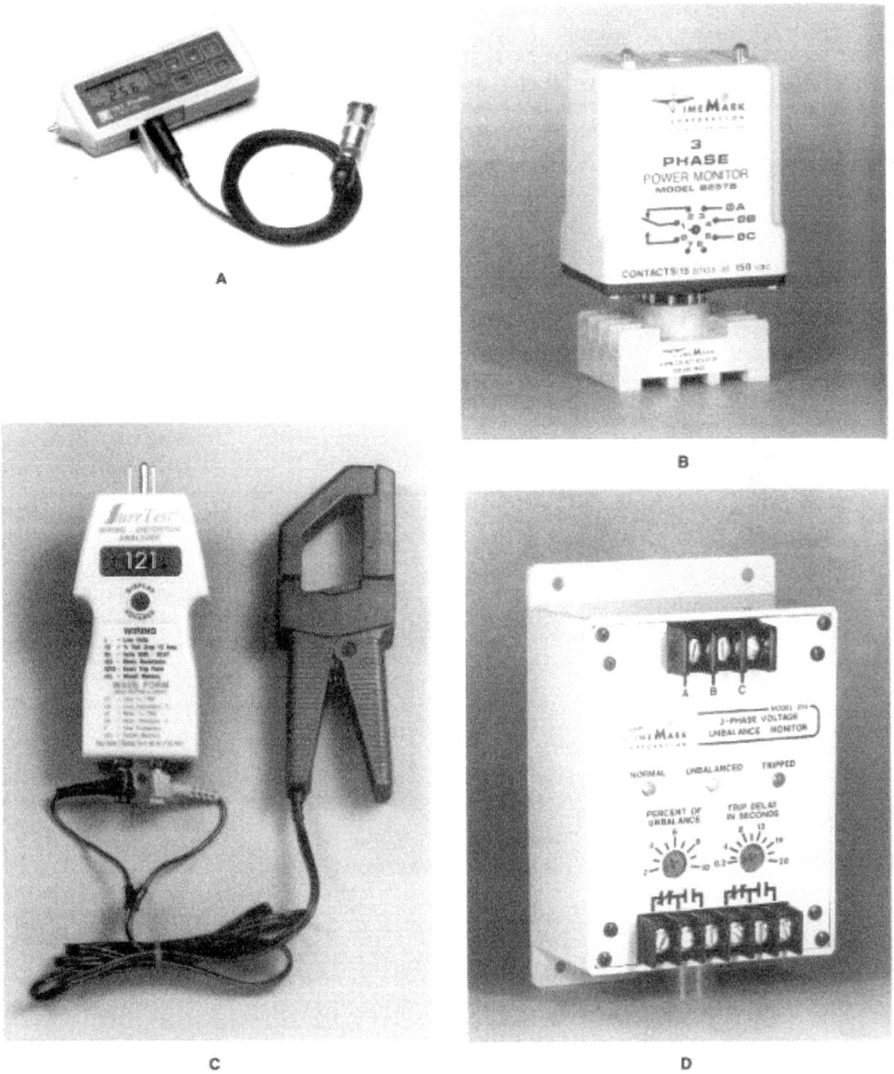

Figure 23.24 A – Portable data collector/analyzer. This device collects and records machinery data and interfaces with equipment software. B – Three-phase power monitor. C – This digital power monitor analyzes branch circuits to determine power quality in terms of ground impedance, false grounds, current capacity, proper wiring, and safety hazards. D – Three-phase voltage unbalance monitor. E – Logic analyzer used for testing complex digital systems such as microprocessor-based equipment. F – Computer diagnostic equipment. This system is used by consumer electronics companies to test circuit boards. (SKF Condition Monitoring, Inc., Time Mark Corp. Industrial Commercial Electronics Co., Time Mark Corp., Teradyne, Inc.)

E F

Figure 23.24 Continued.

Review Questions

1. Identify five areas of technology used by automated industrial systems.

2. List the seven steps involved in servicing automated manufacturing systems.

3. During the initial observation of equipment to be serviced, what should you look for?

4. How does a technician proceed in gathering information about equipment?

5. What are some types of measurements that can be taken to check equipment?

6. What factors should be considered when replacing a component?

7. What information is important to write down in keeping good records when servicing equipment?

8. What is meant by troubleshooting? Which of your senses are used in troubleshooting equipment?

9. List five methods that technicians find helpful when troubleshooting.

10. What is preventive maintenance?

APPENDICES

Appendix A

Calculator Examples

This appendix is devoted to the effective use of electronic calculators as efficient tools for problem solving. The approach to calculator usage is that it can be used in ways that strengthen the achievement of learning. *Understanding* the calculator and *understanding* the problem go hand in hand with getting correct answers and gaining knowledge.

Simple electronic circuit problems involving small, whole numbers are often solved in a few seconds by pencil-and-paper arithmetic. We can determine that the application of 12 V to a 6 – Ω resistor results in a current of 2 A through the resistor. Ohm's law (see Chapter 4) instructs us to divide voltage by resistance to get current. As the electronic values grow larger and use more decimal places, calculations become more difficult and time-consuming. An electronic calculator is a great labor-saving device, which also reduces errors and increases the accuracy of answers. Multidigit numbers can be added, subtracted, multiplied, divided, squared, and more by a few simple strokes on the buttons of a calculator keypad. The fundamental advantage of electronic calculators is to make complicated calculations easier.

A *scientific* calculator has standard arithmetic functions, scientific notation, trigonometric functions, and logarithms. A calculator with these functions is needed in the study of electronics. You may wish to get other calculator features, such as number base conversions and programmable features.

There are many keys that are standard to the typical scientific calculator. The keys for the arithmetic functions of addition, subtraction, multiplication, and division are self-explanatory. The 1/X key provides the reciprocal of the number entered in the display. The reciprocal of a number is one divided by that number. The X^2 key provides the square of the number entered, while the $\sqrt{X}$ key provides the square root of the number. The STO key stores the number from the display in memory while the RCL key will return the number from memory to the display. Some calculators have more than one location for storing data. The manual provided with the calculator describes special functions and features. The EE, or EXP, key allows the entry of numbers in scientific notation.

The calculator procedures in this appendix illustrate how to use a scientific calculator to solve electronic and math problems. There are many calculators on the market. These procedures will work on most scientific calculators with a few changes in key labels and button sequences. One major exception is the type of calculator that uses RPN (reverse Polish notation) logic in the keying sequences, see Figure A.1. If you are using an RPN calculator, it will be necessary to translate the procedures given here.

The calculator sequences are self-explanatory. Calculator function buttons used in the book are summarized here as an aid to understanding the sequences and as a cross-reference with functions for other types of calculators. The specific sequences given here will not solve every problem that may come along. They should, however, provide enough experience for students to be able to "design" the sequences for other problems. The calculator procedures of this appendix are listed in Figure A.2.

The remainder of Appendix A shows calculator procedures for solving various types of problems.

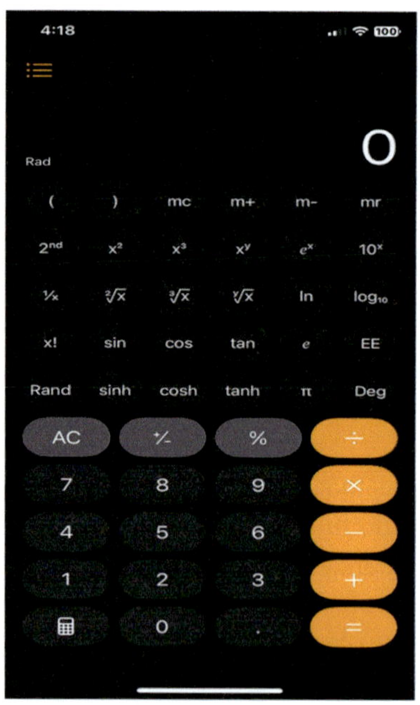

Figure A.1 Typical scientific calculator keys.

Problem	Figure
Addition	A-3
Subtraction	A-4
Multiplication	A-5
Division	A-6
Fractions	A-7
Powers and Roots	A-8
Percentages	A-9
Unit Conversion	A-10
Scientific Notation	A-11
Ohm's Law	A-12
Series Circuits	A-13
Parallel Circuits	A-14
Combination Circuits	A-15
Trigonometric Functions	A-16
Series RC Circuits	A-17

Figure A.2 Calculator procedures.

Example	Operation	Readout
5 + 3 = 8	5 [+] 3 [=]	8.
2.3 + 6.8 = 9.1	2 [•] 3 [+] 6 [•] 8 [=]	9.1
10 + 14 + 5 = 29	10 [+] 14 [+] 5 [=]	29.

Figure A.3 Calculator procedures for addition.

Example	Operation	Readout
42 − 13 = 29	42 [−] 13 [=]	29.
400 − 0.25 = 399.75	400 [−] [•] 25 [=]	399.75

Figure A.4 Calculator procedures for subtraction.

Example	Operation	Readout
$150 \times 50 = 7500$	150 [×] 50 [=]	7500.
$6.5 \times 86 = 559$	6.5 [×] 86 [=]	559.

Figure A.5 Calculator procedures for multiplication.

Example	Operation	Readout
$\frac{600}{3} = 200$	600 [÷] 3 [=]	200.
$\frac{487}{5} = 97.4$	487 [÷] 5 [=]	97.4

Figure A.6 Calculator procedures for division.

Example	Operation	Readout
$\frac{3}{4} = 0.75$	3 [÷] 4 [=]	0.75
$\frac{32}{3} = 10.67$	32 [÷] 3 [=]	10.66666667
$\frac{3}{4} \times \frac{2}{5} = \frac{3}{10}$	3 [÷] 4 [×] 2 [÷] 5 [=]	0.3
$\frac{3}{4} \div 8 = \frac{3}{4} \times \frac{1}{8} = \frac{3}{32}$	3 [÷] 4 [×] 1 [÷] 8 [=]	0.09375
$8 \div \frac{3}{4} = 8 \times \frac{4}{3} = \frac{32}{3} = 10\frac{2}{3}$	8 [×] 4 [÷] 3 [=]	10.66666667
$\frac{3}{4} \div \frac{1}{3} = \frac{3}{4} \times \frac{3}{1} = \frac{9}{4} = 2\frac{1}{4}$	3 [÷] 4 [×] 3 [=]	2.25

Figure A.7 Calculator procedures for fractions.

Example	Operation	Readout
$3^2 = 3 \times 3 = 9$	3 [x^2]	9.
$5^5 = 5 \times 5 \times 5 \times 5 \times 5 = 3125$	5 [y^x] 5 [=]	3125.
$\sqrt{64} = 8$	64 [$\sqrt{\ }$]	8.
$64^{1/2} = 64^{0.5} = 8$	64 [y^x] [•] 5 [=]	8.
$\sqrt[4]{16} = 2$	16 [2nd F] [$\sqrt[x]{y}$] 4 [=]	2.

Figure A.8 Calculator procedures for powers and roots.

Example	Operation	Readout
Percent: $\frac{23}{50} \times 100 = 46\%$	[2ndF] [Δ%]	46.
Percent of: $200 \times 15\% = 30$	200 [×] 15 [2ndF] [Δ%]	30.
Percent increase: An increase from 10 volts to 12 volts $\frac{12-10}{10} \times 100 = 20\%$ increase	12 [−] 10 [2ndF] [Δ%]	20.

Figure A.9 Calculator procedures for percentage.

Amperes to microamperes ($1\ A = 10^6\ \mu A$)

Example	Operation	Readout
$0.0008\ A \times 1{,}000{,}000\ \mu A/A = 800\ \mu A$	[•] 0008 [EXP] 6 [=]	800.

Volts to millivolts ($1\ V = 10^3\ mV$)

Example	Operation	Readout
$0.75\ V \times 10^3\ mV/V = 750\ mV$	[•] 75 [EXP] 3 [=]	750.

Watts to megawatts (1 W = 0.000001 mW)

Example	Operation	Readout
$\frac{50{,}000{,}000\ W}{10^6\ W/mW} = 50\ mW$	50000000 [÷] [EXP] 6 [=]	50.

Figure A.10 Calculator procedures for conversion of electrical units.

Example	Operation	Readout
$0.0000085 = 8.5 \times 10^{-6}$	[•] 0000085 [=] [F↔E]	8.5-06
$52{,}000{,}000 = 5.2 \times 10^7$	52000000 [=] [F↔E]	5.2 07
$4.698 \times 10^5 = 469{,}800$	4 [•] 698 [EXP] 5 [=]	469800.
$2.32 \times 10^{-4} = 0.000232$	2 [•] 32 [EXP] 4 [±] [=]	0.000232
$(4.8 \times 10^3) \times (4.5 \times 10^{-5}) = 0.216$	4 [•] 8 [EXP] 3 [×] 4 [•] 5 [EXP] 5 [±] [=]	0.216
$(9.5 \times 10^4) \div (8 \times 10^{-4}) = 118{,}750{,}000$	9 [•] 5 [EXP] 4 [÷] 8 [EXP] 4 [±] [=]	118750000
(Answer to scientific notation)	118750000	1.1875 08

Figure A.11 Calculator procedures for scientific notation.

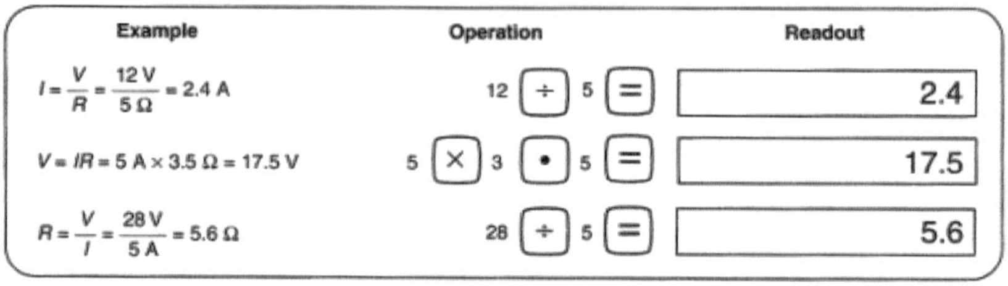

Figure A.12 Calculator procedures for Ohm's law.

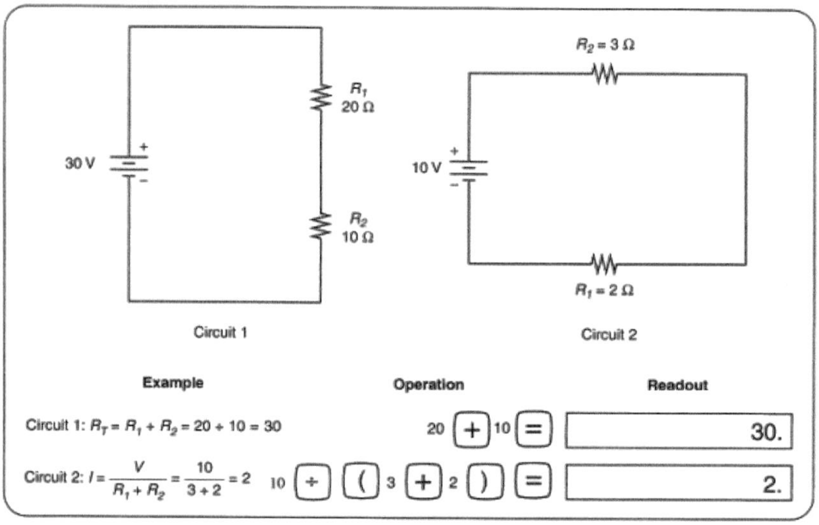

Figure A.13 Calculator procedures for series circuits.

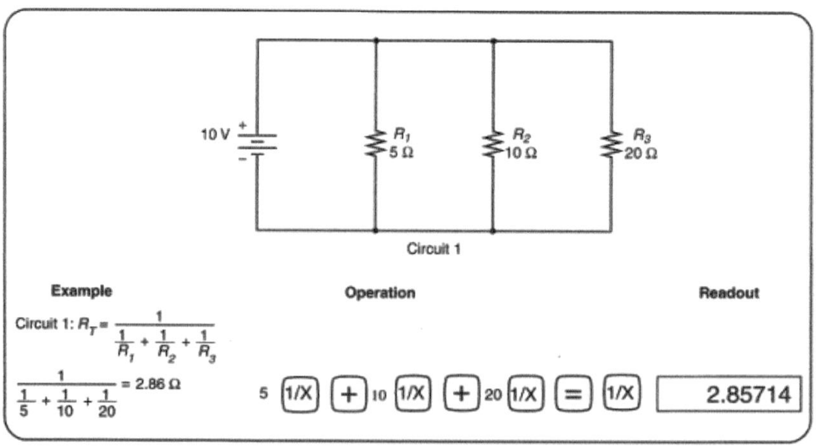

Figure A.14 Calculator procedures for parallel circuits.

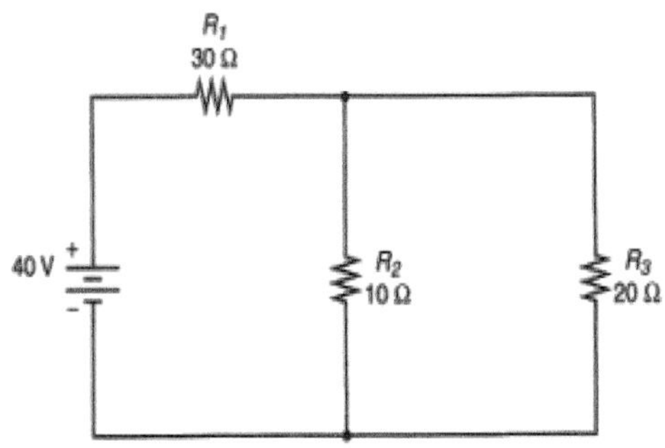

Circuit 1

Example	Operation	Readout
Circuit 1: $R_T = \dfrac{1}{\dfrac{1}{R_2} + \dfrac{1}{R_3}} + R_1$		
$\dfrac{1}{\dfrac{1}{10} + \dfrac{1}{20}} + 30 = 36.67$	10 [1/X] + 20 [1/X] [=] [1/X] [+] 30 [=]	36.6667

Note: In more complicated combination circuits, it may be necessary to solve each parallel portion with parentheses so that reciprocal operations will apply only to that part of the circuit. In the example above, if the order of the expression were reversed, it would be necessary to use parentheses.

$30 + \dfrac{1}{\dfrac{1}{20} + \dfrac{1}{10}} = 36.67$	30 + 1 [÷] [(] 10 [1/X] [+] 20 [1/X] [)] [=]	36.6667

Figure A.15 Calculator procedures for combination circuits.

Example	Operation	Readout
$P = I^2 R = (3\ A)^2 \times 10\ \Omega = 90\ W$	3 [X²] [×] 10 [=]	90.
$P = \dfrac{V^2}{R} = \dfrac{(30\ V)^2}{10\ \Omega} = 90\ W$	30 [X²] [÷] 10 [=]	90.

Figure A.16 Calculator procedures for electric power.

Sines, cosines, tangents, cotangents, secants, cosecants

Example	Operation	Readout
sin 30° = 0.5	30 [sin]	deg 0.5
tan 40° = 0.84	40 [tan]	deg 0.839
cos 150° = –0.866	150 [cos]	deg –0.866
$\cos^{-1} 0.5 = 60°$	[•] 5 [$\cos^{-1}$]	deg 60.
$\sin^{-1} 0.5 = -30°$	[•] 5 [$\sin^{-1}$]	deg –30.
sin1 rad = 0.84	1 [sin]	rad 0.84
$\sin \frac{\pi}{4}$ rad = 0.707	π [÷] 4 [=] [sin]	rad 0.707
$\tan^{-1} 2.5 = 1.19$ rad	2 [•] 5 [$\tan^{-1}$]	rad 1.19
20 × sin 30° = 10	20 [×] 30 [sin] [=]	deg 10.

Figure A.17 Calculator procedures for trigonometric functions.

Capacitive reactance $X_c = \frac{1}{2\pi fc}$

Example	Operation	Readout
$X_c = \frac{1}{2\pi\, 60\, (100 \times 10^{-6})}$	2 [×] [π] [×] 60 [×] 100 [exp] 6 [±] [=] [1/X]	26.53

Impedance $Z = \sqrt{R^2 + X_c^2}$

Example	Operation	Readout
$Z = \sqrt{25^2 + 26.53^2}$	26.53 [x^2] [+] 25 [x^2] [=] [√]	36.45

Note: It is convenient here to proceed from the previous calculation without clearing the calculator.

Total current $I_T = \frac{V}{Z}$

Example	Operation	Readout
$I_T = \frac{10}{36.45}$	36.45 [1/X] [×] 10 [=]	0.274

Resistance voltage $V_R = I_T R$

Example	Operation	Readout
$V_R = 0.274 \times 25$	0.274 [×] 25 [=]	6.85

Voltage across capacitor $V_C = I_T X_C$

Example	Operation	Readout
$V_C = 0.274 \times 26.53$	[•] 274 [×] 26 [•] 53 [=]	7.27

Circuit phase angle $\theta = \sin^{-1} \frac{X_C}{Z}$

Example	Operation	Readout
$\theta = \sin^{-1} \frac{26.53}{36.45}$	26 [•] 53 [÷] 36 [•] 45 [=] [$\sin^{-1}$]	46.7

Figure A.18 Calculator procedures for series RC circuits.

Appendix B

Right Triangles and Trigonometry

Right triangles and trigonometry are very valuable in the study of electricity and electronics. Trigonometry, or "trig," is a form of mathematics that deals with angles and triangles, particularly the right triangle. A right triangle has one angle that is 90 °. An example of a right triangle is shown in Figure B.1. This example illustrates how resistance, capacitive reactance, and impedance are related in an AC series RC circuit. Resistance (*R*) and capacitive reactance (X°) are 90 ° apart. Their angle of intersection forms a right angle. The law of right triangles, known as the Pythagorean theorem, can be used to solve circuit problems. The theorem states that "in a right triangle, the square of the hypotenuse is equal to the sum of the squares of the other two sides." With reference to Figure B.1, the Pythagorean theorem is used as $Z^2 = R^2 + XC$ or $Z = \sqrt{R^2 + XC_2}$.

By using trig relationships, problems dealing with phase angles, power factor, and reactive power in AC circuits can be solved. The three most used trig functions are the sine, the cosine, and the tangent. These functions show the ratios of the sides of a triangle. They determine the size of the angle. Figure B.2 shows how these ratios are expressed. Their values are found with a calculator. The process can be reversed to find the size of an angle when the value of the sides are known. The term "inverse" is used for this process. For example, inverse sine 0.5 = 30 ° means that 30 ° is the angle whose sine is 0.5. When given inverse sine 0.866 = 60 °, find that the angle whose sine is 0.866 is equal to 60 ° (See Figure A.17).

Trig ratios hold true for angles of any size. Angles in the first quadrant of a standard graph are from 0 ° to 90 °. They are used as a reference. All first-quadrant angles have positive values. Angles in the second, third, and fourth quadrants have two negative values and one positive value. See Figure B.3.

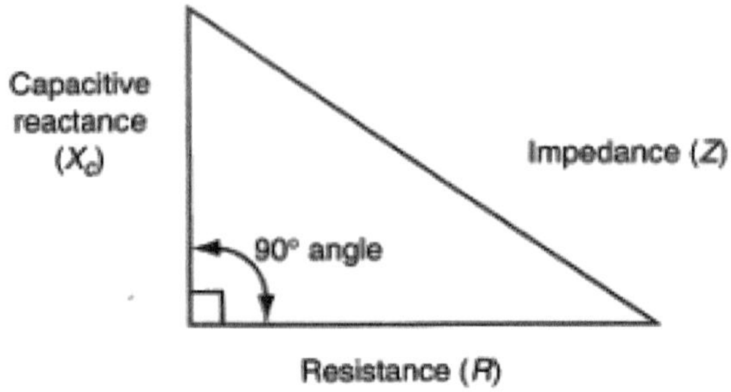

Figure B.1 Right triangle for an AC series RC circuit.

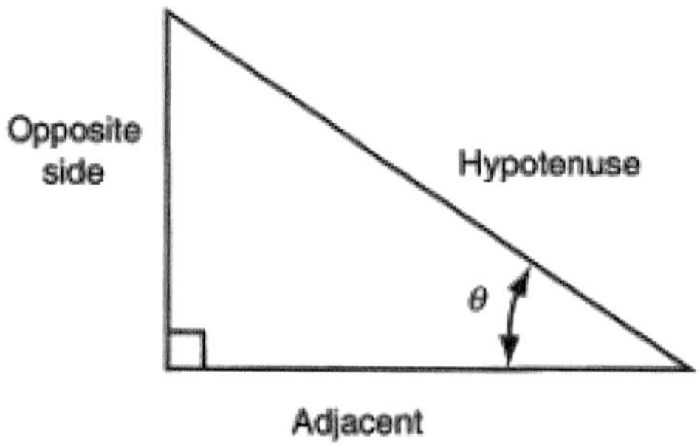

$$\text{sine } \theta = \frac{\text{opposite side}}{\text{hypotenuse}}$$

$$\text{cosine } \theta = \frac{\text{adjacent side}}{\text{hypotenuse}}$$

$$\text{tangent } \theta = \frac{\text{opposite side}}{\text{adjacent side}}$$

Figure B.2 Sine, cosine, and tangent ratios.

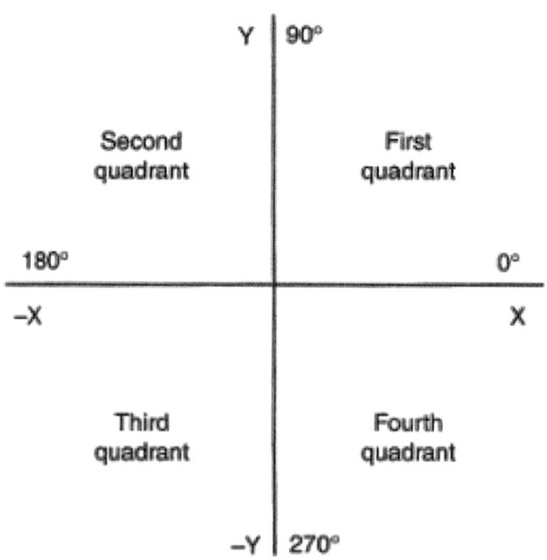

Second quadrant (90° to 180°)

$\sin\theta = \sin(180° - \theta)$

$\cos\theta = -\cos(180° - \theta)$

$\tan\theta = -\tan(180° - \theta)$

Third quadrant (180° to 270°)

$\sin\theta = -\sin(\theta - 180°)$

$\cos\theta = -\cos(\theta - 180°)$

$\tan\theta = \tan(\theta - 180°)$

Fourth quadrant (270° to 360°)

$\sin\theta = -\sin(360° - \theta)$

$\cos\theta = \cos(360° - \theta)$

$\tan\theta = -\tan(360° - \theta)$

Figure B.3 Standard graph and the method used to find sine, cosine, and tangent values.

Appendix C

Basic Electric Symbols

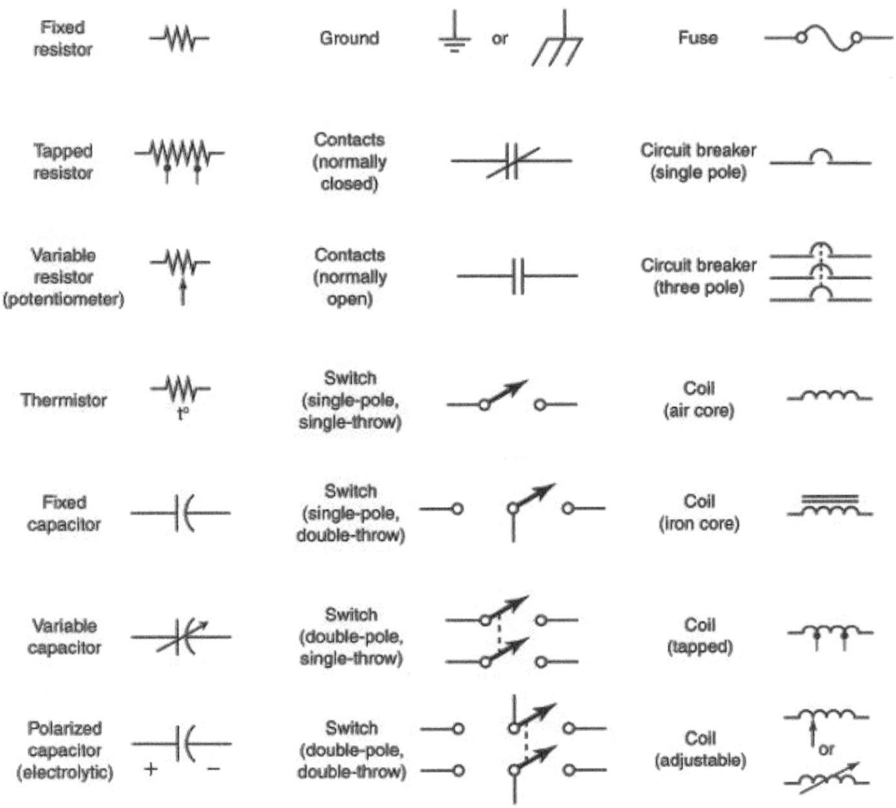

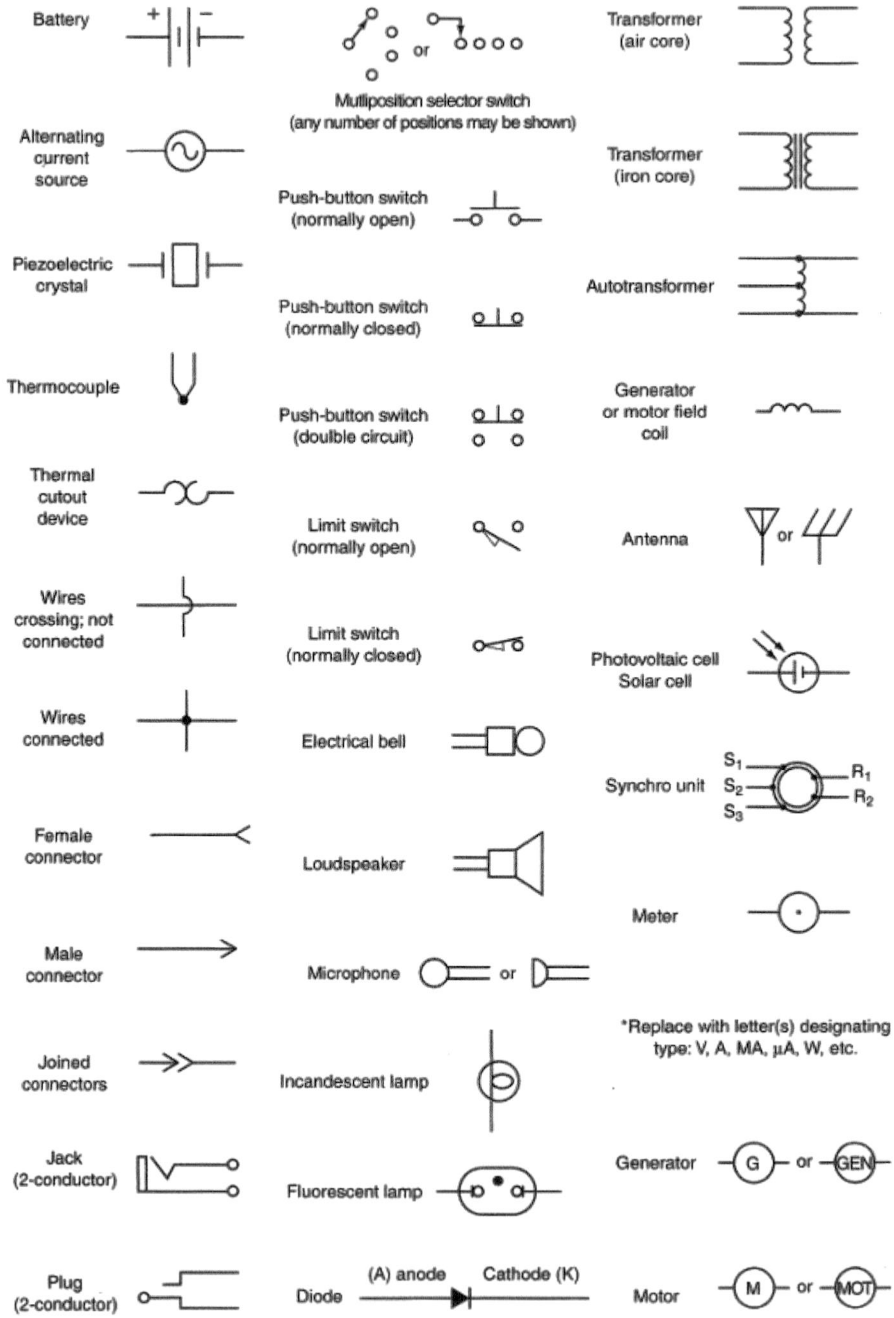
Battery
+
−
Alternating current source
Piezoelectric crystal
Thermocouple
Thermal cutout device
Wires crossing; not connected
Wires connected
Female connector
Male connector
Joined connectors
Jack (2-conductor)
Plug (2-conductor)
or
Mutliposition selector switch (any number of positions may be shown)
Push-button switch (normally open)
Push-button switch (normally closed)
Push-button switch (doulble circuit)
Limit switch (normally open)
Limit switch (normally closed)
Electrical bell
Loudspeaker
Microphone
or
Incandescent lamp
Fluorescent lamp
Diode
(A) anode
Cathode (K)
Transformer (air core)
Transformer (iron core)
Autotransformer
Generator or motor field coil
Antenna
or
Photovoltaic cell Solar cell
Synchro unit
S1
S2
S3
R1
R2
Meter
*Replace with letter(s) designating type: V, A, MA, µA, W, etc.
Generator
G
or
GEN
Motor
M
or
MOT

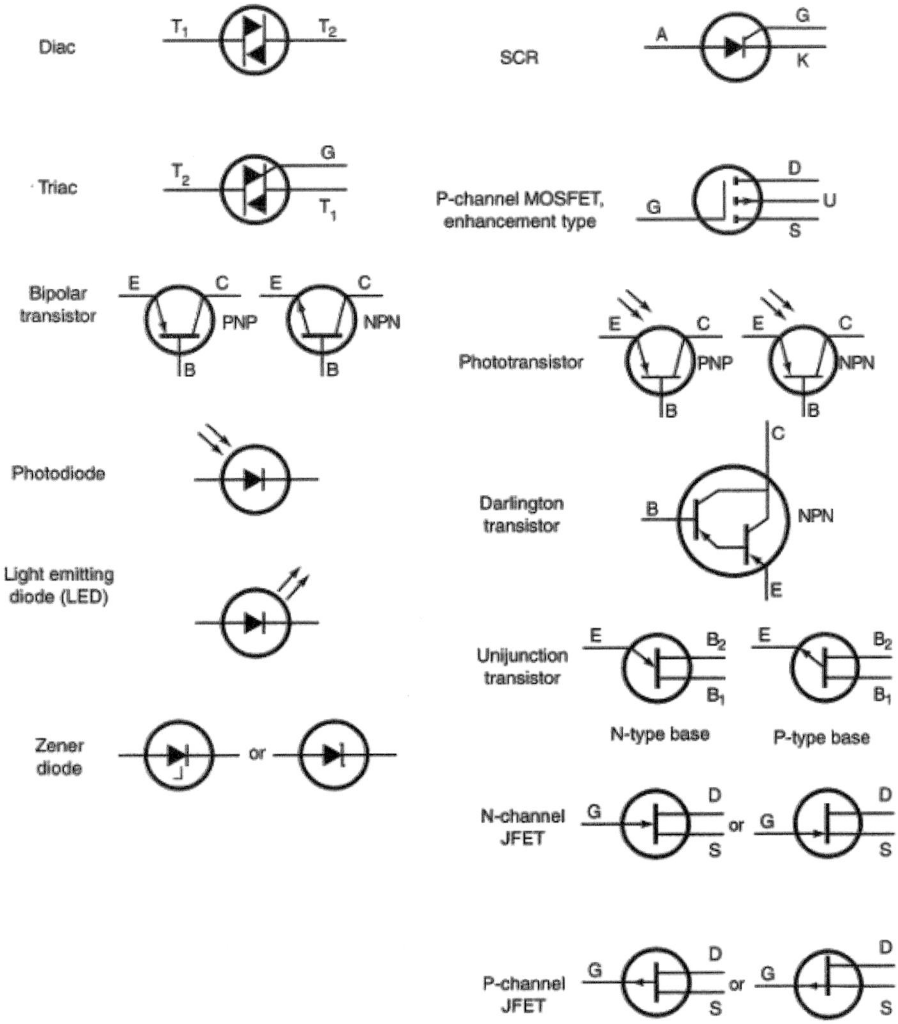

Diac
T1
T2
SCR
A
G
K
Triac
T2
G
T1
P-channel MOSFET, enhancement type
G
D
U
S
Bipolar transistor
E
C
PNP
B
E
C
NPN
B
Phototransistor
E
C
PNP
B
E
C
NPN
B
Photodiode
Darlington transistor
B
C
NPN
E
Light emitting diode (LED)
Unijunction transistor
E
B2
B1
N-type base
E
B2
B1
P-type base
Zener diode
or
N-channel JFET
G
D
S
or
G
D
S
P-channel JFET
G
D
S
or
G
D
S

N-channel MOSFET,
depletion type

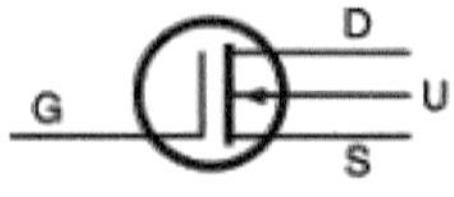

N-channel MOSFET,
enhancement type

P-channel MOSFET,
depletion type

Operational
amplifier

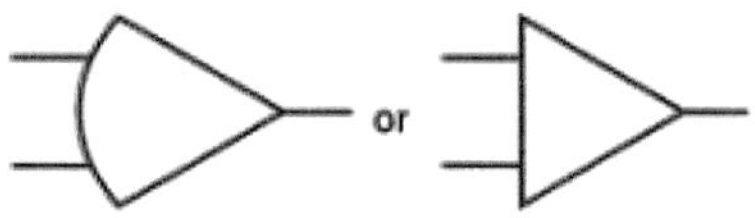

AND
function

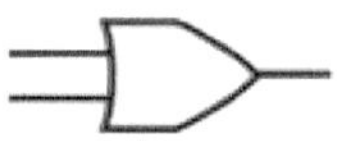

OR
function

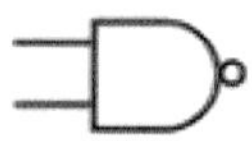

NAND
function

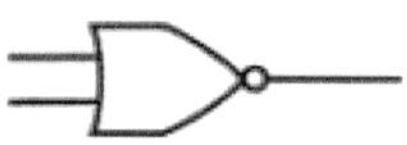

NOR
function

Exclusive
OR
function

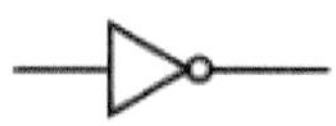

Inverter
(NOT)
function

Flip-flop
(general)

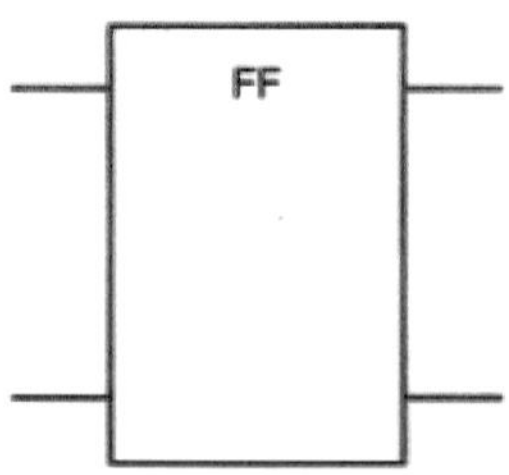

Glossary

A

accumulator: Part of a computer that temporarily stores operands that are to be processed by the ALU. The output of the ALU is also stored in the accumulator after processing.

across-the-line starter: Magnetic contactor motor starting circuit.

active power: In-phase (resistive) component of power that results in power conversion in a circuit. Active power is the true power of the circuit and is measured in watts.

actuating speed: Time it takes for the load to be energized after the control signal has been applied.

actuator: (1) Motor or valve that converts power into robot movement. (2) Part of a limit switch that causes the internal NO or NC contacts to change state.

address: Spot in a computer's memory where certain information is located. With the appropriate command, data can be removed from memory and distributed as output.

address register: Part of a computer that temporarily stores the address of a memory location that is to be accessed.

air-break disconnect switch: Switch that provides a means of switching power lines and disconnecting them for repair.

Allen wrench: Fastening tool commonly used with socket-headed fasteners.

alloy: Mixture of materials.

American National Standards Institute (ANSI): Well-known organization that publishes standards for industry. The National Electrical Code is an American National Standards Institute standard.

American Wire Gage (AWG): System of measurement for round conductors. Wire sizes in the AWG system range from No. 40 (smallest) to No. 0000 (largest). The cross-sectional area of a conductor doubles with each increase of three sizes. The diameter doubles with every increase of six sizes.

ammeter: Electrical instrument used to measure current.

ampacity: Measure of the ability of a conductor to carry electrical current.

ampere (A): Quantity of current equal to one coulomb passing a point on a conductor in one second.

ampere-turn (At or NI): Measure of magnetomotive force. Quantity equal to the current in amperes multiplied by the number of turns of wire ($I \times N$).

analog meter: Meter that gives a reading using a needle pointer on a scale.

analog timer: Timer that responds to physical quantities or changes.

AND gate: Logic gate that requires all inputs to be 1 for an output of 1.

angstrom (Å): Length equal to one-tenth of a nanometer.

angular actuator: Actuator that provides rotation.

anode: Electrode that attracts or gathers in electrons.

anthropomorphic: Humanlike in form.

apparent power: Voltage reading multiplied by the current reading.

arithmetic logic unit (ALU): Part of a computer that acts as a calculator and that performs mathematical and logic operations. It combines the input from the data register and the accumulator. Addition, subtraction, and logic comparisons using binary numbers are its primary operations.

armature reaction: Magnetic resistance to the armature created when the armature passes through the magnetic field.

assembler: Program that automatically translates a program's mnemonics into a form the computer can read.

assembly language: Machine-oriented instructional set for a specific microprocessor.

assignment sheet: Sheet that describes maintenance tasks that must be performed on a regular basis. The sheet identifies the unit to be maintained, the maintenance intervals, and the type of action required.

astable multivibrator: Multivibrator in which each active device alternately conducts and is cut off for periods of time without the use of external triggers.

autotransformer: Transformer that has only one winding, with a common connection between the primary and secondary.

autotransformer starting: Motor starting method that employs one or more autotransformers to control the voltage that is applied to a motor. The autotransformers used are ordinarily tapped to provide a range of starting-current control. When the motor has accelerated to near its normal operating speed, the autotransformer windings are removed from the circuit.

average voltage: Voltage equal to the mathematical average of all instantaneous voltages that occur at each period of time throughout an alternation. The average value is equal to 0.636 times the peak value for one-half cycle (alternation) AC voltage.

B

barrier potential: Potential created across a depletion zone at a PN junction.

base: (1) Element of a bipolar junction transistor sandwiched between the emitter and the collector. Base current controls collector current. (2) Number of digits used in a counting system.

B–H curve: Curve that shows the relationship between a magnetizing force and flux density. Also called a magnetization curve.

bias voltage: An external source of energy applied to a PN junction. Also called biasing.

biasing: An external source of energy applied to a PN junction. Also called bias voltage.

bifilar: Winding construction where two separate wires are wound into the coil slots at the same time. The two wires are small in size, allowing more turns than could be achieved with a larger wire.

bimetal-strip timer: Electrical part of the timer that employs two self-supporting metal strips with electrical contacts. When electrical energy is applied to the heating element, it causes the bimetal strip to bend or flex. The side of the strip near the heating element expands faster than the contact side. As a result, the bending action causes the contacts to be forced together after a short delay period.

binary: Number system uses 2 as its base. The largest value that can be expressed in a specific place is the number 1. As a result, only the numbers 0 and 1 are used.

binary-coded-decimal (BCD): Number system in which a base 10 number is first divided into digits according to place value, and then each digit then is converted to a four-digit binary number.

binary counter: Flip-flops connected so that the Q output of the first circuit drives the trigger, or clock input, of the next circuit.

bipolar: Types of semiconductor devices that conduct with holes and electrons.

bipolar junction transistor (BJT): Three-element semiconductor device that has two PN junctions.

bistable: Any electronic device that can be set in one of two stable operational states by an outside signal.

bit: Contraction of the phrase binary digit. A single pulse of information.

block diagram: Diagram that shows the functions of the subparts of any electrical system.

branch circuit: Conductors that extend beyond the last overcurrent protective equipment of the power system.

breakdown voltage: Reverse bias voltage at which there is a sharp increase in reverse current in a zener diode. Also called zener voltage.

brush: Spring-mounted conductive material used for making contact with slip rings in AC generators or commutators of DC generators.

brushless DC motor: Motor that has neither brushes nor commutator assemblies. Solid-state switching circuits are used.

bus: Distribution path called in a computer.

bus network: Series of registers connected together in a computer.

BX cable: Common type of metal-clad cable.

byte: Group of eight bits.

C

cabling diagram: Wiring diagram that shows the actual location of parts and wires on equipment. Also called wiring diagram.

cache: Localized block of read/write memory that is internally accessible by the chip in which it resides.

candela (cd): Unit of luminous intensity of a standard light source.

capacitance (*C*): Property that allows the storage of an electric charge. The unit of capacitance is the farad (F).

capacitive reactance (X_C): Opposition to the flow of alternating current caused by the electrostatic field that is developed around a capacitor. It is measured in ohms.

capacitor: Devices that have the capability of storing an electrical charge.

capacitor analyzer: Instrument that checks the value of the capacitor and measures the leakage current under the normal working voltage.

capacitor-start, capacitor-run motor: Motor that employs two capacitors. One, of higher value, is in series with the start winding and centrifugal switch. The other, of lower value, is in series with the start winding and remains in the circuit during operation. The larger capacitor is used only to increase starting torque. It is removed from the circuit by the centrifugal switch during normal operation. The smaller capacitor and the entire start winding are part of the operational circuit of the motor. The smaller capacitor helps to produce a more constant running torque as well as quieter operation and improved power factor. Also called two-value capacitor motor.

capacitor-start single-phase induction motor: Motor similar to a split-phase motor except that a capacitor is placed in series with the start winding. The capacitor causes the current in the start winding to lead (rather than lag) the applied voltage.

Cartesian configuration: Arm movement that can be described by three intersecting perpendicular straight lines, referred to as the *X*, *Y*, and *Z* axes.

cathode: Electrode that gives off or emits electrons.

channel: Area in a junction field-effect transistor where electrons flow between the source and drain.

chisel: Tool used with a hammer to cut or chip. Cold chisels are used to cut metal. Wood chisels are made of a softer metal and should only be used for cutting wood.

circuit breaker: Safety device that springs open when too high a current passes through it. Can be turned on or off like a wall switch.

circular mils (cmil): Unit of measure for the area of conductors. Sizes larger than No. 0000 AWG are expressed in thousand circular mil (MCM) units.

closed-loop system: Control system that contains a self-correcting mechanism using information called feedback.

cold solder joint: Bad solder connection.

collector: One of the three elements of a bipolar junction transistor. Collector current is controlled by base current.

commutator assembly: Device used with motors and generators to produce pulsating DC from a sine wave.

comparator: (1) Op-amp that changes states when one of its inputs exceeds the reference voltage. (2) Device that compares a feedback signal to a reference signal.

compass: Object that detects the presence of magnetic fields.

component: Piece of an electrical circuit.

compound-wound DC motor: Motor with two sets of field windings, one set in series with the armature and one in parallel. This motor combines the desirable characteristics of the series and shunt-wound motors.

compressed-air circuit breaker: Circuit breaker in which a stream of compressed air is concentrated on the contacts when the power line is opened. The compressed air aids in extinguishing the arc that is developed when the contacts open.

computer: General term that is used to describe a device that performs automatic arithmetic computations when provided with information. A computer can choose, copy, move, compare, and perform other nonarithmetic operations on alphabetic, numeric, and other symbols that are used to represent different things.

conductor: Material through which current flows.

conduit: Metal tube used in routing cables and wires throughout an industrial or commercial building.

conduit bender: Tool used to bend electrical conduit around corners.

contact current: Maximum current that can safely flow through a contact circuit.

continuity check: Check that is made to see if a circuit is open or closed (a continuous path).

control: Achieved when a system is turned on or off. Control of this type can take place anywhere between the source and the load device. Also called *full control*.

control center: (1) Grouping of motor starters and associated control equipment. (2) Center that provides a means of data collection and recording, system monitoring, frequency control, and signaling.

controller: Part of a robot that coordinates all movements of the mechanical system. It also receives input from the immediate environment through various kinds of sensors.

conventional current flow: System in which an electrical charge is considered to move from a high charge (positive) to a low charge (negative).

convergence: Directing light waves into a single or common point.

copper loss: Loss caused by the resistance in a conductor. Also called *IR* loss.

coulomb (C): The unit of measurement of electrical current. One coulomb contains 6,280,000,000,000,000,000 electrons (6.28×10^{18} in scientific notation).

counterelectromotive force (CEMF): Property of a coil to oppose any change in the instantaneous building of a magnetic field. Also opposes a change in current.

curling: Condition where the corners of a contact material separate from the backing plate.

current: The movement of electrical charges from one point to another. Electrical current is produced when electrons are removed from their atoms. Represented by the letter *I*.

current transformer: Transformer used to reduce a large value of line current to a smaller value for measurement or control purposes.

cylindrical configuration: Robot configuration that consists of two orthogonal (at a 90-degree angle) slides mounted on a rotary axis. Reach is accomplished as the arm of the robot moves in and out. For vertical movement, the carriage moves up and down on a stationary post, or the post can move up and down in the base of the robot. Movement along the three axes traces points on a cylinder.

D

damper winding: Winding that is similar to the conductors of a squirrel-cage rotor. This winding is placed within the laminated iron of the rotor assembly.

data register: Part of a computer that temporarily stores information applied to the data bus.

DC stepper motor: Specialized electrical machine that is used to control automatic industrial equipment.

decade counter: Digital counter that counts in the decimal (base 10) system.

decimal point: Reference point used in the decimal system that gives each digit its position.

degrees of freedom: Term used to describe a robot's dexterity, or freedom of motion. For each degree of freedom, a joint is required. A robot requires six degrees of freedom to be completely versatile.

delay timer: Device used to alter the starting time of a load device or controlled circuit.

delta configuration: Method of winding connection in a three-phase circuit.

depletion mode: One type of MOSFET that responds to the depletion of current carriers in its channel.

depletion zone: Area around a semiconductor junction that is void of majority current carriers. The zone is created by diffusion of current carriers across the junction.

detent torque: Torque that holds the rotor in a position where a pair of north and south pole rotor teeth are aligned with two of the stator poles.

diac: Special diode that can be triggered into conduction by voltage. This device is classified as a bidirectional trigger diode. It can be triggered into conduction in either direction. This device is used primarily to control the gate current of a triac.

diagram: Drawing that shows how components are connected together in a circuit.

dielectric: Insulating material that separates two conductive materials (plates) in a capacitor.

differential compound DC motor: One method of connecting compound-wound DC motors. This method has series and shunt fields that oppose each other.

diffusion: Movement of electrons from an N material across the junction to fill holes in the *P* material.

digital meter: Meter that gives its reading in a digital format.

digital multimeter (DMM): Meter that has a numerical readout.

digital storage oscilloscope (DSO): Type of oscilloscope that can store a waveform and sample it for future analysis.

digital timer: Timer designed to count pulses generated by an electronic clock circuit.
digitizing: Process of sampling an analog signal and assign each sample a number.
diode: Semiconductor device that allows current to flow in only one direction.
direct-drive electric motor: High-torque motor that drives a device directly without the need for reducer gears.
distribution substation: Substation used to receive electrical power from a high-voltage system and convert it to voltage levels suitable for industrial, commercial, or residential use.
divergence: Spreading a light wave out.
DMM: Digital multimeter.
domain theory: Theory that deals with the alignment of domains in materials rather than molecules. A domain is a group of atoms (about 10^{15} atoms). Each domain acts like a tiny magnet.
doping: Adding impurities to a semiconductor material.
drain: One lead on a FET. Current is controlled by the gate. Similar to the collector on a BJT.
drill: Tool used for boring holes.
dropout current: Minimum current that will keep a relay energized.
drum controller switch: Special-purpose switch that ordinarily is used to control a large motor.
duct: Metal channel used in routing cables and wires throughout an industrial or commercial building.
dynamic braking: Breaking method used to bring a motor to a quick stop whenever power is turned off. Motors with wound armatures utilize a resistance connected across the armature as a dynamic braking method.
dynamometer: Instrument that is used to load a motor and measure its power output directly.
dynamotor: Device that converts a DC voltage of one value to a DC voltage of another value. Each time the stator winding polarity is changed by the input logic circuitry, the motor shaft rotates a specific amount.

E

eddy current: Small current induced into the core material of transformers.
effective value: The AC voltage that will do the same amount of work as a DC voltage of the same value.
efficiency: Trait of electrical equipment that expresses the power in divided by the power out.

electrical metallic tubing (EMT): Pipe conduit that can be bent with a special conduit-bending tool. Also called *thinwall conduit.*

electromagnet: Coil that has an iron or steel core inside.

electromagnetic induction: Principle that causes current to flow in a conductor placed inside a magnetic field when there is motion between the conductor and the magnetic field.

electromagnetism: Magnetism that is produced due to the flow of an electrical current.

electromechanical timer: Classification of timing devices that refers to those driven by a motor mechanism. These low-cost timing devices are commonly used to achieve delay, interval, repeat cycle, and reset functions with powerline frequency accuracy.

electromotive force (EMF): Electrical pressure or voltage.

electron current flow: System in which the direction of electron movement is from negative to positive through a circuit.

electronic timer: Broad classification of timers that are used to describe a number of devices employing electronic components.

emitter: One of the three elements of a bipolar junction transistor. The emitter provides current to both the base and collector.

end effector: Device at the end of a robot's arm (hand or tool).

energy: Capacity to do work.

enhancement mode: One type of MOSFET that does not have an interconnecting channel. Current carriers pulled from the substrate form an induced channel. Gate voltage controls the size of the induced channel.

equipment ground: Ground that places operating equipment at ground potential. The conductor that is used for this purpose is either bare wire or a green insulated wire.

expulsion-type fuse: Fuse with an element that will melt and vaporize when it is overloaded. This causes the power line it is connected in series with to open.

F

farad (F): Unit of capacitance. A capacitance of one farad results when a potential of one volt causes an electrical charge of one coulomb to accumulate on a capacitor.

Faraday's law: Law that states that when a coil of wire moves across the lines of force of a magnetic field, electrons flow through the wire in one direction. When the coil of wire moves across the magnetic lines of force in the opposite direction, electrons flow through the wire in the opposite direction.

feedback: Information from the output of a system that is sent back to the input in a closed-loop system.

feeder line: Line that supplies power to branch lines that are connected to it.

feeler gauge: Instrument used in measuring wear allowance.

firmware: Programs that are placed in read-only memory.

fish tape: Tool used to pull wire through electrical conduit.

flange mounting: Method used to attach a motor directly to a machine.

flashover: Arcing between power lines.

flip-flop: Memory device used in digital circuits. It is often the basic logic element for counting, temporary memory, and sequential switching operations. It can be used to hold an output state even when the input is completely removed and changes output based on an appropriate input signal.

float switch: Switch used when it is necessary to control the level of a liquid. Some designs have an operating lever connected to a rod and float assembly. The float assembly is placed into a tank of liquid where the motion of the liquid controls the movement of the operating lever of the float switch. Also called a liquid level switch.

flux density (B): Number of lines of force per unit area of a material. The unit of flux density is the gauss per square centimeter of area.

foot switch: Switch that is controlled by a foot pedal. This type of switch is used for applications where a machine operator has to use both hands during the operation of the machine.

forcer: Moving component in a linear stepper motor.

forward biasing: Applying an external voltage that reduces the barrier potential and causes current carriers to return to the depletion zone.

fractional-horsepower motor: Motor that is less than one hp.

frequency converter: Device used to change an incoming frequency into some other frequency. Frequency converters are motor-generator sets that are connected together.

full-voltage starting: Motor starting method in which full power is applied to the motor initially. Maximum starting torque and minimum acceleration time result with this starting circuit.

function-select switch: Switch on a meter that changes the size or type of measurement to be taken.

fuse: Safety device that blows open if too high a current passes through it.

G

gate: One lead on a FET. Controls current through the drain. Similar to the base on a BJT.

gear-reduction motor: Motor used where reduced speed of operation is required. Most gear-reduction motors have either horizontal shafts or shafts at a right angle (90°) to the motor frame.

general-purpose relay: Relay used for low-power applications. They are relatively inexpensive and small in size.

generator action: Electrical energy produced by the turning of a motor/generator.

GFCI circuit breaker: Combination of ground-fault protection and circuit interruption in the same overcurrent and short-circuit protective equipment. A GFCI circuit breaker fits the same space required by a standard circuit breaker. It provides the same branch-circuit wiring protection as the standard circuit breaker, as well as ground-fault protection. The GFCI sensing system continuously monitors the current balance in the ungrounded (hot) conductor and the grounded (neutral) conductor.

GFCI receptacle: Receptacle that provides ground-fault protection on 120-, 208-, or 240-volt AC systems.

ground-current relaying: Method used to protect equipment from ground faults.

ground-fault circuit interrupter (GFCI): Receptacle device that immediately cuts off power if a shock occurs. Used for bathroom and outdoor power receptacles. These devices are also called ground-fault interrupters (GFIs).

grounding electrode: Actual ground connection is made by welding, brazing, or bolting a conductor to a metal rod or bar. This rod is then physically placed in the earth.

H

half stepping: Sequence in which a motor shaft moves half of its normal step angle for each input pulse applied to the stator.

Hall effect: When a magnetic field is placed at right angles to a current-carrying conductor, a small voltage is produced across the conductor.

hammer: Tool used to drive and remove nails, as well as for striking chisels and punches.

heater: Bimetallic or melting-alloy overload relay in series with motor branch-circuit power lines.

heat sink: Tool that is clamped to a wire between the joint and the device being soldered. A heat sink absorbs heat and protects delicate devices.

henry (H): Unit of measurement for inductance. A circuit has a one-henry inductance if a current changing at a rate of one ampere per second produces an induced counterelectromotive force of one volt.

heterochromatic: Source of radiation that produces a very limited number of different colors.

hierarchical control: Arrangement in which a given level is lower (more elemental) than the one above it and is dependent on the level above it for its instructions.

holding torque: Amount of torque in ounceinches (oz-in.) required to move the motor shaft of a DC stepper motor from a zero revolutions per minute condition.

hole: Space in a semiconductor that is looking for an electron.

hybrid motor: Motor that combines the operating principles of permanent magnet and variable reluctance motors.

hydraulic drive system: Drive system that consists of an electric pump connected to a reservoir tank, control valves, and a hydraulic actuator.

I

image acquisition: Illuminating a workpiece and digitally scanning its image.

image analysis: Process that involves gathering information from the image that will enable a robot to do a task. Analysis is performed by computer software. Using algorithms (special codes), the software identifies and measures features of the digital image. After all of the features are analyzed, the information is interpreted.

image preprocessing: Process in which an analog-to-digital (AD) converter is used to change the analog signal into an equivalent digital signal. The digital signal represents light intensity values over the entire image. These values are stored in memory, permitting the digital image to be analyzed and interpreted.

impedance (Z): Total opposition of an AC circuit to the flow of current. It is measured

in ohms. in-phase: Relationship that exists when the minimum and maximum values of both voltage and current occur at the same time interval.

inching: Momentary operation of a motor to provide small movements of its shaft. Also called jogging.

indicator: Device designed to display certain operating conditions at various points throughout the system.

induced pole: Magnetic pole formed due to the influence of the flux lines.
inductance (*L*): Opposition to a change in current in a circuit. The unit of measurement for inductance is the henry (H).
inductance analyzer: Instrument that measures the inductance of windings.
inductive reactance (X_v): Opposition to current created by inductance when AC is applied. It is measured in ohms.
inductor: An electrical device that produces inductance. A coil of wire.
infrared detector: Device that responds to radiation in the infrared region of the electromagnetic spectrum.
inherent-mode instruction: One-byte instruction that is designed to send data to the accumulator registers of the ALU.
input port: Port at which a robot or computer receives information. Institute of Electrical and Electronics Engineers
(IEEE): Well-known organization that publishes standards for industry.
instruction decoder: Part of a computer that examines a coded word and decides which operation is to be performed by the ALU.
insulator: Material that does not allow electrical current to flow easily.
integral horsepower motor: Motor that is larger than one hp.
interface: Common point at which two or more systems communicate with each other.
interruptible system: Optoelectronic system that includes a source of electromagnetic radiation and a detector or receiver.
interval timer: Timing device used in applications where the load is turned on at the start of an operational cycle and kept on for a specific time period. At the end of this interval, the load is turned off automatically.
inverter: Logic gate that inverts the input signal. Also called a NOT gate.
iron loss: Loss caused by eddy currents in the metallic core material of a transformer.
isolation: Expression of a breakdown voltage that separates an input and load.
I^2R: loss: Loss caused by the resistance in a conductor. Also called copper loss.

J

jogging: Momentary operation of a motor to provide small movements of its shaft. Also called inching.
joint: Part of a robot that connects two segments together and allows them to move relative to one another.
junction field-effect transistor (JFET): Three-element electronic device that has operation based on the conduction of current carriers through a continuous piece of semiconductor material.

junction: Connecting point where P- and N-crystals are joined. A junction diode permits current carriers to flow in one direction and blocks the flow of current in the opposite direction.

K

keeper: Soft-iron piece used to join magnetic poles. The keeper provides the magnetic flux with an easy path between poles.

kilowatt-hour (kWh): Product of power (in watts) and time (in hours). It is the kWh quantity of electrical energy that is used to determine the amount of electrical utility bills. A kilowatt-hour is 1000 watts in 1 hour of time.

kinetic energy: Energy in the form of movement.

L

lamination: Thin, insulated section that is stacked to produce the required core length in transformers. Used to reduce eddy currents.

laser: Letters stand for light amplification by stimulated emission of radiation. A laser produces a tightly-focused beam that can travel long distances with relatively little spreading.

latching relay: Type of relay that has a latching mechanism that holds the contacts in position after the power has been removed from the coil. A latching relay usually has a special type of unlatching coil connected in series with a push-button stop switch.

left-hand rule for polarity: Method used to find the north pole of a coil. Grasp the coil with the left hand. Point the fingers in the direction that the current flows through the coil. The thumb points to the north pole of the coil.

left-hand rule of magnetic flux: Method used to find the direction of the current and the magnetic flux lines around a conductor. If conductor is held in the left hand, the thumb points in the direction of the current from negative to positive. The fingers then encircle the conductor in the direction of the magnetic flux lines.

lens: Object that is used to focus or send light energy to a specific point or in a particular direction.

light-activated silicon-controlled rectifier (LASCR): Semiconductor optoelectronic switch that has a lens that focuses light on its gate. It is triggered into conduction by light rather than by an electrical trigger pulse. They can be used as switching devices in systems for counting, sorting, or similar functions where a circuit must be switched on and off.

light-dependent resistor (LDR): Resistor that has a resistance that varies in inverse proportion to its conductivity.

light-emitting diode (LED): Diode that contains a PN junction that emits light when passing forward current.

lightning arrester: Device used on power lines to conduct to ground excessively high voltages caused by lightning strikes or other system problems.

limit switch: On/off switch that uses a mechanical movement to cause a change in the operation of the electrical control circuit of a motor or other load device.

linear actuator: Actuator that provides motion along a straight line.

liquid fuse: Fuse with liquid filled metal enclosure that contains the fuse element. The liquid acts as an arc suppressing medium. When the fuse element melts due to an excessive current in a power line, the element is immersed in the liquid to extinguish the arc.

liquid level switch: Switch used when it is necessary to control the level of a liquid. Some designs have an operating lever connected to a rod and float assembly. The float assembly is placed into a tank of liquid where the motion of the liquid controls the movement of the operating lever of the float switch. Also called a float switch.

load: Refers to a specific part or number of parts designed to produce some form of work.

logic gate: Circuit with the ability to pass or block certain signals. An IF ... THEN sentence is often used to describe its basic operation.

long-shunt: Way in which series windings are placed in the circuit of a compound-wound DC motor. With this method, the shunt field winding is placed across both the armature and the series field.

lumen (lm): Amount of light falling on a surface of which all points are a unit distance from a uniform point source of one candela.

lux (lx): Unit of illumination equal to the number of lumens per square meter.

M

machine language: Numeric code that a computer control unit can manipulate according to programmed instructions.

magnet: Object that has the property of attracting iron.

magnetic air breaker: Circuit breaker in which the contacts separate, in the air, when the power line is overloaded.

magnetic field: Lines of force that surround a magnet.

magnetic flux (Φ): Magnetic lines of force. The lines never cross one another and they always form individual closed loops around a magnetic material. They have a definite direction from the north to south pole along the outside of a magnet.

magnetic motor contactor: Relay that acts as a control element to start and stop motors. It operates due to electromagnetic relay action.

magnetic saturation: Point at which the maximum alignment of domains takes place in a material.

magnetic starter: Device that relies upon an electromagnetic effect to open or close the power-source circuit of the motor.

magnetization curve: Curve that shows the relationship between a magnetizing force and flux density. Also called a $B - H$ curve.

magnetizing force (*H*): The measure of strength of a magnetic field. Magnetizing force is measured in oersteds.

magnetomotive force (MMF): Force that causes a magnetic flux to be developed. Comparable to voltage in an electrical circuit.

manipulator: Part of a robot that consists of segments that can be jointed and that move about, allowing the robot to do work.

metal oxide semiconductor field-effect transistor (MOSFET): One type of transistor with four terminals: source, gate, drain, and substrate.

metal-clad cable: Electrical cable with a flexible spiral metal covering.

microprocessor unit (MPU): Unit of a computer that combines the arithmetic logic unit and control section of the computer.

microstepping: Reduced step size in stepper motors. It is used to improve low speed application accuracy.

microstepping drive: Drive that subdivides the basic motor step by proportioning the current in the two adjacent windings. In this way, the step size is reduced.

mini-PLC: Compact programmable logic controller that can control simple machine operations and many types of manufacturing processes.

mirror: Object designed to reflect light.

mnemonics: Operation programming code in the form of symbols or words.

modulated: Signal that is interrupted or pulsed on and off several times per second.

momentary contact switch: Switch that accomplishes temporary control.

monochromatic: Source that radiates energy of only one specific wavelength or a very narrow part of the spectrum.

monostable multivibrator: Multivibrator with one stable state and one unstable state. The starting of a monostable multivibrator is based on the application of a trigger pulse.

motor action: Motion produced in a motor due to the interaction of a magnetic field and a set of conductors.

motor-controller enclosure: Enclosure that protects an operator against accidental contact with high voltages that could cause death or shock.

motor-driven cycle timer: Timing device that is used to control one or more different on and off operations on a continual basis.

motor-driven delay timer: Timing device that is used in industrial applications where a specific amount of time is needed before a load device is energized. Timers of this type are used to provide warm-up time for some electrical components or to ensure proper sequencing steps in production-line assembly applications.

motor-driven interval timer: Timing device commonly used in industrial applications where the load is turned on immediately and left on for a certain interval of time. After this interval, the load is automatically turned off. It then remains in the off state until the timer is reset and started again.

motor-starting device: Type of power control used to accelerate a motor from a stopped condition to its normal operating speed.

multimeter: Meter that is used to measure resistance, voltage, or current.

mutual induction: Magnetic coupling that connects inductors.

N

NAND gate: Logic gate that combines a NOT gate with an AND gate. This gate requires all inputs to be 0 for an output of 1.

nanometer (nm): Length equal to 1000 micrometers (μm) (millionths of a meter, more commonly called microns) or 10^{-9} meters.

National Electrical Code (NEC): Code that governs all electrical wiring in buildings.

National Electrical Manufacturers' Association (NEMA): Association that has developed standards for electrical control systems. NEMA standards are used extensively to obtain information about the construction and performance of various electrical power equipment. These standards provide information concerning the voltage, frequency, power, and current ratings for various equipment.

negative logic: Electrical system that assigns a voltage to the value of 0 and no voltage to the number 1.

negative power: Refers to electrical energy that is returned from the load device to the power source without being converted to another form of energy.

noninterruptible system: Optoelectronic system that is designed to respond to a constant flow of light through a slightly unusual medium.

nonmetallic-sheathed cable (NMC): Common type of electrical cable used for interior wiring. Also called Romex cable.

Non-servo robot: Nonintelligent robot.

NOR gate: Logic gate that combines a NOT gate with an OR. This gate produces an output of 0 if one or more of the inputs are 1.

normal excitation: Amount of DC excitation applied to a rotor when the motor operates at a power factor of 1.0.

normally closed (NC): Condition in which a device (such as a contact) is closed when no power is applied and opens when power is applied to activate.

normally open (NO): Condition in which a device (such as a contact) is open when no power is applied and closes when power is applied to activate.

NOT gate: Logic gate that inverts the input signal. Also called an inverter.

NPN transistor: Transistor that has a thin layer of P material placed between two pieces of N material.

N-type crystal: Semiconductor crystal that has an excess of free electrons. Created by adding a pentavalent impurity to a semiconductor.

numerical control (NC) system: System that uses a series of numerical-coded instructions to control the operation of a machine automatically. Duplicate parts can be produced by this process, or specific assembly operations can be performed automatically.

O

oersted: Unit of measure of magnetizing force.

ohm (Ω): Unit of measure of resistance.

ohmmeter: Electrical instrument used to measure resistance.

ohms adjust: Control used to zero the meter when taking a resistance reading.

Ohm's law: Law that explains the relationship of voltage, current, and resistance. Ohm's law states that an increase in voltage increases current if resistance remains the same. Stated another way: an increase in resistance causes a decrease in current if voltage remains the same.

oil-filled circuit breaker: Circuit breaker in which the contacts are immersed in an insulating oil contained in a metal enclosure.

open circuit: Incomplete path for current to flow. open-ground condition: Condition that occurs when a grounded conductor is opened and the ground is no longer functional. This condition can present severe safety problems and cause abnormal system operation.

open-loop system: Control system for a process in which there is no self-correcting mechanism.

operand: Data input supplied to the ALU, normally in the form of an 8-bit binary number (byte).

operational amplifier (op-amp): High-performance amplifier circuit containing several transistor devices. The entire assembly is built on a small silicon substrate and packaged as an integrated circuit. ICs of this type are capable of high-gain signal amplification, from DC to several million hertz.

operational life expectancy: Length of time that a device can be expected to continue functioning properly.

optical encoder: Device that operates by means of a grating that moves between a light source and a detector. When light passes through the transparent areas of the grating, an output is seen from the detector.

optical filter: Object placed in a light path that restricts the passage of specific frequencies.

OR gate: Logic gate that requires at least one input to be a 1 for an output of 1.

output port: Part at which a robot or computer sends out information.

overexcitation: Amount of DC excitation applied to a rotor when a motor operates at a leading power factor (capacitive effect).

overshoot: Amount by which momentum carries a rotor past the stable position in a stepper motor.

overtravel: Distance that an actuator moves beyond the point where the contacts change state.

P

panchromatic: Source that produces radiation that extends over a very large portion of the optical spectrum.

panelboard: Metal cabinet that encloses the main disconnect switch and the branch-circuit protective equipment.

parallel port: Computer port designed for parallel transmission.

parallel transmission: Digital data transmitted as multiple bits at the same time, following side-by-side paths like the lanes of a highway.

part-winding starting: Motor starting method in which the power line voltage is applied across only part of the parallel-connected motor windings, thus reducing starting current. Once the motor has started, the line voltage is placed across all of the motor windings.

partial control: Causes some type of an operational change in a system other than turning it on or off. Changes in electrical current or light intensity are examples of alterations achieved by partial control.

path: Provides a channel for the transfer of energy. It starts with the energy source and continues through the system to the load.

peak-to-peak: Voltage value from peak positive to peak negative.

permanent capacitor motor: Motor that has no centrifugal switch, so its capacitor is permanently connected into the circuit. These motors are used only for very low torque requirements and are made in small, fractional-horsepower sizes.

permanent-magnet DC motor: Motor used for low-torque applications. The DC power supply is connected directly to the armature conductors through the brush/commutator assembly. The magnetic field is produced by permanent magnets mounted on the stator. The rotor is a wound armature.

permeability (μ): Ability of a magnetic material to transfer magnetic flux.

phase angle (θ): Angular separation between voltage and current.

phasor diagram: Diagram that shows the size and phase angle relationships between two quantities (e.g. current and voltage). Also called vector diagram.

photoconductive: Optoelectronic device that varies in conductivity according to fluctuations in light. Electrical resistance decreases when light is more intense and increases when light intensity decreases.

photodiode: Specialized semiconductor diode that has a linear increase in reverse current as light intensity increases. When light of the proper wavelength is focused toward the PN junction, current carriers move across the junction. This causes a current through the external load that is proportional to the amount of light falling on the junction.

photoelectric emission: Device that emits electrons when it is struck by light.

photoemissive: Optoelectronic device that emits electrons in the presence of light.

photo-FET (PFET): Optoelectronic device similar to a conventional junction field-effect transistor, but in its construction has a lens that focuses light on a sensitive photodiode gate. Also called a photo-field-effect transistor.

photo-field-effect transistor (PFET): Optoelectronic device similar to a conventional junction field-effect transistor, but in its construction has a lens that focuses light on a sensitive photodiode gate. Also called a photo-FET.

photometer: Instrument used to measure illumination.

photometric: Source of radiation that produces some form of visible light.

photoresistive: Devices that are designed to change their electrical conductivity when variations of light energy occur.

phototransistor: Amplifying device constructed so that when light is focused on the collector-base junction, base current flows. This base current is amplified by the transistor.

photovoltaic: Optoelectronic device that converts light energy into electrical energy. When light energy falls on a photovoltaic device, it creates an electrical voltage.

pickup current: Minimum current required to energize a relay.

piezoelectric effect: Effect experienced in certain crystals. When they are subjected to mechanical stress, an electrical potential is developed in them.

pilot light: Device used to provide a visual indication. They usually indicate either an on or off condition.

pipe cutter: Tool used by electricians to cut electrical conduit.

pitch: Up-and-down movement of the wrist of a robot.

pixel: Smallest unit used to designate an area on a graphic display.

platen: Fixed part in a linear stepper motor. The platen is equivalent to the rotor in a normal stepper motor.

pliers: Tool that is used for holding, bending, pulling. Many pliers also have cutting edges.

plug-in GFCI receptacle: GFCI receptacle that provides protection by plugging into a standard wall receptacle.

pneumatic drive system: Drive system that uses air-driven actuators.

pneumatic timing relay: Timing relay that is dependent upon the movement of air within a chamber.

PNP transistor: Transistor has a thin layer of N material placed between two pieces of P material.

polarity markings: Markings used so that transformer windings can be connected correctly.

pole: End of a magnet.

positive logic: Electrical system that uses the value of 0 to indicate low-voltage or no voltage, and the number 1 is used to indicate a voltage larger than 0.

positive power: Electrical energy supplied by the source that is used by the load.

potential energy: Stored energy that an object has because of its position.

potentiometer: Variable resistor whose value can be changed by adjusting a rotary shaft.

power: Rate at which work is done.

power factor (PF): Ratio of true power converted in a circuit to apparent power.

power factor-corrective capacitors: Type of capacitor unit that is used to correct the power factor of AC power systems.

precision resistor: Resistors with tolerances lower than 5%.

pressure switch: Switch that has a set of electrical contacts that change states due to a variation in the pressure of air, hydraulic fluid, water, or some other medium.

pretravel: Distance that an actuator must move to change the normally open or normally closed state of the limit-switch contacts.

preventive maintenance: Process of regularly checking equipment, cleaning and maintaining it, and replacing worn parts before breakdowns occur.

primary-reactor starting: Motor starting method similar to primary-resistance starting in which reactors (coils) are used in place of resistors since they consume smaller amounts of power from the AC source. This method is more appropriate for large motors that are rated at over 600 volts.

primary-resistance starting: Motor starting method that uses resistors in series with power lines to reduce the motor-starting current. Starting torque is reduced according to the magnitude of the current.

primary windings: Windings that make up the coil that receives the input energy from an AC source.

prism: Object used to divide a light beam into its component wavelengths.

program: Detailed set of instructions that directs the computer to function in a way that will produce a desired result.

program counter: Memory device that indicates the location in memory of the next instruction to be executed.

programmable logic controller (PLC): Digital solid-state system that can be programmed to control machine operations such as logic, sequencing, timing, counting, and arithmetic.

prony brake assembly: Instrument that can be used to measure torque.

proximity detection: Optoelectronic detection system achieved with a combined light source and detector in a single unit. The light source and detector in proximity detection usually respond to visible light. The object being detected reflects the light signal back to the detector.

proximity sensor: Sensor that detects the absence or presence of an object within a certain distance. Can be used to provide feedback about the distance between the sensor and an object, such as an end effector.

P-type crystal: Semiconductor crystal that has an excess of holes. Created by adding a trivalent impurity to a semiconductor.

pulse stretcher: Monostable multivibrator in which the time duration of the output is altered according to the application. A short-duration input pulse can be used to produce a rather long-duration output pulse.

punch: Tool used with a hammer to create holes.

push-button switch: Switch in which a push button is manually operated to close or open a control circuit. Many motor control applications use push-button switches as a means of starting, stopping, or reversing a motor.

Q

quadrature: Describes how four sets of teeth on a forcer are spaced so that only one set at a time can be aligned with a platen in a linear stepper motor.

R

raceway: Conduit that carries conductors that carry the power to the various pieces of equipment throughout a building.

radial traverse: In-and-out motion relative to the base. This is the extension and retraction of a robot's arm.

radiometric: Source of radiation that generates infrared wavelengths or energy outside of the visible part of the spectrum.

ramping: Gradual sequence of accelerations and decelerations that allows higher speeds to be achieved.

random-access memory (RAM): Memory in a computer that can be written to as well as read.

range sensor: Sensor that determines the precise distance from the sensor to an object. Such devices are useful for locating objects near a workstation.

RC time constant (τ): Formula $\tau = R \times C$. Refers to the time in seconds that it takes a capacitor to build up to 63% of the full charge potential of the applied source.

reactive power: Component of power that results from an inductive or capacitive reactance and is 90° out of phase with the active power. The unit measure of reactive power is the voltampere reactive (VAR).

read-only memory (ROM): Permanently stored information often placed in a computer chip when it is manufactured.

recirculating chopper: Method of current control used in some stepper drives. It provides a feedback voltage proportional to the current in the motor.

rectification: Process of converting alternating current to direct current.

rectification system: System that converts alternating current into a form of direct current.

rectilinear propagation: Motion in a plane that is at right angles to the direction of travel.

reed switch: Switch that can either make or break contact in response to a magnetic field. Two flat metal strips, or reeds, are housed in a hollow glass tube filled with an inert gas. When the reeds are exposed to a magnetic field, they are forced together, completing an electrical circuit. When the magnetic field is turned off, the reeds spring open, breaking the circuit.

reflection factor: Percent of light reflected from a surface expressed as a decimal.

reflex detection: Optoelectronic system that has the light source and detector unit both mounted in the same assembly. The source can be either an incandescent light or a modulated LED. Operation is based on the reflection of an emitted light beam back to the detector. A light beam sent out by the source strikes a retroreflective target and reflects back to the detector.

regenerative chopper: Variation on a recirculating chopper in which the supply voltage is applied across the motor winding in alternating directions.

relative permeability (μ*r*): Comparison of the permeability of a material to the permeability of air (1,0).

relay: Electrical device that relies on magnetism to operate.

relay logic: Programming method for PLCS in which a ladder diagram is entered into the system by means of the keyboard.

reluctance (*Q*): Opposition to the development of a magnetic flux. Comparable to resistance in electrical circuit.

repeatability: Rate or number of times that the load can be switched on and off in a given period.

repeat-cycle timer: Timing device designed to perform an operating cycle continually until the operation is stopped.

repulsion motor: Type of AC induction motor. This motor was once used for many applications, but it is now being replaced by other types of single-phase motors.

reset timer: Type of timer that employs a motor-driven gear assembly that is actuated by a clutch mechanism. At a certain time during a cycle, the clutch mechanism is tripped. This action disengages the drive motor from the gear mechanism and permits it to return to the starting position by spring action. Upon its return to the starting point, this timer can turn off, go into an idling condition, or repeat the cycle again automatically.

residual magnetism: Magnetism left in an electromagnet's core after current stops flowing. Also called "leftover" magnetism.

resilient mounting: Mounting that provides a vibration-reducing insulated contact between the motor and the mounting surface.

resistance (*R*): The opposition to current in electrical circuits. Unit of measure is the ohm (Ω).

resistivity (ρ): Resistance of a conductor for a specified cross-sectional area and length. This measurement is given in ohms per circular mil-feet (cmil-ft).

resistor: Device used to control the flow of electrical current.

resolution: Smallest change of a quantity that the meter can measure. Resolution of a meter is determined by the number of digits in the display.

resolver: Feedback device provides accurate position and velocity feedback. They are resistant to shock and vibration, which are often encountered in industrial settings.

retentivity: Material's ability to retain a magnetic flux after a magnetizing force is removed.

retroreflective: Condition where a target reflects or turns back radiant waves of energy so that the reflected waves are parallel to the original wave.

reverse biasing: Applying an external voltage of the same polarity to the barrier potential and causing an increase in the width of the depletion zone. This action hinders current carriers from entering the depletion zone.

revolute configuration: Jointed-arm configuration for a robot.

rework: Process of fixing parts that do not meet product specifications.

rigid conduit: Solid pipe used in special locations for enclosing electrical conductors.

rigid mounting: Method used to attach a motor directly to a machine.

rms value: Value equal to 0.707 × peak value. The letters rms stand for root mean square.

rocker switch: Simple switch used for on-off control.

roll: Swivel or rotation in the wrist of a robot.

Romex cable: Brand name for nonmetallic sheathed cable.

rotary switch: Switch in which the shaft is attached to sets of moving contacts. These moving contacts touch different sets of stationary contacts that are mounted on ceramic segments when the rotary shaft is turned to different positions.

rotational traverse: Movement about a vertical axis. This is the side-to-side swivel of a robot's arm about its base.
rotor: Rotating shaft and its associated parts in a motor.
run windings: Windings used in a split-phase AC induction motor while the motor is running.
runaway speed: Characteristic of series motors at no load. With no load, the speed of a series motor shaft can become so high as to damage the motor.
running torque: Torque required to keep a rotor turning.

S

safety switch: Type of switch that is mounted in a metal enclosure and operated by means of an external handle. Safety switches are used only to turn a circuit off or on.
saw: Commonly used cutting device.
schematic diagram (or schematic): Diagram used to represent the parts of equipment or circuits. Schematics show how the components in each circuit fit together.
scrap: Part that cannot be fixed.
screwdriver: Tool used in fastening. Screwdrivers come in a wide variety of sizes and lengths and tip shapes.
secondary windings: Windings that make up the coil that receives the energy from the primary windings and deliver it to the load.
semiconductor: Material is not a conductor or an insulator. Semiconductors do not conduct electrical current easily and do not insulate well. Semiconductors have four electrons in their outer orbits.
sequence controller: Part of a computer that maintains the proper sequence of events required to perform a task. After instructions are received and decoded, it issues a signal that starts the proper action.
sequencer: Counting register that is widely used in automatic assembly operations and in the control of production lines.
sequencing: Control of a load device in each of its stepping positions.
serial port: Computer port designed for serial transmission.
serial transmission: Digital data transmitted one bit at a time.
series-wound DC motor: Motor that has the armature and field circuits connected in a series arrangement. The field is wound of relatively few turns of large-diameter wire.
service entrance: Access to a building through which electrical power is brought from overhead power lines or from underground cable.
servomechanism: Special type of AC or DC motor that drives a precision piece of equipment in specific increments. Also called a servo system.

servomotor: Motor that is responsible for producing mechanical changes from an electromagnetic actuating device. A device of this type is normally coupled to the controlled device by a gear train or some mechanical linkage.

servo robot: Intelligent or highly intelligent robot. The primary difference between an intelligent and highly intelligent robot is its level of awareness of its environment.

servo system: Special type of AC or DC motor that drives a precision piece of equipment in specific increments. Also called a servomechanism.

shading: Producing torque by a simulated two-phase method.

short circuit: Occurs when a conductor connects directly across the terminals of an electrical energy source.

short shunt: Way in which series windings are placed in the circuit of a compound-wound DC motor. With this method, the shunt field is placed across the armature.

shunt-wound DC motor: Motor with field coils connected in parallel with its armature. This type of DC motor has field coils that are wound of many turns of small-diameter wire.

silicon-controlled rectifier (SCR): Semiconductor device that is used primarily as a switching device. Power control is achieved by switching the SCR on and off during one alternation of the AC source voltage.

single-phase, three-wire system: Power distribution system used in most new residences. These systems are derived from a three-phase power line.

single-phase, two-wire system: Power distribution system that has a secondary that produces one single-phase voltage, such as 120 or 240 volts. This system has one hot line and one neutral line. In residential distribution systems in the past, this was the most used type to provide 120-volt service.

slip: Difference between the revolving stator speed of an induction motor and the rotor speed.

socket set: Provides a handle and an assortment of sockets that can be used to place or remove nuts and bolts.

software: Programs in which changes can be made. These instructions are stored in read/write memory.

solar cell: Device that is used to convert light energy into electrical energy.

solder: Alloy of tin and lead used to connect electrical components. The solder that is most used is made from 60% tin and 40% lead.

soldering: Method of attaching electrical components using a melted metal (solder).

solenoid: Electrical component that has a plunger that moves when the coil is energized. The back of the plunger is attached to a spring. It causes the plunger to return to its original position when the solenoid de-energizes. The movement of the plunger of the solenoid is used to activate some type of load connected to it.

solid-material fuse: Fuse in which the arc is extinguished in a chamber filled with a solid material.

solid-state relay: Relay that uses solid-state electronics instead of a mechanical switch. These relays are used where improved reliability or a rapid rate of operation is necessary. They are not sensitive to shock, vibration, dust, moisture, or corrosion.

source: (1) Converts energy of one form into a form used by the system; or supplies energy used by the system. (2) One lead on a FET. Provides current to the drain and gate. Similar to the emitter on a BJT.

speed regulation: Measure of how much motor speed varies with varying loads.

spherical configuration: Configuration in which pivot point gives the robot its vertical movement, reach is accomplished through use of a telescoping boom that extends and retracts, and rotary movement occurs around an axis perpendicular to the base. Also called polar configuration.

split-phase AC induction motor: Motor with two sets of stator windings. One set, called the run windings, is connected directly across the AC line. The other set, called the start windings, is also connected across the AC line. The start windings are connected in series with a centrifugal switch mounted on the shaft of the motor. The centrifugal switch is in the closed position when the motor is not rotating.

squirrel-cage rotor: Type of rotor that has large-diameter copper conductors soldered at each end to a circular connecting plate.

stack: Magnet section added to a motor's shaft that is used to increase torque output without reducing acceleration performance.

stadimetry: Determines the distance to an object based on the apparent size of a camera image.

standard: Reference signal that is compared to a feedback signal in a closed-loop system.

start windings: Windings used in a split-phase AC induction motor to start rotation. A centrifugal switch takes them out of the circuit when the motor starts running.

static torque curve: Shows how the restoring torque varies with rotor position as it is moved from its stable point in a hybrid step motor.

stator: Frame and other stationary components in a motor.

step accuracy: Rating that is used to define a position accuracy tolerance. This percentage figure indicates the total possible error of a stepper motor in moving one step.

step angle: Degree of rotation in a stepper motor. The number of steps required for the shaft to rotate 360° or one complete revolution may be found by dividing the step angle into 360°.

step-down transformer: Transformer that has more turns of wire on the primary than on the secondary.

step response time: Indicates the time required for a motor shaft to complete a single step after a DC input signal has been applied to the stator. This time rating is usually in milliseconds and relies on the torque-to-inertia ratio of the motor and the DC input source characteristics. steps per second: Rating of the number of steps a stepper motor shaft rotates in one second.

step-up transformer: Transformer that has fewer turns of wire on the primary than on the secondary.

strain gauge: Device that senses mechanical movement (a change in dimension resulting from strain or stress).

stress sensor: Sensor that produces a signal that indicates the magnitude of the contact made.

substation: Locations encountered along a power distribution system. At these locations, transformers are used to reduce the high transmission voltages to a lower voltage level.

sustained contact switch: Switch that causes a load to remain in an on or off condition until the switch position is manually changed.

switch: Device whose primary function is to turn a circuit on or off. However, more complex functions can be performed using switches.

switchboard: Large enclosure that has several overcurrent protective devices (fuses or circuit breakers).

switchgear: Type of equipment that houses all the necessary control devices for the electrical circuits that are connected.

symbol: Drawing that represents a component in a diagram.

synchronous capacitor: Three-phase synchronous motor when no load is intended to be connected to the shaft. Situation when the motor is being used solely for its power factor correction abilities.

synchronous motor: Constant-speed drive motor. This motor operates from a single-phase AC line. It has stator windings that are connected across the AC line. The rotor is made of permanent-magnetic material. Some external method of starting must be used in order to initiate rotation. Once the rotor is started, it will rotate in synchronism with the revolving stator field since it does not rely on the induction principle.

synchronous speed: Stator speed.

synchro system: Machine that employs rotary motion to control the angular position of one shaft by positioning the shaft of a second device.

system grounding: Involves the actual grounding of a current-carrying conductor (usually called the neutral) of a power distribution system.

T

table: Group of consecutive registers all holding data.

tactile sensor: Sensor that indicates the presence of an object by touch.

temperature switch: Common type of control device used in industry. The control element of a temperature switch contains a specific amount of liquid. The liquid increases in volume when the temperature increases. Thus, changes in temperature can be used to change the position of a set of contacts within the temperature-switch enclosure.

terminal connector: Fastening device that allows wires to be quickly connected and disconnected to machinery.

thermal expansion timer: Timing device with an operating principle based on the expansion characteristic of metal. In this device, metal is made to expand.

thermistor: Temperature-sensitive resistor. The device's resistance decreases as temperature increases (and vice versa).

thermocouple: Device that converts heat energy into electrical energy. The device consists of two dissimilar metal strips fused together at one end.

thermoelectric sensor: Sensors that produce a change in electrical output due to a change in temperature.

thinwall conduit: Pipe conduit that can be bent with a special conduit-bending tool. Also called electrical metallic tubing (EMT).

three-phase, four-wire system: Most widely used three-phase power distribution system. Used to supply 120 / 208 volts and 277/480 volts for industrial or commercial load requirements.

three-phase synchronous motor: Specialized motor that is considered a constant-speed motor. It can be used to correct the power factor of three-phase systems.

three-phase, three-wire system: Power distribution system used to supply motor loads of 240 volts or 480 volts. Its major disadvantage is that it only supplies one voltage as only three hot lines are supplied to the load.

three-phase, three-wire system with neutral: Three-phase, three-wire system power distribution system with a center-tapped winding.

through-beam detection: Optoelectronic system that has the light source and the detector both mounted in independent units. The source sends out a light signal from its location. The detector then responds to this signal at some remote location. Object detection occurs when the light beam is broken.

timing relay: Relay used to turn a load device on or off after a specific period of time.

toggle switch: Simple switch that has two states controlled by a lever, or toggle.

tolerance: Percentage of variation between the marked color-code value and the actual value of a resistor.

torque: The rotating effect produced by the interaction of two magnetic fields.

torque-to-inertia ratio: Motor ratio equal to the holding torque in oz-in. divided by the rotor inertia in oz-in. s^2.

total reactance (X_Y): Quantity found by subtracting the smallest reactance (X_L or X_c) from the largest reactance.

transducer: Device that transforms energy of one form to energy of a different form.

transformer: Electrical device that uses magnetism to link one coil of wire to another. It is capable of controlling massive amounts of power for efficient distribution.

transistor: Three-element device made of semiconductor materials. These devices have one or more PN junctions in their construction.

translator: Logic section of a stepper drive. Its function is to translate the step and direction signals into control waveforms for the switch set.

transmission and distribution system: System used to interconnect electrical power production systems and to provide a means of delivering electrical power from a generating station to its point of utilization.

triac: Semiconductor device that acts as a three-terminal AC switch. Gate current is used to control the conduction time of either alternation of the AC waveform. A triac is the equivalent of two reverse-connected SCRs feeding a common load device.

triangulation: Method that involves measuring angles and the baseline of a triangle to determine an object's position.

troubleshooting: Process of finding out why something does not work properly.

true power: Power that is converted to another form of energy by a circuit.

two-value capacitor motor: Motor that employs two capacitors. One, of higher value, is in series with the start winding and centrifugal switch. The other, of lower value, is in series with the start winding and remains in the circuit during operation. The larger capacitor is used only to increase starting torque. It is removed from the circuit by the centrifugal switch during normal operation. The smaller capacitor and the entire start winding are part of the operational circuit of the motor. The smaller capacitor helps to produce a more constant running torque as well as quieter operation and improved power factor. Also called capacitor-start, capacitor-run motor.

U

underexcitation: Amount of DC excitation applied to a rotor when a motor operates at a lagging power factor (inductive effect).

unijunction transistor (UJT): Semiconductor device that acts as a voltage-controlled diode.

unipolar: Type of semiconductor device that responds to one type of current carrier passing through a channel.

unity power factor: Maximum value of power factor. The factor is equal to 1.0, or 100%. Obtained in a purely resistive circuit.

universal motor: Motor that can be powered by either AC or DC power sources.

V

variable autotransformer: Transformer in which the winding tap can be adjusted along the entire length of the winding to provide a variable AC voltage to a load.

V-curve: Curve that shows the variation of stator current drawn by the synchronous motor as rotor current varies.

vector: Straight lines that have a specific direction and length. They are used to represent voltage or current values.

vector diagram: Diagram that shows the size and phase angle relationships between two quantities (e.g. current and voltage). Also called phasor diagram.

vertical traverse: Provides up-and-down motion.

visual sensor: Sensor that can be used to recognize objects or to measure their characteristics.

V-MOS: Vertical-groove MOS technology. A MOS device that is constructed with a V-shaped groove etched into the substrate. Construction of this type requires less space than a horizontally assembled device. The construction of a V-MOS device permits better heat dissipation and a high density of material in the channel area. Also called -MOSFET.

V-MOSFET: Vertical-groove MOS technology. A MOS device that is constructed with a V-shaped groove etched into the substrate. Construction of this type requires less space than a horizontally assembled device. The construction of a V-MOS device permits better heat dissipation and a high density of material in the channel area. Also called *V*-MOS.

volt (V): Unit of measure for voltage.

voltage (*V* or *E*): Electrical pressure. Measured in volts (V).

voltage regulator: Part of distribution system used to maintain the voltage levels at the proper value.

voltampere (VA): Product of voltage and current. Used to measure apparent power.

voltampere reactive (VAR): Unit of measure of reactive power.

voltmeter: Electrical instrument used to measure voltage.

W

watt (W): Unit of measured power.

watthour meter: Meter that measures the amount of electrical energy used over a period of time.

wattmeter: Meter that monitors both voltage and current in a circuit. It is used to measure the power converted by electrical machines.

watt-second: Product of power (in watts) and time (in seconds). It is used to determine a quantity of electrical energy.

window: Covering or closure through which light passes from a packaged light source to the outside or from the outside to the inside of a packaged detector.

wire strippers: Tool that strips wire. Can also be used to cut wire and bolts and crimp wire connections.

wiring diagram: Wiring diagram that shows the actual location of parts and wires on equipment. Also called cabling diagram.

work: Transferring or transforming of energy. Work is done when a force is exerted to move something over a distance against opposition.

work envelope: Area that a robot's end effector can reach.

wound-rotor induction motor: Specialized type of three-phase motor. This motor can be controlled externally by placing resistances in series with its rotor circuit. The starting torque can be varied by the value of external resistance. The advantages of this type of motor are a lower starting current, a high starting torque, smooth acceleration, and ease of control.

wrench: Tool used in assembling and disassembling many types of threaded fasteners.

wye configuration: Method of winding connection in a three-phase circuit.

wye-delta-starting: Motor starting method that employs a switching arrangement that places the motor windings in a wye configuration during starting and a delta arrangement for running.

Y

yaw: Side-to-side movement in the wrist of a robot.

Z

zener diode: Special diode designed to operate in the reverse direction without being damaged.

zener voltage: Reverse bias voltage at which there is a sharp increase in reverse current in a zener diode. Also called breakdown voltage.

zeroing: Adjusting the needle of the meter so that it is exactly over the zero mark.

Index

D

F

P

About the Authors

Dale R. Patrick was a Professor in the Department of Applied Engineering and Technology at Eastern Kentucky University. His experience includes teaching and organizing laboratory classes in electrical engineering/electronics technology for many years. He completed Bachelor of Science and Graduate Degrees at Indiana State University. His experience also includes research projects, technical teacher training, and energy consulting for business and industry. He and Dr. Fardo have co-authored over 30 textbooks and laboratory manuals.

Stephen W. Fardo is currently a Foundation Professor Emeritus in the Department of Applied Engineering and Technology at Eastern Kentucky University. His experience includes teaching and organizing laboratory classes in Electrical Engineering/ Electronics Technology for many years and coordinating / advising technical teacher programs for the State of Kentucky. He completed Bachelor of Science, Master of Science and Doctoral Degrees at Eastern Kentucky University and University of Kentucky. His experience also includes research projects, technical teacher education, and energy consulting for business and industry. He and Dale Patrick have co-authored over 30 textbooks and laboratory manuals.

Vigyan (Vigs) Chandra is a professor and coordinator of the Cyber Systems Technology-related undergraduate and graduate degree programs offered by the Departments of Computer Science and Information Technology (CSIT) and Engineering Technology at Eastern Kentucky University. He earned a doctoral degree from the University of Kentucky in Electrical and Computer Engineering and a master's degree in Career and Technical Education from Eastern Kentucky University. He holds certifications in various computer networking areas and teaches computer systems and applications, network hardware, communication systems, and digital, analog, and machine-control electronics. He is the recipient of the 2020–2021 Critical Thinking and Reading Teacher Award and the 2013 Golden Apple Award for Teaching Excellence at Eastern. His professional interests include implementing active teaching and learning strategies, metacognition, integrating open-source software/hardware with online control and computer networking technologies.

Brian W. Fardo is an Adjunct Professor in the Department of Applied Engineering and Technology and the Noel Studio for Academic Creativity at Eastern Kentucky University. His experience includes teaching classes in applied creative thinking, electrical systems, materials and processes, transportation, manufacturing, construction, engineering graphics, and quality control. He has a B.Sc. Degree in Industrial Distribution and Logistics, an M.Sc. in Technology Systems from East Carolina University and an M.Sc. in Career and Technical Education Eastern Kentucky University. He has completed coursework for a Doctoral Degree in Engineering/Technology Education at North Carolina State University. He currently owns and operates a telecommunications business.

For Product Safety Concerns and Information please contact our
EU representative GPSR@taylorandfrancis.com Taylor & Francis
Verlag GmbH, Kaufingerstraße 24, 80331 München, Germany